Organic Reactions

Organic Reactions

VOLUME 32

JOHN WILEY & SONS, INC.

New York · Chichester · Brisbane · Toronto · Singapore

Published by John Wiley & Sons, Inc.

Library of Congress Catalog Card Number 42–20265

ISBN 0-471-88101-5

Printed in the United States of America

10 9 8 7 6 5 4 3 2 1

PREFACE TO THE SERIES

In the course of nearly every program of research in organic chemistry the investigator finds it necessary to use several of the better-known synthetic reactions. To discover the optimum conditions for the application of even the most familiar one to a compound not previously subjected to the reaction often requires an extensive search of the literature; even then a series of experiments may be necessary. When the results of the investigation are published, the synthesis, which may have required months of work, is usually described without comment. The background of knowledge and experience gained in the literature search and experimentation is thus lost to those who subsequently have occasion to apply the general method. The student of preparative organic chemistry faces similar difficulties. The textbooks and laboratory manuals furnish numerous examples of the application of various syntheses, but only rarely do they convey an accurate conception of the scope and usefulness of the processes.

For many years American organic chemists have discussed these problems. The plan of compiling critical discussions of the more important reactions thus was evolved. The volumes of *Organic Reactions* are collections of chapters each devoted to a single reaction, or a definite phase of a reaction, of wide applicability. The authors have had experience with the processes surveyed. The subjects are presented from the preparative viewpoint, and particular attention is given to limitations, interfering influences, effects of structure, and the selection of experimental techniques. Each chapter includes several detailed procedures illustrating the significant modifications of the method. Most of these procedures have been found satisfactory by the author or one of the editors, but unlike those in *Organic Syntheses* they have not been subjected to careful testing in two or more laboratories.

Each chapter contains tables that include all the examples of the reaction under consideration that the author has been able to find. It is inevitable, however, that in the search of the literature some examples will be missed, especially when the reaction is used as one step in an extended synthesis. Nevertheless, the investigator will be able to use the tables and their accompanying bibliographies in place of most or all of the literature search so often required.

Because of the systematic arrangement of the material in the chapters and the entries in the tables, users of the books will be able to find information desired by reference to the table of contents of the appropriate chapter. In the interest of economy the entries in the indices have been kept to a minimum, and, in particular, the compounds listed in the tables are not repeated in the indices.

The success of this publication, which will appear periodically, depends upon the cooperation of organic chemists and their willingness to devote time and effort to the preparation of the chapters. They have manifested their interest already by the almost unanimous acceptance of invitations to contribute to the work. The editors will welcome their continued interest and their suggestions for improvements in *Organic Reactions.*

Chemists who are considering the preparation of a manuscript for submission to *Organic Reactions* are urged to write to the secretary before they begin work.

CONTENTS

Organic Reactions

CHAPTER 1

THE INTRAMOLECULAR DIELS–ALDER REACTION

ENGELBERT CIGANEK

Pharmaceuticals Research and Development Division, Du Pont Pharmaceuticals, E. I. du Pont de Nemours & Company, Wilmington, Delaware

CONTENTS

ACKNOWLEDGMENTS

Part of the literature search was carried out by Edward C. Worden and Rita S. Ayers. I also gratefully acknowledge the assistance of the many colleagues who furnished unpublished data or preprints of their work. I thank Professors Robert K. Boeckman, Jr. and William R. Roush for critically reading the entire manuscript, and Professor Barry Trost and Dr. Tadamichi Fukunaga for helpful advice. Dr. Robert M. Joyce provided invaluable assistance with the preparation of the manuscript. The Du Pont Company generously made available the services of its library and graphic arts facilities. I am especially indebted to Dorothy M. Tinker, Jeanette Jordan, and Collette Firmani for drawing all the structures and to Donna R. Weibley for typing the entire manuscript.

INTRODUCTION

The Diels–Alder reaction is one of the most useful reactions available to the synthetic organic chemist.[1] It is thus surprising that the intramolecular version, in which both diene and dienophile are part of the same molecule, remained virtually unexplored for many years. Early examples were often the result of unexpected observations, and the many advantages of this reaction were recognized and put to use in the synthesis of complex polycyclic molecules only recently.

In the intramolecular Diels–Alder reaction, two rings are formed in one step. In addition to the six-membered ring formed by the [4 + 2]cycloaddition, the

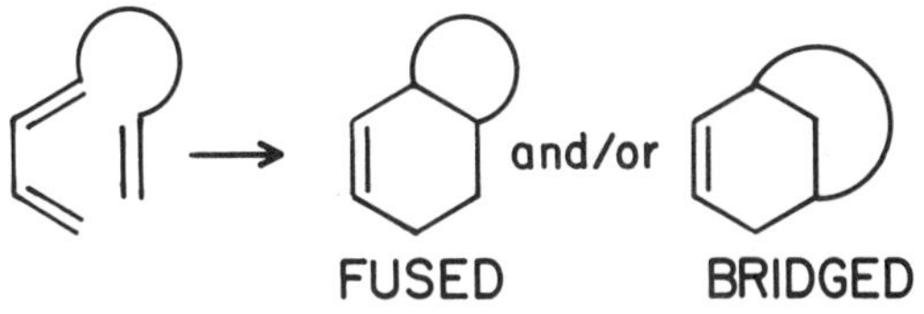

and/or

FUSED BRIDGED

product contains a second ring, the size of which depends on the length of the chain connecting diene and dienophile. Of the two possible regiochemical modes of addition, that leading to the fused product usually predominates to the virtual exclusion of the bridged product. The Diels–Alder reaction proceeds through a highly ordered transition state that is reflected in large negative activation entropies. In the intramolecular version, some of the ordering has been accomplished in advance by making the two reacting functionalities part of the same molecule. This results in less negative activation entropies and increased reaction rates under often surprisingly mild conditions. On the other hand, by using forcing conditions one can sometimes carry out intramolecular Diels–Alder additions that would be doomed to failure in the intermolecular version. Lower reaction temperatures, together with the constraints imposed by the connecting chain, often result in pronounced regio- and stereoselectivity. Side reactions such as dimerization or polymerization can be avoided by using high dilution or low pressures in the gas phase. Examples where the bimolecular Diels–Alder reaction interferes with the intramolecular cyclization are rare.

All of these features suggest that the intramolecular Diels–Alder reaction should be considered for any synthesis of a molecule containing a six-membered ring fused to a second ring, especially if that ring is five- or six-membered. Such a synthetic scheme may be made convergent by building up the diene and dienophile portions separately and connecting them just prior to the actual cyclization. Another potential use of the intramolecular Diels–Alder reaction that has not received much attention to date is as a supplement to the intermolecular Diels–Alder reaction in cases where the latter is unsuccessful or gives the wrong regio- or stereochemistry. By judicious choice of the connecting chain, the reaction may be made to proceed with the desired selectivity; this is followed by cleavage of the chain and conversion of the two ends to the desired functionalities.

Once the feasibility of employing the intramolecular Diels–Alder reaction has been recognized, the main challenge often becomes the preparation of the substrate. A discussion of that aspect is beyond the scope of this chapter; however, by scanning the tables for the particular diene, dienophile, and chain one is interested in, and then referring to the original literature, it will often be possible to obtain some indications on how to proceed.

An attempt has been made to cover the literature of the thermal intramolecular Diels–Alder reaction as completely as possible up to and including the year 1981. Many of the papers that appeared in 1982 and some from 1983 have also been included. More recent references are given in an addendum at the end of the tabular survey. The thermal cyclization of conjugated trienes to 1,3-cyclohexadienes is not covered; however, the thermal $[_{\pi}4_{a} + {}_{\pi}2_{a}]$ cyclization of conjugated trienes to bicyclo[3.1.0]hexenes is included, whereas its photochemical equivalent[2] is not. The few known examples of intramolecular photo Diels–Alder reactions of 1,3,n-trienes ($n > 5$) have been included in the tables under the appropriate diene type. The reverse intramolecular Diels–Alder reaction is discussed briefly, and all examples that could be found in the literature are collected in Tables XXVIII.

The intramolecular Diels–Alder reaction has been reviewed previously;[3–8b] reviews limited to intramolecular Diels–Alder reactions of *o*-quinodimethanes have also appeared.[9–14] The reverse intramolecular Diels–Alder reaction has been reviewed as well.[15]

DEFINITIONS

The following terms and symbols are used throughout this chapter:

1. The terms diene and dienophile have the same meaning as conventionally used for the intermolecular Diels–Alder reaction.
2. The prefixes *cis* and *trans* are used in connection with the diene to denote the stereochemistry of the double bond to which the chain is attached:

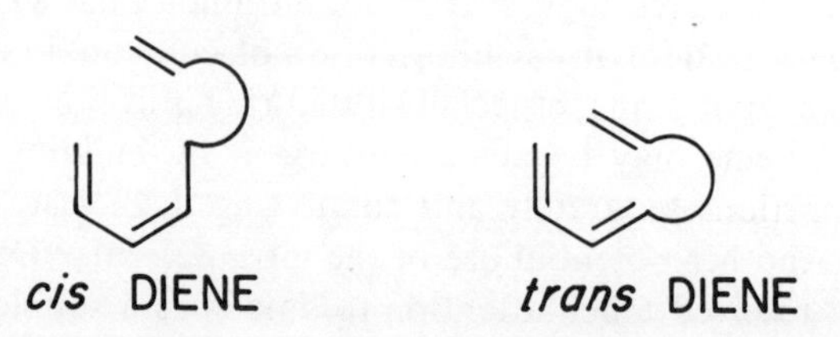

3. The chain is the shortest array of atoms connecting diene and dienophile. By this definition, the chain in the following example consists of three (not four) atoms. (Note, however, that in the tables this substrate is listed under "Cyclohexadienes, 2-Atom Chain Attached to C-5".)

4. Regiochemically, the intramolecular Diels–Alder reaction can give either fused or bridged products:

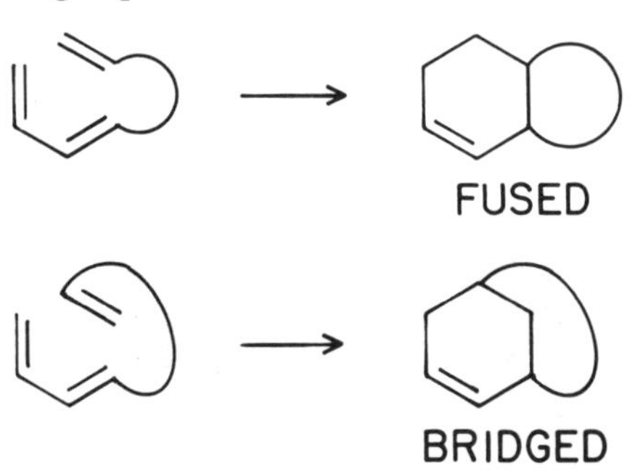

5. Stereochemically, the reaction may proceed by way of either a *syn* or an *anti* transition state; the terms *syn* and *anti* denote the orientation of the dienophile, as defined by its attachment to the chain, relative to the diene. This definition is independent of diene stereochemistry. The terms

H R → H R

syn TRANSITION STATE
(*trans* DIENE)

H R → H R

anti TRANSITION STATE
(*cis* DIENE)

syn and *anti* are used in place of the familiar *endo* and *exo* to avoid the confusion existing in the literature. For instance, the above *anti* transition state is usually called *exo* if R is hydrogen or alkyl but is often designated *endo* if R is a substituent, such as carbalkoxy, that can enter into secondary orbital interactions with the diene.

6. An asterisk (*) is used in place of brackets to denote hypothetical intermediates or intermediates that have not been isolated. This applies to both substrates and products.

7. Frequent reference is made to entries in the tables. Tables are arranged according to type of diene and are numbered in Roman numerals. Each entry within a table is assigned an Arabic entry number.

Throughout this chapter it has been assumed that the cyclization temperatures reported in the literature are those required for the intramolecular Diels–Alder reactions in question to proceed at a reasonable rate. This assumption may not be valid in every case, and some of the mechanistic conclusions drawn from comparisons of reaction temperatures may have to be revised in the future.

MECHANISM

Concerted versus Two-Step Mechanism

There is almost unanimous agreement that the Diels–Alder reaction is a concerted $[_{\pi}4_s + _{\pi}2_s]$ process,[1,16] and there are no compelling reasons to assume a different mechanism for most cases of the intramolecular version. The few instances where the stereochemical integrity of either the diene[17–19] or the dienophile[20] does not remain intact can usually be attributed to isomerization of the substrate prior to cyclization or to epimerization of the product. In an experiment specifically designed to test the concertedness of an intramolecular Diels–Alder reaction, 5-allyl-1,3-cyclohexadiene (**1**) was shown to cyclize exclusively in a concerted manner to give the tricyclononene **2**; intervention of the symmetrical diradical **3** would have led to **2** and **4** in essentially

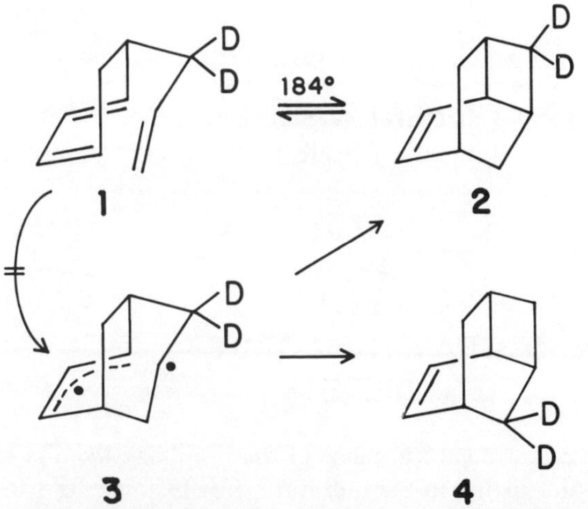

D D 184° D D **1** **2** (Eq. 1) D D D D **3** **4**

equal amounts (Eq. 1).[21] There are, however, a few reactions where a stepwise, diradical or dipolar mechanism has been proved or shown to be a viable alternative. Thus some scrambling of the deuterium labels occurs in the reaction of Eq. 1 at more elevated temperatures, indicating that a diradical mechanism now competes with the concerted cycloaddition. Heating the enamide **5** in refluxing toluene for 90 minutes gives, presumably via the diradical intermediate

6, the cyclobutane derivative **7** in 95% yield. Further heating of **7** in refluxing xylene for 7 hours produces the formal, bridged intramolecular Diels–Alder adduct **8** in 98% yield, again most likely via the diradical **6**.[22]

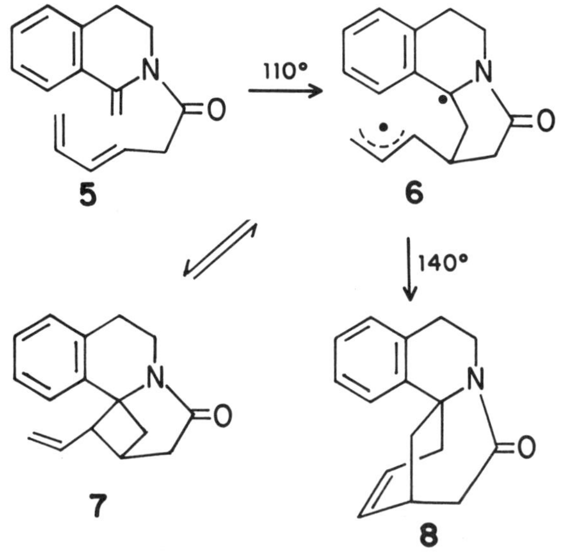

N O 110° N O

5 **6**

140°

N O N O

7 **8**

Diradical intermediates **11** have been postulated for the base-catalyzed cyclization of the diynes **9** to the fused naphthalene derivatives **12**.[23] Other authors consider a concerted intramolecular Diels–Alder reaction of the allenes **10** more likely (see Tables XIV).

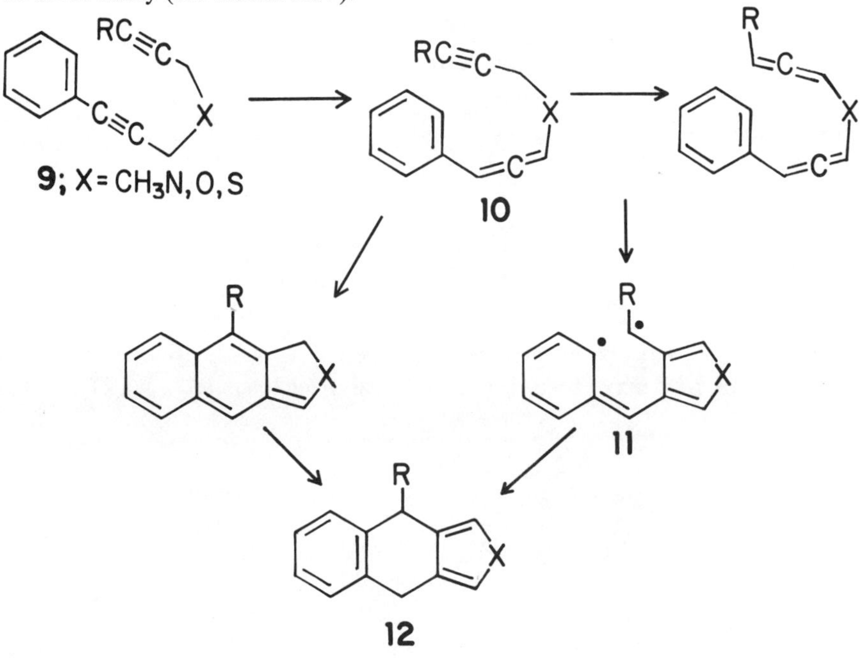

RC≡C X C≡C

9; X = CH_3N, O, S

RC≡C X =C=

10

R =C= X =C=

R X

R X

11

R X

12

Iminium zwitterions of type **14**, formed by Michael addition of the enamine to the acrylate ester, may be intermediates in the cyclization of the dihydrosecodines **13** to the *Vinca* alkaloids **15**;[24–26] however, in view of the pronounced stereoselectivity, this reaction may well be concerted (see Table XV for additional examples).

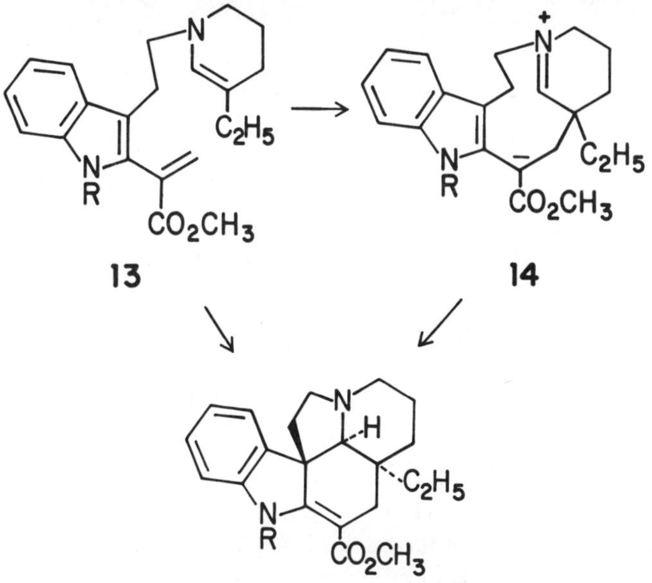

13 → **14**

13 → **15** ← **14**

A special case of an intramolecular Diels–Alder reaction leading to a bridged product is the cyclization of 1,3,5-trienes to bicyclo[3.1.0]hexenes; for instance,

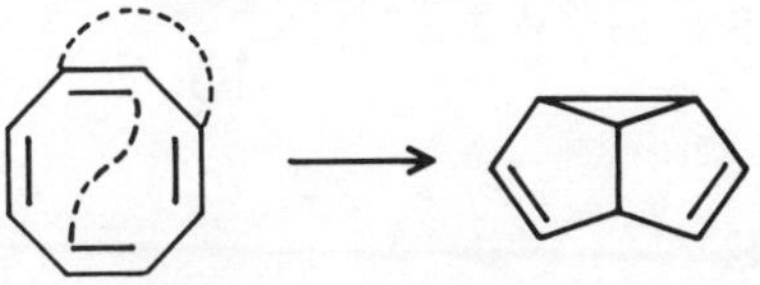

This process may be a symmetry-allowed $[_{\pi}4_{a} + _{\pi}2_{a}]$cycloaddition[27–31] shown in **16**, or it may proceed by a diradical mechanism (e.g., via **17**).[31,32]

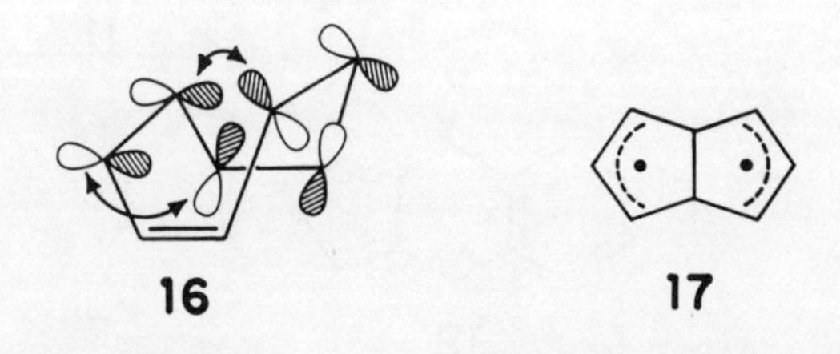

Synchronous or Nonsynchronous Bond Formation

A concerted mechanism for most intramolecular Diels–Alder reactions does not preclude the possibility that the two σ bonds are formed at different rates. Nonsynchronous concerted cycloadditions have been discussed for some time.[33,34] Frontier molecular orbital theory predicts more advanced bond formation between those termini of the diene and dienophile that have the largest coefficients in the highest occupied (HOMO) and lowest unoccupied (LUMO) molecular orbitals, respectively.[35] The constraints imposed by the chain on the attainable transition state geometries of the intramolecular Diels–Alder reaction may further reinforce the trend toward a nonsynchronous mechanism. Models indicate that for connecting chains of three or fewer atoms synchronous bond formation is unlikely.[36] Equation 2 shows that these considerations may have important consequences for the stereochemical course of intramolecular Diels–Alder reactions.[36–39] In the ester **18**, the coefficient of the dienophile LUMO at C-3 is larger than that at C-2. This results in an unsymmetrical transition state in which bond formation between C-3 and C-7 is more advanced than between C-2 and C-10. The development of the five-membered ring has thus progressed significantly and the *syn* transition state **18b**, where the two substituents on the quasi-cyclopentane ring are eclipsed, should be energetically less favored than the *anti* transition state **18a** with staggered substituents.[39] In addition, nonbonded interactions within the chain appear to be more severe in the unsymmetrical *syn* transition state.[38] Experimentally, cyclization of the ester **18** at 150° gives the hexahydroindenes **19a** and **19b** in

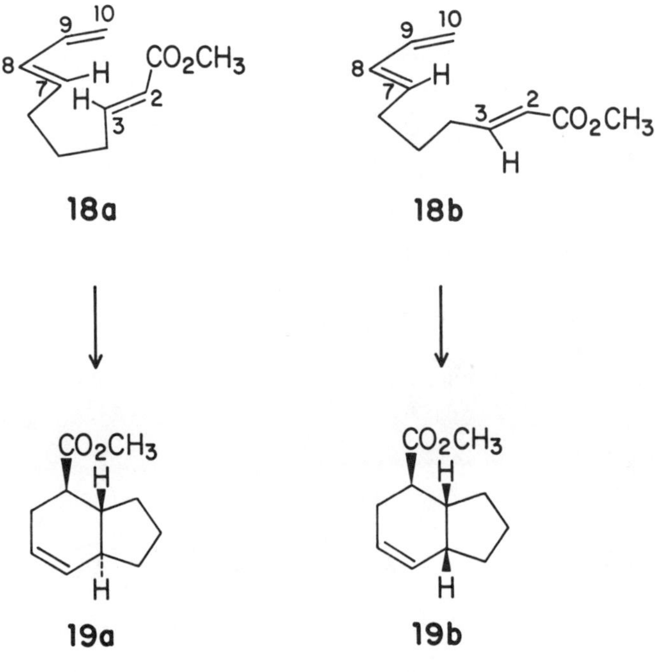

18a **18b**

(Eq. 2)

19a **19b**

39 and 26% yields, respectively.[40,41] Lewis-acid catalysis of the Diels–Alder reaction not only causes rate enhancement, but also leads to more pronounced regioselectivity by increasing the difference in magnitude of the dienophile LUMO coefficients (at C-2 and C-3 in ester **18**).[42] For reasons discussed above, this should lead to increased preference of transition state **18a** over **18b**. Experimentally, cyclization of ester **18** in the presence of menthoxyaluminum dichloride at 23° gives exclusively the *trans*-fused product **19a** in 72% yield.[41] The predominant formation of the *cis*-hexahydroindene **21** from the aldehyde **20**

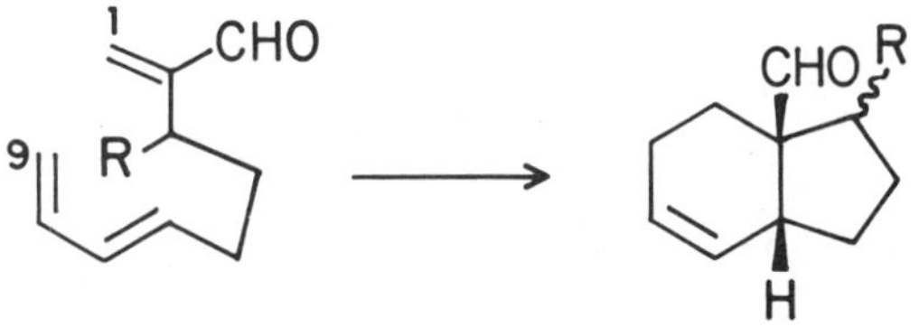

20 $R = C_6H_5CH_2O$ **21**

has been rationalized similarly. In this reaction, bond formation between C-1 and C-9 is more advanced in the transition state and formation of the nine-membered ring is considered to occur more readily via the *syn* transition state **20**.[38] As is seen later, other factors, such as nonbonded interactions between chain and diene or dienophile, and, in certain instances, secondary orbital overlap between the diene and substituents on the dienophile, also play a role in determining the stereochemistry of the products. In fact, enhancement of secondary orbital interactions by the catalyst[42] may well be the dominant factor in the exclusive formation of isomer **19a** in Eq. 2.

Thermodynamic Parameters

The highly ordered transition state of the Diels–Alder reaction is reflected in a large negative activation entropy that is typically in the range of −35 to −45 eu.[1] Linking diene and dienophile with a short chain reduces the degrees of freedom available to the two reacting functionalities, which should result in less negative activation entropies; indeed, this has been found to be the case. Thus, the activation entropies for the cyclization of the amides **22** are in the

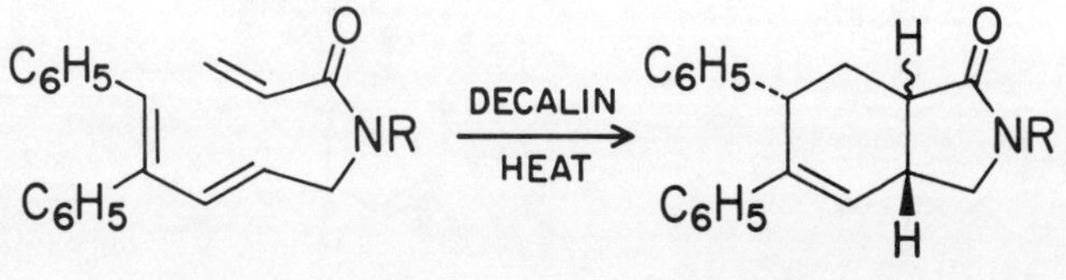

DECALIN HEAT

22 *cis/trans* ≈ 1

range of −13 to −21 eu, depending on the substituent R.[43] Similar activation entropies have been determined for the cyclization of *N*-(5-*p*-anisyl-2,4-pentadienyl)-*N*-methyl-*trans*-cinnamamide (Table IIb/31; −16 eu),[44] *N*-methyl-*N*-propargyl-9-anthracenecarboxamide (Table XVIIc/20; −11 eu),[45] *N*-(2-furylmethyl)-*N*-phenylmaleamic acid (Table XVIIIb/38; −17 eu),[46] and a series of *N*-allyl-*N*-(2,4-pentadienyl)carboxamides (Table IIb/2–6a)[47,48] and quaternary *N*-propargyl-*N*-(2,4-pentadienyl)ammonium salts (Table IIc/1, 6a, 12a, 12b).[49] The activation parameters for a number of *N*-allyl-*N*-(furylmethyl)-carboxamides (Table XVIIIb/11–14, 15a)[48,50] and a series of quaternary *N*-allyl-*N*-(furylmethyl)ammonium salts (Table XVIIIc/10–25)[51] and *N*-propargyl-*N*-(furylmethyl)ammonium salts (Table XVIIIc/1–4, 8)[51] have also been reported. The activation energies for these intramolecular Diels–Alder reactions are in the range of 15–25 kcal/mol and are thus comparable to those encountered in the normal Diels–Alder reaction. The often considerable rate enhancement observed for intramolecular Diels–Alder reactions thus appears to be predominantly the result of a less negative activation entropy.

All examples listed above involve substrates with three-atom chains containing nitrogen, and the measurements were all made in solution. Thermodynamic parameters of intramolecular Diels–Alder reactions as a function of chain length, preferably measured in the gas phase, are still lacking. However, it is likely that activation entropies will become more negative with increasing chain length, reaching the values typical of intermolecular Diels–Alder reactions when the chain contains five or six atoms.

Effect of Solvents and Pressure

The rates of intermolecular Diels–Alder reactions are essentially independent of solvent polarity, varying by a factor of no more than 10 in most cases.[1] No systematic studies of solvent effects in the intramolecular Diels–Alder reaction have been reported, but it is probably equally insensitive to solvent polarity. The ammonium salt **23a** is reported to cyclize in water, 94% ethanol, and

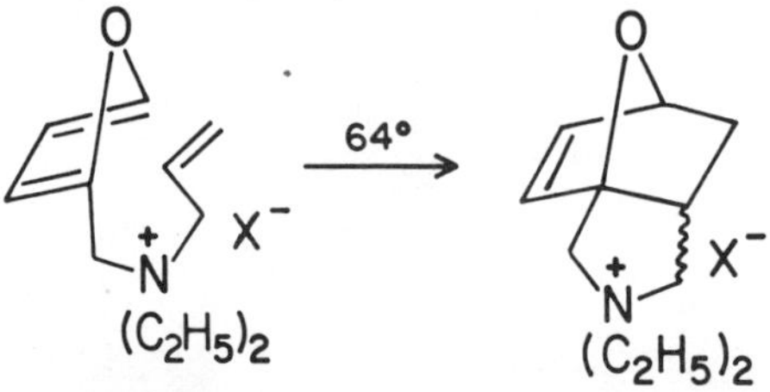

23a

acetonitrile with the relative rates of 6.8, 2.4, and 1, respectively, and this result has been interpreted in terms of frontier molecular orbital theory.[51,52] Alternatively, a hydrophobic effect may operate that, in aqueous solution, would force the molecule to adopt a coiled conformation resembling the transition state.[53] The rate of cyclization of another furan derivative, **23b**, as well as the product ratio, show moderate solvent dependency.[54] Under the conditions shown, the yields (and ratios of β/α-epimeric products) are: 45% (0.5) in water, 63% (1.2) in benzene, and 93% (0.9) in water–ethanol (5:2). β-Cyclodextrin, but not α-cyclodextrin, increases the rate of cyclization of substrate **23b** in water; the product is racemic.[55] The cavities in β-cyclodextrin are large enough to accommodate both diene and dienophile, thus bringing the two reacting functionalities into close proximity. The same would be accomplished by complexation of the dithiane ring, a situation that would in effect place two very bulky geminal substituents on the chain.

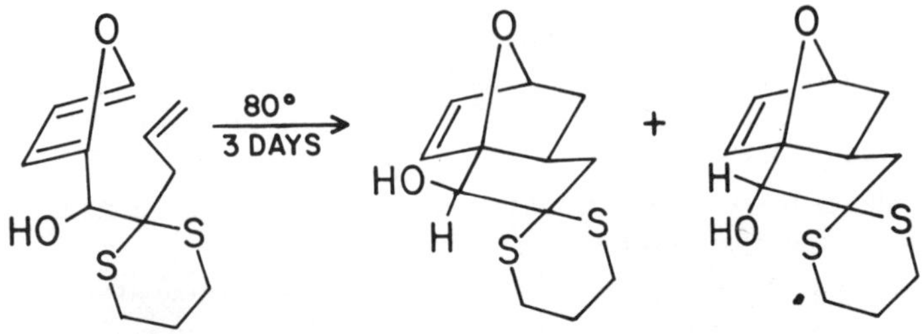

80°
3 DAYS

23b

The rate of the intramolecular Diels–Alder reaction of *N*-propargyl-9-anthracenecarboxamide increases with increased solvent viscosity, but again the effect is small.[56]

Many intramolecular Diels–Alder reactions have been carried out in the gas phase, either under static conditions or in a flow system. However, since data on the corresponding solution reactions are lacking, no comparison is possible.

Intermolecular Diels–Alder reactions have large negative activation volumes and thus can be accelerated dramatically by application of high pressure.[1] The only intramolecular Diels–Alder reaction for which the volume of activation has been measured is the cyclization of amide **24** ($R = C_2H_5$).[46]† As expected,

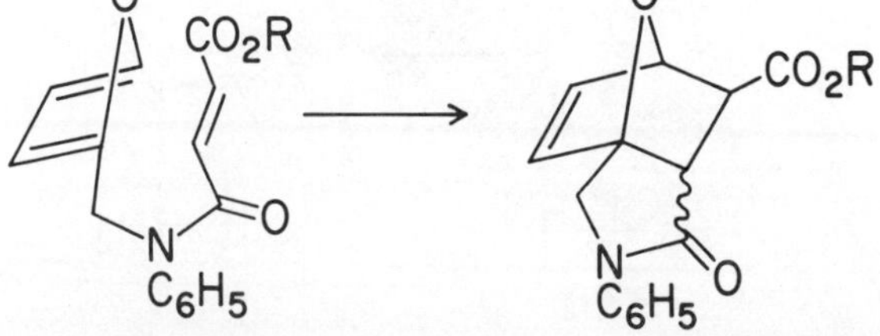

24

† In the original paper, the cyclization product is assigned the *syn* structure.[46] Actually, the stereochemistry of the product has not been determined.[57] In addition, the configuration of the dienophile is in question since the literature procedure that was followed to prepare the acid **24** (R = H) gives the maleic rather than the fumaric acid derivative.[58]

the observed value (−25 mL/mol) is somewhat less negative than that measured for typical intermolecular Diels–Alder reactions (about −30 mL/mol). Since the cyclization of substrate **24** proceeds readily at room temperature under normal conditions, application of high pressure is of no synthetic value in this case. Attempts to force more reluctant cyclizations in this manner have failed so far.[57,57a] Thus no reaction occurs when the ester **25a** (R = $C_2H_5O_2C$) is heated to 110° under 20-kbar pressure, whereas the adduct **25b** is obtained in

$$\mathbf{25a} \xrightarrow[\text{CH}_3\text{CO}_2\text{H},\ 205^\circ]{\text{C}_6\text{H}_6} \mathbf{25b} \qquad \text{(Eq. 3)}$$

50% yield when the reaction is carried out at 205° in benzene in the presence of acetic acid as a catalyst (Eq. 3).[59] No cyclization takes place in the absence of acetic acid. Similarly, the acetate **25a** (R = $CH_3CO_2CH_2$) cannot be forced to cyclize by application of high pressure but does undergo the intramolecular Diels–Alder reaction at 230°.[59] The ester **26**, under super-high pressure, yields only the intermolecular Diels–Alder adduct.[60]

$$\mathbf{26} \xrightarrow[\text{PRESSURE}]{\text{HIGH}} \text{adduct}$$

Catalysis

Some intermolecular Diels–Alder reactions are subject to Lewis-acid catalysis. A necessary (but not always sufficient) precondition is that there be a substituent on either the diene or the dienophile that can complex with the catalyst. Lewis acids not only affect the rates, but also result in more pronounced stereo- and regioselectivity, a finding that has been rationalized in terms of frontier molecular orbital theory.[42] A number of Lewis-acid-catalyzed intramolecular Diels–Alder reactions have been reported. Two examples have already been mentioned (Eqs. 2 and 3), and the effect of catalysts on the regioselectivity, *syn*/*anti* selectivity, and diastereoselectivity is discussed in the sections dealing with these subjects. The main problem with Lewis-acid catalysts is that they also lead to polymerization of the diene; this appears to be particularly true of the classical Lewis acids such as boron trifluoride etherate, titanium tetrachloride, aluminum trichloride, and stannic chloride.[41] Bulky ester groups on the dienophile increase the tendency toward polymerization.[41] Catalysts have occasionally caused epimerization of substituents in the product.[41] The catalysts that seem to offer the best compromise between rate enhancement and suppression of diene polymerization are menthoxyaluminum dichloride, ethylaluminum dichloride, and diethylaluminum chloride.[41] Other catalysts that have been used are tungsten tetrachloride,[41] niobium pentachloride,[41] bornyloxyaluminum dichloride,[41] trifluoroacetic acid,[61] acetic acid,[59] and sulfuric acid.[62] Florisil®[63] and manganese dioxide[64,65] also have been claimed to catalyze intramolecular Diels–Alder reactions, but the rate enhancement with these catalysts is small. The presence of groups, such as ethers, that form strong complexes with Lewis acids usually precludes catalysis of the cyclization.[38,41,66]

The finding that the intramolecular Diels–Alder reaction of the furan derivative **27** (R = H) can be accelerated by formation of the magnesium or

C_6H_6, Reflux

27

zinc salts (e.g., R = MgCl or ZnBr) has been attributed in part to a weak Lewis-acid activation of the dienophile; the other contributing factor is that coordination of the metal with the amide carbonyl group increases the population of the *s-cis* amide conformer required for attaining the transition state geometry.[67–70b]

Successful cyclization of the cyclohexadiene derivative **28** requires the presence of a small amount of potassium *tert*-butoxide. Catalysis by homo-

Decalin, 5% KOC_4H_9-t, 280°

28 (60%)

conjugation of the alcoholate anion with the diene system has been proposed as one of the possible explanations.[71]

Intramolecular Diels–Alder reactions of a number of furan substrates are catalyzed by transition-metal (molybdenum and tungsten) complexes; the rate enhancement is moderate.[71a]

All Lewis-acid-catalyzed intramolecular Diels–Alder reactions that may be found in the tables are as follows: Table IIa/4, 6, 21, 23, 24, 26, and 27; Table IIIa/10a, 10b, 17b, 20a, and 24; Table IV/1, 2c; Table VI/10, 10a; Table XVIIIa/2; Table XIX/7a, 18; Table XXVIc/22–25; and Refs. 71b,c.

SELECTIVITY

Regioselectivity

Intermolecular Diels–Alder reactions between unsymmetrically substituted dienes and dienophiles usually produce mixtures of both regioisomers, although one often predominates. In the intramolecular Diels–Alder reaction, the two regiochemical alternatives lead to either fused or bridged products. In practice, the vast majority of intramolecular Diels–Alder reactions give the fused isomers exclusively, irrespective of the presence, on either the diene or the dienophile, of substituents that exert a strong directing influence in the intermolecular Diels–Alder reaction.

Models indicate that for a *trans* diene a reasonably unstrained transition state leading to the bridged product can be constructed only if the chain is at least five-membered. This applies to both the *anti* and the *syn* stereochemical alternatives:

anti *syn*

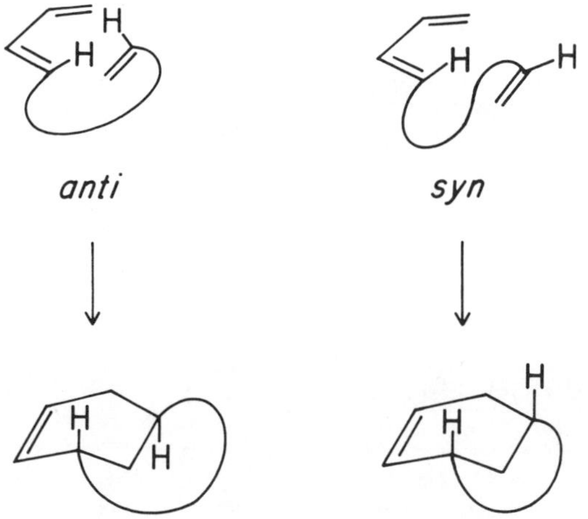

In fact, only three intramolecular Diels–Alder reactions of dienes known with certainty to have the *trans* configuration are reported to lead to bridged products, and in all three cases the chain contains 10 or more atoms. Thus the ester

29 gives the *syn* bridged product in 21% yield and the *anti* bridged product in 5% yield, in addition to the two possible fused products (53% yield; Eq. 4).[72]

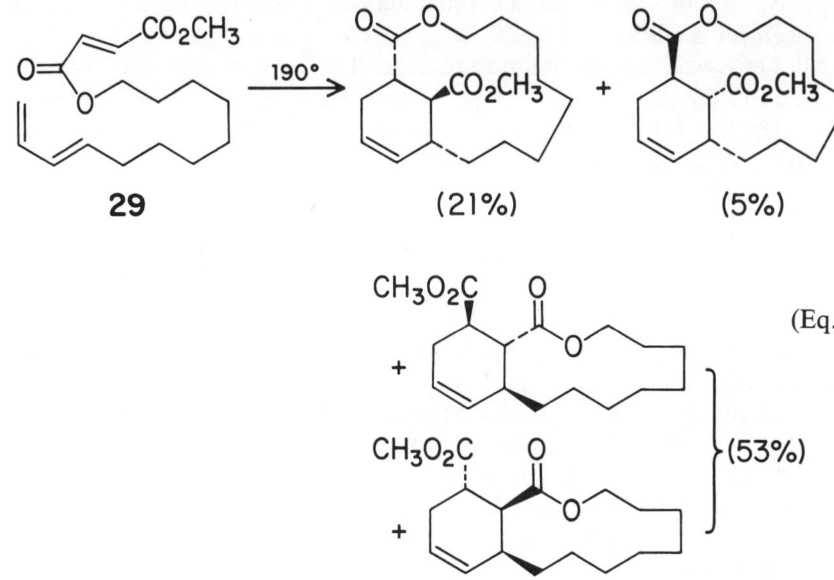

(Eq. 4)

The two other reactions are entries 6 and 8 in Table XXVIc. The formation of a bridged product from *trans* diene **5** having a three-atom bridge as described earlier does not contradict the above statements, since it has been shown to proceed via a diradical intermediate. The *o*-quinodimethane **30**, which has a

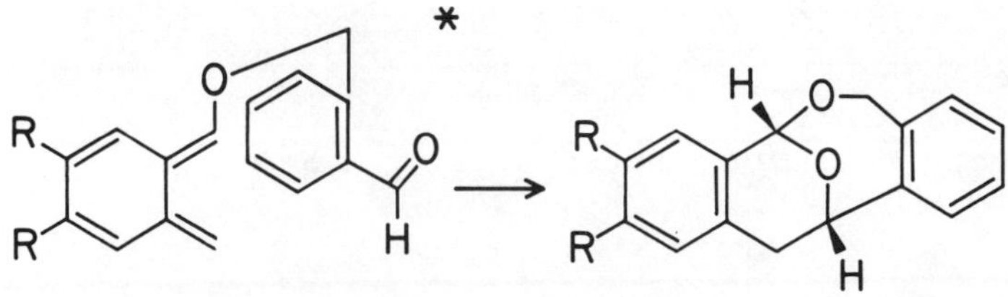

four-membered chain, cyclizes to a single product assigned the bridged structure in 49% yield.[73] The configuration of the diene has been considered to be *trans* as shown, but opening of the benzocyclobutene precursor to the *cis* diene and cyclization via an *anti* transition state cannot be excluded. The exclusive formation of the bridged product in this reaction has been attributed to electronic factors.[73]

The situation is more favorable to formation of bridged products in the case of *cis* dienes provided the stereochemistry is *anti*; an example where the chain consists of a single atom is known (Eq. 5).[74]

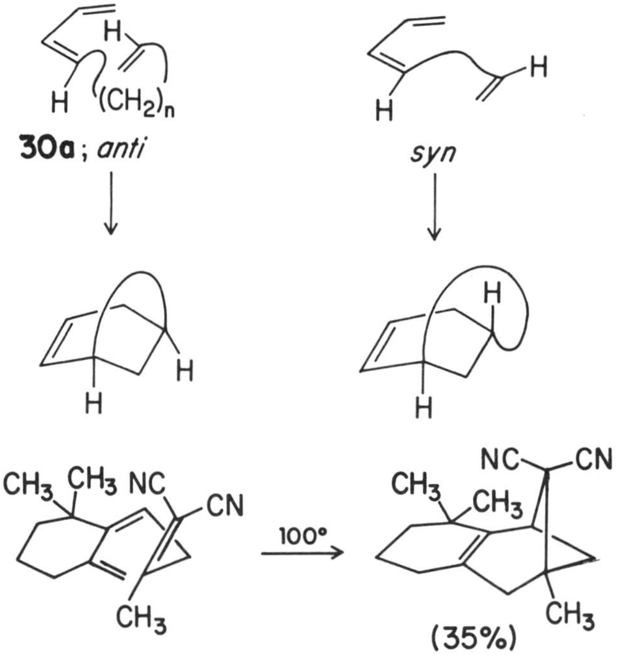

H $(CH_2)_n$
30a; *anti*

H H
syn

H H

H H

CH_3 CH_3 NC CN CH_3 $\xrightarrow{100^\circ}$ NC CN CH_3 CH_3 CH_3 (35%) (Eq. 5)

Molecular mechanics calculations, assuming a symmetrical transition state, indicate that even for substrates with six-atom chains the conformation leading to the bridged product (**30a**) has a considerably higher energy than that leading to the fused product. For all cases calculated (n = 3–6), angle strain is the major reason for this energy difference; torsional strain and nonbonded interactions contribute to a lesser degree.[75]

Cyclohexadienes with chains attached at C-5 are *cis* dienes and sometimes give rise to both regioisomers. Thus, the cyclohexadienone **31**, on heating in refluxing benzene, gives the bridged product **32** and its fused isomer **33** in

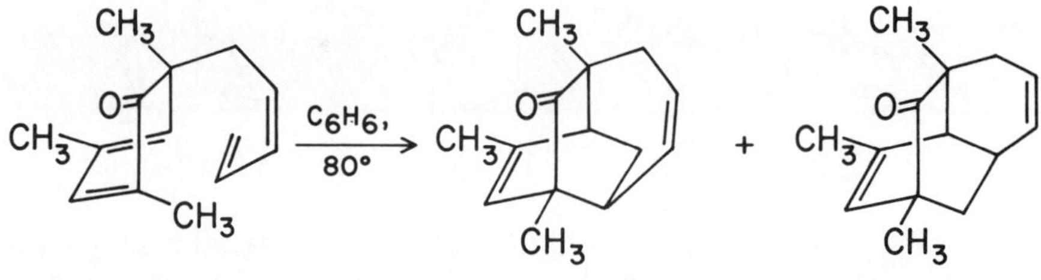

CH_3 O CH_3 CH_3 $\xrightarrow[80^\circ]{C_6H_6,}$ CH_3 O CH_3 CH_3 + CH_3 O CH_3 CH_3

31 **32** (35%) **33** (5%)

35% and 5% yields, respectively.[76] Various explanations for the preponderance of the bridged product in this and similar reactions have been advanced,[76–78] but it should be kept in mind that ketone **32** is the bridged product only by virtue of defining the chain as the shortest array of atoms linking diene and dienophile. If one considers the chain to include the carbonyl group, the ketone **32** becomes the fused product. The regioselectivity observed in the cyclization of the cyclohexadienone **31** is in fact predicted by frontier molecular orbital (FMO) theory.[35] A similar example is encountered in unsymmetrically substituted cyclopentadienes carrying the chain on C-5. The cyclopentadiene **34**, on heating to 250° followed by hydrolysis, thus gives the two regioisomeric ketones **35** and **36** in about equal amounts; by contrast, a single regioisomer is formed from the cyclopentadiene **37**.[79] The observed regioselectivity in this and related reactions is again in accord with FMO theory.[35,79] Another example is the

$(CH_3)_3SiO$
1. 250°
2. CH_3OH
O
O
+
34 **35** (39%) **36** (33%)

$(CH_3)_3SiO$
CO_2CH_3
CH_3
1. 110°
2. CH_3OH
O
CH_3 CO_2CH_3
37 (94%)

cyclization of the cycloheptadienone **38** (generated by a Claisen rearrangement),[80] which represents an intramolecular Diels–Alder reaction with inverse electron demand.[1] FMO theory predicts comparable coefficients on the two carbon atoms of the dienophile, and thus little regioselectivity.

O
*
200°
O
+
O
38 (41%) (41%)

Cyclization of the aldehyde **39** in the presence of a Lewis-acid catalyst gives the bridged product **41** exclusively.[62] The *cis* diene **40** is considered to be a likely intermediate and the exclusive formation of the bridged product is attributed to electronic effects.[62] Thermolysis of aldehyde **39** in the gas phase at 350° gives the fused product **42** exclusively in low yield; at higher temperatures, the bridged product **41** is also formed.[62]

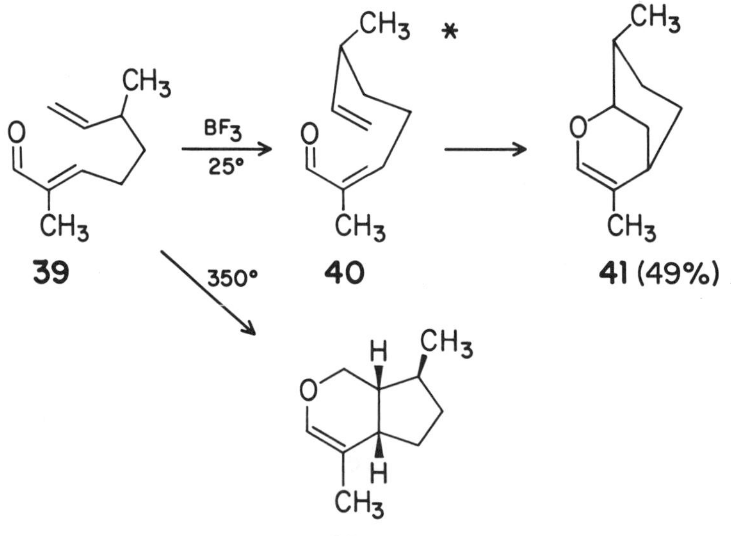

All examples of intramolecular Diels–Alder reactions of 1,3,n-trienes (n > 5) leading to bridged products are listed in Table XXVIc.

The $[_{\pi}4_a + _{\pi}2_a]$cycloaddition of 1,3,5-trienes to produce bicyclo[3.1.0]-hexenes is a special case of a *cis* diene undergoing an intramolecular Diels–Alder reaction with the formation of a bridged product.[80a] Although the reaction is thermally allowed, a diradical mechanism is also possible. As opposed to its very common photochemical equivalent,[2] the thermal cyclization of this type is relatively rare and of limited synthetic potential. By way of illustration, the ketene **44**, generated by photolysis of the cyclohexadienone **43** at −100°, produces the bicyclohexenone **45** in 50% yield on warming to room temperature.[81] Other examples are listed in Tables XXVIa (acyclic trienes) and XXVIb (cyclic trienes).

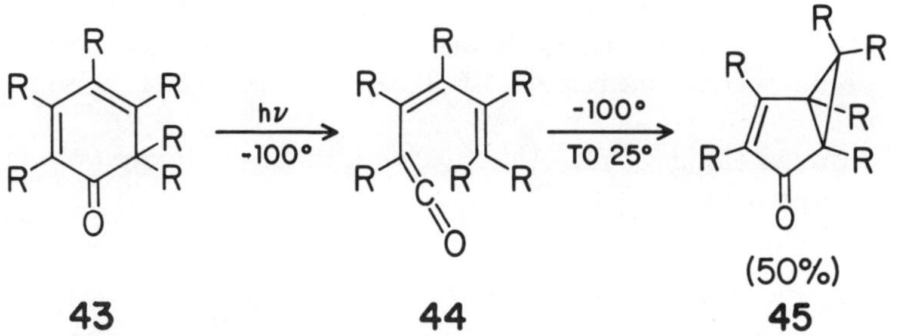

When the chain is attached to the 2-position of the diene, both regioisomeric products are bridged. In the absence of substituents on the termini of the diene, *syn* transition states (as drawn) lead to the same products as the *anti* transition states, but models indicate that for short chains the former are less strained.

meta - BRIDGED

para - BRIDGED

The *meta*-bridged products are obtained exclusively from substrates with chains containing five or fewer members.[82] In the ester **46**, which contains a six-membered chain, the *para*-bridged isomer is formed in low yield, but the *meta*-bridged isomer is still the major product.[82,83] In the intermolecular Diels–Alder

XYLENE 190°

46 (49%) (4%)

reaction of acrylate esters with 2-substituted dienes, the *para* isomer is the major product; the intramolecular Diels–Alder reaction thus offers a means of reversing the regioselectivity.[82] This is, of course, true not only for the case of dienes carrying the chain on C-2, but also for the much more common intramolecular Diels–Alder reactions of dienes where the chain is attached at C-1; fused products are formed in almost all cases irrespective of the substitution pattern.

Since intramolecular Diels–Alder reactions leading to bridged products are so rare, the remainder of the discussion deals exclusively with the formation of fused compounds.

Syn/*Anti* Selectivity

A. Introduction. *Cis* and *trans* dienes can each give either *cis*- or *trans*-fused products depending on the orientation of the dienophile relative to the diene in the transition state (TS). *Trans* dienes give *trans*-fused products via the *anti* TS and *cis*-fused products via the *syn* TS. Conversely, *cis* dienes give *cis*-fused products via the *anti* TS and *trans*-fused products via the *syn* TS. The length of

trans DIENE
anti TS

trans-FUSED

trans DIENE
syn TS

cis-FUSED

the chain, however, limits the number of possibilities. Thus cyclization of *trans* dienes with one- or two-atom chains requires excessively strained transition states, and all known examples of substrates containing such short chains involve *cis* dienes, most of them cyclic (see p. 47 for a list of intramolecular Diels–Alder reactions involving substrates with one- and two-atom chains).

cis·DIENE
anti TS

cis-FUSED

cis DIENE
syn TS

trans-FUSED

Cis dienes with chains containing one to four (and possibly more) atoms can cyclize only via the *anti* transition states to produce *cis*-fused products, since the *syn* transition states are too strained. This is not surprising for substrates containing one or two atoms since the products of *syn* addition would be the highly strained *trans*-fused bicyclo[4.1.0]heptenes and bicyclo[4.2.0]octenes. However, all known *cis* dienes with chains of three or four atoms also cyclize via the *anti* transition states to produce *cis*-fused products (Tables IIa/38, 39; IIb/17, 34, 34a; IIIb/9a; VIIIb/1b–20; IXd/1–32; IXe/1–22; X/15–20; XI/10; XIII/4–15; XIVb/21; XIVf/27; XV/1–15; XVId/4–23; XXI/8–11, 14, 15; XXV/1–4, 7). There appear to be no examples of intramolecular Diels–Alder reactions involving *cis* dienes with chains containing five or more atoms.

If a *cis*-fused product is desired, the simplest way to ensure stereospecificity would thus be to employ *cis* dienes. Acyclic *cis* dienes are readily prepared, such as by partial hydrogenation of vinylacetylenes, and they appear to undergo intramolecular Diels–Alder reactions at rates comparable to those of the *trans* dienes, judging from the few examples where a direct comparison is possible.

CH_3 $OSi(CH_3)_3$ $\xrightarrow{245°}$ CH_3 H H $OSi(CH_3)_3$ (15%) + CH_3 CH_3 $OSi(CH_3)_3$ (60%)

(Eq. 6)

This strategy has not been widely employed, at least for acyclic dienes, primarily because *cis* dienes are prone to undergo 1,5-hydrogen shifts at elevated temperatures (Eq. 6).[84] A possible solution to this problem is to lower the reaction temperatures by using activated dienophiles,[85,86] perhaps in conjunction with Lewis-acid catalysts.

Trans dienes containing chains of three or four atoms constitute the majority of substrates known to undergo the intramolecular Diels–Alder reaction. These, as well as *trans* dienes with longer chains, may cyclize via either the *syn* or the *anti* transition states. Since the intramolecular Diels–Alder reaction often occurs late in a multistep synthesis, it is important to be able to predict the stereochemistry in these cases, and considerable effort has been made to delineate the factors that influence the partition between the two possible transition states. These factors include chain length; substituents on the chain, diene, and dienophile; type of diene; type of dienophile; and catalysts.

B. Acyclic Dienes†

a. Acyclic Dienes with All-Carbon Chains. Acyclic *trans* dienes with three-carbon chains usually give mixtures of the *cis*- and *trans*-fused isomers, with

† This discussion includes semicyclic dienes such as vinylcycloalkenes and vinylheterocycles; *o*-quinodimethanes are discussed in Section E.

the latter predominating slightly. This lack of selectivity is due in part to the high reaction temperatures required in this system. Electron-withdrawing substituents on the terminal carbon atom of the dienophile, which in the intermolecular Diels–Alder reaction favor the *endo* transition state by secondary orbital interaction (Alder rule), have little effect on the stereoselectivity of the uncatalyzed intramolecular Diels–Alder reaction. The ester **18** and its geometric isomer **47** thus both cyclize preferentially via the *anti* transition state.[41] The

CH_3O_2C **18** → CO_2CH_3 H H + CO_2CH_3 H H

150° (39%) (26%)

MENTHOXY $AlCl_2$, 23° (72%) (0%)

CO_2CH_3 **47** → CO_2CH_3 H H + CO_2CH_3 H H

180° (49%) (26%)

$C_2H_5AlCl_2$, 23° (14%) (13%)

cyclization of ester **47** in violation of the Alder *endo* rule is only one of many intramolecular Diels–Alder reactions where this rule breaks down. A possible reason for the preference of the *anti* transition state in both cases has been discussed earlier (Eq. 2) in terms of unsymmetrical transition states anticipating the thermodynamic stabilities of the developing five-membered ring. A nonbonded interaction between the hydrogen atoms on C-3 and C-7, destabilizing the *syn* transition state, has also been advanced as a possible explanation for

syn TS *anti* TS

the *anti* selectivity in this system.[41] A study of system **48** (Table XV/24–33a), which also evaluates the influence of two identical or different activating groups R^1 and R^2 attached to the terminus of the dienophile, leads to the same conclusions.[87]

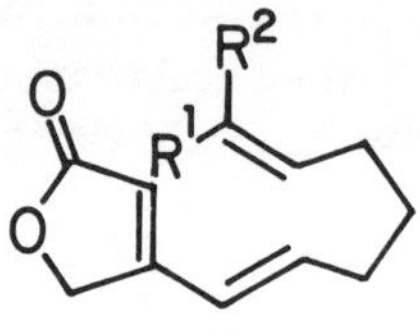

48

Lewis-acid catalysis, which enhances the *endo* selectivity in the intermolecular Diels–Alder reaction,[42] also does so quite dramatically in the ester **18** (p. 11), but fails to have any effect (except on the rate) in the ester **47** where increased secondary orbital interaction is not able to overcome the preference for the *anti* transition state.[41]

There are two important exceptions to the above generalizations. As discussed earlier, the *cis*-fused product predominates (again in violation of the Alder *endo* rule) when the nonterminal carbon atom of the dienophile carries an aldehyde group. Only two examples of this effect are reported (Table IIa/14, 16); both involve aldehyde groups, and both substrates have bulky groups on the first carbon atom of the chain. It remains to be established whether this effect is general and extends to other activating groups. The second exception involves substrates where the dienophile is activated by a carbonyl group that is also part of the chain. For instance, ketone **49** (R = H) gives the *cis*-fused product preferentially by a factor of 2.3;[88] for other examples, see Table IIa/41, 43. The very high reaction temperature (the corresponding substrates with four-carbon chains cyclize at room temperature) indicates that the carbonyl group actually provides very little activation to the dienophile.

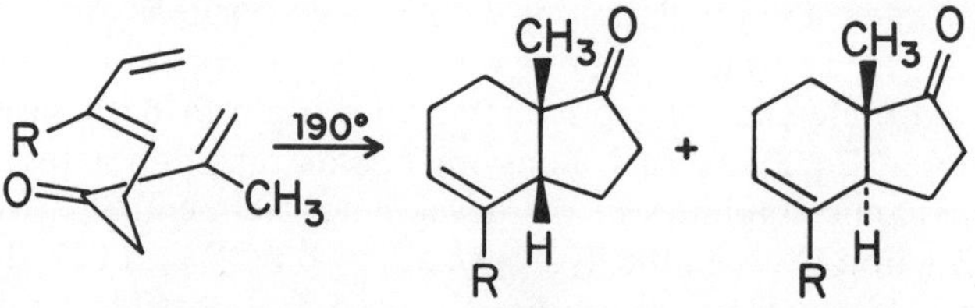

49 —190°→ **50** + **51**

Models show that the carbonyl group is twisted out of the plane of the double bond in both the *syn* and the *anti* transition states, and the preference for the *syn* transition state is thus unlikely to be due to secondary overlap of the carbonyl group with the diene. A possible explanation is that the reaction is thermodynamically controlled.[89] The *trans* isomer **51** is formed exclusively when R is a bulky group, causing a destabilizing interaction with the carbonyl group in the *syn* transition state.[90] For the same reason, ketals of ketone **49** (R = H) give predominantly the *trans*-fused products **51**.[88]

Acyclic *trans* dienes with four-carbon chains, in general, also give mixtures of *cis*- and *trans*-fused products with the former sometimes predominating; as with the lower homologs, π-acceptors on the terminal carbon atoms of the dienophile have no effect on the stereoselectivity of the uncatalyzed reaction (Table IIIa/9–17, 17b, 18),[91–93] but almost exclusive formation of the *trans*-fused product has been achieved using Lewis-acid catalysis (Eq. 7;[92] for other examples, see Table IIIa/10a, 20a). If, on the other hand, the dienophile has the *cis* geometry, Lewis-acid catalysis leads predominantly to the *cis*-fused product by enhancement of the secondary orbital interaction (Table IIIa/10b).[94]

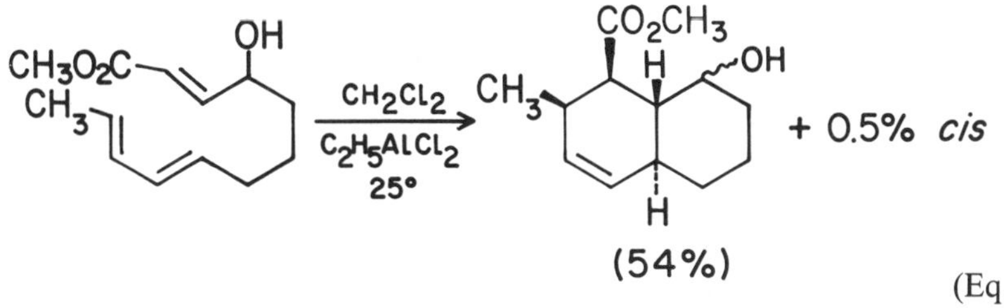

CH_2Cl_2, $C_2H_5AlCl_2$, 25°

(54%) + 0.5% *cis*

(Eq. 7)

Preferential or exclusive formation of the *trans*-fused decahydronaphthalene system can be achieved in a number of other ways. Introduction of a substituent such as a methyl group on C-3 of the diene thus causes a severe nonbonded interaction between that group and the axial hydrogen on C-6 in the *syn*

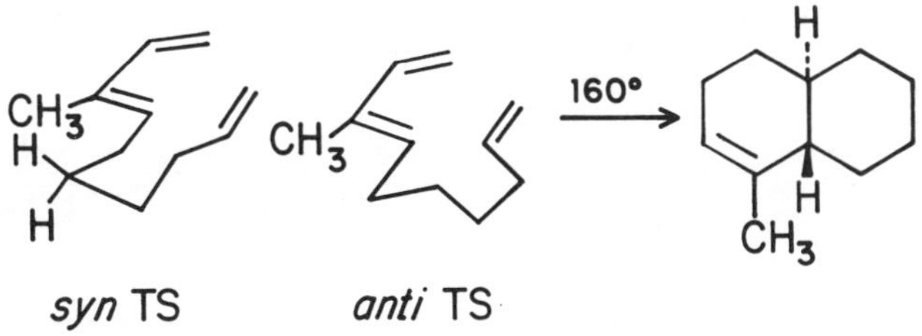

160°

syn TS *anti* TS

(Eq. 8)

transition state, assuming a chair form for the chain. This interaction is absent in the *anti* transition state leading to the *trans*-fused product, which is obtained in 95% yield (Eq. 8;[95] for other examples, see Table IIIa/4–8, 28, 29). However, the presence of a *cis* double bond (in the form of a benzene ring) in the chain reverses the stereoselectivity, and the *cis* products are formed preferentially (Table IIIa/31, 32) or in yields comparable to those of the *trans*-fused isomers (Table IIIa/33). The nonbonded interaction discussed above is no longer of importance since the chain assumes a boat form; alternatively, the reaction may be thermodynamically controlled.[96]

Introduction of a carbonyl group into the chain alpha to the diene also results in stereoselective formation of the *trans*-fused isomers as the primary products in the three cases reported so far (Table IIIa/18–20). In the *anti* transition state leading to the observed *trans* products, the carbonyl group remains coplanar with the diene, whereas it is twisted out of the plane of the diene in the *syn* transition state. However, the carbonyl group also facilitates epimerization to the *cis* isomer (Table IIIa/19)[19] or isomerization of the double bond (Eq. 9).[93]

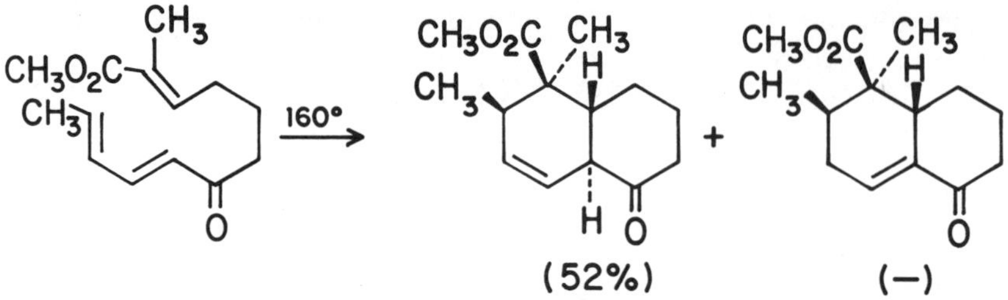

(Eq. 9)

Exclusive formation of the *cis*-decahydronaphthalene system is achieved by introducing a carbonyl group in the chain alpha to the dienophile (Table IIIa/21–27). This structural modification also results in a dramatic lowering of the reaction temperatures required; thus 1,6,8-decatrien-3-one, generated from its silyl enol ether, cyclizes at 0° to give the *cis*-fused product exclusively in 78% yield (Eq. 10).[97]

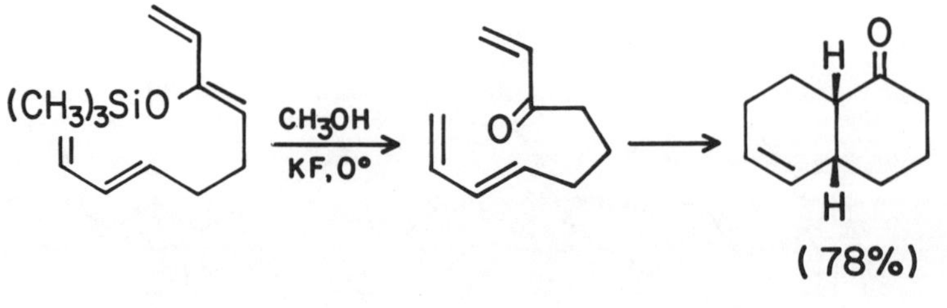

(Eq. 10)

The *cis*-decalones so obtained are easily epimerized by bases or acids[98,99] to the more stable *trans* isomers. The lower free energy of activation and the stereoselectivity of the above intramolecular Diels–Alder reactions are readily explained by the fact that, in contrast to the lower homologs, the *syn* transition state with dienophile and carbonyl group in a coplanar arrangement is strain-free. Since many of the 1,6,8-decatrien-3-one substrates are generated in the presence of Lewis acids, the rate acceleration and the *syn* selectivity may be attributed to acid catalysis. A particularly dramatic example of this effect is the cyclization of the vinylcyclohexene **52** at −78° in the presence of trifluoroacetic acid to give the *cis*-fused adduct in 66% yield in addition to 4% of the *trans* isomer.[61]

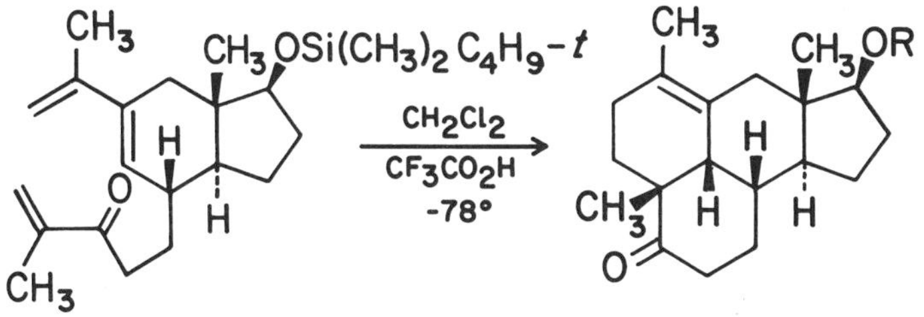

CH_3 CH_3 $OSi(CH_3)_2C_4H_9$-*t* $\xrightarrow[\text{-78°}]{CH_2Cl_2,\ CF_3CO_2H}$ (66%)

52

The *anti* selectivity observed, and the high reaction temperature required, for the cyclization of the vinylallene **52a** (Eq. 11)[100] may be attributed to a strained *syn* transition state and loss of coplanarity between the dienophile and the carbonyl group in the anti transition state.

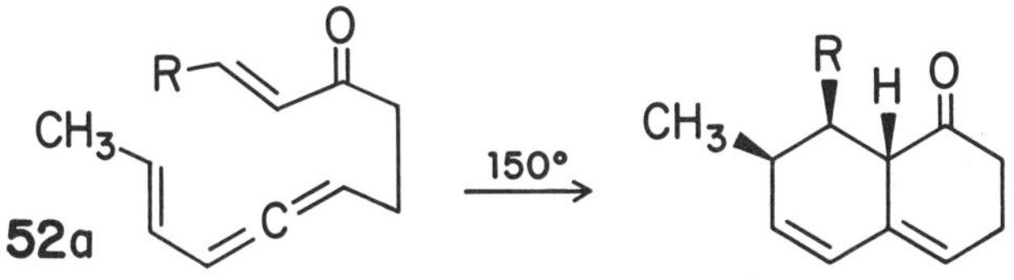

52a $\xrightarrow{150°}$ (75%) (Eq. 11)

$R = C_2H_5OCH(CH_3)O(CH_2)_3$

Only two examples of acyclic *trans* dienes with five-carbon chains undergoing the intramolecular Diels–Alder reaction are known. Both have carbonyl groups alpha to the dienophile and, like their four-carbon homologs, give the *cis*-fused products exclusively (Table IV/1, 1a).

b. Acyclic Dienes with Chains Containing Nitrogen. Only two intramolecular Diels–Alder reactions of acyclic dienes with amine nitrogen in the chain have been reported. The amine **53** ($R = CH_3$) gives the *cis*-fused product predominantly by a factor of 5 (Table IIb/15),[43] whereas the homolog **54**

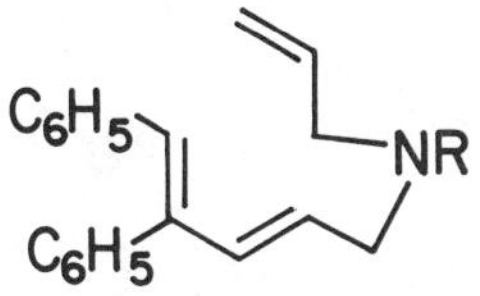

53 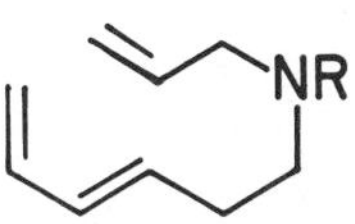 **54**

($R = CH_3$) with a four-membered chain gives essentially equal amounts of the *cis*- and *trans*-fused products (Table IIIb/10).[101] Interestingly, the urethane **54** ($R = CH_3O_2C$) also gives a 1:1 mixture of *cis*- and *trans*-fused products but requires heating to 275°, compared to a cyclization temperature of 140° for the amine **54** ($R = CH_3$).[101] A number of amides of type **53** (R = acyl) undergo the intramolecular Diels–Alder reaction (Table IIb/2–14), but the stereochemistry of the products is not known.

N-Acyldienamines of type **55** with three- or four-membered chains cyclize exclusively through the *syn* transition state to give the *cis*-fused adducts (Tables IIb/1, IIIb/1–6);[102–104] this stereoselectivity has been attributed to the fact

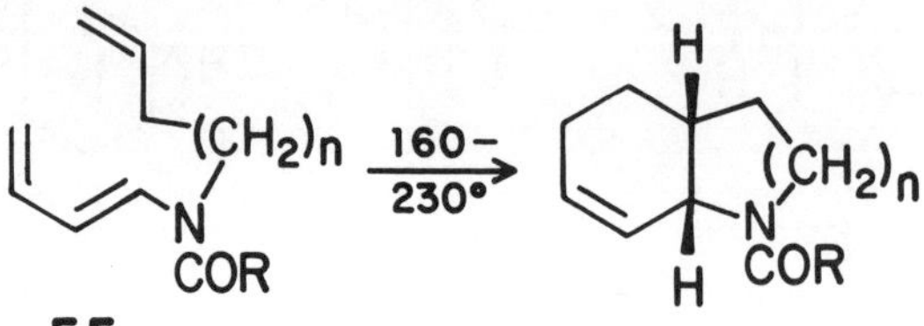

55

n = 1 or 2
R = ALKYL or METHOXY

that the π-orbitals of the diene and the amide nitrogen can overlap only in the *syn* transition state,[102] but it is open to question how important that interaction is for ground-state dienamides at the high cyclization temperature required. The single *N*-alkyldienamine of this type reported to undergo the intramolecular Diels–Alder reaction also gives the *cis*-fused product exclusively (Table XV/18; the reaction may, however, proceed by a Michael addition).

Simple *N*-acyldienamines where the acyl group is part of the chain (e.g., **56**) give mixtures of *cis*- and *trans*-fused products.[102] The enamide **57**, on the other hand, cyclizes to give the *trans*-fused product exclusively (Table XV/16),[105]

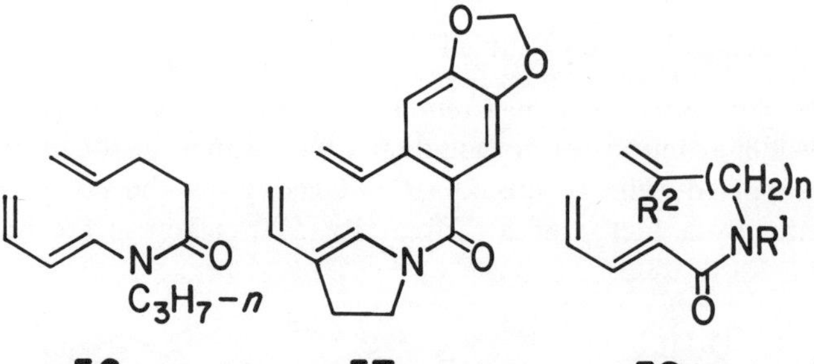

56 **57** **58**

whereas an almost identical substrate that lacks the two-carbon unit forming the pyrroline ring produces the *cis*-fused isomer exclusively (Table IIIb/7).[18]

Butadienecarboxamides **58** (n = 1 or 2; R^2 = H) cyclize to give mixtures of both stereoisomers in comparable yields (Tables IIb/16, IIIb/8). Introduction of a bulky group R^2 results in formation of the *cis* isomer exclusively (Table IIIb/9), possibly because of steric repulsion between R^2 and the diene in the *anti* transition state. The cyclization may also be thermodynamically controlled; alternatively, any *trans*-fused isomer formed may epimerize to the more stable *cis* isomer via the lactam enol. The latter is very likely the case in the formation of the *cis* products from the acylamidines **59** (Table IIb/18–21), which is believed to proceed via the *trans*-fused product **60**; direct formation of the *cis* isomer should have given the epimer of **61** (R^2 α instead of β).[106,107] The reason for the stereoselectivity of this cyclization is not obvious.[106]

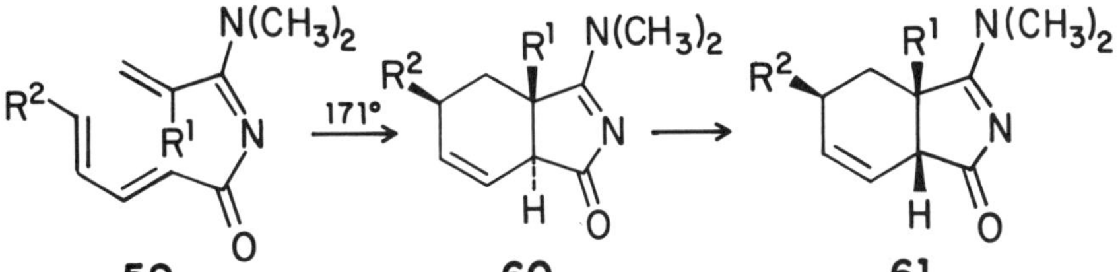

$N(CH_3)_2$ R² R¹ N 171° **59** R¹ $N(CH_3)_2$ R² N H O **60** R¹ $N(CH_3)_2$ R² N H O **61**

Substrates of type **62** where the amide is conjugated with neither the diene nor the dienophile give mixtures of *cis*- and *trans*-fused products in comparable amounts (Table IIIb/11, 13, 14).[101] The same is true of substrates **63** (R^3 = alkyl; n = 1) with acrylamide dienophiles that do not contain additional activating

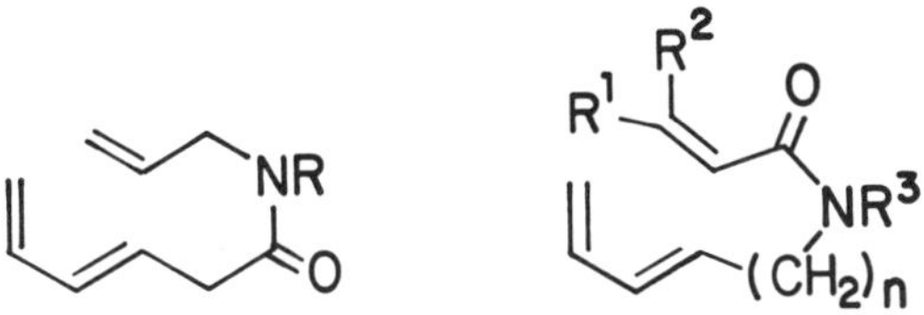

NR O **62** R² R¹ O NR^3 $(CH_2)_n$ **63**

groups R^1 or R^2 (Table IIb/22–27).[43,60] However, the corresponding system with a four-atom chain (**63**, R^1, R^2 = H; $n = 2$; Table IIIb/10a) shows significant *syn* selectivity (ratio of *cis*- to *trans*-fused product 7.3:1),[101] for reasons that are probably the same as for the all-carbon analog discussed earlier; a strain-free *syn* transition state with a coplanar acrylamide system can be achieved that permits secondary overlap of the carbonyl group with the diene.

Introduction of an ester group R^1 *trans* to the carboxamide function in **63** (n = 1) results in the essentially exclusive formation of *trans*-fused products (Table IIb/28–30, 33); when R^1 is phenyl (e.g., **64**) the *trans*-fused isomer is

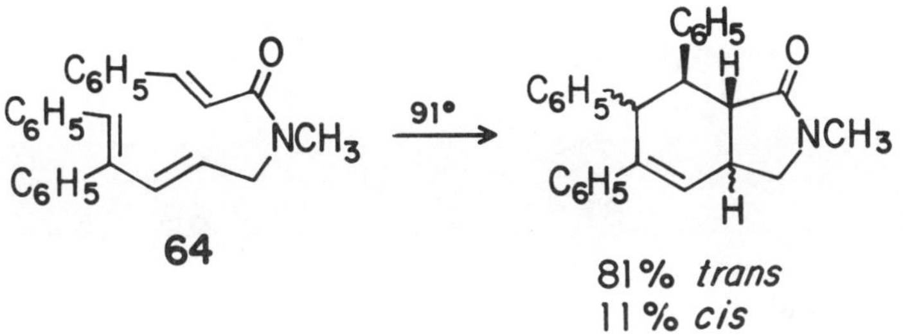

C_6H_5 C_6H_5 C_6H_5 O NCH_3 **64** 91° C_6H_5 H O C_6H_5 NCH_3 C_6H_5 H

81% *trans*
11% *cis*

still obtained predominantly (Table IIb/31, 32).[43,44,108] This is all the more remarkable since all examples reported have a phenyl group on the terminal carbon of the diene system, a situation that might be expected to destabilize the *anti* transition state. The observed stereoselectivity has been attributed to secondary orbital interaction of R^1 with the diene. Since this interaction was shown to be unimportant in the simple acrylate dienophiles (Section A above), it is likely that the presence of a second electron-withdrawing group on the other terminus of the dienophile is crucial in the present example.

Cyclization of a number of substrates **65** and **66** (n = 1 or 2) with enamide dienophiles has been reported (Tables IIb/35–37, 41–43; IIIb/16–19).[20,109–112] Curiously, substrate **65** (R^1 = phenyl) gives only the *cis*-fused product in 45% yield, whereas the closely related compound **65** (R^1 = 3, 4-methylenedioxyphenyl) gives a mixture of both isomers (18% *cis*; 28% *trans*). This result has been explained in terms of the unsymmetrical transition state model discussed earlier (p. 11). Intramolecular Diels–Alder reactions of the acyltetrahydropyridines **66** (n = 2) show little stereoselectivity; on the other hand, the *cis*-fused products are obtained exclusively in the case of the lower homologs **66** (n = 1).

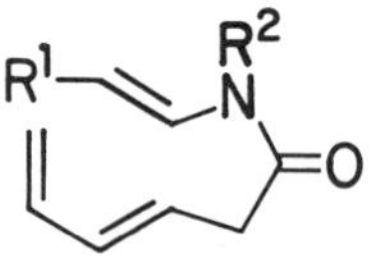

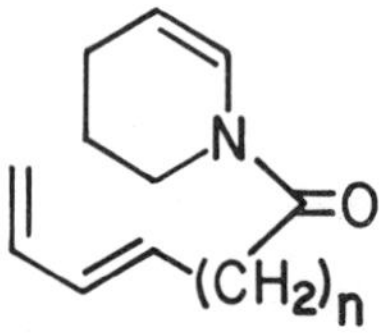

c. Acyclic Dienes with Chains Containing Oxygen. Acyclic dienes with ether oxygen in the chain appear to cyclize at temperatures comparable to those of the all-carbon analogs, but too few examples are known to permit generalizations about the stereochemical course. Ester functions in the connecting chain often have a detrimental effect on the rates of cyclization; nevertheless, a number of examples are known. Ester **67** cyclizes to give the *trans*-fused product predominantly (by a ratio of 4.6:1), whereas the isomer **68** (R = CH_3), in violation of the *endo* rule, gives the thermodynamically less stable *trans* product via the *anti* transition state exclusively (Table IIe/11, 13).[36]

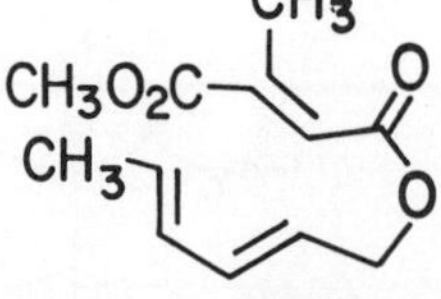

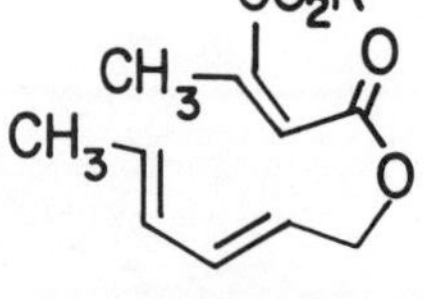

Other related esters show the same stereoselectivity (Table IIe/4a, 10, 12; cf. also Tables IIe/16, XV/20–23), a result that has been rationalized in terms of the unsymmetrical transition state model discussed earlier. By contrast, the acid **68** (R = H) produces only the *cis*-fused product (Table IIe/8) in a cyclization that is possibly thermodynamically controlled. Another example of an ester cyclizing via the *syn* transition state is given in Table VI/12.

The acylimines **69** (n = 1 or 2), generated by an acetate pyrolysis, cyclize to give only products **70**, independent of chain length (Tables IIf/16, 17; IIIb/21, 22).[113,114] Assuming a *trans* arrangement for the imine double bond

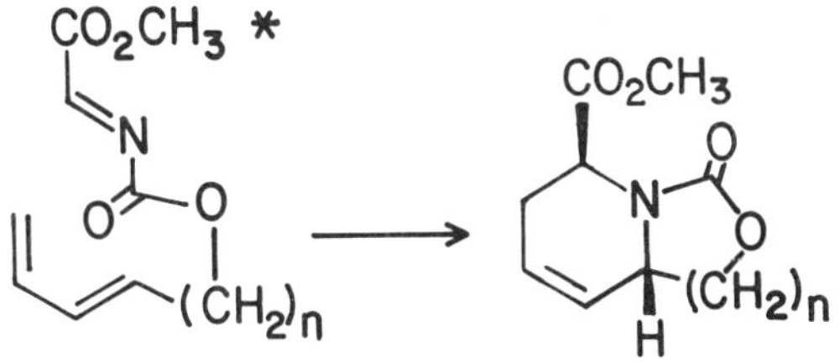

69 **70**

as shown,[113] the cyclization proceeds exclusively via the *syn* transition state; the same stereoselectivity is observed when the oxygen in the chain is replaced by a methylene group (Table IIf/15, 15a, 18–20a). A satisfactory explanation has not yet been proposed.

A few intramolecular Diels–Alder reactions of acyclic diene esters with 10- or 12-membered chains (Table IV/3–5) show the regio- and stereoselectivity of an intermolecular Diels–Alder reaction and are thus not pertinent to this discussion.

C. Cyclic Dienes. Cyclic dienes with chains not directly attached to the terminus of the diene (e.g., **71**) are *cis* dienes and as such can cyclize only through the *anti* transition state (see Section A);† therefore, they require no further discussion.

71

Cyclopentadienes with three- or four-atom chains attached to C-1 cyclize via the *anti* transition state exclusively, at least in those few examples where the stereochemistry of the products is known (Table VIIIa/2–5, 9, 11, 12). The

† This statement may not hold for very long chains or for large-ring dienes where at least one of the two double bonds may have the *trans* configuration; no such reactions have been reported to date.

number of documented examples, however, is too small to permit any generalizations.

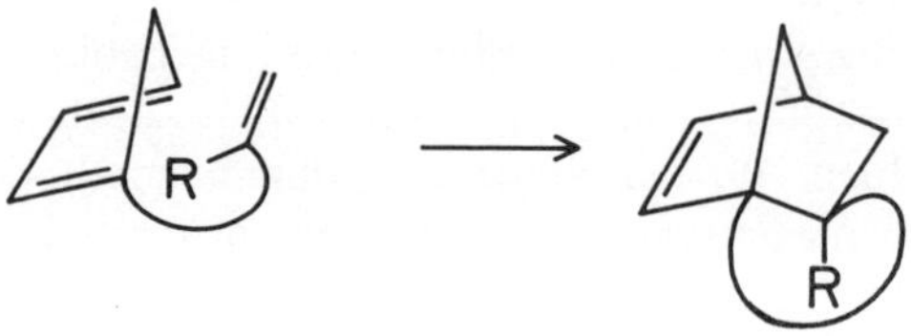

In the cyclization of cyclohexadienes **72** (R = H; X = H_2 or O), the products of *anti* addition predominate by a factor of 3 (Table IXa/1, 2);[77] however, introduction of two methyl groups R in **72** (X = O) causes the

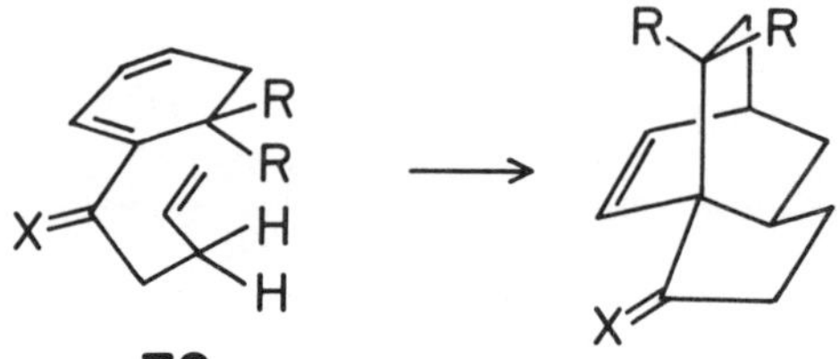

72

cyclization to proceed exclusively via the *syn* transition state (Table IXa/3),[115,116] presumably as a consequence of a nonbonded interaction between the methyl group and the allylic hydrogen. The cyclohexadiene **73** with a four-carbon chain and a dienophile activated by a carbonyl group also cyclizes preferentially via the *syn* transition state; the thermodynamically less stable *syn* product **74** predominates by a factor of 4 (Table IXa/8).[117] It will be

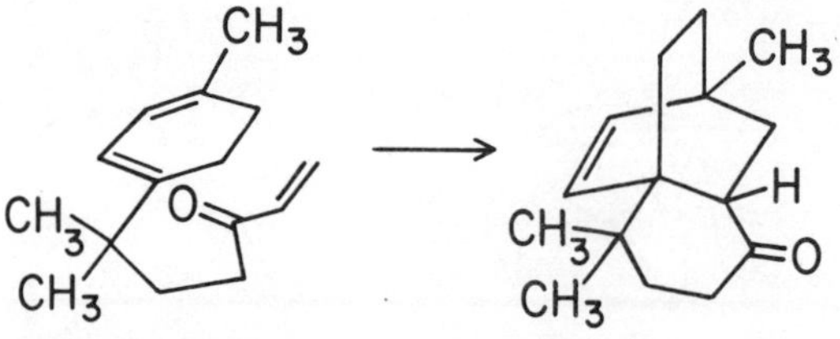

73 **74**

recalled that in the analogous acyclic case, the reaction is stereoselectively *syn*. Two other examples of cyclohexadienes that cyclize preferentially or exclusively via the *syn* transition state are listed in Table IXa/10 and 11.

In the single example where the stereochemistry of the product is known, a cycloheptadiene with a nonactivated dienophile cyclizes exclusively via the *syn* transition state (Table X/3).[118]

D. Vinylaromatics. Information about the stereoselectivity in this system is limited to a few examples. Amides of type **75** ($R^1 = CF_3CO$ or

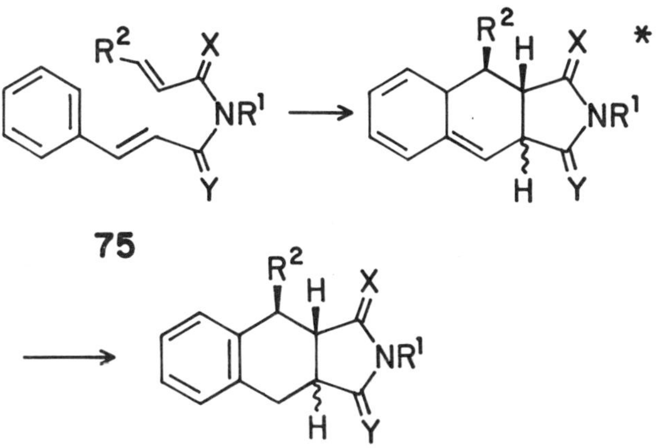

75

p-$CH_3C_6H_4SO_2$; $X = Y = H_2$) give predominantly the *trans*-fused products (Table XIVb/12, 14)[119] unless R^2 is a bulky group (e.g., phenyl), in which case the *cis*-fused products are formed predominantly or exclusively (Table XIVb/13, 15, 16).[119,120] However, it should be noted that the *cis*-cinnamic acid derivatives **76** (X = NH or O), which as *cis* dienes can cyclize only via the *anti* transition

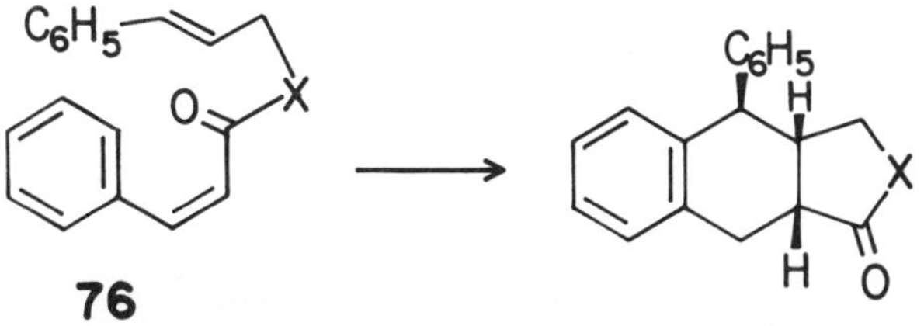

76

state to give the *cis*-fused product, do so in spite of the close proximity of the two phenyl groups (Tables XIVb/21, XIVf/27).[121,122] An example of a *trans* diene with a similar interaction is discussed below.

Amides **75** (R^1 = alkyl; $R^2 = C_2H_5O_2C$; X = O; $Y = H_2$) follow the Alder *endo* rule and cyclize to give the *trans*-fused products exclusively (Table XIVb/18, 19);[123] however, some of the *cis*-fused product is formed when the chain in such amides is four-membered (Table XIVb/26).[123] A number of cinnamamides (**75**, $X = H_2$; Y = O) undergo the intramolecular Diels–Alder reaction; the only reaction where the stereochemistry of the product has been determined involves a substrate with a bulky phenyl group R^2, which again results in exclusive formation of the *cis*-fused product (Table XIVb/25).[119]

Surprisingly, two esters of type **77** cyclize via the more crowded *syn* transition state, even though the *anti* transition state is accessible in this example (Table XIVf/28, 29).[124]

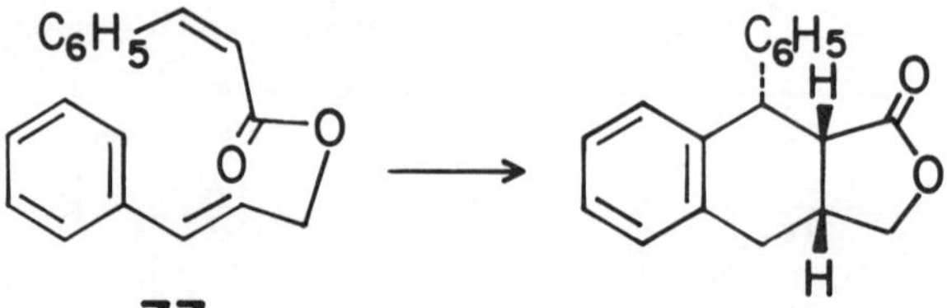

77

E. *o*-Quinodimethanes. *o*-Quinodimethanes are unstable intermediates that can be generated from a variety of precursors. For reasons that are discussed later, it is generally assumed that intramolecular Diels–Alder reactions of *o*-quinodimethanes usually proceed through the *trans* diene. It must be borne in mind, however, that any *cis*-fused product can arise, at least in principle, from the *cis* diene via the *anti* transition state.

cis DIENE; *anti* TS

trans DIENE; *syn* TS

trans DIENE; *anti* TS

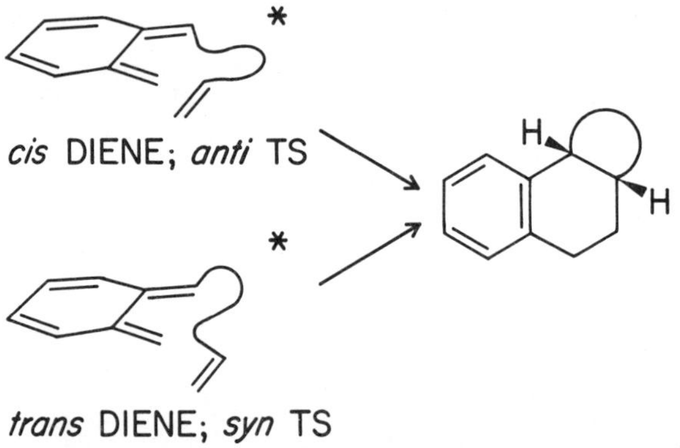

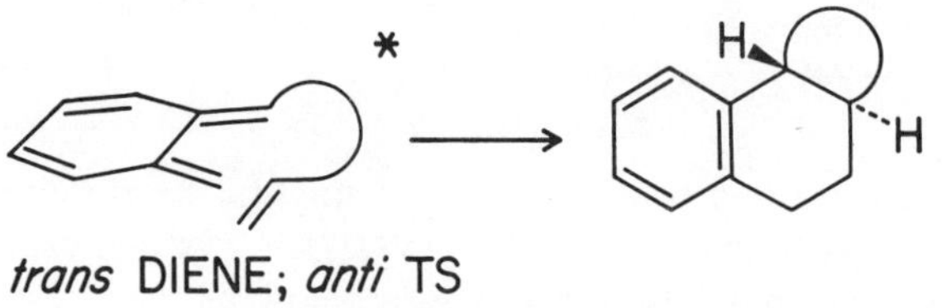

a. o-Quinodimethanes with All-Carbon Chains. *o*-Quinodimethanes with three-carbon chains not containing an sp^2 center appear to cyclize preferentially via the *anti* transition state to give *trans*-fused products. However, the stereochemistry of the products is known in only two examples, and these involve substrates containing additional substituents on the chain and the dienophile (Table XVIa/3, 5).[125] The simplest substrate (**78**, X = H_2) gives two isomers in the ratio of 3:1, but their stereochemistry has not been determined (Table XVIa/1).[126,127] Ketone **78** (X = O) cyclizes exclusively through the *syn* transition state to give the *cis*-fused product (Table XVIa/2; see also Table XVIa/7).[127,128]

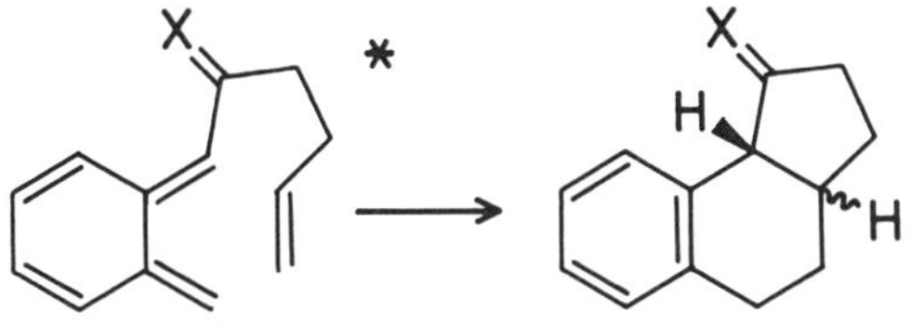

78

The majority of *o*-quinodimethanes with four-carbon bridges cyclize preferentially or exclusively via the *anti* transition state to give the *trans*-fused products, irrespective of any sp^2 centers in the chain or the substitution pattern of the chain or the dienophile (Table XVIa/10, 12–16, 18a, 21–23, 25–41, 43, 44, 48a, 49, 52–60). The selectivity has been attributed to steric repulsion, in the *syn* transition state, between the cyclohexadiene ring and bulky substituents in the chain (e.g., the cyclopentane ring in **79**). However, substrates with no substitution in the chain also give the *trans*-fused products predominantly. There are a few exceptions to the *anti* selectivity. Thus a number of substrates with cyano groups on the diene (**79**, R^1 = CN) furnish only the *cis*-fused

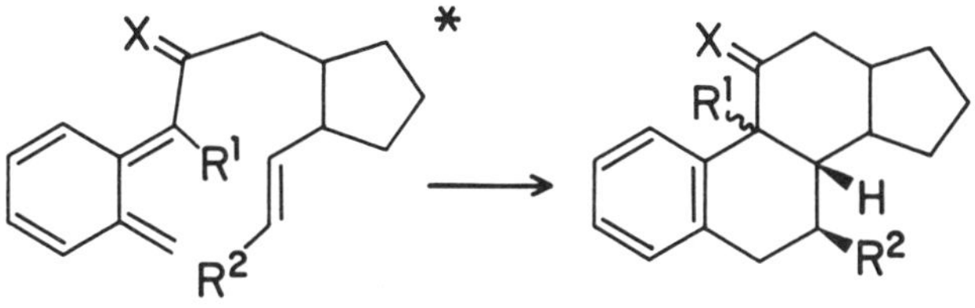

79

products (Table XVIa/19, 20, 24, 50, 51); the reason for this is not clear, and the picture is further confused by the fact that some very similar substrates give only the *trans*-fused products (Table XVIa/57–60). A ketone of type **79** (X = O) gives the *trans*-fused product predominantly (Table XVIa/29).[129] but the corresponding benzyloxime (X = $C_6H_5CH_2ON$) furnishes only the *cis*-fused isomer (Table XVIa/42).[130] Another way of favoring formation of the *cis*-fused product involves introduction of a bulky substituent (e.g., R^2 = *p*-$CH_3C_6H_4SO_2$ in **79**) on the dienophile (Table XVIa/45, 46).[131]

Only one *o*-quinodimethane with a five-carbon chain has been reported to undergo the intramolecular Diels–Alder reaction (Table XVIa/62); thus no generalizations can be made for this system.

b. o-Quinodimethanes with Heteroatom-Containing Chains. *o*-Quinodimethanes with three-membered nitrogen-containing chains appear to cyclize preferentially via the *syn* transition state to give *cis*-fused products regardless of the location of the nitrogen atom or the presence of carbonyl groups in the chain (Table XVIb/1, 6, 11, 12, 14, 15; the single exception is entry 13). The situation is different for substrates with four-atom chains. Amines (e.g., **80**,

X = Y = H_2) seem to give *trans*-fused products preferentially via the *anti* transition state (Table XVIb/22, 23, 26–28, 30, 32), whereas amides (e.g., **80**, X or Y = O) seem to prefer to cyclize via the *syn* transition state (Table XVIb/24,

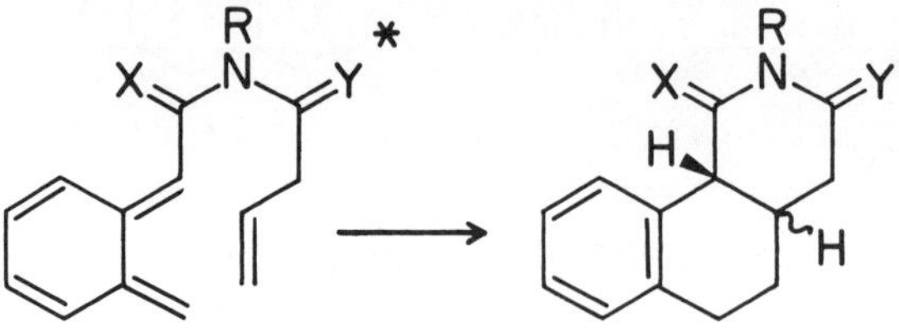

80

25, 29, 33; see also entry 34 for a substrate with a five-atom chain). The situation is reversed when the nitrogen is attached directly to the diene (Table XVIb/19–21).

The number of *o*-quinodimethanes with oxygen-containing chains reported to undergo the intramolecular Diels–Alder reaction (Table XVIc) is too small to permit any generalizations.

F. Furans. The stereoselectivity of many intramolecular Diels–Alder additions involving furan as the diene is unknown, but in all reactions where it has been determined the *anti* products are formed exclusively. Thus ketones **81** (X = $(CH_2)_2$; Table XVIIIa/2), amides **81** (X = NR; Table XVIIIb/34–75), and esters **81** (X = O; Table XVIIId/8–11) all cyclize stereospecifically in violation of the Alder *endo* rule. In view of the well-known propensity of furan adducts to undergo the retro Diels–Alder reaction, the observed selectivity may be a consequence of thermodynamic control.

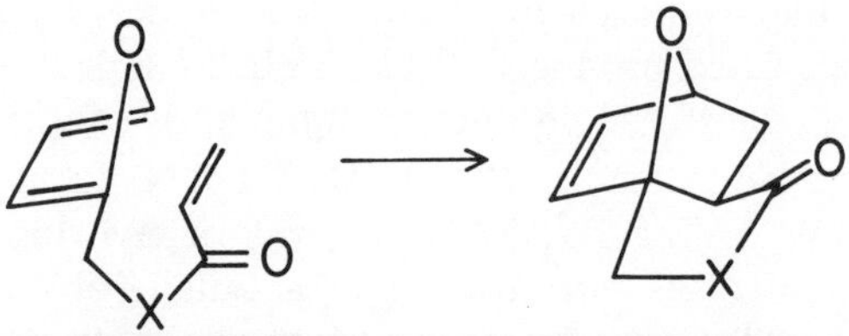

81

G. Conclusions. The above discussion is intended to be of help in predicting the stereoselectivity of a contemplated intramolecular Diels–Alder reaction. It is clear that the stereochemical outcome is often determined by subtle interplay of a number of factors such as nonbonded interactions and electronic effects. Inspection of molecular models can be of some help. A more quantitative insight into the energy differences between the various possible transition states can be gained from molecular mechanics computations.[75]

Another approach to predicting the lowest energy transition state consists of searching the Cambridge Crystallographic Data File for substructures that have the same salient features.[132] If the desired product is the thermodynamically more stable isomer, the obvious solution is to carry along a functionality, such as a carbonyl group, that permits epimerization of the less stable to the more stable isomer. In certain cases the decision as to the stereochemistry of the bicyclic system can be postponed to a later step in a synthetic sequence by proper choice of dienophile or diene. For instance, use of an acetylenic dienophile (Eq. 12) or one of its synthons, such as a vinyl chloride derivative, gives a

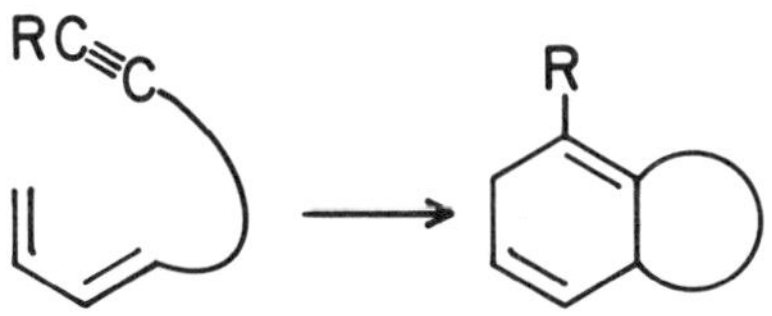

(Eq. 12)

product in which the desired stereochemistry is then introduced by chemical or catalytic reduction.[133] Products with a double bond at the ring junction can also be obtained by using 1,3,4-trienes (vinylallenes) as the dienes (e.g., Eq. 11). Alternatively, choice of a diene such as an α-pyrone (Table XXIV) or a pyridazine derivative permits facile expulsion of a stable molecule (e.g.,

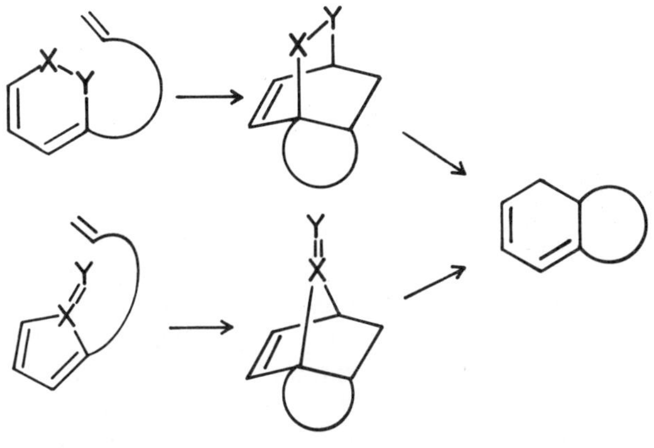

(Eq. 13)

carbon dioxide or nitrogen) from the initial adduct (Eq. 13); the diene so formed is then converted to the bicyclic system with the proper stereochemistry by reduction[134] or other manipulations. An intermediate amenable to this approach can also be obtained by double-bond isomerization (Eq. 14).

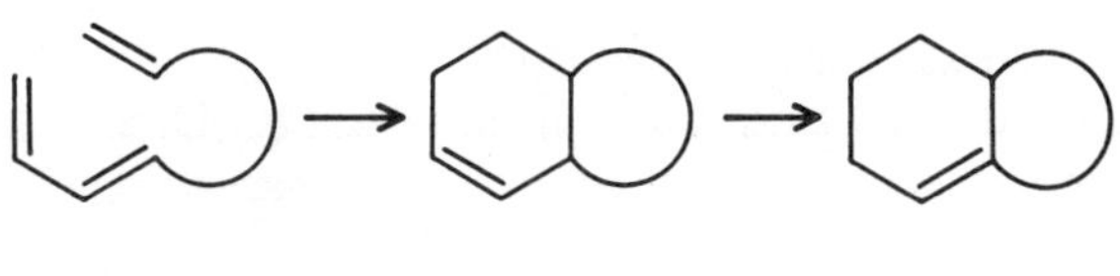

(Eq. 14)

Diastereoselectivity

If the diene or dienophile carries a substituent with one or more asymmetric centers, or if the chain is unsymmetrically substituted, the intramolecular Diels–Alder reaction may lead to mixtures of diastereomeric products. The term "asymmetric induction" has been used to describe cyclizations of chiral substrates that lead to products with two or more new chiral centers. The generation of these new chiral centers is, however, a consequence of a stereo- and diastereoselective cyclization. A true case of asymmetric induction, or enantioselectivity, would be the preferential or exclusive formation of one enantiomer by carrying out the intramolecular Diels–Alder reaction in a chiral solvent or by using a chiral catalyst. Attempts to effect the latter have been unsuccessful so far.[41]

A. Dienes with Asymmetric Centers. Intramolecular Diels–Alder reaction of diene **82**, generated from its SO_2 adduct, gives two diastereomers in a ratio of 5:4;[135,136] the transition state leading to one of the diastereomers is shown in **82**; in the other transition state (both are *syn*), the dienophile adds to the diene from above. Obviously there is little energy difference between the two, and even less would be expected in the *anti* transition state. No other examples of this type of stereoselectivity appear to have been reported.

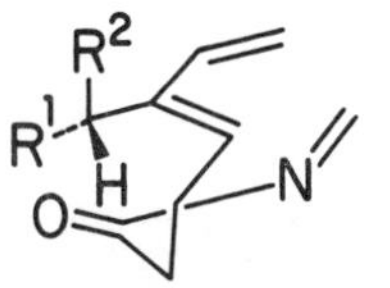

82

$R^1 = n\text{-}C_3H_7$; $R^2 = (CH_3)_3SiO$

B. Dienophiles with Asymmetric Centers. Lewis-acid-catalyzed cyclization of the esters **83**, whose alcohol portion is derived from (−)-phenylmenthol, gives diastereomeric mixtures of *anti* adducts.[41] The ratio **84**:**85** and the overall yield depend on the catalyst used. Cyclization of the unsubstituted diene **83** (R^2 = H) with *l*-bornyloxyaluminum dichloride at 8° for 14 days gives the adducts **84** and **85** (ratio 86:14) in 72% yield. With the use of the same catalyst, the substituted diene **83** ($R^2 = i\text{-}C_3H_7$) is converted to the adducts **84** and **85** (ratio 65:35) in 75% yield; a higher diastereoselectivity (ratio **84**:**85** = 86:14) can be achieved with titanium tetrachloride as the catalyst, but the yield drops to 8%. *d,l*-Menthyloxyaluminum dichloride gives the same ratio of products **84**:**85** as the *l* isomer.[41] The absolute configurations of the major products **84** (R = H or $i\text{-}C_3H_7$) have been determined.[41]

A completely diastereoselective intramolecular Diels–Alder reaction involving a dienophile carrying a chiral center has been reported (Table

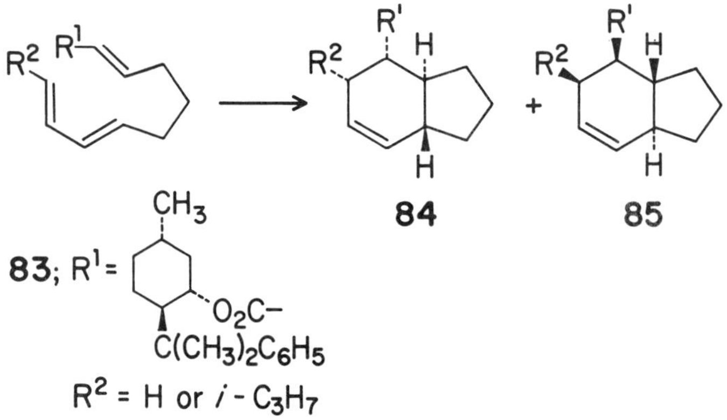

11e/16b);[137] in view of the moderate yield (40%), the possibility exists that the other diastereomer escaped detection.

C. Asymmetric Centers in the Chain. There are many examples of intramolecular Diels–Alder reactions of substrates containing unsymmetrically substituted chains. A simple example is illustrated in Eqs. 15 and 16; only one

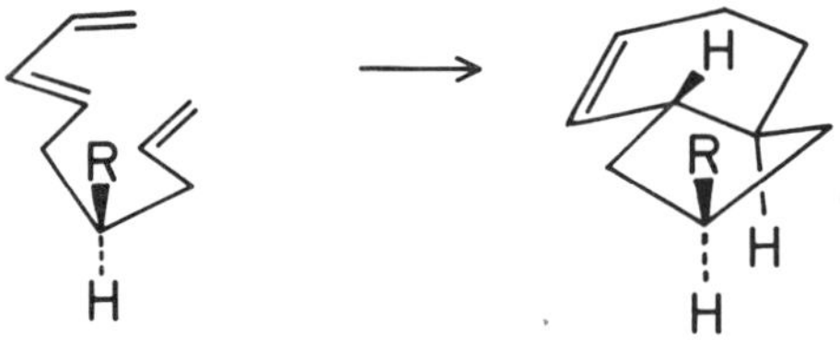

(Eq. 15)

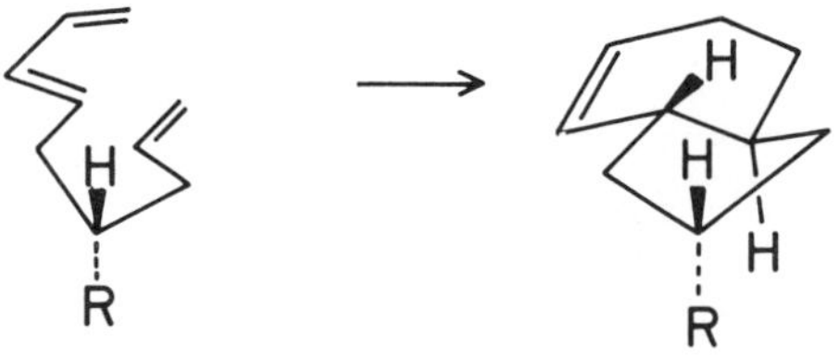

(Eq. 16)

of the two possible enantiomers is shown for each substrate and product. The degree of diastereoselectivity in cyclizations involving such substrates depends on a number of factors such as reaction temperature, chain length, bulk of the substituents R, location of the substituents R relative to the diene and dienophile, and transition state geometry (*syn* or *anti*). Generalizations are difficult to make; the observed diastereoselectivities have usually been rationalized in terms of non-bonded interaction of the substituents R with the diene, dienophile, or the chain as deduced from inspections of models[38,71,86,92,138–141] or molecular mechanics calculations.[75] Bulky substituents do not necessarily ensure diastereoselectivity. Thus ester **86** on heating to 150° furnishes a mixture of all four possible products: the *cis*-fused products **87** (α-isomer, 22%; β-isomer, 29%) and the *trans*-fused products **88** (α-isomer 23%; β-isomer 13%).[93]

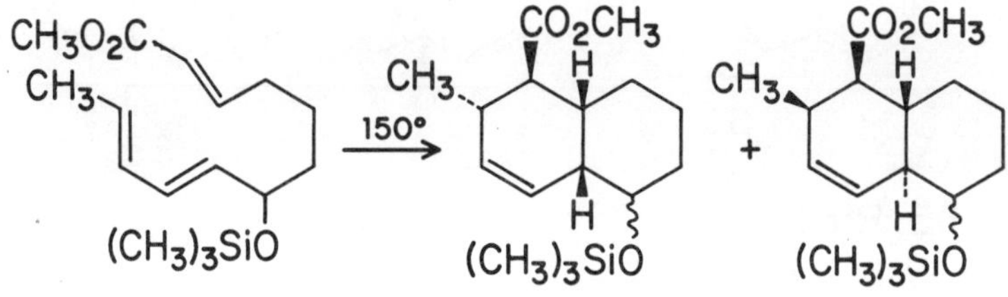

The lack of diastereoselectivity in some cyclizations can be attributed to an early, reactant-like transition state.[142] On the other hand, a bulky substituent in combination with a low reaction temperature can result in pronounced diastereoselectivity. The unprotected hydroxyester **89** (R = H) gives a mixture of diastereomers in both the thermal and catalyzed cyclizations; the same is

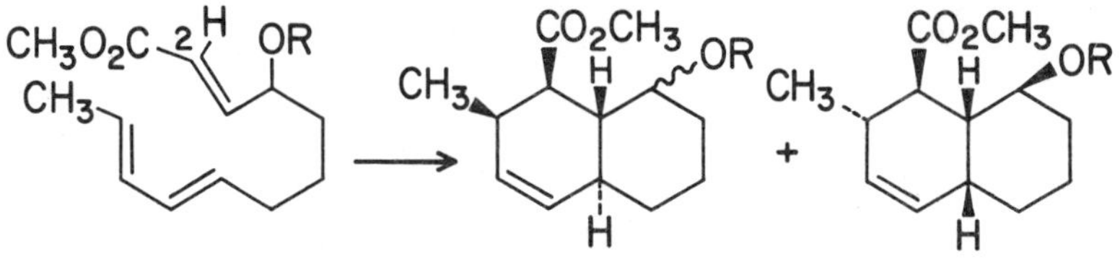

R		90, α-OR	90, β-OR	91
H	155°	6	22	25
H	$C_2H_5AlCl_2$, 25°	30	25	<1
t-$C_4H_9(CH_3)_2Si$	140°	42	8	14
t-$C_4H_9(CH_3)_2Si$	$C_2H_5AlCl_2$, 25°	65–73	0	0

true of the thermal reaction of the siloxy ester **89** [R = t-$C_4H_9(CH_3)_2Si$]; however, the Lewis-acid-catalyzed intramolecular Diels–Alder reaction of the latter is highly diastereo- and stereoselective.[92] This diastereoselectivity has been attributed to an unfavorable steric interaction between the siloxy group and H-2 in the transition state leading to the other diastereomer.[92]

The observation of a moderate solvent dependency of the diastereoselectivity in an intramolecular Diels–Alder reaction was mentioned previously (p. 14).

For thermal intramolecular Diels–Alder reactions that proceed diastereoselectively (ratio 3:1 or better), see Tables IIa/21, 25; IIb/34, 34a; IIIa/17b, 19, 20a, 26, 27 (also refs, 143 and 144); IIIb/5, 6, 9a, 21–23; XV/1, 12; XVIa/5, 19, 20; and XIX/18a–21. Examples exhibiting little or no diastereoselectivity are found in Tables IIa/1, 29; IIf/9–12, 17; IIIa/1, 1a, 5, 6, 14, 16, 17; VIIIa/3–5; IXd/13, 14; XV/13; and XVIa/3. In the following examples, the cyclization is diastereoselective in one stereochemical mode (*syn* or *anti*) but not in the other (see Tables IIa/28, 30, 33–35; IIIa/17a).

Substrates with rings incorporated in the chain usually cyclize to give one diastereomer preferentially or exclusively. Thus the piperidine derivative **92**

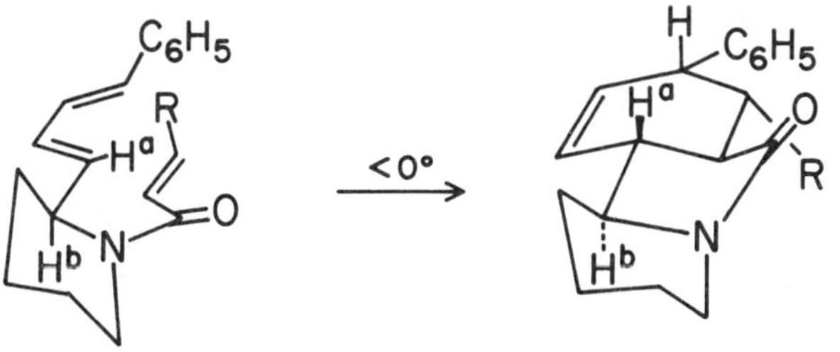

92

R = $C_2H_5O_2C$

reacts below 0° to give the *anti* product predominantly as a single diastereomer in 70% yield.[108] Models indicate that in the transition state leading to the other diastereomer (H^a,H^b *cis*; formed in 1–3% yield), the chain requires considerable distortion from the normal bond angles. For other diastereoselective cyclizations of this type, see Tables IIe/17; IIf/2, 3; IIIa/28; XVIa/22–60; XVIIIa/2a; XVIIId/1a; and XX/6.

Similarly, in certain reactions where the dienophile, or all or part of the diene is incorporated in a ring, only one of the two transition states is strain free, resulting in diastereoselective cyclization. Two examples[145,127] are given in Eqs. 17 and 18; for other examples of this type, see Tables VI/7–12; IXa/11; IXe/12–15; XVIa/6, 61; XVIb/14, 15; XIX/6, 22–24; XX/7–9, 11–24; and

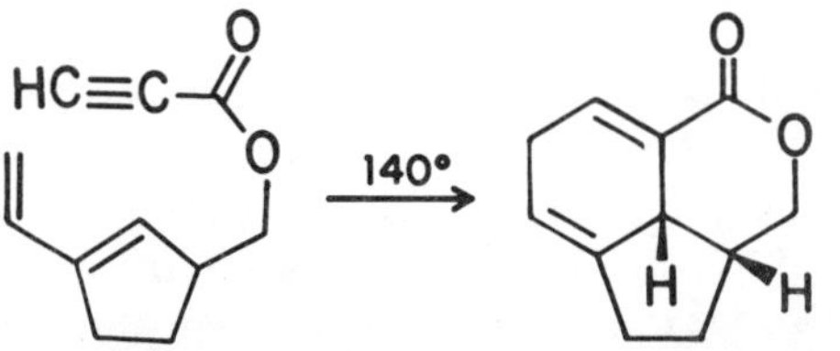

(Eq. 17)

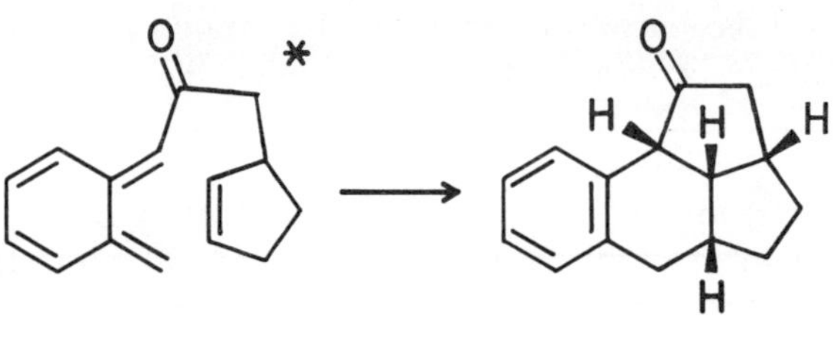

(Eq. 18)

XXV/3 and Ref. 146. All examples listed above involve substrates with three- or four-membered chains; models indicate that for longer chains such reactions may no longer be diastereoselective.

The intramolecular Diels–Alder reaction of the furan derivative **93** (R = H), in which the nitrogen side chain is derived from (−)-phenylglycinol, gives the two diastereomers **94** in equal amounts; however, cyclization of the corresponding magnesium salts **93** (R = MgCl, MgBr, or MgI) leads to mixtures in which diastereomer **94a** predominates (ratios 4:1–7:1). This selectivity has

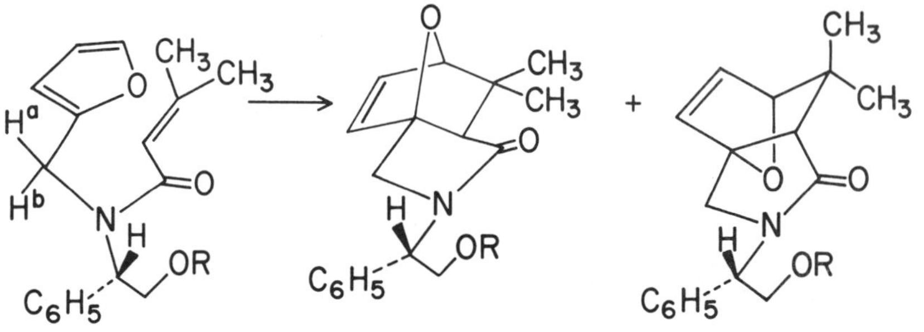

93 → **94a** + **94b**

been attributed to chelation between the magnesium and the carbonyl group, which results in a nonbonded interaction between the phenyl group and the two methylene protons H^a and H^b in one of the two transition states.[69]

SCOPE AND LIMITATIONS

The Chain

A. Chain Length. The length of the chain connecting diene and dienophile has considerable influence on the rate of the intramolecular Diels–Alder reaction. As mentioned earlier, *trans* dienes cyclize only if the chain is three-membered or longer. Entropy considerations alone would suggest that the rates should decrease in the order 3- > 4- > 5-membered chains. Indeed, this occurs in a number of systems; thus in a series of 9-anthracenecarboxamides containing ethynyl dienophiles, the cyclization temperatures for the substrates with three-, four-, and five-membered chains are 110°, 140°, and 220°, respectively (Table XVIIc/20, 44, 46).[45] The temperature required for the last reaction is similar to that of the corresponding intermolecular Diels–Alder reaction. This indicates that the entropy advantage of the intramolecular reaction nearly vanishes in substrates with five-atom chains (unless the degrees of freedom in the chain are reduced, for instance, by bulky substituents or incorporation into a ring). In fact, relatively few reactions involving substrates with five-atom chains have been reported, and none are known that involve substrates with seven- to nine-membered chains. In the only successful cyclization of a substrate with a six-atom chain (Eq. 18a),[45a] the latter is highly rigidized by

$(CH_3)_2AlCl$, C_6H_6, 25°

(86 %)

(Eq. 18a)

incorporation into a bicyclic framework (for attempted cyclizations of substrates with six-atom chains, see Tables IV/2, 2a, 2b; XV/35). The reluctance to form medium-sized (8- to 11-membered) rings is a well-known phenomenon in organic chemistry. A few intramolecular Diels–Alder reactions of substrates with 10- and 12-membered chains are known, but these resemble the intermolecular cyclization in their stereo- and regiochemical outcome.

Substrates containing *cis* dienes can undergo the intramolecular Diels–Alder reaction even when the chains contain only one or two atoms. Indeed, cyclizations of the former type are quite common (see list at the end of this section) and may proceed under very mild conditions in favorable cases (e.g., Eq. 19).[147]

(Eq. 19)

Cyclizations of substrates with two-atom chains, on the other hand, do not proceed readily, as indicated by the example of the cyclohexadienes **95**:[148]

95

n	Temperature	Conversion (%)
0	158°	64
1	158°	0
	225°	44
2	158°	91

The vast majority of known intramolecular Diels–Alder reactions involves substrates with three- and four-atom chains. These two types of substrates often cyclize at comparable temperatures. Although a substrate with a four-membered chain has to overcome a less favorable entropy factor, its cyclohexane-like transition state suffers fewer nonbonded interactions. Thus the esters **96** (n = 3 or 4) cyclize under essentially identical conditions:[40,93]

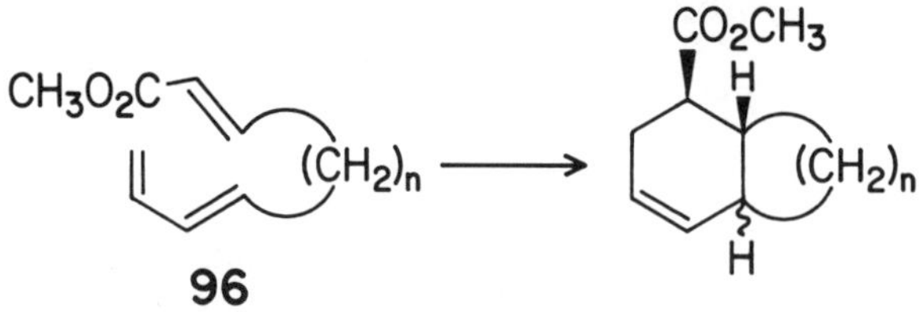

n	Temperature	Time	Yield (%) Cis	Trans
3	150°	24 hr	26	39
4	155°	45 hr	45	47

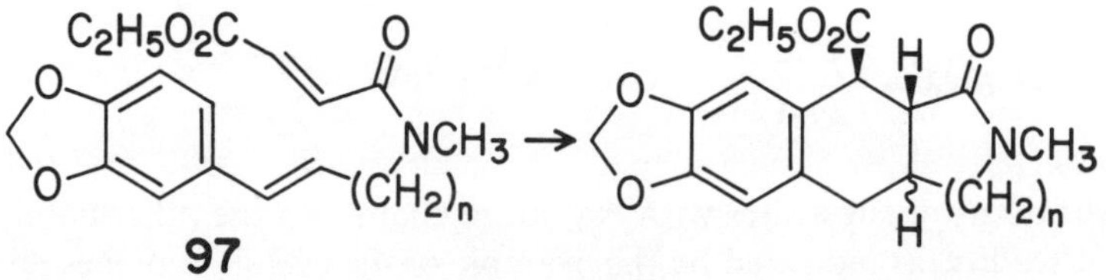

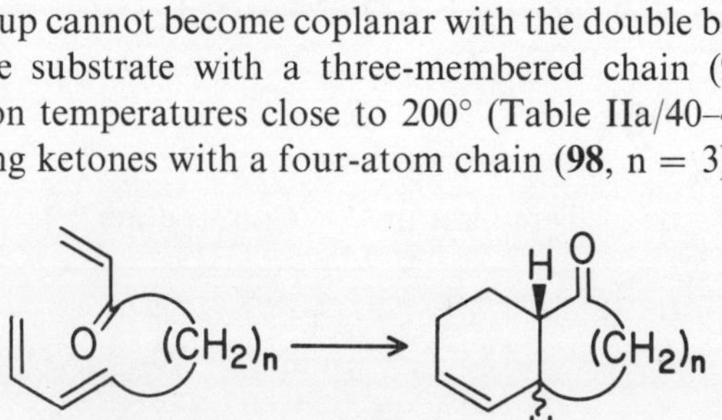

Introduction of one or two sp^2 centers into the chain reduces the nonbonded interactions in a three-atom chain, resulting in a considerable difference in the rate of cyclization. Thus the amides **97** cyclize at 110° (n = 1) and 170° (n = 2), respectively.[123] A similar temperature difference is observed in a series of naphthalenecarboxamides (Table XVIIb/7, 12–15). An important exception is the intramolecular Diels–Alder reaction of vinyl ketones. As discussed earlier, the carbonyl group cannot become coplanar with the double bond in the transition state of the substrate with a three-membered chain (**98**, n = 2), thus requiring reaction temperatures close to 200° (Table IIa/40–45). By contrast, the corresponding ketones with a four-atom chain (**98**, n = 3) cyclize close to, or below, room temperature (Table IIIa/21–27). Even substrates with a five-membered chain (**98**, n = 4) require temperatures of only about 100° (Table IV/1, 1a; the former reaction is Lewis-acid catalyzed).

There are a number of reactions in which substrates with different chain lengths are interrelated by hydride shifts or other sigmatropic rearrangements; the species that undergoes the intramolecular Diels–Alder reaction is the one that has the most favorable chain length, as illustrated in Eq. 20.[60] For other

CH_3 CH_3 CH_3 $CO_2C_2H_5$ 260° CH_3 CH_3 CH_3 $CO_2C_2H_5$

CH_3 CH_3 $C_2H_5O_2C$ H CH_3

(Eq. 20)

examples, see Table IIIb/20a, 20b; VIIIa/2–5, 7–12; VIIIb/2–9a; IXa/3; X/15–17a; XXI/12, 13; and XXII/4c, 19, 20, 25.

The following is a listing of all intramolecular Diels–Alder reactions of substrates containing other than three- or four-membered chains:

1. *Trans* Dienes
   a. Five-atom chains: Tables IV/1, 1a; XV/18; XVIa/62; XVIb/34; XVIc/8; XVIIc/46; XVIIIa/13, 15
   b. Six-atom chains: Table IV/2c
   c. Ten-atom chains: Table IV/3, 4
   d. Twelve-atom chains: Table IV/5 and Ref. 151a
2. *Cis* Dienes
   a. One-atom chains: Tables I/1; VIIIb/1, 1a; IXb/1–37; X/5–12; XI/1–9; XII/1–37; XIII/1; XIX/1–3; XXI/1–3
   b. Two-atom chains: Tables IXc/1–4; X/13, 14; XIII/2, 3, 14; XXI/4–6; XXV/8–14, 31
   c. Five-atom chains: Tables VIIIb/20; XVIIc/47; XXI/16; XXV/2

B. Point of Attachment of the Chain to the Diene. Cyclization of substrates in which the chain is attached to C-2 of the diene leads to products containing bridgehead double bonds. Nevertheless, a number of such intramolecular Diels–Alder reactions are known, with chain lengths (in **99**) varying from three to six. Surprisingly, the cyclization of **99** (n = 3) is exothermic by 18 kcal/mol.[149]

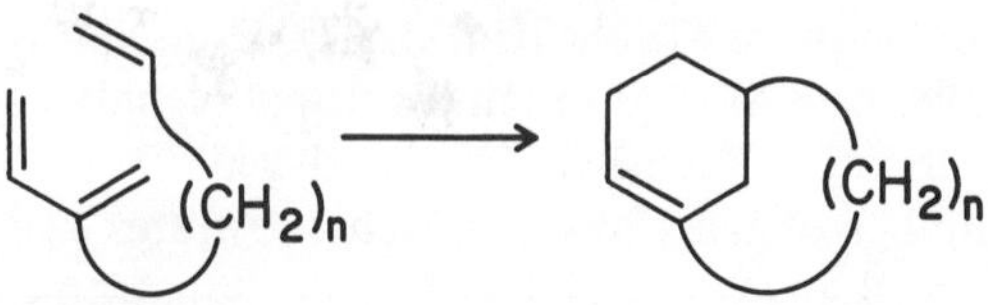

$(CH_2)_n$ $(CH_2)_n$

99

As might be expected, some of these cyclizations are reversible, but respectable yields are obtained in many reactions (Table V/1–17). Another example is the cyclization of the isoquinolinium salt **100**[150,151] (see also Table XXV/5b).

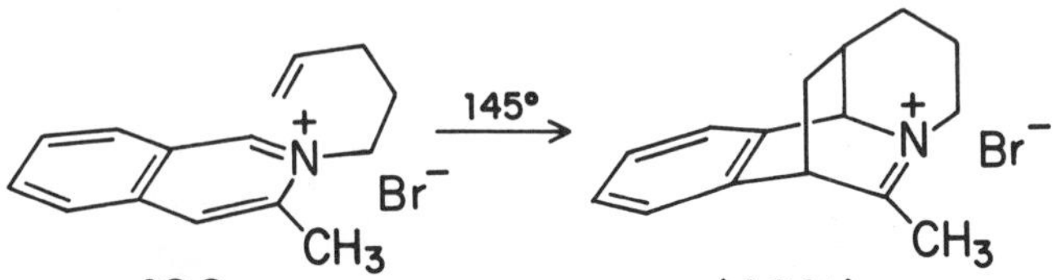

145°

100 (46%)

C. Double Bonds and Rings in the Chain. The chain may contain a double bond as long as it is *cis*, or can become *cis* during the reaction. This requirement does not hold for long chains; successful cyclization of a substrate with a *trans* double bond in a twelve-membered chain is reported.[151a] Thus the cyclohexadienone **101** gives the fused and bridged adducts in 15% and 45% yields, respectively.[152] For other examples, see Tables IIe/19; IXe/3, 4; XXI/14;

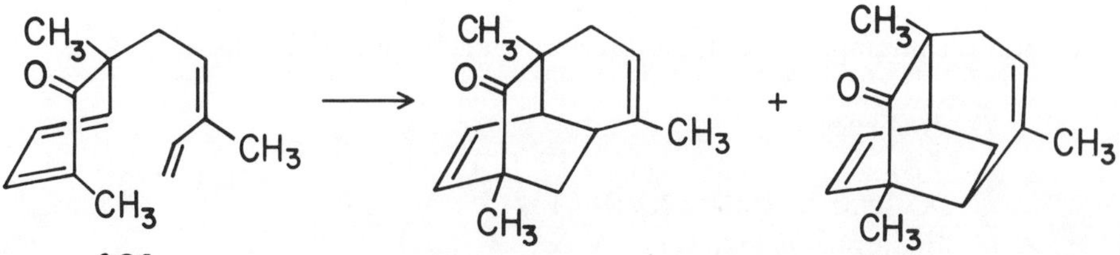

101

and XXV/7. A special case is the cyclization of 1,3,4-trienes, such as the allenylbenzene derivative **102**[153] generated by base-catalyzed isomerization of the corresponding phenylacetylene (see also Eq. 11). For other cyclizations of this type, see Tables IId/1–60; IIe/1; VI/1–6; XIVa/1; XIVb/1; XIVd/1–48; XIVe/1–22; XIVf/1–6, 25, 26; and XIVg/1.

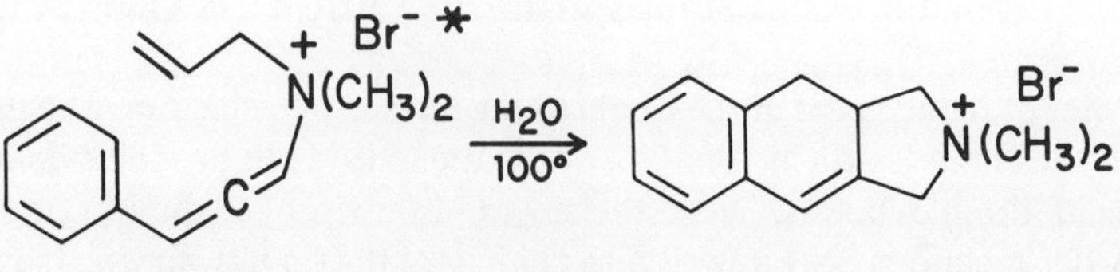

H_2O, 100°

102

The imine **103** presumably exists in the *transoid* form but is able to isomerize to the *cisoid* form, which then cyclizes to give the adduct in 82 % yield.[45] Other substrates with carbon-nitrogen double bonds in the chain are listed in Tables IIb/18–21; and XVIIc/6–8, 22–24.

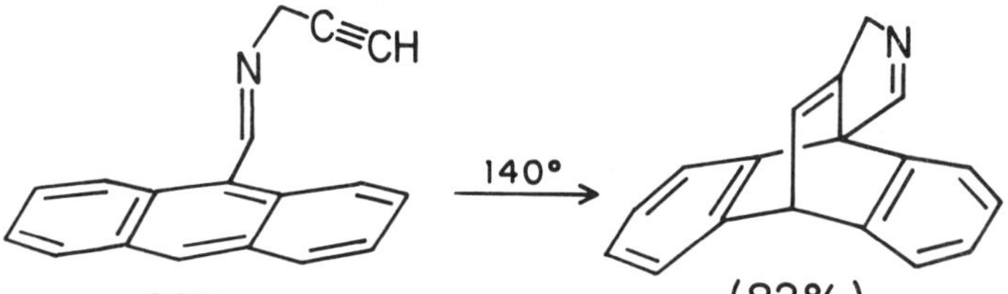

C≡CH
N
140°
N
103
(82%)

Incorporation of rings into the chain can lead to a dramatic rate increase as evidenced by comparison of substrates **104a** and **104b**.[123] In the oxazole

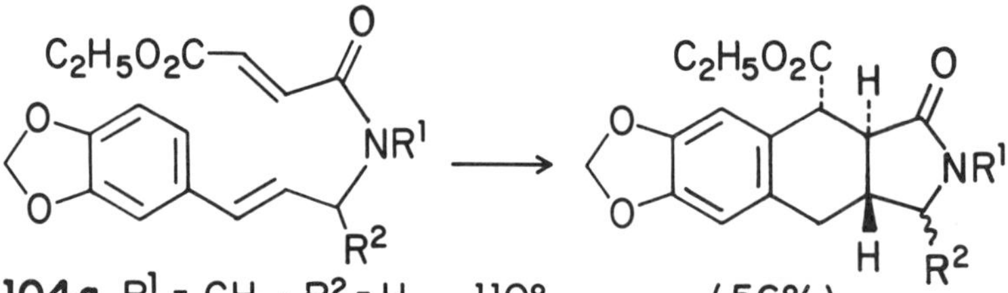

$C_2H_5O_2C$
O
O
O
NR^1
R^2
$C_2H_5O_2C$
H
O
O
O
NR^1
H
R^2

104a $R^1 = CH_3$; $R^2 = H$ 110° (56%)
104b $R^1\text{-}R^2 = (CH_2)_4$ 0° (37%)

derivatives **105**, on the other hand, incorporation of a cyclohexane ring, whether *cis* or *trans*, has no influence on the rate of the intramolecular Diels–Alder reaction (the corresponding substrate lacking the cyclohexane ring also

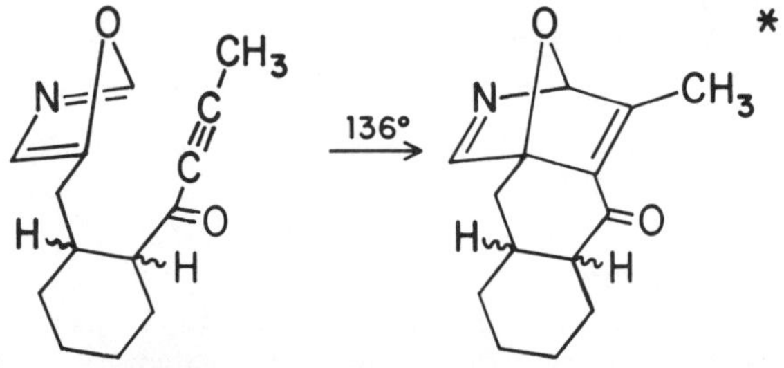

O
N
CH_3
C
C
O
H
H
136°
O
*
N
CH_3
O
H
H
105

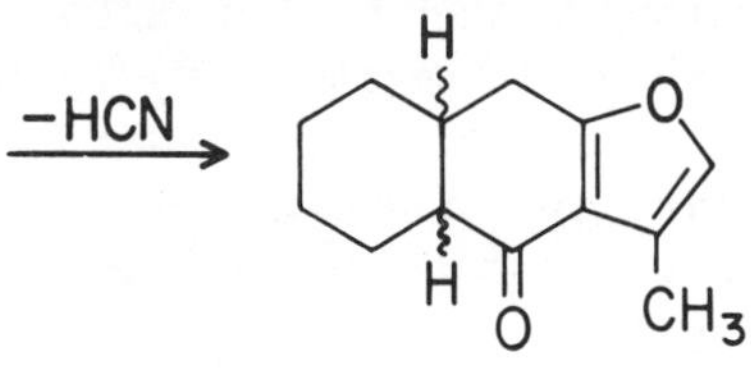

−HCN
H
O
H
O
CH_3

cyclizes at 136°).[154] For other substrates incorporating nonaromatic rings in the chain, see Tables IIa/1d, 2, 44, 45; IIb/33; IIIa/28, 29; VIIIb/18, 19 (epoxides); XVIa/26–46, 48a–60; XVIIc/13, 30; XVIIIa/2a; XVIIId/1a; and XXV/3, 26–29 and Ref. 146. In the following examples, the chains are connected through the *meta* positions of the ring: Tables IIe/17; IV/2c; XVIa/22–26, 47, 48; and Refs. 154a,b.

In the only example where a direct comparison is possible, incorporation of a benzene ring into the chain considerably increases the rate of the intramolecular Diels–Alder reaction. The substrates **106**[155,156] and **107**[96] cyclize

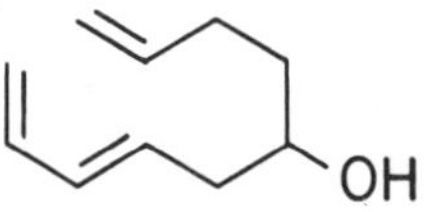

OH

CH_3 OH

106 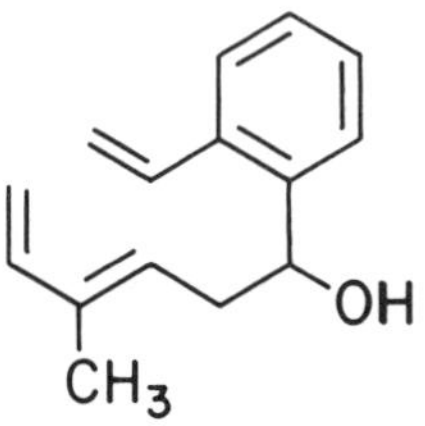 **107**

at 160° and 100°, respectively. Consideration of the other known examples (see below) permits the prediction that incorporation of an aromatic ring in general will not be detrimental to the success of an intramolecular Diels–Alder reaction.

For other substrates with aromatic rings in the chain, see Tables IIf/20b; IIIa/31–35 (the latter contains an indole connected at C-3 and C-4); IIIb/7, 15, 18, 19; IXa/10; IXd/15; XV/16–18 (the latter contains an indole connected at C-2 and C-3); XVIa/8; XVIb/18–21; XVIc/4; XVIIb/16–20; XVIIc/38–42; XVIId/1; XIX/25–31; XXI/16 (indole connected at C-2 and C-3); XXII/4–58; XXIII/12–23; XXV/4b, 25a; and XXVIc/25a, 25b.

D. Heteroatoms in the Chain

a. Amines and Quaternary Ammonium Salts. Substitution of a methylene group in the chain with a tertiary nitrogen does not seem to significantly affect the rate of the intramolecular Diels–Alder reaction. Thus the substrates **108a**[95] and **108b**[101] cyclize at 160° and 140°, respectively (see also Table IIb/15 and the anthracenes in Table XVIIc/2 and 15). Secondary amines appear to pose no problem (Tables XVIb/22, 23, 26–28, 30, 32; XVIIb/16; XVIIc/9, 25). If necessary, they can be protected with the easily removed trifluoroacetyl group.[48,119,120,157] Bulky groups on nitrogen usually increase the rate of the intramolecular Diels–Alder reaction slightly. Protonation and quaternization of a nitrogen atom in the chain also accelerate the cyclization. Hence the tertiary amines **109** do not undergo the intramolecular Diels–Alder reaction, whereas the corresponding protonated amines and the quaternary ammonium salts do.[52,158] Cyclization of the urethane **108c** requires heating to 275°, 135° higher than the tertiary amine **108b**,[101] but this may be an exception since other

108a $X = CH_2$; $R = CH_3$
b $X = NCH_3$; $R = H$
c $X = NCO_2CH_3$; $R = H$

109

urethanes do not exhibit this unusual behavior (Tables IIb/1; XVIb/11, 20; XVId/1, 2, 6, 7, 9–11, 13, 14, 19). There are many examples of the cyclization of *N*-acylamincs (e.g., Tables IIb/2–14; XVIIIb/11–20); the rate in a series of *N*-acyl-*N*-allyl-2,4-pentadienylamines increases in the order of CH_3CO < CF_3CO < $ClCH_2CO$.[48]

b. Amides. There are many examples of intramolecular Diels–Alder reactions of substrates that contain the amide group in the chain because formation of an amide linkage constitutes a convenient method of connecting diene and dienophile. Also, amides are often more tolerant than tertiary and secondary amines of the high reaction temperatures necessary for some cyclizations. Compared to tertiary amines, amides may in some cases require somewhat higher cyclization temperatures. Thus the amine **110a** cyclizes at 140°, whereas amide **110b** must be heated to 185°; the fact that the amide **110c** cyclizes at 80° can be attributed to activation of the dienophile by the carbonyl

110a $X = Y = H_2$
b $X = H_2$; $Y = O$
c $X = O$; $Y = H_2$

111a $R^1 = H$; $R^2 = C_2H_5O_2C$
b $R^1 = CH_3$; $R^2 = C_2H_5O_2C$
c $R^1 = CH_3$; $R^2 = H$
d $R^1 = t\text{-}C_4H_9$; $R^2 = H$

group (Table IIIb/10, 10a, 11).[101] Secondary amides usually require higher reaction temperatures than tertiary amides. For instance, the cyclization temperatures of the amides **111a** and **111b** are 80° and 0°, respectively.[43] This rate difference has been attributed to intermolecular hydrogen bonding in the secondary amide (in nonpolar solvents), which decreases the population of the *cisoid* amide rotamer; the latter is the only rotamer that can undergo the intramolecular Diels–Alder reaction. The prediction that secondary amides would cyclize more rapidly in polar solvents such as dimethylformamide[43] has not been tested. For examples where secondary amides, unlike the corresponding tertiary ones, completely fail to undergo the intramolecular Diels–Alder reaction, see Table XVIIIb/31–33 and 36.

Bulky groups on nitrogen lower the activation energy for the cyclization of the amides **111**; the energies of activation range from 25.3 kcal/mol for the *tert*-butyl derivative **111d** to 28.7 kcal/mol for the methyl derivative **111c**. This finding has been interpreted in terms of an increased population of the *cisoid* amide rotamer with increasing size of the substituent on nitrogen.[43] The same effect has been invoked to explain the observation that *N*-acylation[121] and chelation[67] also increase the rate of cyclization.

c. Ethers and Thioethers. Many examples of substrates with ether linkages in the chain, and a few with thioether groups, are known. The presence of these heteroatoms does not seem to adversely affect the success of the intramolecular Diels–Alder reactions.

For ethers, see Tables IIe/1, 17; IIIb/20c, 20d; VI/7–9; VII/1, 2; VIIIa/6 (*ortho* ester); VIIIb/1b; 4a, 4b, 5, 9a; IXd/16, 28; IXe/18; X/15; XIVf/1, 5, 6, 25, 26; XVIc/3, 4, 6–8; XVIIa/12, 13; XVIIb/17–20; XVIIc/31, 34 (acetal), 35, 36 (acetal); XVIIIa/10, 11 (phenyl ethers); XIX/25, 26, 30 (phenyl ethers); XXII/4–58 (phenyl ethers); XXIII/12–23 (phenyl ethers); and XXVIc/25a, 25b (phenyl ethers). For thioethers, see Tables IIe/19; IXd/29, 32; IXe/19; XIVg/1; XVIIc/32; and XIX/27 (phenyl thioether).

d. Esters. Incorporation of an ester linkage into the chain often has an adverse effect on the intramolecular Diels–Alder reaction. This is evidenced by the requirement of higher reaction temperatures, or, in some cases, even failure to obtain any cyclized product. The reduced reactivity of esters has been attributed to preference for the *transoid* form and a relatively high barrier for interconversion of the two rotamers,[66] or, in some instances, loss of ester resonance in the transition state.[159] Esters thus resemble secondary amides, except that the population of the *cisoid* rotamer cannot be increased by use of bulky substituents. The problem is compounded by the tendency of esters to undergo elimination (ester pyrolysis) at high temperatures.[60] Nevertheless, many successful examples are known, especially those where the ester carbonyl also activates the dienophile. A second activating group on the dienophile may be necessary; thus the fumarate **112b** gives the intramolecular Diels–Alder adduct in 73% yield when heated to 200°; the acrylate **112a** fails to cyclize under these conditions.[160] Esters of propiolic acid, or substituted propiolic acids, also appear to cyclize more readily: heating the butynoic ester **113b** to 140° gives the intramolecular cycloadduct in 96% yield;[36] the propargyl ester **113a**, on the other hand, does not cyclize on heating.[161]

For other examples of successful cyclizations of substrates with ester linkages, see Tables IIe/4a, 4b, 8, 10–13, 16, 18; IV/3–5; V/11–13, 15–17; VI/11, 12; IXd/17–27, 30, 31; XIVf/2–4, 7–24, 27–29; XV/20–23, 34, 36, 37; XVIc/1, 2, 5; XVIIa/10; XVIIc/33; XVIIId/8, 9, 11; XIX/8–13; XXIV/20–24; and XXVIc/6–9b.

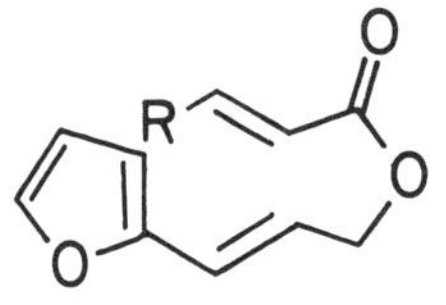

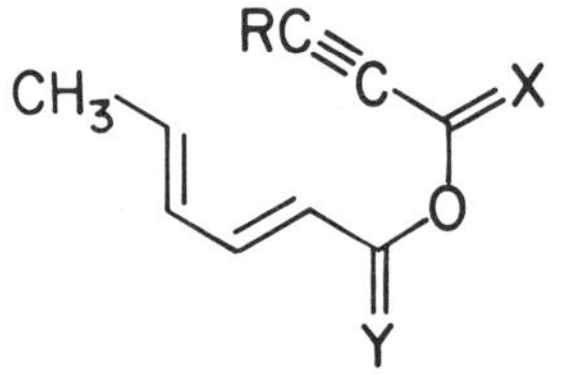

112a R = H
112b R = CH_3O_2C

113a R = H; X = H_2; Y = O
113b R = CH_3; X = O; Y = H_2

E. Substituents on the Chain. A wide variety of substituents may be present on the chain, but some of the more sensitive ones may have to be protected if the cyclization temperature is high. Hydroxyl groups, for instance, may undergo elimination or Michael addition to an activated dienophile; they are most commonly protected as the trialkyl silyl ethers. Protection, cyclization, and deprotection is often carried out without isolation of the intermediates. Other protecting groups that have been employed include benzyl and tetrahydropyranyl. Conversion of the alcohols to the lithium[162] and potassium[71] salts before cyclization has also been reported.

No systematic study of the effect of substituents in the chain on the rate of the intramolecular Diels–Alder reaction has been reported. Introduction of two geminal methyl groups (R = CH_3) in substrate **114a** increases the cyclization rate by a factor of 4.[87] The furan derivative **114b** cyclizes on heating to 80°; no reaction occurs when the analogous substrate lacking the dithiane ring is

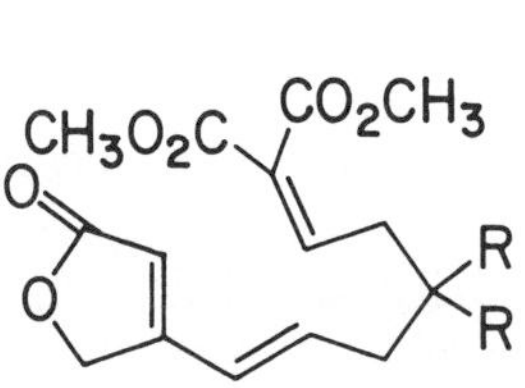

114a

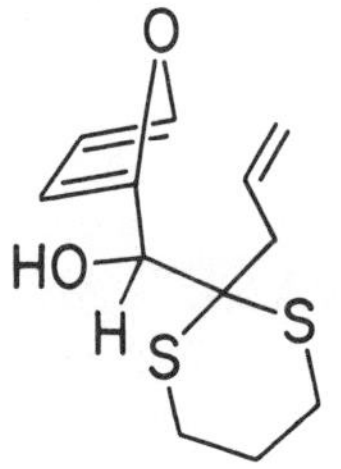

114b

subjected to the same conditions.[54] The fact that substituents in the chain may influence the stereochemical course of the intramolecular Diels–Alder reaction has been discussed earlier.

The Dienophile

A. Nonactivated Dienophiles. One important feature of the intramolecular Diels–Alder reaction is the fact that additions involving nonactivated dienophiles often occur under very mild conditions. Thus the allylamine **115**[163] and the acetylenic amine hydrochloride **116**[45] cyclize at room temperature. By comparison, intermolecular additions of nonactivated dienophiles to furan

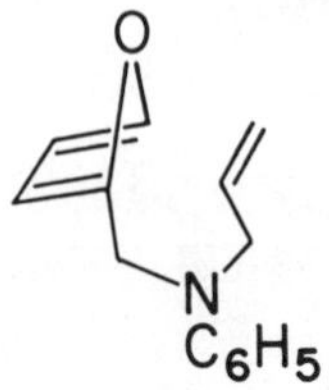

115 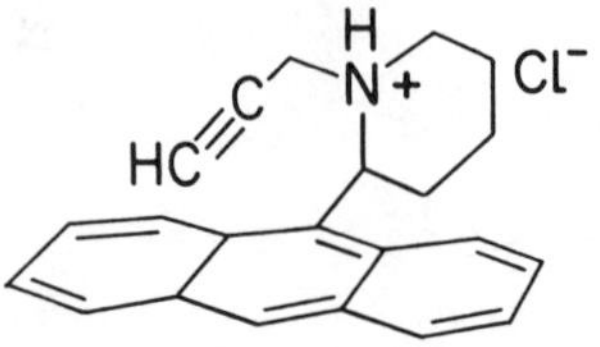 116

usually proceed only under ultrahigh pressures, and addition of acetylene to anthracene requires heating to 250°. Activation of the dienophile can increase the rate of the intramolecular Diels–Alder reaction, but this does not always occur; thus the ester **117a**[93] and the olefin **117b**[95] cyclize at essentially identical temperatures (155° and 160°, respectively) to give the intramolecular Diels–Alder adducts in high yields. The acetylenic ketone **118a** and the corresponding alcohol **118b** also require identical cyclization temperatures.[164]

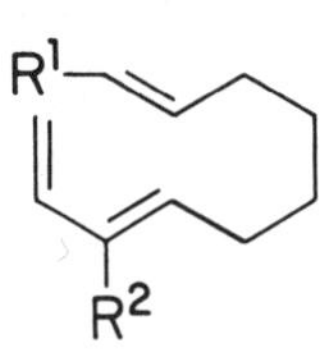

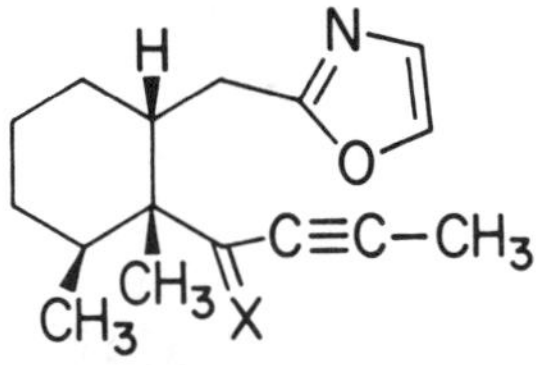

117a: $R^1 = CH_3O_2C$; $R^2 = H$ **118a**: X = O

117b: $R^1 = H$; $R^2 = CH_3$ **118b**: X = H, OH

B. Dienophiles with Electron-Withdrawing Substituents. An activating group can be attached to an olefinic dienophile in four different ways illustrated by the examples **119–122**. Although the different activating groups and the presence of other substituents makes a quantitative comparison impossible, it is clear that the mode of attachment does not greatly influence the cyclization temperatures (given in parentheses). In fact, the term "activating groups" is a misnomer here since the nonactivated olefin **123** cyclizes at a comparable temperature. It should be noted, however, that the cyclization of the esters **119** and **120** (but not of the aldehyde **121**)[38] can be catalyzed by Lewis acids and that ketones analogous to **122** that contain four- or five-atom chains require much lower cyclization temperatures. The diene type may have some influence on the activating effect of electron-withdrawing groups on the dienophile; thus in the cyclopentadiene **124** (a *cis* diene), replacement of the hydrogen (R) by a carbomethoxy group (either *cis* or *trans*) reduces the cyclization temperature from 250 to 110°.[79] A second activating group on the terminal carbon atom of the dienophile, in the single system studied, does not alter the reactivity to any great degree—the diester **125b** cyclizes three times faster than the monoester **125a**; however, not unexpectedly, the dinitrile **125c** cyclizes nine times faster

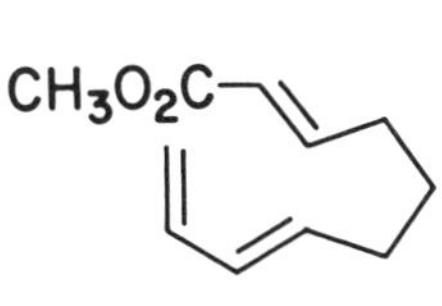

119 (150°; ref. 41)

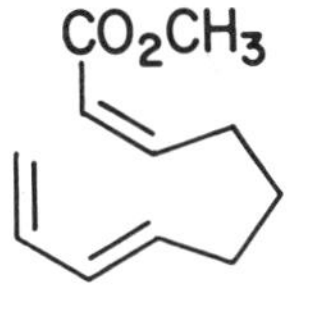

120 (180°; ref. 41)

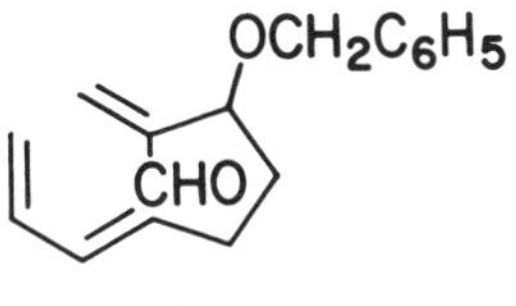

121 (150°; ref. 38)

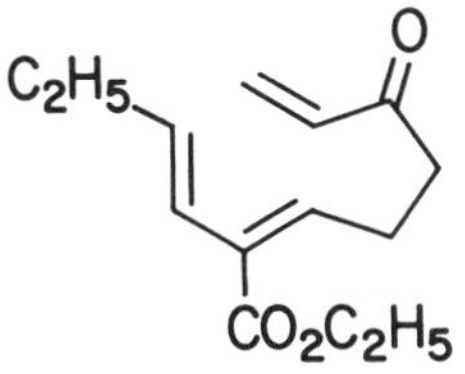

122 (180°; ref. 165)

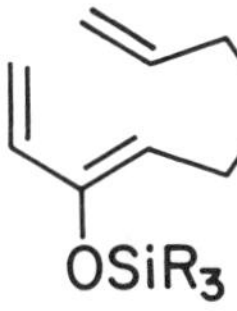

123 (160°; ref. 97)

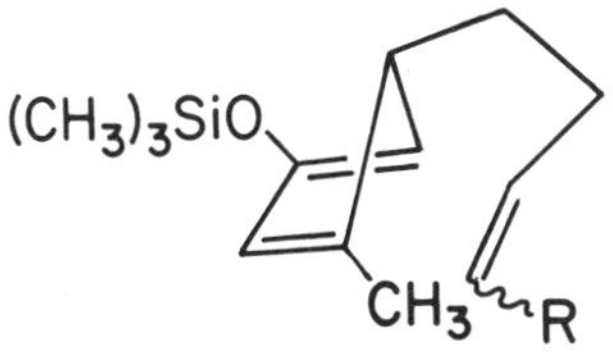

124

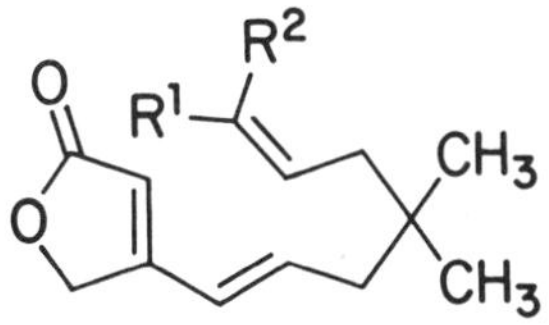

125a $R^1 = CH_3O_2C$; $R^2 = H$
125b $R^1, R^2 = CH_3O_2C$
125c $R^1, R^2 = CN$

than does the diester **125b**.[87] Placing activating groups on both the terminal and nonterminal carbon atoms of the dienophile, on the other hand, does seem to enhance the reactivity significantly as illustrated by the substrates **126** (the cyclization temperatures are given in parentheses).[43] For another example, see Table XV/19, 20.

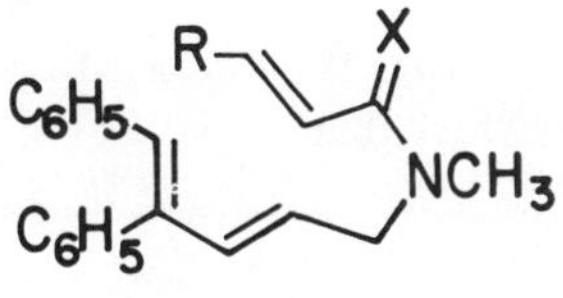

126a R = H; $X = H_2$ (140°)
b R = H; X = O (110°)
c $R = C_2H_5O_2C$; X = O (<25°)

C. Miscellaneous Substituents on the Dienophile. Alkyl or aryl groups on the dienophile may slow down the rate of the intramolecular Diels–Alder reaction, especially when they are attached to the nonterminal carbon atom. For instance, the cyclization temperatures of the ethers **127a** and **127b** are 120 and 310°, respectively.[166] In the *N*-acylcinnamylamines **128**, on the other

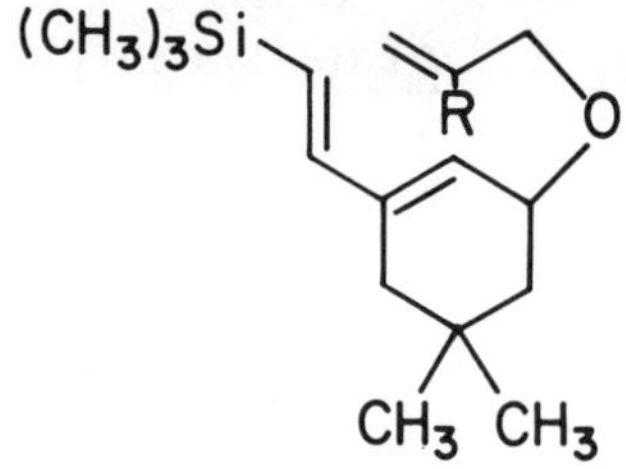

127a R = H
127b R = CH_3

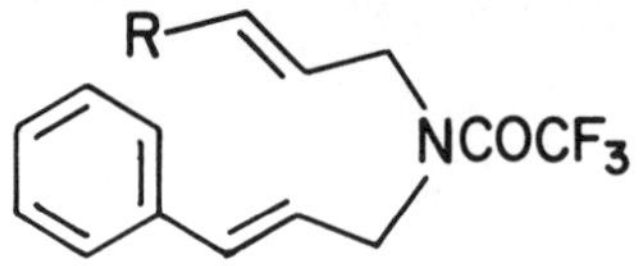

128a R = H
128b R = C_6H_5

hand, the derivative **128b** with a phenyl group on the terminal carbon of the dienophile cyclizes somewhat more readily than does the unsubstituted analog **128b** (180° vs. 235°).[119] Additional substitution of the dienophile usually decreases the cyclization rate further, but cases of intramolecular Diels–Alder reactions involving tri- and even tetra-substituted dienophiles are known. An example of the latter is the benzofuran derivative **129**, which cyclizes at 215°;[167] for another example, see Table XVIa/9.

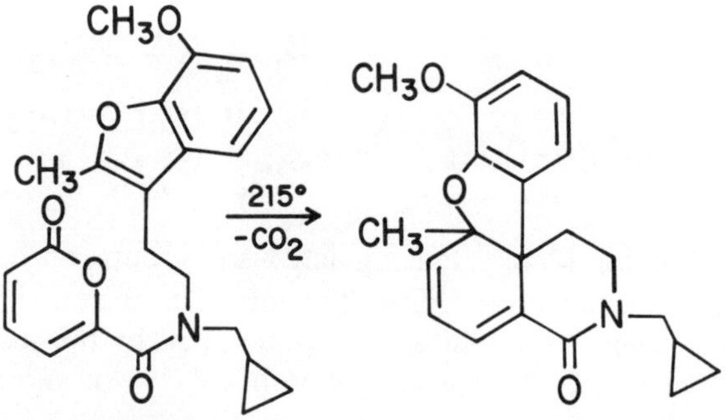

129

As evidenced by the following list, a variety of substituted olefins can act as dienophiles in intramolecular Diels–Alder reactions:

Enamines	Table XV/1–15
Enamides (*N*-acylenamines)	Tables IIb/35–37; 41–43; IIIb/16–19
Vinyl ethers	Tables IIe/17; XVIIa/12, 13; XVIIc/35; XX/10
Vinyl esters	Tables IV/5; IXd/30, 31; XVIa/25, 26

Vinyl thioethers	Tables IXd/32; XVIa/22–24, 47
Bromoolefins	Table IXd/17, 19, 20
Chloroolefins	Tables IId/50, 55; VI/12; XIII/11; XIVc/9–13; XIVe/20–22; XVIb/7; XVIIIc/13, 21; XXV/31
Nitroolefins	Table XVIb/21

D. Cyclic Dienophiles. Intramolecular Diels–Alder reactions involving a wide variety of cyclic dienophiles have been reported. A complete list is found at the end of this section. Some of the more unusual examples are illustrated in Eqs. 21,[168] 22,[169] and 23.[170]

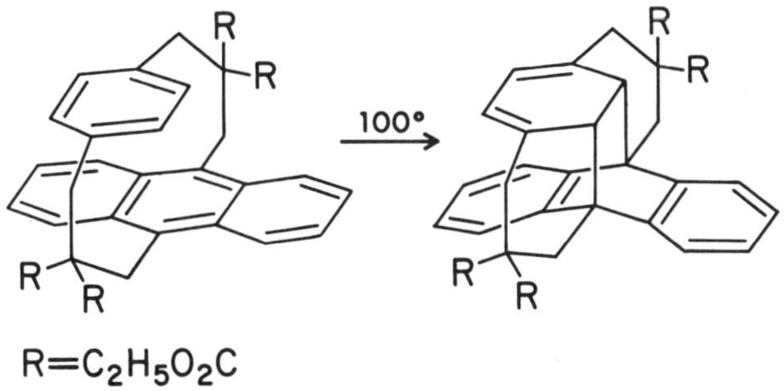

100°

$R = C_2H_5O_2C$

(Eq. 21)

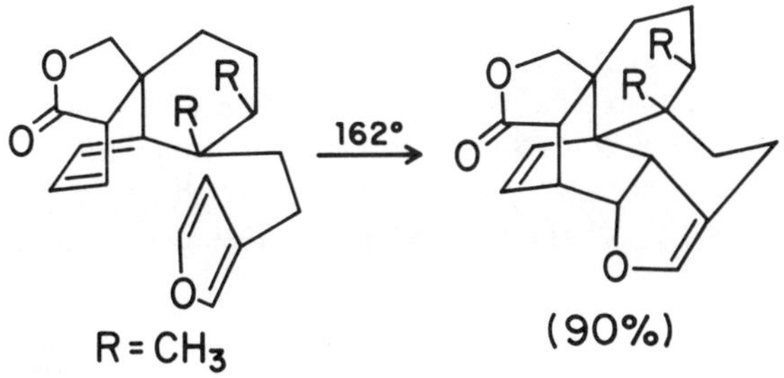

162°

$R = CH_3$

(90%)

(Eq. 22)

*

25°

C_6H_5

(37%)

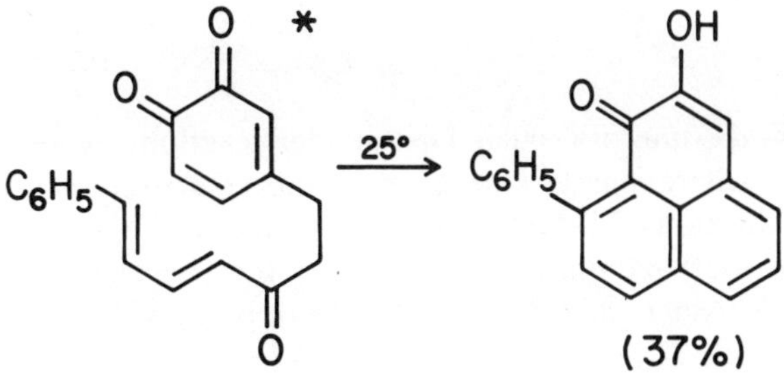

(Eq. 23)

1. Carbocyclic dienophiles
   a. Cyclopropenes: Tables VIIIb/16; IXd/15c–g; XIX/2, 3; XXI/6a; XXII/1–3d
   b. Cyclopentenes: Tables IIb/33a–p; IIe/16; VIIIa/13; VIIIb/20; X/18; XVIa/6, 7; XVIb/14; XVId/15; XVIIc/24; XVIII/69; XIX/6; and XXIV/10 and Ref. 171
   c. Cyclohexenes: Tables IIb/34, 34a; IId/56–60; IIIa/30; VI/6; XIII/2; XVIb/15; XVId/17; XVIIIb/70; and XXIV/11 and Ref. 171
   d. Cycloheptenes: Tables VIIIb/17; IXe/5; and XIII/14 and Ref. 172
   e. Cyclooctene: Table XIII/13
2. Aromatic dienophiles (see also arynes in the section on acetylenic dienophiles)
   a. Benzenes: Table XVIIc/4, 37
   b. Naphthalene: Table XVIIb/4
   c. Anthracenes: Refs. 172a,b
   d. Phenanthrene: Table XXIV/7
3. Heterocyclic dienophiles
   a. Five-membered heterocycles: Tables IIe/16; IIIb/9, 18; IV/5; IXa/11; XV/1; XVIIId/5; XXIV/8, 9, 12–19; and Refs. 172c (imidazolines) and 172d,e (thiazolines).
   b. Six-membered heterocycles: Tables IIb/35–37; IIIb/16, 17, 19; XV/2–14; XVIb/13; XXI/15; and Ref. 172f
4. Bicyclic dienophiles: Tables VIIIb/10–15; X/19, 20; XVIIb/2, 3; XVIIc/47; XVIIId/6, 7; and XXV/4a and Ref. 172

E. Cumulenes. A few examples of intramolecular Diels–Alder reactions where an allene group acts as the dienophile are known. The substrates are often formed by Claisen rearrangement of propargyl vinyl ethers, as illustrated in Eq. 24.[173] For other examples of allenic dienophiles, see Tables VIIIb/1, 1a; IXb/29–37; X/6–12; XVIIa/4–9; XXI/1–3. Ketenes normally do not act as

(Eq. 24)

dienophiles in the intermolecular Diels–Alder reaction. There are, however, a few examples where simple ketenes are believed to undergo intramolecular Diels–Alder additions. The ketene **130**, generated from the corresponding acid chloride with triethylamine, is converted to the tricyclic ketone **132** via the proposed Diels–Alder adduct **131**.[174] For examples where ketenes give bridged adducts, see Table XXVIa/1–11, 19 and XXVIc/1–3.

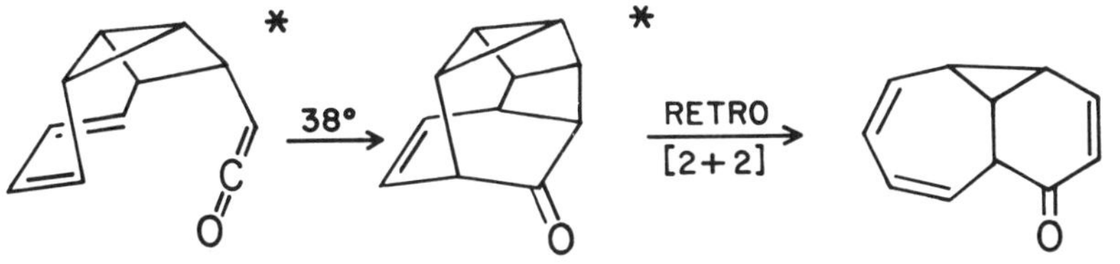

130 → (38°) **131** → (RETRO [2+2]) **132** (70%)

F. Acetylenes. In additions to *trans* dienes, acetylenes are usually somewhat more reactive than the corresponding olefins. For instance, heating the *N*-allyl-*N*-propargylamine **133** to 110° results in preferential addition of the

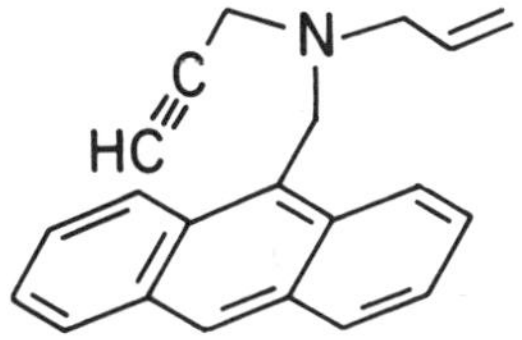

133 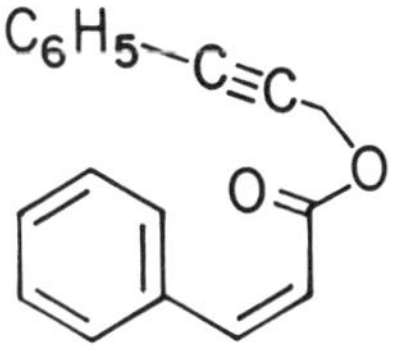 **134**

acetylene by a factor of 3.[45] Incorporation of an activating group on the terminal carbon atom of the acetylene lowers the cyclization temperature. Thus the acetylenic ester **135b** cyclizes at 110°, whereas substrate **135a** lacking the

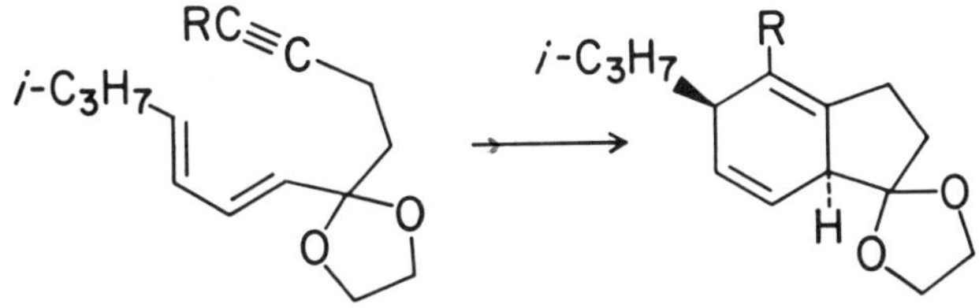

135a R = H
135b R = CH_3O_2C

ester group requires heating to 155°.[175] An example of an acetylenic ester that undergoes the intramolecular Diels–Alder reaction at room temperature is given in Table IIa/1c.[38] Acetylenic substrates with four-atom bridges (Table IIIa/1, 1a) require more rigorous conditions since the chain must adopt a boat-like conformation, whereas the transition state is relatively strain-free in substrates with a three-atom bridge.[133] As mentioned earlier, use of acetylenes has the advantage of allowing introduction of the desired stereochemistry of the ring junction at a later stage.

Cis dienes connected through a short chain (three or fewer members) to an acetylenic dienophile cannot attain a strain-free transition state; thus the ester

134 fails to cyclize at 150°, whereas the corresponding olefin (*trans*-cinnamyl-*cis*-cinnamate) gives the adduct to the styrene diene system in 31% yield under these conditions.[122]

Arynes are among the most powerful dienophiles in the intermolecular Diels–Alder reaction, and, not surprisingly, this holds true for the intramolecular version as well. Thus the benzyne derivative **136**, generated by treatment of the corresponding dibromo derivative with butyllithium, cyclizes at or below 25° to give the adduct **137** in 90% yield.[176] For other additions involving benzynes, see Tables XVIIa/1–3, 11; XVIIb/5; and XVIIIa/10, 11.

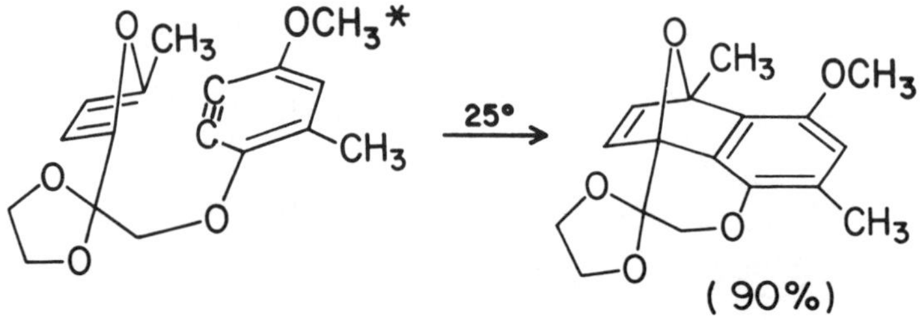

OCH_3* CH_3 25° CH_3 OCH_3 CH_3 CH_3

(90%)

136 **137**

G. Dienophiles Containing Heteroatoms. A variety of dienophiles containing heteroatoms have been employed successfully in the intramolecular Diels–Alder reaction. Examples are given in Eqs. 25,[136] 26,[177] 27,[177] and 28.[178] A complete list of these dienophiles is presented at the end of this section.

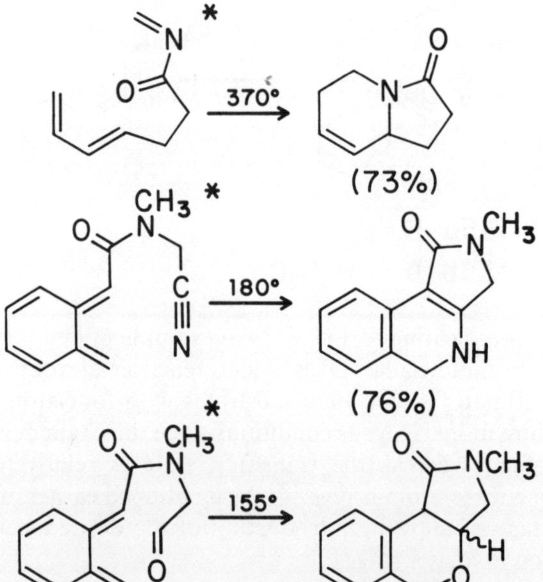

370° (73%) (Eq. 25)

180° (76%) (Eq. 26)

155° (25%) (Eq. 27)

*

-20°

(55%)

(Eq. 28)

Imines	Tables IIf/15–20b; IIIb/21–23; XIVa/2; XXVIa/12–18
Oximes	Tables XVIb/17; XVIc/4
Nitriles	Tables IIf/1–3, 4†; XVIb/3–5; XVIIIa/3†; XVIIIc/9†; XXIII/5; XXIV/2
Aldehydes	Tables XVIb/16, 17a, 17b; XXVIc/20
Ketones	Tables IIf/14†; IIIb/9a; XVIIIc/26†, 28†; XXVIc/22, 23
Thioaldehydes	Table IIf/21 and Ref. 178a
Esters (carbonyl)	Tables IIf/14a†; XVIIIc/27†
Amide (carbonyl)	Table XXVIc/27
Nitroso compounds	Table IIf/5–13
N-Sulfinylimine	Ref. 178b

H. *In Situ* Generation of the Dienophile. In the synthesis of substrates for the intramolecular Diels–Alder reaction, sensitive dienophilic functions can be carried along in protected form and generated *in situ* in the cyclization step. For instance, vinyl ketones have been masked as Diels–Alder adducts with dimethylfulvene (Table IIa/42)[90] or in the form of an aldol (Table IIIa/23, 25).[98,179] They have also been formed by oxidation of the corresponding vinylcarbinol just prior to cyclization. Olefinic dienophiles have been generated *in situ* by *N*-oxide pyrolysis (Table IXe/1)[180] or acetate elimination (Table XVIIc/38).[181] *N*-Acylimines have been generated by acetate pyrolysis (Table IIf/15–20; IIIb/21, 22; XIVa/2)[113,114,136,182] and by reverse ene reaction (Table IIf/20b).[183] Acylnitroso compounds have been masked as the Diels–Alder adduct with 9,10-dimethylanthracene (Table IIf/9–13);[142,184] they have also been generated by periodate oxidation of hydroxamic acids (Table IIf/5–8).[178]

The Diene

A. *Cis* vs. *Trans* Dienes. The *cis* and *trans* dienes **138** and **139** both cyclize at 140°;[185] an isomeric pair of 5-trimethylsiloxy-1,3,8-decatrienes also requires identical reaction temperatures (Table IIa/5, 39),[84] as does a pair of isomeric 2-phenylpentadienoic acid *N*-allylamides (Table IIb/17).[167] A rate factor of 40·

† The structure of the product has not been established with certainty.

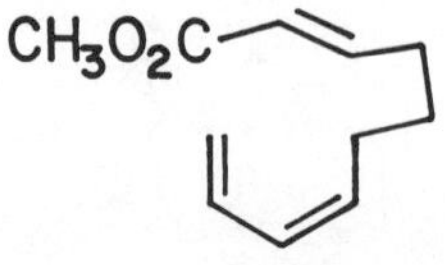

CH_3O_2C — 138 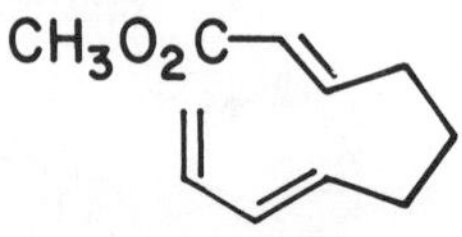 — 139

in favor of the *trans* diene has been measured in a pair of vinylfuranones (Table XV/32, 33a).[85] The *cis* dienes **140** [R^2 = H or $(CH_3)_3Si(CH_2)_2O$] cyclize at 110°; for reasons that are not clear, the corresponding *trans* isomers fail to react under these conditions.[86] The reactivity of *cis* dienes in the intramolecular Diels–Alder reaction is in pronounced contrast to the reluctance of such dienes (e.g., *cis*-piperylene **141**) to undergo the intermolecular Diels–Alder reaction.[186]

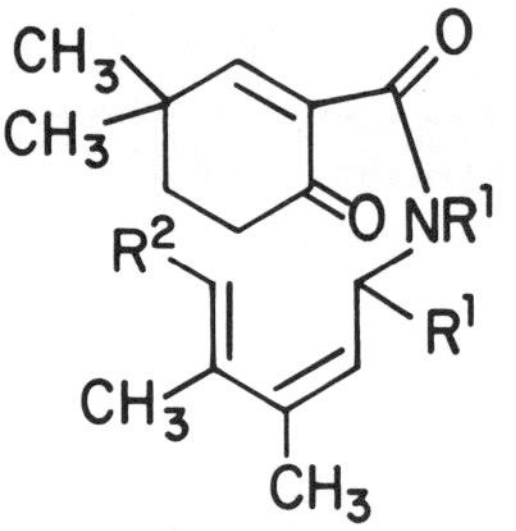

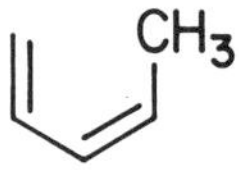

140
$R^1 = C_6H_5CH_2$

141

As mentioned earlier (p. 24), the utility of *cis* dienes in the intramolecular Diels–Alder reaction is somewhat limited by their tendency to isomerize by 1,5-hydrogen shifts.

The comparable reactivity of *cis* and *trans* dienes also extends to cyclic dienes; thus the *cis* diene **142** and the *trans* diene **143** cyclize at 168° and 208°, respectively (Table IXa/1; IXd/1).[77,148] A similar situation exists for the corresponding cyclopentadienes (Table VIIIa/2; VIIIb/2).[187]

142  **143**

Dienes usually preserve their stereochemical integrity, even at fairly high temperatures; only two exceptions to this rule have been reported to date (Table IIa/29[17] and IIIb/7[18]).

B. Substituents on the Diene. With the exception noted below, substituents on the diene do not appear to inhibit the intramolecular Diels–Alder reaction. Thus dienes **144a** and **144b**, differing by an isopropyl group at the terminus, cyclize at the same temperature (150°)[41] as do the corresponding substrates **145a** and **145b** containing a *cis* dienophile (180°).[41] However, if the isopropyl

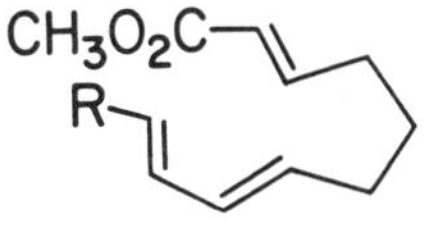

144a R = H
b R = *i*-C_3H_7

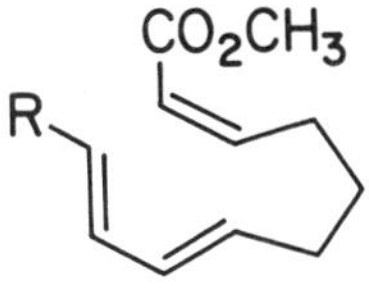

145a R = H
b R = *i*-C_3H_7

group is attached in a *cis* fashion (substrate **146a**), no cyclization takes place at 150°, a temperature at which the *trans* isomer **146b** undergoes the intramolecular Diels–Alder reaction in 70% yield.[188] As in the intermolecular Diels–Alder reaction, incorporation of the offending *cis* substituent in a ring removes the steric inhibition: cyclic dienes readily undergo the intramolecular Diels–Alder reaction.

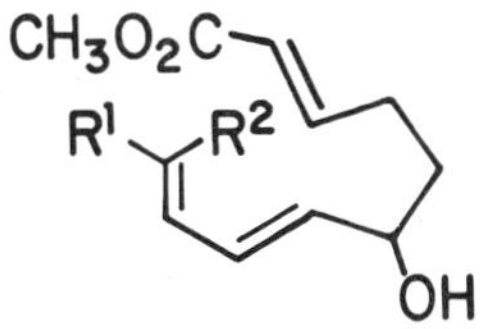

146a R^1 = H; R^2 = *i*-C_3H_7
b R^1 = *i*-C_3H_7; R^2 = H

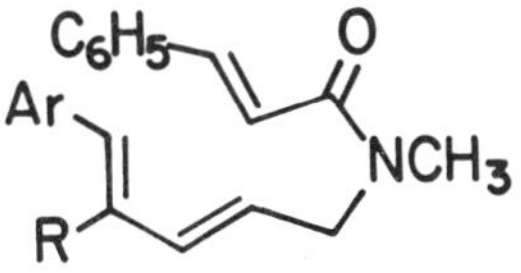

147a R = H; Ar = *p*-$CH_3OC_6H_4$
b R, Ar = C_6H_5

Introduction of a second phenyl group at C-2 of diene **147** has little effect on the rate: diene **147a** cyclizes at 85°,[44] compared to 91° for diene **147b**.[43] Substituents on C-4 of a *trans*-diene surprisingly also do not seem to inhibit the intramolecular Diels–Alder reaction; thus, the dienamide **148** cyclizes slowly at 120°;[167] substrate **149**, the closest comparison compound, cyclizes at 156°.[189]

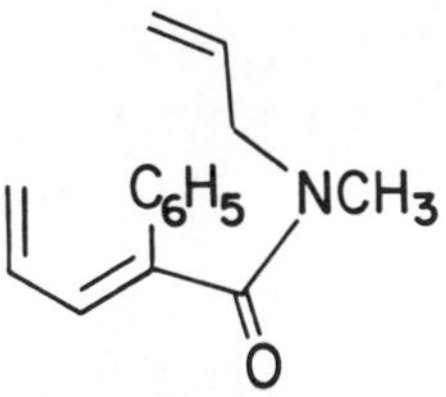

148

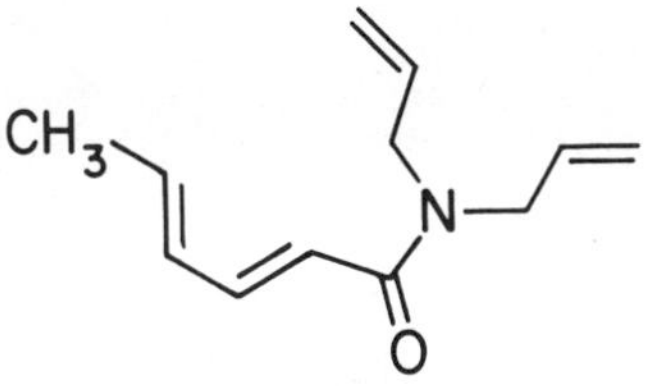

149

Hydroxyl groups can be carried along on the diene as silyl, tetrahydropyranyl, or benzyl ethers; a phosphonate ester has also been used as a protecting group (Table IXd/13).[190] There is one example of a diene bearing an unprotected hydroxyl group.[191] Little is known about the effect of electron-withdrawing substituents attached to the termini of the diene. If they form part of the chain (such as in substrates **148** and **149**), they can exert an activating effect only if the chain is four-membered; the carbonyl group is twisted out of the plane of the diene in substrates with three-atom chains. However, even if conjugation is not inhibited, electron-withdrawing substituents may not provide much rate enhancement; the doubly activated diene **150**, containing a four-atom chain, requires heating to 250°,[19] even though the dienophile is electron-rich (Diels–Alder reaction with inverse electron demand).

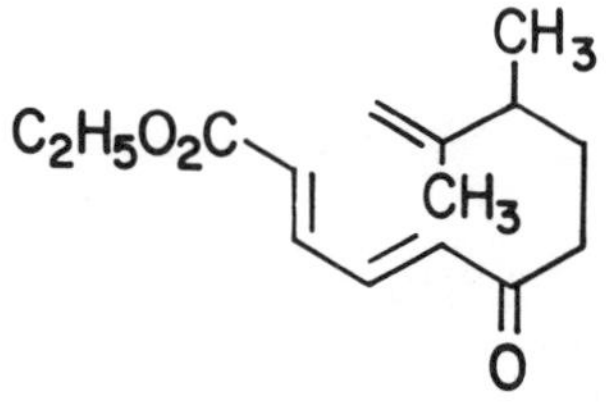

$C_2H_5O_2C$ CH_3 CH_3 O

150

C. Acyclic Dienes (Tables I–V). Many examples of acyclic dienes undergoing the intramolecular Diels–Alder reaction have been given in the preceding sections, and it should be obvious that this diene type is very useful in the construction of bicyclic ring systems. All known examples are listed in Tables I (one- and two-atom chains), II (three-atom chains), III (four-atom chains), IV (chains of five or more atoms), and V (chains attached at C-2).

1,3,4-Trienes of type **151**, generated by base-catalyzed isomerization of the appropriate 4-penten-2-ynylammonium salts, cyclize at room temperature with formation of dihydroisoindolinium salts **152**.[192] For other examples, see

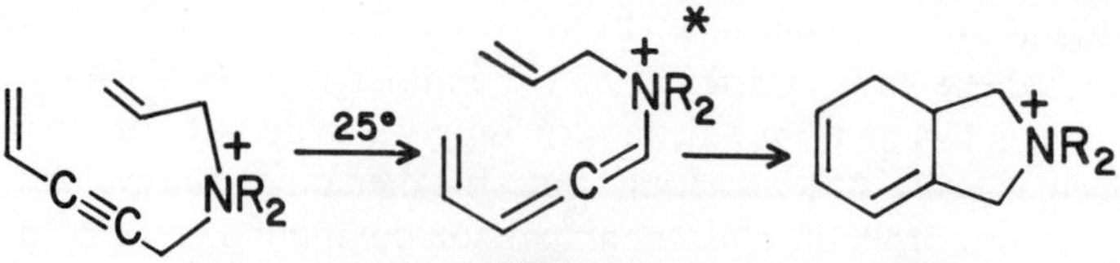

$\overset{+}{N}R_2$ 25° * $\overset{+}{N}R_2$ $\overset{+}{N}R_2$

151 **152**

Table IId/24–49 and 56–60 (see also Table VI/1–6). If the diene or dienophile contains a chlorine substituent, or if the dienophile is an acetylene, isoindolinium salts are formed (Table IId/1–23, 50–55). The cyclization of a vinylallene with a four-carbon chain has been mentioned in the discussion involving Eq. 11.

Sensitive acyclic dienes can be generated *in situ* from precursors in which the diene is masked in various ways. Examples include the electrocyclic ring opening of a cyclobutene (Table IIa/42),[90] the cheletropic extrusion of sulfur dioxide from dihydrothiophene dioxides (Tables IIb/36, 37, 43; XXVIc/4, 5),[20,22,109] and the retro cleavage of an intermolecular Diels–Alder dimer (Table IIIa/35).[193]

D. Vinylcycloalkenes (Table VI). The few known examples of this diene type indicate that vinylcycloalkenes are useful substrates for the construction of tricyclic systems as illustrated in Eq. 29.[166]

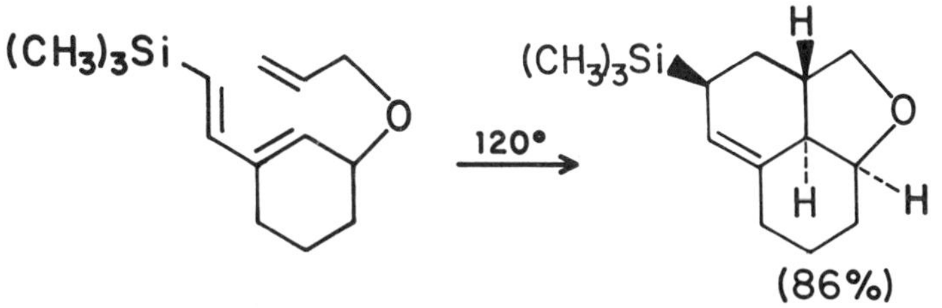

$(CH_3)_3Si$ ... O $\xrightarrow{120°}$ $(CH_3)_3Si$ ... O, H, H, H (86%)

(Eq. 29)

E. Cyclobutadienes (Table VII). In the only two examples involving this reactive diene, the cyclobutadiene is generated by oxidation of the iron tricarbonyl complex; the resulting fused Dewar benzenes aromatize under the reaction conditions (Eq. 30).[194]

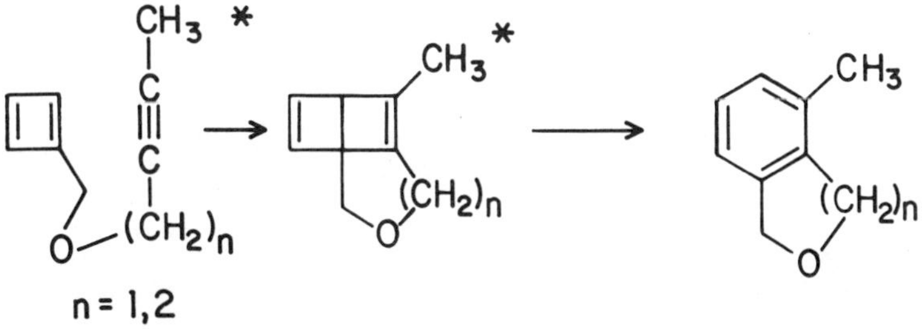

CH_3 * C≡C $(CH_2)_n$ O, n = 1,2 → CH_3 * $(CH_2)_n$ O → CH_3 $(CH_2)_n$ O

(Eq. 30)

F. Cyclopentadienes (Tables VIIIa and VIIIb). Substrates of this diene type are usually prepared by alkylation of metal cyclopentadienides, which often results in mixtures of all three possible isomers. However, since the 1-, 2-, and 5-alkylcyclopentadienes are, at elevated temperatures, interrelated by reversible 1,5-hydrogen shifts (Eq. 31), single products are usually obtained in

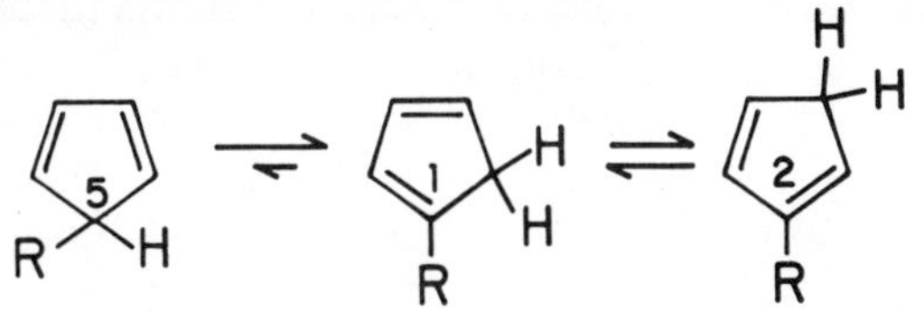

5, R, H ⇌ 1, R, H, H ⇌ 2, R, H, H (Eq. 31)

the subsequent cyclizations. The products are derived from either the 1- or 5-substituted isomer; intramolecular Diels–Alder reactions involving cyclopentadienes with chains attached at C-2 have not been reported to date, even though the corresponding reaction of acyclic dienes is known (Table V). Which isomer reacts appears to depend on the chain length, and substrates with three-atom chains are favored. Thus the equilibrating mixture of pentenylcyclopentadienes **153a** gives rise to adduct **155a** in 67–96% yield when heated to 200°; the isomer **154a** with a four-atom chain is formed, if at all, in only small amounts.[195] On the other hand, heating the butenylcyclopentadienes **153b** to 180° results in exclusive formation of isomer **154b** in quantitative yield.[187]

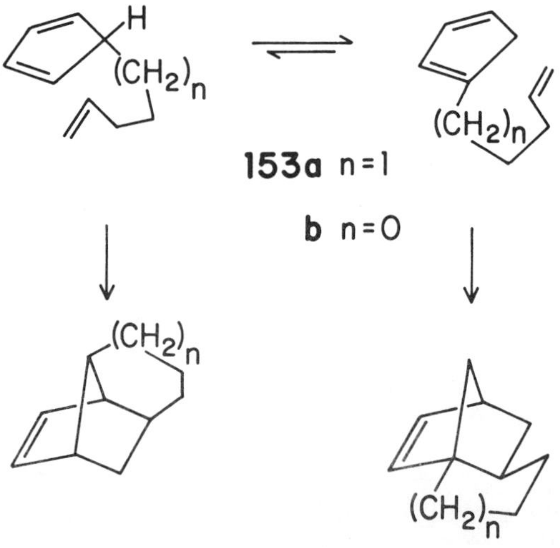

153a n = 1
b n = 0

154 **155**

Cyclopentadienes tend to dimerize, but elaboration of the chain and dienophile can be carried out on either the dimer (e.g., Table VIIIb/18, 19)[196] or a mixed dimer with cyclopentadiene (e.g., Table VIIIa/4, 5);[197] the monomer is then generated *in situ* by retro Diels–Alder reaction.

G. Cyclohexadienes (Tables IXa–d). Alkylcyclohexadienes are also interrelated by a reversible 1,5-hydrogen shift, and the points brought out for cyclopentadienes hold for the cyclohexadienes as well. Thus 5-pentenylcyclohexadiene (**156**), when heated to 210° in the gas phase, gives the rearranged products **159** (mixture of *syn* and *anti* adducts) in 27% yield in addition to isomer **158** (73%); however, when 1-pentenylcyclohexadiene (**157**) is heated to 208°, isomers **159** are formed exclusively.[77] This indicates that cyclization of **156** to **158** is somewhat faster than the 1,5-hydrogen shift (**156** ⇌ **157**). The butenyl 5-cyclohexadienyl ketone **160** rearranges completely to the conjugated dienone **161** before cyclization.[115] An analog of substrate **161** that lacks the two geminal

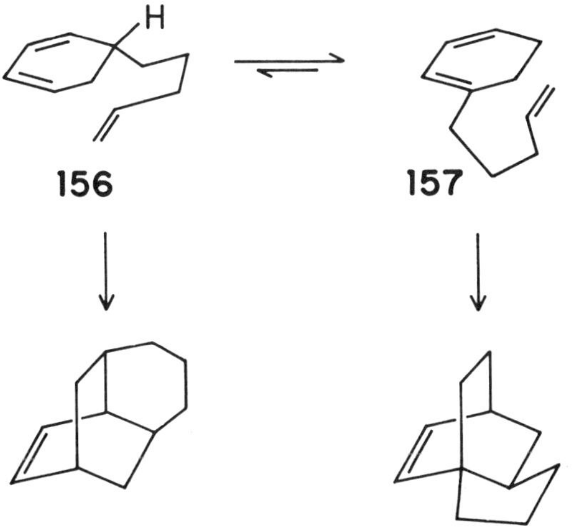

156 157

158 159

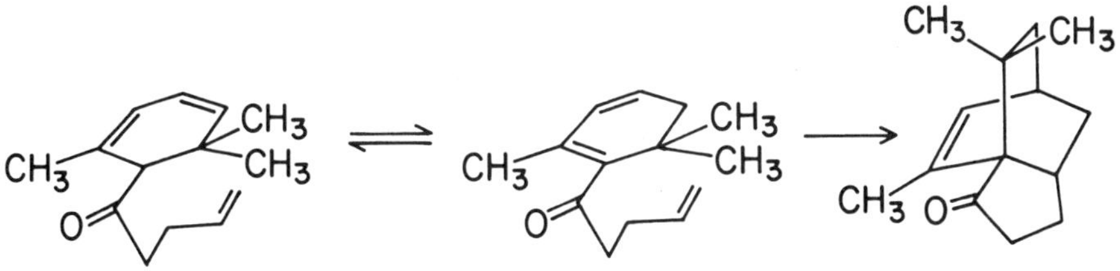

160 161

methyl groups aromatizes on heating, but conversion to a ketal permits the intramolecular Diels–Alder reaction to be carried out successfully at 250° (Table IXa/4).[116]

5-Vinylcyclohexadienes, which can be generated by Diels–Alder addition of pyrones to butadienes, readily cyclize as illustrated for substrate **162**.[198]

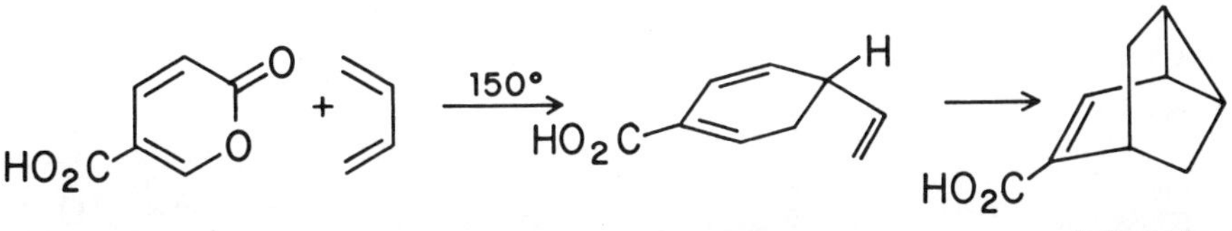

162 (71%)

5-Allenylcyclohexadienes (e.g., **163**)[199] generated by Claisen rearrangement of the appropriate phenyl propargyl ethers also undergo intramolecular Diels–Alder addition, but the initial adducts can be isolated only in certain cases (see Table IXb/29–37). Claisen rearrangements have been used to generate other cyclohexadienone substrates with longer chains (Table IXc/2–4, 10, 11).

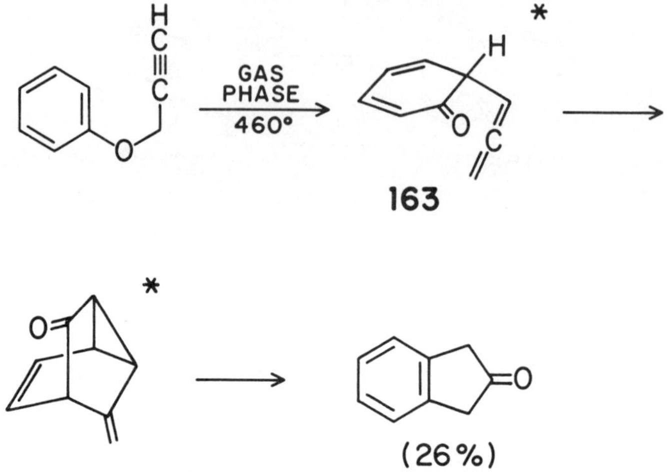

GAS PHASE 460°

163

(26%)

Oxidation of phenols with lead tetraacetate in the presence of acrylic acids gives dienone acrylates (e.g., **164**); unlike other esters, these cyclize under very mild conditions.[200,201]

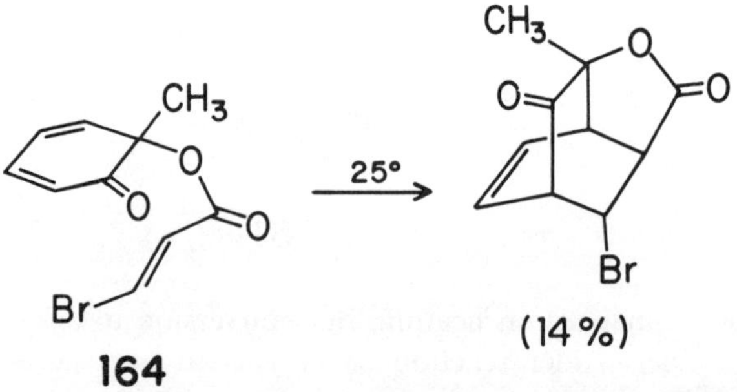

25°

164

(14%)

The cyclooctatrienone **165** isomerizes to the bicyclooctadienone **166** prior to undergoing intramolecular Diels–Alder reaction[202] (see also Table IXd/15a, 15b). Preferred cycloaddition to one of two valence isomers is common in the intermolecular Diels–Alder reaction. Thus cycloheptatrienes usually react as

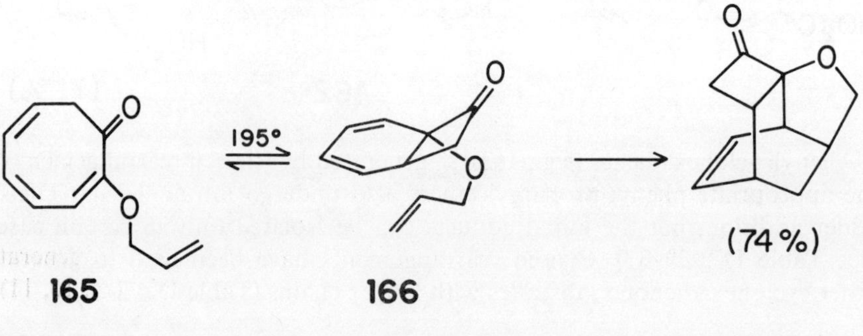

195°

165

166

(74%)

their norcaradiene valence isomers. An intramolecular addition to such a system is known (Table IXd/32), but in 7-butenyl- and 7-allyloxycycloheptatriene, addition occurs to the triene system (Table X/15, 16; see also entries 19, 20).

H. Cycloheptadienes and Cycloheptatrienes (Table X). As in the lower homologs, 1,5-hydrogen shifts occur in cycloheptadienes and cycloheptatrienes, and again the isomer with a three-atom chain appears to cyclize preferentially. From a mixture of isomeric amides, therefore, only the 1-substituted cycloheptadiene undergoes intramolecular Diels–Alder reaction (Eq. 32).[189] In some systems, Claisen or Cope rearrangements occur prior to cyclization (see Table X/15, 16).

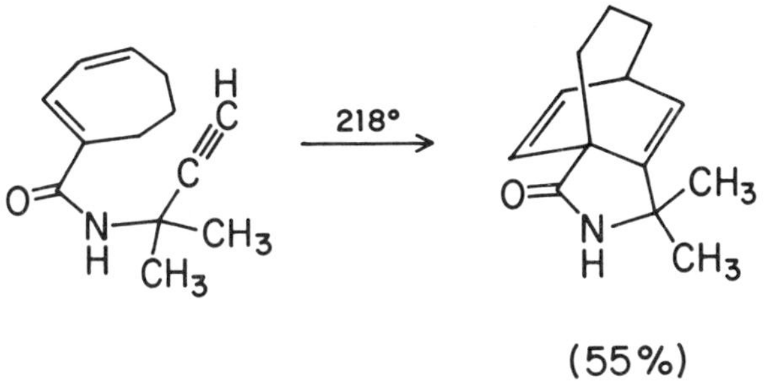

218°

(55%)

(Eq. 32)

I. Cyclic 1,3,n-Trienes (Table XI) and Bridged 1,3,n-Cyclic Trienes (Tables XII and XIII). Although many examples are known, Diels–Alder reactions involving these diene types are of only rather specialized synthetic interest. Examples are given in Eqs. 33[203] and 34.[204]

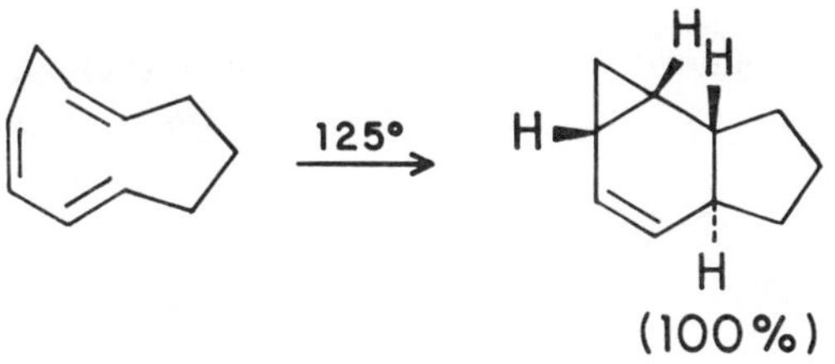

125°

(100%)

(Eq. 33)

200°

(62%)

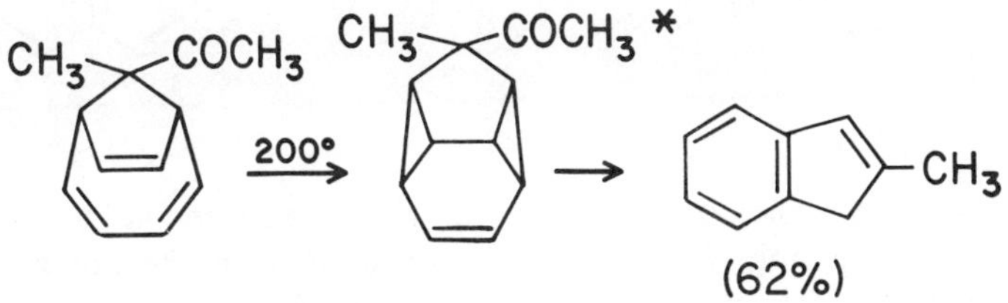

(Eq. 34)

J. Vinylaromatics (Tables XIVa–g). Styrenes are usually poor dienes in the intermolecular Diels–Alder reaction because the initial step entails loss of aromatic resonance stabilization. Intramolecular cyclizations involving these dienes frequently require high reaction temperatures, but many proceed under quite mild conditions. Thus the amide **167** cyclizes at 110°; an analog containing a ring in the chain cyclizes at 0° (Table XIVb/19).[123] The intermediate **168** aromatizes by a 1,3-hydrogen shift to give the lactam **169** in 56% yield. If atmospheric oxygen is not excluded, **168** undergoes an ene-like reaction with triplet oxygen to give eventually the alcohol **170**. This side reaction may account

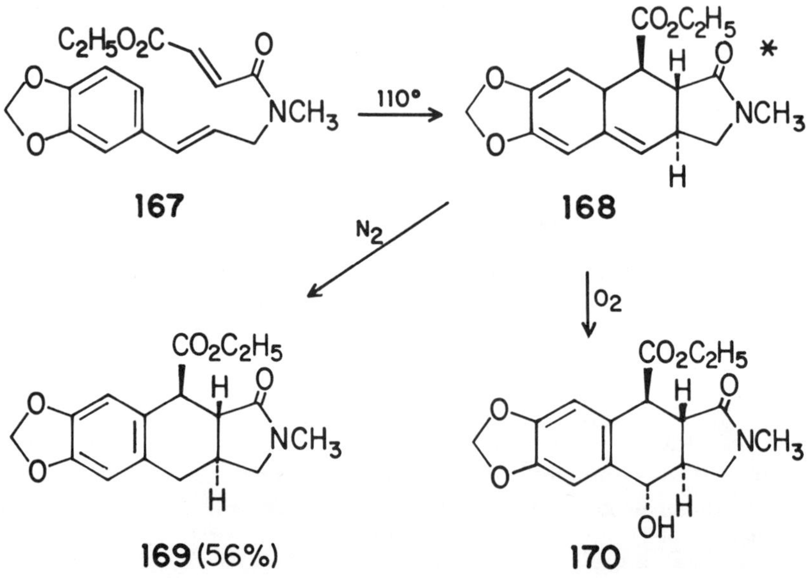

for the low yields obtained in other intramolecular Diels–Alder reactions of vinylaromatics. With acetylenic dienophiles, it is similarly advisable to exclude oxygen in order to avoid aromatization of both six-membered rings (Eq. 35).[205]

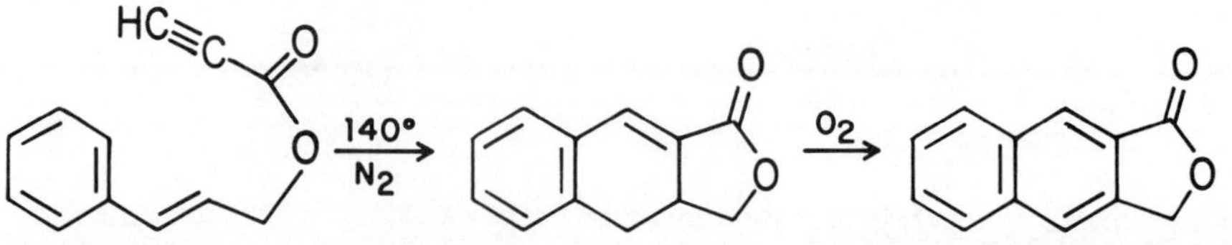

(Eq. 35)

α-Substituted styrenes can give rise to two isomers in the aromatization step. In the only example reported, the main product from the amide **171** is the thermodynamically less stable α-isomer **172a**, formed by a formal suprafacial 1,3-hydrogen shift.[119]

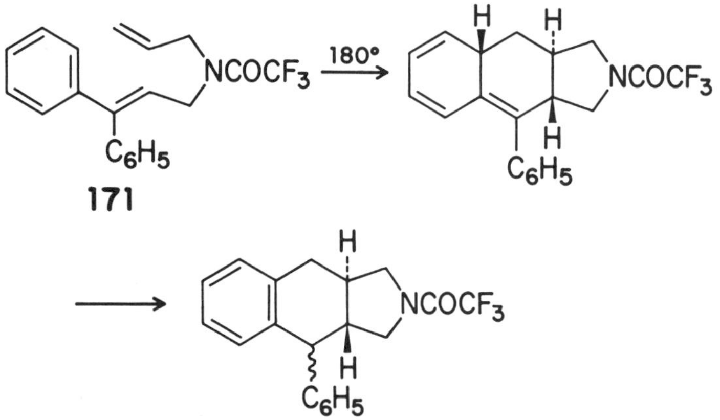

171

172a C_6H_5 α (38%)
172b C_6H_5 β (10%)

A [2 + 2]cycloaddition sometimes occurs as a competing side reaction (Eq. 36).[119]

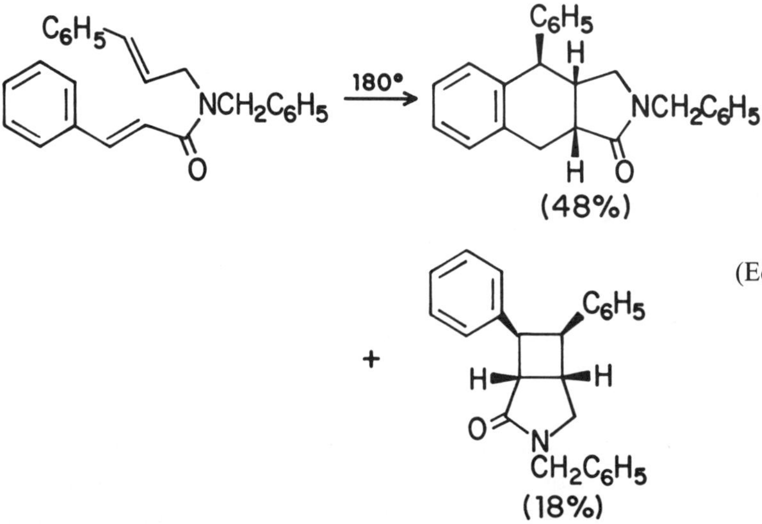

(48%)

(18%)

(Eq. 36)

Allenylbenzenes, generated by base-catalyzed isomerization of suitable phenylpropargyl compounds, undergo intramolecular Diels–Alder reactions with both ethylenic and acetylenic dienophiles. Heating the quaternary ammonium salt **173** in aqueous solution in the presence of a catalytic amount of sodium hydroxide thus furnishes the dihydrobenzoisoindolinium salt **174**.[153,192] With acetylenic dienophiles, benzoisoquinolinium salts are obtained. These reactions, which have also been observed with substrates having chains containing other heteroatoms (or carbon only), may not be concerted.[23,153]

173

174 (76%)

Most known intramolecular Diels–Alder reactions of the type discussed in this section involve styrenes as the diene. However, other vinylaromatics, such as vinylnaphthalenes, may be even better substrates. One example of a 9-vinylphenanthrene derivative undergoing the intramolecular Diels–Alder reaction has been reported (Table XIVa/2).[136]

K. Vinylheterocycles (Table XV). Cyclizations involving this diene type include the intramolecular Diels–Alder reactions of 2-vinylindoles, 3-vinyldihydropyrroles, vinyldihydropyridines, vinylfurans, 3-vinyldihydrofuranones, and vinylthiophenes. An example of the last is illustrated in Eq. 37.[206]

(Eq. 37)

(24 %)

L. *o*-Quinodimethanes (Tables XVIa–c)†. *o*-Quinodimethanes are highly reactive intermediates that can be generated in various ways. The most widely used method is the reversible electrocyclic ring opening of benzocyclobutenes.[207] Substituents R strongly influence the rate of ring opening; the order of reactivity is R = NH_2(25°) > OH(80°) > NHCOR(110°) > COR(150°) > CH_2(180°) > H(200°) (approximate reaction temperatures are given in parentheses).[9] It is generally assumed that the species involved in the intramolecular Diels–Alder reactions are the *trans* dienes **175b**. This assumption is based on the premise that the *cis* dienes **175a** are sterically more crowded and,

† For specialized reviews of intramolecular Diels–Alder reactions involving this diene type, see refs. 9–14.

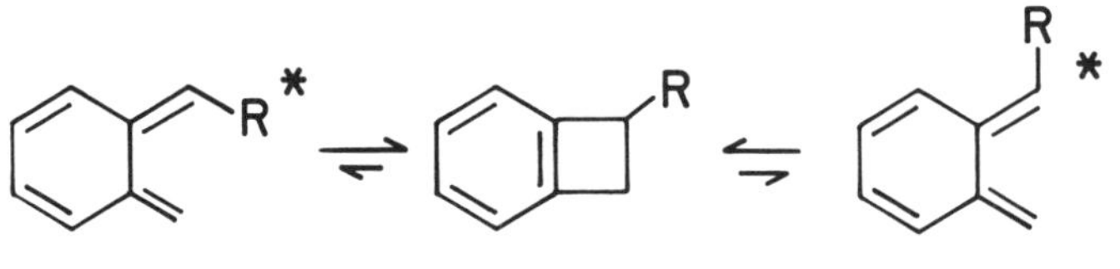

175a **175b**

if formed, undergo side reactions such as electrocyclic ring closure (when R = COR^1; Eq. 38)[127] or 1,5-hydrogen shifts (when R = CHR^1R^2).[126,127]

(Eq. 38)

Such 1,5-hydrogen shifts can occur to the total exclusion of the intramolecular cyclization, as illustrated in Eq. 39.[208]

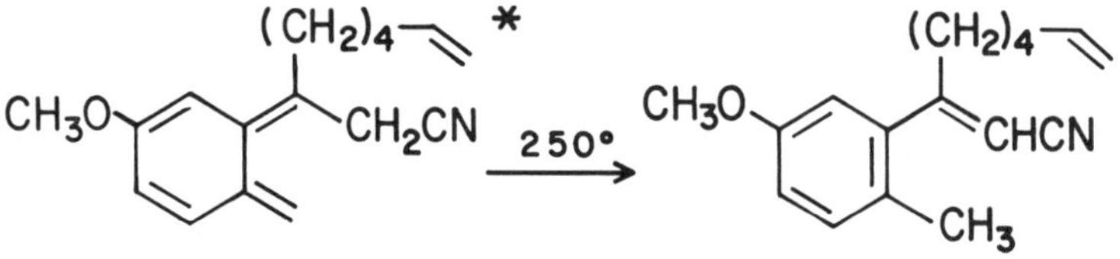

(Eq. 39)

A further argument for invoking the intermediacy of the *trans* dienes **175b** is discussed in the section on stereochemistry: for steric reasons, *trans*-fused products cannot be formed from *cis* dienes (such as **175a**) when the substrate contains three- or four-membered chains.

Other methods for the generation of *o*-quinodimethanes as substrates for intramolecular Diels–Alder reactions include the following: thermolysis of 1,3-dihydroisothionaphthene-2,2-dioxides (Eq. 40)[126,127] (1,5-hydrogen shifts

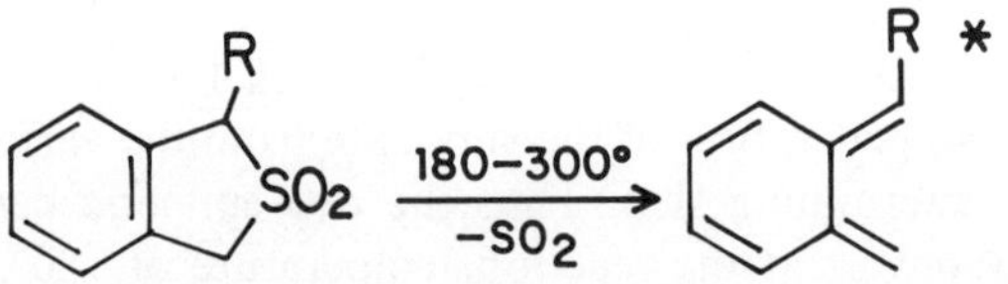

(Eq. 40)

can be minimized by keeping the temperature as low as possible);[209] thermolysis of isochromanones (Eq. 41);[210] photolysis of *o*-methylacylbenzenes (Eqs. 42 and 43);[211,212] (generation of *o*-quinodimethanes by the corresponding photochemical reaction of *o*-methylstyrenes appears to be quite limited in scope);[213] and base-induced 1,4 eliminations (Eqs. 44a[214] and 44b[215,216]).

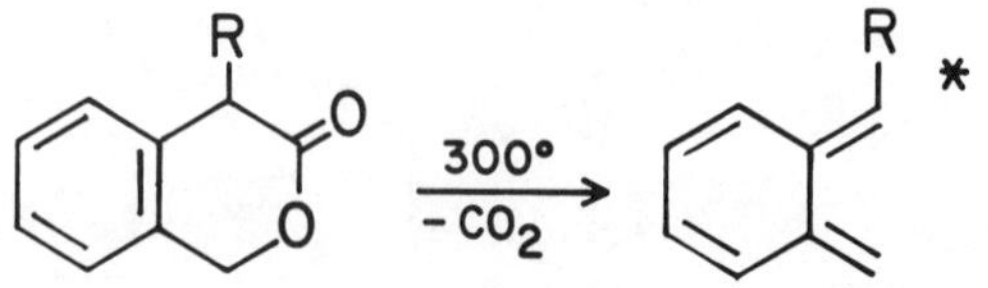

R, O, O; $\xrightarrow[-CO_2]{300°}$ R * (Eq. 41)

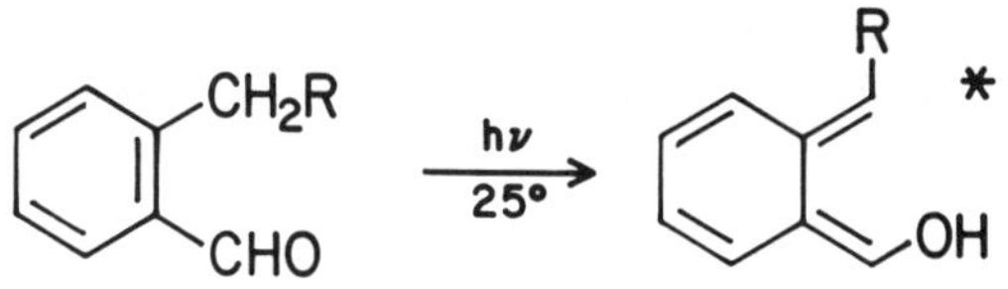

CH_2R, CHO; $\xrightarrow[25°]{h\nu}$ R, OH * (Eq. 42)

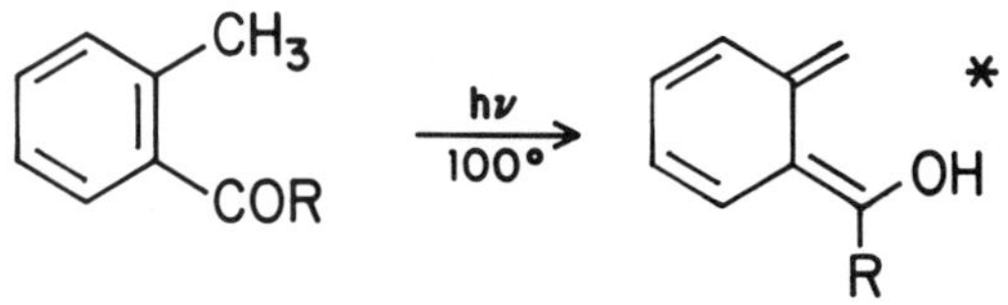

CH_3, COR; $\xrightarrow[100°]{h\nu}$ OH, R * (Eq. 43)

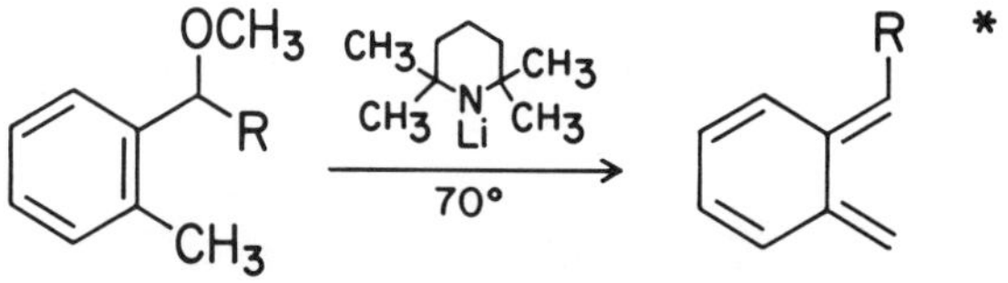

OCH_3, R, CH_3; CH_3, CH_3, N, Li, CH_3, CH_3; $\xrightarrow{70°}$ R * (Eq. 44a)

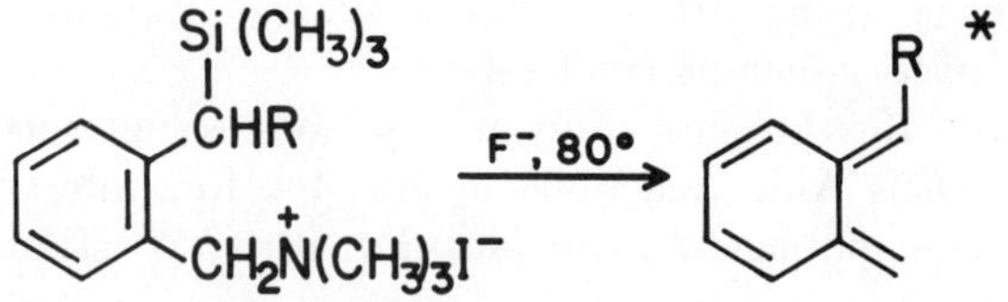

$Si(CH_3)_3$, CHR, $CH_2\overset{+}{N}(CH_3)_3I^-$; $\xrightarrow{F^-, 80°}$ R * (Eq. 44b)

o-Quinodimethanes are very reactive dienes in the intramolecular Diels–Alder reaction, and generation of the substrate from the various precursors is usually the rate-determining step. Thus the two epimeric benzocyclobutenes **176** do not interconvert at the reaction temperature of 180°, indicating that cycloaddition is more rapid than reversal to the substrate.[217,218] This appears to hold true even if the temperature at which the *o*-quinodimethane is generated is much lower, as in the case of substrate **177**.[216]

On the other hand, the *o*-quinodimethane **178**, which lacks the rate-enhancing ring in the chain, reverts to the benzocyclobutene precursor at a rate comparable to that of the cycloaddition.[9,207]

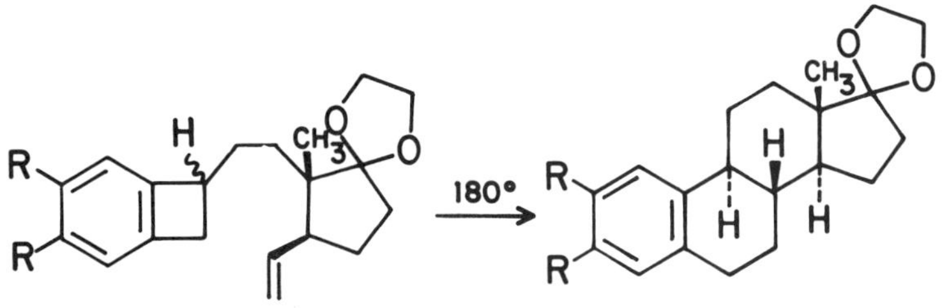

Intramolecular Diels–Alder additions of *o*-quinodimethanes with chains of one or two† members have not been reported, even though such substrates could, at least in principle, cyclize in the *cis* form (**175a**). There are many examples of substrates with three- and four-atom chains, and a few with five-membered chains. In one of the examples involving a five-atom chain, dimerization to a cyclooctane derivative competes with the intramolecular cycloaddition, even when the reaction is carried out in dilute solution (Eq. 45).[219]

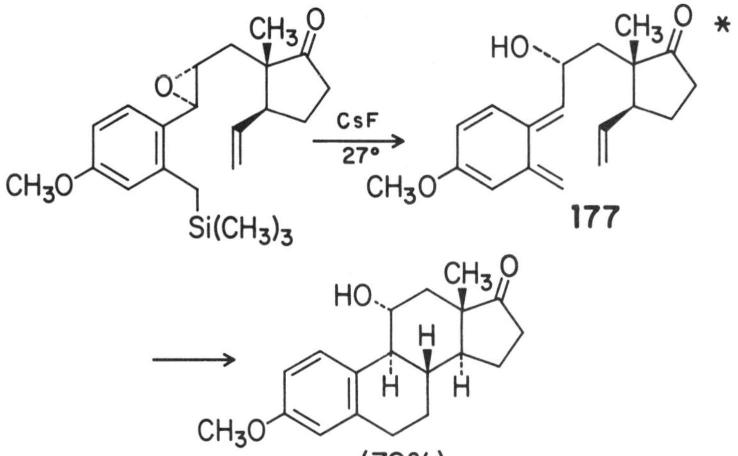

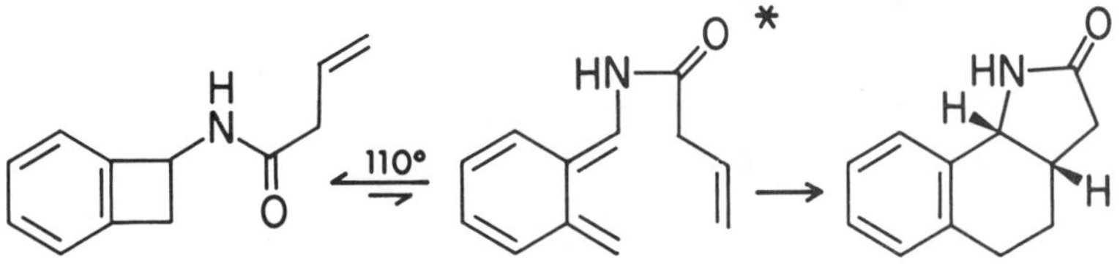

† See ref. 214 for an unsuccessful attempt to carry out such a cycloaddition.

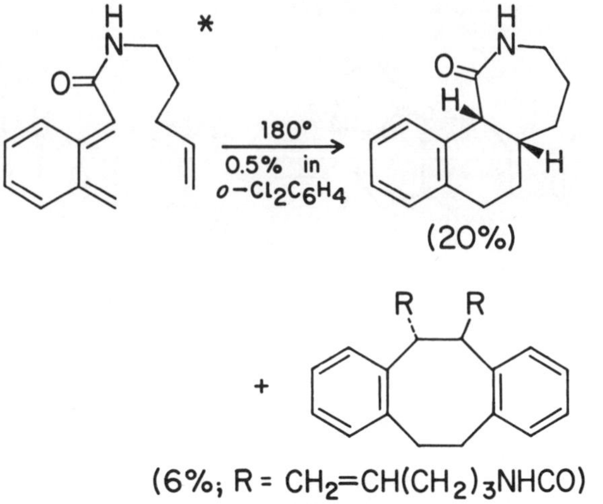

180°, 0.5% in $o\text{-}Cl_2C_6H_4$

(20%)

(Eq. 45)

+ (6%; R = $CH_2{=}CH(CH_2)_3NHCO$)

M. *o*-Dimethyleneheterocycles (Table XVId). In most of the known examples of this type, the diene is generated by acylation of a 2-methyl-3-indolemethylenimine as shown in Eq. 46. The *cis* diene is believed to lead to the less strained transition state in this case.[220]

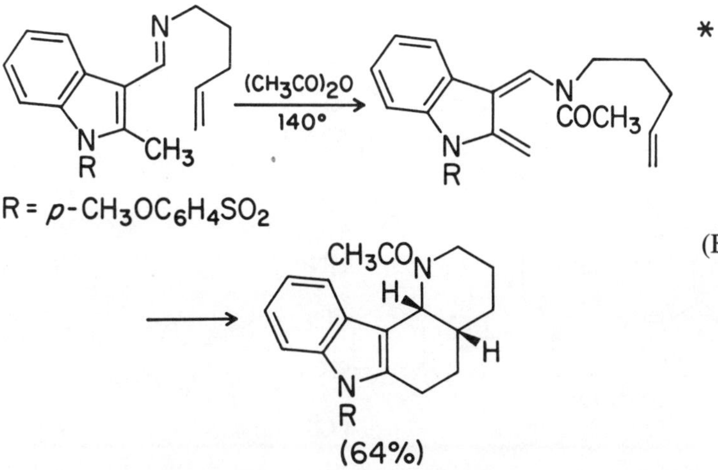

$(CH_3CO)_2O$, 140°

R = $p\text{-}CH_3OC_6H_4SO_2$

(Eq. 46)

(64%)

N. Aromatics (Tables XVIIa–d). As in the intermolecular Diels–Alder reaction, the reactivity of aromatic dienes increases in the order benzenes < naphthalenes < anthracenes. Thermal intramolecular Diels–Alder reactions involving benzene derivatives have been observed only with very powerful dienophiles such as arynes or activated acetylenes and allenes. Examples are the cyclization of the benzyne **179a** (generated from 5-bromo[3.3]paracyclophane and potassium *tert*-butoxide)[221] and the remarkably facile reaction of the allenecarboxamide **179b**.[222]

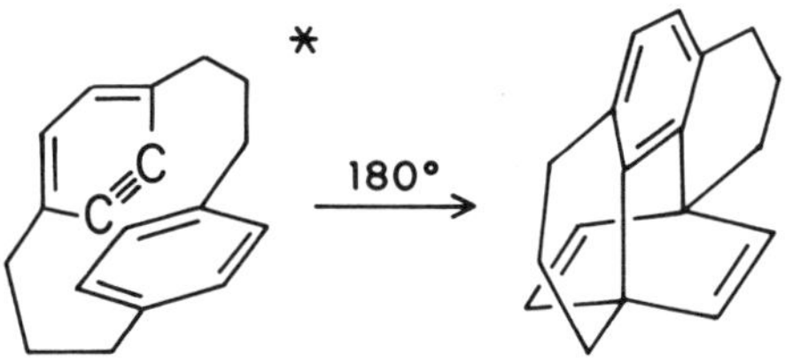

*

180°

179a (66%)

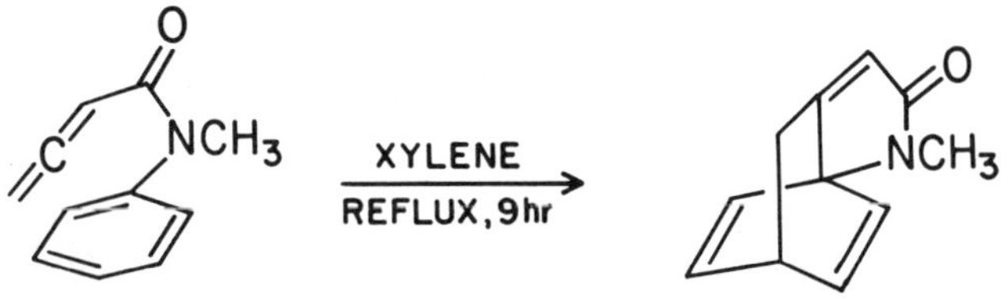

XYLENE
REFLUX, 9hr

179b (90%)

With naphthalenes, nonactivated dienophiles can often be used, although high reaction temperatures are normally required. Thus the propargylamide **179c** cyclizes at 190°; at a slightly higher temperature, acetylene is eliminated in a retro Diels–Alder reaction.[223]

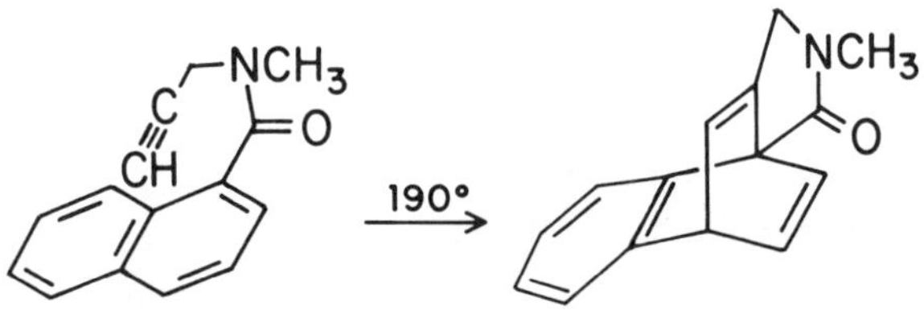

190°

179c (65%)

220°
(– HC≡CH)

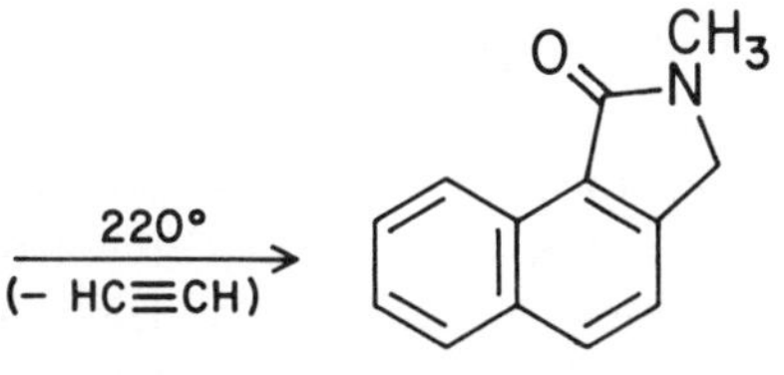

Anthracenes are excellent dienes in the intramolecular Diels–Alder reaction as long as the chain has more than two and less than six members. An unusual addition to the central ring of a phenanthrene derivative is given in Table XVIId/1.

O. Furans (Tables XVIIIa–d). Furans are equivalents of acyclic dienes at a higher oxidation state and would thus be useful in the synthesis of functionalized bicyclic ring systems. However, their utility is severely restricted by the tendency of Diels–Alder adducts of furan to revert to the starting substrates. The many instances of unsuccessful intramolecular Diels–Alder reactions of furan derivatives may simply be a consequence of an unfavorable equilibrium. Activation of the dienophile is often necessary; thus the amide **180a** is in equilibrium with the cycloadduct **181a** (ratio **181a**:**180a** = 0.9 at 120° in benzene, 1.6 at 80° in ethanol). When the amine **181b** is generated by reduction of the lactam **181a**, it reverts completely to substrate **180b**.[224] Lactam **181a** can be converted to the functionalized hexahydroisoindole **182** in two steps.[224]

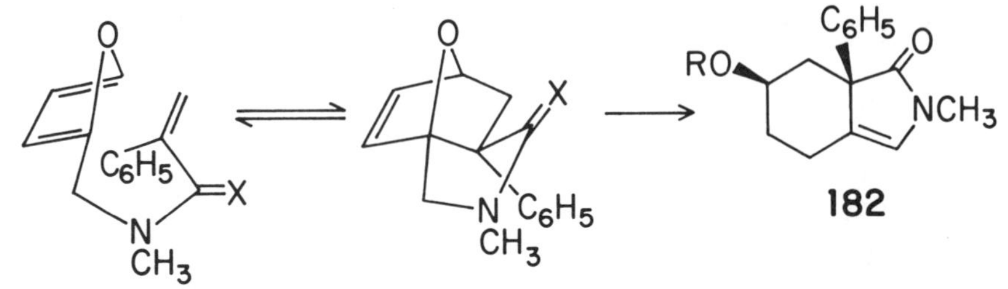

180a X = O
b X = H_2

181

182

There are a number of examples involving unactivated dienophiles, such as the formation of product **183**.[163] Adducts of type **183** are readily aromatized to

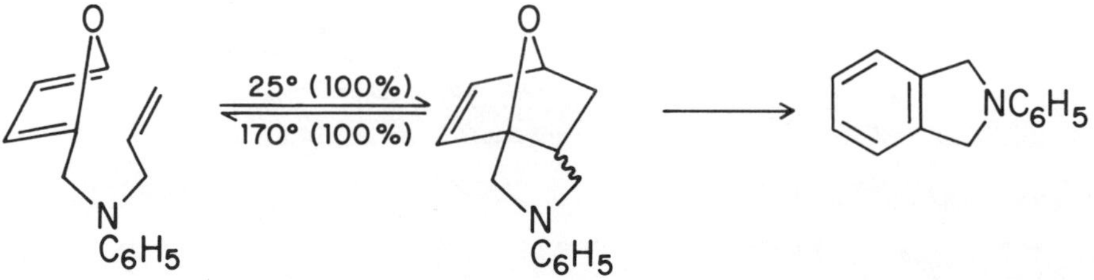

25° (100%)
170° (100%)

183

isoindolines.[163] Geminal substitution on the chain is often beneficial to the success of intramolecular Diels–Alder reactions involving furans (Table XVIIId/1a, 2d–g, 3). Most known intramolecular Diels–Alder reactions of furan derivatives involve substrates with three-atom chains (Tables XVIIIb–d); there are, however, a few examples with four-atom chains (Table XVIIIa/2–11) and, surprisingly, even with five-atom chains (Table XVIIIa/13, 15).

P. Acyclic and Semicyclic Dienes Containing Heteroatoms (Table XIX). Dienes of this type include α,β-unsaturated ketones, aza-, diaza-, and triazabutadienes; and α-methylenecyclohexanones. Two examples are shown in Eqs. 47[225] and 48. Substrate **184** is generated from 1,3-cyclohexanedione and citronellal.[226,227]

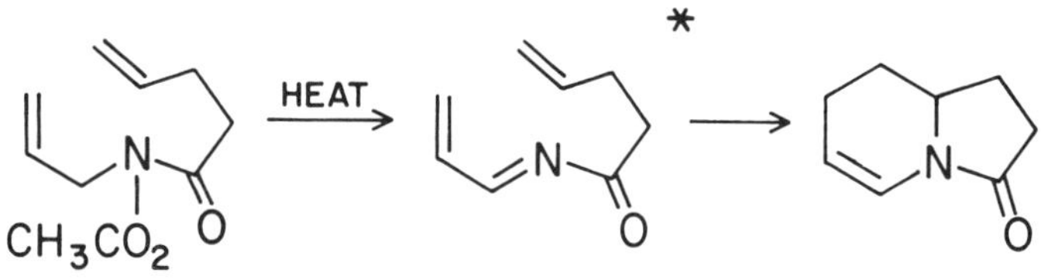

HEAT

CH_3CO_2

(73%)

(Eq. 47)

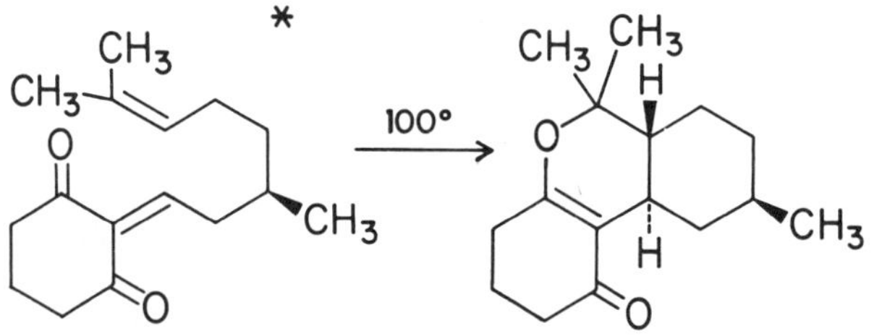

100°

184

(64 %)

(Eq. 48)

Q. *o*-Quinodimethanes Containing Heteroatoms in the Diene System (Table XX). *o*-Quinodimethanes containing nitrogen have been generated by silicon-assisted eliminations (Eq. 49)[228] and by flash vacuum pyrolysis of dihydrobenzoxazinones.[229] They appear to be as reactive as their all-carbon analogs.

The oxygen analogs have been obtained by 1,5-hydrogen shifts (Eq. 50)[230] or by dehydration (Eq. 51).[231]

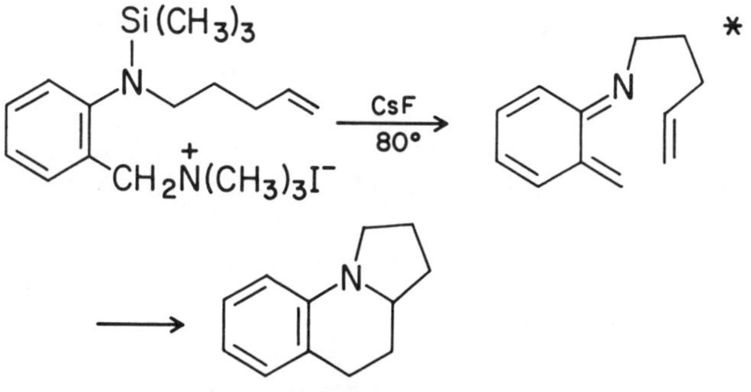

$Si(CH_3)_3$

$CH_2\overset{+}{N}(CH_3)_3I^-$

CsF 80°

(53%)

(Eq. 49)

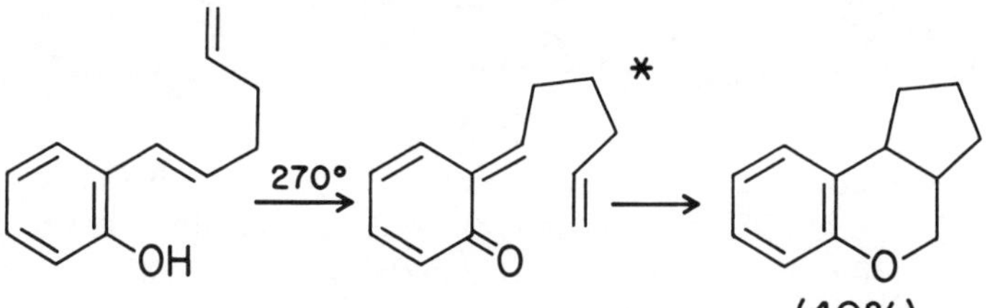

270°

(40%)

(Eq. 50)

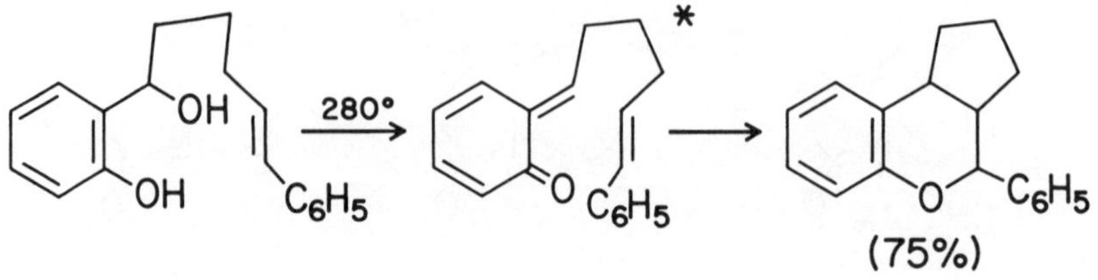

(Eq. 51)

A number of condensations of polyhydroxybenzenes with aldehydes such as citral or farnesal have been reinterpreted as proceeding by intramolecular Diels–Alder reactions because of the pronounced stereoselectivity observed.[232] An example is the reaction of phloroglucinol with citral (Eq. 52).[233,234]

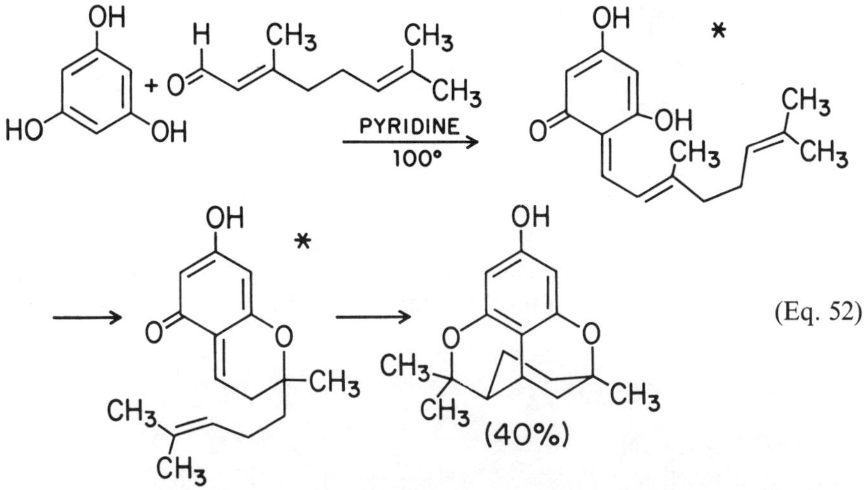

(Eq. 52)

R. Dihydropyridines and Pyridones (Table XXI). 1,2-Dihydropyridines of type **185**, generated by thermolysis of the Diels–Alder dimers, cyclize readily; the yields are lower when the chain is four-membered.[235] An intramolecular

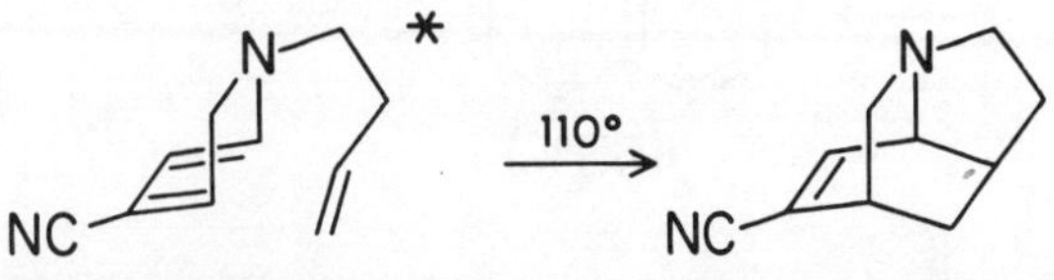

185 (48 %)

Diels–Alder reaction of a 3,4-dihydropyridine derivative is also known (Table XXI/12).[236] Successful intramolecular Diels–Alder reactions have been reported for a number of 3- and 4-pyridone derivatives (e.g., **186**) generated by a Claisen rearrangement.[237]

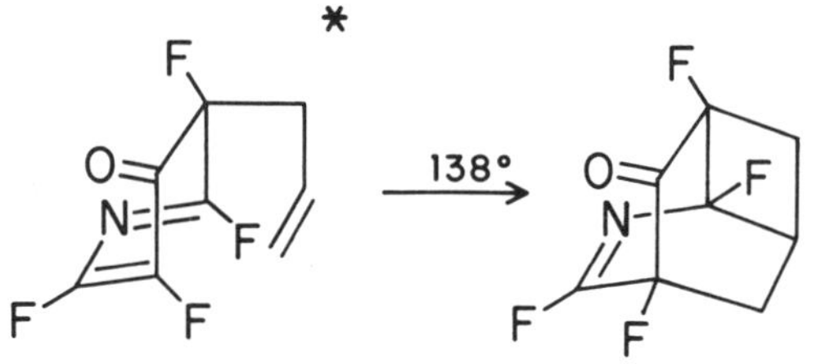

138°

186 (16%)

S. Pyridazines (Table XXII). Apart from a few dihydropyridazines (diazanorcaradienes), the only reported intramolecular Diels–Alder reactions of pyridazines involve ethers of type **187**. The initial adducts lose nitrogen to give dihydroxanthenes; when R is hydrogen, xanthenes formed by loss of hydrogen chloride are the only products isolated.[238–241]

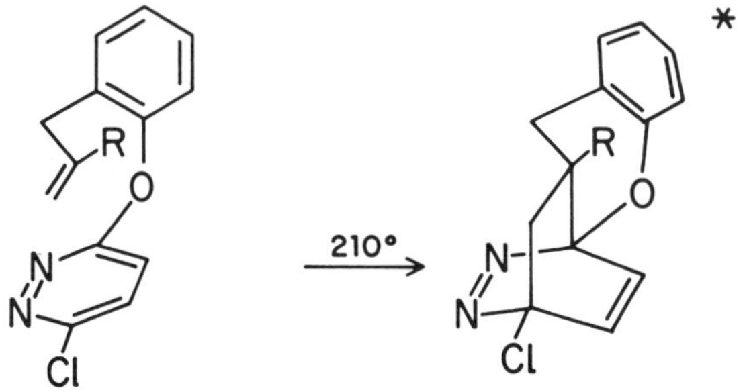

210°

187; R = CH_3

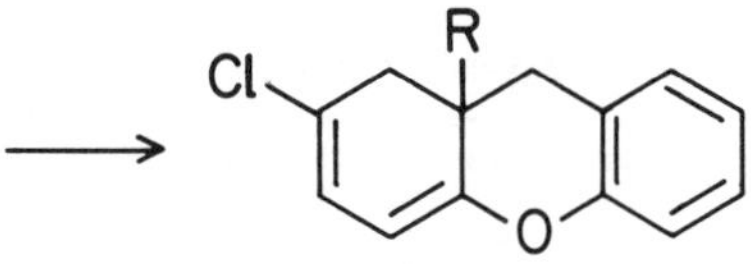

(99%)

T. Pyrimidines and Pyrimidones (Table XXIII). Pyrimidine ethers analogous to the pyridazines **187** cyclize at somewhat lower temperatures (140–150°). The initial adducts can be isolated in some reactions; in others, aromatization occurs with loss of hydrogen cyanide or acetonitrile (Table XXIII/12–22).[242] Similar eliminations are often observed in intramolecular Diels–Alder reactions of pyrimidones, as illustrated in Eq. 53.[243] With olefinic dienophiles, the initial adducts can be isolated in some reactions (Table XXIII/6, 7, 10, 11).[244,245]

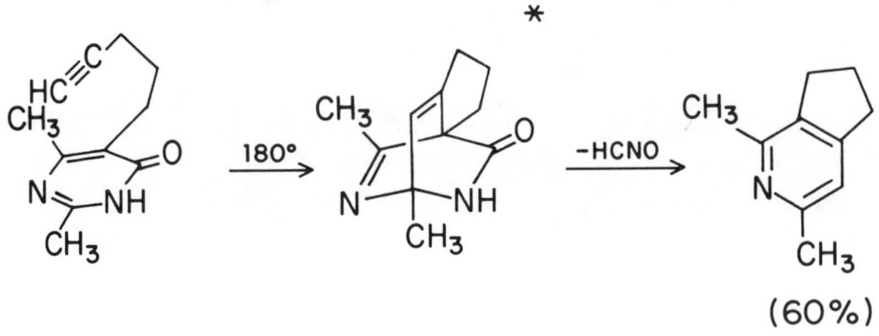

(Eq. 53)

U. α-Pyrones (Table XXIV). *N*-Allyl-*N*-methyl-α-pyrone-6-carboxamide (**188**) cyclizes at room temperature to give a single adduct of unknown stereochemistry, which loses carbon dioxide on heating to 70°.[167] Derivatives of α-pyrone-2-carboxylic acid appear to be poorer dienes; the isomeric amide **189** requires heating to 100°.[167] α-Pyrones are thus equivalents of acyclic dienes at a higher oxidation state. They are more reactive dienes than the corresponding butadienes, and the cyclohexadienes that are formed can be reduced to introduce the desired stereochemistry at the ring junction.[134,167] The cyclohexadiene products can also be used in inter- or intramolecular Diels–Alder reactions; an example of the latter is given in Table XXIV/19.[167] With acetylenic dienophiles, aromatic compounds are obtained (Eq. 54).[167]

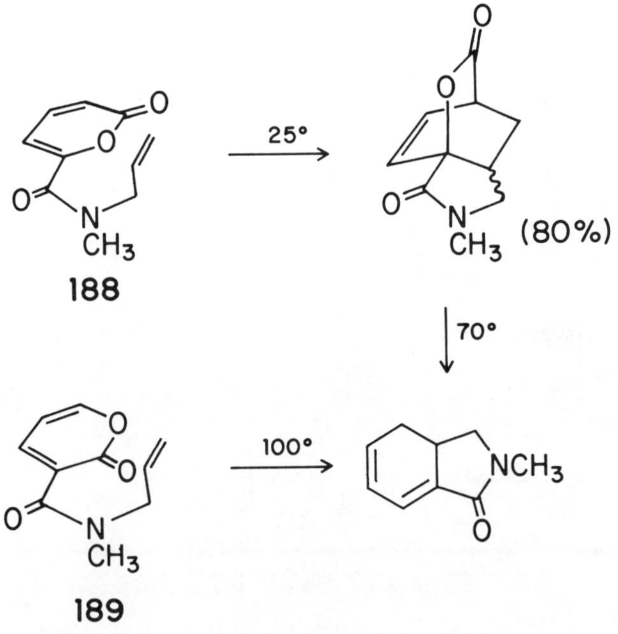

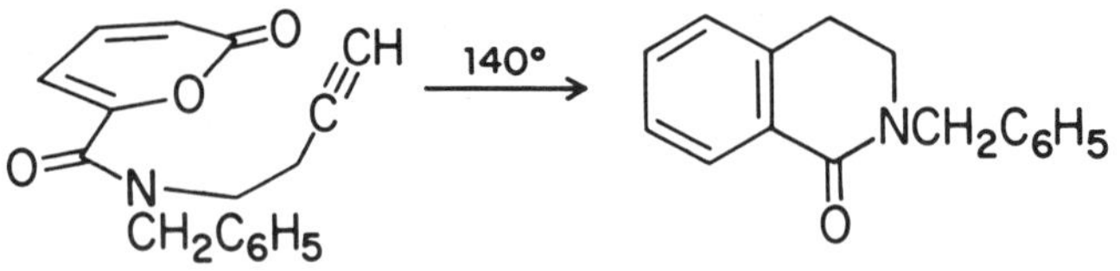

(Eq. 54)

Benzopyrones also undergo intramolecular Diels–Alder reactions but, expectedly, require much higher reaction temperatures (Eq. 55).[246,247]

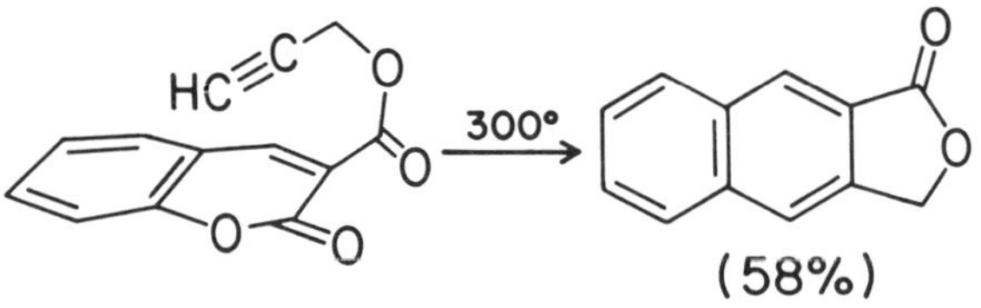

(Eq. 55)

V. Miscellaneous Heterocycles (Table XXV). The dienes listed under this heading include dihydroazepines, indolenines, isoindoles, isobenzofurans, isoquinolines, acridines, phosphacycloalkadienes, thiophenes, metallocycles, oxazoles, and triazines. Metallocycles of type **190**, generated by reaction of suitable enediynes with cyclopentadienylcobalt dicarbonyl, cyclize to give adducts **191**, from which the free ligands can be obtained by treatment with cupric chloride. The cyclization may not be concerted.[248]

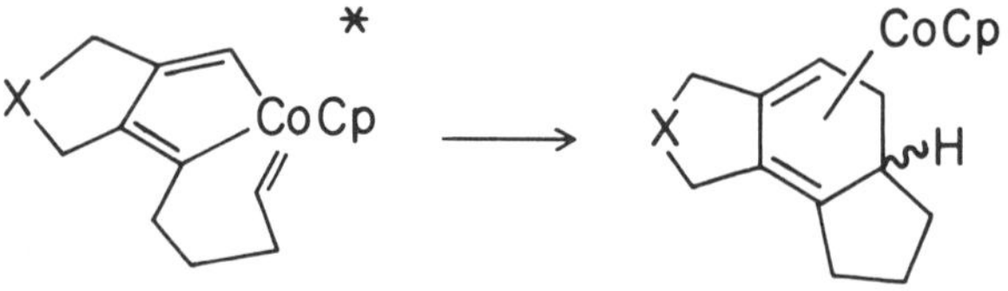

Oxazoles are useful in the construction of fused furans (Eq. 56).[154,164,249]

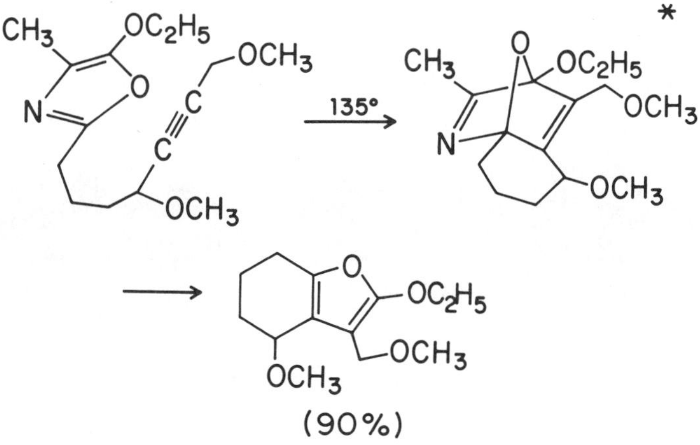

(Eq. 56)

The Intramolecular Homo Diels–Alder Reaction (Table XXVII)

The reaction of dienophiles with certain 1,4 dienes—for instance, the addition of dicyanoacetylene to barrellene (Eq. 57)[250]—has been termed the *homo* Diels–Alder reaction. There has been one unsuccessful attempt to

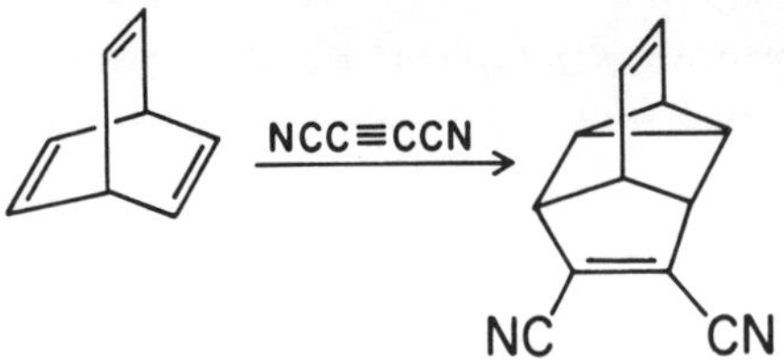

NCC≡CCN → NC, CN (Eq. 57)

demonstrate the intramolecular equivalent: substrate **192** either polymerizes or undergoes intramolecular ene reactions.[251] A few examples of reverse intramolecular homo Diels–Alder reactions are known (Table XXVIIIm).

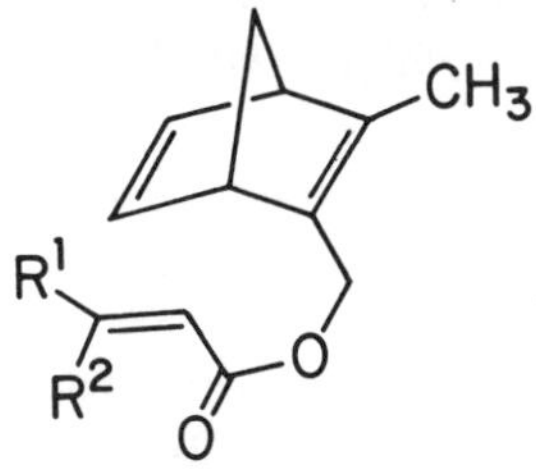

CH_3, R^1, R^2, O, O

192

The Reverse Intramolecular Diels–Alder Reaction (Tables XXVIIIa–m)

For most classes of dienes, the reverse reaction does not seem to occur readily and it is thus of little importance in limiting the utility of the intramolecular Diels–Alder reaction. The adducts to furan are the only major exception; their propensity for reversing to the substrates is discussed earlier.

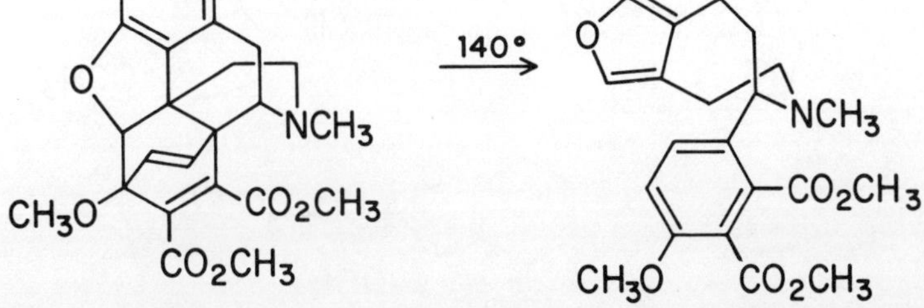

CH_3O, O, NCH_3, CH_3O, CO_2CH_3, CO_2CH_3 **193** —140°→ CH_3O, O, NCH_3, CO_2CH_3, CH_3O, CO_2CH_3 (90%)

Tables XXVIII show that reverse intramolecular Diels–Alder reactions often require fairly high temperatures, even when the resulting diene is a benzene derivative. Thus thermolysis of the adduct of thebaine with dimethyl acetylenedicarboxylate (**193**) occurs only at 140°.[252] When substrates and adducts are in equilibrium, advantage can sometimes be taken of the usually lower boiling point of the adduct to drive the reaction in the desired direction. For instance, *N*-(4-pentenyl)isoindole and its adduct are in equilibrium with an equilibrium constant of unity at 150° (Eq. 58). The pure adduct can be obtained in 80% yield by slow vacuum distillation of the mixture at 160–180° bath temperature.[253]

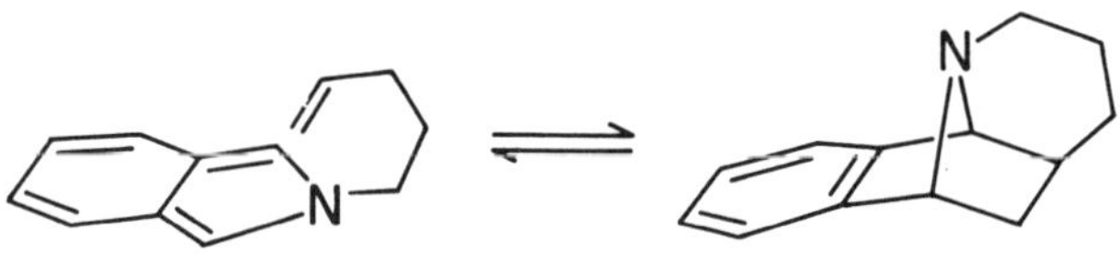

(Eq. 58)

SYNTHETIC UTILITY

Natural Product Syntheses

The following is a survey of natural products total syntheses that involve intramolecular Diels–Alder reactions. Some biosyntheses proposed to proceed via such reactions are also mentioned.† Examples are given for each diene type; these are followed by a listing of other natural products that have been prepared using that diene type.

A. Terpenoids. The examples illustrated here show the syntheses of selina-3,7(11)diene, gibberellic acid, patchouli alcohol, alnusenol, farnesiferol, and evodone.

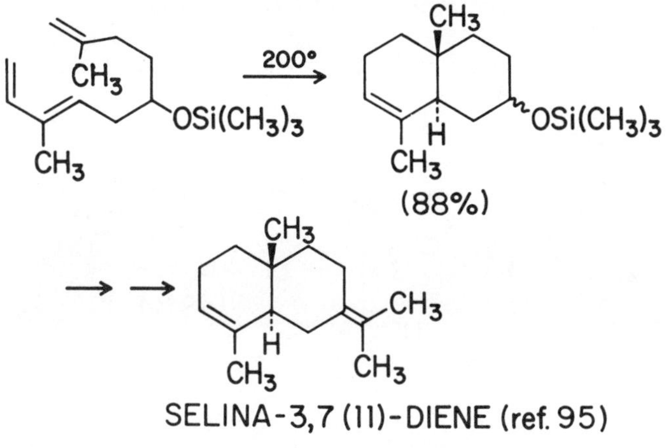

200°

$OSi(CH_3)_3$

(88%)

SELINA-3,7 (11)-DIENE (ref. 95)

† For a discussion of this topic see ref. 170.

Other terpene syntheses using acyclic dienes are γ-cadinene,[254] khusitene,[65] eudesmol,[144] epizonarene,[156] δ-ambrinol,[255] torreyol,[141] α-himachalene,[256,257] fichtelite,[258] *eremophilane* and *valencane* sesquiterpenes,[19] and the abietane ring system.[96] The biosynthesis of ircinianin has been proposed to involve an intramolecular Diels–Alder reaction.[259] A vinyldihydrofuranone is used in the synthesis of marasmic acid.[37]

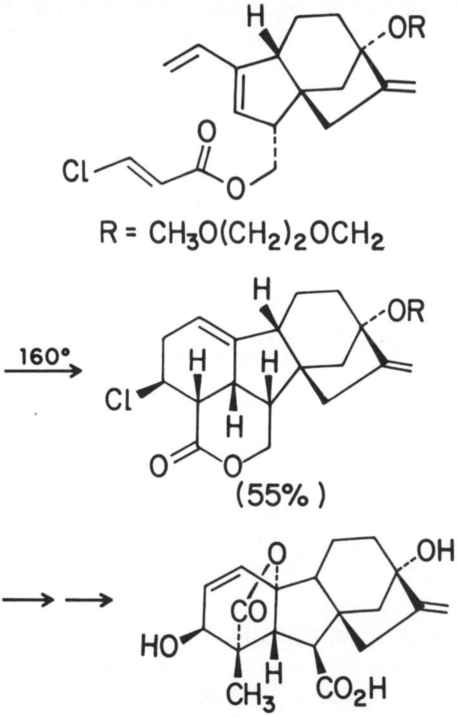

H
OR
O
Cl
O

R = $CH_3O(CH_2)_2OCH_2$

H
OR
160°
H H
Cl
H
O O
(55%)

O
OH
CO
HO
H
CH_3 CO_2H

GIBBERELLIC ACID (ref. 260)

Terpene syntheses using cyclopentadienes are cedrol and cedrene,[261] cedranediol,[197] and sativene.[79]

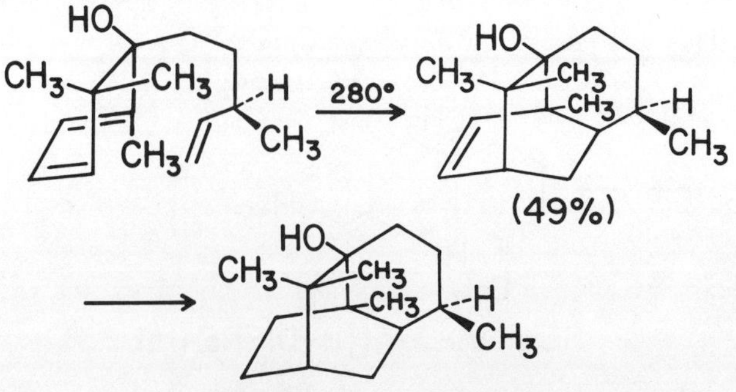

HO
CH_3 CH_3 H
280°
CH_3 CH_3

HO
CH_3 CH_3
CH_3 H
CH_3

(49%)

HO
CH_3 CH_3
CH_3 H
CH_3

PATCHOULI ALCOHOL (ref. 71)

Other terpene syntheses using cyclohexadienes are norpatchoulenol,[262–264] patchouli alcohol analogs,[162,196,265] seychellene,[152] norseychellanone,[180] 2-pupukeanone,[266] 9-pupukeanone,[190,267] 9-isocyanopupukeanone,[268] khusimone,[116] and the major acetylation product of thujopsene.[117]

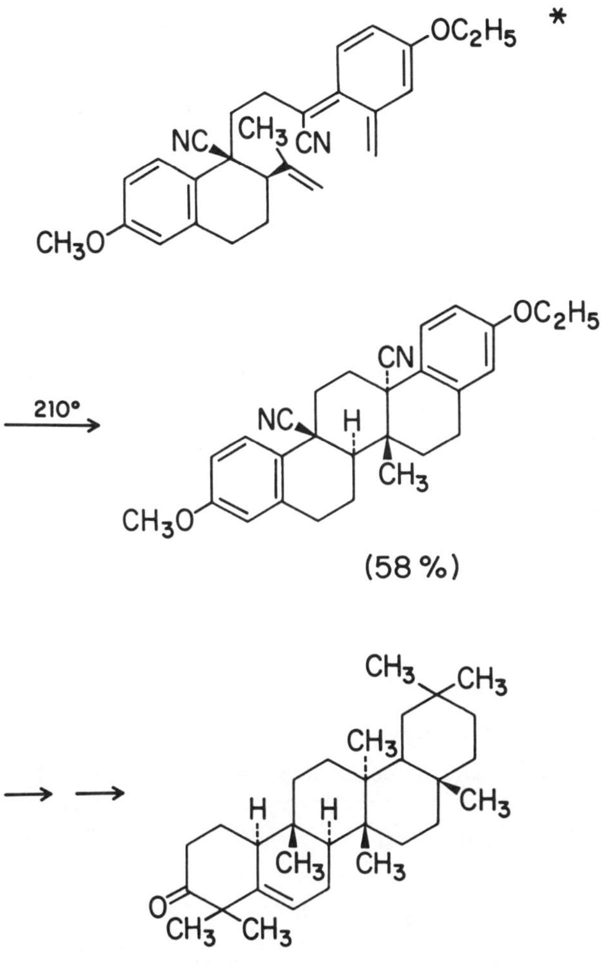

*

210° →

(58 %)

→ →

ALNUSENONE (ref. 269)

Other terpene syntheses using *o*-quinodimethanes are friedeline (using the same intermediate as for alnusenone)[269] hibaol,[270] quassinoids,[271] atisine,[272] garryine,[272] veatchine,[272] the aphidicolan ring system,[273] and gibberellin A_{15}.[272] An α-methylenecyclohexadienone is the substrate for a deoxybruceol synthesis (see Table XX/25).[274,275]

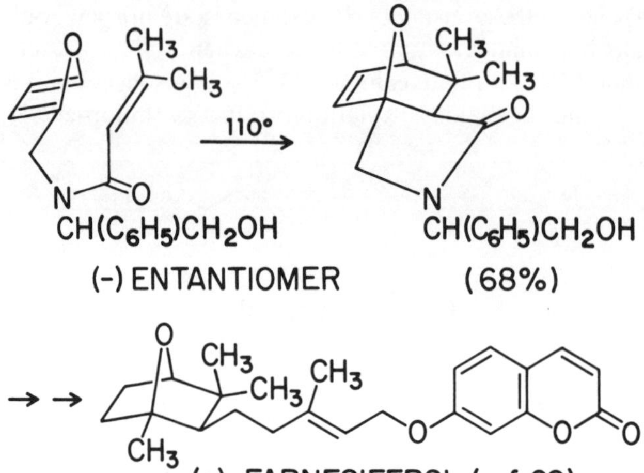

110°

$CH(C_6H_5)CH_2OH$

(-) ENTANTIOMER

(68%)

(+)-FARNESIFEROL (ref. 69)

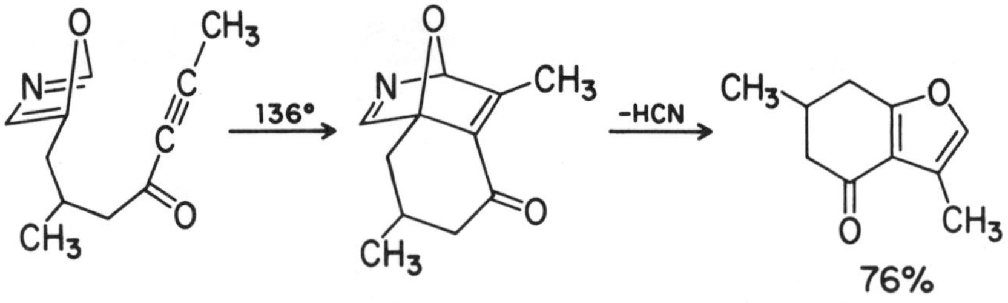

136°

-HCN

76%

EVODONE (ref. 154)

Other terpene synthesis using isoxazoles are ligularone[164] and petasalbine.[164]

B. Alkaloids. The syntheses of lysergic acid, coniceine, heliotridine and retronecine, vincadifformine, chelidonine, and actinidine are illustrated.

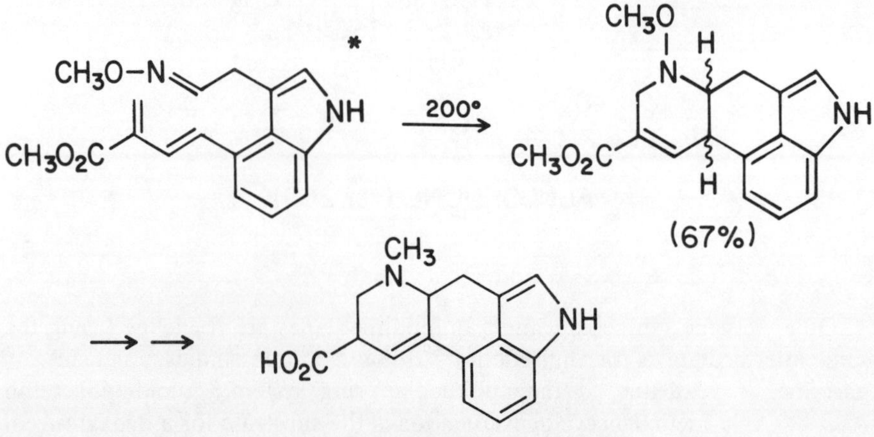

200°

(67%)

LYSERGIC ACID (ref. 193)

Other alkaloid syntheses using acyclic dienes are lycorine,[20,110] dihydrolycoricidine,[18] pumiliotoxin C,[84,103,104] aspidospermine,[109] slaframine,[276] cyclostachine A and B,[91] *epi*-lupinine,[277] dendrobine,[188,278] *epi*-dendrobine,[17] and *epi*-galanthan.[111] The lycorine skeleton has also been made using a vinylpyrroline.[105]

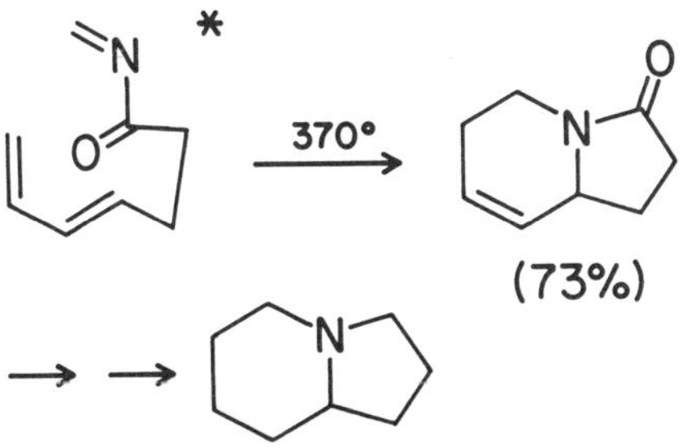

370°

(73%)

CONICEINE (ref. 136)

Other alkaloid syntheses using acylimines are tylophorine[136] and elaeokanine A and B.[136]

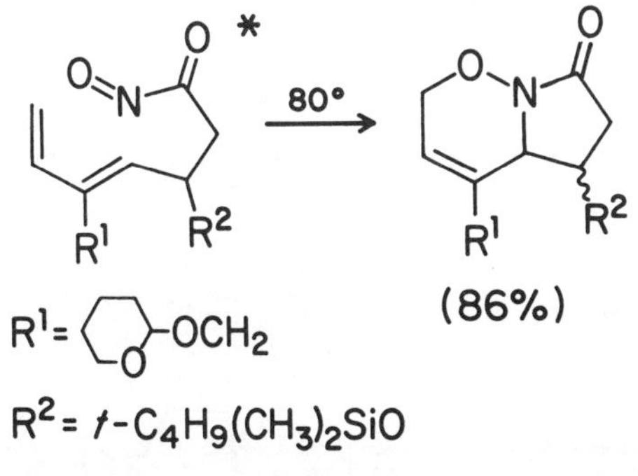

80°

(86%)

R^1 = OCH_2

R^2 = t-$C_4H_9(CH_3)_2SiO$

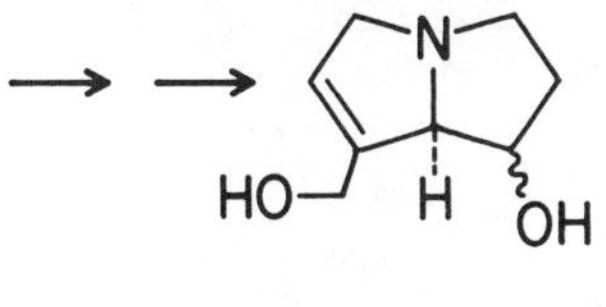

HELIOTRIDINE (α-OH) AND RETRONECINE (β-OH) (ref. 142)

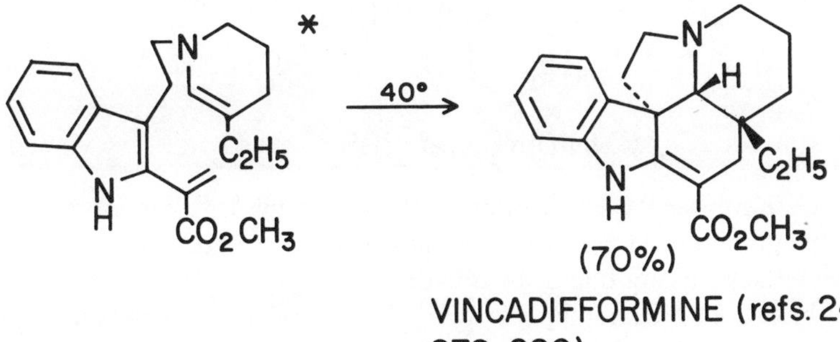

40°

(70%)

VINCADIFFORMINE (refs. 24-26, 279, 280)

The cyclization of a vinylindole has also been used in the synthesis of andranginine.[281] This approach mimics the proposed biosynthesis of *aspidosperma* alkaloids. Intramolecular Diels–Alder reaction of the corresponding dihydropyridine (diene) to the acrylate ester (dienophile) has been suggested to lead to the *iboga* alkaloids;[282] an example where such a cyclization proceeds at room temperature is known (Table XXI/16;[283] see also Ref. 284).

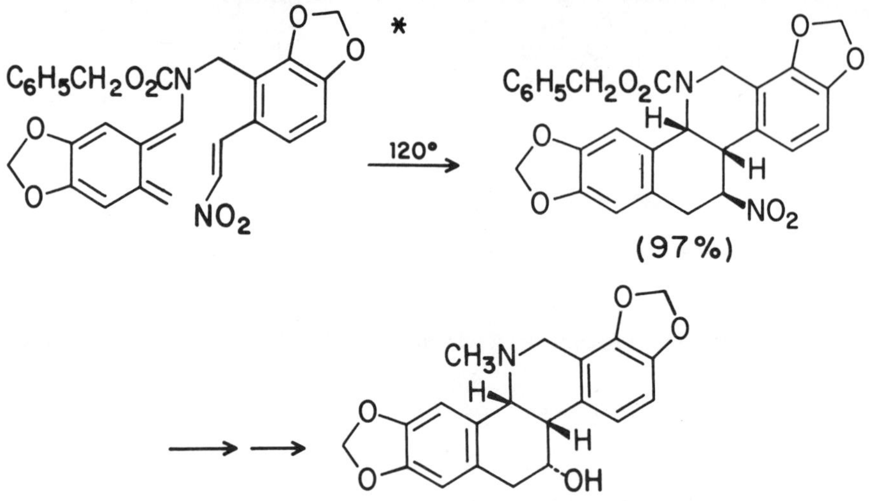

$C_6H_5CH_2O_2CN$ → 120° → $C_6H_5CH_2O_2CN$, H, H, NO_2 (97%)

→ → CH_3N, H, H, OH

CHELIDONINE (refs. 285, 286)

Other alkaloid syntheses using *o*-quinodimethanes are aspidospermine,[287] and precursors for diterpene alkaloids[288] and indole alkaloids.[220]

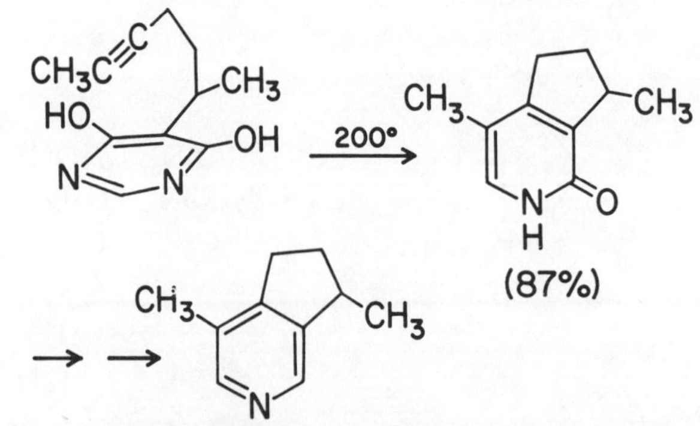

$CH_3C\equiv C$, CH_3, HO, OH, N, N → 200° → CH_3, CH_3, N, H, O (87%)

→ → CH_3, CH_3, N

ACTINIDINE (ref. 289)

α-Pyrones have been used to prepare the morphine fragments octahydrobenzofuroisoquinoline (lacking ring B of the alkaloid)[134] and 4a-aryldecahydroisoquinoline (lacking ring B and the dihydrofuran ring).[167] 4a-Aryloctahydroisoindoles have been prepared by this method[167] as well as by an intramolecular Diels–Alder reaction involving a furan as the diene.[224]

C. Lignans. The examples show the syntheses of γ-apopicropodophyllin and carpanone. Cyclization of vinylaromatics has also been used in the preparation of attenuol,[294] collinusin,[295] and justicidin B.[295]

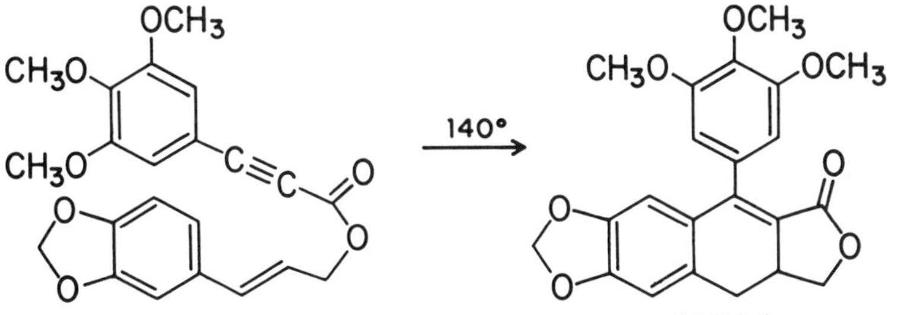

140°

(29%)
γ-APOPICROPODOPHYLLIN
(refs. 290, 291; *cf.* also refs. 292, 293)

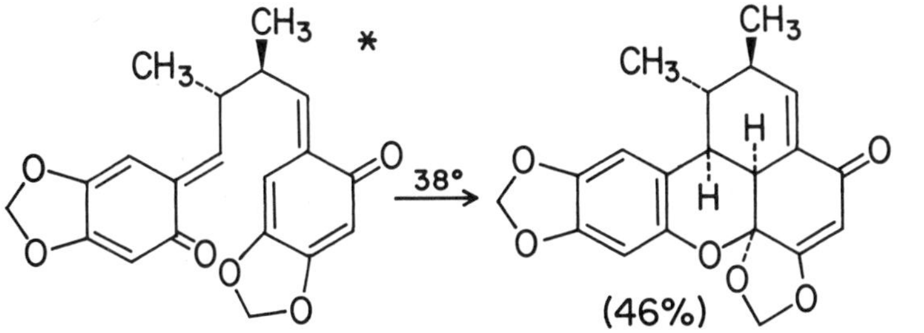

38°

(46%)
CARPANONE (ref. 296)

D. Steroids. The examples show the syntheses of estrone methyl ether and 11-ketotestosterone. With the exception of the latter, and one approach using an acyclic diene,[146] all steroid syntheses reported so far involve *o*-quinodimethanes as the diene. They differ in the approach to the precursor and the method of generating the *o*-quinodimethanes.

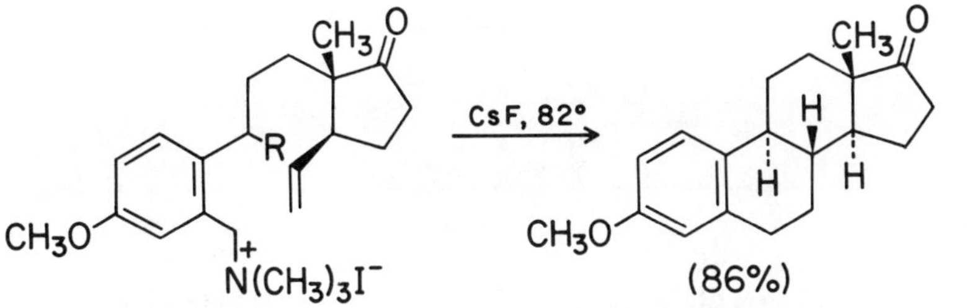

CsF, 82°

R = $(CH_3)_3Si$

(86%)
ESTRONE METHYL ETHER
(ref. 297)

Other syntheses of estrone and estrone types are given in Refs. 126, 129, 131, 212, 216–218, and 298–304. Homoestrone types are given in Ref. 305, pregnane types in Ref. 306, homoandrostanes in Ref. 307, ring D aromatic steroids in Ref. 308, cortisone in Ref. 309, and chenodeoxycholic acid in Ref. 310.

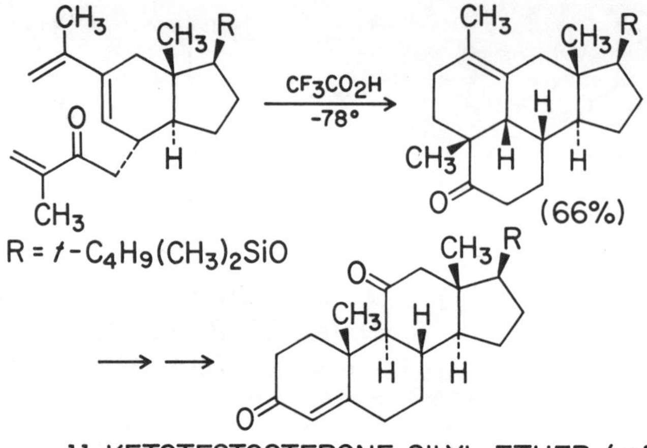

CF_3CO_2H, -78° (66%)

R = t-$C_4H_9(CH_3)_2SiO$

11-KETOTESTOSTERONE SILYL ETHER (refs. 61, 311)

For the synthesis of an aza- and diazasteroid skeleton, see Ref. 108.

E. Miscellaneous Natural Products. The examples show the synthesis of lachnanthocarpone and the ionophore antibiotic X-14547A.

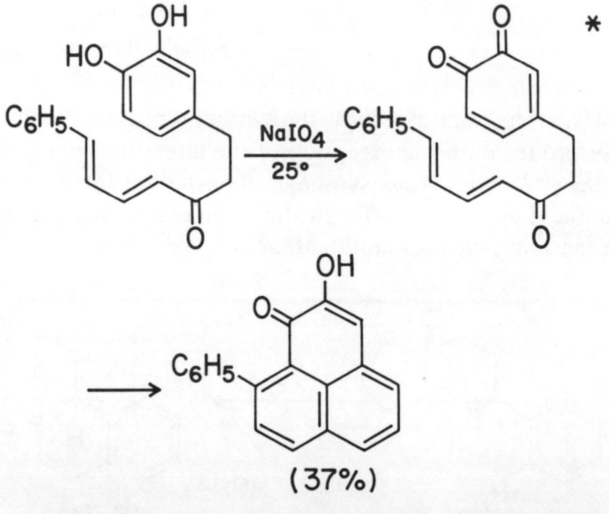

$NaIO_4$, 25° (37%)

LACHNANTHOCARPONE (THIS SYNTHESIS MIMICS THE PROPOSED BIOSYNTHESIS) (ref. 170)

$R = (C_6H_5)_2C_4H_9SiO$ (70%)

X-14547A
(IONOPHORE ANTIBIOTIC; ref. 138;
cf. also refs. 139, 312)

Other syntheses using acyclic dienes are coronafacic acid,[90,165] the indole nucleus of cytochalasins[313] (see also refs. 86, 137, 172, and 314), part of the aglycone of chlorothricin,[133] compactin,[100] and endiandric acid.[315–317] The latter synthesis follows the proposed biogenetic path.[318]

The biosynthesis of ikarugamycin has been proposed[319] to proceed via intramolecular Diels–Alder reaction of a 1,3,8-cyclododecatriene derivative.[320,320a]

The synthesis of a cannabinoid by intramolecular Diels–Alder reaction of an α-methylenecyclohexanone was discussed earlier. For other syntheses of cannabinoids, see refs. 321 and 322; see also Table XX/7–9 and 11–25.

The synthesis of mansonone E, an *o*-naphthoquinone, involves an intramolecular Diels–Alder reaction of a benzyne with a furan.[176] Resistomycin is obtained by cyclization of an isobenzofuran.[323] The key step in a tetradeoxydaunomycinone synthesis is the intramolecular Diels–Alder reaction of a benzopyrone.[247]

Miscellaneous Syntheses

The syntheses of iceane and triquinacene are illustrated. For the preparation of anti-Bredt's rule bridgehead alkenes, see Table V. For a triquinacene

(21%)

ICEANE (ref. 324
cf. also ref. 325)

synthesis by reverse intramolecular homo Diels–Alder reaction, see ref. 327. A synthesis of semibullvalenes using the intramolecular Diels–Alder reaction is reported in ref. 328 (see Table XXII/1–3a). A general synthesis of isoindolines using intramolecular Diels–Alder reactions of furans was discussed

-78 TO 25°

(6%)

TRIQUINACENE
(ref. 326)

earlier. For other syntheses of isoindole derivatives using this approach, among others, see refs. 58, 158, and 329–343. Dihydroisobenzofuranones have been prepared by a similar route.[344] For syntheses of rigid dopamine analogs, see ref. 345.

Net Formation of One Ring

The net formation of one ring can be accomplished by cleavage of either of the two rings generated in the intramolecular Diels–Alder reaction. This approach can be used to force the formation of regioisomers that are disfavored in the intermolecular Diels–Alder reaction. In the cyclization of substrates **194** and **195**, the isomers obtained exclusively are the minor products of the corresponding intermolecular Diels–Alder reaction.

194

25°

(35%)

(refs. 200, 346)

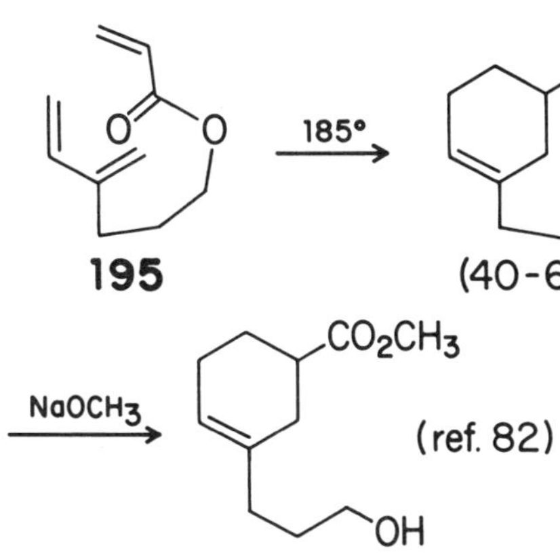

185°
195
(40-60%)
NaOCH3
CO2CH3
OH
(ref. 82)

Cyclization of substrate **196** followed by cleavage of the six-membered ring in the product results in net formation of a five-membered ring. (Another example of cleavage of the six-membered ring is mentioned in the section on steroid synthesis.)[61,311]

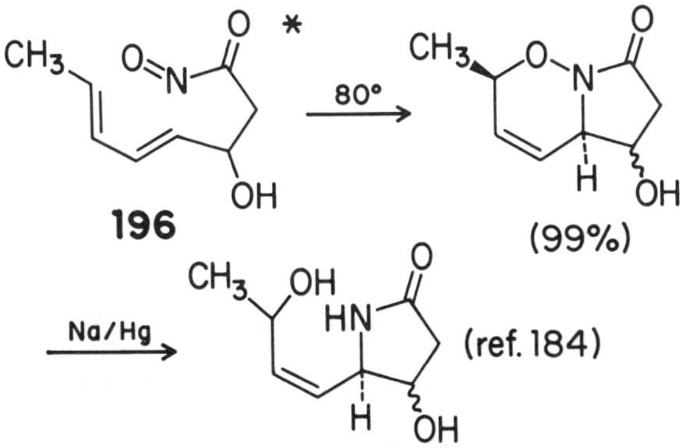

80°
196
(99%)
Na/Hg
(ref. 184)

Another approach, termed *quasi*-intramolecular Diels–Alder reaction, employs a Lewis acid in the chain that can complex with the dienophile and thus provide the regiochemical and stereochemical control usually associated with a true intramolecular Diels–Alder reaction (Eq. 59).[347]

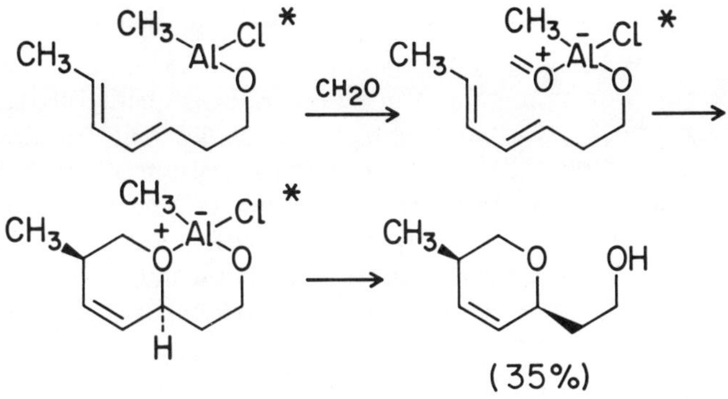

CH2O
(Eq. 59)
(35%)

This strategy has been used in a synthesis of pseudomonic acids A and C.[348]

Cleavage of *both* rings of an intramolecular Diels–Alder adduct has been used to generate acyclic systems with complete stereocontrol.[178b]

EXPERIMENTAL CONDITIONS

Many intramolecular Diels–Alder reactions require no more than heating the substrate in a solvent that boils at the cyclization temperature. The temperature at which the reaction proceeds at a convenient rate should be determined on a small scale; these scouting experiments can be carried out directly in an NMR tube. Since the rates of most intramolecular Diels–Alder reactions are essentially independent of solvent polarity, any inert solvent will give satisfactory results. Aliphatic and aromatic hydrocarbons, chlorinated aliphatic and aromatic hydrocarbons, and aromatic ethers are used most frequently; others include dimethylformamide, aliphatic and aromatic nitriles, esters of aromatic acids (including dialkyl phthalates), and dialkylanilines. Some intramolecular Diels–Alder reactions can be carried out without a solvent. Where dimerization or other bimolecular reactions of substrate or product are a problem, substrate concentrations should be kept low. In one example, the yield is almost doubled by reducing the substrate concentration from 10% to 0.6%.[19] High-dilution techniques, such as slow addition of the substrate solution by means of a motor-driven syringe to a large volume of refluxing solvent, may be required, especially with substrates having chains containing more than four atoms. An example is given in the Experimental Procedures section.

Reactions that require very high temperatures are best carried out using sealed glass tubes or autoclaves. The solvent should have a boiling point low enough to permit ready removal, but not so low as to cause excessive pressure buildup at the reaction temperature; toluene and xylene are suitable. Sealed-tube techniques also permit complete removal of atmospheric oxygen by a series of freeze-thaw cycles (see Experimental Procedures section). Use of a protective atmosphere such as nitrogen or argon may be necessary to avoid peroxide formation,[123] aromatization,[205] or other oxygen-induced side reactions.

Cis-trans isomerizations or other undesired reactions catalyzed by acid can be minimized by washing the glass apparatus with a base such as ammonium hydroxide solution or by pretreatment with *N,O*-bis(trimethylsilyl)acetamide. Acid scavengers such as tertiary amines or epoxides may be added to the reaction medium. Free-radical-initiated polymerization of the substrate is suppressed by addition of an inhibitor; among those used are methylene blue,[39,112,258] 3-*tert*-butyl-4-hydroxy-5-methylphenyl sulfide,[105] 2,6-di-*tert*-butyl-*p*-cresol,[72] 4,4′-thiobis-(6-*tert*-butyl-3-methylphenol),[110,145] and hydroquinone.[249] There is, however, at least one report of charring in the presence of hydroquinone.[246]

Intramolecular Diels–Alder reactions can also be carried out in the gas phase. Either stationary or flow systems can be used; examples are given in the Experimental Procedures section. A modification of a flow system involves passing substrate solutions through vertical tubes packed with glass helices and heated to temperatures of up to 600°.[109,136]

Lewis acids catalyze some intramolecular Diels–Alder reactions. A list of catalysts is given in the section dealing with that subject (p. 17). The solvents of choice for catalyzed reactions appear to be chlorinated hydrocarbons such as methylene chloride or carbon tetrachloride. One equivalent of catalyst is required when the product is a stronger Lewis base than the substrate.[41]

EXPERIMENTAL PROCEDURES

Deoxygenation—Sealed Tube Technique. The neck of a heavy-walled glass tube is constricted, and the substrate solution is introduced using a disposable pipette reaching beyond the constriction in order to avoid contamination of the glass seal. The tube should not be filled to more than 75% of capacity. The tube is then connected through a stopcock to a source of high vacuum, preferably a mercury or oil diffusion pump. Ideally, the connection is made by glass, but in practice heavy-walled, flexible vacuum tubing is usually satisfactory. With the stopcock closed,† the tube is cooled with liquid nitrogen. It is then evacuated by opening the stopcock. After full vacuum has been reached, the stopcock is closed and the tube is allowed to warm and kept at room temperature until the solvent has melted. It is best to avoid solvents that melt above 0°, such as benzene or *p*-xylene; toluene (mp −95°) or *m*-xylene (mp −50°) are preferable. The tube is then cooled with liquid nitrogen and evacuated again. This completes the freeze–thaw cycle. Several such cycles may be necessary to remove all gases. At the end of the last cycle, the tube is sealed at the constriction using a torch; during this operation the tube is still immersed in liquid nitrogen and the stopcock is open. The sealed tube is transferred into a heavy steel jacket that is then heated to the desired temperature in an explosion-proof, shielded oven. The tube should not be removed from the oven until it has cooled to room temperature. If formation of gaseous products is suspected, the tube should be cooled in dry ice–acetone or liquid nitrogen before opening.

4,5,6,7,10a,13,14,14a-Octahydro-2,9-benzodioxacyclododecin-1,8(3*H*, 10*H*)-dione (High-Dilution Technique) (Table IV/3).[72] A solution of 22 mg (0.1 mmol) of 2,6-di-*tert*-butyl-*p*-cresol in 25 mL of dry benzonitrile (distilled from calcium hydride) was heated to reflux (190°) and a solution of 479 mg (1.90 mmol) of 2,4-pentadienyl-6-[(1-oxo-2-propenyl)oxy]hexanoate in 20 mL of dry benzonitrile was added over a period of 30 hours, under argon, using a

† **Caution:** Never cool a tube that is open to the atmosphere with liquid nitrogen since oxygen from the air will be condensed in it. This will lead to an explosion when the sealed tube is warmed to room temperature.

mechanically driven syringe. Heating under reflux was continued for another 2 hours, the solvent was removed, and the residue was chromatographed on 80 g of silica gel (elution with 9:1 benzene-ethyl acetate) to give 368 mg (77% yield) of three intramolecular Diels–Alder adducts (*cis*- and *trans*-fused isomers of the title compound and the bridged isomer).

1,2,3,4,4a,5,6,8a-Octahydro-1,6-methanonaphthalene (Gas-Phase Reaction in a Stationary System) (Table IXe/6).[77] Three 250-mL Pyrex tubes were washed with acetone and water, rinsed with dilute ammonium hydroxide solution, and dried at 500°. Degassed samples of 60–90 μL of 5-(4-pentenyl)-1,3-cyclohexadiene were transferred by bulb-to-bulb distillation into the tubes at liquid nitrogen temperature, and the tubes were sealed and heated to 208° for 21 hours. The tips of the tubes were cooled with liquid nitrogen and opened, and the condensed liquid was subjected to gas-liquid chromatography (GLC) to give the title compound, mp 41–43°, in 73% yield.

Bicyclo[4.3.1]dec-6-ene (Gas-Phase Reaction in a Flow System) (Table V/6).[149,349] The apparatus consisted of a three-necked flask connected to a horizontal quartz tube (300 × 10 mm), heated by a cylindrical tube furnace. A sample of 3-methylene-1,8-nonadiene, placed in the flask, was swept through the hot zone by a stream of dry nitrogen at atmospheric pressure. The products were collected in a trap cooled with dry ice–acetone and analyzed by GLC. The contact time was estimated from the flow rate. At a tube temperature of 455° and a contact time of 5 seconds, the yield of bicyclo[4.3.1]dec-6-ene was 55%.

5,6,6a,7,9a,9b-Hexahydro-4*H*-pyrrolo[3,2,1-*ij*]quinolin-2(1*H*)-one (Gas-Phase Reaction in a Modified Flow System) (Table IIb/35).[109,350] The apparatus consisted of an addition funnel attached to the top of a vertical quartz tube that was packed with glass beads and heated to 600°. The lower end of the tube was connected to a two-necked flask that was cooled in dry ice–isopropyl alcohol and protected by a drying tube. A stream of dry nitrogen was passed through the apparatus at a rate of 3 mL/min. A solution of 1-(1-oxo-3,5-hexadienyl)-1,2,3,4-tetrahydropyridine in toluene was allowed to enter the hot tube at a rate of 15 drops/min. The contents of the receiver flask were concentrated under vacuum and the residue was purified by high-performance liquid chromatography to give the title compound in 45% yield.

Methyl 5β-Isopropyl-2,3,3aβ,4,5,7aα-hexahydroindene-4β-carboxylate (Lewis-Acid-Catalyzed Reaction) (Table IIa/6).[41] A solution of 1.50 g (6.76 mmol) of methyl (*E,E,E*)-11-methyldodeca-2,7,9-trienoate in 8 mL of dry carbon tetrachloride was treated under nitrogen with 4.5 mL (6.1 mmol) of a 1.36 *M* solution of diethylaluminum chloride in hexane. After the reaction mixture had been stirred at room temperature for 24 hours an additional 2.3 mL (3.2 mmol) of the catalyst solution was added. After a total of 48 hours, an excess of 1 *N* HCl was added, and the aqueous phase was extracted several times with ether. The ether extracts were washed with saturated sodium bicarbonate solution and dried. Removal of the solvent and chromatography of the residue (silica gel, elution with 95:5 hexane: ether) gave 1.27 g (85%) of the title compound.

TABULAR SURVEY

An attempt has been made to cover the literature completely through December 1981; many references to papers published in 1982 and some from 1983 are also included. More recent references are given in an addendum at the end of the tables. Since only a small fraction of the known intramolecular Diels–Alder reactions are indexed under that heading in *Chemical Abstracts*, omissions are inevitable.

The tables are arranged according to diene type. For a number of dienes, the tables are subdivided according to length of chain, point of attachment of the chain to the diene, type of heteroatom in the chain, and, in some instances, type of dienophile. Within each table, entries are usually arranged according to the following order of priorities, which differs from that normally used in *Organic Reactions*:

1. Chain length; entries are ordered according to increasing chain length.
2. Heteroatoms; substrates containing heteroatoms in the diene, dienophile, or chain are listed in the alphabetical order of the heteroatoms (i.e., N, O, S, . . .); in substrates with heteroatoms in the chain, the listing starts with the substrate having the heteroatom attached directly to the diene and ends with the substrate having the heteroatom attached directly to the dienophile.
3. Dienophile; dienophiles containing triple bonds take precedence over those containing double bonds; dienophiles containing carbon only are listed before those containing heteroatoms.

When priority items are equal, entries are arranged according to increasing number of carbon atoms. However, in order to save space, entries are combined in subtables whenever feasible. This has made strict adherence to the above order of priorities impossible.

All intramolecular Diels–Alder reactions involving equilibria are listed under the forward reactions in Tables I–XXVII. These equilibria do not appear again in Tables XXVIII (Reverse Intramolecular Diels–Alder Reactions), but cross-references to them are given in footnotes.

Elemental formulas and carbon atom numbers have been omitted. Instead, each reaction is assigned an "entry number" for the purpose of cross-referencing.

The structural formulas of the substrates are drawn to approximate the transition state that leads to the major product. Where this is impractical, that transition state is denoted below the formula by "*syn* TS" or "*anti* TS."

Substrates that have not been isolated, as well as hypothetical substrates, are marked with an asterisk (*); the precursors to these substrates are usually given in a footnote, although in some reactions their formulas appear in the body of tables preceding those of the substrates. The reaction conditions listed for the cyclizations of these substrates obviously are those necessary to generate the substrate from the precursors; the actual intramolecular Diels–Alder reactions would thus proceed at lower temperatures, or with shorter reaction times. Yields of these reactions are based on the substrate precursors.

Products of intramolecular Diels–Alder reactions that have not been isolated are also denoted by an asterisk (*); formulas for these intermediates have been omitted in those cases where their structures are obvious from the structures of the final products.

Only products arising from intramolecular Diels–Alder reactions are listed. In some entries, major products arising from other pathways are given in footnotes. Unsuccessful intramolecular Diels–Alder reactions are included.

A dash (—) in the column listing the conditions signifies that no conditions were reported. A dash (—) in the yield column means that the product was isolated but the yield was not reported. A zero (0) means that the reaction did not proceed under the conditions listed. Where a reaction has been reported in more than one publication, the conditions producing the highest yield are given and the reference to that paper is listed first. The following abbreviations have been used:

Cp	Cyclopentadienyl (as ligand)
Diglyme	Bis(2-methoxyethyl) ether
DMF	Dimethylformamide
Et_2O	Diethyl ether
Me_2CO	Acetone
Me_2SO	Dimethyl sulfoxide
TLC	Thin-layer chromatography

TABLE I. ACYCLIC DIENES; 1- AND 2-ATOM CHAINS

Entry No.	Substrate	Conditions	Product(s) and Yield(s) (%)	Refs.
1	 R *[a] R = CH_3O_2C	—	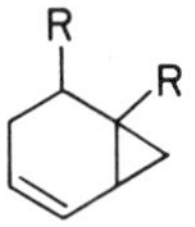 R, R (—)[b]	351
2	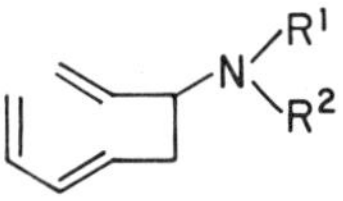 N R^1 R^2 $R^1, R^2 = CH_3, C_2H_5$, $-(CH_2)_2O(CH_2)_2-$, $-(CH_2)_5-$	DMF, 100°	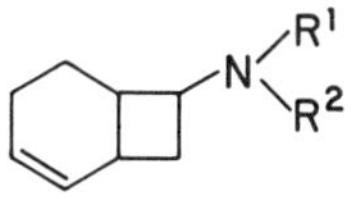 N R^1 R^2 (0)	342
3	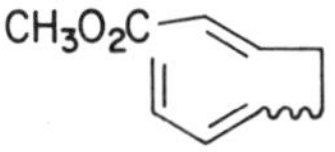 CH_3O_2C [c]	210°, 4.5 hr or *hv* (Corex)	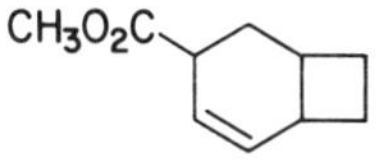 CH_3O_2C (0)	185
4	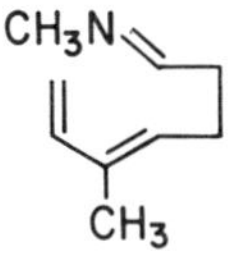 CH_3N, CH_3	Heat	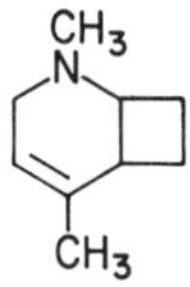 CH_3, N, CH_3 (0)	183

[a] The substrate was generated by an ene reaction between 1,4-pentadiene and dimethyl acetylene-dicarboxylate.

[b] Although the stereochemistry was not discussed, it is likely that both the ene and intramolecular Diels–Alder reactions proceed in a *cis* fashion.

[c] Neither *cis* nor *trans* dienes cyclized under these conditions.

*This intermediate is hypothetical, or has not been isolated.

TABLE IIa. ACYCLIC DIENES; 3-ATOM CHAIN CONTAININ

Entry No.	Substrate				
	R^1	R^2	R^3	R^4	R^5
1	H	H	HO	H	i-C_3H_7
1a	H	H	$-O(CH_2)_2O-$		i-C_3H_7
1b	CH_3O_2C	H	$-O(CH_2)_2O-$		i-C_3H_7
1c	CH_3O_2C	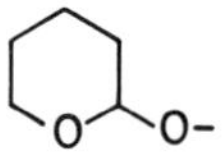 (tetrahydropyran-2-yl)oxy	H	H	H
1d	R = t-$C_4H_9(CH_3)_2SiO$ *a				

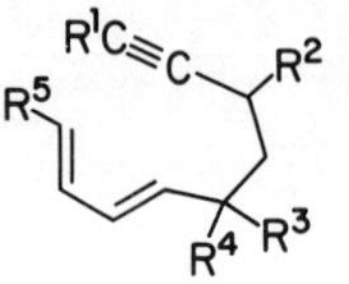

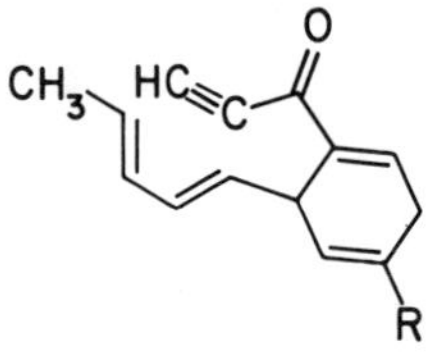

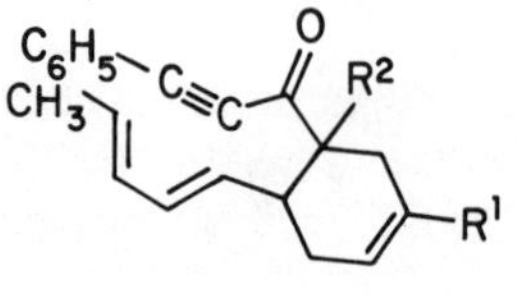

Entry No.	R^1	R^2
2a	t-$C_4H_9(CH_3)_2SiO$	H
2b	"	CH_3

ARBON ONLY; ETHYNYL AND VINYL DIENOPHILES

try o.	Conditions	Product(s) and Yield(s)(%)	Refs.
		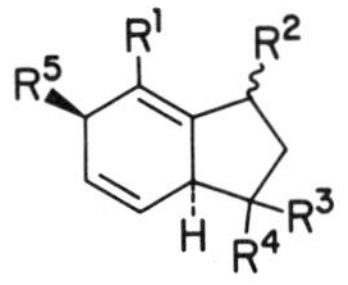 R1, R2, R3, R4, R5, H	
	150°, 5 hr	(68)[u]	175
	", "	(65)	175
	110°, "	(60)	175
	CH_2Cl_2, 23°, 110 hr	(50)	38
	CCl_4, warm, 16 hr	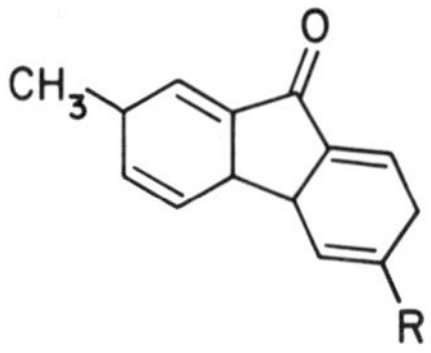 CH_3, O, R (34)	352
	Toluene, 240°, 3–5 hr	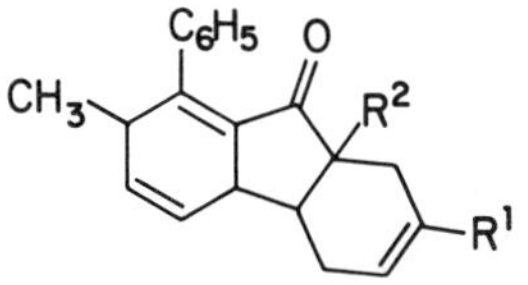 C_6H_5, O, CH_3, R2, R1	
		(30)	352
		(28)	352

TABLE IIa. ACYCLIC DIENES; 3-ATOM CHAIN CONTAININ

Entry No.	Substrate								
	R^1	R^2	R^3	R^4	R^5	R^6	R^7	R^8	R^9
3	CH_3	H	H	H	HO	H	H	H	H
4	CH_3O_2C	H	H	H	H	H	H	H	H
5	CH_3	H	H	H	$(CH_3)_3SiO$	H	H	H	H
6	CH_3O_2C	H	H	H	H	H	H	i-C_3H_7	H
7	CH_3O_2C	H	H	H	HO[e]	H	H	i-C_3H_7	H
8	CH_3O_2C	H	H	H	HO	H	H	H	i-C
9	H	H	H	H	H	H	$(C_2H_5)_3SiO$	H	H
10	H	CH_3	O O	H	H	H	H	H	H
11	CH_3O_2C	H	H	H	CH_3CO_2	H	H	i-C_3H_7	H
12	CH_3O_2C	H	H	H	$-O(CH_2)_2O-$		H	i-C_3H_7	H
13	H	CH_3	t-$C_4H_9(CH_3)_2SiO$	H	H	H	H	H	H
14	H	HCO	$C_6H_5CH_2O$	H	H	H	H	H	H
15	N H CO	H	H	H	C_2H_5	H	H	$HOCH_2$	H
16	H	HCO	$C_6H_5CH_2O$	H	H	H	CH_3	H	H
17	CH_3O_2C	H	H	H	$(CH_3)_3SiO$	H	H	i-C_3H_7	H
18	$C_2H_5O_2C$	H	H	H	C_2H_5	H	H	$CH_3O(CH_2)_2OCH_2$	H
19	CH_3O_2C	CH_3	$C_6H_5CH_2O$	H	H	H	H	H	H
20	$C_2H_5O_2C$	H	H	CH_3[i]	C_2H_5[i]	H	H	O [j]	H
21	N H CO	H	H	H	C_2H_5	H	H	CH_3O_2C	H

ARBON ONLY; ETHYNYL AND VINYL DIENOPHILES (*Continued*)

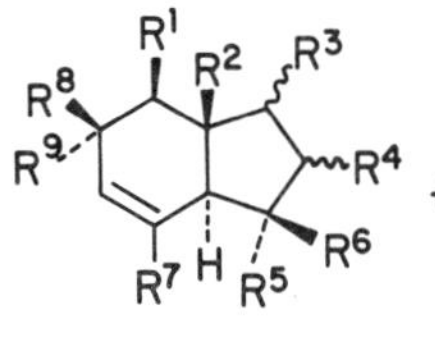

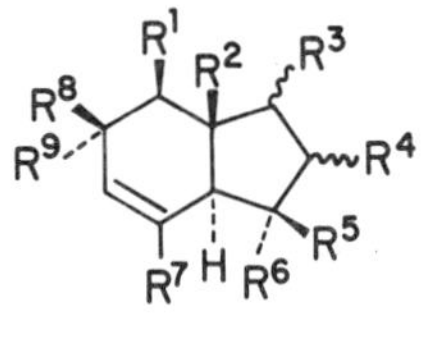

try).	Conditions	Product(s) and Yield(s)(%)				Refs.
		A	B	C	D	
	Toluene, 250°		(10)[b]			84
	", 150°, 24 hr	(39)		(26)		41,40,185
	Toluene–CH_2Cl_2, menthoxy-$AlCl_2$, 23°, 48 hr	(72)		(0)		41,353
	Toluene, 245–250°, 16 hr	(17)[c]		(34)[c]		84
	", 150°, 40 hr	(52)		(20)		41,40
	Toluene, CH_2Cl_2, menthoxy-$AlCl_2$[d], 23°, 48 hr	(75–84)		(0)		41,353
	Toluene, 150°, 18 hr	(23)	(26)	(3)	(18)	41,188,354
	", "	(0)	(0)	(0)	(0)	188
	", 160°, 17 hr		(84)[f]			97
	", 220°, 30 hr	(35)[m]		(35)[m]		38
	", 150°, 18 hr	(21)	(32)	(0)	(18)	354
	", ", 15 hr	(62)		(13)		41,278
	", 200°, 20 hr	(30)[m]		(30)[m]		355
	CCl_4, 150°, 18 hr	(19)		(63)		38
	Toluene, reflux, 17 hr	(50)[g]				139
	", 110°, 18 hr	(10)[h]		(45)[h]		39
	Toluene, 150°, 15 hr	(40)	(26)	(3)	(14)	41,188,354
	", 110°, 36–70 hr	(>90)	(0)	(0)	(0)	312
	", 250°, 6 hr	(52)[n]		(23)[o]		38
	o-$Cl_2C_6H_4$, 175°, 44 hr	(20)[j]	(20)[j]	(0)	(0)	320
	CH_2Cl_2, $C_2H_5AlCl_2$, 0 to 25°, 1.5 hr	(71)[k]				139

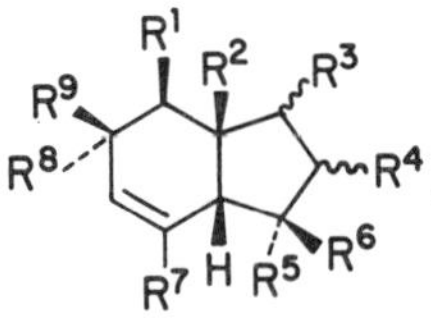

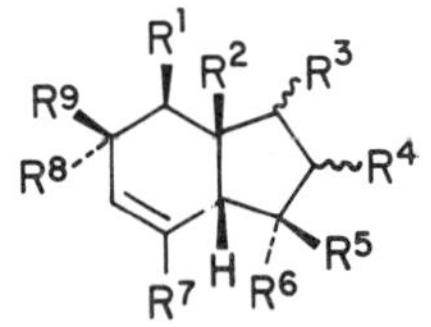

TABLE IIa. ACYCLIC DIENES; 3-ATOM CHAIN CONTAININ

Entry No.	Substrate								
	R^1	R^2	R^3	R^4	R^5	R^6	R^7	R^8	R^9
22	CH_3O_2C	H	H	H	$C_6H_5CH_2O$	H	H	i-C_3H_7	H
23	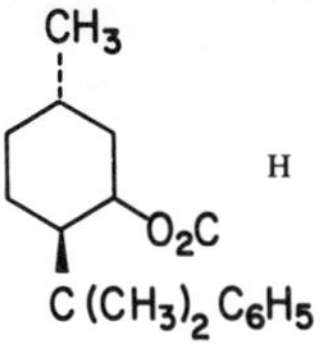 CH_3, O_2C, $C(CH_3)_2C_6H_5$	H	H	H	H	H	H	H	H
24	"	H	H	H	H	H	H	i-C_3H_7	H
25	$C_2H_5O_2C$	H	H	H	C_2H_5	H	H	$(C_6H_5)_2(C_4H_9)SiOCH_2$	H

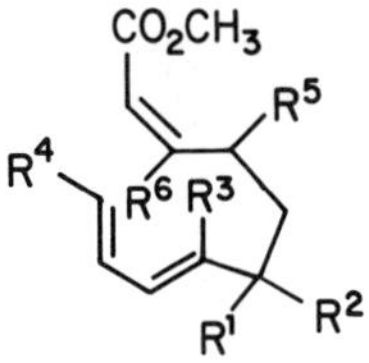

Entry No.	R^1	R^2	R^3	R^4	R^5	R^6
26	H	H	H	H	H	H
27	H	H	H	i-C_3H_7	H	H
28	HO	H	H	i-C_3H_7	H	H
29	NC	H	CH_3	i-C_3H_7	H	H[p]
30	CH_3CO_2	H	H	i-C_3H_7	H	H
31	H	H	H	H	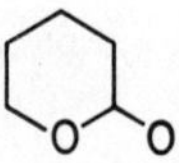 O O	CH_3
32	H	H	H	H	$C_6H_5CH_2O$	CH_3
33	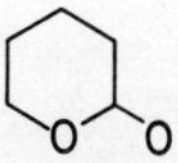 O O	H	H	i-C_3H_7	H	H
34	$(CH_3)_3SiO$	H	H	i-C_3H_7	H	H
35	$C_6H_5CH_2O$	H	H	i-C_3H_7	H	H
36	$—O(CH_2)_2O—$		H	i-C_3H_7	H	H
37	$—O(CH_2)_3O—$		H	i-C_3H_7	H	H

ARBON ONLY; ETHYNYL AND VINYL DIENOPHILES (*Continued*)

try o.	Conditions	Product(s) and Yield(s)(%)				Refs.
		A	B	C	D	
	Toluene, 115°, 110 hr	(19)	(34)	(3)	(8)	41,188,354
	CH_2Cl_2, various catalysts, 25°	(64–82)			(0)	41
	", ", "	(8–75)			(0)	41
	Toluene, 130°, 48 hr	(70)[l]	(—)	(—)	(—)	138

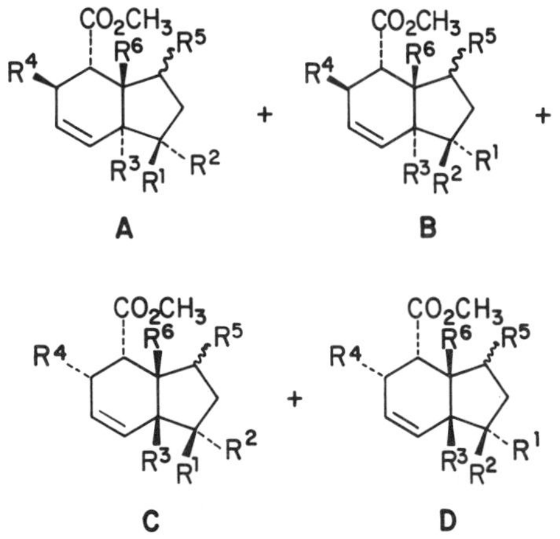

A + B + C + D

try o.	Conditions	Product(s) and Yield(s)(%)				Refs.
		A	B	C	D	
	Toluene, 180°, 5 hr;	(49)		(26)		40,41
	CH_2Cl_2, $C_2H_5AlCl_2$, 23°, 40 hr	(14)		(13)		41,353
	Toluene, 180°, 5 hr;	(50)		(25)		40,41
	CH_2Cl_2, $C_2H_5AlCl_2$, 23°, 40 hr	(37)		(22)		41,353
	Toluene, 150°, 6 hr	(27)	(19)	(14)	(0)	41,188,354
	o-$Cl_2C_6H_4$, reflux, 3 days	(0)	(0)	(24)	(25)	17
	Toluene, 150°, 6 hr	(15)	(26)	(17)	((0)	41,354
	", 220°, 10 hr	(62)[q]		(16)[q]		38
	", 210°, 11 hr	(68)[q]		(10)[q]		38
	", ", 15 hr	(24)	(23)	(18)	(0)	41,354
	", 180°, 2 hr	(38)	(22)	(18)	(0)	41,188,354
	", 115°, 44 hr	(34)	(27)	(31)	(0)	41,345
	", 180°, 4 hr	(58)		(15)		41,278
	", ", 0.5 hr	(75)		(15)		41,278

TABLE IIa. ACYCLIC DIENES; 3-ATOM CHAIN CONTAININ

Entry No.	Substrate		
	R^1	R^2	
38	CH_3O_2C	H	
39	CH_3	$(CH_3)_3SiO$	
	R^1	R^2	R^3
40	H	H	CH_3[r]
41	H	CH_3	CH_3
42	C_2H_5	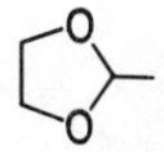 1,3-dioxolan-2-yl	H*[s]
43	C_2H_5	$C_2H_5O_2C$	H
	R^1	R^2	
44	CH_3	CH_3	
45	C_6H_5	t-C_4H_9	

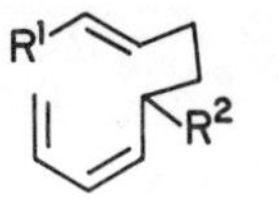

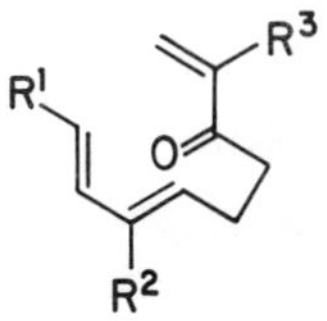

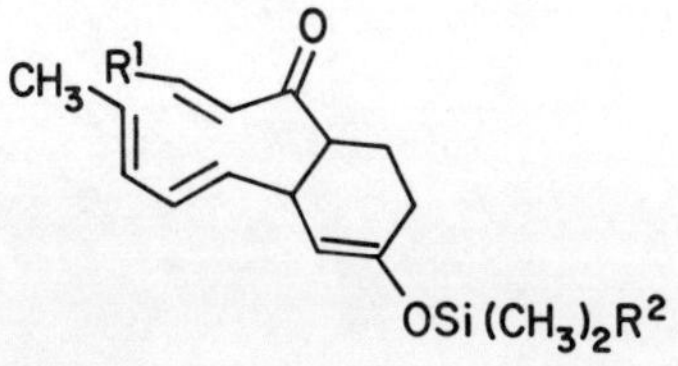

RBON ONLY; ETHYNYL AND VINYL DIENOPHILES (*Continued*)

ry

Conditions	Product(s) and Yield(s) (%)	Refs.
	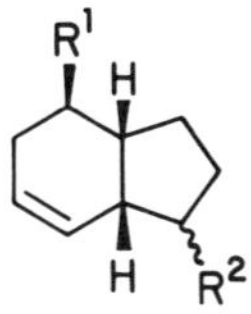	
Xylene, reflux, 92 hr	(—)	185
Toluene, 245–250°, 16 hr	(15)	84
	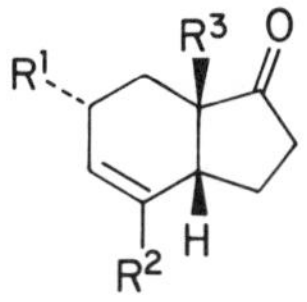 A	
	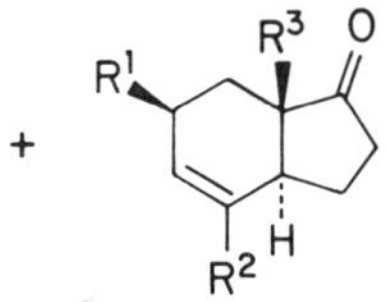 B	
	A B	
Benzene, 190°, 13 hr	(70) (30)	88,89
Toluene, 190°, 18 hr	(21) (10)	39
", 170–185°, 3 hr	(0) (92)	90
", 180°, 4 hr	(58) (38)	165
Toluene, 240°, 3–5 hr	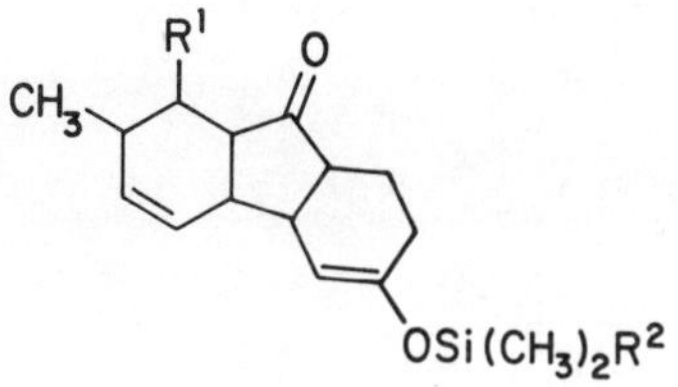	
	(35)^{f}	352
	(32)	352

TABLE IIa. ACYCLIC DIENES; 3-ATOM CHAIN CONTAININ

Entry No.	Substrate
	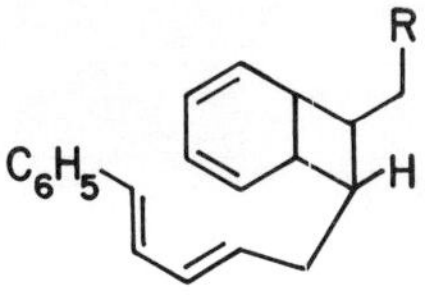
	R
46	CH_3O_2C
47	$CH_3O_2CCH{=}CH$
48	$t\text{-}C_4H_9(C_6H_5)_2SiO$

[a] The substrate was generated by an intermolecular Diels–Alder reaction of 2-(dimethyl-*tert*-butylsiloxy)-1,3,5,7-nonatriene and 1,4-pentadiyn one.

[b] The product was a mixture of *cis* and *trans* isomers in unspecified ratio.

[c] The yields are of perhydroindanones obtained after desilylation, catalytic hydrogenation, and oxidation.

[d] The yields were comparable with $AlCl_3$, $(C_2H_5)_2AlCl$, or $C_2H_5AlCl_2$, but lower with $SnCl_4$, $TiCl_4$, or $BF_3 \cdot (C_2H_5)_2O$.

[e] The same result was obtained with $R^5 = (CH_3)_3SiO$ or $C_6H_5CH_2O$.

[f] The initially formed adduct isomerized to the more stable 3a,4-unsaturated hexahydroindene.

[g] Other isomers were formed in 17% yield.

[h] The *cis–trans* ratio was established in the methyl ketones obtained by Wolff–Kishner reduction of the aldehyde, debenzylation, and oxidat

[i] R^4, R^5 are *cis*.

[j] The substrate was a single diastereomer.

[k] Other isomers were formed in < 5% yield.

[l] Other isomers were formed in 15% yield.

ARBON ONLY; ETHYNYL AND VINYL DIENOPHILES (*Continued*)

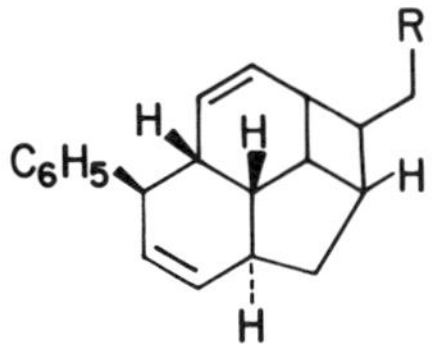

try).	Conditions	Product(s) and Yield(s)(%)	Refs.
6	Toluene, 70°	(—)	315
7	", "	(—)	315
8	", 100°, 5 hr	(100)	316

[m] The numbers are yields after deprotection and oxidation to the ketones.

[n] The products were 22% $R^3\alpha$, 30% $R^3\beta$.

[o] The products were 10% $R^3\alpha$, 13% $R^3\beta$.

[p] The actual starting material was the corresponding *cis* diene, which isomerized to the *trans* diene under the reaction conditions; the *trans* diene ıld be isolated.

[q] R^5 is β.

[r] The dimethyl and diethyl ketals produced **A/B** in the approximate ratio 30:70; for the propylene ketal, the ratio was about 1:3.[355]

[s] The substrate was generated from a precursor in which the diene was masked as the cyclobutene and the dienophile as the Diels–Alder adduct h dimethylfulvene.

[t] The product was isolated as the diketone.

[u] A mixture of hydroxyl epimers (ratio 1:1.2) was obtained.

* This intermediate is hypothetical, or has not been isolated.

TABLE IIb. ACYCLIC DIENES; 3-ATOM CHAIN CONTAINING UNCHARGED NITROGEN; VINYL DIENOPHILES

Entry No.	Substrate		Conditions	Product(s) and Yield(s)(%)	Refs.
1	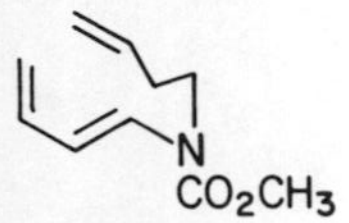 N		160°, 16 hr	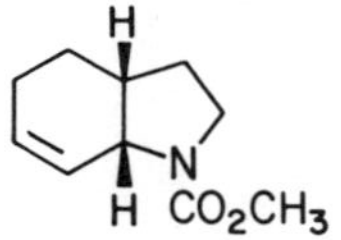 (38)	102
	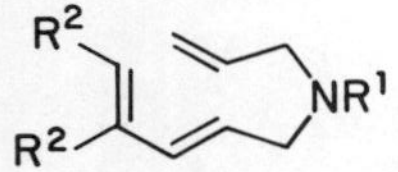 R^1	R^2		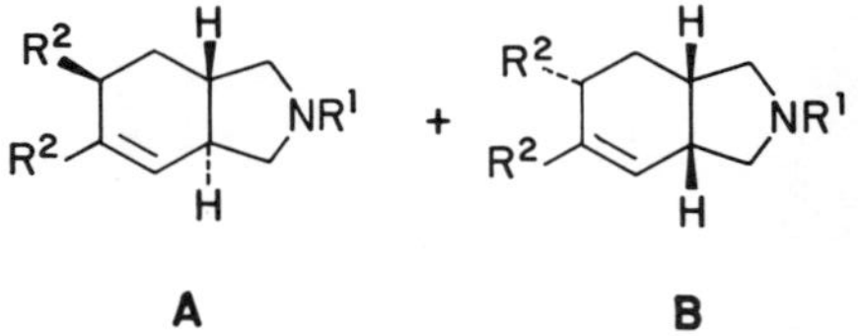 A B	
2	CH_3CO	H	Neat or in C_2H_5OH, 90–95°, 25 hr	(95)[a]	356,48,50,357
3	$ClCH_2CO$	H	", ", "	(85)[a]	356,50
4	Cl_3CCO	H	Neat or in C_2H_5OH, 90–95°, 25 hr	(85)[a]	356,50
5	F_3CCO	H	90–95°	(85)[a]	48
6	C_2H_5CO	H	Neat or in C_2H_5OH, 90–95°, 25 hr	(85)[a]	356,50
6a	$(CH_3)_3CCO$	H	90–95°	(88)[a]	48
7	2-Furoyl	H	Neat or in C_2H_5OH, heat, 10–12 hr	(90)[a]	358
8	3-Pyridoyl	H	", ", ", "	(90)[a]	358
9	$C_6H_5SO_2$	H	", ", 90–95°, 50–60 hr	(90)[a]	358

No.	Substrate		Conditions	Product(s) and Yield(s) (%)	Refs.
10	p-$ClC_6H_4SO_2$	H	", ", ", "	(95)[a]	358
11	C_6H_5CO	H	", ", 20°, 4–6 hr	(90)[a]	358
12	p-BrC_6H_4CO	H*[b]	", ", "	(95)[a]	358
13	p-$O_2NC_6H_4CO$	H*[b]	", ", "	(95)[a]	358
14	p-$CH_3C_6H_4SO_2$	H	", ", 90–95°, 50–60 hr	(95)[a]	358
15	CH_3	C_6H_5	C_6D_6, 140–142°, 12 hr	(16) (84)	43
16	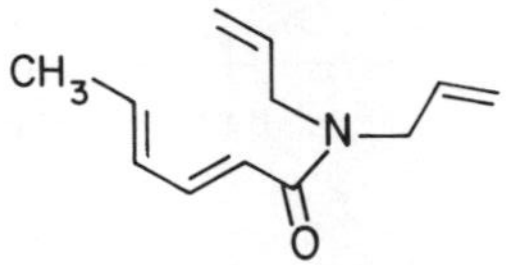		DMF, 156°, 3 hr	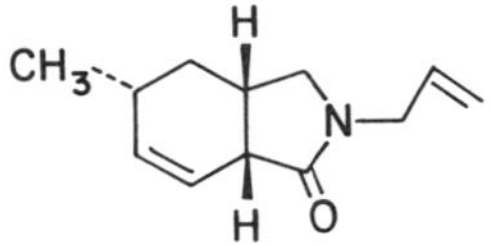 A (44) 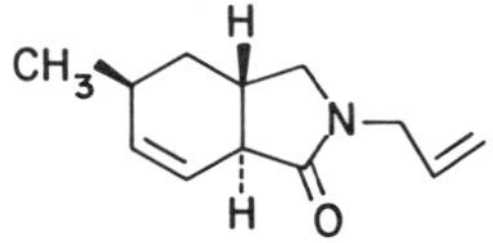 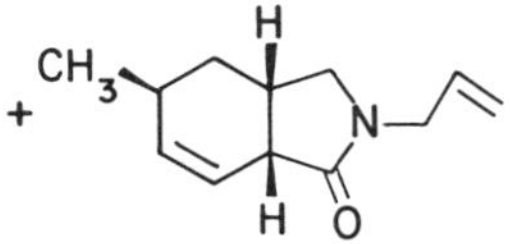 + B (32)[c] + C (4)[c]	189
17	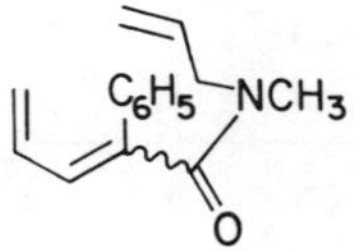 [d]		Toluene, 200°, 8 hr[e]	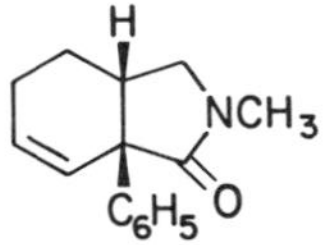 (—)	167

TABLE IIb. ACYCLIC DIENES; 3-ATOM CHAIN CONTAINING UNCHARGED NITROGEN; VINYL DIENOPHILES (*Continued*)

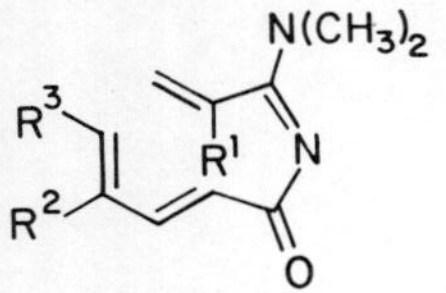

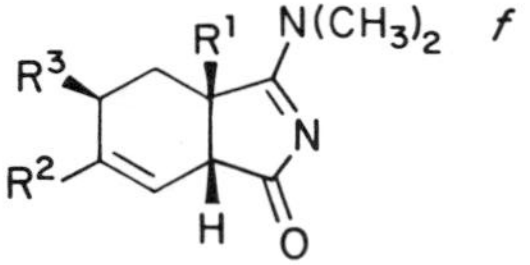

Entry No.	Substrate			Conditions	Product(s) and Yield(s)(%)	Refs.
	R^1	R^2	R^3	CH_3CN, 171°, 3–4 hr	[f]	
18	CH_3	H	H		(70)	107
19	CH_3	H	CH_3		(78)	106,107
20	CH_3	CH_3	H		(63)	107
21	C_6H_5	H	CH_3		(66)	107

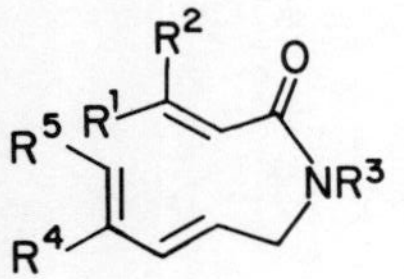

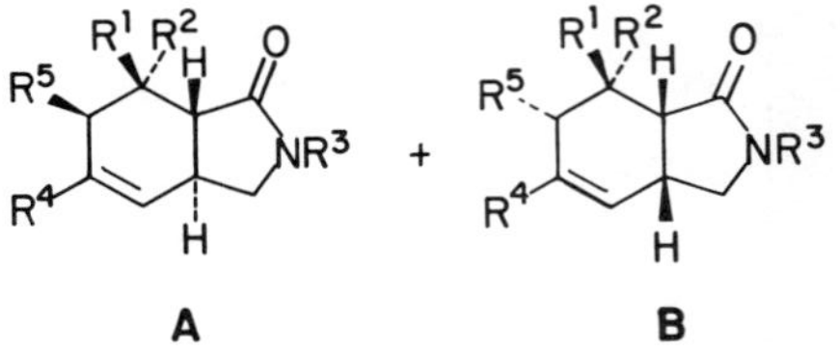

Entry No.	R^1	R^2	R^3	R^4	R^5	Conditions	A	B	Refs.
22	CH_3	CH_3	CH_3	H	H	Benzene, 220°, 2 hr	(70)[d]		60
23	CH_3	CH_3	t-C_4H_9	H	H	", ", "	(70)[d]		60
24	H	H	CH_3	C_6H_5	C_6H_5	Toluene, 110°, 8 hr	(42)	(40)	43
25	H	H	C_2H_5	C_6H_5	C_6H_5	Decalin, 90–109°	Ratio ≈ 1:1		43
26	H	H	i-C_3H_7	C_6H_5	C_6H_5	", 69–90°	"		43
27	H	H	t-C_4H_9	C_6H_5	C_6H_5	", 45–61°	"		43
28	$C_2H_5O_2C$	H	CH_3	H	p-$CH_3OC_6H_4$*[g]	Pyridine, <0°	(77)	(trace)	44
29	$C_2H_5O_2C$	H	H	C_6H_5	C_6H_5	Benzene, reflux, 5 hr	(58)	(0)	43

32	C_6H_5	H	CH_3	C_6H_5	C_6H_5	C_6D_6, 91–92°, 9 hr	(81)	(11)	43
33	$R^1 = C_2H_5O_2C$, $R^2 = C_6H_5$ *g*					CH_2Cl_2, <0°, 2 hr	(70) + (1–3)		108

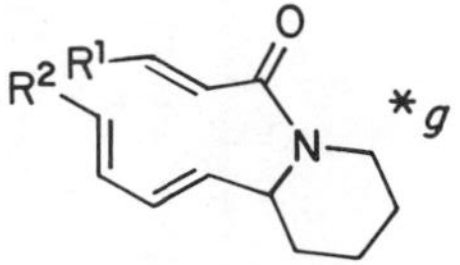

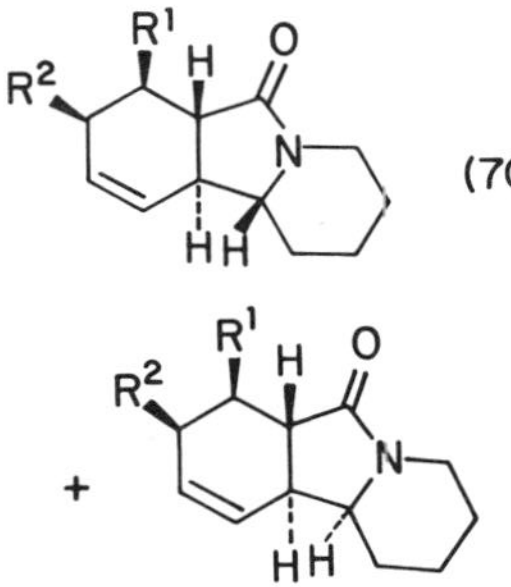

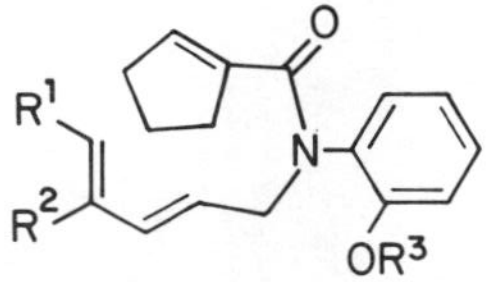

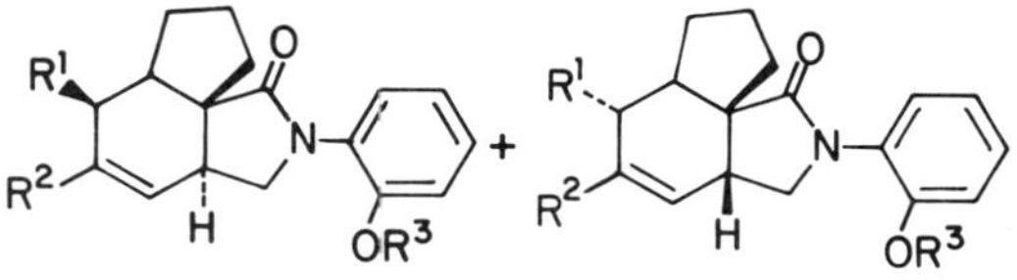

	R^1	R^2	R^3		A	B	
33a	H	H	H	Toluene, reflux, 6 hr	(12)	(17)	70
33b	H	H	MgCl	", ", 4 hr	(56)	(35)	70
33c	H	H	MgBr	", ", "	(53)	(33)	70
33d	H	H	MgI	", ", "	(49)	(32)	70
33e	CH_3	H	H	", ", 17 hr	(23)	(23)	70
33f	CH_3	H	MgCl	", ", 10 hr	(68)	(12)	70
33g	CH_3	H	MgBr	Toluene, reflux, 10 hr	(62)	(16)	70
33h	CH_3	H	MgI	", ", "	(65)	(13)	70
33i	H	CH_3	H	", ", 20 hr	(13)	(17)	70
33j	H	CH_3	MgCl	", ", 10 hr	(43)	(28)	70
33k	H	CH_3	MgBr	", ", "	(49)	(26)	70

TABLE IIb. ACYCLIC DIENES; 3-ATOM CHAIN CONTAINING UNCHARGED NITROGEN; VINYL DIENOPHILES (*Continued*)

Entry No.	Substrate	Conditions	Product(s) and Yield(s)(%)	Refs.
	R^1 R^2 R^3			
33l	H CH_3 MgI	Toluene, reflux, 10 hr	(46) (29)	70
33m	i-C_3H_7 H H	", ", 11 hr	(15) (10)	70
33n	i-C_3H_7 H MgCl	", ", 10 hr	(56) (18)	70
33o	i-C_3H_7 H MgBr	", ", "	(56) (14)	70
33p	i-C_3H_7 H MgI	", ", "	(62) (8)	70
	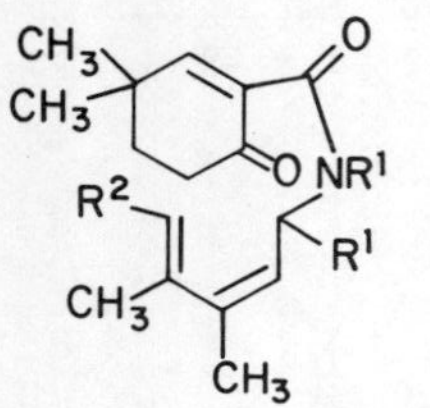 *h*		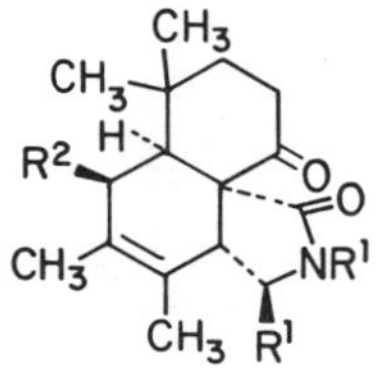	
	R^1 R^2			
34	$C_6H_5CH_2$ H	Toluene, reflux, 40 hr	(95)[i]	86,359
34a	$C_6H_5CH_2$ $(CH_3)_3Si(CH_2)_2O$	", ", 15 hr	(79)[j, k]	86
	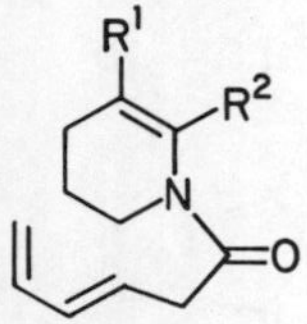		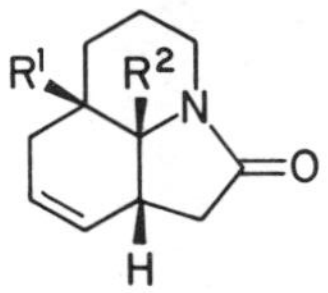	
	R^1 R^2			
35	H H	Toluene, 260°, 20 hr or gas phase, 600°	(45–55)	109

36	H	CH_3*[l]	", ", "	(67)	109
37	C_2H_5	H*[l]	", "	(58)	109

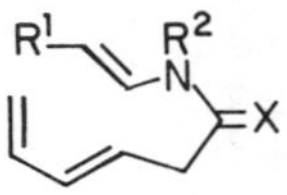

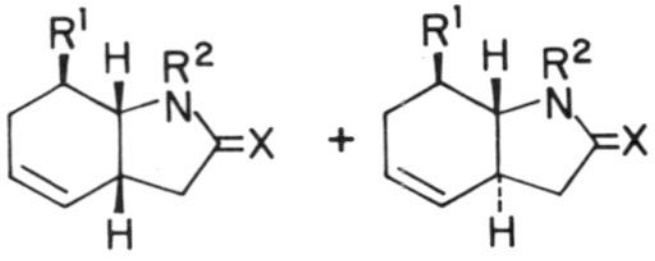

	R^1	R^2	X		A	B	
38	C_6H_5	CH_3CO	H_2	To 600°	(0)	(0)	20
39	C_6H_5	Cl_3CCO	H_2	" "	(0)	(0)	20
40	C_6H_5	CH_3O_2C	H_2	" "	(0)	(0)	20
41	C_6H_5	$C_6H_5CH_2$	O	Xylene, reflux, 18 hr	(45)	(0)	20
42	3,4-$CH_2O_2C_6H_3$	$C_6H_5CH_2$	O	", ", "	(18)	(28)[m]	20
43	3,4-$CH_2O_2C_6H_3$	p-$CH_3OC_6H_4CH_2$	O*[l]	", ", "	(20)	(27)[m]	20,110

[a] The stereochemistry was not determined; the structures were assigned on the basis of elemental analysis and IR and/or UV spectra.
[b] The starting material cyclized too rapidly to be isolated.
[c] Isomer **B** is converted quantitatively into the more stable *cis* isomer **C** in refluxing DMF; isomer **C** is thus a secondary product.
[d] The substrate was a mixture of *cis* and *trans* isomers.
[e] The half-life at 120° is 16 hr.
[f] The reaction is believed to proceed via the *anti* transition state to give initially the *trans*-fused product, which epimerizes to the *cis*-fused isomer via the enol.[106]
[g] The substrate was generated from the secondary amine and the acid chloride.
[h] The chiral substrate was derived from L(−)-phenylalanine.
[i] The corresponding *trans* diene failed to cyclize under these conditions.
[j] The adduct involving the cyclohexenone carbonyl group as the dienophile was formed in 15% yield (see Table IIIb/9a).
[k] The corresponding *trans* diene failed to cyclize at 168° for 48 hr.
[l] The diene was generated by elimination of SO_2 from the corresponding dihydrothiophene-1,1-dioxide.
[m] An isomer formed by intramolecular Diels–Alder addition of the internal double bond of the diene to the styrene diene system was obtained in approximately 2% yield.
* This intermediate is hypothetical, or has not been isolated.

TABLE IIc. ACYCLIC DIENES; 3-ATOM CHAIN CONTAINING QUATERNARY NITROGEN; ETHYNYL AND VINYL DIENOPHILES

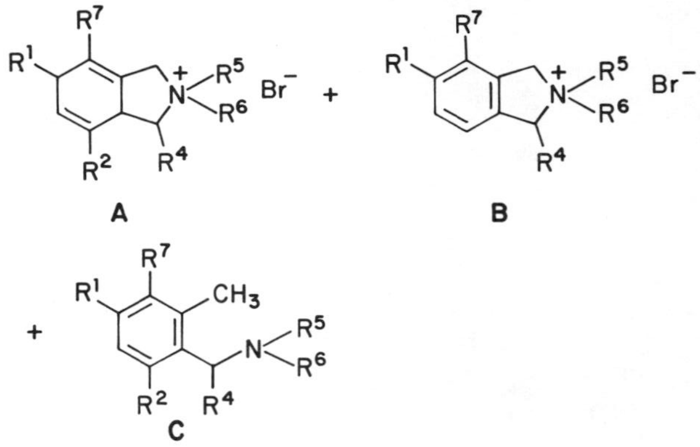

Entry No.	R¹	R²	R³	R⁴	R⁵	R⁶	R⁷	Conditions	A	B	C	Refs.
1	H	H	H	H	CH_3	CH_3	H	DMF, 100°, 12–15 hr	(≈100)[a–c]	(0)	(0)	360,49,338
2	H	Cl	H	H	CH_3	CH_3	H	H_2O, KOH, 100–110°, 3 hr	(0)	(37)	(17)	337,361,362
3	H	CH_3	H	H	CH_3	CH_3	H	DMF, 100°, 50–60 hr	(≈100)[a,c]	(0)	(0)	343
4	H	H	H	CH_3	CH_3	CH_3	H	", ", "	(≈100)[a,c]	(0)	(0)	343
5	H	H	H	H	CH_3	CH_3	CH_3	DMF or CH_3CN, 100°, 12–15 hr or H_2O, 100°, 90–100 hr	(≈100)[a]	(0)	(0)	338,343
6	CH_3	Cl	H	H	CH_3	CH_3	H	H_2O, KOH, 100–110°, 3 hr	(0)	(20)	(23)	337,361,362
6a	H	H	H	H	C_2H_5	C_2H_5	H	", 85°	(–)	(–)	(–)	49
7	H	Cl	H	H	C_2H_5	C_2H_5	H	", KOH, 100–110°, 3 hr	(0)	(6)	(37)	337,361,362
8	H	Cl	Cl	H	C_2H_5	C_2H_5	H	", 100″, 14 hr	(0)	(90)[d]	(0)	332
9	CH_3	Cl	H	H	C_2H_5	C_2H_5	H	", KOH, 100–110°, 3 hr	(0)	(30)	(27)	337,361,362
10	H	Cl	H	H	—$(CH_2)_5$—		H	", ", ", "	(0)	(18)	(50)	337

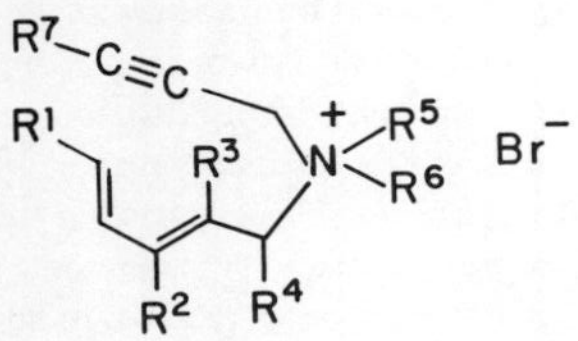

No.								Conditions				Refs.
11	H	H	H	CH_3	$-(CH_2)_5-$		H	H_2O, 85°	(—)	(—)	(—)	363
12	H	H	H	$CH_2{=}CHCH_2$	C_2H_5	C_2H_5	H	DMF, 100°, 90 hr	(0)	(0)	(70)	342
12a	H	H	H	H	$-(CH_2)_2O(CH_2)_2-$		H	H_2O, 85°	(—)	(—)	(—)	49
12b	H	H	H	H	$-(CH_2)_5-$		H	", "	(—)	(—)	(—)	49
13	H	H	H	H	$-(CH_2)_5-$		C_6H_5	DMF, 90–95°	(90)[a]	(0)	(0)	364, 49

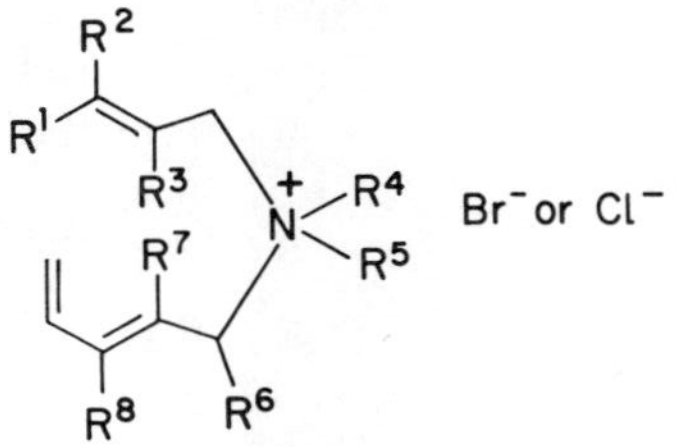

Br$^-$ or Cl$^-$ 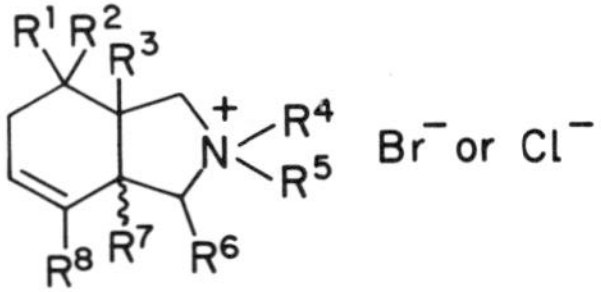 Br$^-$ or Cl$^-$

No.	R^1	R^2	R^3	R^4	R^5	R^6	R^7	R^8	Conditions	Yield	Refs.
14	H	H	H	CH_3	CH_3	H	H	H	DMF or CH_3CN, 100°, 12–15 hr; or H_2O, 100°, 90–100 hr	(≈100)[a, e]	338, 49
15	H	H	CH_3	CH_3	CH_3	H	H	H	DMF, 90–95°	(85)[a]	364
16	H	H	H	CH_3	CH_3	CH_3	H	H	DMF, 100°, 50–60 hr	(≈100)[a, e]	343
17	H	H	H	CH_3	CH_3	H	H	CH_3	", ", "	(≈100)[a, e]	343
18	H	H	H	C_2H_5	C_2H_5	H	Cl	Cl	H_2O, KOH, 25°, then HBr	(88)*[a, f]	332
19	H	H	CH_3	C_2H_5	C_2H_5	H	H	H	DMF, 90–95°	(90)[a]	364
20	H	H	H	$-(CH_2)_5-$		H	H	H	H_2O, 85°	(—)	49
21	CH_3	H	H	$-(CH_2)_5-$		H	H	H	DMF, 90–95°	(90)[a]	364,49,52
22	H	H	CH_3	$-(CH_2)_5-$		H	H	H	DMF, 90–95°	(90)[a]	364,49,52
23	H	H	H	$-(CH_2)_5-$		CH_3	H	H	H_2O, 85°	(—)	363
24	CH_3	CH_3	H	$-(CH_2)_5-$		H	H	H	DMF, 90–95°	(90)[a]	364,49
25	$CH_3O(CH_2)_2$	H	H	$-(CH_2)_5-$		H	H	H	", "	(80)[a]	364
26	C_6H_5	H	H	$-(CH_2)_5-$		H	H	H	", "	(90)[a]	364,49
27	H	H	C_6H_5	$-(CH_2)_5-$		H	H	H	H_2O, 85°	(—)	49

TABLE IIc. ACYCLIC DIENES; 3-ATOM CHAIN CONTAINING QUATERNARY NITROGEN; ETHYNYL AND VINYL DIENOPHILES (*Continued*)

Entry No.	Substrate 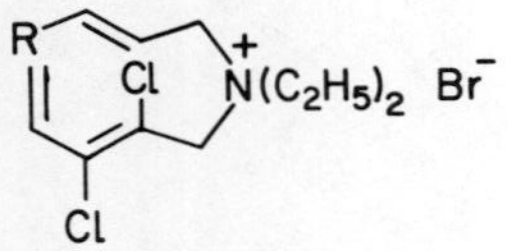	Conditions	Product(s) and Yield(s)(%) 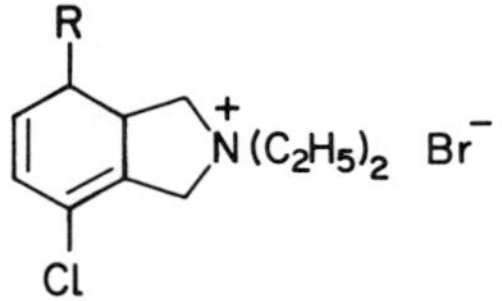	Refs.
	R			
28	H	H_2O, KOH, 25°, then HBr	(88)[a]	332
29	CH_3	H_2O, KOH(cat.), 100°, 2 hr	(71)[a]	333
	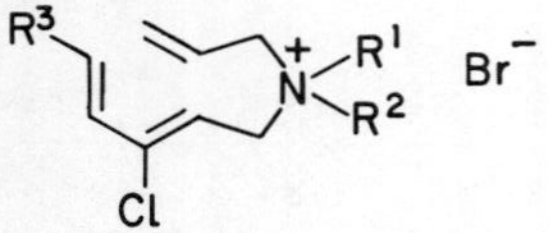	H_2O, 20% KOH, 100–110°, 3 hr	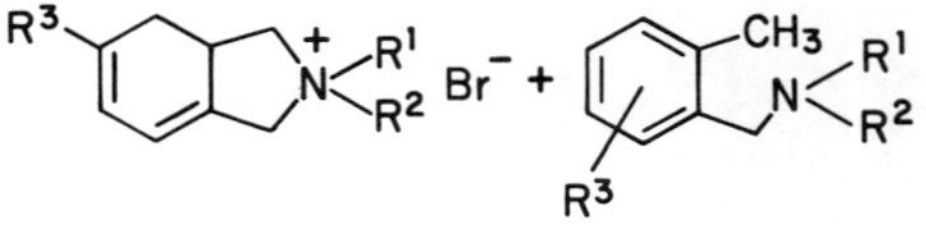 A + B	
	R^1 / R^2 / R^3		A / B	
30	CH_3 / CH_3 / H		(27)[a] / (27)	337,361,362
31	CH_3 / CH_3 / CH_3		(31)[a] / (34)	337,361,362
32	C_2H_5 / C_2H_5 / H		(12)[a] / (35)	337,361,362
33	C_2H_5 / C_2H_5 / CH_3		(0) / (40)	337,361,362
34	$-(CH_2)_5-$ / CH_3		(42)[a] / (8)	337,361,362

[a] The structure is supported only by elemental analysis and IR and/or UV spectra.
[b] The product gave *N,N*-dimethyl-2-methylbenzylamine on treatment with base.
[c] The reaction was also carried out with $R^5 = R^6 = C_2H_5$ and with $R^5, R^6 = -(CH_2)_2O(CH_2)_2-$ or $-(CH_2)_5-$.
[d] The product was the 4-chloroisoquinolinium salt.
[e] The reaction was also carried out with $R^4 = R^5 = C_2H_5$ and with $R^4, R^5 = -(CH_2)_2O(CH_2)_2-$ or $-(CH_2)_5-$.
[f] The product was the elimination product, 7-chloro-2,2-diethyl-3a,4-dihydroisoquinolinium bromide.

TABLE IId. Acyclic 1,3,4-Trienes; 3-Atom Chain Containing Quaternary Nitrogen; Ethynyl and Vinyl Dienophiles

Substrate: $R^1—C{\equiv}C—CH_2—N^+(R^2)(R^3)—CH{=}C{=}CH—C(R^4){=}CH_2$ Br^- or Cl^- [*a]

Product: 4-R^1-6-R^4-2,2-(R^2,R^3)-isoindolinium Br^- or Cl^-

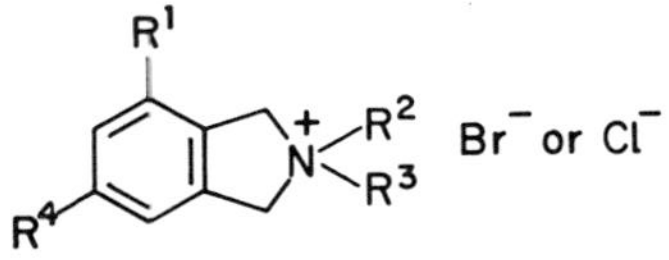

Entry No.	R^1	R^2	R^3	R^4	Conditions	Product(s) and Yield(s) (%)	Refs.
1	H	CH_3	CH_3	H	H_2O, 25°	(85)	329,365
2	H	CH_3	CH_3	Cl	H_2O, 100°, 2 hr	(82)	333
3	CH_3	CH_3	CH_3	H	H_2O, 25°, 4 days	(81)	366,335,367
4	H	CH_3	CH_3	CH_3	H_2O, 25°	(97)	329,365
5	H	C_2H_5	C_2H_5	H	H_2O, 25°	(86)	329,365
6	H	$—(CH_2)_4—$		H	H_2O, 25°, 2 hr	(83)	368
7	CH_3	CH_3	CH_3	CH_3	H_2O, 25°, 70 hr	(78)	369
8	$CH_2{=}CH$	CH_3	CH_3	H	H_2O, 25°, 1 hr	(90)	331
9	H	$—(CH_2)_2O(CH_2)_2—$		H	H_2O, 25°, 2 hr	(89)	368
10	CH_3	C_2H_5	C_2H_5	H	H_2O, 25°, 4 days	(71)	369,491
11	CH_3	C_2H_5	C_2H_5	H	H_2O, 25°, 70 hr	(84)	369
12	$CH_2{=}CH$	C_2H_5	C_2H_5	H	H_2O, 25°, 1 hr	(87)	331
13	$CH_2{=}C(CH_3)$	CH_3	CH_3	CH_3	H_2O, 25°, 1 hr	(92)	331
14	H	$C_6H_5CH_2$	CH_3	H	H_2O, 25°, 2 hr	(71)	368
15	C_6H_5	CH_3	CH_3	H	H_2O, 25°, 2 hr	(98)[b]	370
16	$CH_2{=}C(CH_3)$	C_2H_5	C_2H_5	H	H_2O, 25°, 1 hr	(84)	331
17	C_6H_5	CH_3	CH_3	CH_3	H_2O, 25°, 2 hr	(85)[b]	370

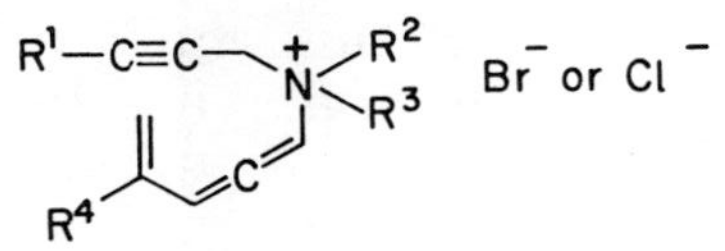

TABLE IId. ACYCLIC 1,3,4-TRIENES; 3-ATOM CHAIN CONTAINING QUATERNARY NITROGEN; ETHYNYL AND VINYL DIENOPHILES (*Continued*)

Entry No.	Substrate R^1	R^2	R^3	R^4	Conditions	Product(s) and Yield(s)(%)	Refs.
18	$C_6H_5OCH_2$	CH_3	CH_3	H	H_2O, 25°	(80)	371
19	$C_6H_5OCH_2$	CH_3	CH_3	CH_3	H_2O, 25°	(65)	371
20	C_6H_5	$-(CH_2)_2O(CH_2)_2-$		H	H_2O, 25°, 2 hr	(93)[b]	370
21	C_6H_5	$-(CH_2)_2O(CH_2)_2-$		CH_3	H_2O, 25°, 2 hr	(92)[b]	370
22	$C_6H_5OCH_2$	C_2H_5	C_2H_5	H	H_2O, 25°	(60)	371
23	$C_6H_5OCH_2$	C_2H_5	C_2H_5	CH_3	H_2O, 25°	(50)	371

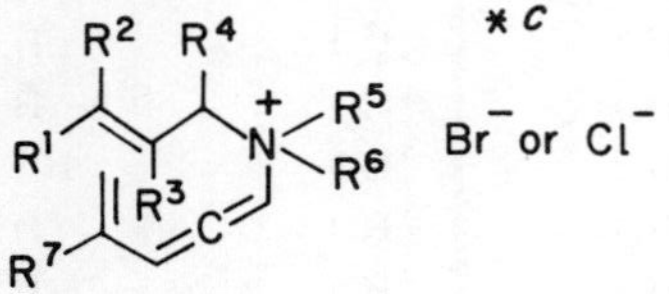

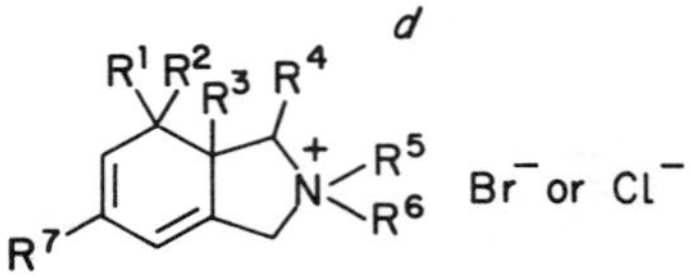

*c

d

Entry No.	R^1	R^2	R^3	R^4	R^5	R^6	R^7	Conditions	Product(s) and Yield(s)(%)	Refs.
24	H	H	H	H	CH_3	CH_3	H	H_2O, 25°	(85)	192,372,373
25	CH_3	H	H	H	CH_3	CH_3	H	H_2O, 25°, then heat	(49)*[e]	373
26	H	H	H	H	CH_3	CH_3	CH_3	", ", "	(73)*[e]	373
27	H	H	CH_3	H	CH_3	CH_3	H	H_2O, 25°, then 40–50°	(85)	374,192,375
28	H	H	H	H	C_2H_5	C_2H_5	H	H_2O, 25°, then heat	(54)*[e]	373,376
29	CH_3	CH_3	H	H	CH_3	CH_3	H	H_2O, 40°, then heat	(26)*[f]	377
30	H	H	CH_3	H	CH_3	CH_3	CH_3	H_2O, 25°	(92)	192, 375
31	H	H	H	H	$-(CH_2)_4-$		H	H_2O, 25°	(94)	378
32	H	H	H	H	$-(CH_2)_2O(CH_2)_2-$		H	H_2O, 25°, then 40–50°	(90)	374, 378
33	CH_3	H	H	H	C_2H_5	C_2H_5	H	H_2O, 25°, then heat	(35)*[e]	373,376

34	CH_3	H	H	CH_3	CH_3	CH_3	CH_3	H_2O, 110–120°, 4 hr	(48)*[e]	336
35	H	H	CH_3	H	C_2H_5	C_2H_5	H	H_2O, 25°, then 40–50°	(90)	374
36	H	H	CH_3	H	—$(CH_2)_2O(CH_2)_2$—		H	H_2O, 25°, then 40–50°	(90)	374
37	H	H	H	H	—$(CH_2)_5$—		H	H_2O, 25°, then 40–50°	(90)	374
38	CH_3	CH_3	H	H	C_2H_5	C_2H_5	H	H_2O, 40°, then heat	(56)*[f]	377
39	H	H	CH_3	H	—$(CH_2)_5$—		H	H_2O, 25°, then 40–50°	(87)	374
40	H	H	H	H	—$(CH_2)_5$—		H	H_2O, 110–120°, 4 hr	(57)*[e]	336
41	H	H	H	H	$C_6H_5CH_2$	CH_3	H	H_2O, 25°	(80)	378
42	C_6H_5	H	H	H	CH_3	CH_3	H	H_2O, 25°	(71–82)	371
43	H	H	C_6H_5	H	CH_3	CH_3	H	H_2O, 40–50°, many hours	(100)	379
44	H	H	C_6H_5	H	CH_3	CH_3	CH_3	", ", "	(100)	379
45	C_6H_5	H	H	H	C_2H_5	C_2H_5	H	H_2O, 25°	(77–79)	371
46	H	H	C_6H_5	H	C_2H_5	C_2H_5	H	H_2O, 40–50°, many hours	(100)	379
47	H	H	C_6H_5	H	—$(CH_2)_2O(CH_2)_2$—		H	", ", "	(100)	379
48	C_6H_5	H	H	H	—$(CH_2)_5$—		H	H_2O, 25°	(73–94)	371
49	H	H	C_6H_5	H	—$(CH_2)_5$—		H	H_2O, 40–50°, many hours	(100)	379

*[a] Br^- or Cl^-

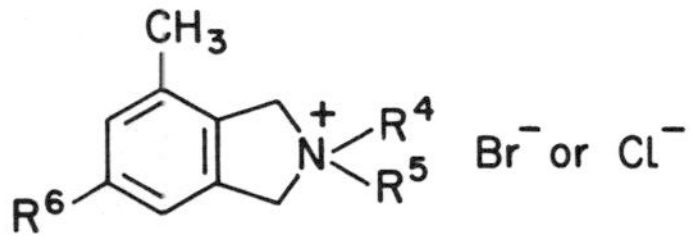

Br^- or Cl^-

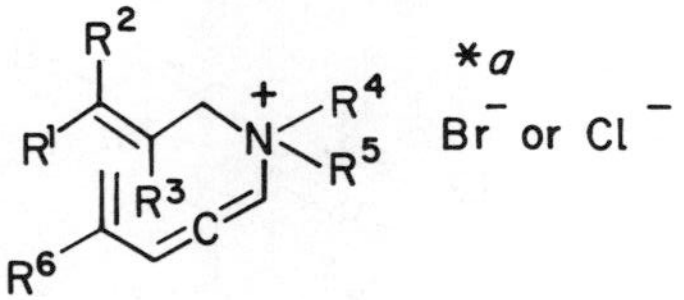

	R^1	R^2	R^3	R^4	R^5	R^6			
50	CH_3	Cl	H	CH_3	CH_3	H	H_2O, 25°, 4 days	(55)	366,335,367
51	Cl	CH_3	H	CH_3	CH_3	H	", ", "	(96)	335,367
52	CH_3	H	Cl	CH_3	CH_3	H	", ", 70 hr	(30)	369
53	CH_3	Cl	H	C_2H_5	C_2H_5	H	H_2O, 25°, 4 days	(25)	366
54	CH_3	Cl	H	—$(CH_2)_5$—		H	", ", "	(51)	366

TABLE IId. ACYCLIC 1,3,4-TRIENES; 3-ATOM CHAIN CONTAINING QUATERNARY NITROGEN; ETHYNYL AND VINYL DIENOPHILES (*Continued*)

Entry No.	Substrate	Conditions	Product(s) and Yield(s) (%)	Refs.
	R^1 R^2 R^3 R^4 R^5 R^6			
55	CH_3 Cl H C_2H_5 C_2H_5 CH_3	H_2O, 25°, 70 hr	(54)	369
	[c]* 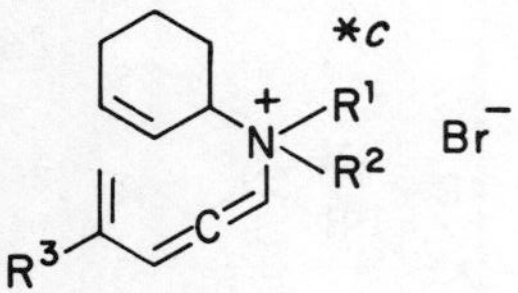 Br^-	H_2O, 40°, 60 hr	[d] 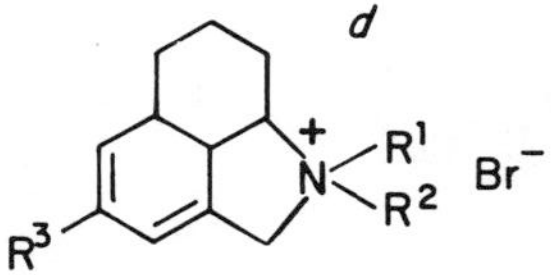 Br^-	
	R^1 R^2 R^3			
56	CH_3 CH_3 H		(—)	380
57	CH_3 CH_3 CH_3		(—)	380
58	C_2H_5 C_2H_5 H		(—)	380
59	$-(CH_2)_2O(CH_2)_2-$ H		(—)	380
60	$-(CH_2)_5-$ H		(—)	380

[a] The substrate was generated by KOH-catalyzed isomerization of the corresponding 4-penten-2-ynylammonium salt.

[b] The product was a mixture containing unspecified amounts of the isomer resulting from addition of the 4-penten-2-ynyl group (dienophile) to the phenylallenyl system (diene).

[c] The substrate was generated by KOH- or $(C_2H_5)_3N$-catalyzed isomerization of the corresponding 4-penten-2-ynylammonium salt.

[d] The structures are based only on elemental analyses and IR and/or UV spectra.

[e] The product was one or more benzylamines formed by Hofmann degradation.

[f] The product was a 5-methylene-1,3-cyclohexadiene-4-methylamine formed by Hofmann degradation.

* This intermediate is hypothetical, or has not been isolated.

TABLE IIe. ACYCLIC DIENES; 3-ATOM CHAIN CONTAINING OXYGEN OR SULFUR; ETHYNYL AND VINYL DIENOPHILES

Entry No.	Substrate	Conditions	Product(s) and Yield(s)(%)	Refs.
1	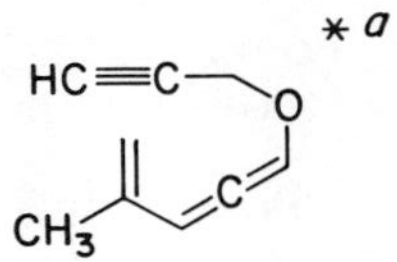 HC≡C—O, CH_3 * [a]	t-C_4H_9OH, 25°, 4 days	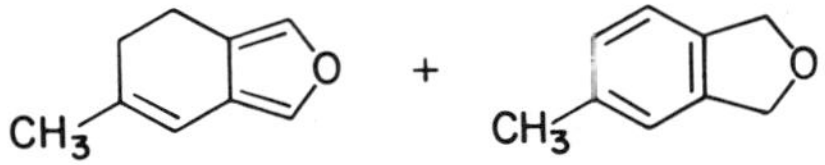 CH_3 (45) + CH_3 (15)	381
	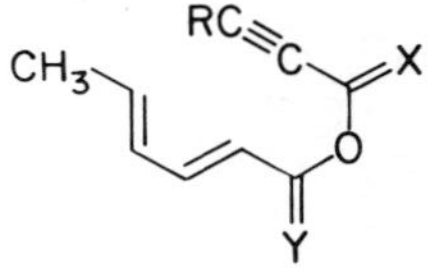 CH_3, RC≡C, X, O, Y		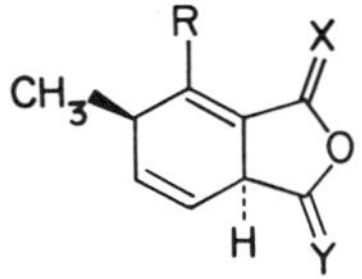 CH_3, R, X, O, H, Y	
	R X Y			
2	H H_2 O	Heat	(0)	161
3	CH_3 O H_2	Xylene, reflux, 24 hr	(96)	36
	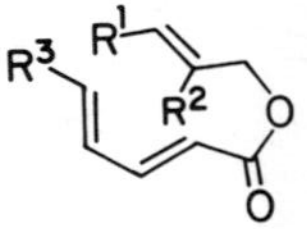 R^3, R^1, R^2, O		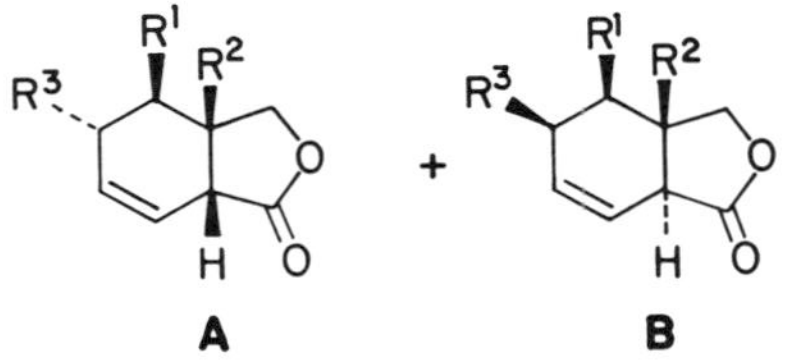 R^3, R^1, R^2, O, H, A + R^3, R^1, R^2, O, H, B	
	R^1 R^2 R^3		A B	
4	H H H	Heat	(0)	185,382
4a	$C_2H_5O_2C$ H CH_3	135°, 15.5 hr	(8) (71)	159
4b	H CH_3O_2C CH_3	250°, 4 hr	(55)[b] (45)[b]	159

TABLE IIe. ACYCLIC DIENES; 3-ATOM CHAIN CONTAINING OXYGEN OR SULFUR; ETHYNYL AND VINYL DIENOPHILES (*Continued*)

Entry No.	R^1	R^2	R^3	R^4	R^5	Conditions	A	B	Refs.
5	H	H	H	H	H	25°[c]	(0)	(0)	185
6	CH_3	CH_3	H	H	H	Benzene, 220°	(0)[d, e]	(0)	60
7	CH_3	CH_3	H	H	CH_3	Heat to 300°	(0)	(0)	383
8	CH_3	HO_2C	H	H	CH_3	Xylene, reflux, 15 hr	(0)	(25)	383,384
9	H	HO_2C	CH_3	H	CH_3	", "	(0)	(0)	384
10	CH_3	CH_3O_2C	H	H	H	Xylene, reflux, 32 hr	(55)	(0)	36,384
11	CH_3O_2C	CH_3	H	H	CH_3	", ", 5 days	(41)	(9)	36
12	CH_3O_2C	H	CH_3	H	CH_3	", ", 11 days	(35)[f]	(7)[f]	36
13	CH_3	CH_3O_2C	H	H	CH_3	Xylene, reflux, 24 hr	(40)	(0)	36,384

14	CH_3O_2C	H	CH_3	H	CH_3	180°		(0)	(0)	384
15	H	H	H	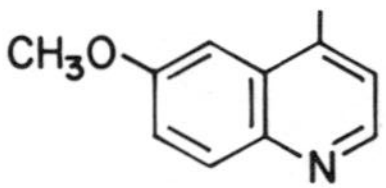	H	C_6D_6, 120°		(0)	(0)	385
16	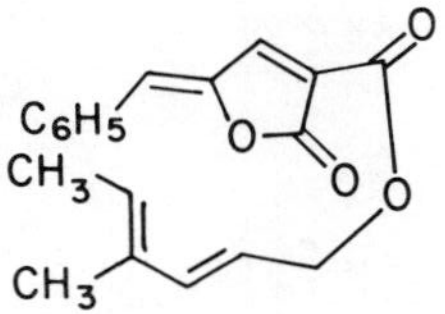					o-$Cl_2C_6H_4$, 180°, 2.5 hr	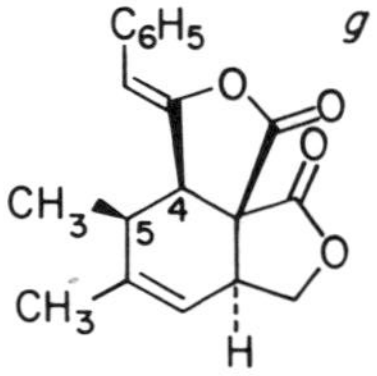 [g]	(—)		313
17	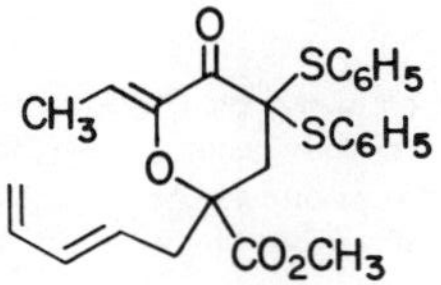					Toluene, 130°, 3.5 hr	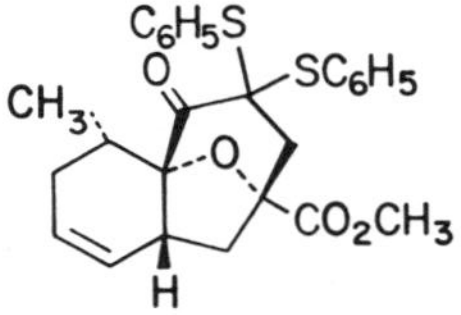		(79)	386
18	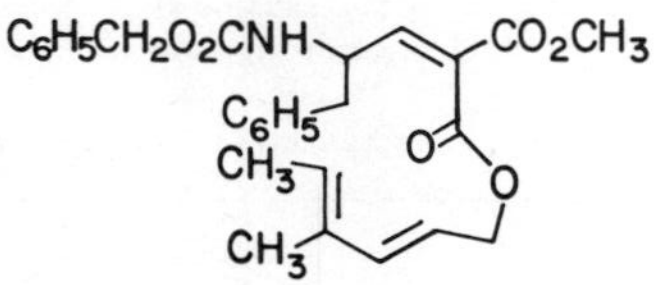					Toluene, reflux, 7 days	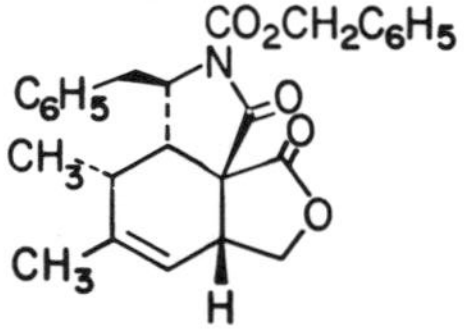		(40)	137

TABLE IIe. Acyclic Dienes; 3-Atom Chain Containing Oxygen or Sulfur; Ethynyl and Vinyl Dienophiles (*Continued*)

Entry No.	Substrate	Conditions	Product(s) and Yield(s) (%)	Refs.
19	O O_2CCH_3 C S *[h]	520°	O O_2CCH_3 S * → HO S (28)[i]	387

[a] The substrate was generated by t-C_4H_9OK-catalyzed isomerization of propargyl 4-methyl-4-penten-2-ynyl ether; the corresponding 2-methylallyl, 3,3-dimethylallyl, and cinnamyl ethers could not be cyclized.

[b] The numbers are the ratios **A**/**B**.

[c] The substrate polymerized at 25°, even in the presence of radical inhibitors.

[d] The product was 3-methyl-2-butenoic acid.

[e] On heating under super-high pressure, the intermolecular Diels–Alder adduct was formed.

[f] The structure assignment is tentative.

[g] The stereochemical assignment is based on ^{1}H NMR, $J_{H\text{-}4,H\text{-}5} = 7$ Hz.

[h] The substrate was generated from

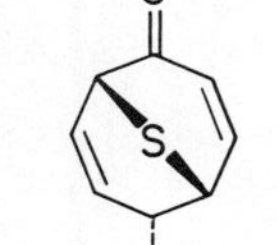

[i] The bridged product was formed in 19% yield; see Table XXVIc/2.

* This intermediate is hypothetical, or has not been isolated.

TABLE IIf. ACYCLIC DIENES; 3-ATOM CHAIN; DIENOPHILES CONTAINING HETEROATOMS

Entry No.	Substrate		Conditions	Product(s) and Yield(s)(%)	Refs.
1			500°	CH_3, CH_3 (—)	388
	R^1	R^2			
2	$C_2H_5O_2C$	CH_3, CH_3, CH_3	C_6H_{12}, reflux, 2 hr	(73)	389
3	$CH_3CO_2CH_2$	"	", ", 3 hr	(77)	389

TABLE IIf. ACYCLIC DIENES; 3-ATOM CHAIN; DIENOPHILES CONTAINING HETEROATOMS (*Continued*)

Entry No.	Substrate				Conditions	Product(s) and Yield(s) (%)		Refs.
4	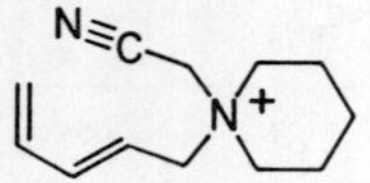 N≡C, N+, Cl⁻				DMF, 90–95°	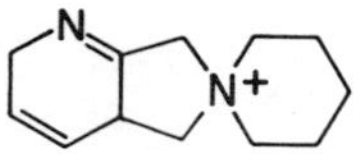 N, N+, Cl⁻ (90)[a]		364
	R3, O=N, O, X, R2, R1 *					R3, O, N, O, X, R2, H, R1 **A** + R3, O, N, O, X, R2, H, R1 **B**		
	R^1	R^2	R^3	X		A	B	
5	H	H	H	O[b]	−20°	(55–75)		178
6	H	H	CH_3	O[b]	−20°	(55–75)		178
7	H	H	H	CH_2[b]	−20°	(55–75)		178
8	H	H	CH_3	CH_2[b]	−20°	(55–75)		178
9	HO	H	CH_3	CH_2[c]	Benzene, reflux, 5 hr	(66)	(33)	184
10	DO	H	CH_3	CH_2[c]	", ", "	(66)[d]	(33)[d]	184
11	CH_3CO_2	H	CH_3	CH_2[c]	", ", "	(43)[d]	(57)[d]	184
12	$t\text{-}C_4H_9(CH_3)_2SiO$	H	CH_3	CH_2[c]	", ", "	(41)[d]	(59)[d]	184
13	"	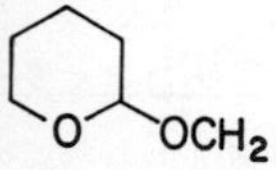 O, OCH_2	H	CH_2[c]	", ", 4.5 hr	(86)[e]		142

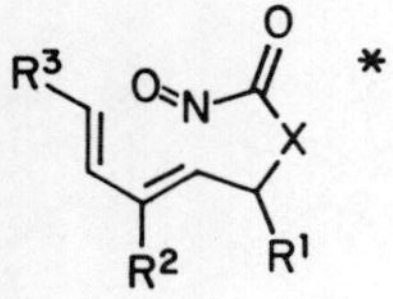

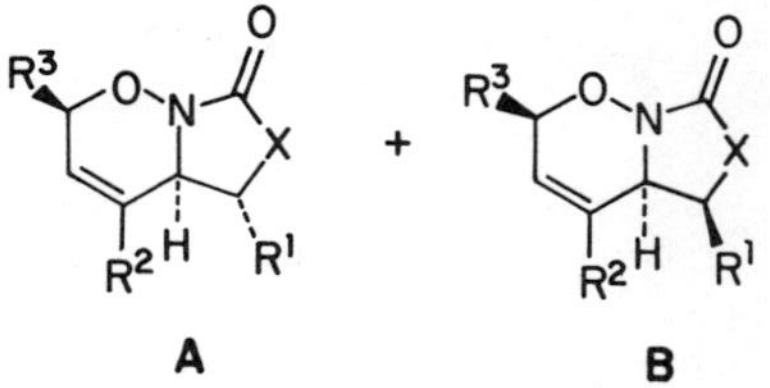

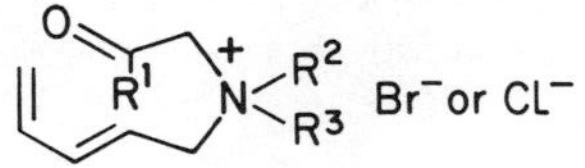

No.	Reactant					Conditions	Product(s) and Yield(s) (%)	Refs.
	Br⁻ or Cl⁻					DMF, 90–95°	Br⁻ or Cl[a]	
	R^1	R^2	R^3					
14	CH_3	—$(CH_2)_5$—					(90)	364
14a	CH_3O	C_2H_5	C_2H_5				(90)	364
	*						A + B	
	R^1	R^2	R^3	R^4	X		A B	
15	H	H	H	H	CH_2[f]	370–390°	(73)	182, 136
15a	H	H	CH_3	H	CH_2[g]	10^{-3} torr, 800°	(<20)	183
16	CH_3O_2C	H	H	H	O[f]	Toluene, 215°, 2 hr	(30)	113
16a	CH_3	H	CH_3	H	CH_2[g]	10^{-3} torr, 800°	(<20)	183
17	CH_3O_2C	H	H	H	O[f]	C_6H_5Br, 230–240°, 2.5 hr	(32) (27)	113
18	H	H	n-C_3H_7CO	H	CH_2[f,h]	Heat	(poor)	136, 135
19	H	H	n-$C_3H_7CH(OH)$	H	CH_2[f,h]	370–390°	(low)	135
20	H	H	n-$C_3H_7CH[OSi(CH_3)_3]$	H	CH_2[f,h]	"	(68)[i]	136, 135
20a	H	t-$C_4H_9(CH_3)_2SiO$	H	CH_3O_2C	CH_2[f]	o-$Cl_2C_6H_4$, reflux, 4 hr	(33) (57)	276
20b	*[g]					10^{-3} torr, 800°	(<20)	183

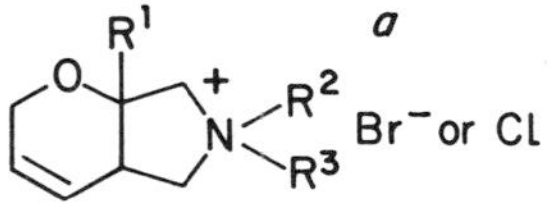

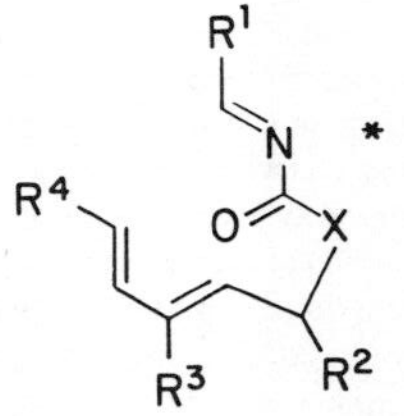

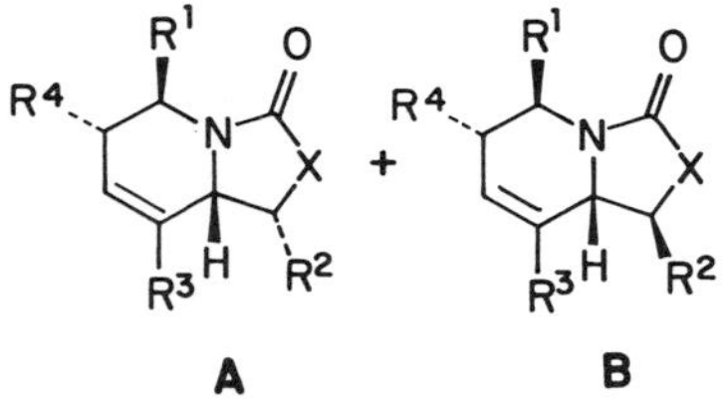

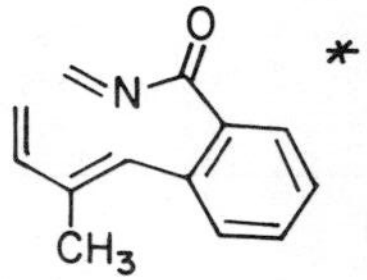

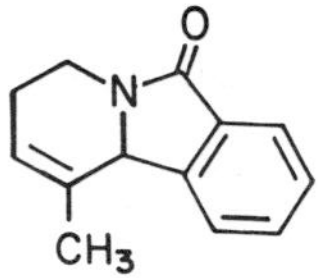

TABLE IIf. ACYCLIC DIENES; 3-ATOM CHAIN; DIENOPHILES CONTAINING HETEROATOMS (*Continued*)

Entry No.	Substrate	Conditions	Product(s) and Yield(s)(%)	Refs.
20c	$CH_3N{=}$ … CH_3	Heat	N-CH_3 … CH_3 (0)	183
21	O=C=… S= … O_2CCH_3 *[j]	Benzene, hν	O= S … O_2CCH_3 (28)[k]	387

[a] The structure assignment is based only on elemental analysis and IR and UV spectra.
[b] The substrate was generated by oxidation of the corresponding hydroxamic acid with tetraethylammonium periodate.
[c] The substrate was generated by thermolysis of the adduct of the nitrosoacylderivative to 9,10-dimethylanthracene.
[d] The numbers are the ratios of products.
[e] The product was a mixture of isomers of unassigned stereochemistry in a ratio of 57:43.
[f] The substrate was generated from the corresponding acetoxymethylamide.
[g] The substrate was generated by reverse ene reaction from a butadienylpyrrolidone.
[h] The diene was masked as the adduct with SO_2.
[i] The product was a mixture of diastereomers in a ratio of 5:4.
[j] The substrate was generated from

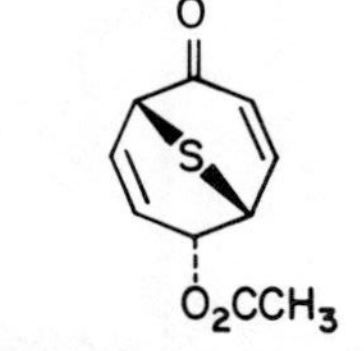

[k] The product was a ≈1:1 mixture of the two epimers; the stereochemistry of the ring junction was not specified.
*This intermediate is hypothetical, or has not been isolated.

TABLE IIIa. ACYCLIC DIENES; 4-ATOM CHAIN CONTAINING CARBON ONLY

Entry No.	Substrate			Conditions	Product(s) and Yield(s)(%)		Refs.
	R^1	R^2	R^3		A	B	
1	$Cl_3CCH_2O_2C$	$CH_3O(CH_2)_2OCH_2O$	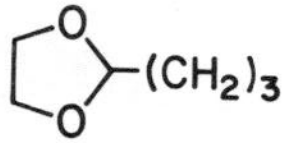 (1,3-dioxolan-2-yl)$(CH_2)_3$	Toluene, 165°, 50 hr	(28)	(67)	133
1a	$Cl_3CCH_2O_2C$	t-$C_4H_9(CH_3)_2SiO$	$C_6H_5CH_2O(CH_2)_3$	", 160°, 61 hr	(37)	(63)	133
1b	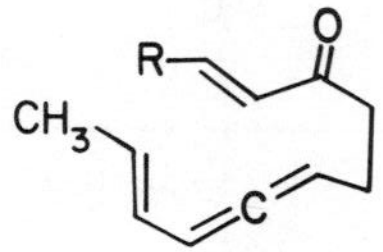 R = $C_2H_5OCH(CH_3)O(CH_2)_3$			Benzene, 150°, 2 hr	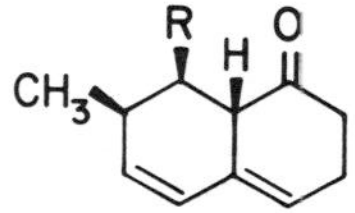 (75)		100

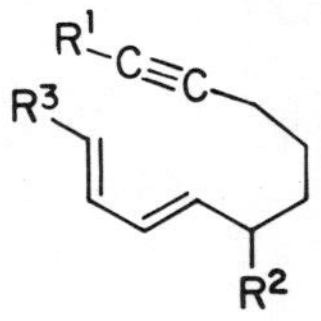

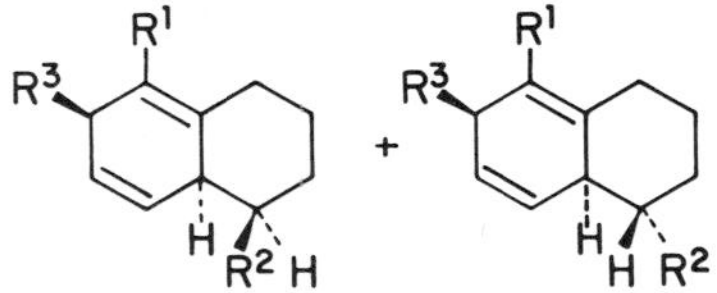

TABLE IIIa. ACYCLIC DIENES; 4-ATOM CHAIN CONTAINING CARBON ONLY (*Continued*)

Entry No.	Substrate			Conditions	Product(s) and Yield(s)(%)			Refs.
	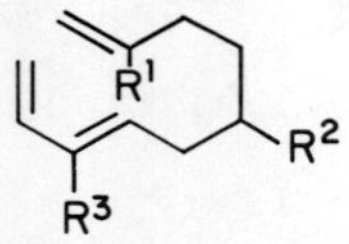				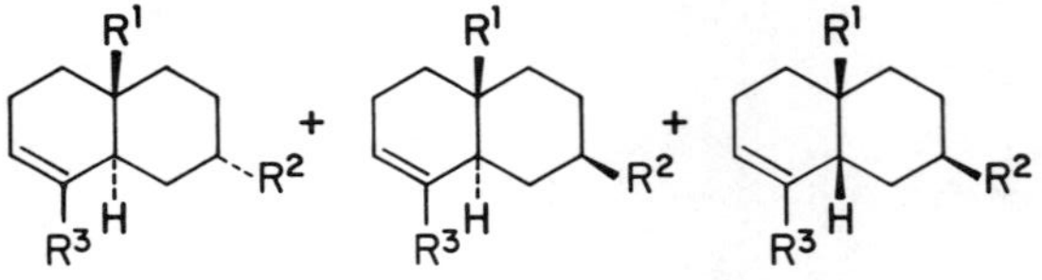 A	B	C	
	R^1	R^2	R^3		A	B	C	
2	H	OH	H	160°	(55)[a]		(45)[a]	155
3	H	H	CH_3	”	(95)		(0)	95
4	CH_3	H	CH_3	190°	(30)		(0)	95
5	H	OH	CH_3	160°	(24)	(33)	(0)	95
6	CH_3	$(CH_3)_3SiO$	CH_3	Toluene, 200°, 100 hr	(34)	(54)	(5)	95
6a	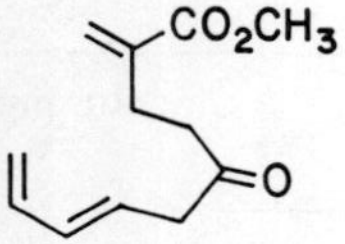			170°, 22 hr	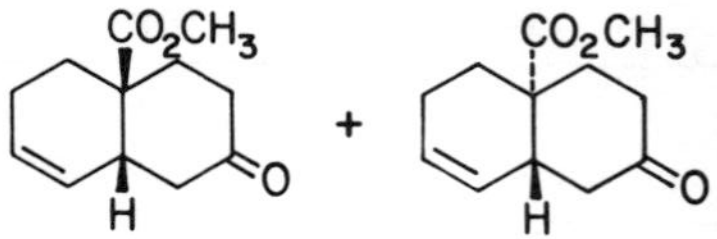 (38) + (12)			
	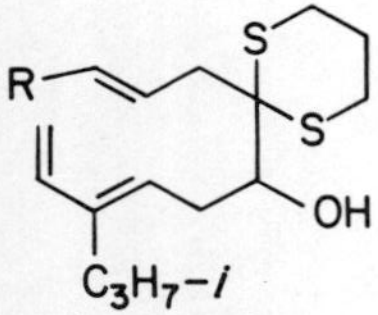			Benzene, 180°, 20 hr	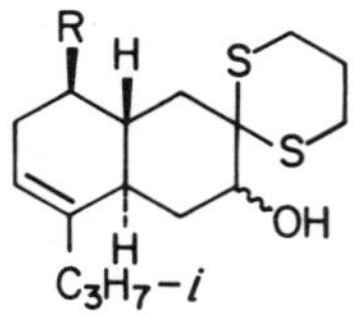			159

	R			
7	H		(85)	156
8	CH_3		(91)	156

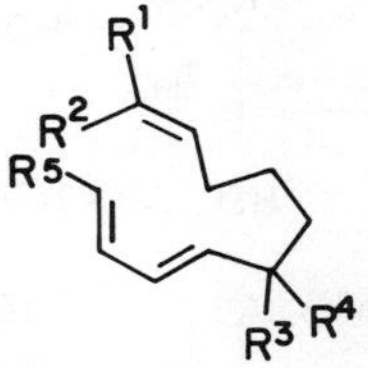

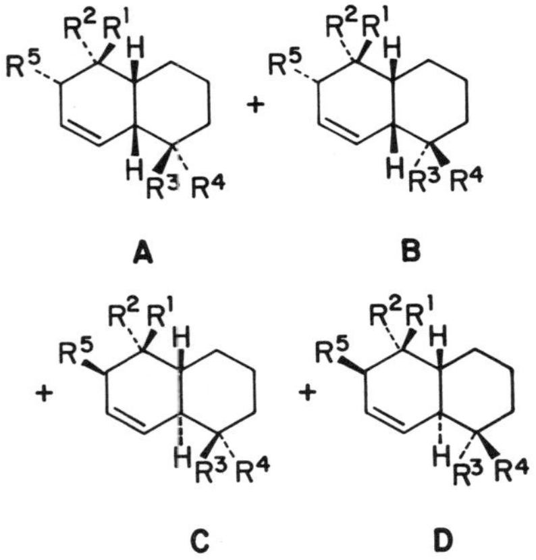

A B C D

	R^1	R^2	R^3	R^4	R^5		A B	C D	
9	CH_3O_2C	H	H	H	H	Toluene, 155°, 45 hr	(45)	(47)	93
10	H	CH_3O_2C	H	H	H	", ", "	(44)	(46)	93
10a	CH_3O_2C	H	H	H	i-C_3H_7	", 160°, 4 hr	(35)	(35)	94
						$C_2H_5AlCl_2$, CH_2Cl_2, 23°, 3 hr	(8)	(60)	94

TABLE IIIa. Acyclic Dienes; 4-Atom Chain Containing Carbon Only (*Continued*)

Entry No.	Substrate R^1	R^2	R^3	R^4	R^5	Conditions	Product(s) and Yield(s)(%) A	B	C	D	Refs.
10b	H	CH_3O_2C	H	H	i-C_3H_7	Toluene, 180°, 3 hr	(32)		(39)		94
						$C_2H_5AlCl_2$, CH_2Cl_2, 23°, 4 hr	(71)		(6)		94
11	CH_3O_2C	H	$—O(CH_2)_2O—$		CH_3	Toluene, 150°, 18 hr	(63)		(17)		93
12	H	CH_3O_2C	$—O(CH_2)_2O—$		CH_3	Toluene, 150°, 19 hr	(54)		(17)		93
13	CH_3O_2C	CH_3	$—O(CH_2)_2O—$		CH_3	", 140°, 16 hr	(61)		(14)		93
14	CH_3O_2C	H	$(CH_3)_3SiO$	H	CH_3	", 150°, 19 hr	$(29)^b$	$(22)^b$	$(13)^b$	$(23)^b$	93
15	CH_3O_2C	H	H	H	3,4-methylenedioxyphenyl	Xylene, reflux, 3 hr	(37)		(19)		91
16	H	CH_3O_2C	t-$C_4H_9(CH_3)_2SiO$	H	CH_3	Toluene, 150°, 17 hr	(34)	(16)	(18)	(13)	93
17	CH_3O_2C	CH_3	"	H	CH_3	$Cl_2C{=}CCl_2$, 150°, 3 hr	$(27)^b$	$(14)^b$	$(3)^b$	$(7)^b$	93
17a						Toluene, Na_2CO_3, 220°, 20 hr, then HCl	*trans* (40), *cis* (3) + *trans* (13), *cis* (7)				255

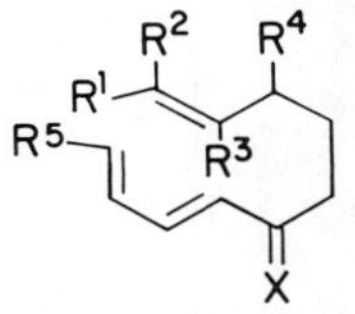

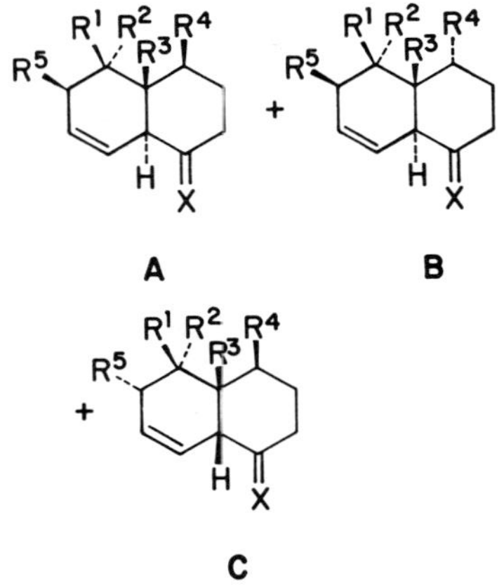

	R^1	R^2	R^3	R^4	R^5	X		A	B	C	
17b	CH_3O_2C	H	H	HO	CH_3	H_2	C_6H_5Br, reflux, 60 hr	(22)	(6)	(25)	92
							CH_2Cl_2, $C_2H_5AlCl_2$, 25°	(24)	(30)	(0.5)	92
18	CH_3O_2C	CH_3	H	H	CH_3	O	Toluene, 160–165°, 138 hr	(52)[d]		(0)	93
19	H	H	CH_3	CH_3	$C_2H_5O_2C$	O	", 250°, 6 hr	(46)	(0)	(46)[c]	19
20	CH_3O_2C	CH_3	H	H	$CH_3O_2C(CH_2)_2$	O	", 200°, 102 hr	(—)[d]		(0)	93
20a	CH_3O_2C	H	H	$t\text{-}C_4H_9(CH_3)_2SiO$	CH_3	H_2	Xylene, reflux, 120 hr	(8)	(42)	(14)	92
							CH_2Cl_2, $C_2H_5AlCl_2$, 25°, 72 hr	(0)	(65–73)	(0)	92

TABLE IIIa. ACYCLIC DIENES; 4-ATOM CHAIN CONTAINING CARBON ONLY (*Continued*)

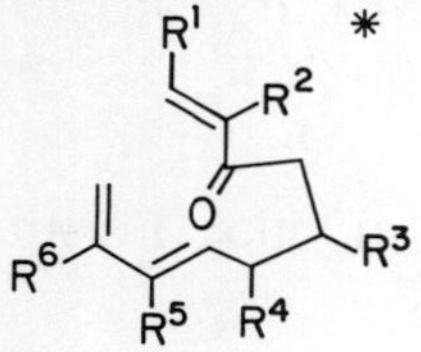

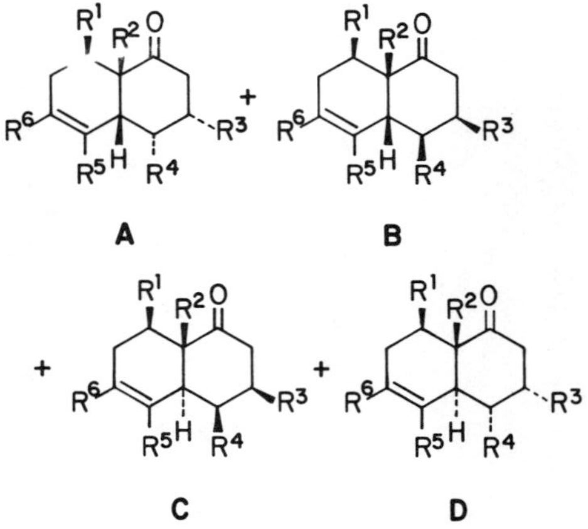

Entry No.	Substrate						Conditions	Product(s) and Yield(s)(%)		Refs.
	*							A + B + C + D		
	R^1	R^2	R^3	R^4	R^5	R^6		A B	C D	
21	H	H	H	H	H	H[e]	$CHCl_3$, 22°, 4 hr;	(9.5)[f]	(0.5)[f]	98
							", ", 5 days	(1)[f]	(9)[f]	98
22	H	H	H	H	H	H[k]	CH_3OH, 0°, 1 hr	(78)	(0)	97
22a	H	H	CH_3S	H	H	H[k]	", ", "	(63)[h]	(0)	390
23	i-C_3H_7	H	H	H	H	H[g]	THF	(85)	(0)	98
24	H	H	H	C_2H_5	H	CH_3[e]	$CHCl_3$, 25°, 70 hr	(80)[h]	(0)	65

No.							Conditions					Refs.
25	i-C_3H_7	H	H	H	CH_3	H^l	Benzene, 42°, 1 hr		(84)		(0)	179
26	H	H	H	i-C_3H_7	H	$CH_3^{\,i}$	Et_2O, 0°, 25 min	(11)	(76)	(0)	(0)	141,254, 390a
27	H	CH_3	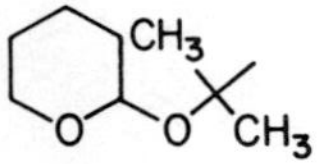	H	CH_3	$H^{e,j}$	—	(42)	(0)	(17)	(4)	99
28	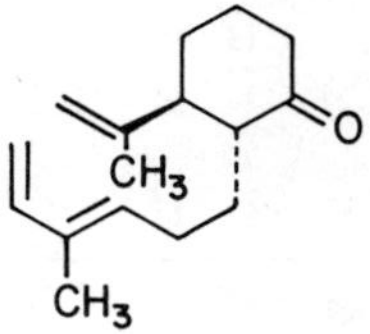						Toluene, 170°, 68 hr	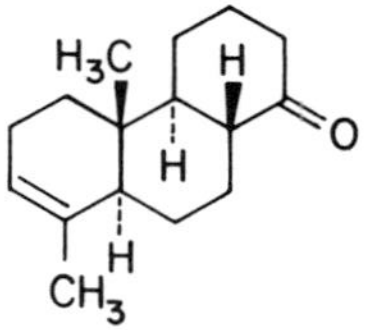 (73)				258
29	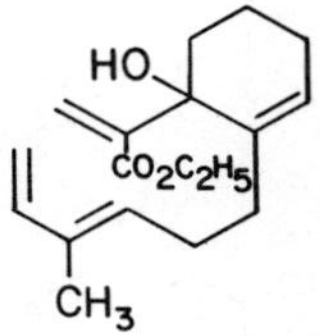						110°, 18 hr	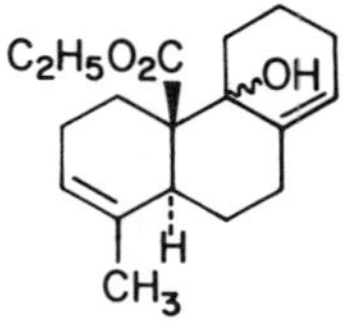 (—)				258
30	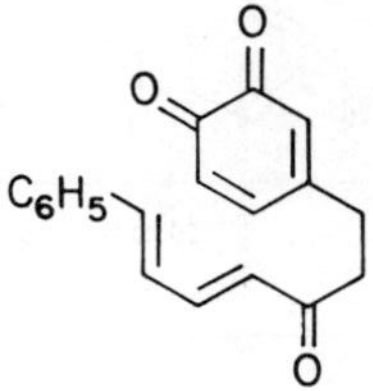 *m						$CHCl_3$, 25°, 16 hr	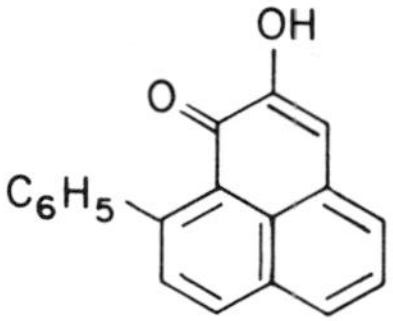 (37)				170

TABLE IIIa. ACYCLIC DIENES; 4-ATOM CHAIN CONTAINING CARBON ONLY (*Continued*)

Entry No.	Substrate			Conditions	Product(s) and Yield(s) (%)		Refs.
	R¹, R², OR³, CH_3				R¹, R², OR³, H, CH_3 **A** + R¹, R², OR³, H, CH_3 **B**		
	R^1	R^2	R^3		**A**	**B**	
31	H	H	H	Neat, 100°, $t_{1/2}$ = 2 hr	(58)	(14)	96
32	CH_3	H	$(CH_3)_3Si$	180°, 24 hr	(29)[n]	(14)[n]	96
33	CH_3	i-C_3H_7	"	180°, 22 hr	(39)[n]	(39)[n]	96
34	$(CH_3)_3SiO$, $OSi(C_2H_5)_3$			Xylene, reflux, 3.5 hr	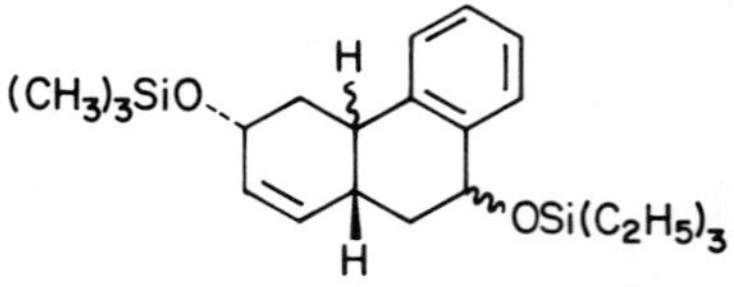 $(CH_3)_3SiO$, H, H, $OSi(C_2H_5)_3$ (70; 1:1 mixture of *cis* and *trans* isomers)[o]		391

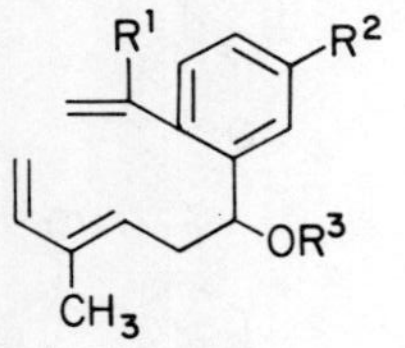

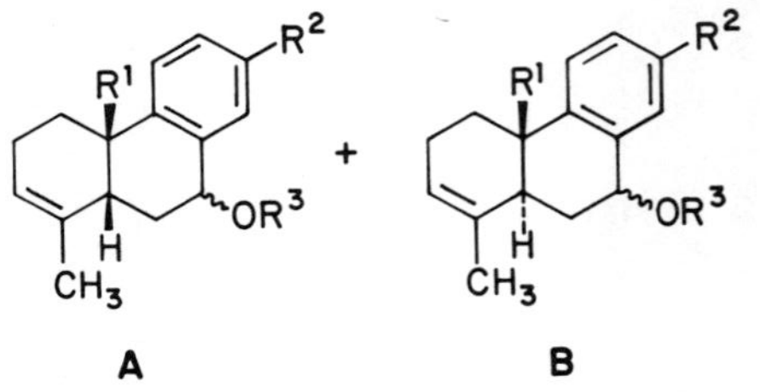

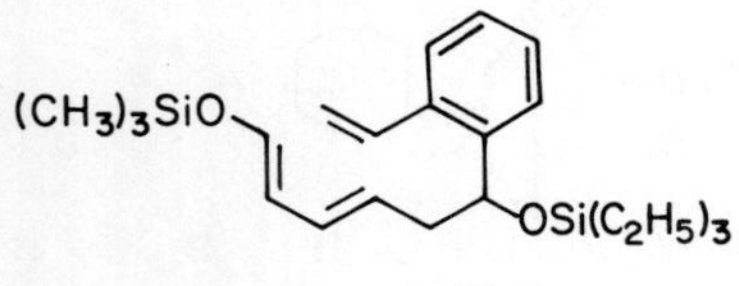

No. of C Atoms	Substrate	Conditions	Product(s) and Yield(s) (%)	Refs.
35	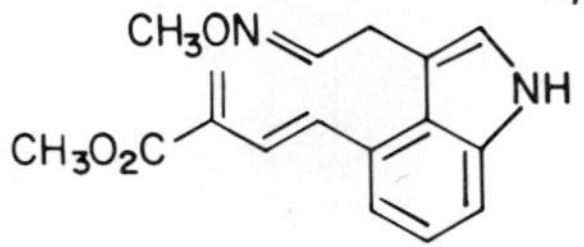 CH_3O_2C, CH_3ON … NH *[p]	1,2,4-$Cl_3C_6H_3$, 200°, 5 hr	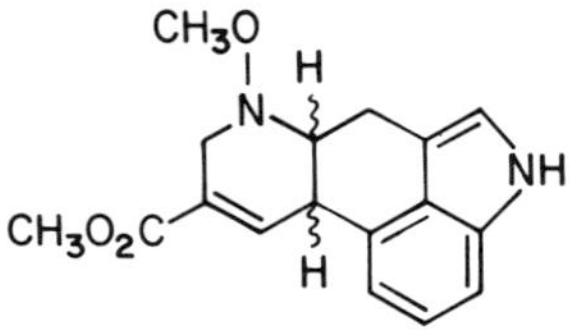 CH_3O, H, N, CH_3O_2C, H, NH (67)[q]	193

[a] The numbers are the ratio of *trans*:*cis* isomers; the stereochemistry of R^2 was not reported.
[b] The yield is of the alcohol obtained after desilylation.
[c] R^5 is β; this was a secondary product derived from **A**.
[d] In addition, conjugated isomers were formed in unspecified yields.
[e] The substrate was generated by oxidation of the corresponding alcohol with MnO_2.
[f] The numbers are the ratios of *cis*:*trans* isomers; the yields were 70–85%. The reaction proceeded stereoselectively by *syn* addition to give the less stable *cis* isomer; isomerization of the *cis* to the *trans* isomer was catalyzed by acid (e.g., MnO_2).
[g] The substrate was generated from $CH_2{=}CHCH{=}CH(CH_2)_3COCH_2CHOHC_3H_7$-*i* with methanesulfonyl chloride and triethylamine.
[h] The ratio of R^3 or R^4 epimers was not reported.
[i] The substrate was generated by oxidation of the corresponding alcohol with CrO_3.
[j] The yield and ratio of products depended on the oxidizing agent used.
[k] The substrate was generated by KF-catalyzed desilylation of the corresponding 3-(triethylsilyloxy)-1,3,7,9-decatetraene.
[l] The substrate was generated by elimination of H_2O from the corresponding aldol.
[m] The substrate was generated by sodium periodate oxidation of the corresponding pyrocatechol derivative.
[n] The yield is after desilylation and oxidation to the ketone.
[o] The yield and composition were determined after desilylation and oxidation of the benzylic hydroxyl group.
[p] The substrate was generated from the intermolecular Diels–Alder dimer (mixture of three isomers).
[q] The product was a mixture of two isomers in 3:2 ratio.
* This intermediate is hypothetical, or has not been isolated.

TABLE IIIb. ACYCLIC DIENES; 4-ATOM CHAIN CONTAINING NITROGEN OR OXYGEN

Entry No.	Substrate				Conditions	Product(s) and Yield(s)(%)		Refs.
	R^1	R^2	R^3	R^4		A	B	
1	CH_3CO	H	H	H	Toluene, 190°, 16 hr	(59)	(0)	102
2	CH_3O_2C	H	H	H	", ", 24 hr	(84)	(0)	102
3	n-C_3H_7	O=		H	—	(25)	(37)	102
4	CH_3O_2C	H	H	CH_3	Toluene, 205°, 22 hr	(36)	(0)	102
4a	$C_2H_5O_2C$	H	H	CH_3O_2C	", 200–210°, 14 hr	(65)	(5)	392
5	CH_3O_2C	n-C_3H_7	H	CH_3	Benzene, 215°, 20 hr	(25)	(0)	103
6	i-C_3H_7CO	n-C_3H_7	H	CH_3[a]	Toluene, 230°, 16 hr	(60)[b]	(0)	104
7	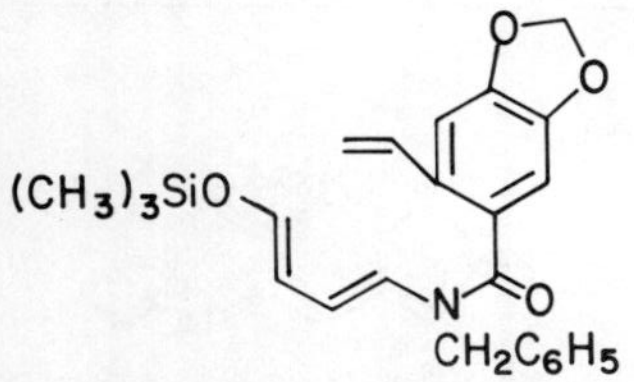 $(CH_3)_3SiO$ ... $CH_2C_6H_5$ *[c]				DMF, 160°, 12 hr, then HCl	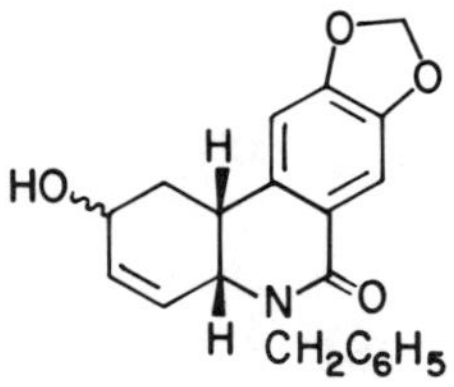 HO ... $CH_2C_6H_5$ (60–65)[d]		18

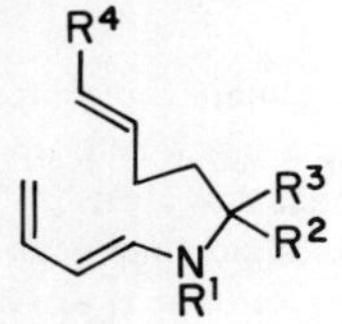

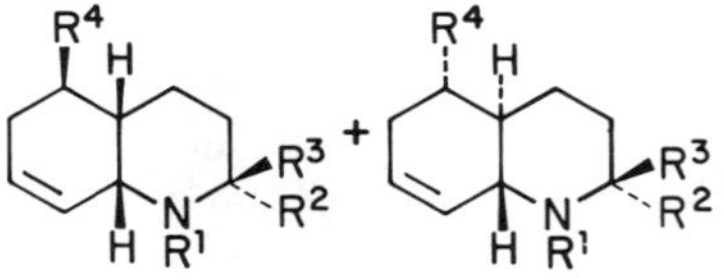

8	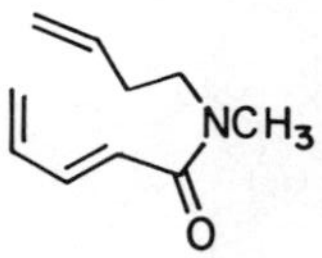 NCH3	138°, 72 hr	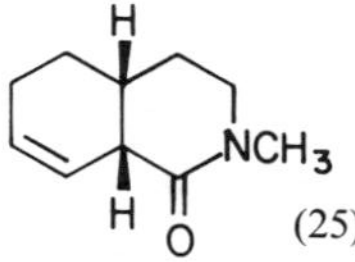 NCH3 + NCH3 (25) 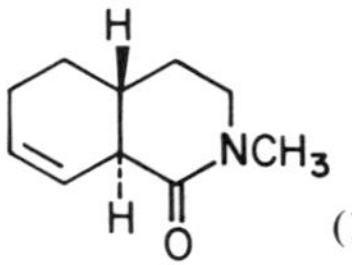 (15)	101
9	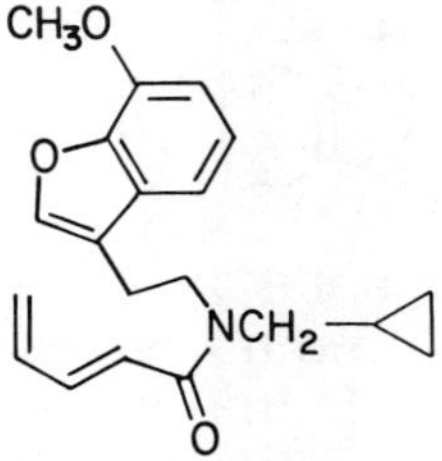 CH3O, NCH2	Toluene, 240°, 10 hr	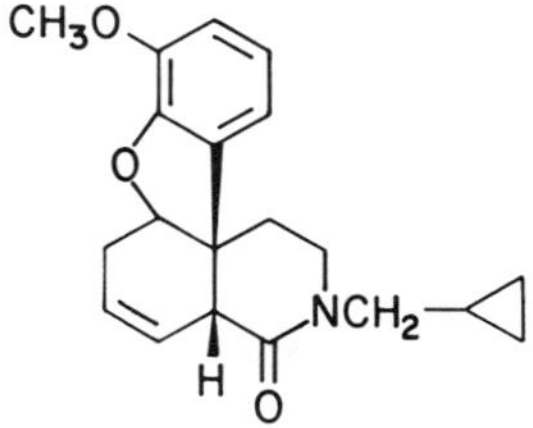 CH3O, NCH2 (10)	134
9a	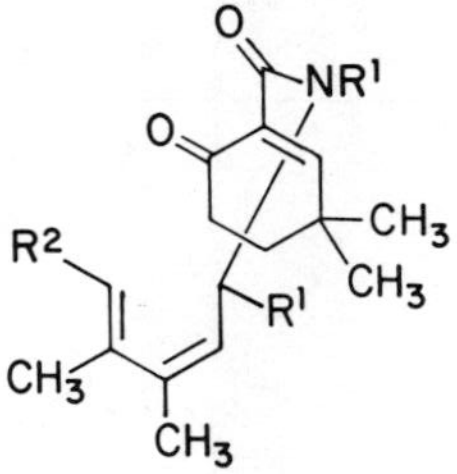 NR1, CH3, CH3, R2, R1, CH3, CH3; R^1 = $C_6H_5CH_2$; R^2 = $(CH_3)_3Si(CH_2)_2O$	Toluene, reflux, 15 hr	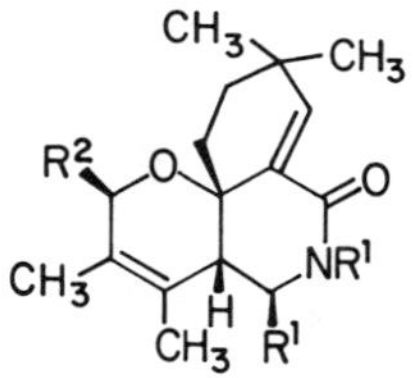 CH3, CH3, R2, NR1, CH3, CH3, R1 (15)[e]	86

TABLE IIIb. ACYCLIC DIENES, 4-ATOM CHAIN CONTAINING NITROGEN OR OXYGEN (*Continued*)

Entry No.	Substrate			Conditions	Product(s) and Yield(s)(%)		Refs.
	X	Y	R		A	B	
10	H_2	H_2	CH_3	Xylene, 140°, 1 day	(36)	(33)	101
10a	H_2	O	CH_3	", 80°, 3 hr	(73)	(10)	101
11	O	H_2	CH_3	", 185°, 2.5 days	(54)	(26)	101
12	H_2	H_2	CH_3O_2C	", 275°, 2 days	(38)	(34)	101
13	O	H_2	$C_6H_5(CH_2)_2$	", 185°, 2.5 days	(47)	(23)	101
14	O	H_2 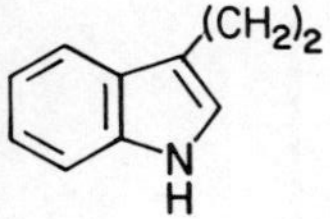	2-(indol-3-yl)ethyl, $(CH_2)_2$	", 160°, 4 days	(37)	(34)	101
15	HO_2C, O, NH			Toluene, reflux, 1.5 hr	HO_2C, H, O, NH, H (44)		393

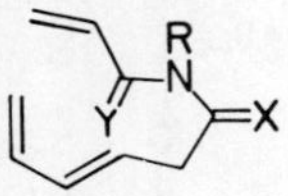

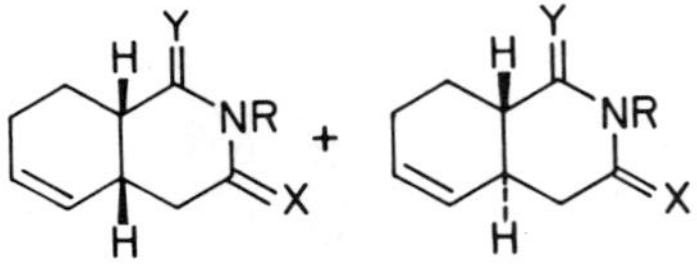

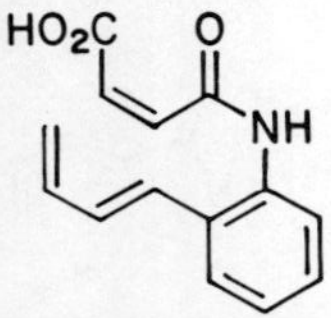

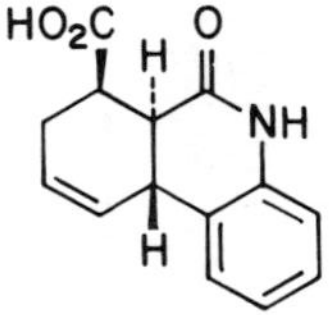

No.	Reactant	Conditions	Product(s) and Yield(s) (%)			Refs.
	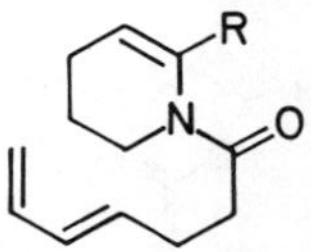	Gas phase, 600°	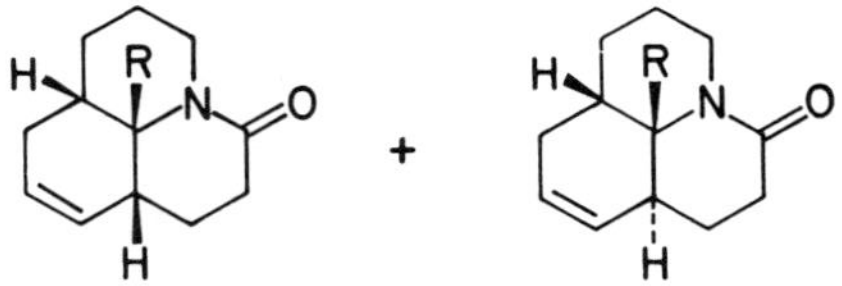 A + B			
	R			A	B	
16	H			(32)	(16)	109
17	CH_3			(22)	(22)	109
	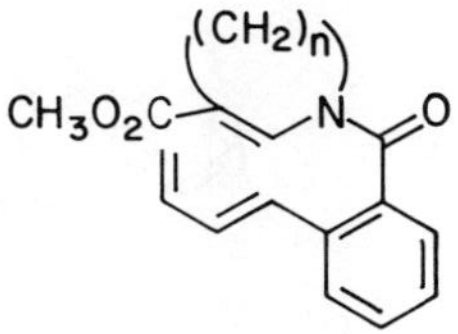		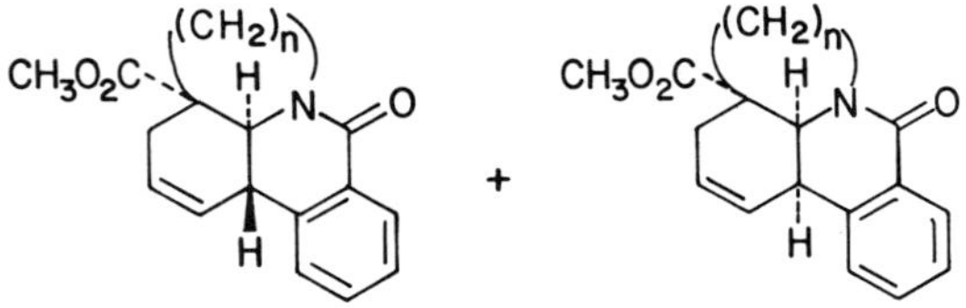 A + B			
	n			A	B	
18	2	o-$Cl_2C_6H_4$, reflux, 2.5 hr		(38)	(46)	111,112
19	3	", ", 3.5 hr		(49)	(25)	111,112
20	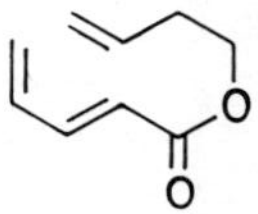	Heat	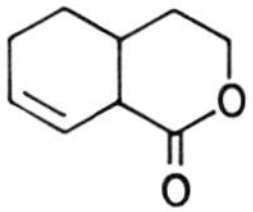 (0)			382

TABLE IIIb. ACYCLIC DIENES, 4-ATOM CHAIN CONTAINING NITROGEN OR OXYGEN (*Continued*)

Entry No.	Substrate 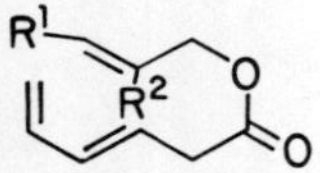		Conditions	Product(s) and Yield(s)(%) 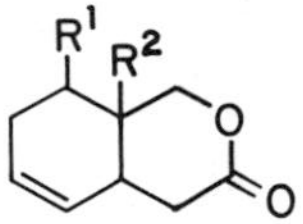		Refs.
	R^1	R^2				
20a	$C_2H_5O_2C$	H	200°	(0)[f]		159
20b	H 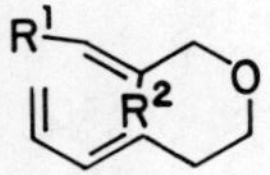	CH_3O_2C	220°	(0)[f] 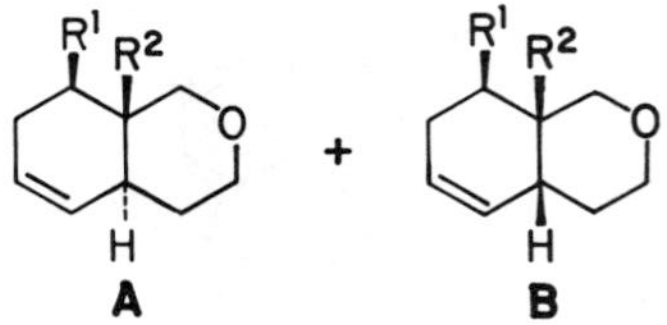		159
	R^1	R^2		A	B	
20c	$C_2H_5O_2C$	H	170°, 18–22 hr	(52)	(34)	159
20d	H	CH_3O_2C	", "	(15)	(35)	159
	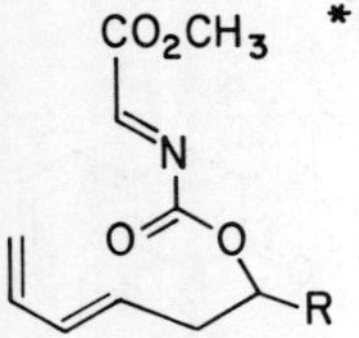 CO_2CH_3 *g		Toluene, 210°, 2 hr	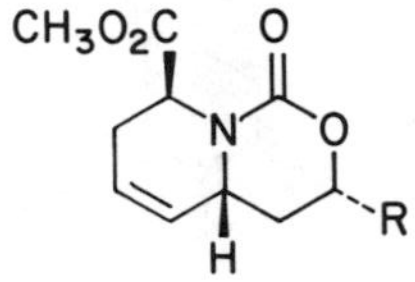 CH_3O_2C O		

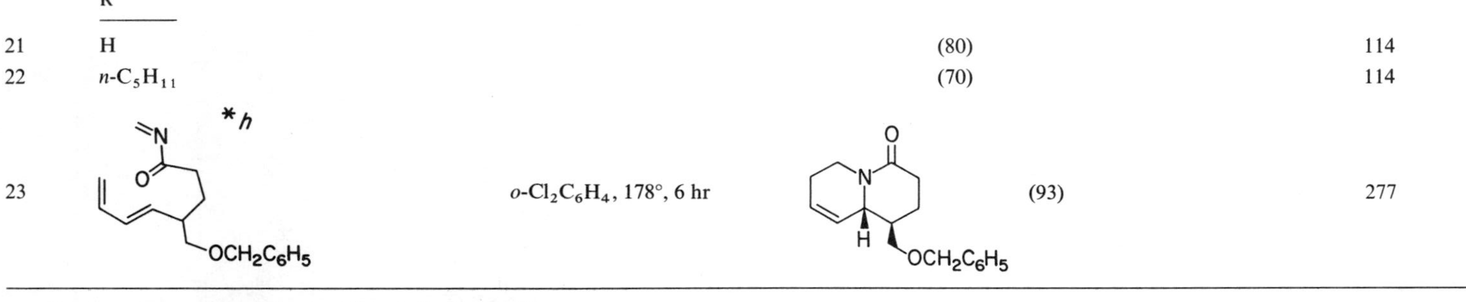

	R				
21	H			(80)	114
22	n-C_5H_{11}			(70)	114
23	*[h]	o-$Cl_2C_6H_4$, 178°, 6 hr		(93)	277

[a] The reaction was carried out with both the *R* and *S* isomers.

[b] The yield includes minor amounts of unspecified isomers.

[c] The substrate was generated by silylation of the α,β-unsaturated aldehyde with $(C_2H_5)_3N/(CH_3)_3SiCl$.

[d] The product comprised α and β epimers in a ratio of 3:1.

[e] The major adduct (79%) involved the carbon-carbon double bond of the cyclohexenone as the dienophile; see Table IIb/34a.

[f] At higher temperatures the substrate isomerized to the conjugated diene ester, which then underwent an intramolecular Diels–Alder reaction to give the products shown in Table IIe/4a and 4b, respectively.

[g] The substrate was generated from the $—CONHCH(O_2CCH_3)CO_2CH_3$ derivative.

[h] The substrate was generated from the $—CONHCH_2O_2CCH_3$ derivative.

*This intermediate is hypothetical, or has not been isolated.

TABLE IV. ACYCLIC DIENES, CHAINS 5 ATOMS AND LONGER

Entry No.	Substrate	Conditions	Product(s) and Yield(s)(%)	Re
	R			
1	H	Toluene, $AlCl_3$, reflux, 2 hr	(—)	25
1a	$OP(O)(OC_2H_5)_2$	Benzene, reflux, 18 hr	(90)	25
2	n = 3,4,5	Heat	(0)	38
2a	R = H, $C_2H_5O_2C$	Toluene, 250°, 24 hr	(0)	39
2b	R = H, $C_2H_5O_2C$	Benzene, 250°, 24 hr	(0)	3
			cis *trans*	
2c		Benzene, $(CH_3)_2AlCl$, 25°	(86) (0)	4
		Toluene, 160°	(9) (70)	4

TABLE IV. ACYCLIC DIENES, CHAINS 5 ATOMS AND LONGER (*Continued*)

Entry No.	Substrate	Conditions	Product(s) and Yield(s)(%)	Refs.
	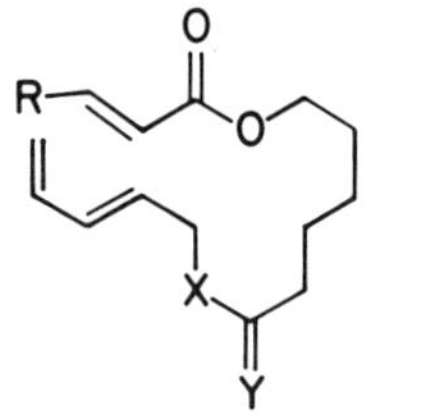	High dilution (slow addition to C_6H_5CN, 190°, 30 hr, argon)	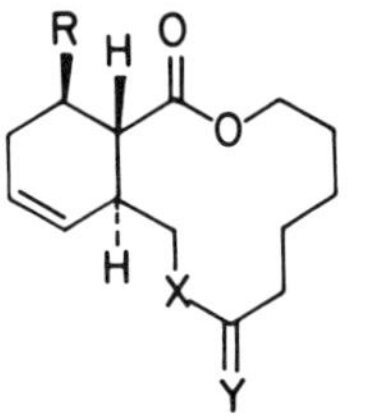 + 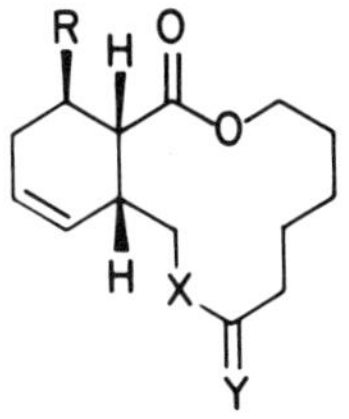 A B	
	X Y R		A B	
	O O H		(37) (34)[a]	72
	CH_2 H_2 CH_3O_2C		(53)[b,c]	72
	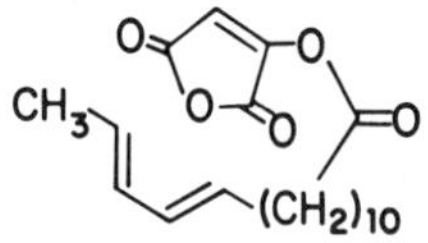	0.1% in toluene, reflux, 96 hr	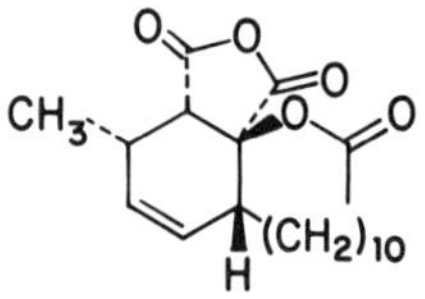 (27)[d]	314

[a] A bridged cycloadduct was formed in 6% yield (see Table XXVIc/6).

[b] Two fused isomers were formed in 37% and 16% yields, respectively. Stereochemical assignments were not made.

[c] Two bridged cycloadducts were also formed in 21% and 4% yields (see Table XXVIc/7).

[d] A bridged product was formed in 5% yield (see Table XXVIc/8).

TABLE V. ACYCLIC DIENES, CHAIN ATTACHED AT C-2

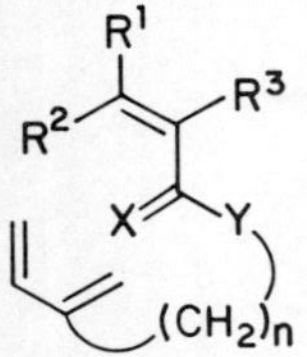

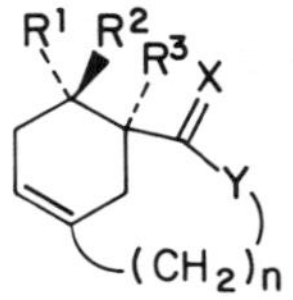

Entry No.	Substrate n	R^1	R^2	R^3	X	Y	Conditions	Product(s) and Yield(s)(%)	Refs.
1	1	H	H	H	H_2	CH_2	Gas phase, flow system, 420°, 23 sec	(32)[a]	395,349
2	1	H	H	H	O	CH_2	", ", 395°, 18 sec	(72)[a]	395
3	1	H	H	H	CH_2	CH_2	", ", 400°	(9)[a]	349
4	1	CH_3O_2C	H	H	H_2	CH_2	", ", 318°, 18 sec	(76)[a]	395
5	1	H	H	$C_2H_5O_2C$	H_2	CH_2	", ", 390°, 12 sec	(30)[a]	395
5a	2	H	H	H	O	O	Heat	(0)	83
6	2	H	H	H	H_2	CH_2	Gas phase, flow system, 455°, 8 sec	(55)[a]	395,349
7	2	H	H	H	O	CH_2	Gas phase, flow system, 398°, 8 sec	(85)[a]	395
							Xylene, 201°, 15 min	(57)	395
7a	2	CH_3O_2C	H	H	O	O	Heat	(0)	83
7b	2	H	CH_3O_2C	H	O	O	"	(0)	83
8	2	$C_2H_5O_2C$	H	H	H_2	CH_2	Gas phase, flow system, 420°, 8 sec;	(80)[a]	395
							Xylene, 206°, 2 hr	(91)	395
9	2	H	$C_2H_5O_2C$	H	H_2	CH_2	", 232°, 4 hr	(65)[a]	395
10	2	H	H	$C_2H_5O_2C$	H_2	CH_2	", 220°, 3 hr	(63)	395
11	3	H	H	H	O	O	Benzene, 190°, 9 hr	(59)	83,82

No.	n						Conditions	Product(s) and Yield(s) (%)	Refs.
12	3	CH_3O_2C	H	H	O	O	Xylene, 130°, 5–6 hr	(40–60)	82
13	3	H	CH_3O_2C	H	O	O	Benzene, 190°, 9 hr	(52)	83,82
14	3	H	H	H	H_2	CH_2	Gas phase, flow system, 510°	(28)[a]	149
15	4	H	H	H	O	O	Benzene, 190°, 9 hr	(49)[b]	83,82
16	4	CH_3O_2C	H	H	O	O	Xylene, 130°, 5–6 hr	(40–60)[b]	82
17	$C_2H_5O_2C$, CH_3, O, O						Benzene, 185°, 18 hr	$CO_2C_2H_5$, H, CH_3, O, O (78)	396

[a] The reaction was reversible; the remainder was unreacted starting material.
[b] The *para*-bridged isomer was also formed; see Table XXVIc/9a and 9b, respectively.

TABLE VI. VINYLCYCLOALKENES

Entry No.	Substrate	Conditions	Product(s) and Yield(s)(%)	Refs.
	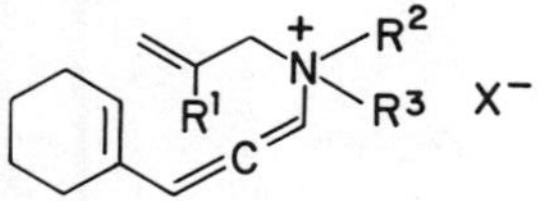 [structure] X^- * [a]		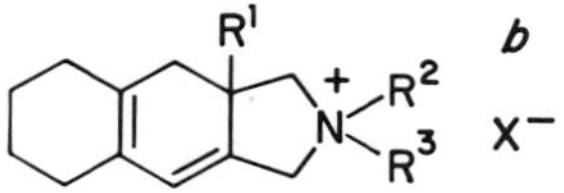 [structure] X^- [b]	
	R^1 / R^2 / R^3 / X^-			
1	H / CH_3 / CH_3 / Br	H_2O, 100°, 1–2 hr	(80)	192,336,372
2	H / $-(CH_2)_5-$ / Br	H_2O, 25°, 16 hr	(96)	372,336
3	CH_3 / CH_3 / CH_3 / Cl	H_2O, 25–40°	(90)	374,192
4	CH_3 / C_2H_5 / C_2H_5 / Cl	", "	(90)	374
5	CH_3 / $-(CH_2)_5-$ / Cl	", "	(95)	374
6	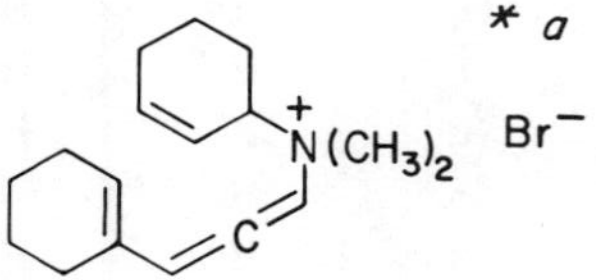 [structure] $\overset{+}{N}(CH_3)_2$ Br^- * [a]	H_2O, 40°, 60 hr	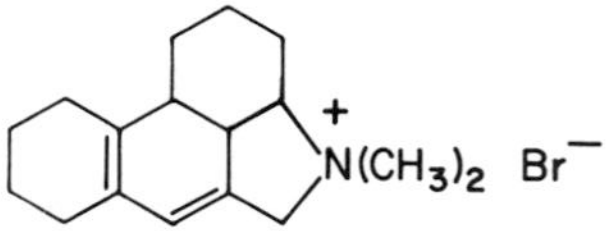 [structure] $\overset{+}{N}(CH_3)_2$ Br^- (—)[b]	380

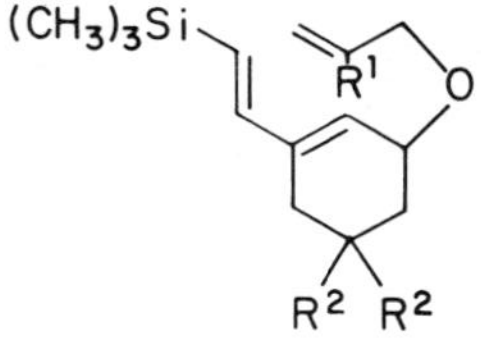

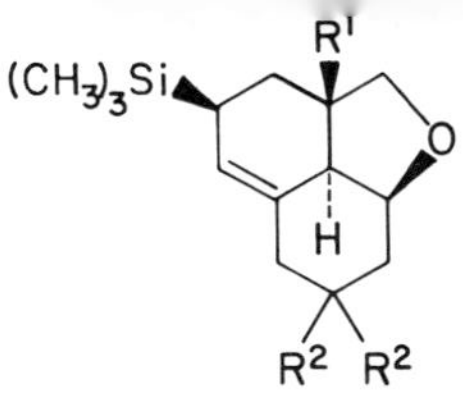

(CH3)3Si … R¹ … R² R²   (CH3)3Si … R¹ … H … R² R²

	R^1	R^2			
7	H	H	120°, 67 hr	(86)	166
8	H	CH_3	", 50 hr	(80)	166
9	CH_3	CH_3	310°, 20 min	(30)	166

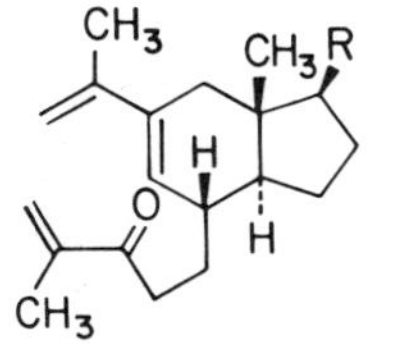

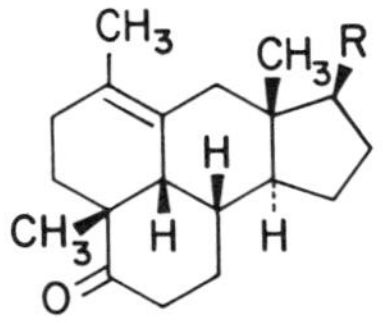

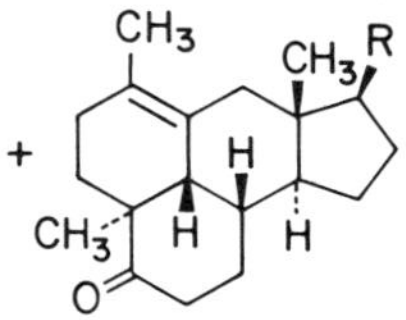

	R		A	B	
10	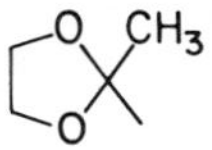 (2-methyl-1,3-dioxolan-2-yl)	C_2H_5OH, reflux	(95) (R = CH_3CO)	(0)	311
		Silica gel, CH_2Cl_2, reflux, 1 hr	(90)	(0)	311
10a	t-$C_4H_9(CH_3)_2SiO$	CH_2Cl_2, CF_3CO_2H, −78°, 4 hr	(66)	(4)	61

TABLE VI. VINYLCYCLOALKENES (*Continued*)

Entry No.	Substrate	Conditions	Product(s) and Yield(s)(%)	Refs.
11	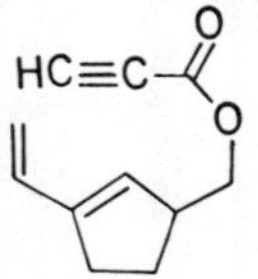	$(CH_3CO)_2O$, reflux, 20 hr	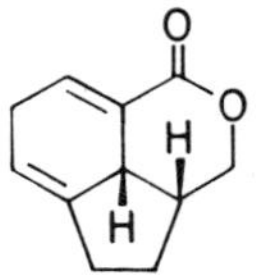 (70)	145
12	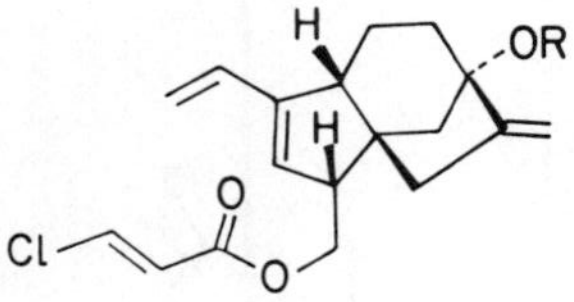 R = $CH_3O(CH_2)_2OCH_2$	Benzene, 160°, 45 hr[c]	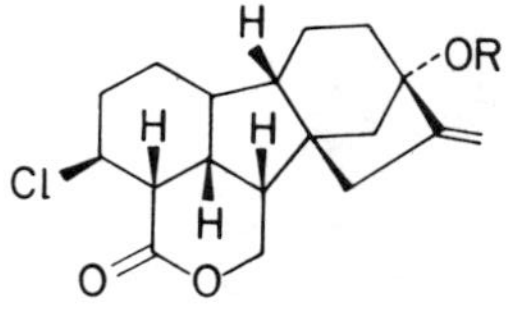 (55)	260
13	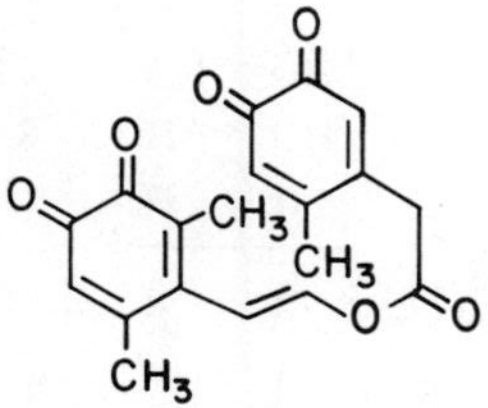	Heat to 150° or Lewis-acid catalysts at 25°	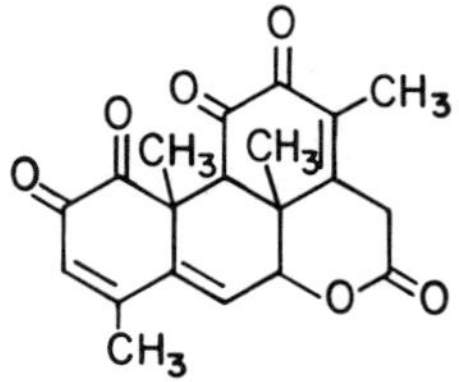 (0)	397

[a] The substrate was generated by base-catalyzed alkyne–allene rearrangement.
[b] The structures were assigned on the basis of elemental analyses and IR and UV spectra.
[c] Propylene oxide (100 eq) was added as an acid scavenger.
* This intermediate is hypothetical, or has not been isolated.

TABLE VII. CYCLOBUTADIENES

Entry No.	Substrate	Conditions	Product(s) and Yield(s) (%)	Refs.
	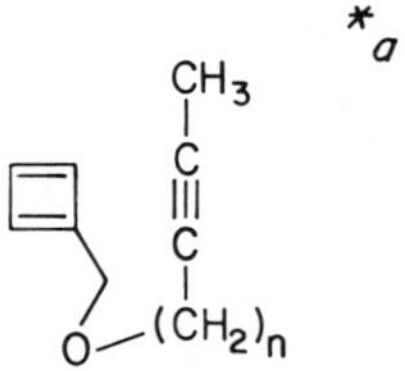 CH_3–C≡C–$(CH_2)_n$–O–CH_2–cyclobutadiene *[a]	Me_2CO	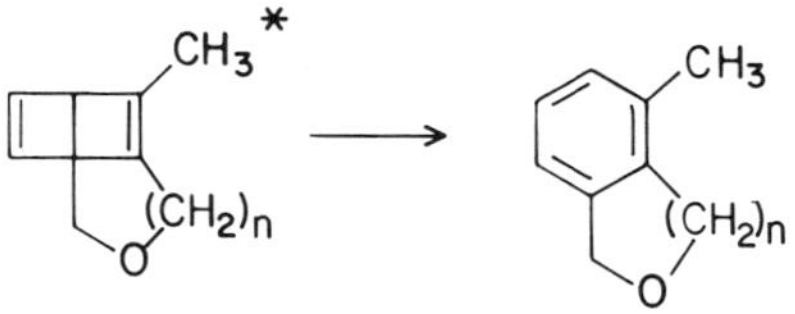 CH_3 *, $(CH_2)_n$, O (bicyclic cyclobutadiene) → CH_3, $(CH_2)_n$, O (benzo-fused)	
	n			
1	1		(50)	194
2	2		(83)	194

[a] The substrate was generated from the iron carbonyl complex with Ce(IV).

* This intermediate is hypothetical, or has not been isolated.

TABLE VIIIa. CYCLOPENTADIENES; CHAIN ATTACHED AT C-1

Entry No.	Substrate				Conditions	Product(s) and Yield(s)(%)	Refs.
1					$(n\text{-}C_4H_9)_3N$, 200°, 45 min	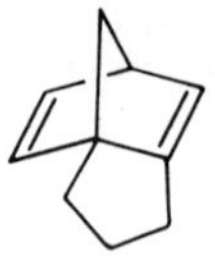 (80)	398
	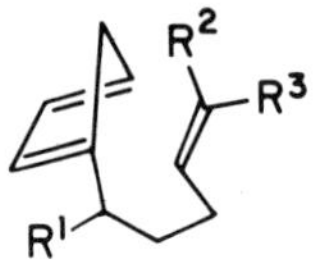					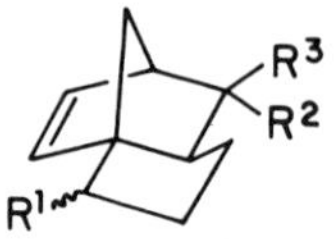	
	R^1	R^2	R^3				
2	H	H	H[a]		$(n\text{-}C_4H_9)_3N$, 200°, 1 hr	(67–96)[b]	195
					Benzene, 180°, 4 hr	(100)[c]	187
3	CH_3	CH_3	CH_3[a]		$[(CH_3)_2N]_3PO$ or $(n\text{-}C_4H_9)_3N$, 205°, 7 hr	(36)[d]	261
4	$CH_3OCH_2OCH_2$	CH_3O_2C	CH_3*[e]		Toluene, 180°, 24 hr	(80)[f]	197
5	$(CH_3)_3SiOCH_2$	CH_3O_2C	H*[e]		", ", 48 hr	(60)[f]	197
6	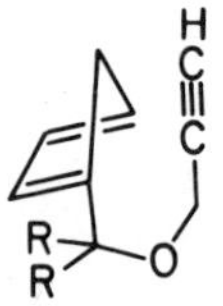 R = HC≡CCH$_2$O				$CDCl_3$, 25°, 7 days	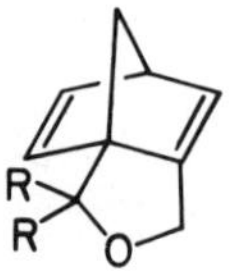 (100)	399
	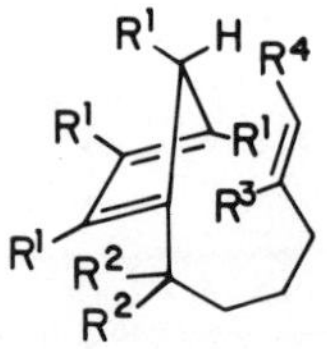					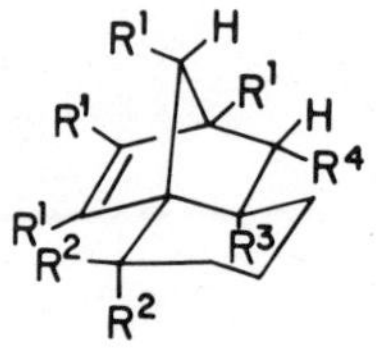	
	R^1	R^2	R^3	R^4			
7	H	H	H	H[a]	Benzene, 180°, 4 hr	(73)[c]	187,8
8	Cl	CH_3	CH_3	$HOCH_2$*[g]	Decalin, reflux, 5 hr	(0)	400
9	Cl	CH_3	CH_3	CH_3OCH_2[h]	", ", "	(70–80)	400
10	Cl	CH_3	CH_3	$CH_3CO_2CH_2$*[g]	", ", "	(0)	400
11	Cl	CH_3	CH_3	$C_6H_5CH_2OCH_2$[h]	", ", "	(41)	400
12	H	CH_3	CH_3	$HOCH_2$[a]	Pseudocumene, 176°, 48 hr	(90)	401

TABLE VIIIa. CYCLOPENTADIENES; CHAIN ATTACHED AT C-1 (*Continued*)

Entry No.	Substrate	Conditions	Product(s) and Yield(s)(%)	Refs.
13	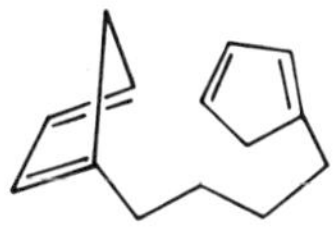 [a]	Benzene, reflux, 3 hr	 (—)	402

[a] The substrate was an equilibrating mixture of 1- and 2-substituted cyclopentadienes.

[b] An isomer of unknown structure was formed in 3–5% yield.

[c] The stereochemistry was not determined.

[d] The product was a 1:1 mixture of R^1 epimers.

[e] The substrate was generated by retro Diels–Alder reaction from the corresponding adduct with cyclopentadiene.

[f] The product was a mixture of R^1 epimers.

[g] The substrate was generated by pyrolysis of the corresponding pentachlorocyclopentadiene.

[h] The substrate was a mixture of isomeric tetrachlorocyclopentadienes.

* This intermediate is hypothetical, or has not been isolated.

TABLE VIIIb. CYCLOPENTADIENES; CHAIN ATTACHED AT C-5

Entry No.	Substrate	Conditions	Product(s) and Yield(s)(%)	Refs.
1	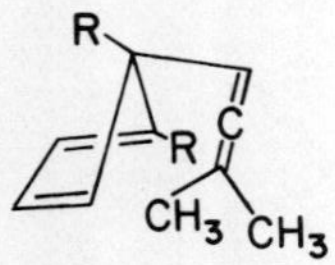 R = CH_3O_2C	CCl_4, reflux, 20 hr	* a 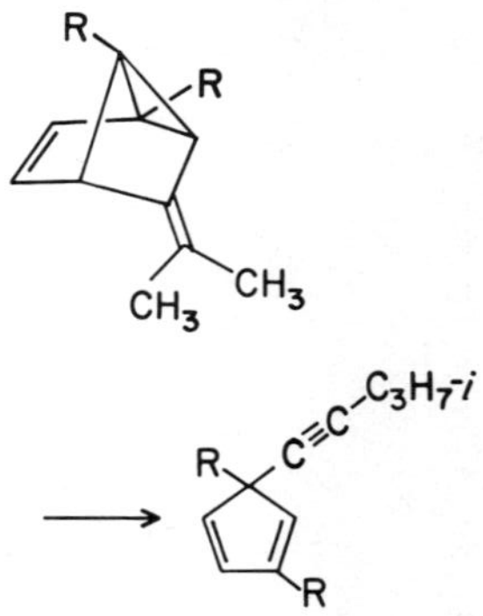 → (38)	403
1a	"	Et_2O, $h\nu$ (Pyrex), 10 hr	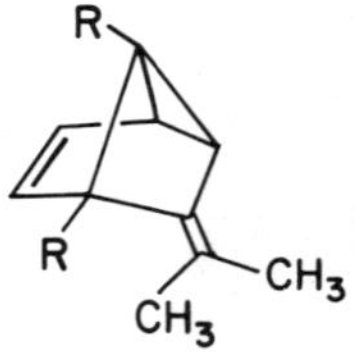 (80)	403

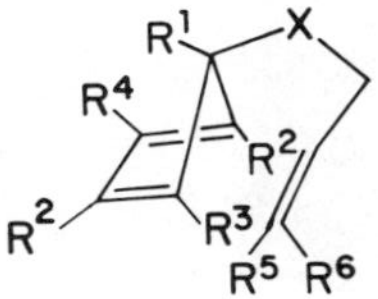

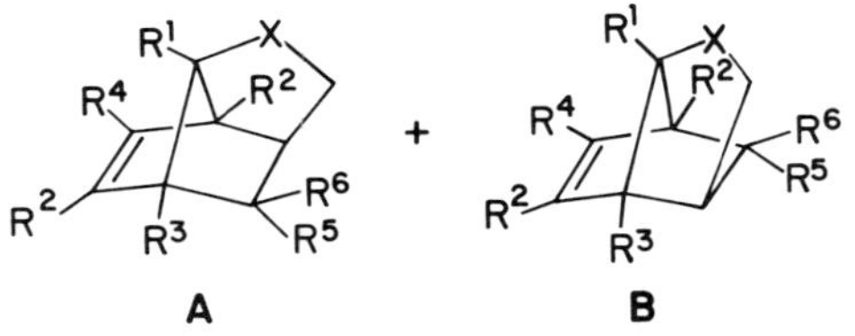

	R^1	R^2	R^3	R^4	R^5	R^6	X	Conditions	A	B	Ref.
1b	Cl	Cl	Cl	Cl	H	H	O*[c]	$CH_2{=}CHCH_2OH$, 25°, 3 hr	(45)		404,405
2	H	H	H	H	H	H	CH_2[b]	Benzene, 180°, 4 hr	(100)		187
3	H	H	CH_3	H	H	H	CH_2[b]	", 250°, 24 hr	(22)	(8)	79
4	H	H	H	H	CH_3	H	CH_2[b]	$(n\text{-}C_4H_9)_3N$, heat	(—)		406
4a	Cl	Cl	Cl	Cl	CH_3	CH_3	O*[c]	$(CH_3)_2C{=}CHCH_2OH$, 25°, 9 hr	(12)		404
4b	$CH_2{=}CHCH_2O$	Cl	Cl	Cl	H	H	O*[c]	$CH_2{=}CHCH_2OH$, 25°, 3 hr	(15)		404
5	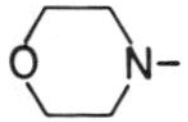 morpholino (O⟨ ⟩N–)	Cl	Cl	Cl	H	H	O*[c]	Benzene, 25°, 28 days	(42)		407
6	H	H	H	$(CH_3)_3SiO$	H	H	CH_2[b]	", 250°, 24 hr	(39)[d]	(33)[d]	79
7	H	H	CH_3	$(CH_3)_3SiO$	H	H	CH_2[b]	Benzene, 250°, 24 hr	(60)[d]	(17)[d]	79
8	H	H	CH_3	"	CH_3O_2C	H	CH_2*[b,e]	Toluene, 110°, 3 days	(94)[d]	(0)	79
9	H	H	CH_3	"	H	CH_3O_2C	CH_2*[b,e]	", ", "	(—)	(0)	79
9a	$(CH_3)_2C{=}CHCH_2O$	Cl	Cl	Cl	CH_3	CH_3	O	Benzene, 80°, 2 hr	(—)		404

*

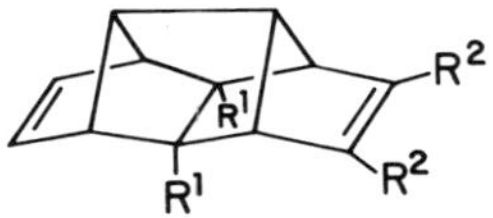

	R^1	R^2	Conditions	Yield	Ref.
10	H	H[f]	Diglyme, 150°, 3–4 hr	(8–12)	408
11	H	CH_3O_2C[g]	$CDCl_3$, 0–5°	(20)	409,326
12	CH_3O_2C	H[g]	$CDCl_3$, 0–5°	(20)	409,326 409a

TABLE VIIIb. CYCLOPENTADIENES; CHAIN ATTACHED AT C-5 (*Continued*)

Entry No.	Substrate	Conditions	Product(s) and Yield(s)(%)	Refs.
	R3, R3, R3, R3, R3, R3, R3, R3, N–R1, N–R2		R3, R3, R3, R3, R3, R3, R3, R3, N–R1, N–R2	
	R^1 / R^2 / R^3			
13	$C_2H_5O_2C$ / $C_2H_5O_2C$ / H*[h]	THF, −78 to 25°	(5.5)	326,410
14	—CON(CH_3)CO— / Cl	$CH_3CO_2C_2H_5$, reflux, 3 days	(63)	326
15	—CON(C_6H_5)CO— / H*[i]	THF, −78 to 25°	(1.1–3.0)	326,410
16	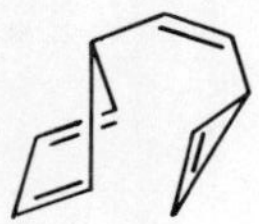 *[j]	(—)	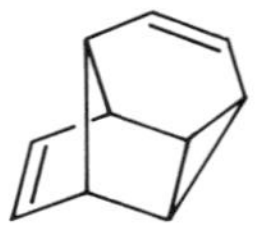 (—)	411
17	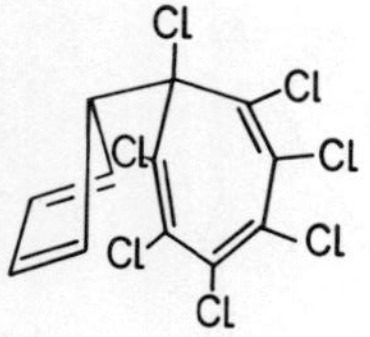 *[k] Cl, Cl, Cl, Cl, Cl, Cl, Cl	Et_2O, 25°, 6 hr	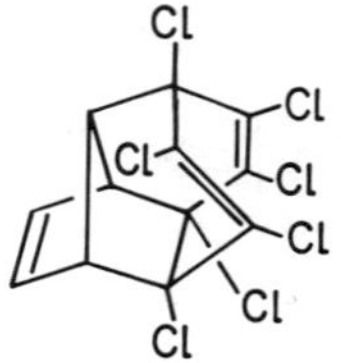 Cl, Cl, Cl, Cl, Cl, Cl, Cl (60)	412

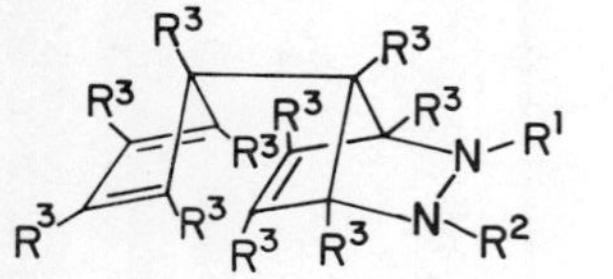

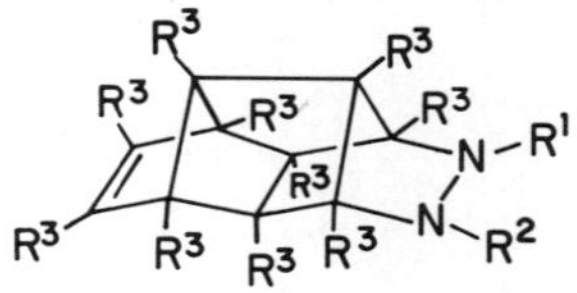

Entry	Substrate	Conditions	Product (Yield, %)	Refs.
	*[l] (R^1, R^2, n)		* → (R^1, R^2, n)	
18	R^1 = H, R^2 = CH_3, n = 1	Toluene, 270–280°, 2 hr	(55)	196
19	R^1 = CH_3, R^2 = H, n = 2	", ", "	(71)	196
20	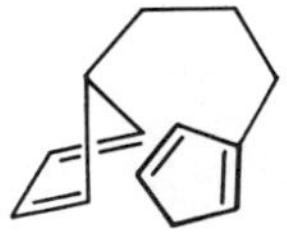 [m]	Benzene, 190°, 9 hr	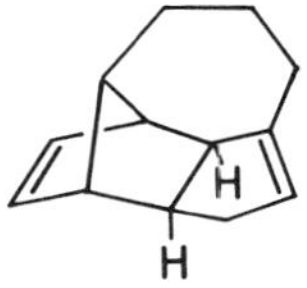 (100)	402

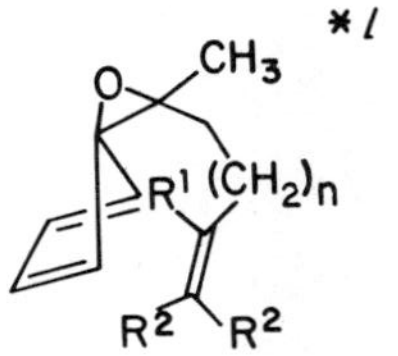

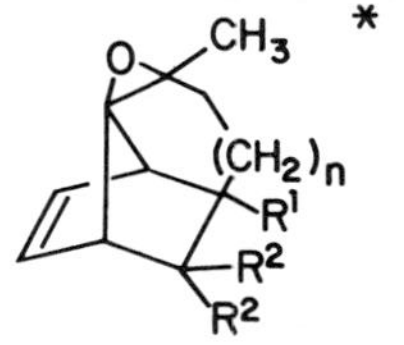

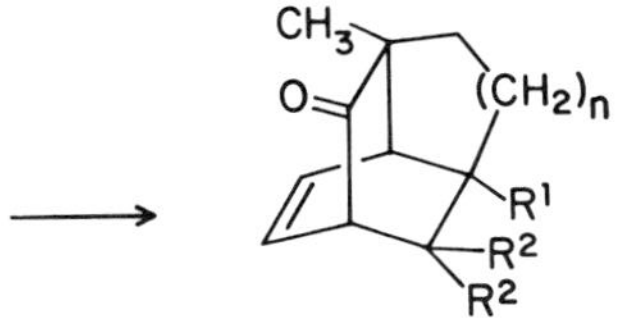

[a] This intermediate was converted into the final product by a 1,3 shift followed by reverse intramolecular Diels–Alder reaction.

[b] The substrate was an equilibrating mixture of 1-, 2-, and 5-substituted cyclopentadienes.

[c] The substrate was generated from hexachlorocyclopentadiene and allyl alcohol (or 3,3-dimethylallyl alcohol) and KOH, or (entry 5) from 5,5-bis(1-morpholino)-1,2,3,4-tetrachlorocyclopentadiene and allyl alcohol.

[d] The yields are of ketones obtained after hydrolysis of the silyl enol ethers.

[e] The substrate was generated from the corresponding cyclopentenone with $(CH_3)_3SiCl$ and $(C_2H_5)_3N$.

[f] The substrate was generated by coupling thallium cyclopentadienide with 7-chloronorbornadiene.

[g] The substrate was generated from dimethyl acetylenedicarboxylate and 9,10-dihydrofulvalene.

[h] The substrate was generated from diethyl azodicarboxylate and 9,10-dihydrofulvalene.

[i] The substrate was generated from *N*-phenyltriazolinedione and 9,10-dihydrofulvalene.

[j] The substrate was generated by irradiation of bullvalene through Vycor.

[k] The substrate was generated from nickelocene and heptachlorocycloheptatriene.

[l] The substrate was generated from the corresponding cyclopentadiene dimer.

[m] The substrate was an equilibrating mixture of isomers.

* This intermediate is hypothetical, or has not been isolated.

TABLE IXa. CYCLOHEXADIENES; CHAIN ATTACHED AT C-1

Entry No.	Substrate					Conditions	Product(s) and Yield(s)(%)		Refs.
	R^1	R^2	R^3	R^4	R^5		A	B	
1	H	H	H	H	H	Gas phase, 208°	(78)[a,b]	(22)[a,b]	77
2	H	H	H	O=		", 227°, 21 hr	(79)[a,b]	(21)[a,b]	77
3	CH_3	CH_3	H	O=*[c]		—	(0)	(70)	115
4	H	H	CH_3	—$OCH_2CH(CH_3)O$—[d]		Mesitylene, 250°, 24 hr	(14)[e]	(41)[e]	116
5	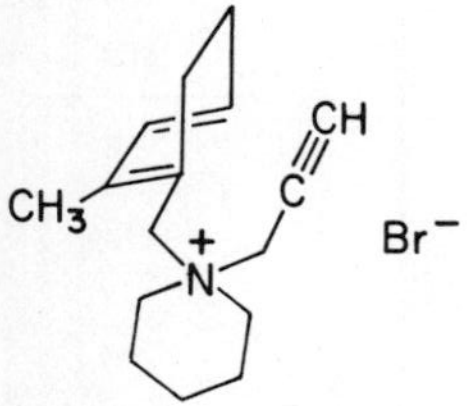 Br^-					DMF, 90°, 60–75 hr	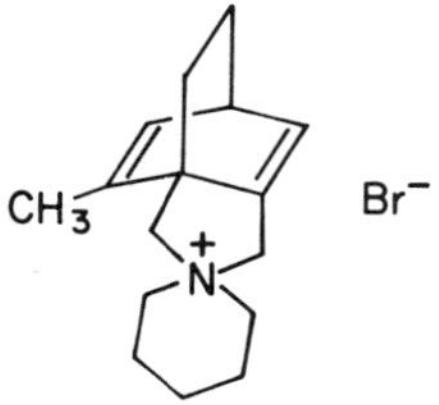 Br^- (100)[f]		413
6	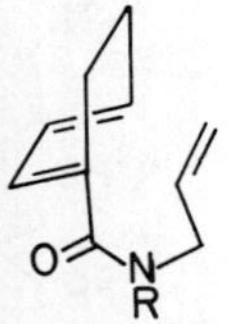 R = $CH_2CH=CH_2$					DMF, 156°, 1.5 hr	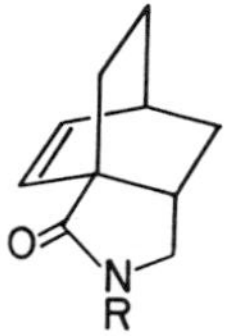 (24)	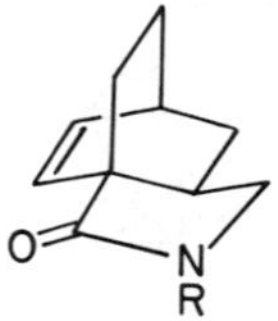 (56)	189

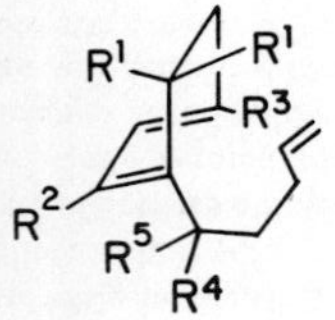

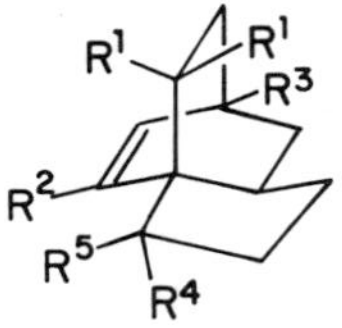

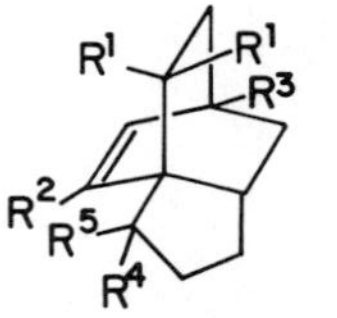

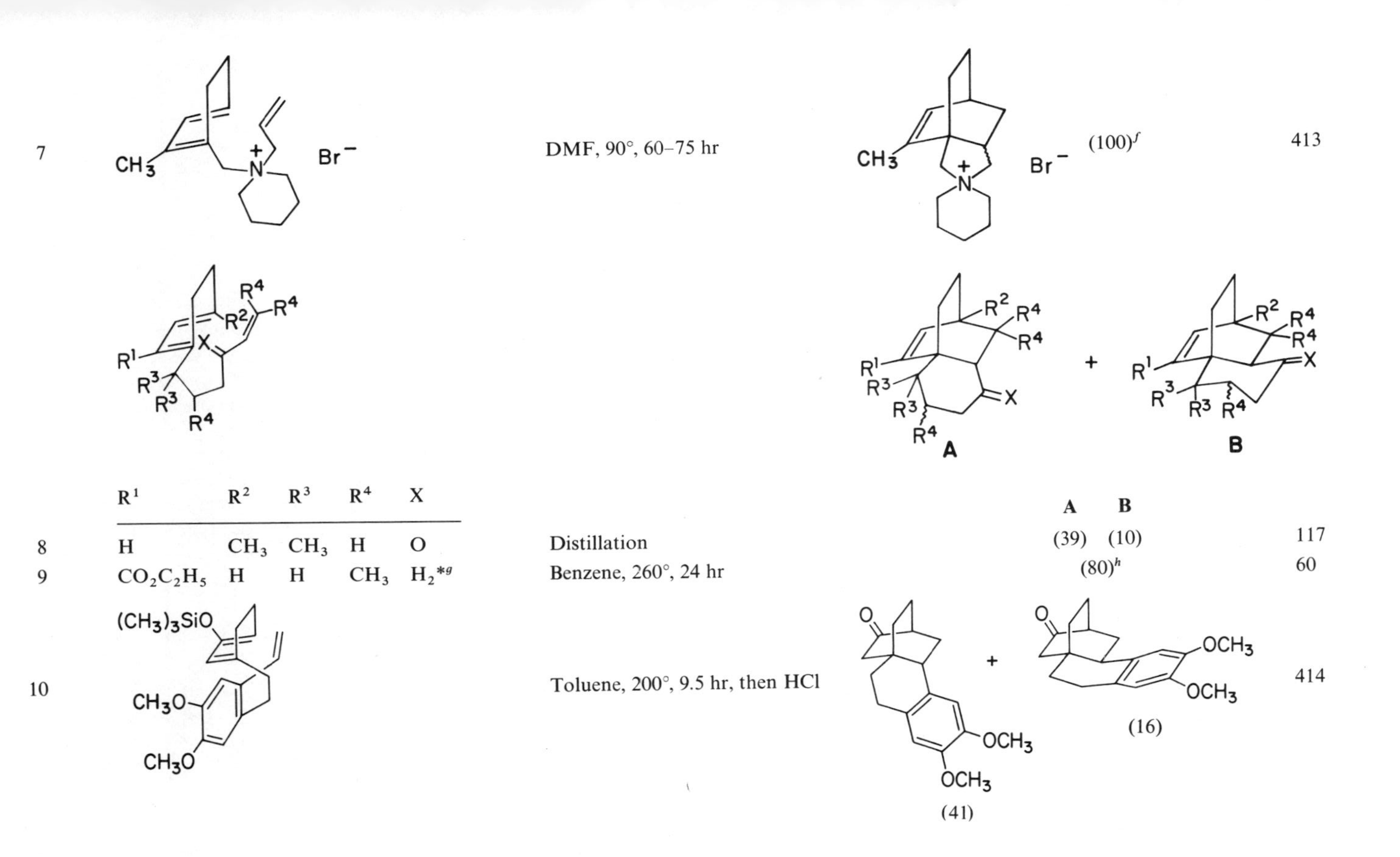

No.	Conditions	Product(s) and Yield(s) (%)	Refs.
7	DMF, 90°, 60–75 hr	Br^- (100)[f]	413

No.	R^1	R^2	R^3	R^4	X	Conditions	A	B	Refs.
8	H	CH_3	CH_3	H	O	Distillation	(39)	(10)	117
9	$CO_2C_2H_5$	H	H	CH_3	H_2*[g]	Benzene, 260°, 24 hr	(80)[h]		60

No.	Conditions	Product(s) and Yield(s) (%)	Refs.
10	Toluene, 200°, 9.5 hr, then HCl	(41) + (16)	414

TABLE IXa. CYCLOHEXADIENES; CHAIN ATTACHED AT C-1 (*Continued*)

Entry No.	Substrate	Conditions	Product(s) and Yield(s) (%)	Refs.
11	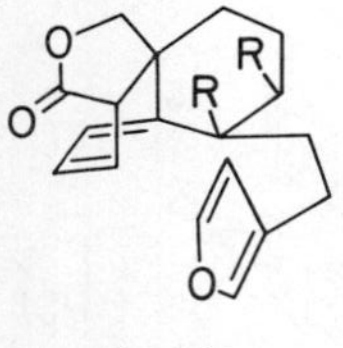 R = CH_3	Diglyme, reflux, 32 hr	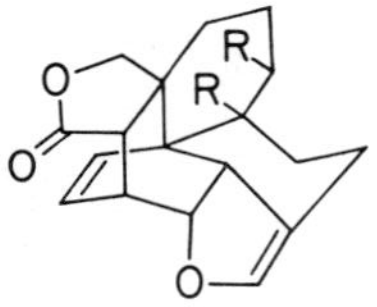 (90)	169

[a] The numbers are product ratios.
[b] The assignment of stereochemistry is tentative.
[c] The substrate was generated from 5-(4,6,6-trimethyl-1,3-cyclohexadienyl) allyl ketone by 1,5 hydrogen shift.
[d] Pyrolysis of the corresponding ketone resulted mainly in aromatization of the cyclohexadiene ring.
[e] The yields are of ketones obtained on hydrolysis.
[f] The structure was assigned on the basis of elemental analysis and IR and/or UV spectra.
[g] The substrate was generated from ethyl 6-(2,6-dimethyl-5-heptenyl)-1,3-cyclohexadiene-1-carboxylate by 1,5-hydrogen shift.
[h] The product was a mixture of *exo* and *endo* isomers.
* This intermediate is hypothetical, or has not been isolated.

TABLE IXb. CYCLOHEXADIENES; 0-ATOM CHAIN ATTACHED AT C-5

Entry No.	Substrate 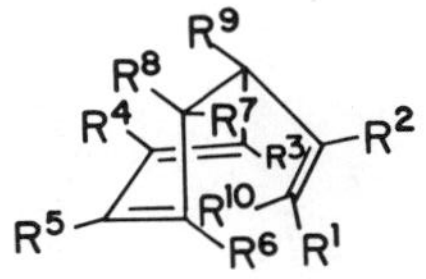	Conditions	Product(s) and Yield(s)(%) 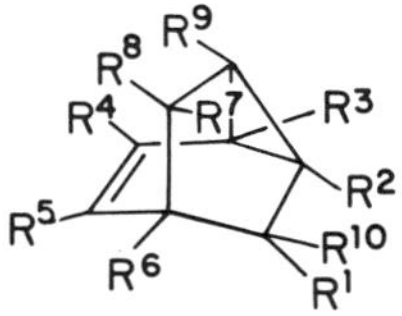	Refs.
	All substituents R = hydrogen except as noted below.			
1		Gas phase, 225°, 24 hr	(—)[r]	148,415,416
2	$R^5 = HO_2C$*[a]	Benzene, 150°, 20 hr	(71)	198
3	$R^3 = CH_3O_2C$*[b,c]	", ", 18 hr	(33)[b], (27)[c]	417
4	$R^4 = CH_3O_2C$*[d]	", ", 24 hr	(24)	417
5	$R^5 = CH_3O_2C$*[d]	", ", "	(14)	417
	" *[e]	", 170°, 40 hr	(50)	418
6	$R^6 = CH_3O_2C$*[c]	Sulfolene, 150°, 13 days	(8)	417
7	$R^3 = CH_3O_2C$; $R^5 = HO_2C$*[a]	CH_3OH, 100°, 6 hr	(70)	198
8	$R^2, R^9 = CH_3$; $R^5 = HO_2C$*[f]	", 100–110°, 30 hr	(15)	198
9	$R^2 = CH_3$; $R^5 = CH_3O_2C$*[g]	Benzene, 100°, 17 hr	(28)	418
10	$R^1 = CH_3$; $R^3 = CH_3O_2C$; $R^5 = HO_2C$*[m]	CH_3OH, 100°, 6 hr	(21)	198
11	$R^2 = CH_3$; $R^3 = CH_3O_2C$; $R^5 = HO_2C$*[h]	", ", "	(23)	198
12	$R^2, R^9 = CH_3$; $R^5 = CH_3O_2C$*[i]	Benzene, 100°, 40 hr	(8)	418
13	$R^2, R^9 = CH_3$; $R^3 = CH_3O_2C$*[j]	", 150°, 6 days	(7)	417
14	$R^2, R^9 = CH_3$; $R^6 = CH_3O_2C$*[j]	", ", "	(12)	417
15	$R^1, R^2 = CH_3O_2C$*[k]	Reflux[l], 48 hr	(major product)	419,420
16	$R^3, R^4, R^5, R^6 = CH_3$	180°	(100)	421
17	$R^3, R^6 = CH_3O_2C$; $R^4, R^5 = CH_3$	"	(100)	421

TABLE IXb. CYCLOHEXADIENES; 0-ATOM CHAIN ATTACHED AT C-5 (*Continued*)

Entry No.	Substrate	Conditions	Product(s) and Yield(s)(%)	Refs.
18	$R^3, R^5 = CH_3O_2C$; $R^4, R^6 = CH_3$	180°	(100)	421
19	$R^3, R^4, R^5, R^6 = CH_3OCH_2$	"	(100)	421
20	$R^1, R^2, R^5, R^6 = CH_3O_2C$; $R^7 = Cl$*[o]	HCO_2H, reflux, 60 hr	(12)[n]	422
21	$R^3, R^5 = C_2H_5O_2C$; $R^4, R^6 = n\text{-}C_4H_9$	180°	(100)	421
22	$R^1, R^8 = C_6H_5$*[p]	175°, 7 hr	(20)	423
23	$R^8, R^{10} = C_6H_5$*[q]	", "	(26)	423
24	$R^3, R^6 = C_2H_5O_2C$; $R^4, R^5 = n\text{-}C_4H_9$	180°	(100)	421
25	$R^3, R^5 = CH_3O_2C$; $R^4, R^6 = C_6H_5$	"	(100)	421
26	$R^3, R^6 = CH_3O_2C$; $R^4, R^5 = C_6H_5$	"	(100)	421
27	$R^3, R^6 = C_6H_5$; $R^4, R^5 = CH_3O_2C$	"	(100)	421
28	$R^3, R^5 = C_6H_5$; $R^4, R^6 = CH_3O_2C$	"	(100)	421

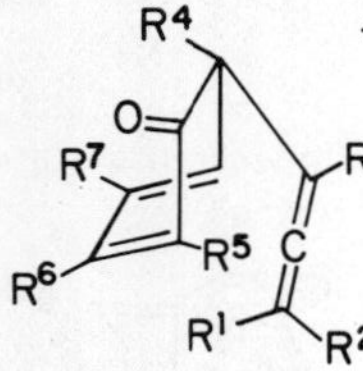

*[s]

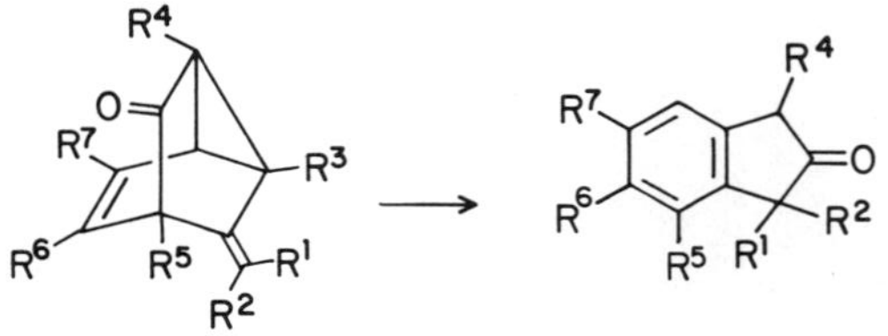

A → B[t]

Entry No.	R^1	R^2	R^3	R^4	R^5	R^6	R^7	Conditions	A	B	Refs.
29	H	H	H	H	H	H	H	Gas phase, 10^{-}–10^{-5} torr, 460°	(0)	(26)	199
30	H	H	H	CH_3	H	H	H	", ", "	(0)	(24)	424

31	H	H	H	H	CH_3	H	H	", ", "	(0)	(28)	424
32	H	H	H	H	H	CH_3	H	", ", "	(0)	(18)	424
33	CH_3	CH_3	H	H	H	H	H	", ", 400°	(0)	(61)	199
34	H	H	H	CH_3	CH_3	H	H	", ", 410°	(0)	(65)	199
									(—)	(0)	425
35	H	CH_3	H	CH_3	CH_3	H	H	—	(—)[u]	(0)	425
36	H	H	CH_3	CH_3	CH_3	H	H	—	(—)	(0)	425
37	H	H	H	CH_3	CH_3	H	CH_3	—	(—)	(0)	425

[a] The substrate was generated from α-pyrone-5-carboxylic acid and butadiene.
[b] The substrate was generated from methyl α-pyrone-3-carboxylate and butadiene.
[c] The substrate was generated from methyl α-pyrone-6-carboxylate and butadiene.
[d] The substrate was generated from methyl α-pyrone-4-carboxylate and butadiene.
[e] The substrate was generated from methyl α-pyrone-5-carboxylate and butadiene.
[f] The substrate was generated from α-pyrone-5-carboxylic acid and 2,3-dimethylbutadiene.
[g] The substrate was generated from methyl α-pyrone-5-carboxylate and 2-methylbutadiene.
[h] The substrate was generated from α-pyrone-5-carboxylic acid and 2-methylbutadiene.
[i] The substrate was generated from methyl α-pyrone-5-carboxylate and 2,3-dimethylbutadiene.
[j] The substrate was generated from methyl α-pyrone-6-carboxylate and 2,3-dimethylbutadiene.
[k] The substrate is the hypothetical product of the ene reaction between 1,4-cyclohexadiene and dimethyl acetylenedicarboxylate.
[l] The reflux temperature was that of a 1:1 mixture of 1,4-cyclohexadiene and dimethyl acetylenedicarboxylate.
[m] The substrate was generated from α-pyrone-5-carboxylic acid and 1,3-pentadiene.
[n] The stereochemistry of R^1 was not determined.
[o] The substrate was generated from $Cl_2CHCH{=}C(CO_2CH_3)CH_2CO_2CH_3$ and sodium formate.
[p] The substrate was generated by cyclization of (*E,Z,Z,E*)-1,8-diphenyl-1,3,5,7-octatetraene.
[q] The substrate was generated by isomerization of (*E,Z,Z,E*)- to (*E,Z,Z,Z*)-1,8-diphenyl-1,3,5,7-octatetraene followed by cyclization.
[r] The reaction is reversible.
[s] The substrate was generated by Claisen rearrangement of the appropriate phenyl propargyl ether.
[t] The product was formed by reverse intramolecular Diels–Alder reaction followed by cyclization of the ketene intermediate.
[u] The stereochemistry of R^2 was not reported.
* This intermediate is hypothetical, or has not been isolated.

TABLE IXc. Cyclohexadienes; 1-Atom Chain Attached at C-5

Entry No.	Substrate	Conditions	Product(s) and Yield(s) (%)	Refs.
1	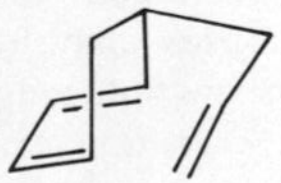	Gas phase, 225°, 8 hr	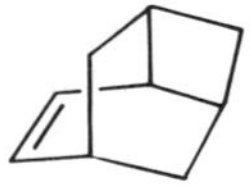 (44)[a]	148,21
2	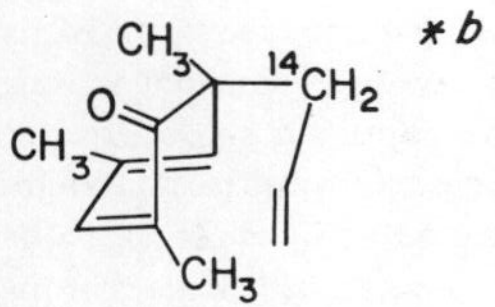 *b	$C_6H_5N(C_2H_5)_2$, 170°	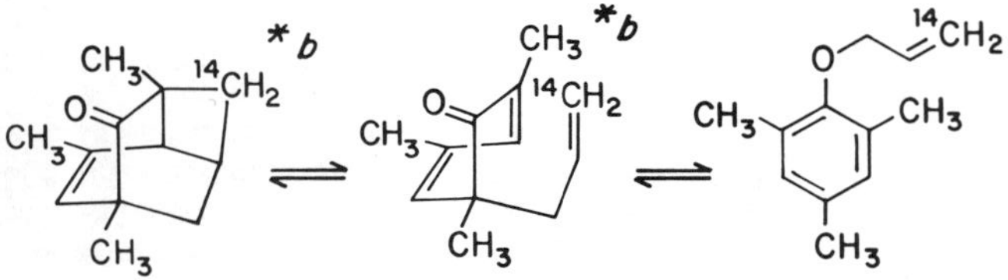 *b ⇌ *b ⇌ (—)	78,426
	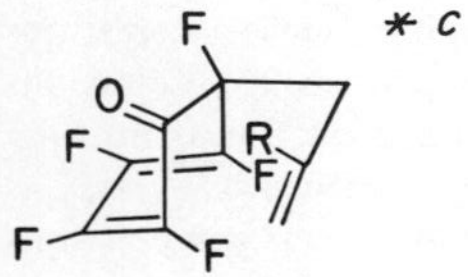 *c		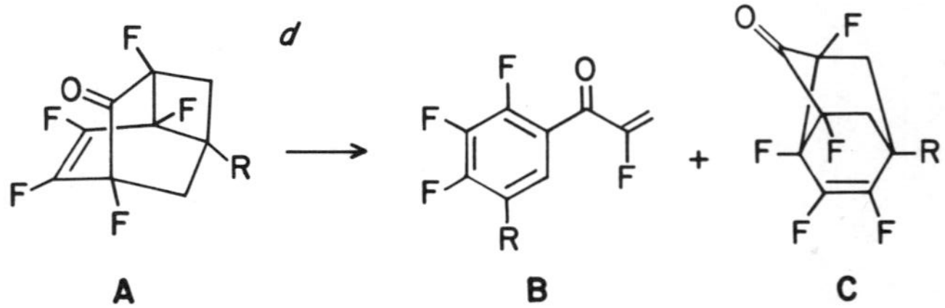 A → B + C [d]	
	R		A B C	
3	H	Gas phase, 137–141°, 13 days	(19) (0) (—)	427,428
		", 440–480°	(0) (—) (0)	429,430
4	CH_3	", 160°, 15 days	(0) (0) (—)	428
		", (0.05–0.1 torr), 410°	(0) (30) (0)	431

[a] The number is the conversion; the reaction is reversible; $K = 1.38$ at 203°.

[b] The structures are hypothetical intermediates in the degenerate rearrangement of 2,4,6-trimethylphenyl allyl ether.

[c] The substrate was generated by Claisen rearrangement of pentafluorophenyl 2-R-allyl ether.

[d] The bridged product was also formed; see Table XXVIc/12, 13.

* This intermediate is hypothetical, or has not been isolated.

TABLE IXd. CYCLOHEXADIENES; 2-ATOM CHAIN ATTACHED AT C-5

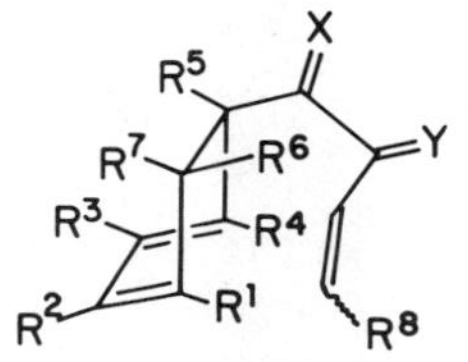

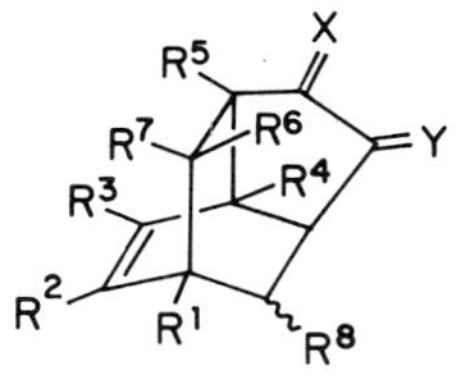

Entry No.	R^1	R^2	R^3	R^4	R^5	R^6	R^7	R^8	X	Y	Conditions	Product(s) and Yield(s)(%)	Refs.
1	H	H	H	H	H	H	H	H	H_2	H_2	Gas phase, 168°, 4 hr	(92)	148
2	H	Cl	Cl	H	H	H	H	H	H_2	H_2*[a]	Benzene, reflux, 16 hr	(17)	432
3	Cl	Cl	Cl	Cl	H	H	H	H	H_2	H_2*[a]	80°, 3 hr	(50)	432
4	Br	Br	Br	Br	H	H	H	H	H_2	H_2*[a]	$Cl(CH_2)_2Cl$, reflux, 4 hr	(68)	432
5	Cl	Cl	Cl	Cl	H	H	H	CH_3	H_2	H_2*[a]	80°, 3 hr	(48)	432
6	Cl	Cl	Cl	Cl	H	CH_3	H	CH_3	H_2	H_2*[a]	100°, 18 hr	(57)[b]	432
7	CH_3	H	H	H	CH_3	O=		H	H_2	H_2*[c]	Toluene, reflux, 11 days	(8)	78
8	H	H	H	CH_3	H	CH_3	CH_3	H	O	H_2	Xylene, reflux, 1 hr	(70)	433
9	CH_3	CH_3O	H	H	CH_3	H	H	H	H_2	O	Benzene, 60°, 1.5 hr	(87)	267
10	CH_3	H	CH_3	H	CH_3	O=		H	H_2	CH_2*[d]	", reflux, 24 hr	(12)[e]	78
11	CH_3	H	CH_3O	H	CH_3	O=		H	H_2	CH_2*[d]	", ", 36 hr	(9)[f]	78
12	H	CH_3	H	CH_3	LiO	CH_3	CH_3	H	H_2	H_2	", 200°, 6 hr	(—)	162
13	CH_3	RO[g]	H	H	CH_3	H	H	H	H_2	H, OH[h]	Xylene, reflux, 4.5 hr	(89)[h]	190
14	CH_3	RO[i]	H	H	CH_3	H	H	H	H_2	H, OR[j]	Benzene, 160°, 30 min	(100)[k]	268
14a	CH_3	H	H	H	CH_3	O=		H	H_2	$(CH_3)_2C$	Benzene, 80°, 6 hr	(42)[l]	266

TABLE IXd. CYCLOHEXADIENES; 2-ATOM CHAIN ATTACHED AT C-5 (*Continued*)

Entry No.	Substrate	Conditions	Product(s) and Yield(s)(%)	Refs.
15	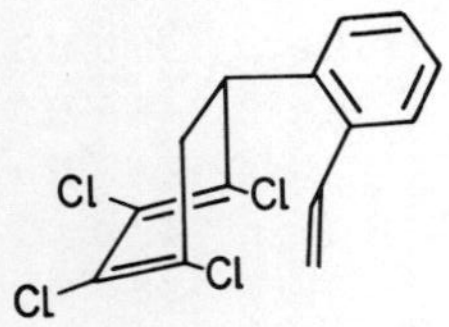 * /	Benzene, 100°, 30 min	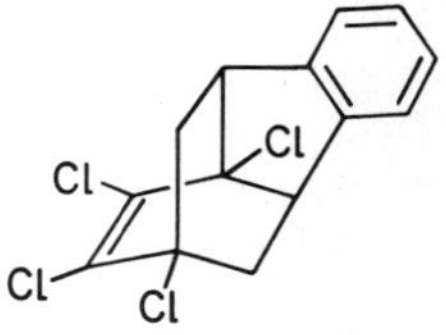 (72)	432
	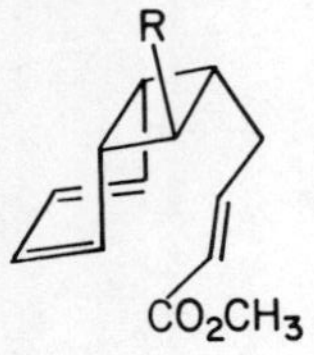		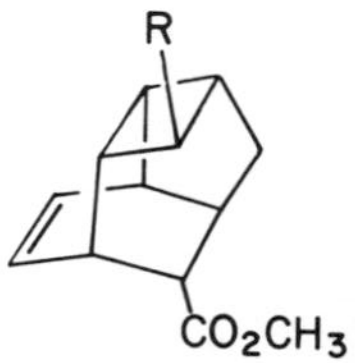	
	R			
15a	$C_6H_5(CH{=}CH)_2CH_2$	Toluene, 70°	(—)	315
15b	$t\text{-}C_4H_9(C_6H_5)_2SiOCH_2$	", 110°, 12 hr	(92)	317
	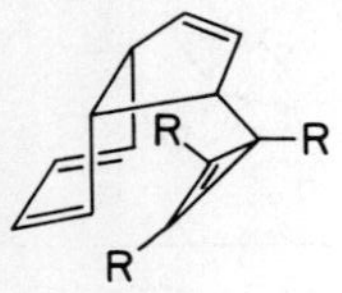		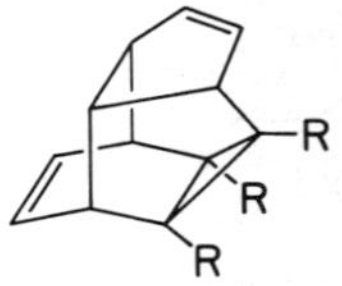	
	R			
15c	H*[q]	Flow system, 300°	(3)	434
15d	Cl*[r]	THF, 25°, 16 hr	(2)	435

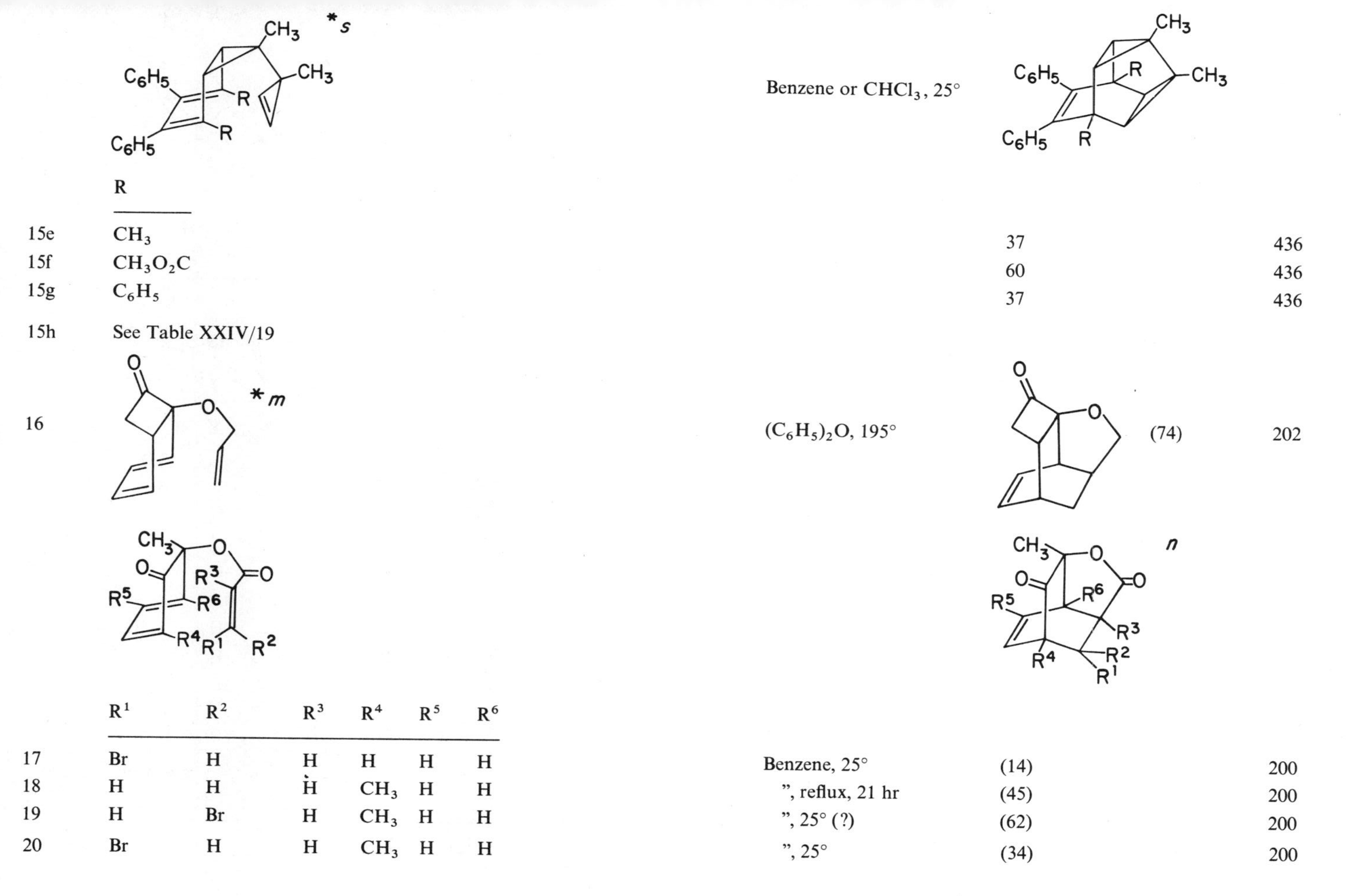

No.	Reactant	Conditions	Product	Yield (%)	Refs.
	R	Benzene or $CHCl_3$, 25°			
15e	CH_3			37	436
15f	CH_3O_2C			60	436
15g	C_6H_5			37	436
15h	See Table XXIV/19				
16		$(C_6H_5)_2O$, 195°		(74)	202

No.	R^1	R^2	R^3	R^4	R^5	R^6	Conditions	Yield (%)	Refs.
17	Br	H	H	H	H	H	Benzene, 25°	(14)	200
18	H	H	H	CH_3	H	H	", reflux, 21 hr	(45)	200
19	H	Br	H	CH_3	H	H	", 25° (?)	(62)	200
20	Br	H	H	CH_3	H	H	", 25°	(34)	200

TABLE IXd. CYCLOHEXADIENES; 2-ATOM CHAIN ATTACHED AT C-5 (*Continued*)

Entry No.	Substrate R^1	R^2	R^3	R^4	R^5	R^6	Conditions	Product(s) and Yield(s)(%)	Refs.
21	H	CH_3O_2C	H	H	H	H	Benzene, CH_3OH, 25°	(35)	200
22	H	H	H	CH_3	CH_3	H	n-C_6H_{14}, reflux, 16 hr	(41)	201
23	CH_3	H	CH_3	CH_3	H	H	Benzene, 25°	(13)	200
24	H	CH_3O_2C	H	CH_3	H	H	Benzene, 25°	(41)	200
25	CH_3O_2C	H	H	CH_3	H	H	", "	(30)	200
26	C_6H_5	H	H	CH_3	H	H	", "	(59)	200
27	H	H	H	H	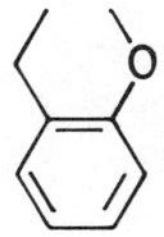		", reflux, 20 hr	(23)	201

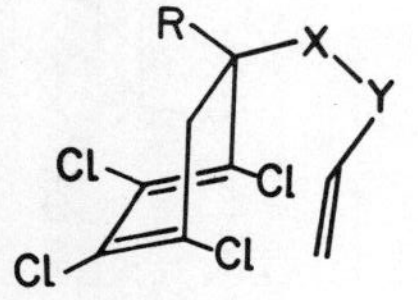

*o

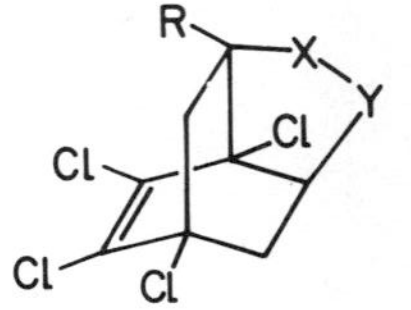

Entry No.	X	Y	R	Conditions	Product(s) and Yield(s)(%)	Refs.
28	O	CH_2	H	100°, 1 hr	(80)	432
29	S	CH_2	H	100°/1 hr; 150°/10 min	(75)	432
30	CO	O	H	100°, 16 hr	(56)	432
31	CO	O	CH_3	", 3 hr	(35)	432

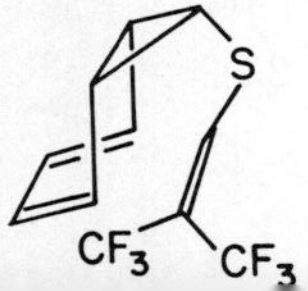

*p

Entry No.	Substrate	Conditions	Product(s) and Yield(s)(%)	Refs.
32		< 25°	(51)	437

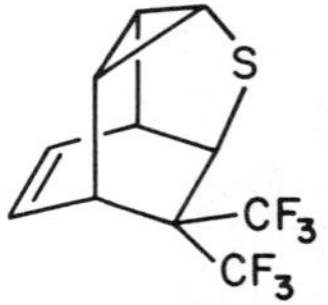

[a] The substrate was generated from $R^8CH{=}CH(CH_2)_2C(R^5){=}CR^6R^7$ and 2,5-R^1R^4-3,4-R^2R^3-thiophene dioxide.
[b] The stereochemistry of R^6 was not determined.
[c] The substrate was generated from sodium 2,6-dimethylphenoxide and 4-bromobutene.
[d] The substrate was generated by Claisen rearrangement from the corresponding phenyl allyl ether.
[e] The bridged product was formed in 44% yield (see Table XXVIc/14).
[f] The bridged product was formed in 35% yield (see Table XXVIc/15).
[g] $R = (C_2H_5O)_2P(O)$.
[h] The substrate and product were 1:1 mixtures of hydroxy epimers at Y.
[i] $R = (CH_3)_3Si$.
[j] R = tetrahydropyranyl.
[k] The yield is of ketoalcohols (mixture of hydroxy epimers) after hydrolytic removal of the two protecting groups.
[l] The substrate was generated from *o*-divinylbenzene and tetrachlorothiophene dioxide.
[m] The substrate was generated from the corresponding cyclooctatrienone valence isomer.

[n] The yields are based on phenols (OH; CH_3; R^4; R^5; R^6) that on lead tetraacetate oxidation, in the presence of acids $R^1R^2C{=}C(R^3)CO_2H$, give the substrates for the intramolecular Diels–Alder reaction.
[o] The substrate was generated from $CH_2{=}CHYXCR{=}CH_2$ and 2,3,4,5-tetrachlorothiophene dioxide.
[p] The substrate was generated by an ene reaction of cycloheptatriene and bis(trifluoromethyl)thioketene followed by valence isomerization.
[q] The substrate was generated from 9-(2-cyclopropenyl)bicyclo[6.1.0]nona-2,4,6-triene.
[r] The substrate was generated from lithium cyclononatetraenide and tetrachlorocyclopropene.
[s] The substrate was generated by addition of 3-methyl-3-(3′-methyl-3′-cyclopropenyl)cyclopropene to the appropriate 3,4-diphenylcyclopentadienone.
[t] The bridged isomer was formed in 14% yield; see Table XXVIc/15a.
* This intermediate is hypothetical, or has not been isolated.

TABLE IXe. CYCLOHEXADIENES; 3-ATOM CHAIN ATTACHED AT C-5

Entry No.	Substrate	Conditions	Product(s) and Yield(s)(%)	Refs.
	* R¹ R² R³ R⁴			
1	H H CH_3 CH_3[a]	150°, reduced pressure	(13)	180
2	CH_3 CH_3 H H[b]	Benzene, reflux, 24 hr	(7)	76
	R¹ R²			
3	CH_3 H	Benzene, reflux, 48 hr	(5)[c]	76
4	H CH_3	", ", 6 hr	(15)[d]	152,76
5	*e	Benzene, reflux, 24 hr	(26)	76

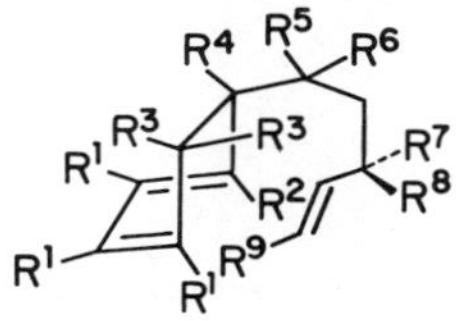

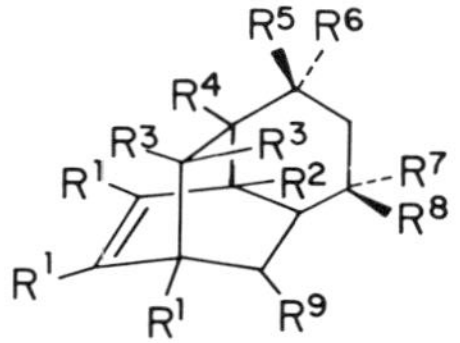

No.	R^1	R^2	R^3	R^4	R^5	R^6	R^7	R^8	R^9	Conditions	Yield (%)	Refs.
6	H	H	H	H	H	H	H	H	H	Gas phase, 210°, 21 hr	(73)[f]	77
7	H	H	H	H	O=		H	H	H*[g]	", 230°, 50 hr	(37)[h]	77
8	Cl	Cl	H	H	H	H	H	H	H	170°, 1 hr	(26)	432
9	H	$HOCH_2$	H	H	CH_3	H	H	H	H or CH_3	Heat or high pressure	(0)	59
10	H	HO_2C	H	H	CH_3	H	H	H	H	Benzene, 190°, 24 hr	(52)[i, j]	59
11	H	CH_3	CH_3	OH	H	H	H	H	H	Decalin, t-C_4H_9OK, 280°, 24 hr	(—)	265
12	H	CH_3	CH_3	OH	H	H	H	CH_3	H	", ", ", "	(49)	71
13	H	CH_3	CH_3	OH	H	H	CH_3	H	H	", ", ", "	(0)	71
14	H	$C_2H_5O_2C$	H	H	CH_3	H	H	H	H	Benzene, CH_3CO_2H[k], 205°, 3 hr	(50)[i]	59
15	H	$CH_3CO_2CH_2$	H	H	CH_3	H	H	H	H	", 230°, 24 hr	(40)[i]	59
16	H	$C_2H_5O_2C$	H	H	CH_3	H	H	H	CH_3	", CH_3CO_2H, 220°	(0)	59
17	H	CH_3	CH_3	RO[l, m]	H	H	O=		H	", 230°, 12 hr	(76)	264,262, 263

*[n]

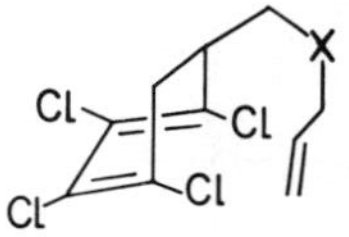

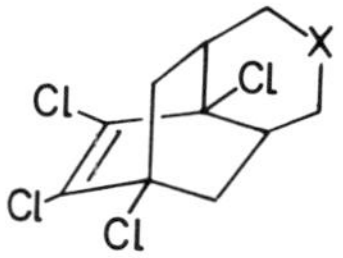

No.	X	Conditions	Yield (%)	Refs.
18	O	170°, 1 hr	(65)	432
19	S	170°, 3 hr	(66)	432

TABLE IXe. Cyclohexadienes; 3-Atom Chain Attached at C-5 (*Continued*)

Entry No.	Substrate R^1 R^2	R^3	R^4	R^5	R^6	R^7	R^8	R^9	Conditions	Product(s) and Yield(s)(%)	Refs.
20	NCN								170°, 1 hr	(56)	432
21	$NCCH_2N$								170°, 10 min	(75)	432
22	CH_3CON								170°, 30 min	(59)	432

[a] The dienophile was generated by pyrolysis of the terminal *N*-oxide.
[b] The substrate was generated from sodium 2,4,6-trimethylphenoxide and 1-bromo-4-pentene.
[c] The bridged isomer was formed in 35% yield; see Table XXVIc/16.
[d] The bridged isomer was formed in 44% yield; see Table XXVIc/17.
[e] The substrate was generated from sodium 2,4,6-trimethylphenoxide and 5-bromo-1,3-cycloheptadiene.
[f] Two minor products were formed by initial 1,5-hydrogen shift followed by intramolecular Diels–Alder reaction of the resulting 1-(4-pentenyl)-1,3-cyclohexadiene; see Table IXa/1.
[g] The substrate was generated from 1,6-heptadien-3-one and α-pyrone followed by loss of CO_2.
[h] The major products were formed by initial 1,5-hydrogen shift followed by intramolecular Diels–Alder reaction of the resulting 1-(2-oxo-4-pentenyl)-1,3-cyclohexadiene; see Table IXa/2.
[i] The stereochemistry of the methyl group was assigned on the basis of transition-state models.
[j] The product was isolated as the methyl ester.
[k] No reaction occurred in the absence of one equivalent of acetic acid or at 110°/20-kbar pressure.
[l] $R = (CH_3)_3Si$.
[m] The free alcohol did not cyclize at 160° and decomposed at higher temperatures.
[n] The substrate was generated from $X(CH_2CH{=}CH_2)_2$ and tetrachlorothiophene dioxide.
*This intermediate is hypothetical, or has not been isolated.

TABLE X. Cycloheptadienes and Cycloheptatrienes[a]

Entry No.	Substrate	Conditions	Product(s) and Yield(s)(%)	Refs.
1	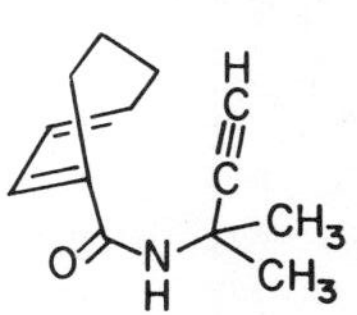 [b]	$C_6H_5N(C_2H_5)_2$, 218°, 3 hr	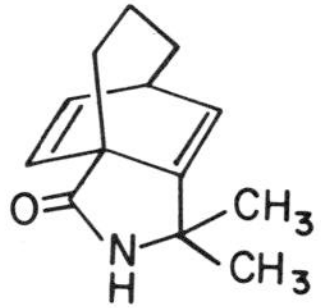 (55)	189
2	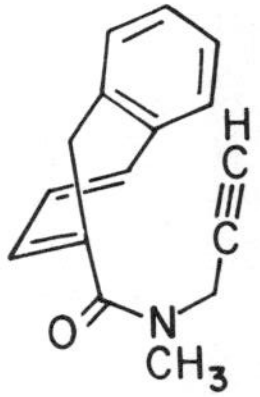	Toluene, 185°, 3 hr	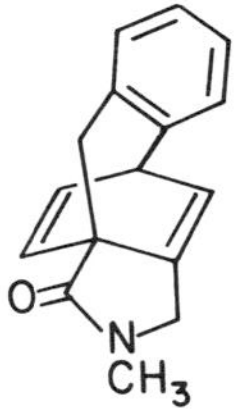 (78)	167
3	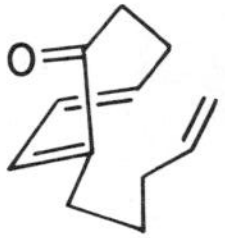 [b]	C_7H_{16}, 200–205°, 24 hr	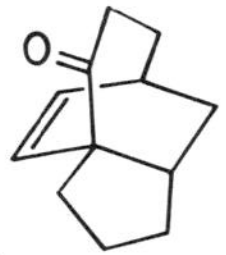 (77)	118
4		Toluene, 200°, 4 hr	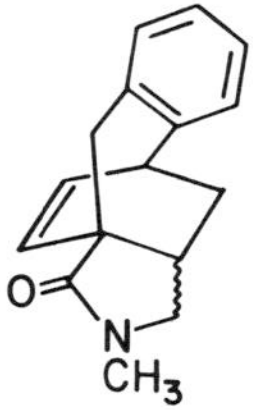 [c] (83)	167

TABLE X. CYCLOHEPTADIENES AND CYCLOHEPTATRIENES (*Continued*)

Entry No.	Substrate			Conditions	Product(s) and Yield(s)(%)	Refs.
5	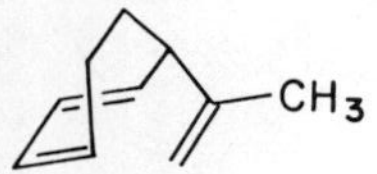 CH_3			Gas phase, 350°	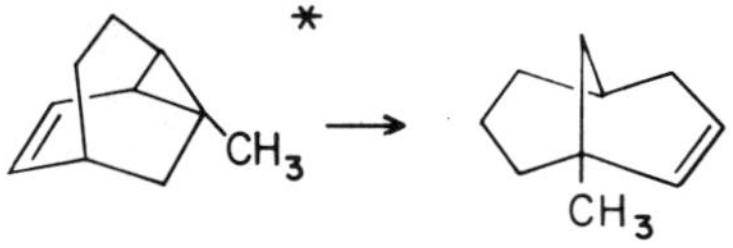 * CH_3 → CH_3 (—)	438
	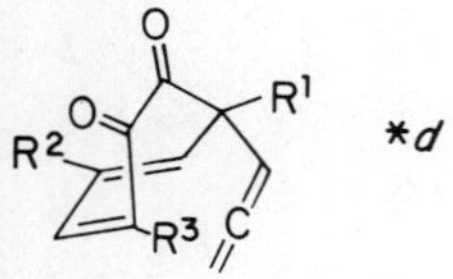 R^1, R^2, R^3, O, O, C *d				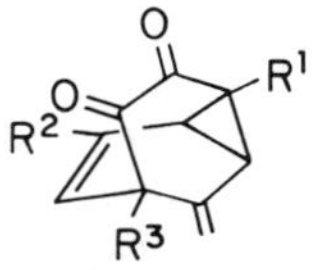 R^1, R^2, R^3, O, O	
	R^1	R^2	R^3			
6	H	H	H	Mesitylene, 170°	(0)[e]	439
7	Br	H	H	*p*-Cymene, reflux, 2 hr	(0)	173
8	H	H	CH_3	", ", "	(4)[f]	173
9	CH_3	H	H	", ", "	(81)	173
10	*n*-C_3H_7	H	H	", ", "	(69)	173
11	CH_3	CH_3	CH_3	*p*-Cymene, reflux, 1 hr	(80)	440
12	C_6H_5	H	H	$C_{10}H_{22}$, reflux, 4.5 hr	(53)	441
13	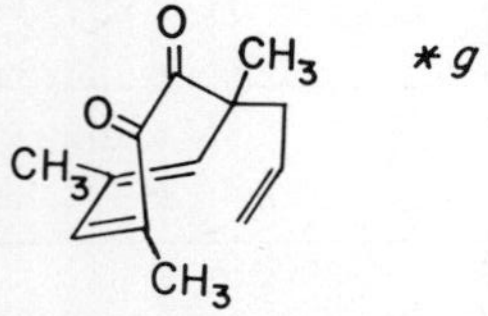 O, O, CH_3, CH_3, CH_3 *g			140°	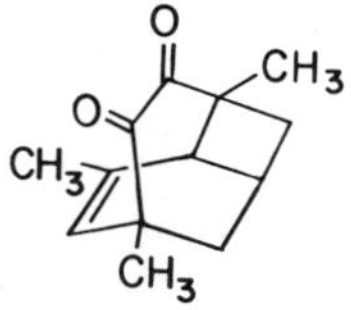 O, O, CH_3, CH_3, CH_3 (—)[h]	440

14	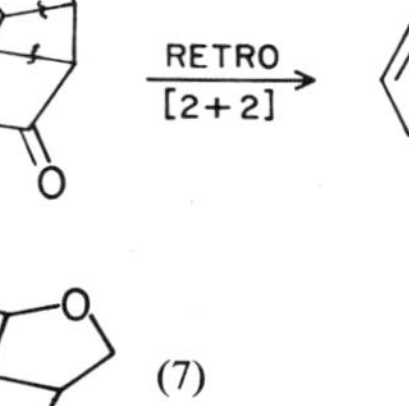 $*^{i}$	Et_2O, reflux	$*$ RETRO [2+2] → (70)	174
15	^{j}	  →	(7)	80
	Three 1,5-hydrogen shifts followed by Claisen rearrangement ↓	Neat, 200°, 24 hr		
15a	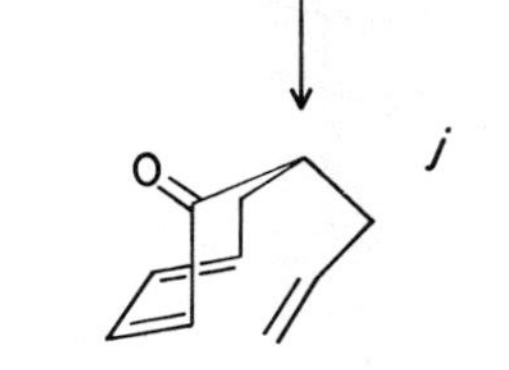 ^{j}	→	(41) + (41)	80

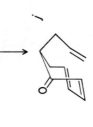

TABLE X. CYCLOHEPTADIENES AND CYCLOHEPTATRIENES (*Continued*)

Entry No.	Substrate	Conditions	Product(s) and Yield(s)(%)	Refs.
16	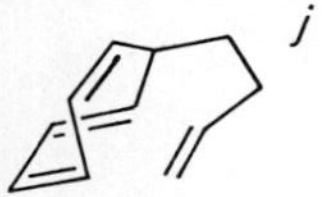 [j]	→	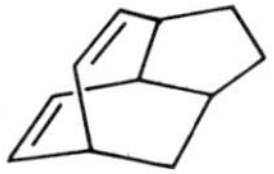 (9)	442
	↓ Two 1,5 shifts, then Cope rearrangement	n-C_7H_{16}, 215°, 4 days		
16a	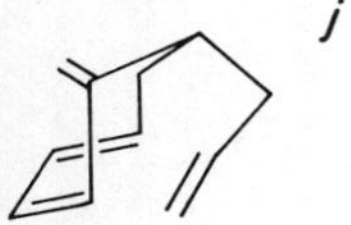 [j]	→	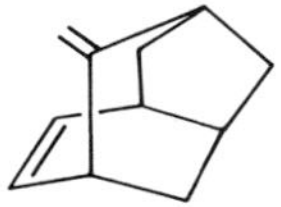 (12) +  (37)	442
17	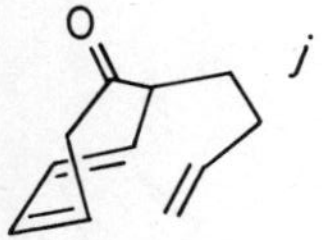 [j]	→	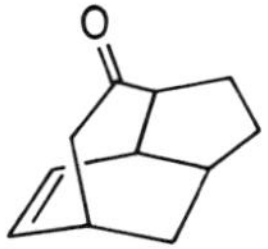 (29)	443
	⇅	n-C_7H_{16} or neat, 200–210°		
17a	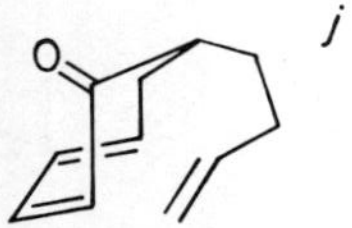 [j]	→	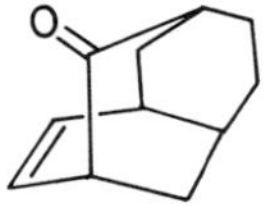 (22) + 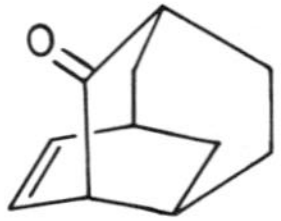 (10)	443

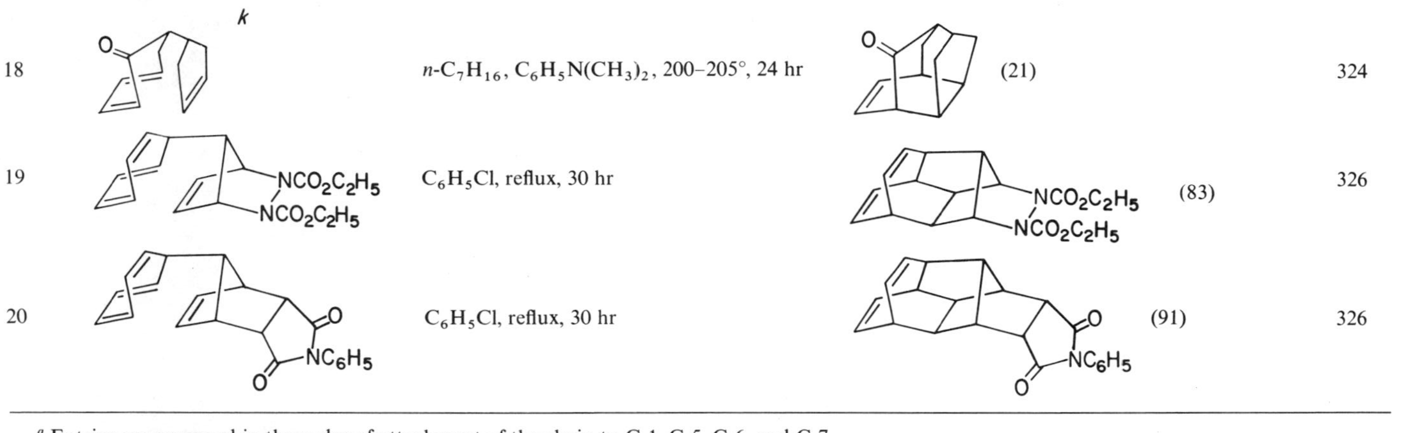

No.	Reactant	Conditions	Product(s) and Yield(s) (%)	Refs.
18	[k]	n-C_7H_{16}, $C_6H_5N(CH_3)_2$, 200–205°, 24 hr	(21)	324
19		C_6H_5Cl, reflux, 30 hr	(83)	326
20		C_6H_5Cl, reflux, 30 hr	(91)	326

[a] Entries are arranged in the order of attachment of the chain to C-1, C-5, C-6, and C-7.
[b] The substrate was a mixture of isomers interrelated by 1,5-hydrogen shifts.
[c] The product was a single isomer of undetermined stereochemistry.
[d] The substrate was generated by Claisen rearrangement from the corresponding tropolone propargyl ether.
[e] The intramolecular ene adduct was formed in 81% yield.
[f] The product was not completely characterized.
[g] The substrate was generated by Claisen rearrangement from the corresponding tropolone allyl ether.
[h] The bridged product was also formed (see Table XXVIc/18).
[i] The substrate was generated from the corresponding acid chloride and triethylamine.
[j] Both reactions occurred simultaneously under the conditions shown.
[k] The substrate was a mixture of cycloheptadienone isomers.
* This intermediate is hypothetical, or has not been isolated.

TABLE XI. CYCLIC 1,3,n-TRIENES

Entry No.	Substrate					Conditions	Product(s) and Yield(s) (%)	Refs.
1						CH_3OH, $h\nu$	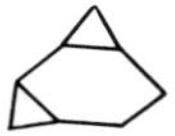 (10)	416,444
	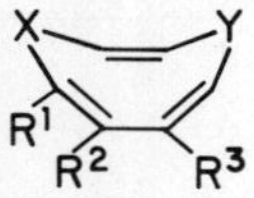 A					⇌	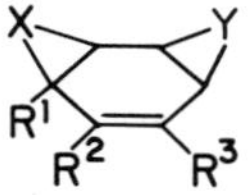 B	
	X	Y	R[1]	R[2]	R[3]		Ratio **A/B**	
2	O	O	H	H	H	Benzene, 60°	95/5	445
3	O	O	H	Br	H	CCl_4, 77°	65/35	446
4	CH_2	O	CH_3O_2C	CH_3O_2C	H	25°	40/60	147
						110°	70/30	147
5	CH_2	O	CH_3O_2C	CH_3O_2C	Br	Toluene, 100°, 4 hr	87/13	147
6	CH_2	O	CH_3O_2C	CH_3O_2C	CH_3	Benzene, 120°	95/5	147
7	CH_3O_2CCH	CH_3O_2CCH	H	H	H	$CDCl_3$, 25°	20/80	447

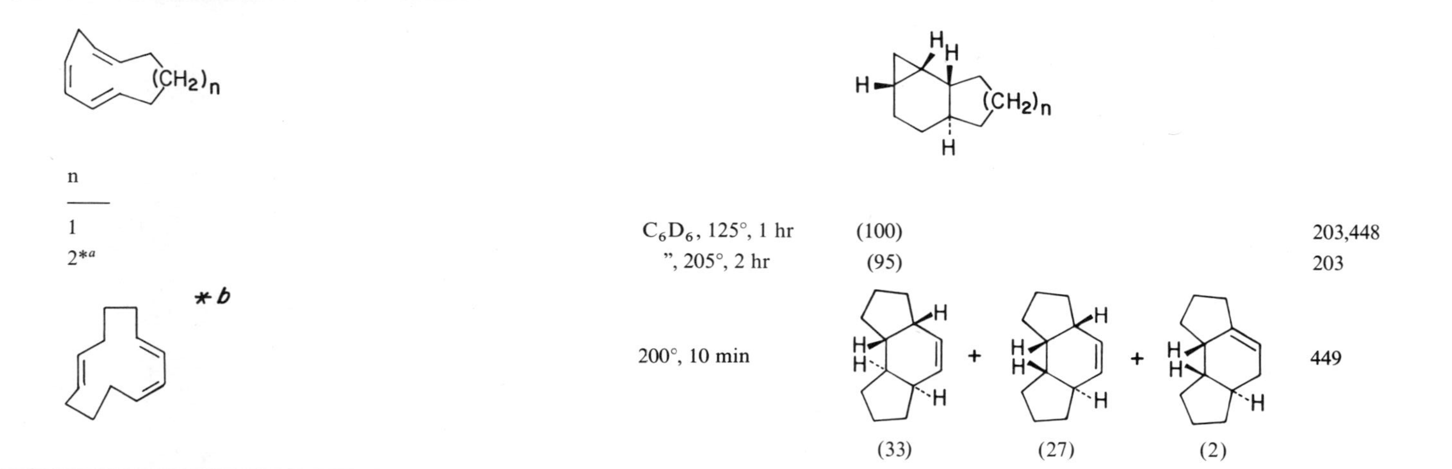

	n				
8	1		C_6D_6, 125°, 1 hr	(100)	203,448
9	2*[a]		", 205°, 2 hr	(95)	203
10			200°, 10 min	(33) + (27) + (2)	449

[a] The substrate was generated from *cis,trans,cis*-cycloundeca-1,3,5-triene or *cis*-bicyclo[7.2.0]undeca-3,10-diene.

[b] The substrate was generated from *cis,trans,cis*-1,5,9-cyclododecatriene on treatment with $Cp_2TiCl_2/LiAlH_4$. The authors do not comment on the mechanism of this reaction; the *trans,cis,trans*-1,3,8-cyclododecatriene shown would give the major observed product on intramolecular Diels–Alder reaction.

* This intermediate is hypothetical, or has not been isolated.

TABLE XII. BRIDGED[a] 1,3,6-CYCLOOCTATRIENES

Entry No.	Substrate	Conditions	Product(s) and Yield(s) (%)	Refs.
	*b; R^1 R^2		* ⇌ *	
1	CH_3 H	Gas phase (18 mm), 395–500°	(0–29)[c]	450,451
2	H CH_3	", 405–590°	(—)[c]	452,451
3	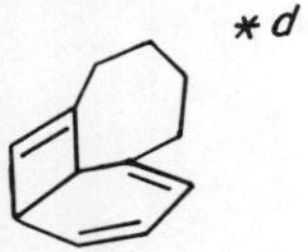 *d	Gas phase, flow system, 500°	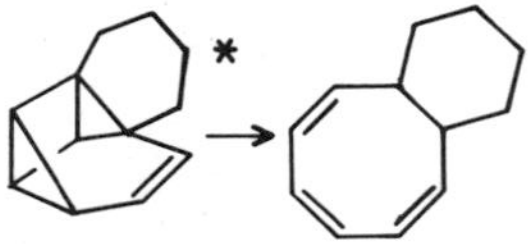 * → (69)	453
	n R		* → * →	
4	1 H	Gas phase, 615°	(low)	454
5	2 H	", 470°	(60)	454

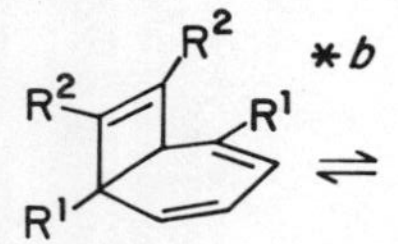

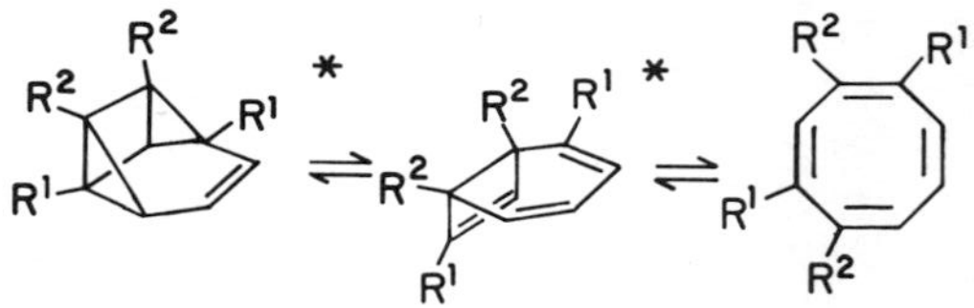

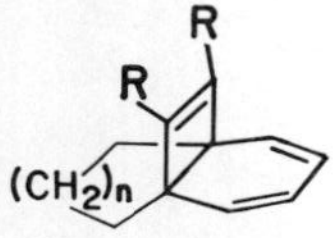

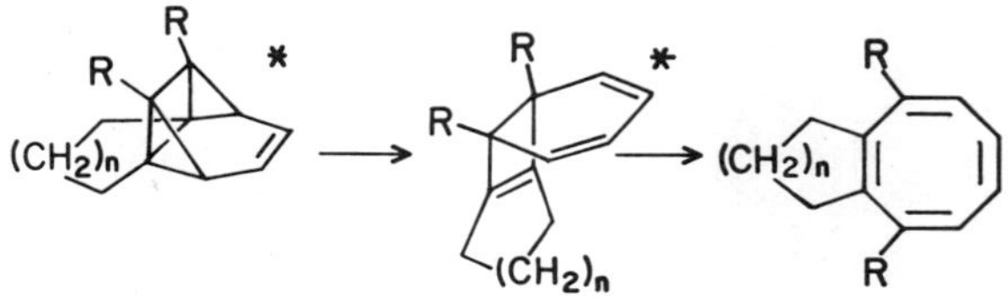

			Conditions	Products (Yield %)	Ref.
6	3	H	", "	(47)	454
7	2	CH_3	", "	(50)[e]	454
8			Gas phase, 470°	(53)	454

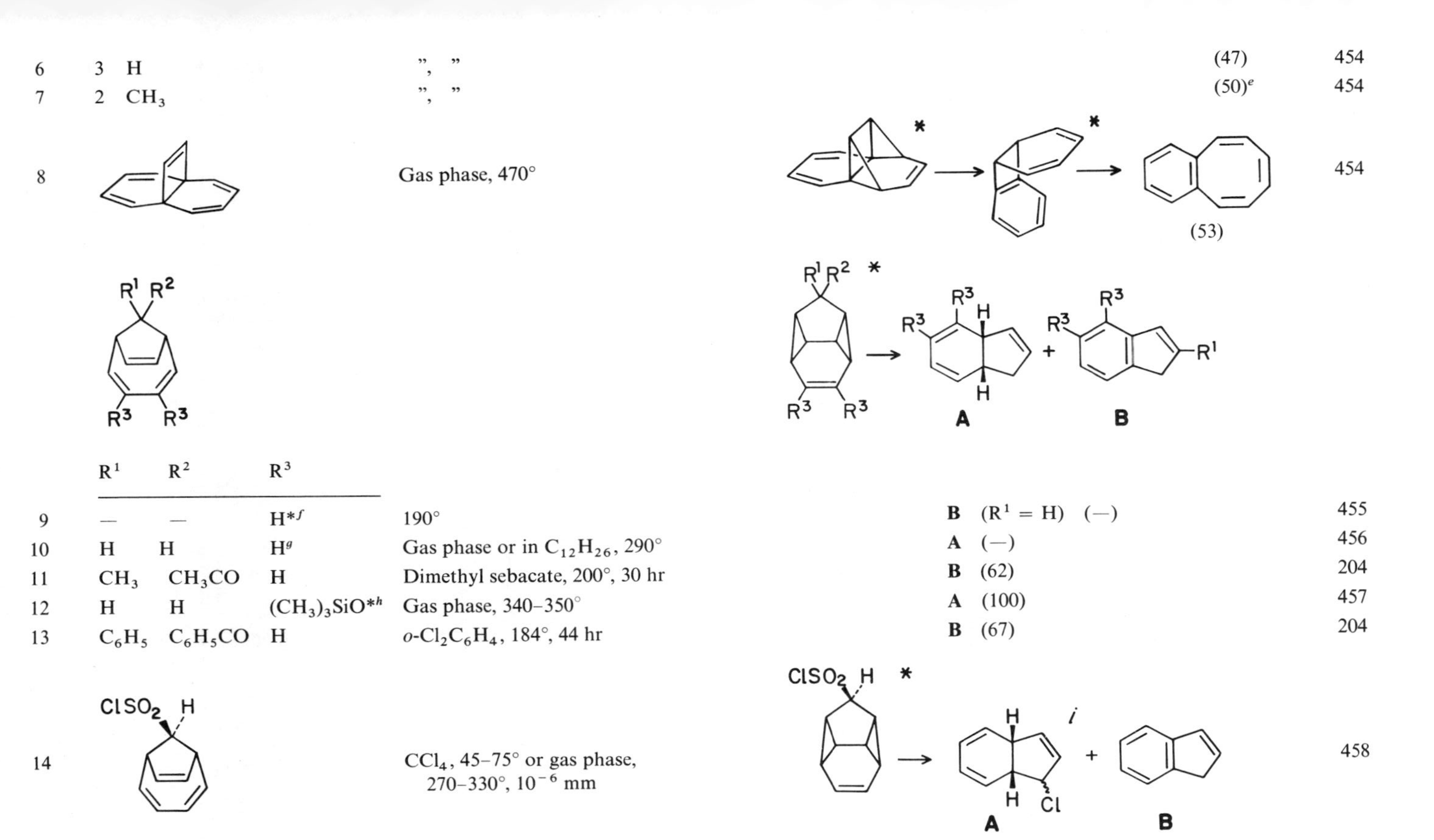

	R^1	R^2	R^3	Conditions	Products (Yield %)	Ref.
9	—	—	H*[f]	190°	**B** (R^1 = H) (—)	455
10	H	H	H[g]	Gas phase or in $C_{12}H_{26}$, 290°	**A** (—)	456
11	CH_3	CH_3CO	H	Dimethyl sebacate, 200°, 30 hr	**B** (62)	204
12	H	H	$(CH_3)_3SiO$*[h]	Gas phase, 340–350°	**A** (100)	457
13	C_6H_5	C_6H_5CO	H	o-$Cl_2C_6H_4$, 184°, 44 hr	**B** (67)	204
14				CCl_4, 45–75° or gas phase, 270–330°, 10^{-6} mm	[i] Ratio **A/B** 90/10 (in CCl_4); 95/5 (gas phase)	458

TABLE XII. BRIDGED 1,3,6-CYCLOOCTATRIENES (*Continued*)

Entry No.	Substrate	Conditions	Product(s) and Yield(s)(%)	Refs.
15	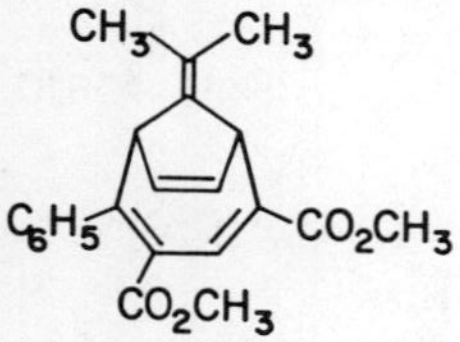 [structure]	180°	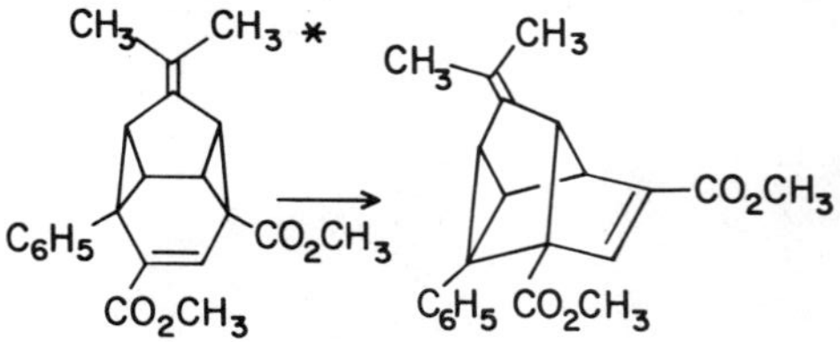 [structures] * (90)	459

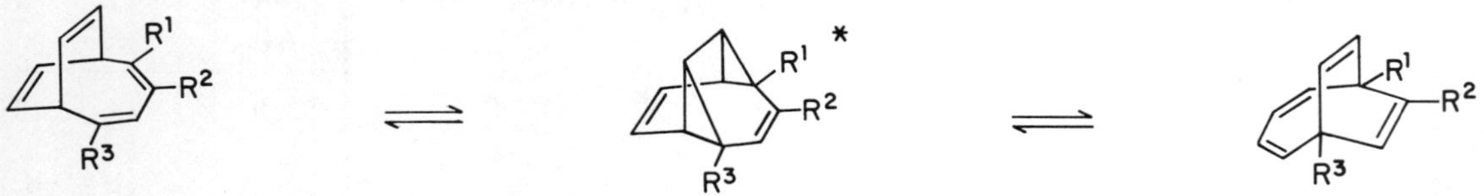

A ⇌ B * ⇌ C

Entry No.	R^1	R^2	R^3	Conditions	Ratio A/C	Refs.
16	H	H	H	—	—[j,k]	460–464
17	Cl	H	H	35°	75/25	465
18	Br	H	H	"	70/30	465
19	H	Br	H	"	$<5/>95$	465
20	Cl	H	Cl	"	$>95/<5$	465
21	CH_3	H	H	"	85/15	465
22	H	CH_3	H	"	$<5/>95$	465
23	CH_3	H	CH_3	"	$>95/<5$	465
24	CH_3O_2C	H	H	"	$>95/<5$	465

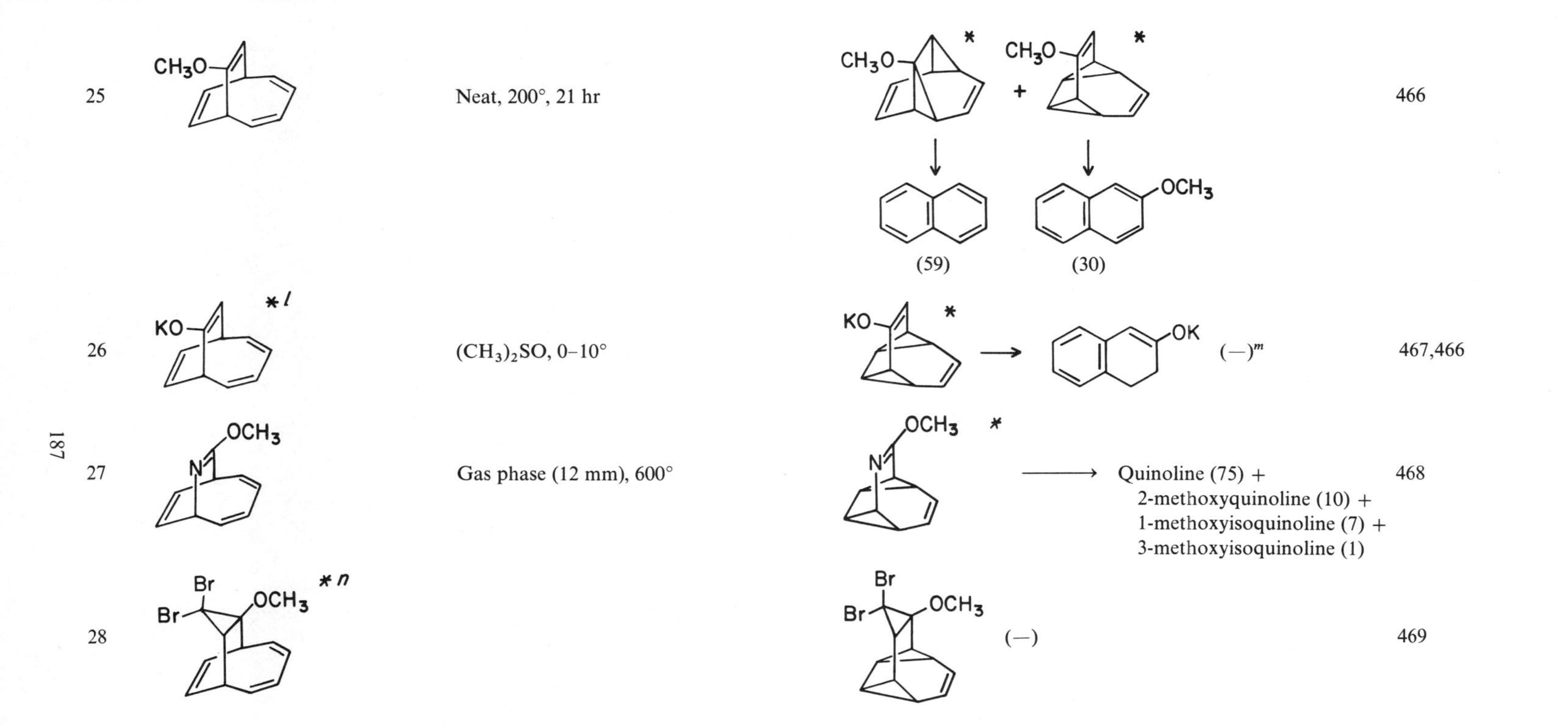

No.	Conditions	Products (%)	Refs.
25	Neat, 200°, 21 hr	(59) + (30)	466
26	$(CH_3)_2SO$, 0–10°	(—)[m]	467,466
27	Gas phase (12 mm), 600°	Quinoline (75) + 2-methoxyquinoline (10) + 1-methoxyisoquinoline (7) + 3-methoxyisoquinoline (1)	468
28		(—)	469

TABLE XII. BRIDGED 1,3,6-CYCLOOCTATRIENES (*Continued*)

Entry No.	Substrate			Conditions	Product(s) and Yield(s) (%)	Refs.
	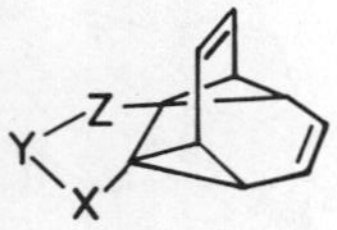 A				 B ⇌ C ⇌ D	
	X	Y	Z			
29	CH_2	O	CH_2	CH_3OH or $CHCl_3$, 26° ($t_{1/2}$ = 110 min)	**D** (81.5)	470
30	CH_2	O	CO	−40°	**B** (32); **C** (45); **D** (23)	471,470
31	CO	O	CO	CH_3OH, 25°, 40 min	**D** (40)	470,462,472
32	CO	O	CO	0°, $(Et_2O)_2O^a$	**B** (—)	462
33	CO	NH	CO	CH_3OH, reflux, 3 hr	**D** (80)	470
34	CO 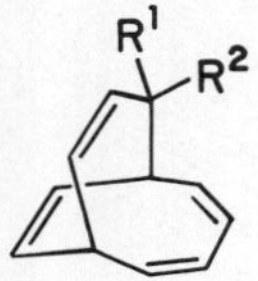	NCH_3	CO	$CDCl_3$, 59°	**D** (86) 	470
	R^1		R^2		COPE REARRANGEMENT	
35	HO		H	$Cl_2C{=}CCl_2$, 100°	(69)	473,474
36		O=		", "	(84)	473,474a
37	H		p-$O_2NC_6H_4CO_2$	Me_2CO/H_2O, 80°	(—)	475

[a] For this table, a single bond is considered to be a bridge.
[b] The substrate was generated by valence isomerization of the corresponding cyclooctatetraene; only one of the possible valence isomers is shown.
[c] The product is in equilibrium with the starting material.
[d] The substrate was generated from 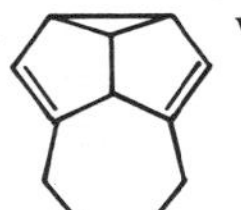 via a number of proposed intermediates.

[e] The number is the conversion.
[f] The carbene was generated from the diazoalkane; the analogous intramolecular Diels–Alder reaction is also postulated for the corresponding carbonium ion.
[g] The mechanism was inferred from deuterium labeling.

[h] The substrate was generated by thermolysis of 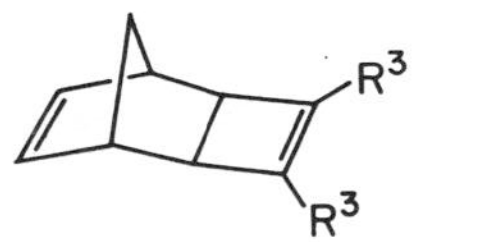

[i] The product was a mixture of epimers.
[j] Intermediate **B** has also been generated by photolysis of *cis*-9,10-dihydronaphthalene in pentane[460] or tetrahydrofuran;[464] yields of **A**/**C** were 26% and 100%, respectively.
[k] The equilibrium was demonstrated by deuterium labeling.[462]
[l] The substrate was generated from the ketone with potassium *tert*-butoxide.
[m] The product was isolated as the methyl ether.
[n] The substrate was obtained by dibromocyclopropanation of the corresponding enol ether (conditions not reported); see Table **XXVIIId**/19 for an equilibrium involving the monobromo analog.
[o] The first-order rearrangement of **B** to **D** was measured in $CDCl_3$ between 17° and 25°; $k = 5.5 \times 10^{13} \exp[(-24.5 \pm 2.7)\ \text{kcal}/RT]\ \text{sec}^{-1}$.
* This intermediate is hypothetical, or has not been isolated.

TABLE XIII. BRIDGED[a] 1,3,n-CYCLIC TRIENES OTHER THAN CYCLOOCTATRIENES

Entry No.	Substrate	Conditions	Product(s) and Yield(s)(%)	Refs.
1	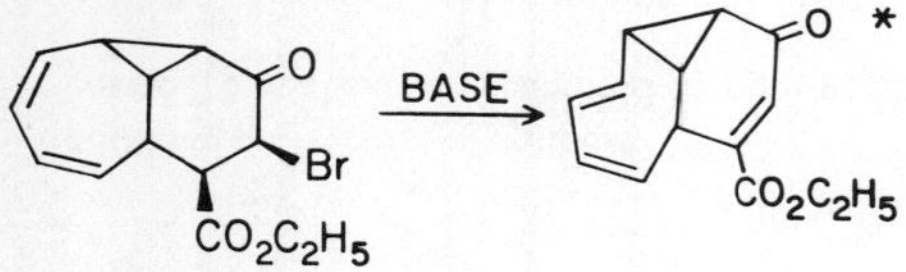 Br, $CO_2C_2H_5$ —BASE→ $CO_2C_2H_5$ *	(—)	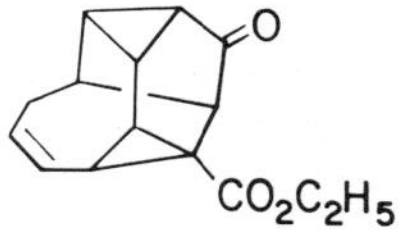 $CO_2C_2H_5$ (75)	476
2	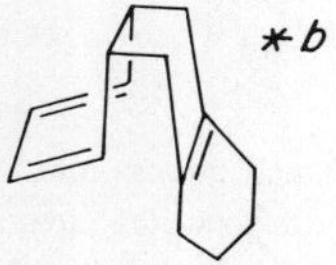 *b	Neat, 180°, 4.5 hr	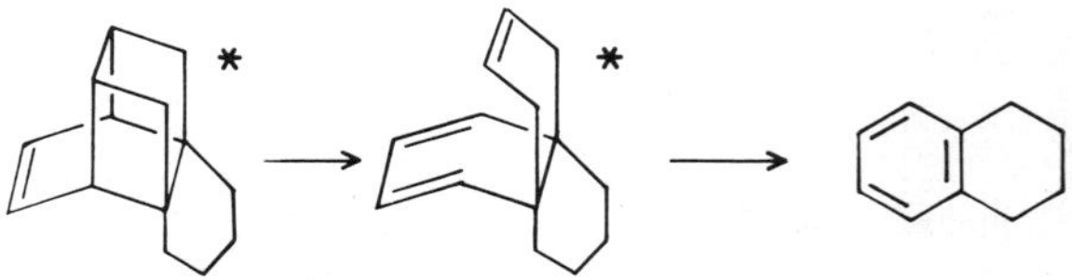 * → * → (52) + butadiene (—)	477
3	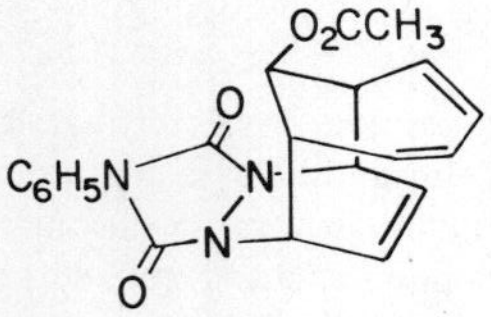 O_2CCH_3, C_6H_5N	(—)	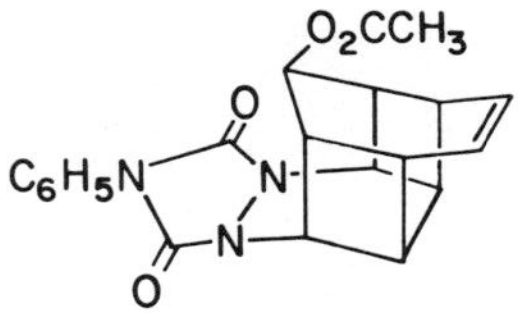 O_2CCH_3, C_6H_5N (—)	478

No.	Reactant	R¹	R²	Conditions	Product (Yield, %)	Refs.
	*					
4		H	H[c]	150°, 15 hr	(35)	479
5		Cl	H[d]	Benzene, 100°, 15 hr	(67)	432
6		Cl	Cl[d–f]	100°, 1 hr	(90)	432,479
7		Br	Br[d]	”	(56)	432
8		C_6H_5	C_6H_5[g]	150°, 96 hr	(78)	479
9	*			Flow system, 480°	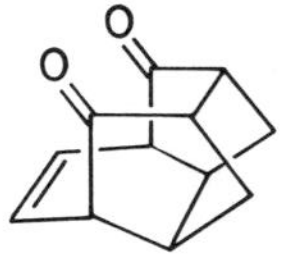 (43)	480
10	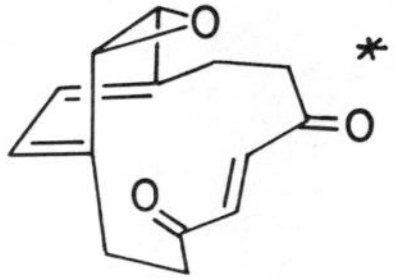 * [h]			CH_3OH, 25°	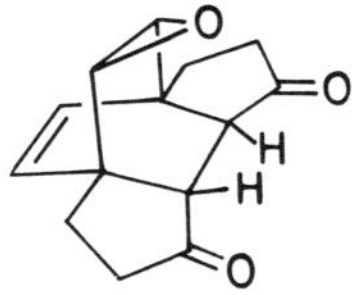 (5)	481
11	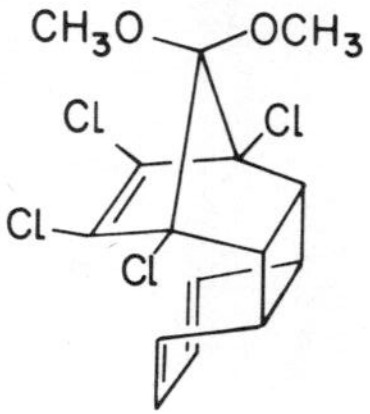 * [i]			*p*-Cymene, reflux, 16 hr	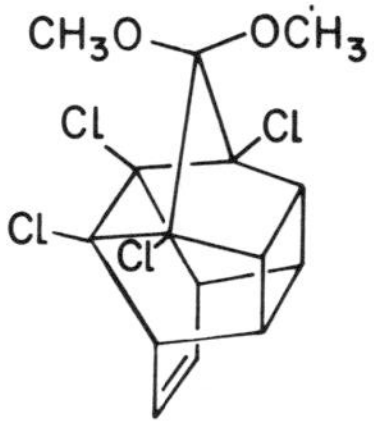 (25)	31

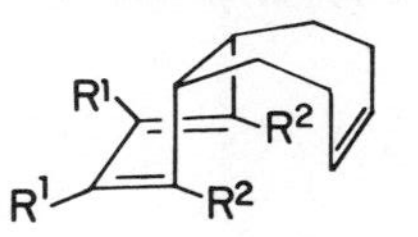

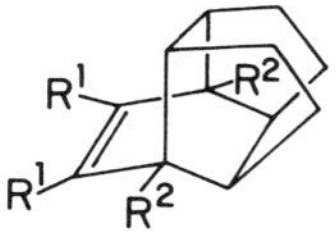

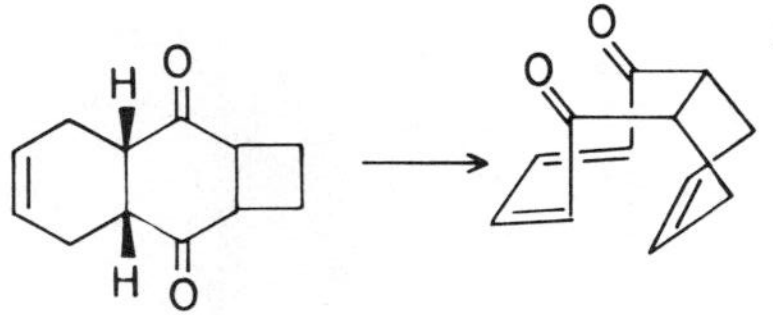

TABLE XIII. BRIDGED[a] 1,3,n-CYCLIC TRIENES OTHER THAN CYCLOOCTATRIENES (*Continued*)

Entry No.	Substrate	Conditions	Product(s) and Yield(s) (%)	Refs.
12	CH_3N … * j	160°	CH_3N … (—)	482
13	Cl, Cl, Cl, Cl * k	125°, 5 hr	Cl, Cl, Cl, Cl (57)	432
14	* l,m		**A** (90)[n, o]	
	III	Benzene, $(C_2H_5)_2AlCl$, $TiCl_4$, 40°, 6 hr		
14a			**B** (10)[b]	483

15	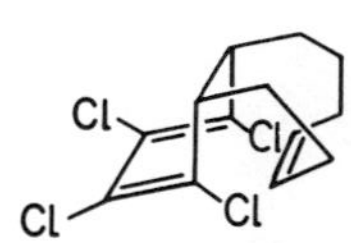 Cl, Cl, Cl, Cl *[p]	150–190°, 10 min	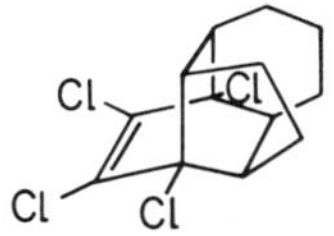 Cl, Cl, Cl, Cl (58)	432

[a] For this table, a single bond is considered to be a bridge.

[b] The substrate was generated from

[c] The substrate was generated from α-pyrone and 1,5-cyclooctadiene.

[d] The substrate was generated from 2,5-di(R^1)-3,4-di(R^2)thiophene dioxide and 1,5-cyclooctadiene.

[e] The substrate was generated by debromination of 5,6-dibromo-1′,2′,3′,4′-tetrachlorobenzocyclooctane.

[f] The substrate was generated from tetrachloro-α-pyrone and 1,5-cyclooctadiene.

[g] The substrate was generated from tetraphenylcyclopentadienone and 1,5-cyclooctadiene.

[h] The substrate was generated from [2.2]furanoparacyclophane and oxygen by methylene-blue-sensitized photolysis.

[i] The substrate was generated from the corresponding *anti*-bicyclo[4.2.0]octadiene.

[j] The substrate is a hypothetical intermediate formed on heating the corresponding *anti* isomer.

[k] The substrate was generated from dibenzocyclooctatetraene and tetrachlorothiophene dioxide.

[l] The substrate was generated by catalyzed dimerization of cycloheptatriene.

[m] Both reactions occurred simultaneously.

[n] The number is the conversion.

[o] **A** was converted quantitatively into **B** on heating to 150° for 12 hours, indicating that the initial cycloaddition is reversible.

[p] The substrate was generated from 1,5-cyclononadiene and tetrachlorothiophene dioxide.

* This intermediate is hypothetical, or has not been isolated.

TABLE XIVa. VINYL- AND ALLENYLAROMATICS; CHAIN CONTAINING CARBON ONLY

Entry No.	Substrate	Conditions	Product(s) and Yield(s) (%)	Refs.
1	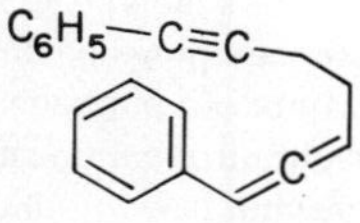 $C_6H_5-C\equiv C-$ … $*$[a]	t-C_4H_9OH, 62–63°, 2.5 hr	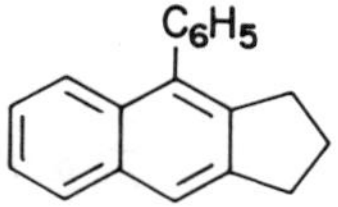 C_6H_5 (35)	484
2	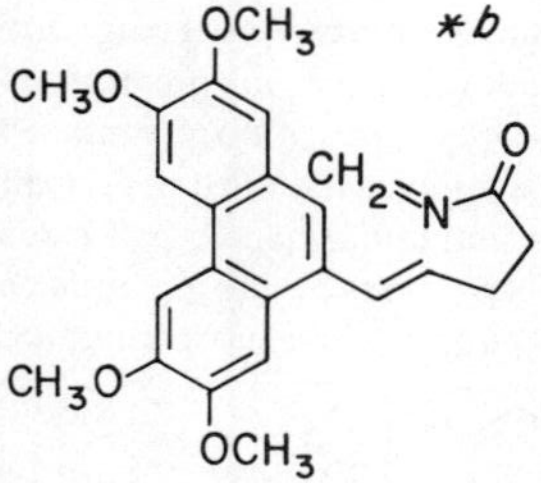 CH_3O, OCH_3, CH_3O, OCH_3, $CH_2{=}N$, O $*$[b]	C_6H_5Br, 220°, 5 hr	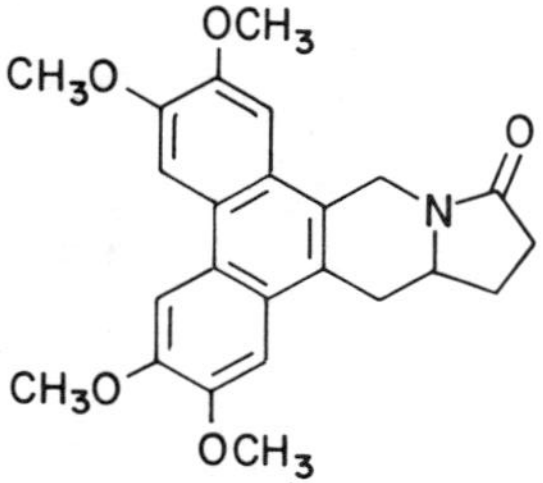 CH_3O, OCH_3, CH_3O, OCH_3, N, O (50)	136,182

[a] The substrate was generated by t-C_4H_9OK-catalyzed isomerization of 1,7-diphenyl-1,6-heptadiyne.
[b] The substrate was generated by acetate pyrolysis of the $CH_3CO_2CH_2NH—$ derivative.
* This intermediate is hypothetical, or has not been isolated.

TABLE XIVb. VINYL- AND ALLENYLAROMATICS; CHAIN CONTAINING UNCHARGED NITROGEN

Entry No.	Substrate	Conditions	Product(s) and Yield(s)(%)	Refs.
1	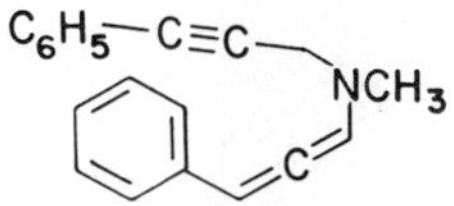 *[a]		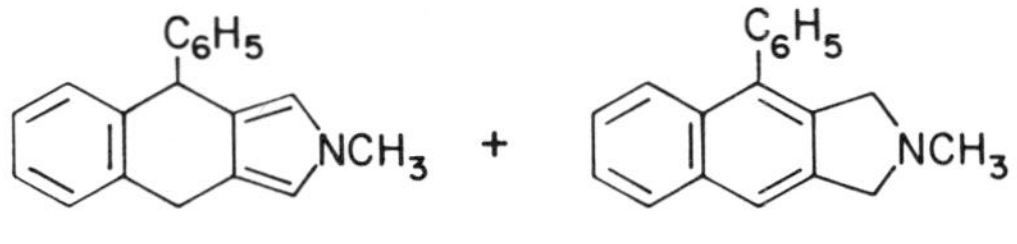 A + B	
			A B	
		Benzene, 20°, 6 hr	(12) (24)	23[b]
		t-C_4H_9OH, 62–63°, 75 min	(0) (85)	484
2	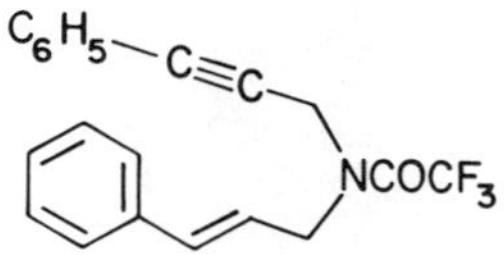	o-$Cl_2C_6H_4$, 180°, 5 hr	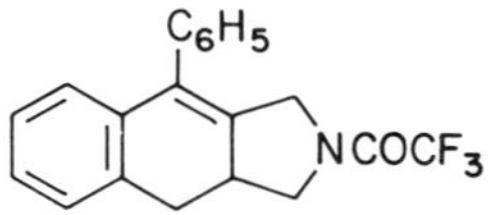 (—)	120

TABLE XIVb. VINYL- AND ALLENYLAROMATICS; CHAIN CONTAINING UNCHARGED NITROGEN (*Continued*)

Entry No.	Substrate 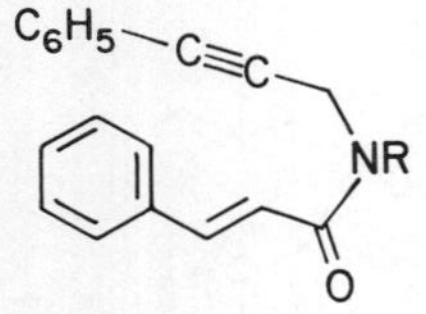	Conditions	Product(s) and Yield(s)(%) 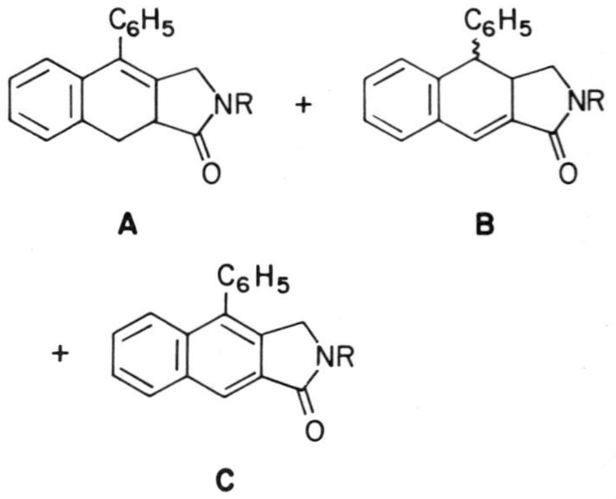	Refs.
	R		A B C	
3	H	$(CH_3CO)_2O$, reflux	(0) (60) (R = CH_3CO)[c,d] (15) (R = CH_3CO)[d]	121
4	$C_6H_5CH_2$ 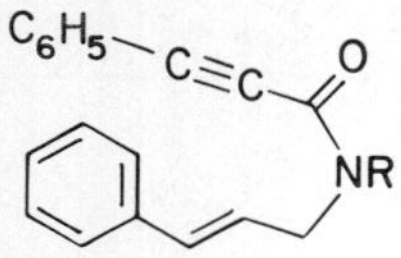	", "	(16) (0) (8) 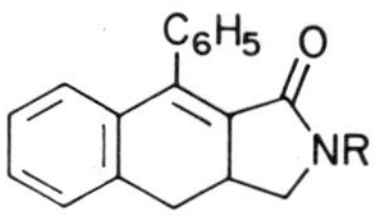	485
	R			
5	H	DMF, reflux, 6 hr	(55)	121
6	H	$(CH_3CO)_2O$, reflux, 4 hr	(74)[e]	121

No.	Reactant	Conditions	Product(s) and Yield(s) (%)	Refs.
7	CH_3CO*[f]	", 25°, 2 hr	(37)	121
8	$C_6H_5CH_2$	", reflux, 5 hr	(100)	485
9	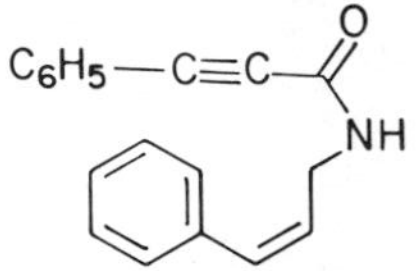 C_6H_5—C≡C—CO—NH—	$(CH_3CO)_2O$, reflux	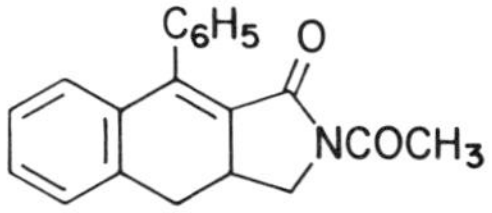 C_6H_5, O, $NCOCH_3$ (32)	121
	p-RC_6H_4—C≡C—, NH, O	$(CH_3CO)_2O$, reflux, 6 hr	C_6H_4R-p, NH + C_6H_4R-p, NH, O; **A**, **B**	
	R		**A** **B**	
	—			
10	H		$(18)^d$ $(5)^d$	486
11	Br		$(24)^d$ $(3)^d$	486
	R^4, R^2, NR^1, R^3		R^4, R^2, H, NR^1, R^3, H + R^4, R^2, H, NR^1, R^3, H; **A**, **B**	

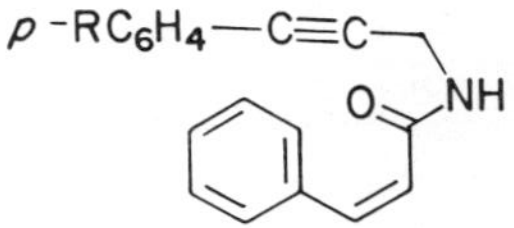

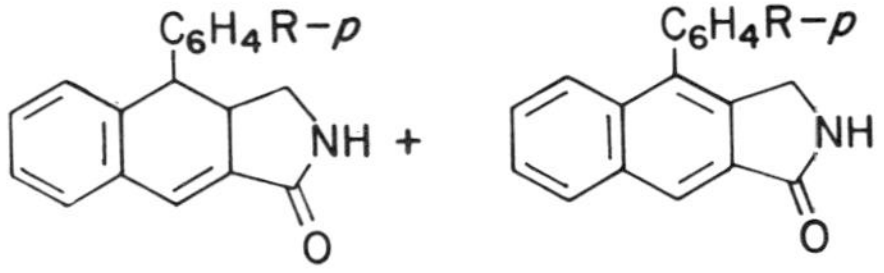

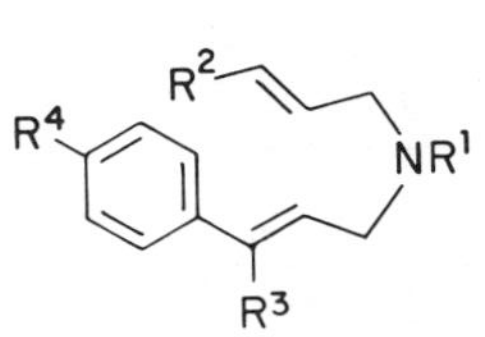

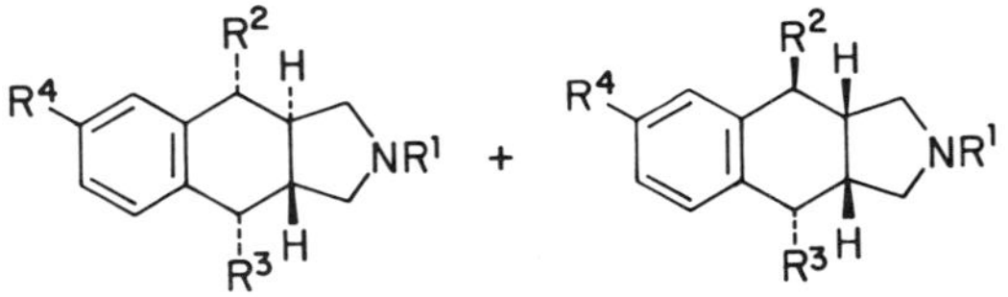

	R^1	R^2	R^3	R^4	Conditions	A	B	Refs.
12	CF_3CO	H	H	H	Toluene, 235°, 20 hr	(41)	(20)	119,157
13	CF_3CO	C_6H_5	H	H	o-$Cl_2C_6H_4$, 180°, 20 hr	(0.6)	(78)	119,120

TABLE XIVb. Vinyl- and Allenylaromatics; Chain Containing Uncharged Nitrogen (*Continued*)

Entry No.	Substrate					Conditions	Product(s) and Yield(s) (%)	Refs.
	R^1	R^2	R^3	R^4				
14	CF_3CO	H	C_6H_5	H		o-$Cl_2C_6H_4$, 180° 6 days	(48)[g] (6)	119
15	p-$CH_3C_6H_4SO_2$	C_6H_5	H	H		", ", 48 hr	(0) (—)	120
16	CF_3CO	p-ClC_6H_4	H	Cl		", ", 16 hr	(0) (18)[h]	119
	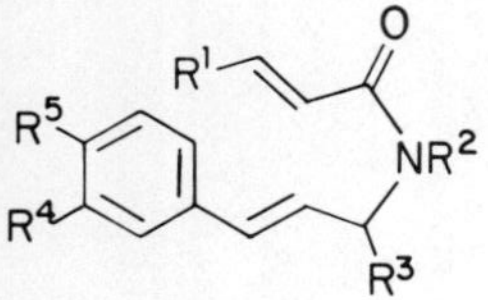						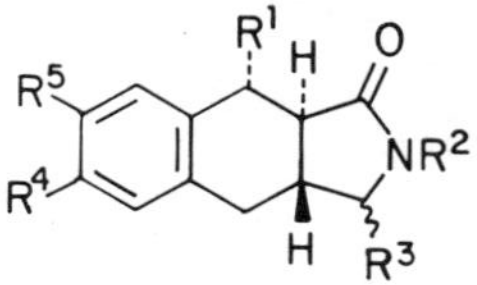	
	R^1	R^2	R^3	R^4	R^5			
17	C_6H_5	H	H	H	H	$(CH_3CO)_2O$, reflux	(0)[i]	121
18	$C_2H_5O_2C$	CH_3	H	—OCH_2O—		Toluene, reflux, 8 hr	(56)	123
19	$C_2H_5O_2C$	—$(CH_2)_4$—		—OCH_2O—		CH_3CN, 0°	(37)	123
20	C_6H_5	$C_6H_5CH_2$	H	H	H	$(CH_3CO)_2O$, reflux	(0)	485
21	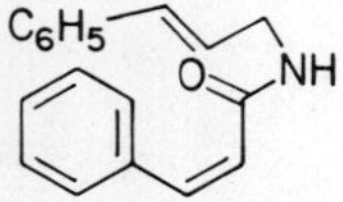					$(CH_3CO)_2O$, reflux	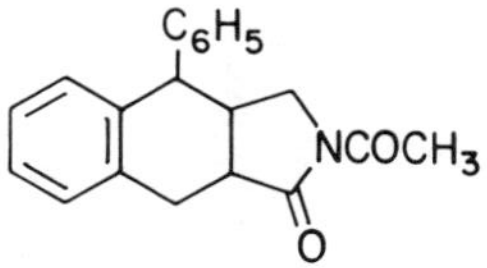 (25)[j]	127

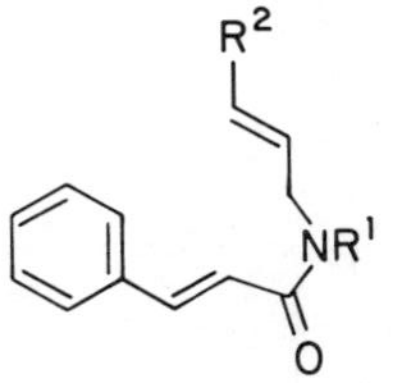

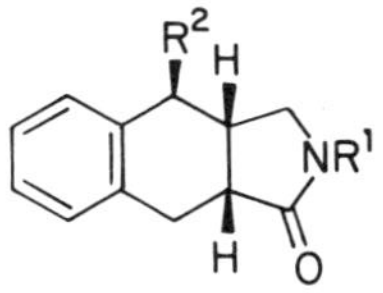

No.	Substrate		Conditions	Product(s) and Yield(s) (%)	Refs.
	R^1	R^2			
22	H	H	DMF, 140°, 30 hr	(60)[j]	47
23	C_6H_5	H	", ", 12 hr	(85)[j]	47
24	p-$O_2NC_6H_4$	H	", ", 8 hr	(85)[j]	47
25	$C_6H_5CH_2$	C_6H_5	o-$Cl_2C_6H_4$, 180°, 6 days	(48)[k]	119
26	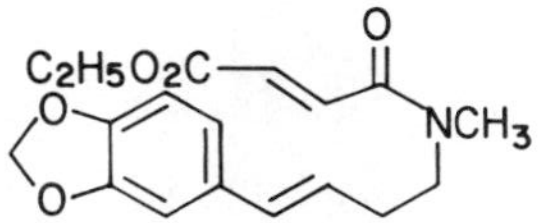		o-$Cl_2C_6H_4$, reflux	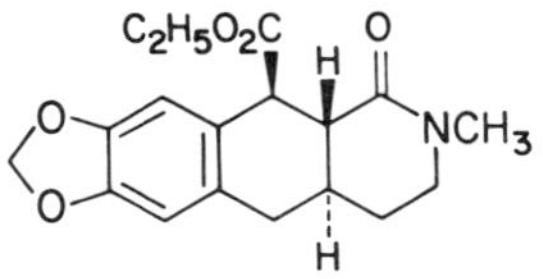 (27) + 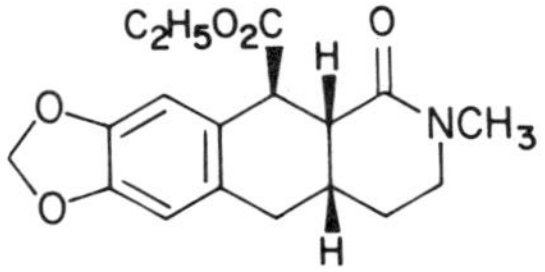 (18)	123

[a] The substrate was generated by base-catalyzed isomerization of N-methyl-N,N-bis(phenylpropargyl)amine.
[b] The authors prefer a biradical mechanism.
[c] Isomerization of the double bond was ascribed to the presence of acetic acid.
[d] The products were not separated.
[e] The product was the N-acetyl derivative ($R = CH_3CO$); acetylation occurred prior to cyclization.
[f] The substrate was generated from the secondary amide by treatment with sodium hydride and acetic anhydride.
[g] The product was a mixture with $R^3\alpha$ as shown (38%) and $R^3\beta$ (10%).
[h] An intramolecular [2 + 2] adduct was obtained in unspecified yield.
[i] The substrate polymerized.
[j] The stereochemistry was not established.
[k] The intramolecular [2 + 2]cycloadduct was obtained in 18% yield.
* This intermediate is hypothetical, or has not been isolated.

TABLE XIVc. VINYLAROMATICS; CHAIN CONTAINING QUATERNARY NITROGEN

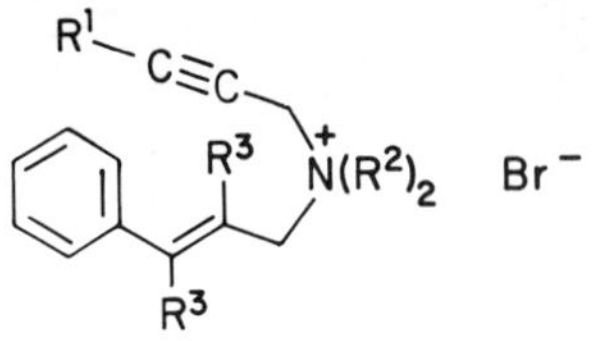

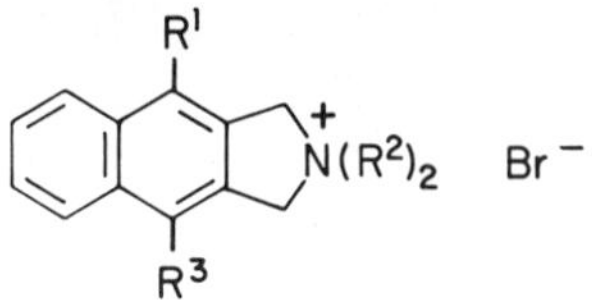

Entry No.	Substrate			Conditions	Product(s) and Yield(s)(%)	Refs.
	R^1	R^2	R^3			
1	H	CH_3	Cl	H_2O, 25° or heat	(88)	372,330
2	H	C_2H_5	Cl	H_2O, KOH, 95°, 1 hr	(85)	330
3	CH_3	CH_3	H	H_2O, 25°, 4 days	(70)	366
4	CH_3	C_2H_5	H	", ", "	(75)	366
5	$CH_2{=}CH$	CH_3	Cl	H_2O, KOH, 100°, 2 hr	(80)	333
6	$CH_2{=}C(CH_3)$	CH_3	Cl	", ", "	(60)	333
7	C_6H_5	CH_3	Cl	", ", "	(87)	333
8	C_6H_5	C_2H_5	Cl	", ", "	(60)	333
	R^1	R^2	X^-			
9	CH_3	CH_3	Br	KOH, H_2O, 25°, 4 days	(74)	366
10	C_2H_5	CH_3	Br	", ", ", "	(83)	366
11	CH_3	$(CH_3)_2NCH_2$	Br	", ", ", 5 days	(61)	366
12	CH_3	$CH_3O(CH_2)_2$	Cl	", ", ", 4 days	(23)	366
13	CH_3	t-$C_4H_9CH_2$	Br	", ", 100°, 4 days	(25)	366

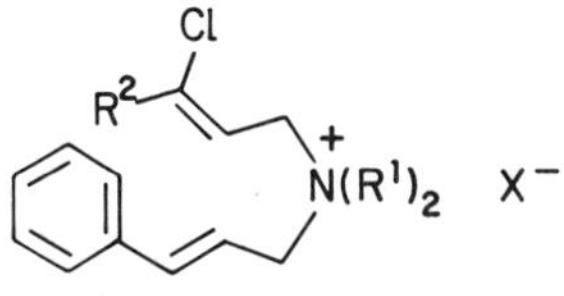

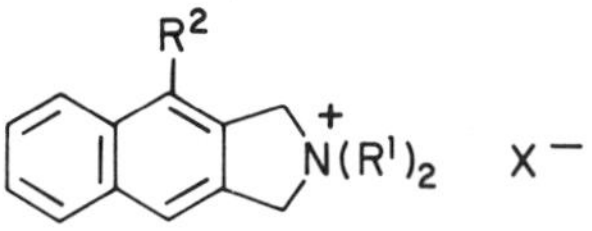

TABLE XIVd. ALLENYLAROMATICS; CHAIN CONTAINING QUATERNARY NITROGEN; ETHYNYL DIENOPHILES

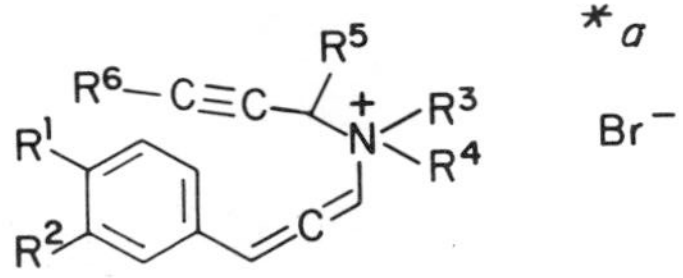

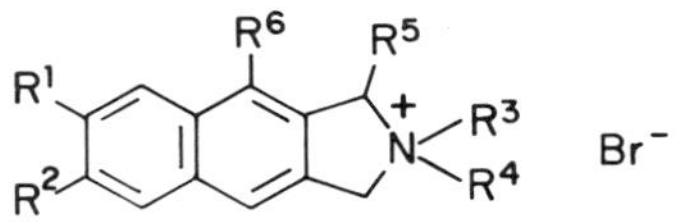

Entry No.	R^1	R^2	R^3	R^4	R^5	R^6	Conditions	Product(s) and Yield(s)(%)	Refs.
1	H	H	CH_3	CH_3	H	H	H_2O, 25°, 1 hr	(95)[b]	329,365,372,487,488
2	Cl	H	CH_3	CH_3	H	H	", ", 2 hr	(83)	489
3	H	Cl	CH_3	CH_3	H	H	", "	(45)[c, d]	334
4	CH_3	H	CH_3	CH_3	H	H	", ", 1 hr	(92)	339
5	H	CH_3	CH_3	CH_3	H	H	", "	(60)[d, e]	334
6	H	H	CH_3	CH_3	H	CH_3	", ", 3 days	(70)	335
7	H	H	CH_3	CH_3	CH_3	H	", "	(75)	334
8	H	H	C_2H_5	C_2H_5	H	H	H_2O, 25°, 1 hr	(86)	329,365,372,490
9	H	H	—$(CH_2)_4$—		H	H	", ", 2 hr	(95)	368
10	H	H	CH_3	$CH_2{=}CHCH_2$	H	H	", ", "	(80)	368
11	H	H	—$(CH_2)_2O(CH_2)_2$—		H	H	", ", "	(94)	368,488
12	Cl	H	C_2H_5	C_2H_5	H	H	", ", "	(95)	489
13	Cl	H	—$(CH_2)_4$—		H	H	", " "	(98)	489
14	H	Cl	C_2H_5	C_2H_5	H	H	", "	(0)[d, f]	334
15	Cl	H	—$(CH_2)_2O(CH_2)_2$—		H	H	", ", 2 hr	(90)	489
16	H	H	CH_3	CH_3	H	$CH_2{=}CH$	", ", "	(98)[g]	370
17	CH_3	H	C_2H_5	C_2H_5	H	H	", ", 1 hr	(82)	339

TABLE XIVd. ALLENYLAROMATICS; CHAIN CONTAINING QUATERNARY NITROGEN; ETHYNYL DIENOPHILES (*Continued*)

Entry No.	Substrate R^1	R^2	R^3	R^4	R^5	R^6	Conditions	Product(s) and Yield(s)(%)	Refs.
18	H	CH_3	C_2H_5	C_2H_5	H	H	H_2O, 25°	(82)	334
19	H	H	$—(CH_2)_5—$		H	H	", ", 1 hr	(94)	329,365,372
20	H	H	C_2H_5	C_2H_5	H	CH_3	H_2O, 38° → 100°	(35)[h]	491
21	H	H	CH_3	CH_3	H	$CH_2{=}C(CH_3)$	H_2O, 25°, 2 hr	(85)[g]	370
22	Cl	H	$—(CH_2)_5—$		H	H	", ", "	(90)	489
23	H	Cl	$—(CH_2)_5—$		H	H	", "	(60)[d, i]	334
24	CH_3O	H	C_2H_5	C_2H_5	H	H	", ", 1 hr	(91)	339
25	CH_3	H	$—(CH_2)_5—$		H	H	", ", "	(87)	339
26	H	CH_3	$—(CH_2)_5—$		H	H	", "	(30)[d, j]	334
27	H	H	C_2H_5	C_2H_5	H	$CH_2{=}CH$	", ", 2 hr	(—)	370
28	H	H	$—(CH_2)_2O(CH_2)_2—$		H	$CH_2{=}CH$	", ", "	(93)[g]	370
29	H	H	C_2H_5	C_2H_5	H	$CH_2{=}C(CH_3)$	", ", "	(94)	370
30	H	H	$—(CH_2)_2O(CH_2)_2—$		H	$CH_2{=}C(CH_3)$	", ", "	(92)[g]	370
31	H	H	n-C_4H_9	n-C_4H_9	H	H	", ", "	(96)	368
32	H	H	CH_3	$C_6H_5CH_2$	H	H	DMF, heat	(35)	368
33	H	H	CH_3	CH_3	H	C_6H_5	H_2O, 25°, 1 hr	(94)	331,153,492,493
34	H	H	CH_3	CH_3	H	$C_6H_5OCH_2$	", "	(95)	371
35	H	H	C_2H_5	C_2H_5	H	n-$C_4H_9OCH_2$	", "	(89)	371
36	H	H	C_2H_5	C_2H_5	H	C_6H_5	", ", 1 hr	(94)	331,492,493
37	H	H	CH_3	$CH_2{=}CHCH_2$	H	C_6H_5	", ", 2 hr	(80)	370
38	H	H	$—(CH_2)_2O(CH_2)_2—$		H	C_6H_5	", ", "	(92)	370
39	H	H	$—(CH_2)_5—$		H	C_6H_5	", ", "	(95)	370,331,492,493

No.							Conditions	Yield (%)	Ref.
40	H	H	C_2H_5	C_2H_5	H	$C_6H_5OCH_2$	", "	(90)	371
41	CH_3	H	C_2H_5	C_2H_5	H	p-$CH_3C_6H_4$	", ", 1 hr	(83)	331
	* k						H_2O, 25°	$2Br^-$	
	R^1	R^2							
42	CH_3	CH_3						(81)	341
43	$-(CH_2)_4-$							(80)	341
44	$-(CH_2)_5-$							(90)	341
	* m						H_2O, C_2H_5OH, 25°, 1 day	$2 Br^-$	
	R^1	R^2							
45	CH_3	CH_3						(65)	494
46	$-(CH_2)_4-$							(72)	494

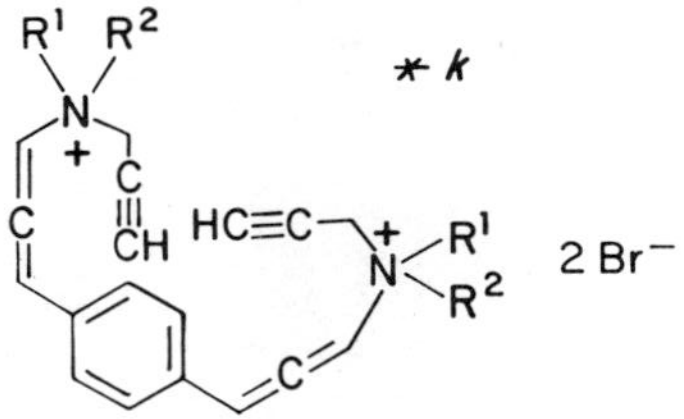

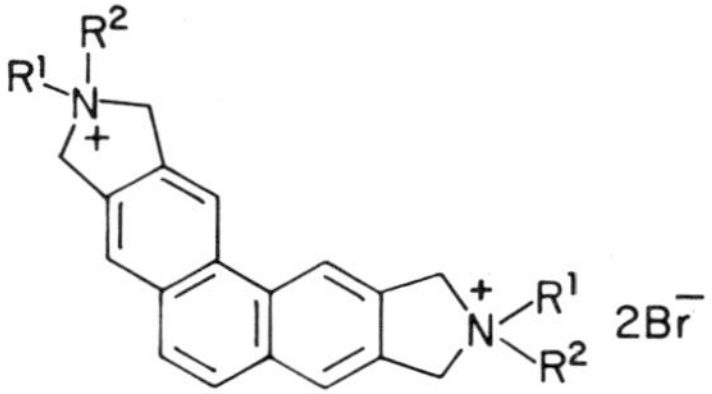

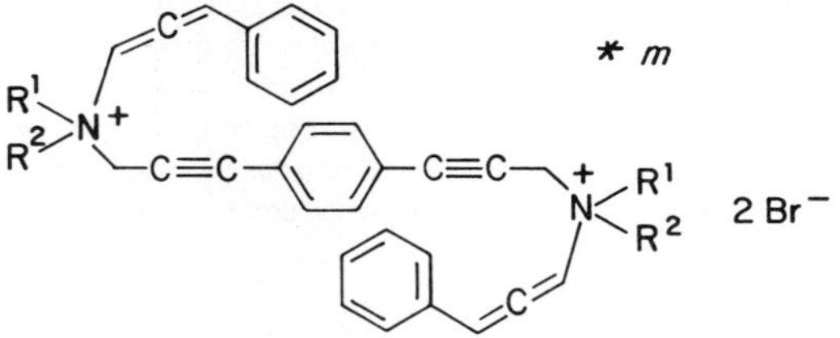

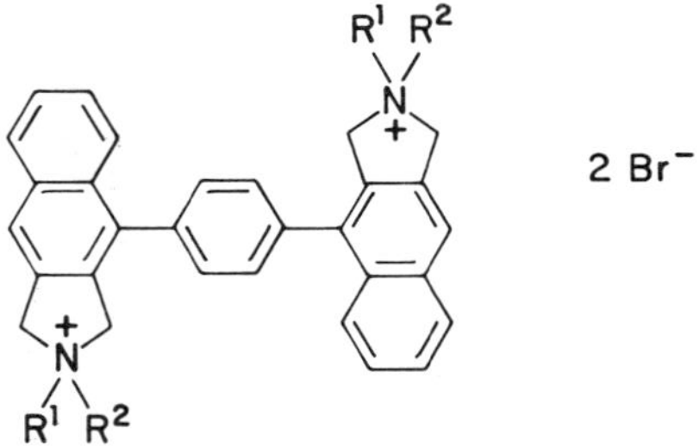

TABLE XIVd. ALLENYLAROMATICS; CHAIN CONTAINING QUATERNARY NITROGEN; ETHYNYL DIENOPHILES (*Continued*)

Entry No.	Substrate	Conditions	Product(s) and Yield(s)(%)	Refs.
	R^1 R^2			
47	$—(CH_2)_2O(CH_2)_2—$		(71)	494
48	$—(CH_2)_5—$		(74)	494

[a] The substrate was generated by base-catalyzed isomerization of the corresponding phenylpropargylammonium salt.
[b] The reaction with sodium ethoxide in refluxing ethanol gave the demethylated tertiary amine.[487]
[c] The isomer with R^2 in position 5 was formed in 30% yield.
[d] The isomer structures were assigned on the basis of IR spectra.
[e] The isomer with R^2 in position 5 was formed in 20% yield.
[f] The isomer with R^2 in position 5 was the sole product (80% yield).
[g] The product includes an isomer formed by addition of the phenylpropargyl moiety to the pentatrienyl group.
[h] The product was the tertiary amine.
[i] The isomer with R^2 in position 5 was formed in 15% yield.
[j] The isomer with R^2 in position 5 was formed in 20% yield.
[k] The substrate was generated by base-catalyzed isomerization of the corresponding aryl bis(propargylammonium) salt.
[l] The absence of the anthracene isomer was based on the UV spectrum.
[m] The substrate was generated by base-catalyzed isomerization of the corresponding bis(phenylpropargylammonium) salt.
* This intermediate is hypothetical, or has not been isolated.

TABLE XIVe. ALLENYLAROMATICS; CHAIN CONTAINING QUATERNARY NITROGEN; VINYL DIENOPHILES

Entry No.	Substrate R[1]	R^2	R^3	R^4	R^5	R^6	Conditions	Product(s) and Yield(s) (%)	Refs.
	(substrate structure: Br^- or Cl^- *[a])							(product structure: Br^- or Cl^-)	
1	H	CH_3	CH_3	H	H	H	H_2O, 100°, 1–2 hr	(76)	192,153,495
2	Cl	CH_3	CH_3	H	H	H	", 92°, 2–3 hr	(64)	489
3	CH_3	CH_3	CH_3	H	H	H	", 90°, 2–3 hr	(71)	339
4	H	CH_3	CH_3	CH_3	H	H	", reflux, 12 hr	(0)	153,495
5	H	CH_3	CH_3	H	CH_3	H	", 100°, 12 hr	(80)[b]	153,495
6	H	CH_3	CH_3	H	CH_3	CH_3	", "	(45)	153,340,495
7	H	$-(CH_2)_4-$		H	H	H	", "	(57)	378
8	H	C_2H_5	C_2H_5	H	H	H	", 100°, 1–2 hr	(84)	192
9	H	$-(CH_2)_2O(CH_2)_2-$		H	H	H	", 100°	(50)	378
10	Cl	C_2H_5	C_2H_5	H	H	H	", 90–92°, 2–3 hr	(90)	489
11	Cl	$-(CH_2)_2O(CH_2)_2-$		H	H	H	", ", "	(60)	489
12	H	$-(CH_2)_5-$		H	H	H	", 100°, 1–2 hr	(70)	192
13	CH_3	C_2H_5	C_2H_5	H	H	H	", 90°, 2–3 hr	(78)	339
14	H	C_2H_5	C_2H_5	H	CH_3	H	", 100°, 1–2 hr	(83)	192
15	Cl	$-(CH_2)_5-$		H	H	H	", 90–92°, 2–3 hr	(62)	489
16	CH_3	$-(CH_2)_5-$		H	H	H	", 90°, 2–3 hr	(76)	339
17	H	CH_3	CH_3	C_6H_5	H	H	", 100°	(0)	153,495

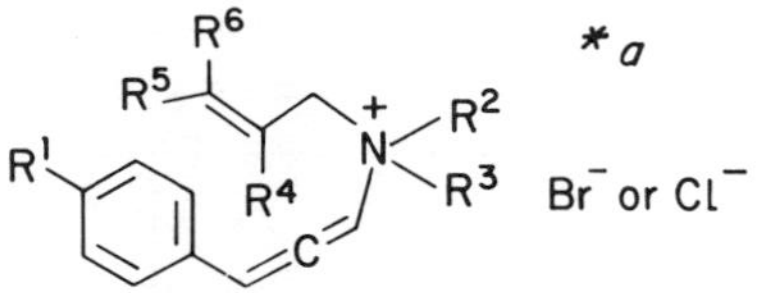

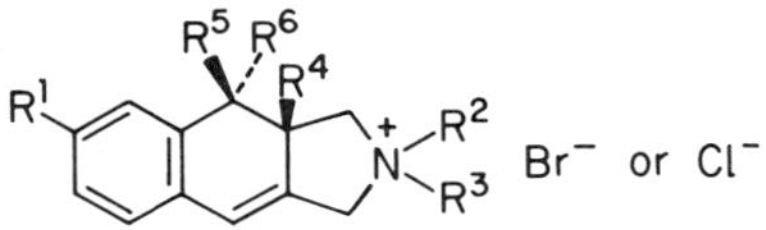

TABLE XIVe. ALLENYLAROMATICS; CHAIN CONTAINING QUATERNARY NITROGEN; VINYL DIENOPHILES (*Continued*)

Entry No.	Substrate						Conditions	Product(s) and Yield(s)(%)	Refs.
	R^1	R^2	R^3	R^4	R^5	R^6			
18	H	CH_3	CH_3	H	C_6H_5	H	H_2O, 100°, 12 hr	(100)	153,495
19	H	CH_3	$C_6H_5CH_2$	H	H	H	", 100°	(0)	378
	CH_3, R^3, R^2, $\overset{+}{N}(R^1)_2$ Br⁻ or Cl⁻ *[a]							CH_3, $\overset{+}{N}(R^1)_2$ Br⁻ or Cl⁻	
	R^1	R^2	R^3						
20	CH_3	Cl	H				H_2O, 25°, 70 hr	(68)	369
21	CH_3	H	Cl				", ", 3 days, then heat	(62)	367,335
22	C_2H_5	H	Cl				", 38–40°, 3 hr, then heat	(30)[c]	491

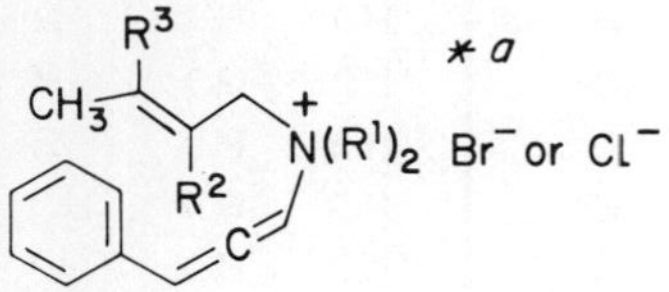

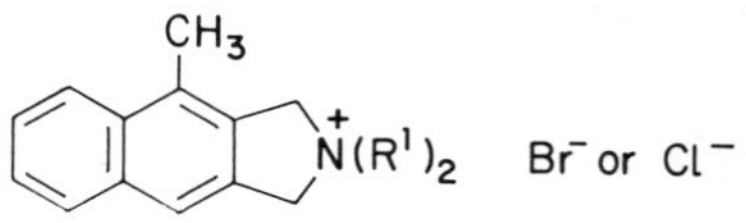

[a] The substrate was generated by base-catalyzed isomerization of the corresponding phenylpropargylammonium salt.
[b] The product was a mixture of isomers.
[c] The product was the tertiary amine.
*This intermediate is hypothetical, or has not been isolated.

TABLE XIVf. VINYL- AND ALLENYLAROMATICS; CHAIN CONTAINING OXYGEN

Entry No.	Substrate: R^1	R^2	R^3	R^4	X	Conditions	Product(s) and Yield(s) (%): A	B	Refs.
	*a								
1	H	H	H	H	H_2	THF, 20°, 4 hr	(20)	(0)	23[b]
						t-C_4H_9OH, 25°, 12 hr	(0)	(77)	381
2	H	H	H	H	O	$(CH_3CO)_2O$, reflux	(0)	(2)	205
3	—OCH_2O—		H	H	O	", "	(0)	(3)	205
4	CH_3O	H	CH_3O	H	O	", "	(0)	(10)	205
5	H	H	H	$CH_2{=}C(CH_3)$	H_2	t-C_4H_9OH, 25°, 16 hr	(0)	(55)[c]	381
6	H	H	H	C_6H_5	H_2	THF, 0°, 15 min	(54)	(0)	23[b]
						t-C_4H_9OH, 62–63°	(0)	(97)	484

TABLE XIVf. VINYL- AND ALLENYLAROMATICS; CHAIN CONTAINING OXYGEN (*Continued*)

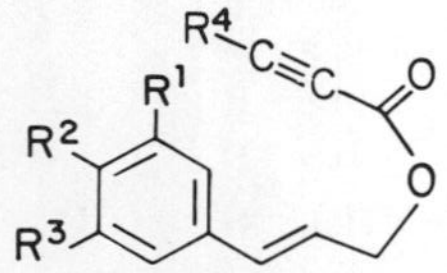

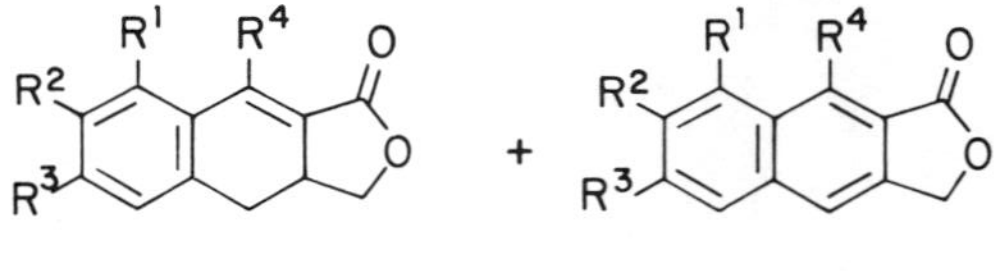

Entry No.	Substrate						Conditions	Product(s) and Yield(s)(%)		Refs.
	R^1	R^2	R^3				$(CH_3CO)_2O$, reflux	A	B	
7	H	H	H					(8)[d]	(0)	205
8	H	—OCH_2O—						(0)	(13)	205
9	CH_3O	H	CH_3O					(0)	(5)	205
	R^1	R^2	R^3	R^4	R^5	R^6		A	B	
10	H	H	H	H	H	H*[e]	DMF, reflux, 5 hr	(51)	(0)	290,496
11	H	H	H	H	CH_3O	H	$(CH_3CO)_2O$, reflux, 5–6 hr	(32)	(0)	124

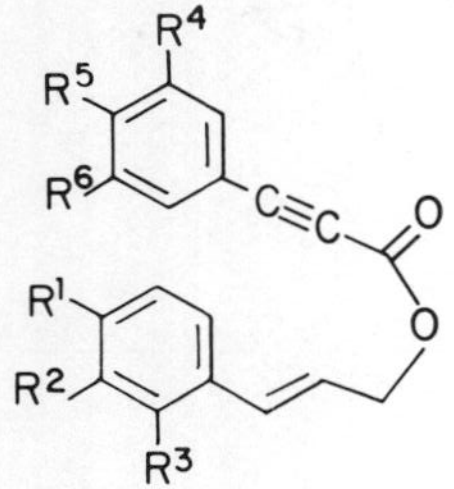

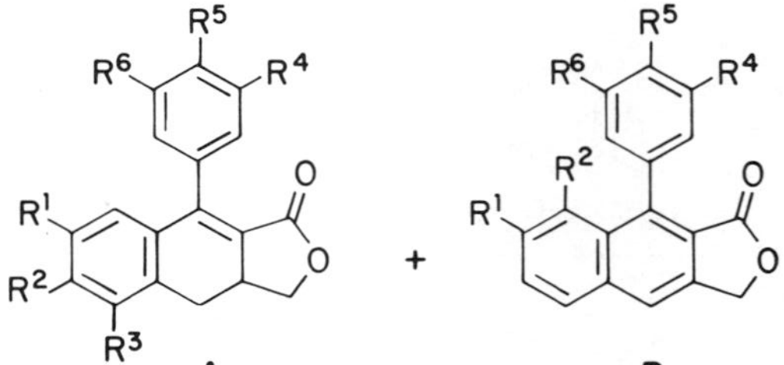

12	CH_3O	CH_3O	H	H	H	H	", ", "	(22)	(0)	124
13	H	H	H	CH_3O	CH_3O	H	", ", "	(43)	(0)	124
14	—OCH_2O—		H	CH_3O	CH_3O	H	$(CH_3CO)_2O$, reflux, 5–6 hr	(14)	(29)[f]	124,291
15	CH_3O	CH_3O	H	—OCH_2O—		H	$(CH_3CO)_2O$, reflux, 11 hr	(9)	(5)	295
16	—OCH_2O—		H	CH_3O	CH_3O	CH_3O	$(CH_3CO)_2O$, reflux, 5–6 hr	(49)	(7)[f]	124,291
17	CH_3O	CH_3O	H	CH_3O	CH_3O	H*[e]	DMF, reflux, 5 hr	(18)	(0)	290
18	—OCH_2O—		H	CH_3O	CH_3O	CH_3O*[e]	", ", "	(29)	(0)	290
19	—OCH_2O—		CH_3O	CH_3O	CH_3O	CH_3O	$(CH_3CO)_2O$, reflux, 4.5 hr	(20)	(0)	293
20	H	H	H	CH_3O	$C_6H_5CH_2O$	H	$(CH_3CO)_2O$, reflux, 10 hr	(33)	(0)	292
21	—OCH_2O—		H	H	$C_6H_5CH_2O$	H	DMF, reflux, 8 hr	(31)	(0)	294
22	$C_6H_5CH_2O$	CH_3O	H	CH_3O	CH_3O	H	$(CH_3CO)_2O$, reflux, 10 hr	(35)	(0)	292
23	$C_6H_5CH_2O$	CH_3O	H	CH_3O	CH_3O	CH_3O	$(CH_3CO)_2O$, reflux 10 hr	(39)	(0)	292
24	$C_6H_5CH_2O$	CH_3O	H	CH_3O	$C_6H_5CH_2O$	H	$(CH_3CO)_2O$, reflux, 10 hr	(23)	(0)	292

TABLE XIVf. VINYL- AND ALLENYLAROMATICS; CHAIN CONTAINING OXYGEN (*Continued*)

Entry No.	Substrate	Conditions	Product(s) and Yield(s)(%)	Refs.
	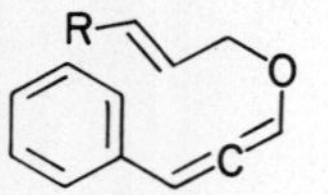 *g		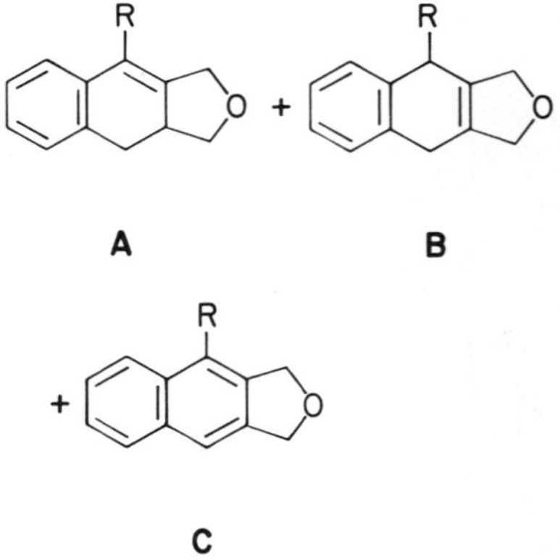	
	R		A B C	
25	H	t-C_4H_9OH, 55°, 22 hr	(55) (28) (0)	381
26	C_6H_5	", ", 16 hr	(52) (0) (30)	381
27	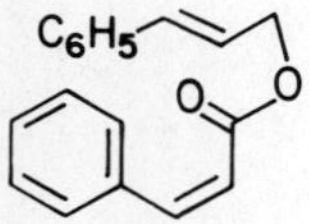 * h,i	DMF, reflux, 35 hr	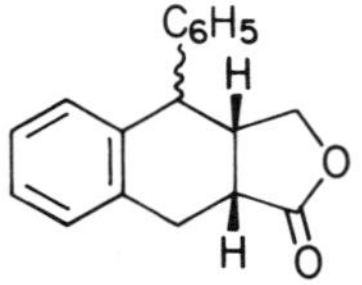 j (31)	122

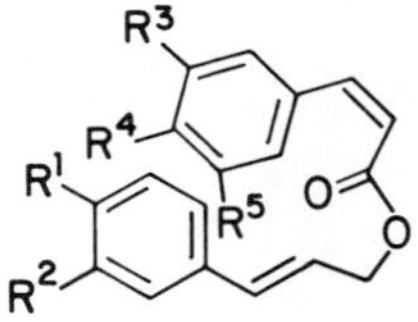

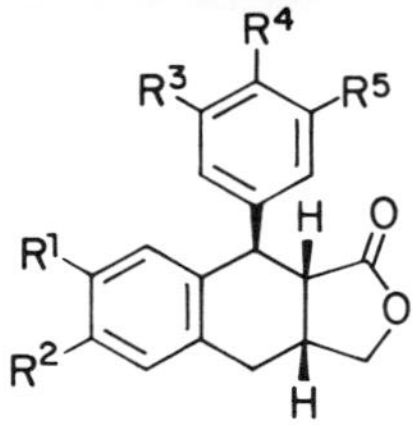

No.	R^1	R^2	R^3	R^4	R^5	Conditions	Product (Yield, %)	Refs.
28	$—OCH_2O—$		CH_3O	CH_3O	H	$(CH_3CO)_2O$, reflux, 24 hr	(43)	124
29	$—OCH_2O—$		CH_3O	CH_3O	CH_3O	$(CH_3CO)_2O$, reflux, 24 hr	(20)	124
30							(0)	497

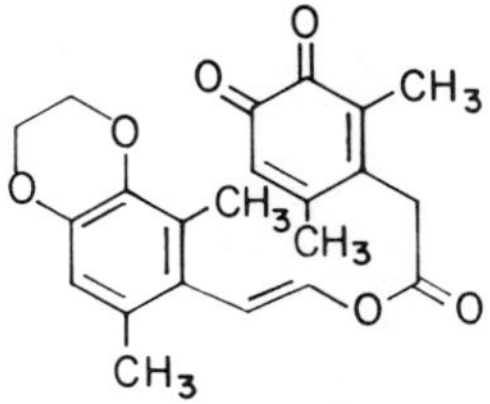

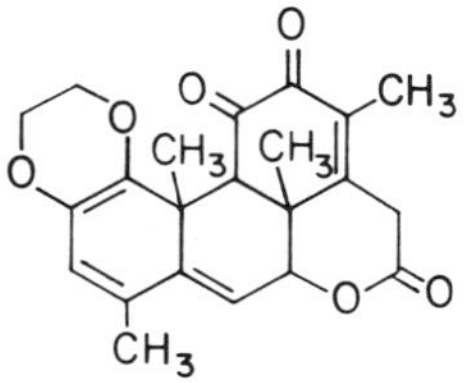

[a] The substrate was generated by base-catalyzed rearrangement of the corresponding phenylacetylene derivative.

[b] The authors consider a stepwise, biradical mechanism to be more likely.

[c] A minor product (11%), tentatively identified as 5-methyl-7-phenyldihydroisobenzofuran, arose from addition of the phenylethynyl group (dienophile) to the isomerized isopropenylethynyl group (diene).

[d] The reaction was carried out under nitrogen.

[e] The substrate was generated *in situ* from the appropriate cinnamyl chloride and sodium arylpropiolate.

[f] The structure assignment is tentative.

[g] The substrate was generated by base-catalyzed isomerization of the corresponding phenylpropargyl ether.

[h] The substrate was generated *in situ* from sodium *cis*-cinnamate and *trans*-cinnamyl chloride.

[i] The corresponding *trans*-cinnamoyl ester did not undergo the intramolecular Diels–Alder reaction.[122,496]

[j] The stereochemistry of the pendant phenyl group was reported to be uncertain.

* This intermediate is hypothetical, or has not been isolated.

TABLE XIVg. ALLENYLAROMATICS; CHAIN CONTAINING SULFUR

Entry No.	Substrate	Conditions	Product(s) and Yield(s)(%)	Refs.
	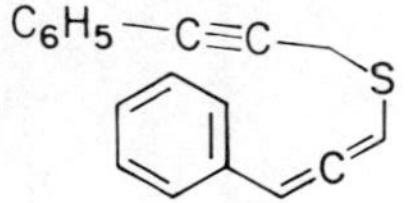 $C_6H_5-C\equiv C-$ *[a]		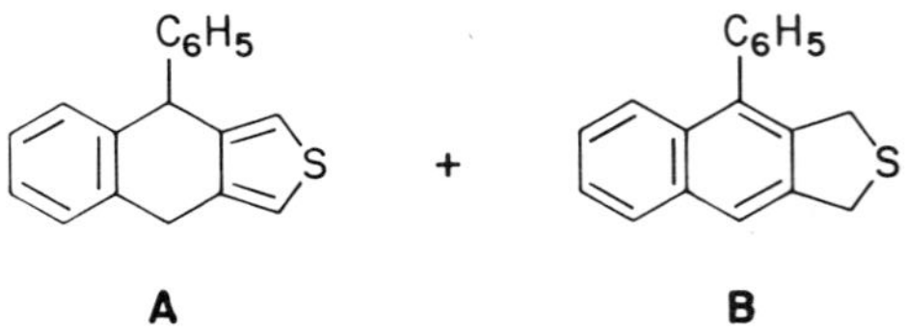 A + B	
			A B	
1		t-C_4H_9OH, 20°, 90 min	(73) (0)	23[b]
		", 62–63°, 4 hr	(0) (40)	484

[a] The substrate was generated by base-catalyzed isomerization of the corresponding phenylpropargyl thioether.
[b] The authors suggest a biradical mechanism for the cyclization.
* This intermediate is hypothetical, or has not been isolated.

TABLE XV. VINYLHETEROCYCLES[a]

Entry No.	Substrate			Conditions	Product(s) and Yield(s) (%)	Refs.
1	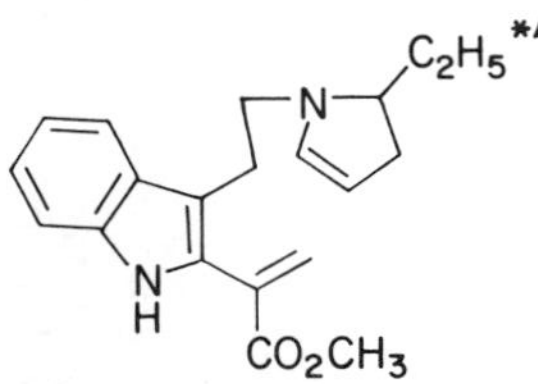 *b			THF, 60°	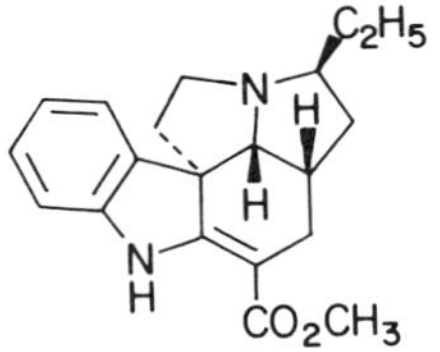 (98)	498
	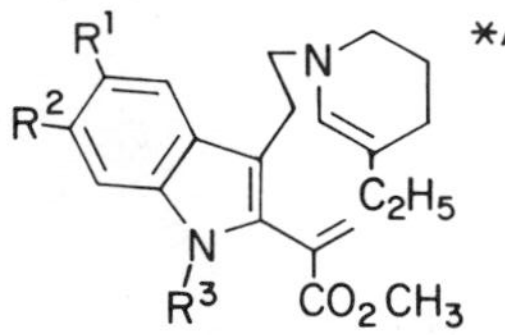 *b				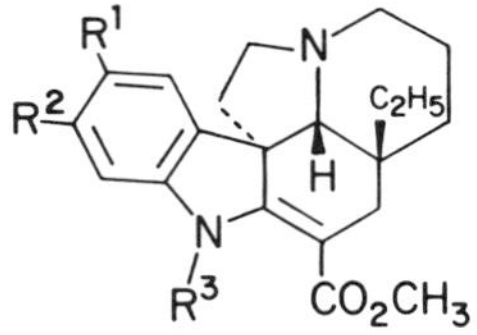	
	R^1	R^2	R^3			
2	H	H	H	Toluene, reflux	(84)	25
3	H	H	H	CH_3OH, $(C_2H_5)_3N$, 40°, 12 hr	(70)	24,26
4	H	H	H	Benzene, 45°, 51 hr	(26)	24
5	H	H	H	DMF, 25°	(70)	24
6	Cl	H	H	CH_3OH, $(C_2H_5)_3N$, 40°, 20 hr	(70)	26
7	Br	H	H	", ", ", "	(51)	26
8	H	H	CH_3	Toluene, reflux	(56)	25
9	CH_3O	H	H	CH_3OH, $(C_2H_5)_3N$, 40°, 20 hr	(82)	26
10	H	CH_3O	H	", ", ", "	(59)	26
11	H	CH_3O	H	Toluene, reflux	(92)	25

TABLE XV. VINYLHETEROCYCLES[a] (*Continued*)

Entry No.	Substrate	Conditions	Product(s) and Yield(s)(%)	Refs.
	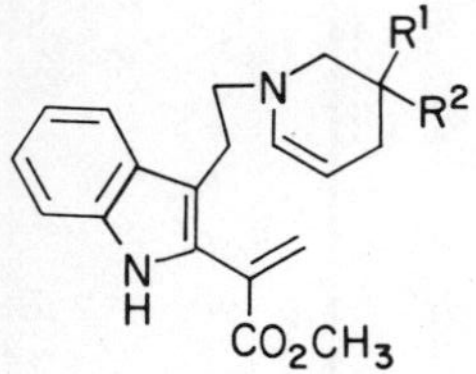 *b (structure: R^1, R^2, CO_2CH_3)		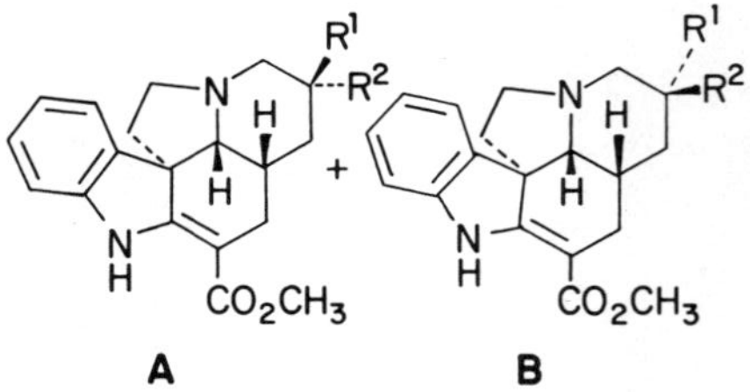 A + B (structures: R^1, R^2, H, CO_2CH_3)	
	R^1 R^2		A B	
12	H C_2H_5	CH_3OH, 40°, 16 hr	(8) (33)	279
13	HO C_2H_5	CH_3OH, 25°, 10 hr, then reflux 1 hr	(32) (32)	279
14	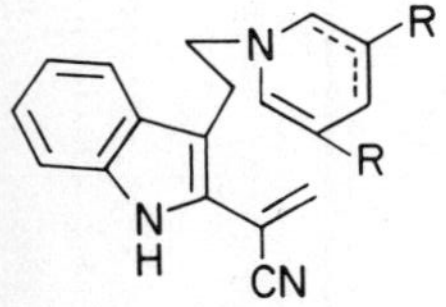 (structure: R, CN) R = CH_3O_2C	CH_2Cl_2, 50°	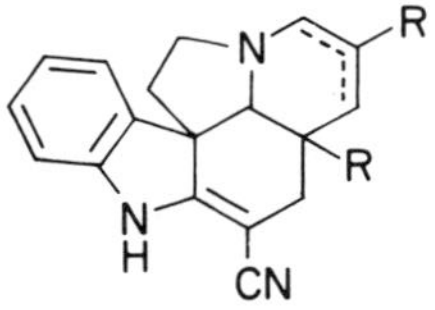 (structure: R, CN) (0)	284

15	[NCH$_2$C$_6$H$_5$, R, NH, CO$_2$CH$_3$ structure] *b R = 2-ethyl-1,3-dioxolan-2-yl-CH_2– (C_2H_5)	CH_3OH, $(C_2H_5)_3N$, reflux, 2 hr	[NCH$_2$C$_6$H$_5$, R, H, NH, CO$_2$CH$_3$ structure]	(82–85)	498
16		C_6H_5Cl, 140°, 24 hr	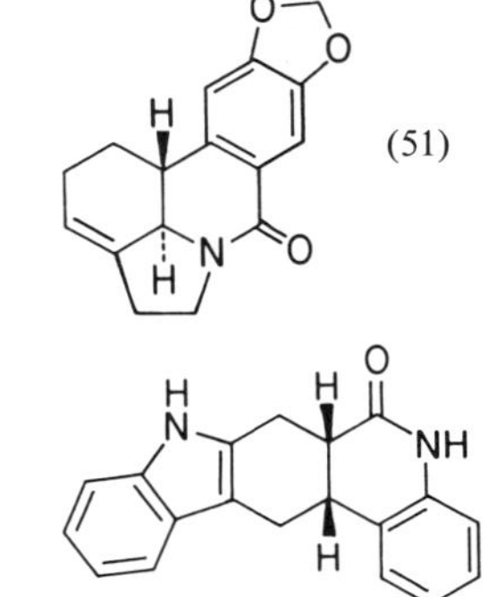	(51)	105
17		210–230°, 1 torr, 15 min (—CO_2)	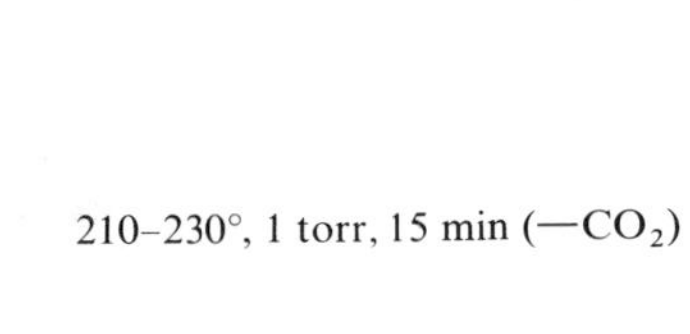	(10)	393

TABLE XV. VINYLHETEROCYCLES[a] (*Continued*)

Entry No.	Substrate	Conditions	Product(s) and Yield(s)(%)	Refs.
18	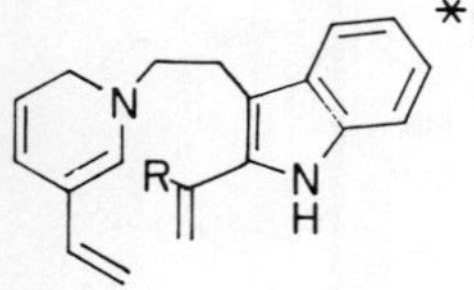 *c R = CH_3O_2C syn TS	$CH_3CO_2C_2H_5$, 100°	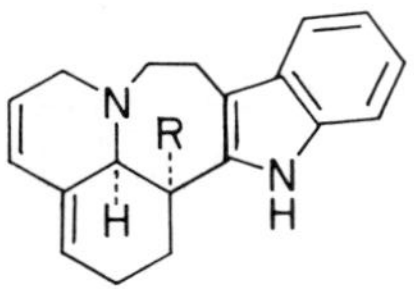 (28)	281
	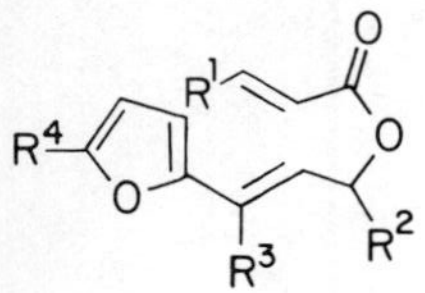 R¹ R² R³ R⁴	Toluene, 200°, 20 hr	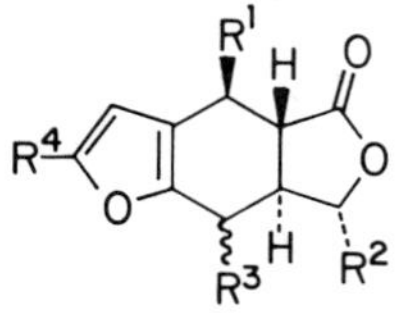	
19	H H H H		(0)	160
20	CH_3O_2C H H H		(73)	160
21	CH_3O_2C CH_3 H H		(56)	160
22	CH_3O_2C H CH_3 H		(46)	160
23	CH_3O_2C H H CH_3		(57)	160

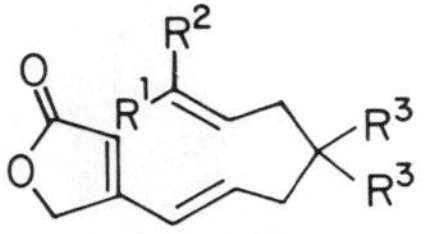

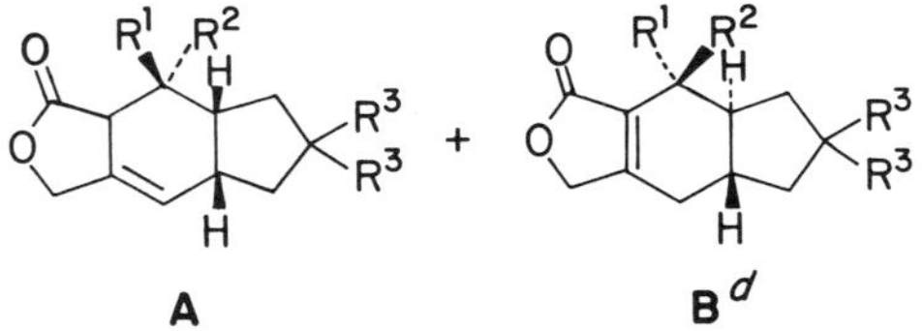

	R^1	R^2	R^3	Conditions	A	B	Ref.
24	NC	H	CH_3	Benzene, 200°, 5 hr	(0)	(—)	87
25	CH_3O_2C	H	H	Toluene, 110°	(0)	(—)	87
26	NC	NC	CH_3	", ", 2.5 hr	(0)	(80)	87
27	CH_3O_2C	CH_3O_2C	H	", "	(0)	(—)	87
28	CH_3O_2C	H	CH_3	Benzene, 155°, 1 hr	(0)	(85)	87
29	CH_3O_2C	NC	CH_3	Toluene, 110°, 4 hr	(0)	(75)	87
30	$t\text{-}C_4H_9O_2C$	NC	H	", "	(0)	(—)	87
31	CH_3O_2C	CH_3O_2C	CH_3	Benzene, 150°, 15 min	(0)	(90)	87
32	$CH_3CO_2CH_2$	CH_3O_2C	CH_3	Toluene, 200°, 0.5 hr	(46)	(46)[e]	37
33	$t\text{-}C_4H_9O_2C$	NC	CH_3	", 110°, 4 hr	(0)	(63)	87
33a	CH_3CO_2 CO_2CH_3 ... CH_3 CH_3				(80)		85

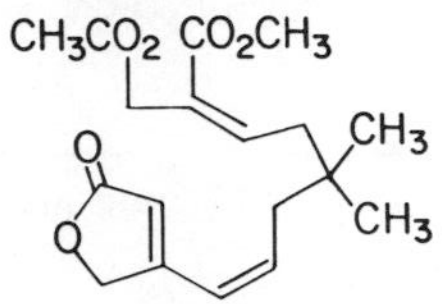

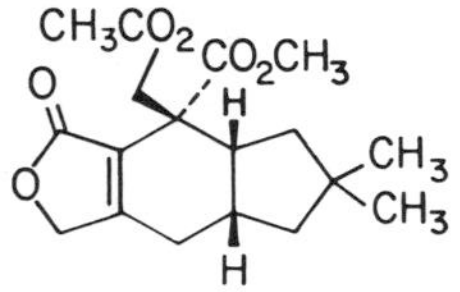

A + B[d]

TABLE XV. VINYLHETEROCYCLES[a] (*Continued*)

Entry No.	Substrate	Conditions	Product(s) and Yield(s)(%)	Refs.
34	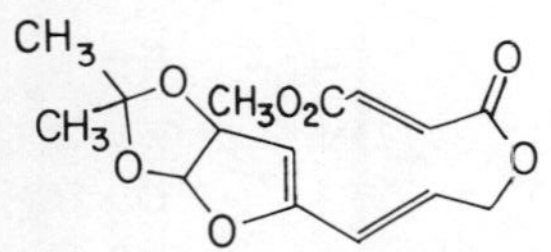 CH_3, CH_3, O, O, CH_3O_2C, O, O, O	Toluene, heat, 3 days	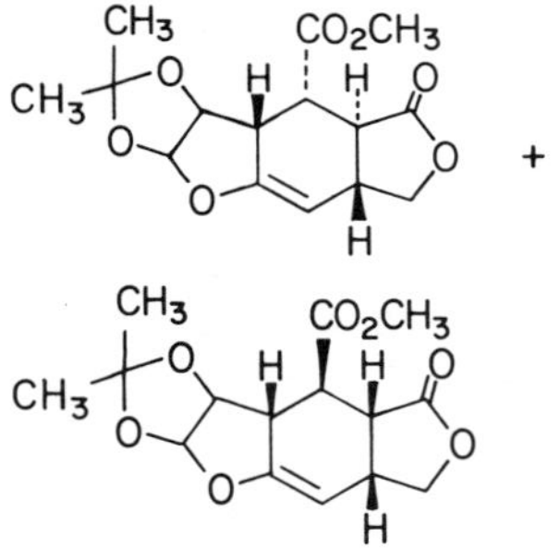 CH_3, CH_3, O, O, O, H, CO_2CH_3, H, O, O, H + (30–40)[f]; CH_3, CH_3, O, O, O, H, CO_2CH_3, H, O, O, H	499
35	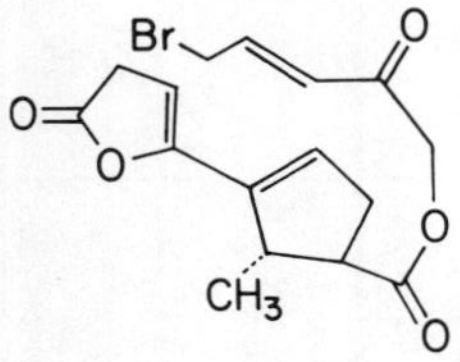 Br, O, O, O, O, CH_3, O	o-$Cl_2C_6H_4$, 190°	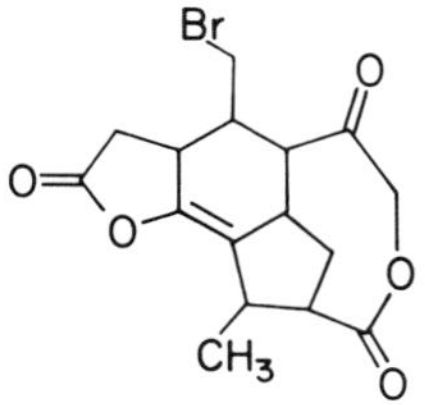 Br, O, O, O, O, O, CH_3, O (0)	500
36	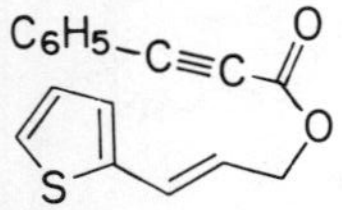 C_6H_5–C≡C, O, O, S	$(CH_3CO)_2O$, reflux, 6 hr	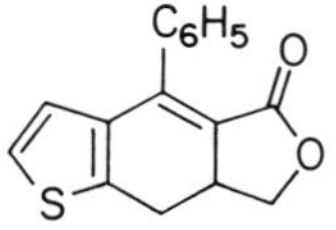 C_6H_5, O, O, S (24)	206

37	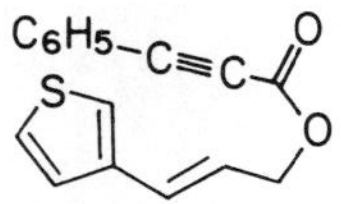 $C_6H_5-C\equiv C-$ (thienyl ester structure)	$(CH_3CO)_2O$, reflux, 6 hr	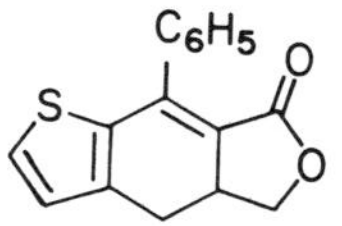 C_6H_5 (thieno-fused lactone structure) (10)	206

[a] Entries are arranged in the order nitrogen-, oxygen-, and sulfur-containing heterocycles.
[b] See original work for the precursor.
[c] The substrate was generated from percondylocarpine acetate; a mechanism involving a Michael-type reaction was also considered.
[d] See original paper for the relative rates of formation of these products.
[e] The initial *trans* adduct was detected spectroscopically, but it isomerized to **B** on attempted isolation.
[f] The combined yield is shown; the ratio of products varied.
* This intermediate is hypothetical, or has not been isolated.

TABLE XVIa. *o*-QUINODIMETHANE

Entry No.	Substrate						
	R^1	R^2	R^3	R^4	R^5	R^6	X
1	H	H	H	H	H	H	H_2 [a, c, aa]
2	H	H	H	H	H	H	O [a]
3	H	CH_3O	H	H	HO	CH_3	H_2 [c]
4	CH_3O	H	NC	CH_3	CH_3	H	H_2 [c]
5	H	CH_3O	H	H	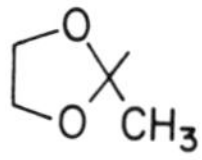 (2-methyl-1,3-dioxolan-2-yl) CH_3	CH_3	H_2 [c]

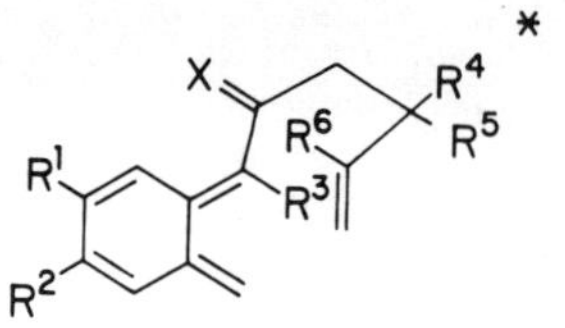

*a

Entry No.	X
6	H_2
7	O

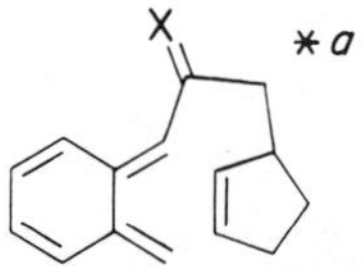

8 *a

9 *c

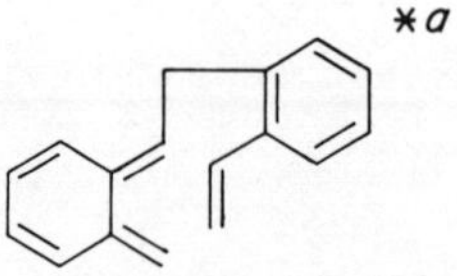

CH_3

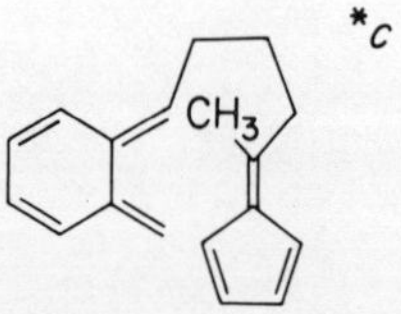

9a See Table XVIIb/4

HAIN CONTAINING CARBON ONLY

ntry o.	Conditions	Product(s) and Yield(s)(%)	Refs.
		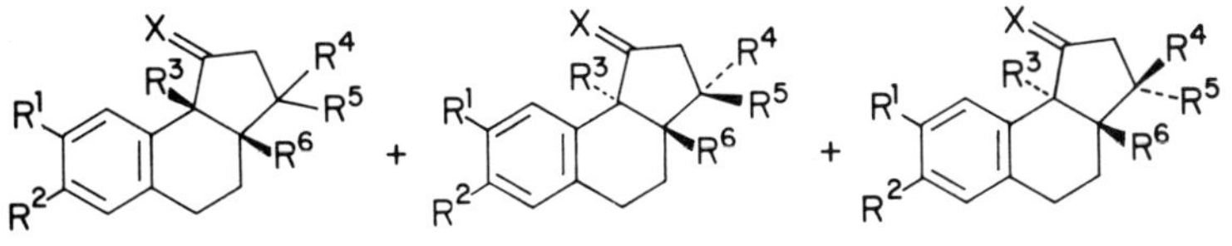 A B C	
		A B C	
	Quartz tube, 300°	(80)[b]	126,127,214,501
	$Cl_3C_6H_3$, 220°	(70) (0) (0)	128
	Toluene, 180°, 6 hr	(0) (43) (43)	125,309
	o-$Cl_2C_6H_4$, 170°, 5 hr	(81)[d]	502
	Toluene, 190–200°, 7 hr	(0) (58)[e] (6)[e]	125
		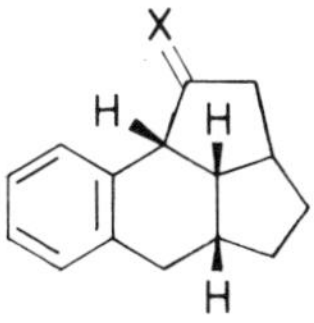	
	240°, 3 hr	(67)[d]	127
	1,2,4-$Cl_3C_6H_3$, 213°, 4 hr	(45)	127
	Diethyl phthalate, 235°, 4 hr	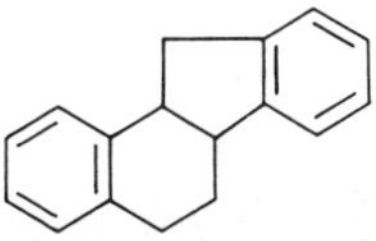 (85)[d]	127
	Toluene, 195°, 12 hr	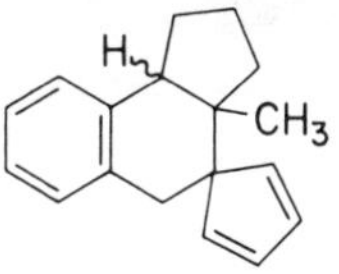 (26)[y]	503

TABLE XVIa. *o*-QUINODIMETHANES

Entry No.	Substrate						
	R^1	R^2	R^3	R^4	R^5	R^6	X
10	H	H	H	H	H	H	H_2 [a, c, f, i, aa]
11	H	H	H	H	H	H	O [a]
11a	H	H	CH_3	H	H	H	H_2 [z]
12	CH_3O	H	H	H	H	H	H_2 [f]
13	H	CH_3O	H	H	H	H	H_2 [a, f]
14	H	NC	H	H	H	H	H_2 [a]
15	$-OCH_2O-$		H	H	H	H	H_2 [f]
16	CH_3O	CH_3O	H	H	H	H	H_2 [f]
17	CH_3O	H	NC	H	H	H	H_2 [c]
18	H	$(C_2H_5)_2NSO_2$	H	H	H	H	H_2 [a]
18a	H	H	H	H	H	$C_2H_5O_2C$	H_2 [a]
19	CH_3O	H	NC	CH_3O_2C	CH_3	H	H_2 [c]
20	H	CH_3O	NC	CH_3O_2C	CH_3	H	H_2 [c]
21	$(CH_3)_3Si$	$(CH_3)_3Si$	H	H	H	H	H_2 [c]

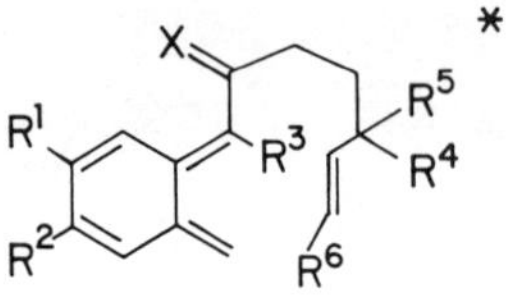

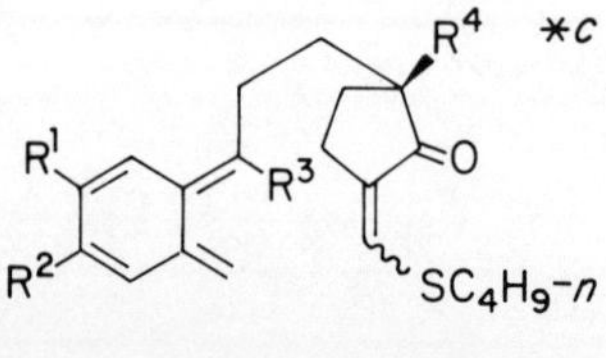

*c

Entry No.	R^1	R^2	R^3	R^4
22	H	CH_3O	H	CH_3
23	CH_3O	H	H	CH_3
24	H	CH_3O	NC	H

CHAIN CONTAINING CARBON ONLY (*Continued*)

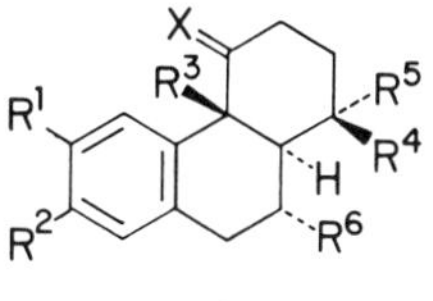

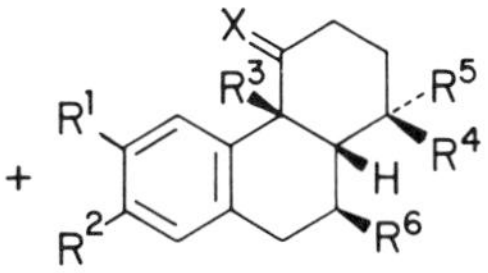

Entry No.	Conditions	Product(s) and Yield(s)(%)		Refs.
		A	B	
	Quartz tube, 300°[a]	(60)	(5)	126,73,127,214,215,501
	$Cl_3C_6H_3$, 220°	(80)[d]		128
a	Benzene, 7°	(24)	(0)	213
	Diethyl phthalate, 300°, 1 hr	(75)	(0)	504,210
	180°	(90)	(0)	210
	1,2,4-$Cl_3C_6H_3$, 213°, 3 hr	(95)	(0)	301
	Diethyl phthalate, 300°	(52)	(0)	210
	", ", "	(60)	(0)	210
	o-$Cl_2C_6H_4$, reflux, 6 hr	(80)[d]		208
	$Cl_3C_6H_3$, 213°, 3 hr	(98)[d]		505
a	Quartz tube, 300°	(69)	(6)	126
	Toluene, 180–230°, 3 hr	(0)	(40–50)	272,288
	", ", "	(0)	(52)	272
	", 230°, 8 hr	(14)[g]	(47)[h]	506
	Decane, reflux, 30 hr	(97)	(<5)	73
	o-$Cl_2C_6H_4$, 180°, 13 hr	(65) ($R^3\alpha^j$)		270
	", ", 6 hr	(78) ($R^3\alpha^j$)		270
	", ", "	(71) ($R^3\beta^j$)		507

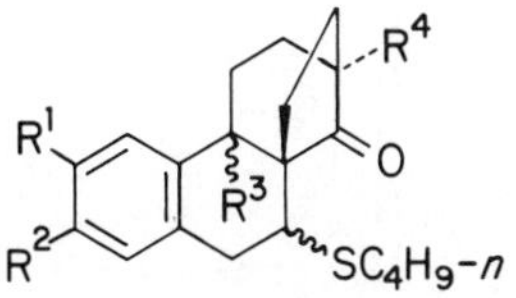

TABLE XVIa. *o*-QUINODIMETHANE

Entry No.	Substrate 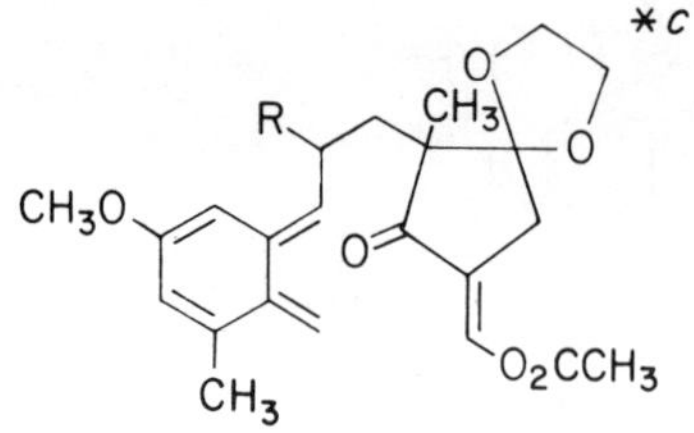

*c

anti TS

Entry No.	R
25	H
26	$C_6H_5CH_2O$

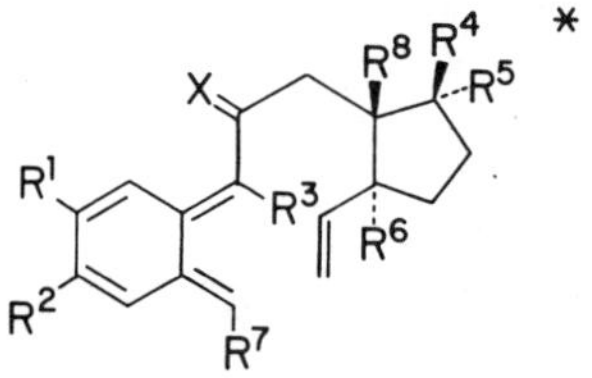

*

Entry No.	R^1	R^2	R^3	R^4	R^5	R^6	R^7	R^8	X
27	H	H	H	O=		H	H	CH_3	H_2[a]
28	H	CH_3O	H	O=		H	H	CH_3	H_2[m]
29	H	CH_3O	H	O=		H	H	CH_3	O[c,k]
30	H	CH_3O	H	O=		H	H	CH_3	H, HO[n]
31	H	CH_3O	H	O=		HO	H	CH_3	H_2[c]
32	H	CH_3O	HO	O=		H	H	CH_3	H_2[l]
33	H	CH_3O	H	HO	H	H	H	CH_3	H_2[c]
34	H	H	H	O=		H	CH_3	CH_3	H_2[m]
35	CH_3O	$(CH_3)_3Si$	H	O=		H	H	CH_3	H_2[c]
36	H	CH_3O	H	$—O(CH_2)_2O—$		H	H	$HOCH_2$	H_2[c]
37	$(CH_3)_3Si$	CH_3O	H	O=		H	H	CH_3	H_2[c]
38	H	CH_3O	H	t-C_4H_9O	H	H	H	CH_3	H_2[c,k]
39	H	CH_3O	H	$—O(CH_2)_2O—$		H	H	CH_3O_2C	H_2[c]
40	$(CH_3)_3Si$	$(CH_3)_3Si$	H	O=		H	H	CH_3	H_2[c]
41	H	NC	H	t-C_4H_9—$(CH_3)_2SiO$	H	H	H	CH_3	H_2[a,k]

HAIN CONTAINING CARBON ONLY (*Continued*)

ntry o.	Conditions	Product(s) and Yield(s)(%)		Refs.
	o-$Cl_2C_6H_4$, 180°, 15 hr	(43)		271
	", 190°, 15 hr	(41)		271
		A	B	
	Dibutyl phthalate, 210°, 8 hr	(85)	(0)	126
	CH_3CN, reflux, 1.5 hr	(86)	(0)	297
	Decane, 150°, 12 hr	(56)	(5)	129
	Diglyme, 27°, 20 hr	(70)[o]	(0)	216
	o-$Cl_2C_6H_4$, 180°, 4 hr	(45)	(0)	299
	Methylcyclohexane, 98°	(major)	(minor)	212,302[k]
	o-$Cl_2C_6H_4$, 200°, 7.5 hr	(78)	(0)	303
	CH_3CN, reflux, 1.5 hr	(95)	(0)	297
	Decane, reflux	(11)	(0)	217,218
	o-$Cl_2C_6H_4$, reflux	(75)	(0)	304
	Decane, reflux	(22)	(0)	218,508
	o-$Cl_2C_6H_4$, 180°, 3 hr	(84)	(0)	298
	—	(60)	(15)	304
	Decane, reflux, 20 hr	(95)	(0)	218,508
	1,2,4-$Cl_3C_6H_3$, 213°, 3 hr	(80)	(0)	301

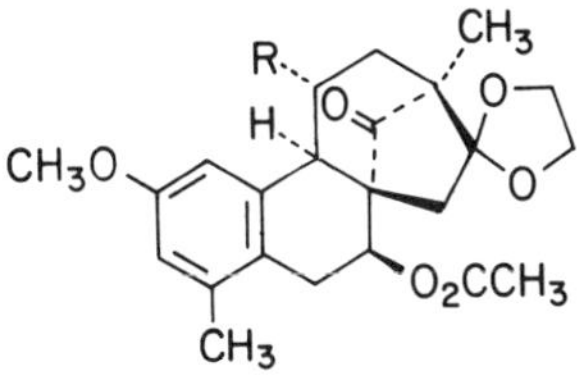

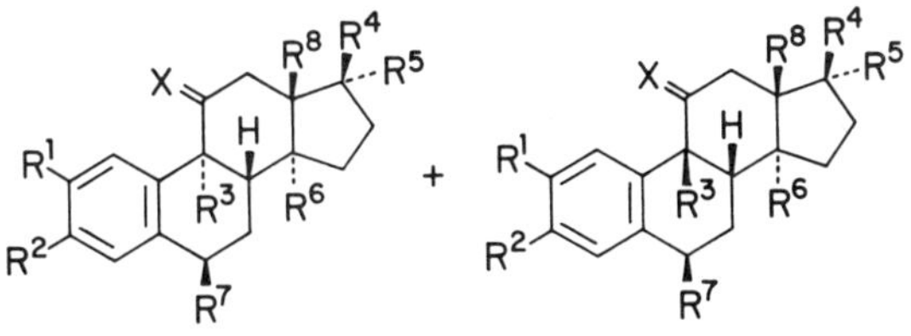

TABLE XVIa. *o*-QUINODIMETHANES

Entry No.	Substrate								
	R^1	R^2	R^3	R^4	R^5	R^6	R^7	R^8	X
42	H	CH_3O	H	O=		H	H	CH_3	$C_6H_5CH_2ON$
43	$(CH_3)_3Si$	$(CH_3)_3Si$	H	—$O(CH_2)_2O$—		H	H	CH_3	H_2[c]
44	H	CH_3O	H	O=		HO	H	CH_3	$C_6H_5CH_2O$, H

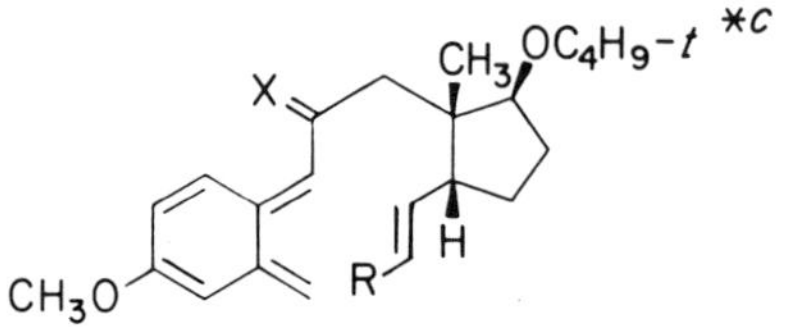

R = *p*-$CH_3C_6H_4SO_3$

syn TS

Entry No.	X
45	H_2
46	O

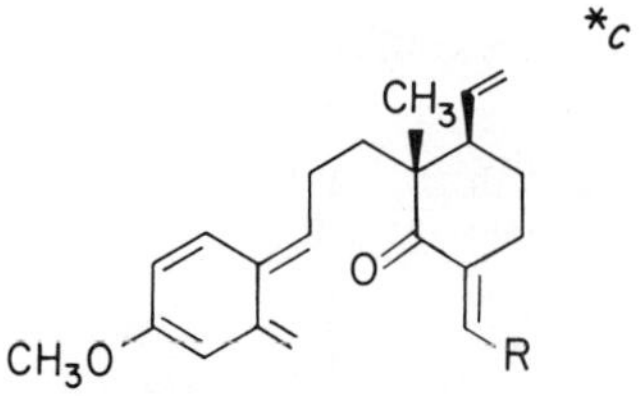

Entry No.	R
47	*n*-C_4H_9S
48	C_6H_5

48a

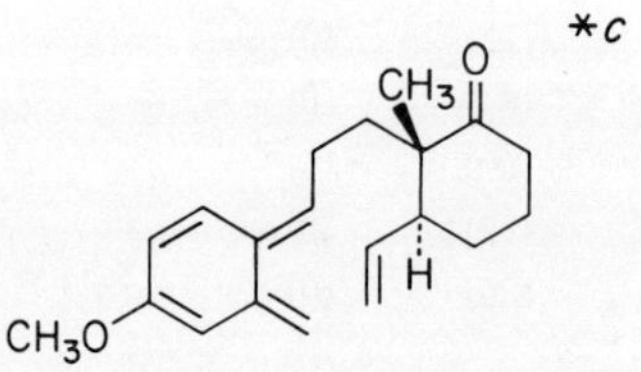

HAIN CONTAINING CARBON ONLY (*Continued*)

ntry No.	Conditions	Product(s) and Yield(s)(%)		Refs.
2	o-$Cl_2C_6H_4$, 140°, 6 hr	(0)	(50)	130
3	Decane, 180°, 20 hr	(81)	(4)[e]	217,218
4	o-$Cl_2C_6H_4$, 180°, 6 hr	(37)[p]	(0)	300
	o-$Cl_2C_6H_4$, reflux, 6 hr			
5		(62)		131
6		(—)		131
7	o-$Cl_2C_6H_4$, 180°, 13 hr	(38)		305
8	", ", "	(17)		305
8a	o-$Cl_2C_6H_4$, 180°, 4 hr	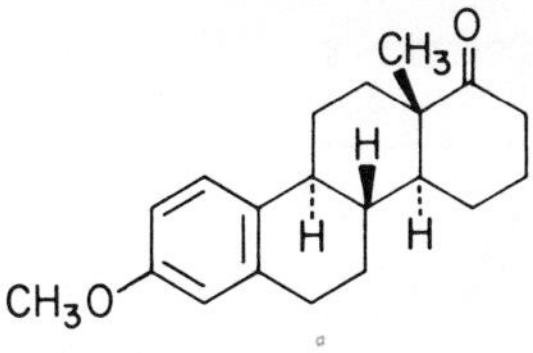 (95)		305

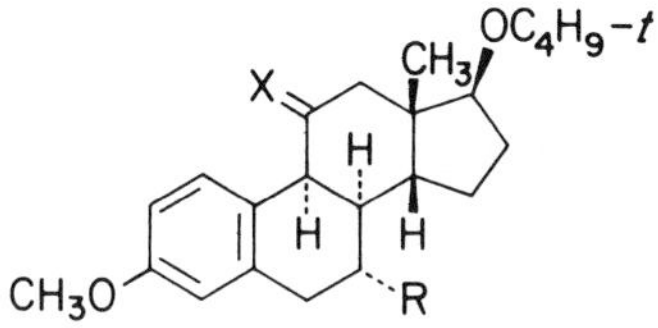

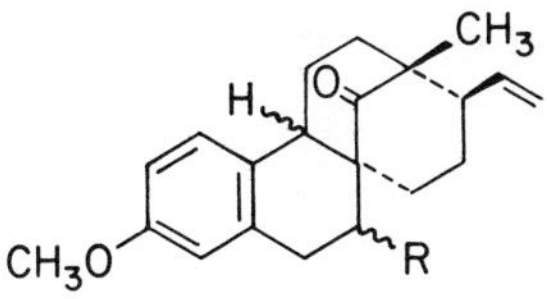

TABLE XVIa. *o*-QUINODIMETHANE

Entry No.	Substrate
	*c
	R^1 / R^2 / R^3 / R^4 / R^5
49	H / H / HO / H / H[q]
50	NC / O_2N / H / CH_3CO_2 / H
51	NC / O_2N / H / —$O(CH_2)O$—
	*c, q
	R^1 / R^2 / X
52	HO / H / O
53	O= / O
54	CH_3CO_2 / H / O[r]
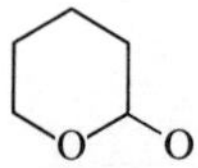 55	(tetrahydropyran-2-yl)O / H / O
56	$C_6H_5CH_2O$ / H / H, $C_6H_5CH_2O$
	*c
	R^1 / R^2 / R^3 / R^4
57	CH_3 / CH_3 / H[v] / C_2H_5
58	C_2H_5 / CH_3 / H[v] / CH_3
59	CH_3 / H / CH_3[w] / C_2H_5
60	C_2H_5 / H / CH_3[w] / CH_3

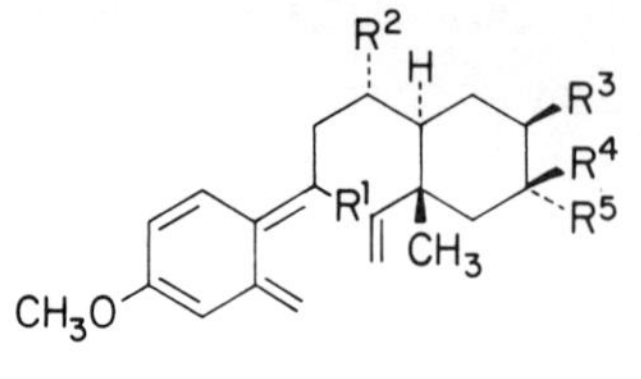

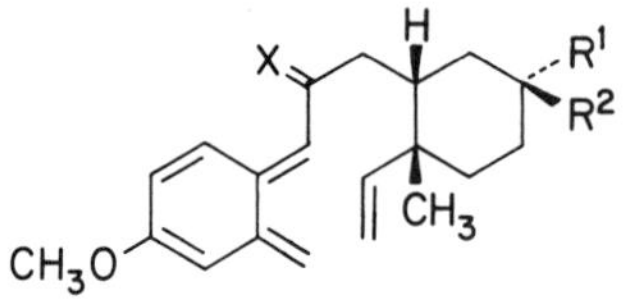

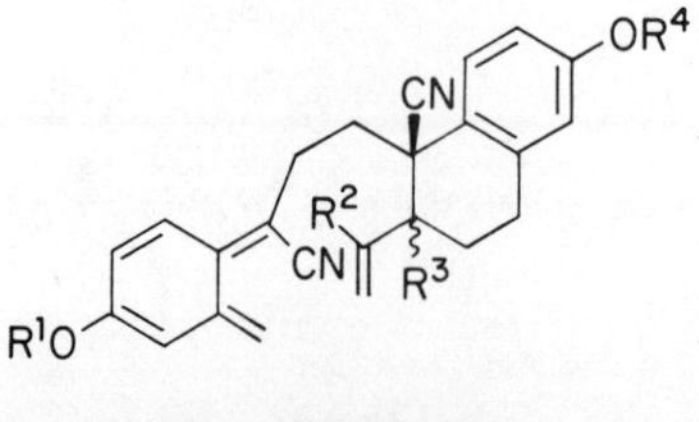

HAIN CONTAINING CARBON ONLY (*Continued*)

try o.	Conditions	Product(s) and Yield(s)(%)		Refs.
		A	B	
	o-$Cl_2C_6H_4$, 195°, 4.5 hr	(93)	(0)	306
	", 180°, 2 hr	(0)	(94)	307,308
	", ", "	(0)	(96)	307
	o-$Cl_2C_6H_4$, reflux, 1 hr	(0)[r]		310
	", ", 0.5 hr	(46)[s]		310
	", ", 0.75 hr	(43)		310
	", ", 1 hr	(0)[r]		310
	", ", 4 hr	(65)[u]		310
	Toluene, 210–215°, 3 hr	(60)[v]		269
		(58)[v]		269
		(82)[w]		269
		(59)[w]		269

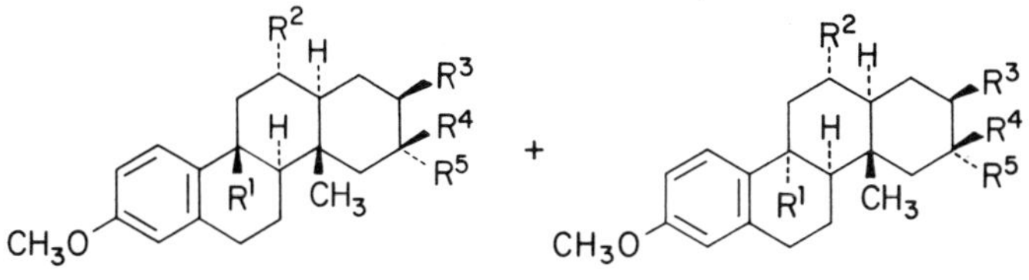

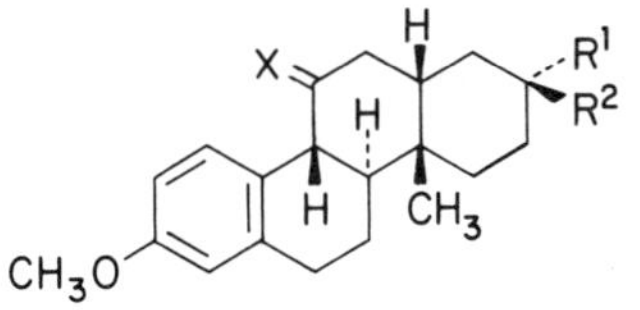

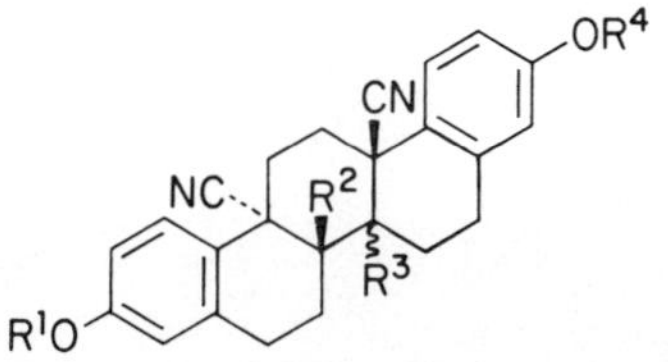

TABLE XVIa. *o*-QUINODIMETHANES

Entry No.	Substrate
61	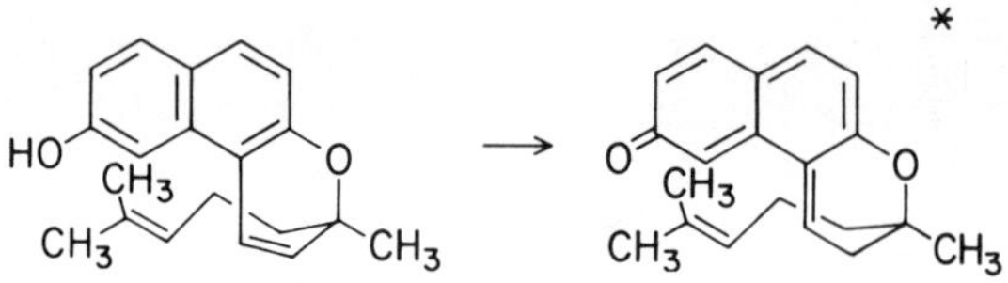 *
62	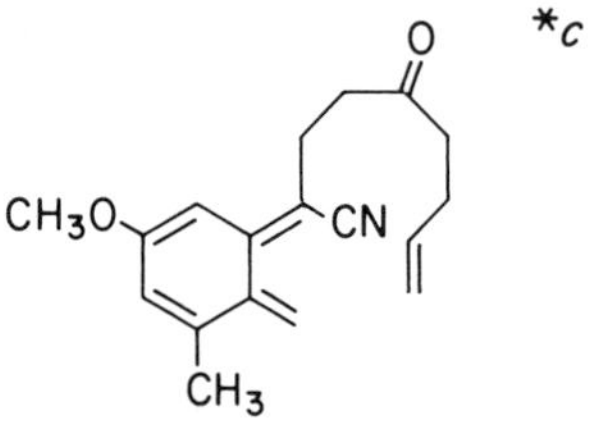 *[c]

[a] The substrate was generated by elimination of SO_2 from the corresponding 1,3-dihydroisothionaphthene-2,2-dioxide.

[b] The product was a mixture of two isomers in a ratio of 3:1; the stereochemistry was not determined.

[c] The substrate was generated from the benzocyclobutene.

[d] The stereochemistry was not reported.

[e] The structure assignment is tentative.

[f] The substrate was generated by elimination of CO_2 from the corresponding 4-(5-hexenyl)isochroman-3-one.

[g] Isomer **A** with R^4/R^5 interchanged was formed in 3% yield.

[h] Isomer **B** with R^4/R^5 interchanged was formed in 34% yield.

[i] The substrate was generated from [structure] and $(n\text{-}C_4H_9)_4NF$; **A** was formed exclusively in 70% yield.

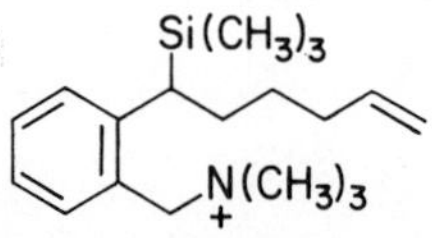

[j] The stereochemistry of the $n\text{-}C_4H_9S$ group was not reported.

[k] The reaction was carried out with a chiral cyclopentane ring corresponding to the natural steroid.

[l] The substrate was generated by irradiation (>340 nm) of the corresponding 3-methyl-4-acylanisole.

[m] The substrate was generated by fluoride-catalyzed 1,4 elimination from the corresponding *o*-(α-trimethylsilylalkyl)benzyltrimethylammonium iodide.

HAIN CONTAINING CARBON ONLY (*Continued*)

ntry o.	Conditions	Product(s) and Yield(s)(%)	Refs.
	230°, 1 hr	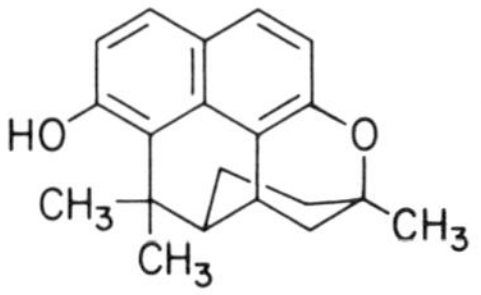 HO, O, CH_3, CH_3, CH_3 (27)	509
	o-$Cl_2C_6H_4$, reflux, 6 hr	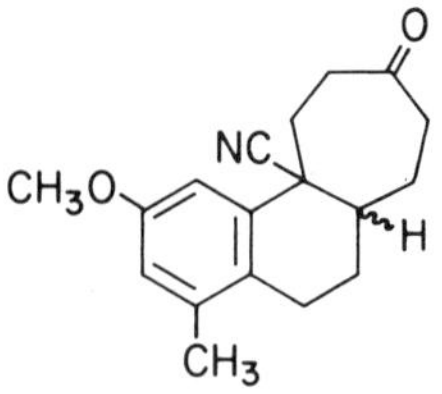 O, NC, CH_3O, H, CH_3 (55)[x]	273

[n] See original work for the precursor,
[o] The hydroxyl group in X is α.
[p] The benzyl group is α.
[q] The substrate contained a chiral cyclohexane ring.
[r] The hemiketal was formed in high yield.
[s] The *cis*-fused product was formed in 28% yield.
[t] The corresponding ethylene ketal failed to undergo the intramolecular Diels–Alder reaction.
[u] The $C_6H_5CH_2O$ group in X is β.
[v] R^3 is α.
[w] The substrate and product were mixtures of R^3 α- and β-isomers.
[x] The stereochemistry of the product was not specified but appears to be *cis*, judged from the structure of a bsequent product.
[y] A [6 + 4]cycloadduct was formed in 44% yield.
[z] The substrate was generated by photolysis of 2-(2-methylphenyl)-1,7-octadiene.
[aa] The substrate was generated from the appropriate o-methylbenzyl ether by lithium tetramethylpiperidide-duced 1,4 elimination.
*This intermediate is hypothetical, or has not been isolated.

TABLE XVIb. *o*-QUINODIMETHANES; CHAIN CONTAINING NITROGEN

Entry No.	Substrate	Conditions	Product(s) and Yield(s) (%)	Refs.
1	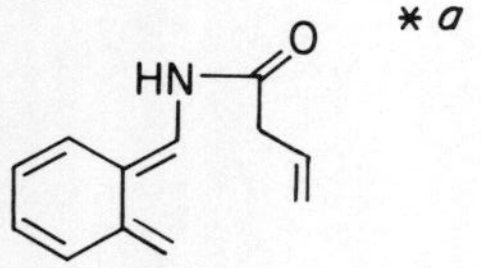 *a	Toluene, reflux, 16 hr	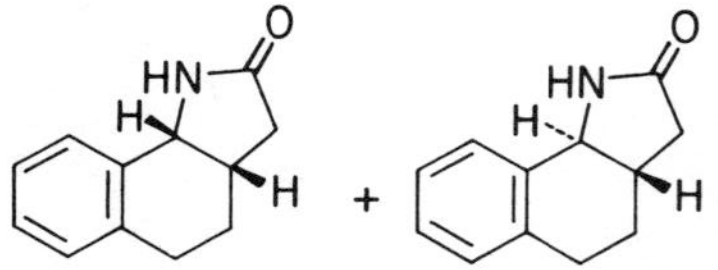 (80) + (17)	207,9
2	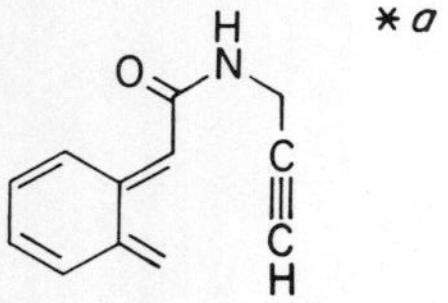 *a	C_6H_5Br, reflux, 16 hr	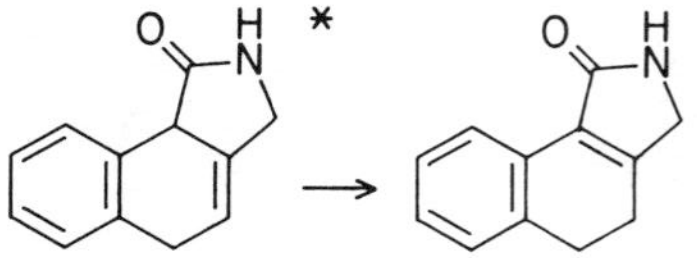 * → (95)	219
3	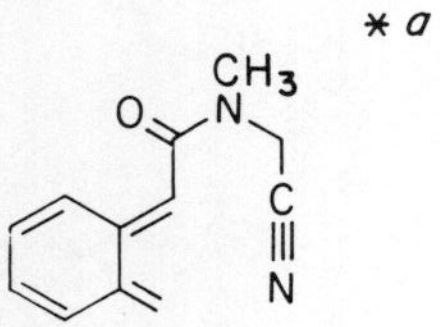 *a	Toluene, 180°, 16 hr	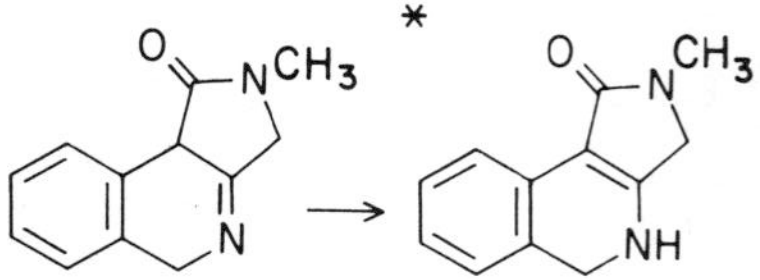 * → (76)	177
4	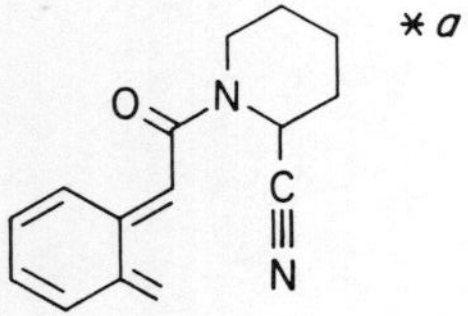 *a	155°	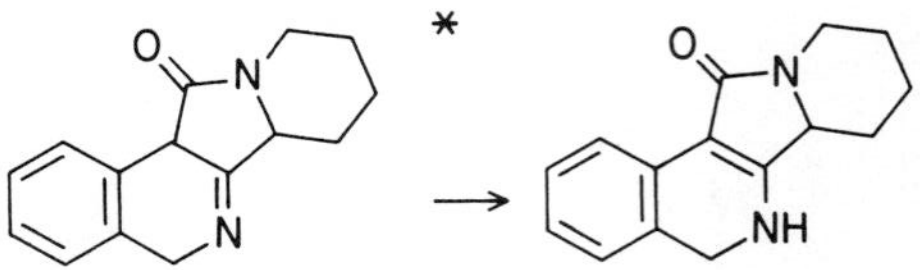 * →	177

No.	Reactant	Conditions	Product(s) (Yield, %)	Refs.
5	* a; R = $C_2H_5O_2C$	155°	* → (82)	177
	*		A + B	

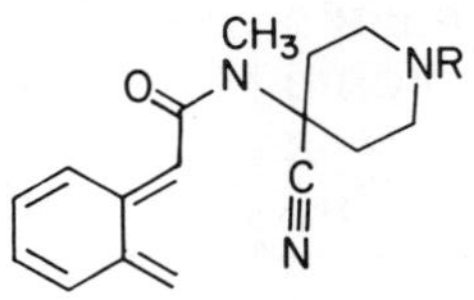

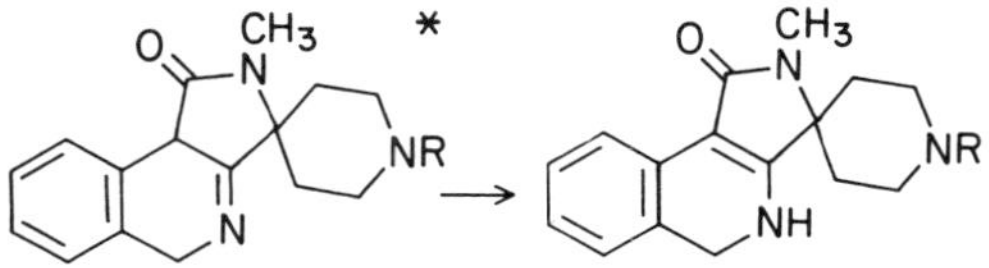

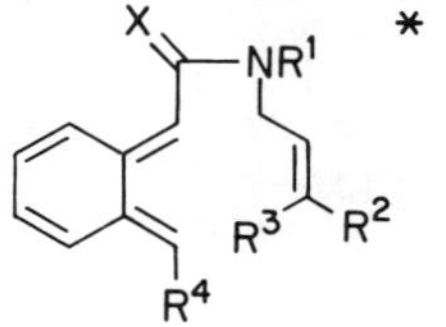

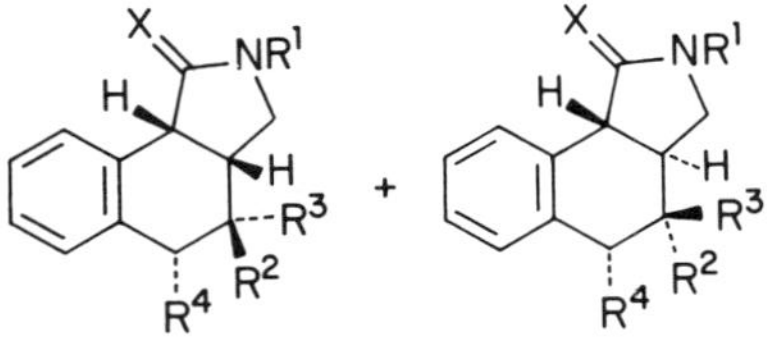

	X	R^1	R^2	R^3	R^4	Conditions	A	B	Refs.
6	O	H	H	H	H[a]	Toluene, 190°, 16 hr	(85)	(—)	219,177,510
7	O	H	Cl	H	H[a]	C_6H_5Br, reflux	(77)[b]		219,9
8	O	H	H	Cl	H[a]	C_6H_5Br, reflux	(73)[b]		219
9	O	CH_3	Cl	H	H[a]	", ", 16 hr	(—)[b]		510
10	O	CH_3	H	Cl	H[a]	", ", 14 hr	(—)[b]		510
11	H_2	$C_2H_5O_2C$	H	H	HO[c]	Toluene, 25°, 4 hr	(52)	(17)	211
12	O	CH_3	C_6H_5	H	H[a]	C_6H_5Br, reflux, 17 hr	(56)	(9)	510,9

No.	Reactant	Conditions	Product(s) (Yield, %)	Refs.
13	*; R = $C_6H_5CH_2$	Toluene, 250°, 8 hr	(64)	219

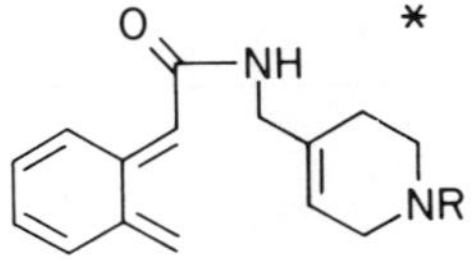

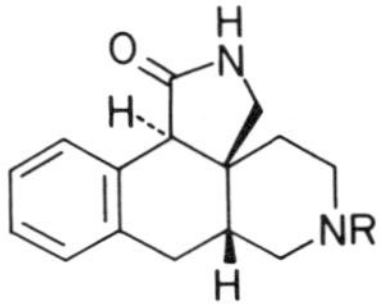

TABLE XVIb. *o*-QUINODIMETHANES; CHAIN CONTAINING NITROGEN (*Continued*)

Entry No.	Substrate				Conditions	Product(s) and Yield(s)(%)	Refs.
	*a						
	n						
14	1				o-$Cl_2C_6H_4$, p-$CH_3C_6H_4SO_3H$, reflux, 16 hr	(80)	9,219,510
15	2				155–180°	(65)	9,219,510
	*a						
	R^1	R^2	X	Y		*cis* *trans*	
16	H	CH_3	O	O	C_6H_5Br, reflux, 23 hr	(25)[b]	177,9
17	H	H	CH_3ON	O	", ", 20 hr	(58)[b]	9,177
17a	CH_3O	CH_3O_2C	O	H_2	", ", 24 hr	(0) (41)	511
17b	$C_6H_5CH_2O$	$C_6H_5(CH_2)_2$	O	O	", ", 16 hr	(12) (70)	511

No.	Substrate	Conditions	Product(s) and Yield(s) (%)	Refs.
18	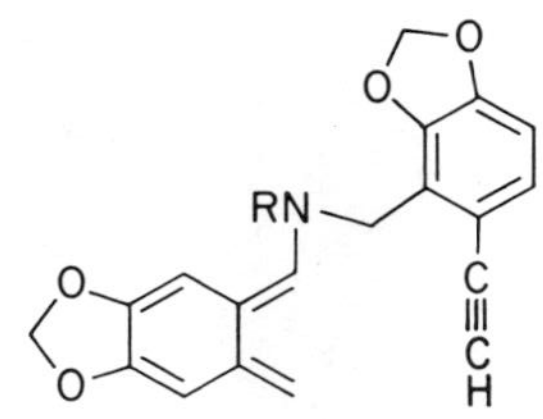 *a; $R = C_6H_5CH_2O_2C$	o-Xylene, 120°, 1 hr	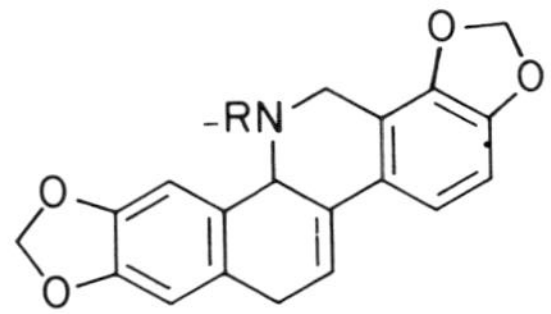 (73)	286
	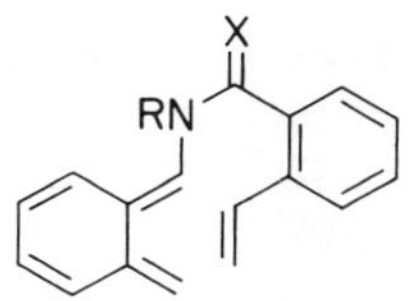 *a		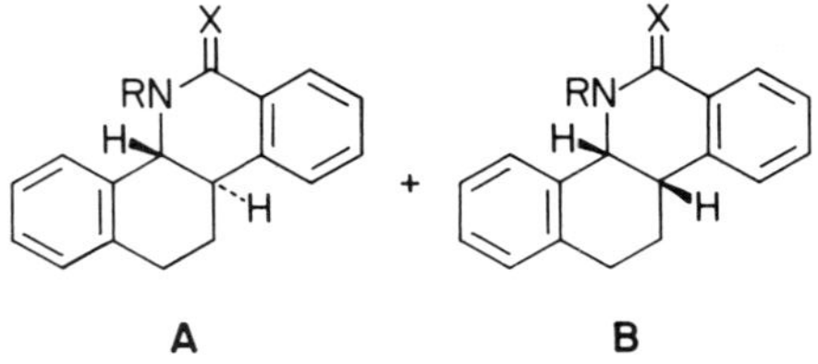 A + B	
	R / X		A / B	
19	H / O	Toluene, 160°, 16 hr	(94) / (0)	9,286,510
20	CH_3O_2C / H_2	C_6H_5Br, reflux, 16 hr	(0) / (78)	9,286,510
21	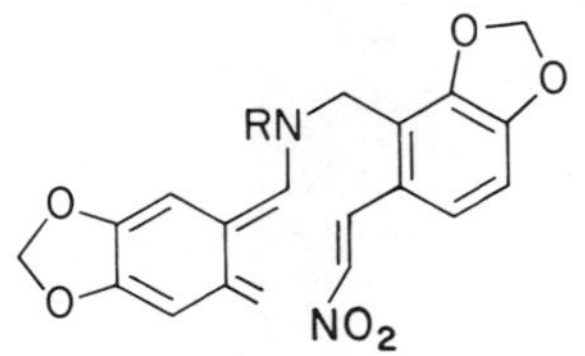 *a; *syn* TS; $R = C_6H_5CH_2O_2C$	Xylene, 120°, 2 hr	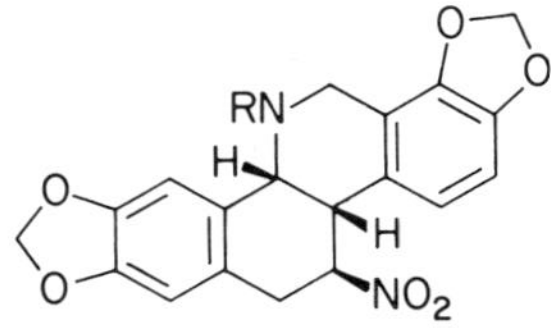 (97)	285

TABLE XVIb. *o*-QUINODIMETHANES; CHAIN CONTAINING NITROGEN (*Continued*)

Entry No.	Substrate					Conditions	Product(s) and Yield(s)(%)		Refs.
	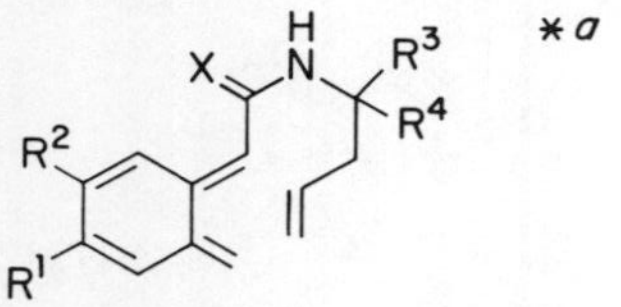 *a						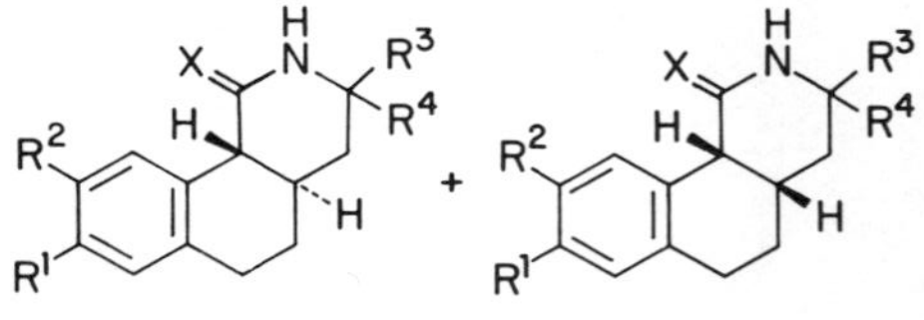 A + B		
	R^1	R^2	R^3	R^4	X		A	B	
22	H	H	H	H	H_2	o-$Cl_2C_6H_4$, reflux, 16 hr	(72)	(27)	512,9
23	H	Cl	H	H	H_2	180°, 16 hr	(98)	(0)	9
24	H	H	H	H	O	o-$Cl_2C_6H_4$, reflux, 16 hr	(0)	(85)	510,9,219,512
25	H	H	O=		H_2	", 180°, 0.5 hr	(31)	(64)	512
26	H	CH_3O	H	H	H_2	180°, 16 hr	(54)	(0)	9
27	H	Cl	CH_3	H	H_2	", "	(75)	(0)	9
28	CH_3O	CH_3O	H	H	H_2	o-$Cl_2C_6H_4$, reflux, 18 hr	(59)	(12)	345,9
29	CH_3O	CH_3O	O=		H_2	", ", "	(0)	(57)	354
30	$C_6H_5CH_2O$	H	H	H	H_2	180°, 16 hr	(75)	(0)	9
31	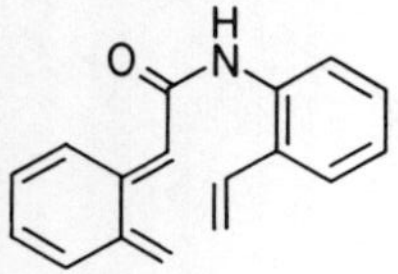 *a					C_6H_5Br, reflux, 12 hr	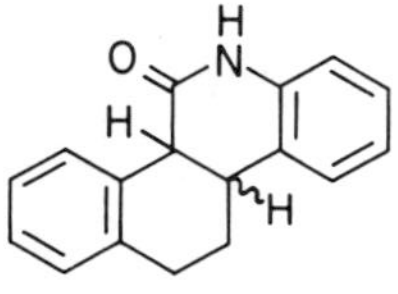 (—)[d]		510

	X	Conditions	Product(s) (%)	Ref.
	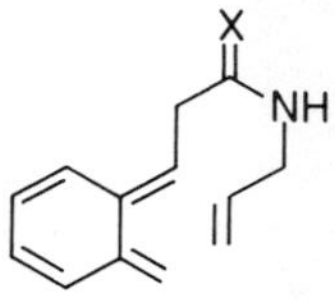 *a		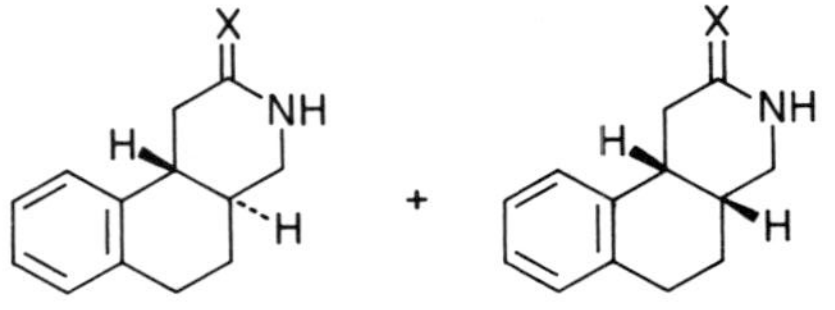 A + B	
	X		A B	
32	H_2	o-$Cl_2C_6H_4$, reflux, 16 hr	(87) (12)	512
33	O	", ", "	(13) (79)	512
34	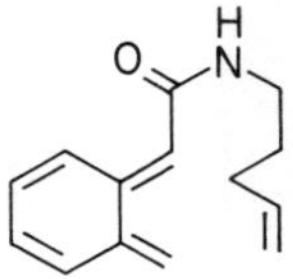 *a	o-$Cl_2C_6H_4$, reflux, 16 hr	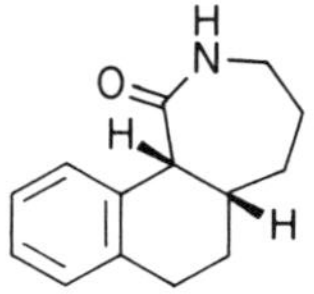 (20)	219

[a] The substrate was generated from the benzocyclobutene.
[b] The product was a mixture of *cis*- and *trans*-fused isomers; the ratio was not determined.
[c] The substrate was generated by irradiation (> 300 nm) of 2-(N-allyl-N-carbethoxy)aminoethylbenzaldehyde.
*This intermediate is hypothetical, or has not been isolated.

TABLE XVIc. *o*-QUINODIMETHANES; CHAIN CONTAINING OXYGEN

Entry No.	Substrate	Conditions	Product(s) and Yield(s)(%)	Refs.
1	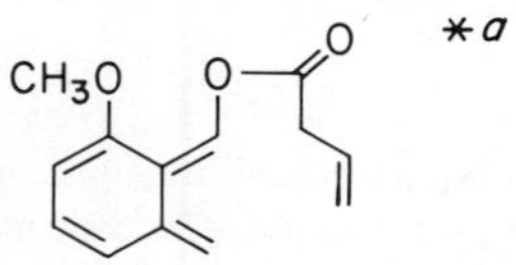 CH_3O, O, O *a	190–200°, 48 hr	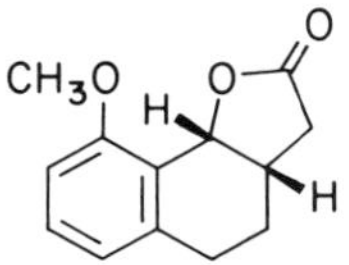 CH_3O, H, O, O, H (61)[b]	513
2	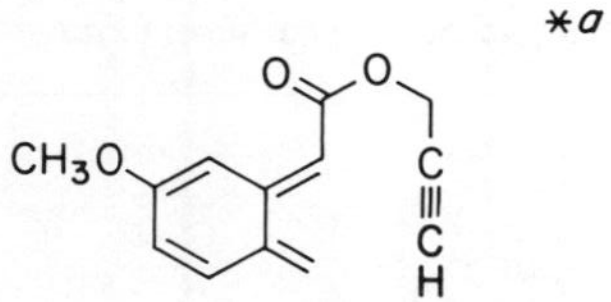 CH_3O, O, O, C≡C, H *a	Toluene, 190–200°, 40 hr	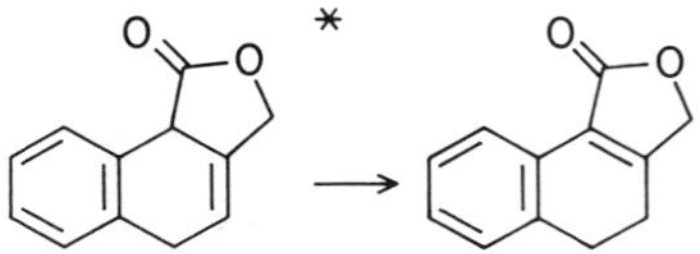 O, O * → O, O (42)	513
3	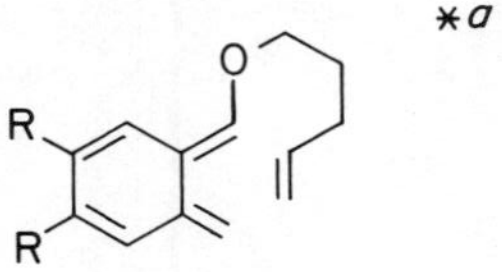 R, R, O *a R = $(CH_3)_3Si$	RC≡CR, reflux	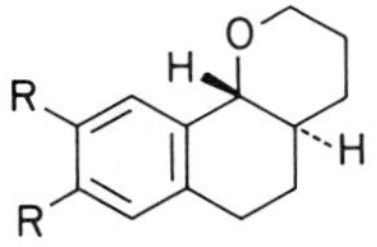 R, R, H, O, H (60)	73
4	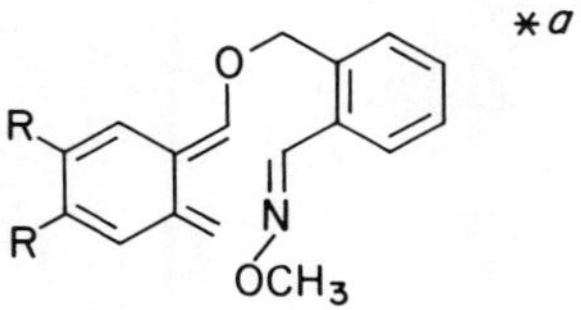 R, R, O, N, OCH_3 *a R = $(CH_3)_3Si$	RC≡CR, reflux	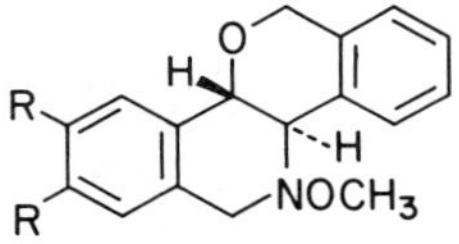 R, R, H, O, H, $NOCH_3$ (45)	73

No.	Substrate	Conditions	Product(s) and Yield(s) (%)	Refs.
5	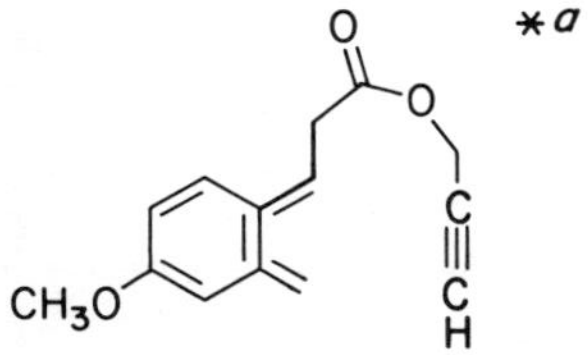 CH_3O (structure) *[a]	Toluene, 190–200°, 40 hr	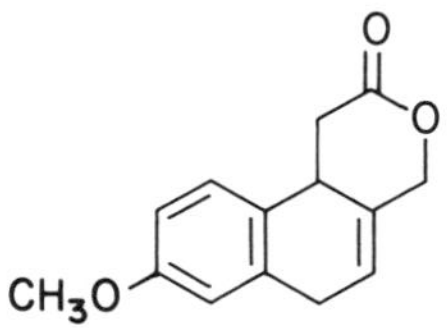 CH_3O (structure) (77)	513
	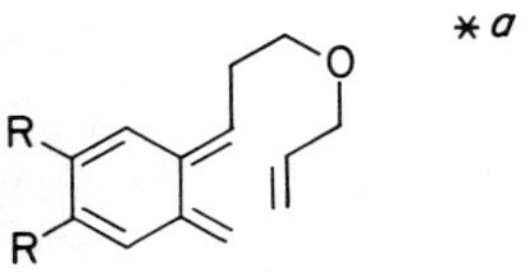 R, R (structure) *[a]		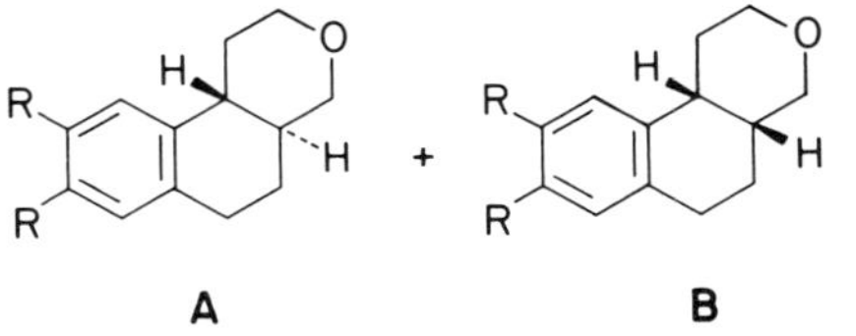 A + B (structures)	
	R		**A** **B**	
6	H	Decane, reflux, 30 hr	(—)[c] (—)[c]	73
7	$(CH_3)_3Si$	", ", "	(74) (19)	73
8	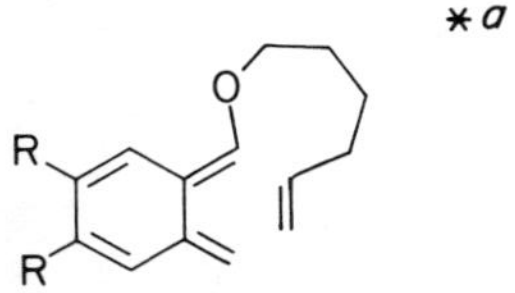 R, R (structure) *[a] R = $(CH_3)_3Si$	RC≡CR, reflux	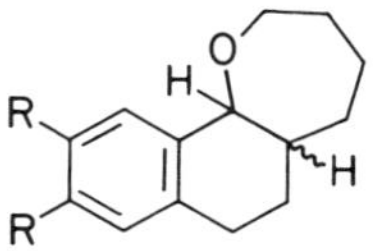 (structure) (65)[d]	73

[a] The substrate was generated from the benzocyclobutene.
[b] The stereochemistry of the product was assigned tentatively.
[c] The isomer ratio was the same as for R = $Si(CH_3)_3$.
[d] The product was a mixture of fused and bridged isomers; see Table XXVIc/19.
* This intermediate is hypothetical, or has not been isolated.

TABLE XVId. *o*-DIMETHYLENEHETEROCYCLES

Entry No.	Substrate	Conditions	Product(s) and Yield(s) (%)	Refs.
	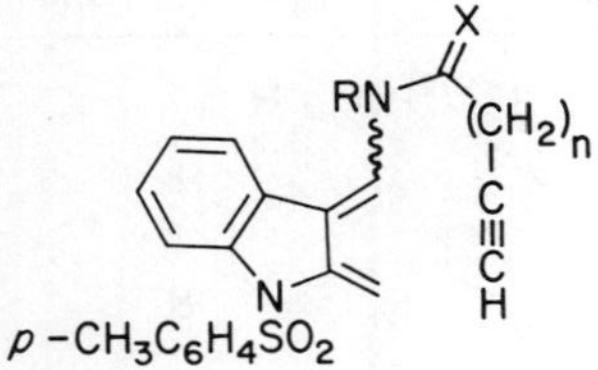 *a; p-$CH_3C_6H_4SO_2$ 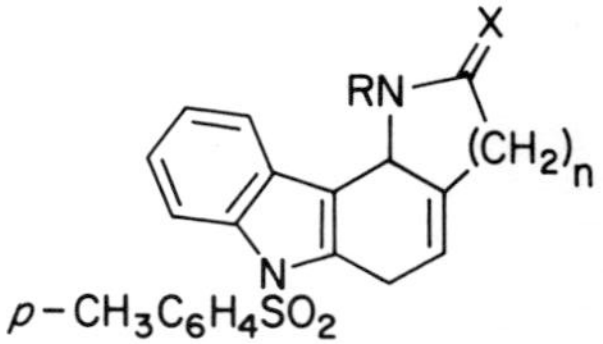		p-$CH_3C_6H_4SO_2$	
	R / X / n			
1	CH_3O_2C / H_2 / 1	C_6H_5Cl, 130°, 1 hr	(17)	514
2	$Cl(CH_2)_2O_2C$ / H_2 / 1	80°, 1 hr	(>95)	514
3	$C_6H_5S(CH_2)_2$ / O / 2	C_6H_5Cl, 130°, 2 hr	(31)	514
	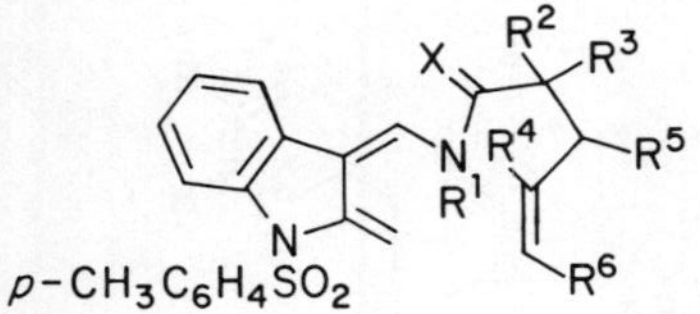 *a, b; p-$CH_3C_6H_4SO_2$ 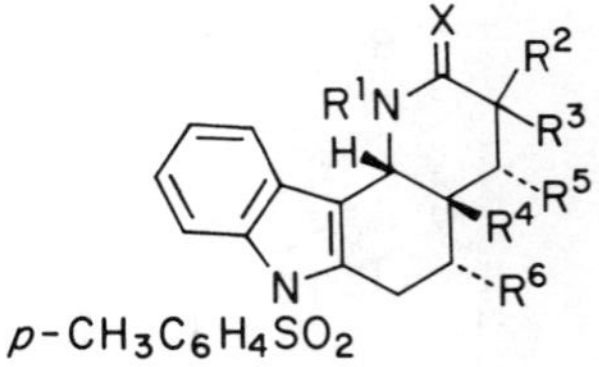		p-$CH_3C_6H_4SO_2$	
	R^1 / R^2 / R^3 / R^4 / R^5 / R^6 / X			
4	CH_3 / H / H / H / H / H / O	C_6H_5Cl, reflux, 24 hr	(40)	515
5	CH_3CO / H / H / H / H / H / H_2	$(CH_3CO)_2O$, 140°, 4 hr	(64)	220
6	CH_3O_2C / H / H / H / H / H / H_2	C_6H_5Cl, 130°, 3.5 hr	(88)	220
7	$C_2H_5O_2C$ / H / H / H / H / H / H_2	Xylene, reflux, 3 hr	(43)	220

No.	R¹	R²	R³	R⁴	R⁵	R⁶	X	Conditions	Product(s) and Yield(s) (%)	Refs.
8	$CH_3O(CH_2)_2$	H	H	H	H	H	O	C_6H_5Cl, reflux, 5 hr	(55)	515
9	$Cl(CH_2)_2O_2C$	H	H	H	H	H	H_2	", 135°, 40 min	(92)	515
10	$Cl_3CCH_2O_2C$	H	H	H	H	H	H_2	", ", 1.25 hr	(79)	515
11	CH_3O_2C	H	H	C_2H_5	H	H	H_2	", ", 3 hr	(54)	515
12	$(CH_3O)_2CHCH_2$	H	H	H	H	H	O	", ", 3.5 hr	(58)	515
13	$Cl(CH_2)_2O_2C$	H	H	C_2H_5	H	H	H_2	", ", 1 hr	(70)	515
14	$Cl_3CCH_2O_2C$	H	H	C_2H_5	H	H	H_2	", ", 10 hr	(46)	515
15	$CH_3O(CH_2)_2$	H	H	H	$-(CH_2)_3-$		O	", ", 5 hr	(48)[c]	515
16	$(CH_3O)_2CHCH_2$	H	H	C_2H_5	H	H	O	", ", 2 hr	(11)	515
17	$CH_3O(CH_2)_2$	H	H	H	$-(CH_2)_4-$		O	", ", 5 hr	(31)[c, d]	515
18	$C_6H_5CH_2$	H	H	H	H	H	O	", reflux, 6 hr	(56)	515
19	$C_6H_5O_2C$	H	H	H	H	H	H_2	Xylene, reflux, 3 hr	(68)	220
20	$CH_3O(CH_2)_2$	$-(CH_2)_5-$		H	H	H	O	C_6H_5Cl, 135°, 4 hr	(38)	515
21	$C_6H_5S(CH_2)_2$	H	H	H	H	H	O	", ", 3 hr	(60)	515
22	$C_6H_5Se(CH_2)_2$	H	H	H	H	H	O	", reflux, 3.5 hr	(22)	515
23	$C_6H_5S(CH_2)_2$	H	H	C_2H_5	H	H	O	", 140°, 2.75 hr	(33)	287,515
24	[structure] *[e]							o-$Cl_2C_6H_4$, reflux, 6 hr	[structure] (100)	516

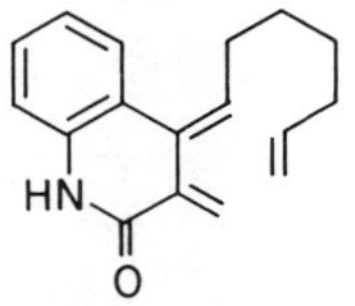

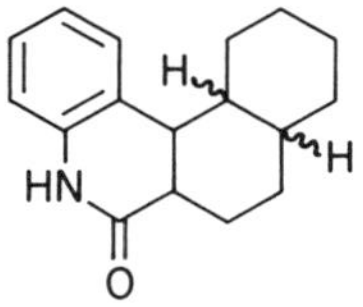

[a] The substrate was generated from a 2-methyl-1H-indole-3-methylenimine and a mixed anhydride or an acid chloride; see original work for details.

[b] See original work for a discussion of possible transition states.

[c] The stereochemical structure assignment was tentative.

[d] R^5 is β.

[e] The substrate was generated from the corresponding cyclobutaquinoline.

* This intermediate is hypothetical, or has not been isolated.

TABLE XVIIa. BENZENES

Entry No.	Substrate	Conditions	Product(s) and Yield(s)(%)	Refs.
	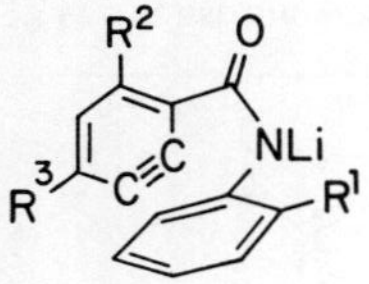 *a	C_6H_{14}, THF, −60° to −20°, then HCl	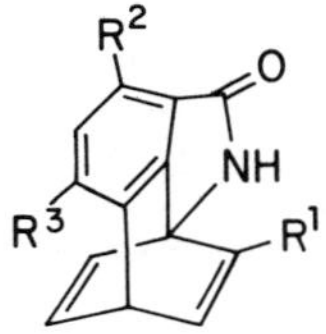	
	R^1 R^2 R^3			
1	H Cl H		(4)	517
2	CH_3 Cl H		(25)	517
3	CH_3 H Cl		(9)	517
	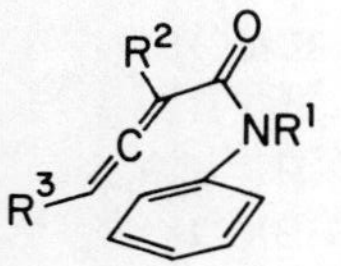 b		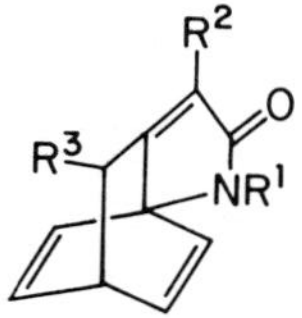	
	R^1 R^2 R^3			
4	CH_3 H H	Xylene, reflux, 9 hr	(90)	222
5	CH_3 CH_3 H	", ", 2.5 hr	(76)	222
6	CH_3 $(CH_3)_3Si$ H	", ", 5 hr	(87)	222
7	C_6H_5 H H	", ", 4 hr	(47)	222
8	C_6H_5 $(CH_3)_3Si$ H	", ", 4 hr	(61)	222
9	CH_3 CH_3 $C_6H_5(CH_3)_2Si$	", ", 4.5 hr	(21)	222

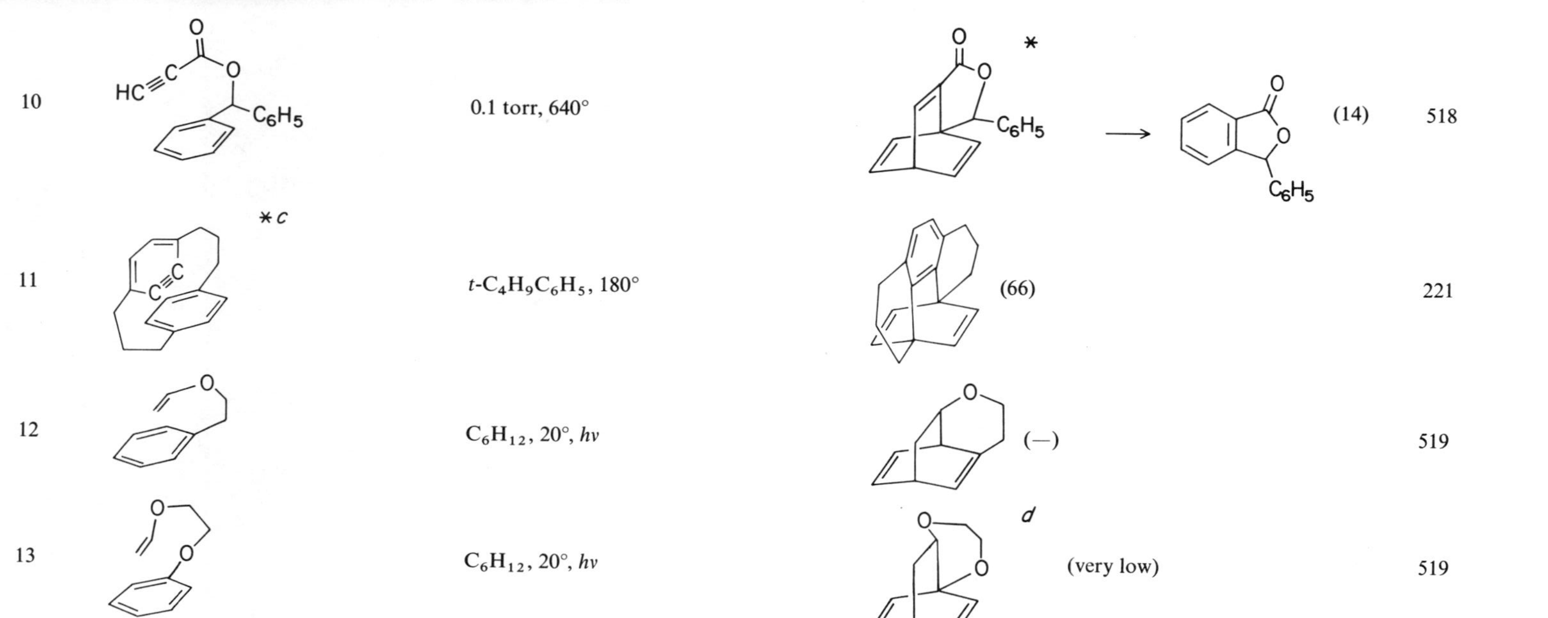

Entry	Conditions	Yield (%)	Ref.
10	0.1 torr, 640°	(14)	518
11	t-$C_4H_9C_6H_5$, 180°	(66)	221
12	C_6H_{12}, 20°, $h\nu$	(—)	519
13	C_6H_{12}, 20°, $h\nu$	(very low)	519

[a] The aryne was generated from the corresponding 2-chlorobenzamide and n-C_4H_9Li.
[b] The reaction is reversible.
[c] The benzyne intermediate was generated from 5-bromo[3.3]paracyclophane and t-C_4H_9OK.
[d] The structure assignment was tentative.
* This intermediate is hypothetical, or has not been isolated.

TABLE XVIIb. NAPHTHALENES

Entry No.	Substrate	Conditions	Product(s) and Yield(s)(%)	Refs.
1	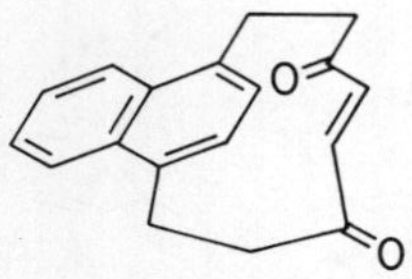 *a	CH_3OH, 0°	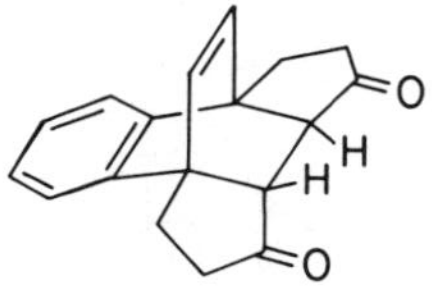 (—)	520
2	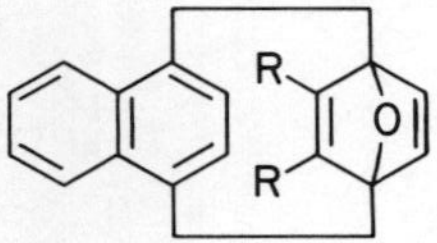 *b R = CH_3O_2C	RC≡CR, 100°, 20 hr	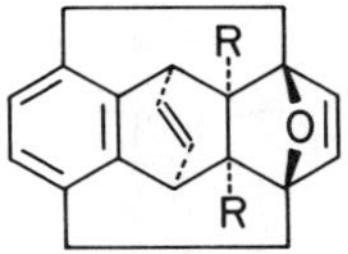 (30)	521
3	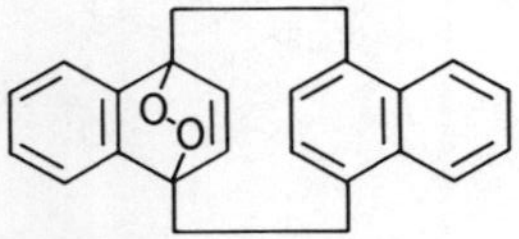 *c	CH_3OH, *hv*, 58°, 10 days	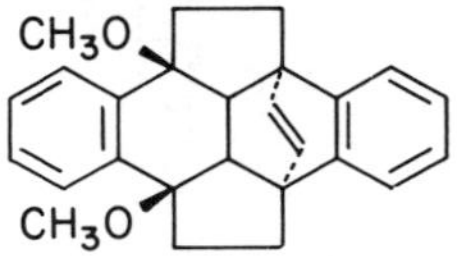 CH_3O, CH_3O (20)	522

No.	Reactant	Conditions	Product(s) and Yield(s) (%)	Refs.
4	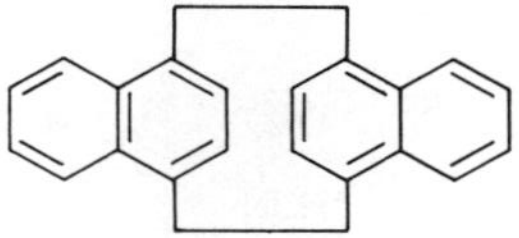	Benzene, CH_3OH, $h\nu$, 10 days	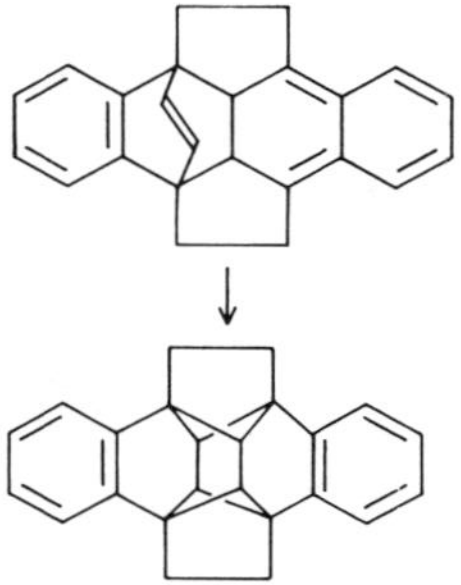 * (50)	523
5	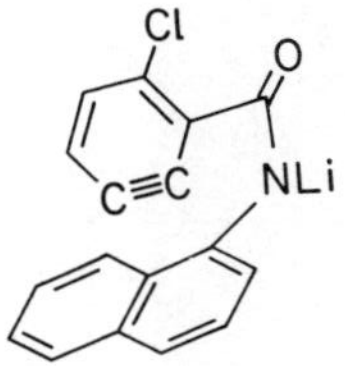 *d 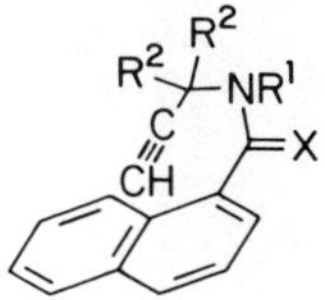	C_6H_{14}, THF, $-60°$ to $-30°$, then HCl	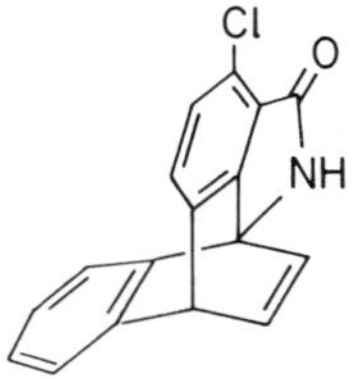 (41) 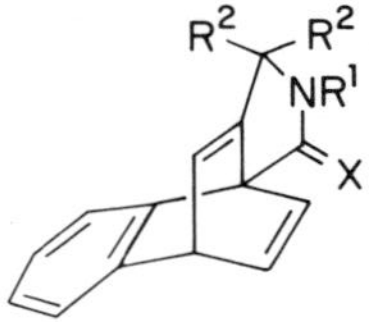	517
	R^1 R^2 X			
6	CH_3 H H_2	Toluene, 210°	(low)	223
7	CH_3 H O	Toluene, 190°, 17 hr	(65)	223
8	H CH_3 O	$C_6H_5N(C_2H_5)_2$, 218°, 3.5 hr	(55)	189

TABLE XVIIb. NAPHTHALENES (*Continued*)

Entry No.	Substrate	Conditions	Product(s) and Yield(s)(%)	Refs.
	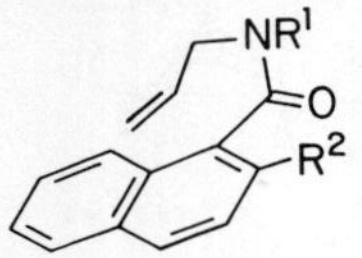 NR[1], O, R[2]		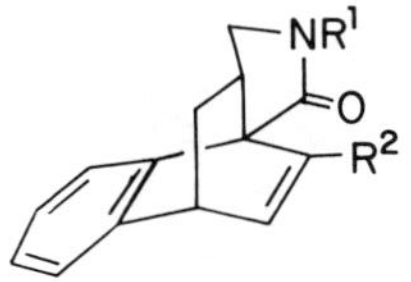 NR[1], O, R[2]	
	R^1 / R^2			
9	CH_3 / H	Toluene, 300°	(0)	223
10	$CH_2{=}CHCH_2$ / HO	$C_6H_5N(C_2H_5)_2$, 218°, 315 hr	$(15)^{e,f}$	189
11	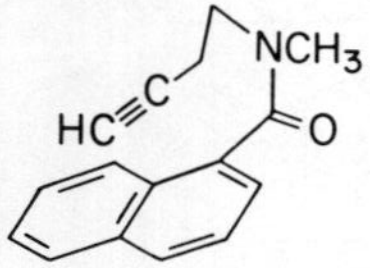 HC≡C, NCH_3, O	Toluene, 270°, 5 hr	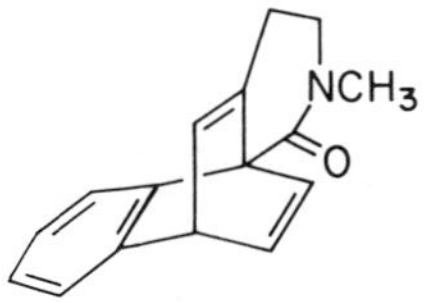 NCH_3, O (19)	223

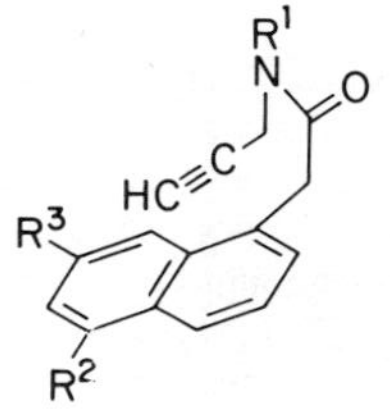

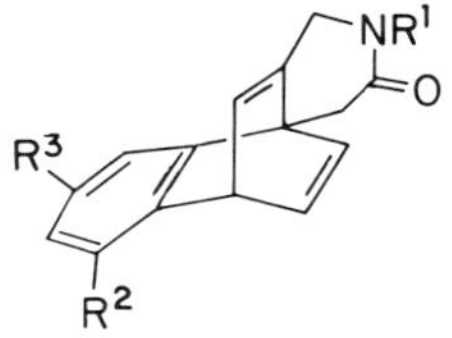

No.	R^1	R^2	R^3	Conditions	Product (Yield, %)	Refs.
12	CH_3	H	H	Toluene, 250°, 6.5 hr	(37)	223
13	CH_3	CH_3O	H	", ", 7 hr	(32)	223
14	CH_3	H	CH_3O	", ", "	(64)	223
15	$C_6H_5CH_2$	H	H	", ", 5 hr	(≈100)	223
16				$CH_3NH(CH_2)_2OH$, H_2O, 230°, 96 hr; or mesitylene, dioxane, H_2O, 230°, 48 hr	(28)[g] + (10)[g]	524

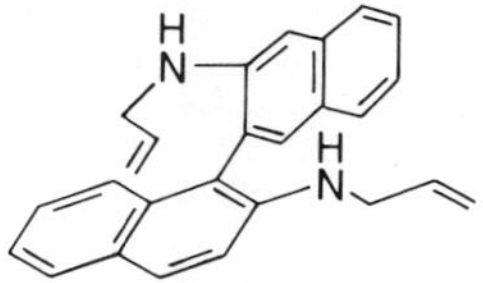

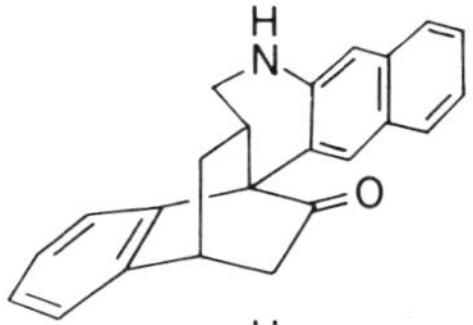

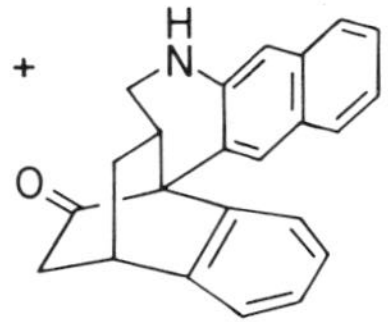

TABLE XVIIb. NAPHTHALENES (*Continued*)

Entry No.	Substrate		Conditions	Product(s) and Yield(s)(%)		Refs.
	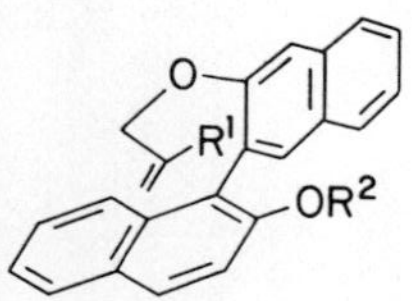			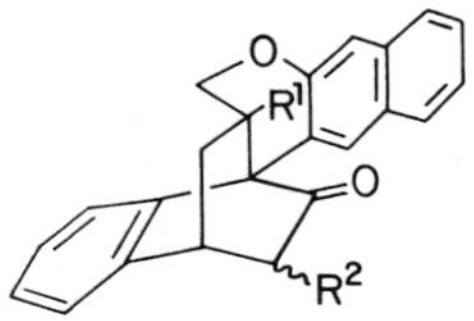 +	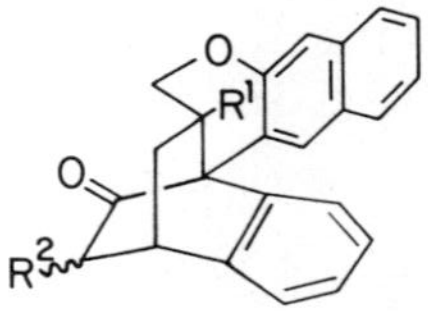	
				A	**B**	
	R^1	R^2		A	B	
17	H	H	Mesitylene, 210°, 16 hr	(11)	(0)	525
18	CH_3	H	", ", 96 hr	(2)	(0)	525
19	H	$CH_2{=}CHCH_2$	", 215°, 6 hr	(47)[h, i]	(19)[h]	525
20	CH_3	$CH_2{=}C(CH_3)CH_2$	", 210°, 46 hr	(34.5)[h, j]	(0)	525

[a] The substrate was generated by bromination of *anti*-[2.2]naphthalenofurophane in methanol.
[b] The substrate was generated from naphthalenofurophane and RC≡CR.
[c] The substrate was generated by addition of singlet oxygen to *anti*-[2.2]paracyclonaphthane.
[d] The aryne intermediate was generated from the corresponding 2-chlorobenzamide and n-C_4H_9Li.
[e] The product isolated exists in the keto form.
[f] The main product (33% yield) was that of an intramolecular ene reaction.
[g] The ketones resulted from hydrolysis of the secondary eneamines formed initially; only resinous products were formed in the absence of water.
[h] The initial cycloadducts underwent Claisen rearrangement to give products **A** and **B**.
[i] The product was a mixture of R^2 *exo* (44%) and R^2 *endo* (3%) isomers.
[j] The product was a mixture of R^2 *exo* (4.5%) and R^2 *endo* (30%) isomers.
* This intermediate is hypothetical, or has not been isolated.

TABLE XVIIc. ANTHRACENES

Entry No.	Substrate	Conditions	Product(s) and Yield(s) (%)	Refs.
	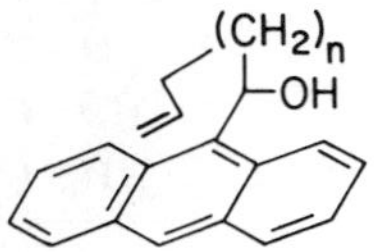 (CH$_2$)$_n$ OH		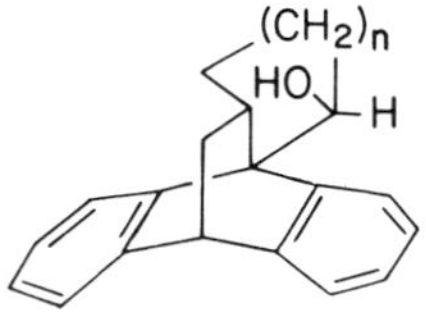 (CH$_2$)$_n$ HO H	
	n			
1	0	Toluene, 180°, 12 hr	(0)	45,167
2	1	*p*-Xylene, reflux, 26 hr	(20)[a,b]	45
3	2	", ", 17 hr	(40)[a,b]	45
4	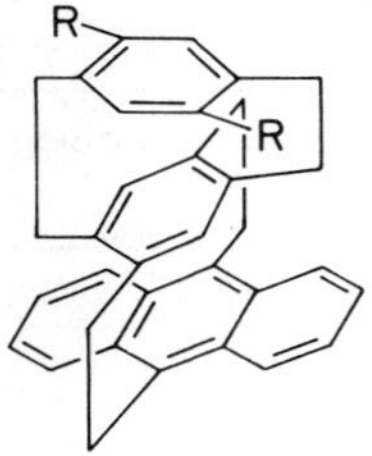 R R * c, d	Xylene, reflux	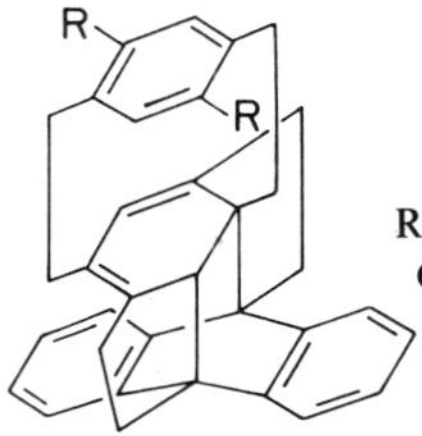 R R; R = H: 8.7, CH_3: 1.5	526

TABLE XVIIc. ANTHRACENES (*Continued*)

Entry No.	Substrate 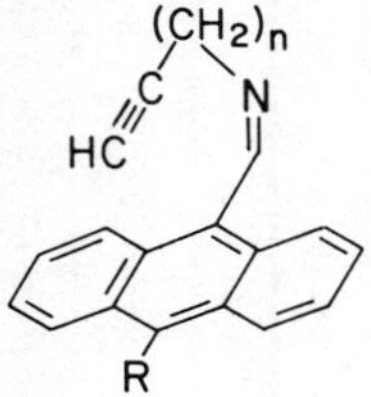	Conditions	Product(s) and Yield(s)(%) 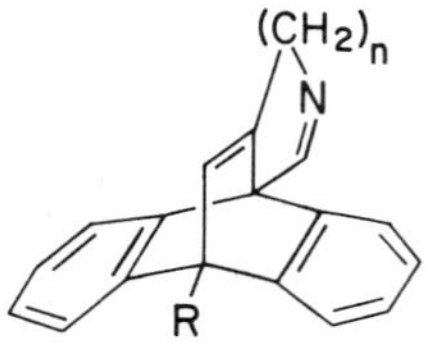	Refs.
	n R			
5	1 H	*p*-Xylene, reflux, 3 hr	(82)	45
6	1 CH_3	Xylene, reflux, 16 hr	(—)	527
7	2 H	Toluene, 200°, 24 hr	(42)	45
8	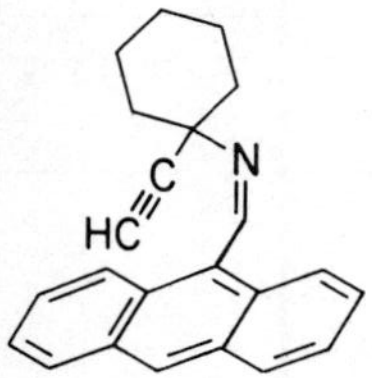	*p*-Xylene, reflux, 3 hr	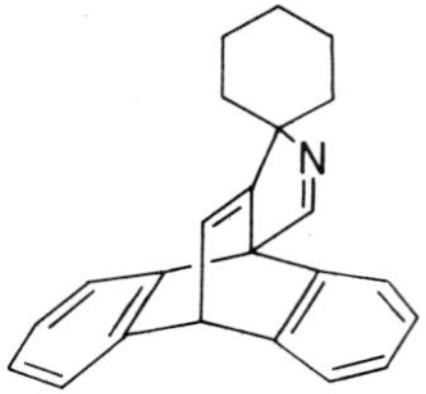 (84)	45

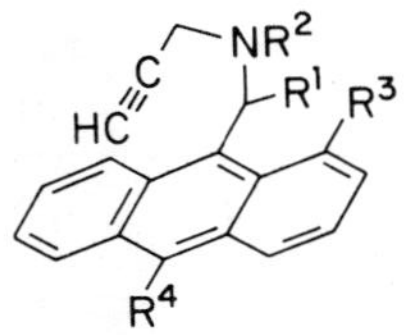

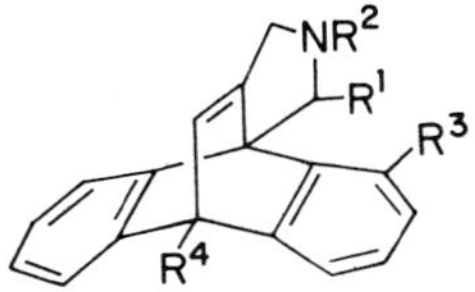

	R^1	R^2	R^3	R^4			
9	CH_3	H	H	H	Toluene, reflux, 7 hr	(≈80)	527
10	H	CH_3	H	CH_3	*p*-Xylene, reflux, 16 hr	(—)	527
11	H	$CH_2{=}CHCH_2$	H	H	Toluene, reflux, 2.2 hr	(—)[e]	45
12	H	c-C_3H_5	H	H	—	(—)	528
13	—$(CH_2)_4$—		H	H	CH_3OH, HCl, 25°, 24 hr	(34)	45
14	H	c-C_6H_{11}	H	H	—	(—)	528
15	H	$C_6H_5CH_2$	H	H	Toluene, reflux, 2.2 hr	(>76)	45
16	H	c-$C_5H_9CH_2$	H	H	Toluene, reflux, 125 hr	(70)	528
17	H	c-$C_6H_{11}CH_2$	Cl	H	—	(—)	528
18	H	c-$C_6H_{11}CH_2$	H	CH_3	*p*-Xylene, reflux, 16 hr	(43)	527

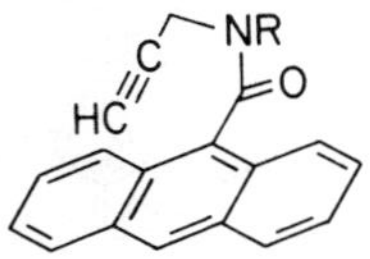

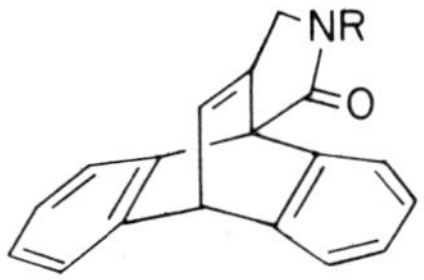

	R			
19	H	*p*-Xylene, reflux, 16 hr	(47)	45
20	CH_3	Toluene, reflux, 4 hr	(70)	45
21	c-$C_6H_{11}CH_2$	*p*-Xylene, reflux, 3 hr	(45)	529

TABLE XVIIc. ANTHRACENES (*Continued*)

Entry No.	Substrate		Conditions	Product(s) and Yield(s)(%)	Refs.
	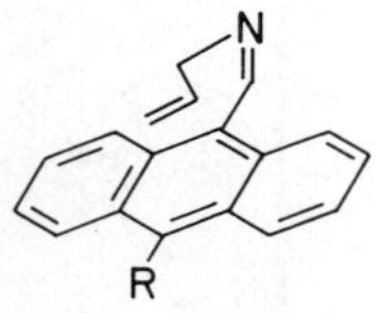			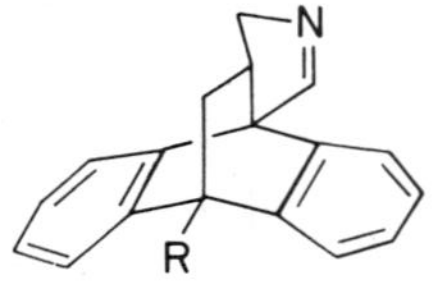	
	R				
22	H		*p*-Xylene, reflux, 24 hr	(85)	45
23	Cl		*p*-Xylene, reflux, 3.5 hr	(34)[f]	45
24	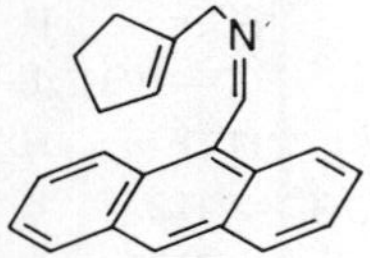		Toluene, 200°, 12 hr	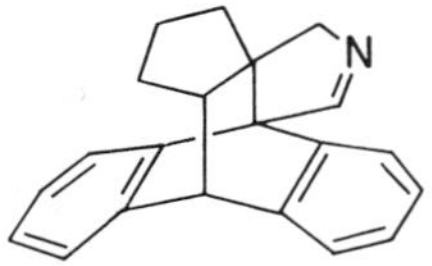 (38)[g]	45
	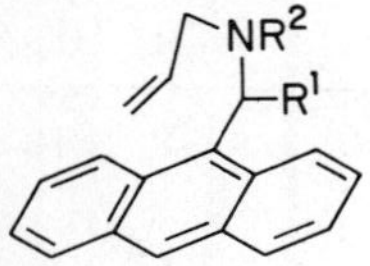			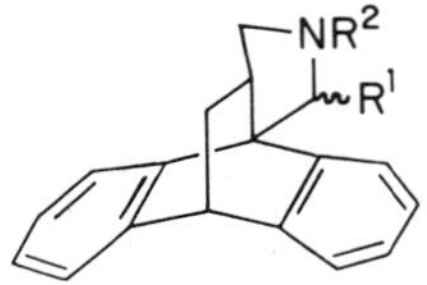	
	R^1	R^2			
25	H	H	*p*-Xylene, reflux, 16 hr	(80)	167,530[h]
26	CN	CH_3	", ", 5 hr	(>38)	45
27	CF_3CO_2	CF_3CO*[i]	CH_2Cl_2, 25–39°, 24 hr	(25)	45

	R			
28	H	p-Xylene, reflux, 5 hr	(97)	45
29	CH_3	", 125°, 1 hr	(77)	45
30		p-Xylene, reflux, 20 hr	(29)	45

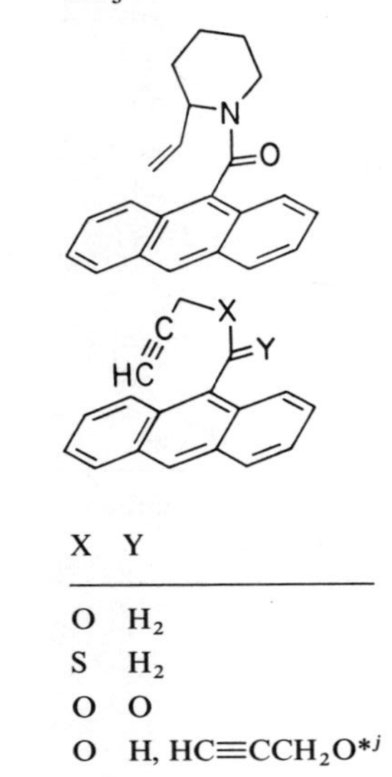

	X	Y			
31	O	H_2	p-Xylene, reflux, 3 hr	(68)	45
32	S	H_2	", ", 3.3 hr	(74)	45
33	O	O	", ", 45 hr	(89)	45
34	O	H, $HC{\equiv}CCH_2O^{*j}$	Benzene, reflux, 20 hr	(31)	45

TABLE XVIIc. ANTHRACENES (*Continued*)

Entry No.	Substrate	Conditions	Product(s) and Yield(s)(%)	Refs.
35	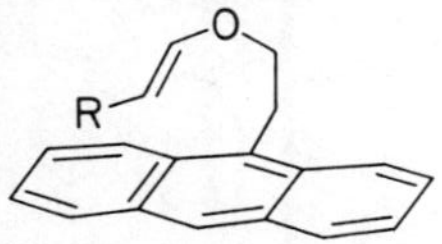 R-group: R=CH_3O_2C	Toluene, 185°, 24 hr	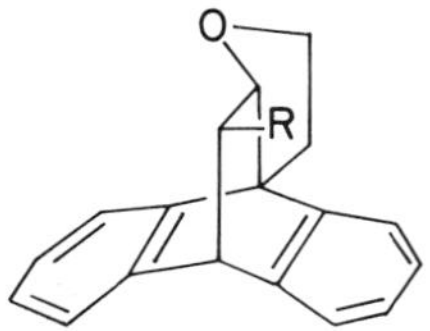 (79)	45
36	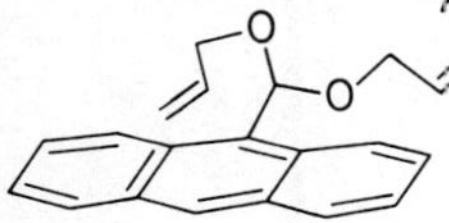 *[k,l]	CH_2=$CHCH_2OH$, 93°, 3 hr[k] (23 hr[l])	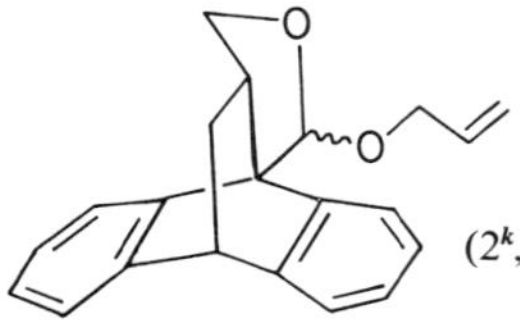 (2[k], 19[l])	531
37	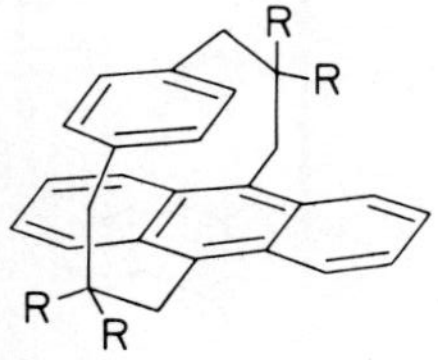 R=$C_2H_5O_2C$	100°	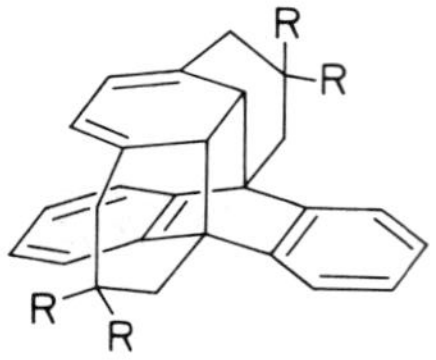 (—)[m,n]	168

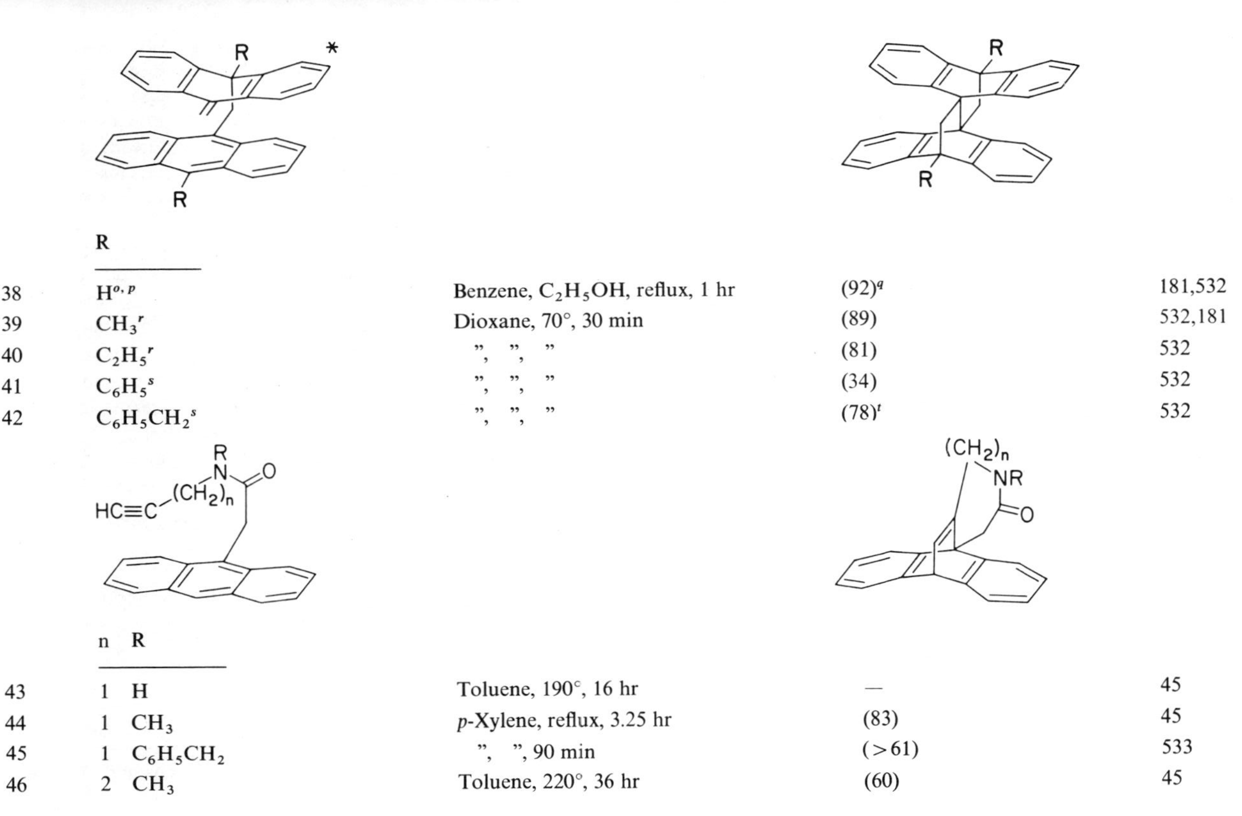

No.	R	Conditions	Product (Yield, %)	Refs.
38	H[o,p]	Benzene, C_2H_5OH, reflux, 1 hr	(92)[q]	181,532
39	CH_3[r]	Dioxane, 70°, 30 min	(89)	532,181
40	C_2H_5[r]	", ", "	(81)	532
41	C_6H_5[s]	", ", "	(34)	532
42	$C_6H_5CH_2$[s]	", ", "	(78)[t]	532

No.	n	R	Conditions	Product (Yield, %)	Refs.
43	1	H	Toluene, 190°, 16 hr	—	45
44	1	CH_3	*p*-Xylene, reflux, 3.25 hr	(83)	45
45	1	$C_6H_5CH_2$	", ", 90 min	(>61)	533
46	2	CH_3	Toluene, 220°, 36 hr	(60)	45

TABLE XVIIc. ANTHRACENES (*Continued*)

Entry No.	Substrate	Conditions	Product(s) and Yield(s)(%)	Refs.
47	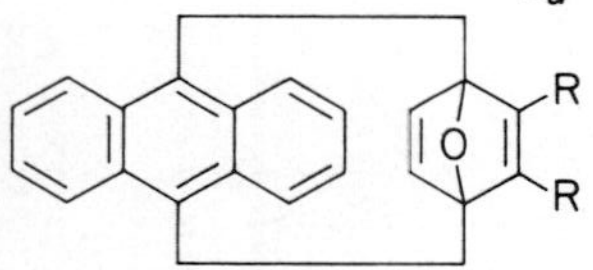 [u] R=CH_3O_2C	RC≡CR, heat	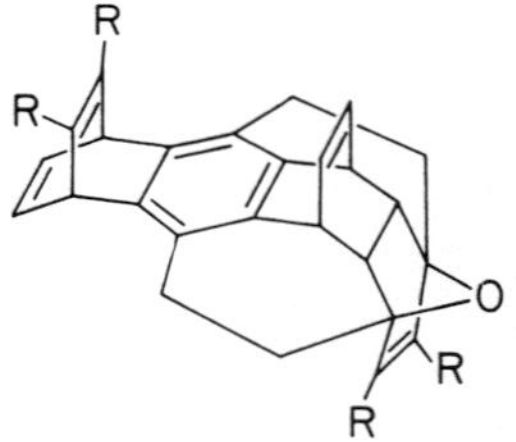 (36)	534

[a] The stereochemistry of the hydroxyl group was assumed on the basis of transition-state models.
[b] The yield includes the preparation of the precursor.
[c] The substrate was the presumed intermediate formed by mixed [6 + 6]cycloaddition of two *p*-quinodimethanes.
[d] Note that the corresponding *double*-layered [2.2]cyclophane did not undergo an intramolecular Diels–Alder reaction.[535]
[e] The product shown, and the adduct involving the allyl group, were formed in a ratio of 3:1.
[f] The product was isolated as the reductive methylation product, 5-chloro-2-methyl-1,2,3,3a,4,5-hexahydro-5,9b-*o*-benzenobenz[*e*]-isoindole.
[g] The product was isolated as the reductive methylation product, 2-methyl-1,3,4,5,6,6a-hexahydro-7*H*-7,11b-*o*-benzenobenzo[*e*]cy-clopent[*h*]isoindole.
[h] The authors report that no intramolecular Diels–Alder reaction occurred under these conditions.
[i] The substrate was generated from 9-anthraldehyde *N*-allylimine and trifluoroacetic anhydride.
[j] The substrate was generated from 9-anthraldehyde and propargyl alcohol in the presence of *p*-toluenesulfonic acid.
[k] The substrate was generated from 9-anthraldehyde and allyl alcohol in the presence of HCl.
[l] The substrate was generated by alcohol exchange from 9-anthraldehyde dimethyl acetal and allyl
[m] The reaction is reversible; heating the adduct at 190° resulted in a 60/40 mixture of cyclized and uncyclized products.
[n] Note that the corresponding [2.2]cyclophane (R = H) did not undergo intramolecular Diels–Alder reaction.[535]
[o] The substrate was generated by base-induced elimination from the corresponding acetate.

[p] See ref. 532 for cyclization of a number of aromatic ring-chlorinated derivatives.

[q] The product is in equilibrium with a small amount of starting material; the equilibrium constants determined in toluene were $K = 810$ at 21° and $K = 258$ at 49°.[647]

[r] The substrate was generated by coupling 9-chloromethyl-10-R-anthracene in the presence of $SnCl_2$.

[s] The substrate was generated by coupling 9-bromomethyl-10-R-anthracene in the presence of $SnCl_2$.

[t] The equilibrium constant in toluene at 25° is 3×10^5.

[u] The substrate was generated from anthracenofurophane and RC≡CR.

* This intermediate is hypothetical, or has not been isolated.

TABLE XVIId. PHENANTHRENES

Entry No.	Substrate	Conditions	Product(s) and Yield(s)(%)	Refs.
1	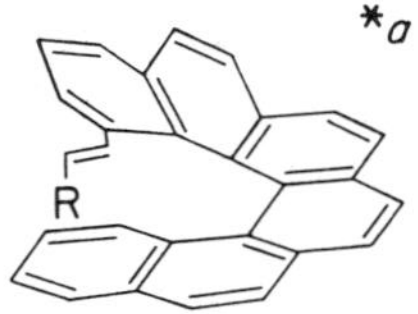 *[a] R = $C_2H_5O_2C$	Benzene, 25°, 14 hr	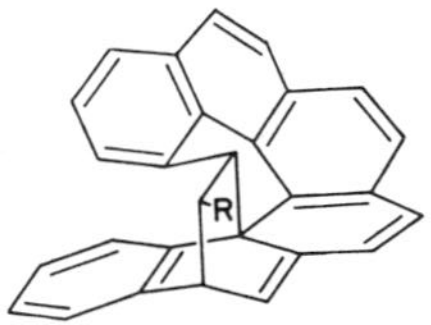 (90)[b]	536

[a] The substrate was generated from *dl*-1-formyl[6]helicene and $[(C_2H_5)_2P(O)CHCO_2C_2H_5]^-$.

[b] The *cis* ester (12%) was obtained in addition to the *trans* ester (80%) when the reaction was carried out at 80°.

TABLE XVIIIa. FURANS; 2-,4-, AND 5-ATOM CHAINS

Entry No.	Substrate	Conditions	Product(s) and Yield(s)(%)	Refs.
1	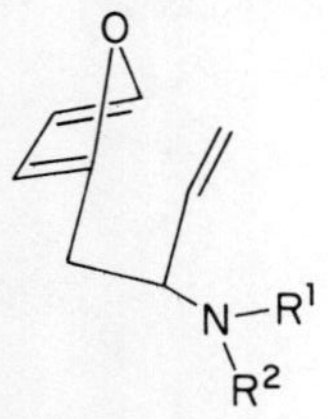 $R^1, R^2 = CH_3, C_2H_5, —(CH_2)_2O(CH_2)_2—, —(CH_2)_5—$ *a*	DMF, 100°	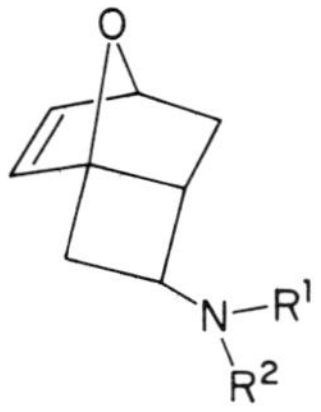 (0)	342
2	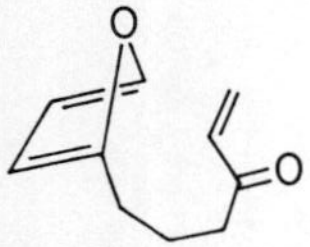 **A**	⇌	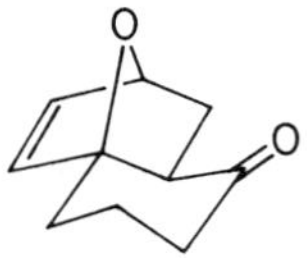 **B**	
			Ratio **A**/**B**	
		CH_2Cl_2, 25°, 6 days	45/55	63
		CH_2Cl_2, reflux, 6 days	18/82 (74)	63
		CH_2Cl_2, Florisil, reflux, 6 days	7/93 (72)	63
		C_2H_5OH, reflux, 6 days	36/64	63
		Benzene, reflux, 6 days	56/44	63

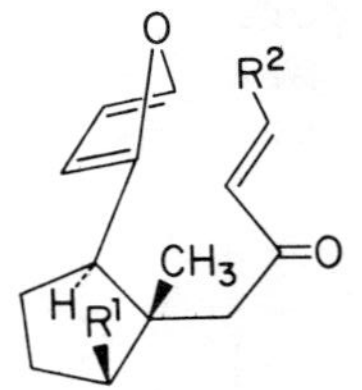

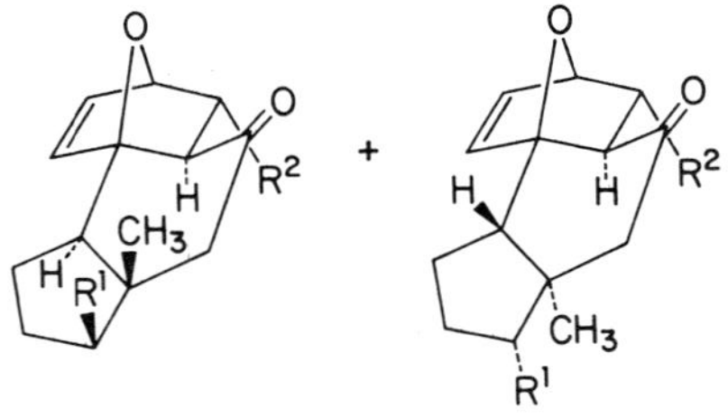

	R^1	R^2	Conditions	Product(s) and Yield(s) (%) A	B	Refs.
2a	t-$C_4H_9(CH_3)_2SiO$	H	CH_2Cl_2, 25°, 6 days	(7)[b]	(54)[b]	537
			Benzene, 80°, 6 days	(45)[c]	(5)[c]	537
2b	t-$C_4H_9(CH_3)_2SiO$	CH_3	", ", "	(0)	(0)	537
3			H_2O, 100°, 60 hr	[d] (85)		158
3a			H_2O, 100°, 60 hr	[d] (90)		158

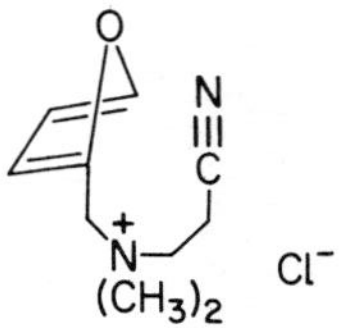

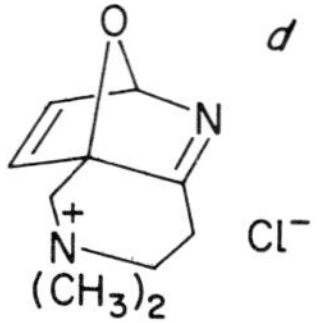

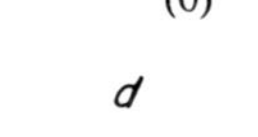

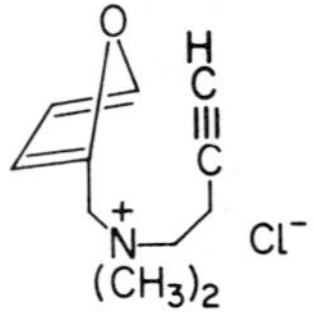

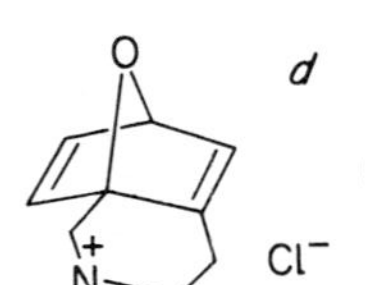

TABLE XVIIIa. FURANS; 2-,4-, AND 5-ATOM CHAINS (*Continued*)

Entry No.	Substrate 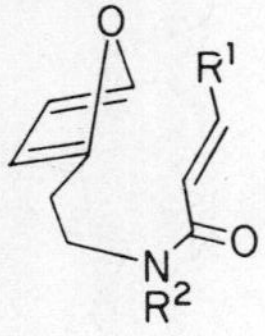	Conditions	Product(s) and Yield(s)(%) 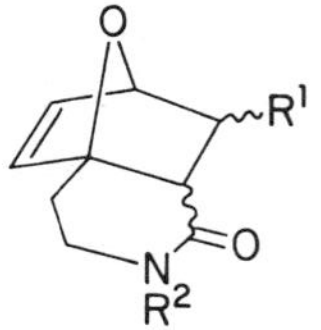	Refs.
	R^1 R^2			
4	H H	Benzene, reflux	(0)	66
5	H CH_3	", ", 4 days	(100)	66
6	CH_3 CH_3	", "	(12)	66
	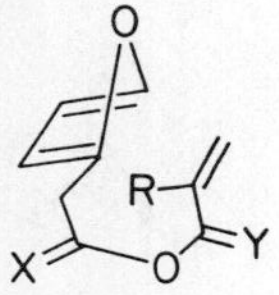	Benzene, reflux	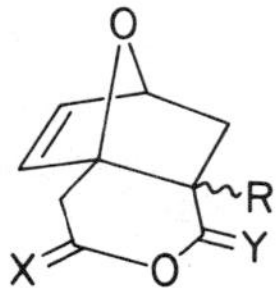	66
7	R = H; X = O, Y = H_2		(0)	
8	R = H; X = H_2; Y = O		(0)	
9	R = $C_2H_5O_2C$; X = O; Y = H_2[e]		(0)	

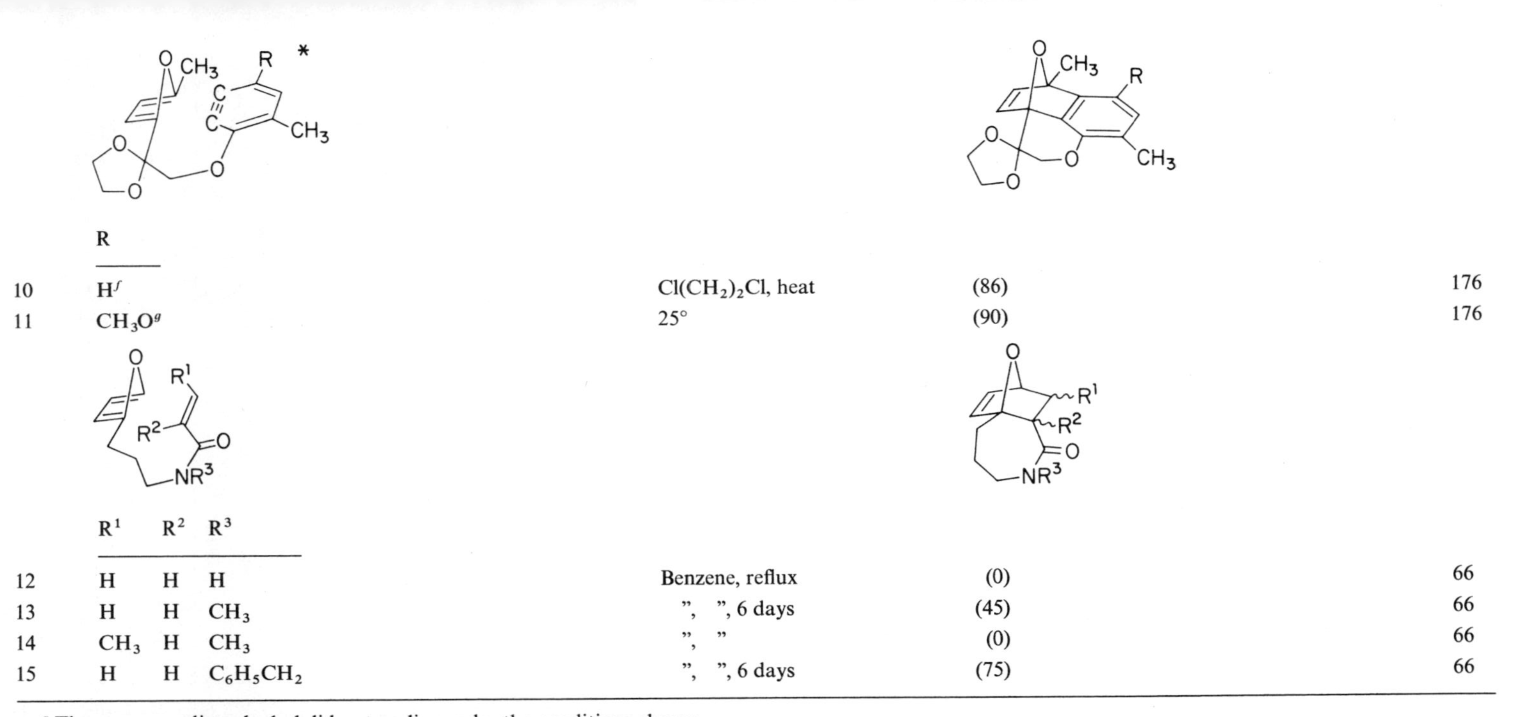

No.	Reactant			Conditions	Product (Yield %)	Ref.
	R					
10	H[f]			$Cl(CH_2)_2Cl$, heat	(86)	176
11	CH_3O[g]			25°	(90)	176
	R^1	R^2	R^3			
12	H	H	H	Benzene, reflux	(0)	66
13	H	H	CH_3	", ", 6 days	(45)	66
14	CH_3	H	CH_3	", "	(0)	66
15	H	H	$C_6H_5CH_2$	", ", 6 days	(75)	66

[a] The corresponding alcohol did not cyclize under the conditions shown.
[b] The numbers are conversions.
[c] The product was an equilibrium mixture; the remainder was starting material. The equilibrium was approached from either side.
[d] The structure was assigned on the basis of elemental analysis and IR spectrum.
[e] The reaction also failed in refluxing ether, ethanol, or dimethylformamide, and in methylene chloride in the presence of aluminum chloride.
[f] The benzyne was generated by diazotization of the corresponding anthranilic acid.
[g] The benzyne intermediate was generated by treatment of the dibromo derivative with n-C_4H_9Li at $-78°$.
*This intermediate is hypothetical, or has not been isolated.

TABLE XVIIIb. FURANS; 3-ATOM CHAIN CONTAINING UNCHARGED NITROGEN

Entry No.	Substrate R¹	R²	Conditions	Product(s) and Yield(s)(%)	Refs.
	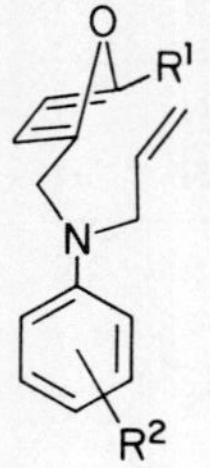			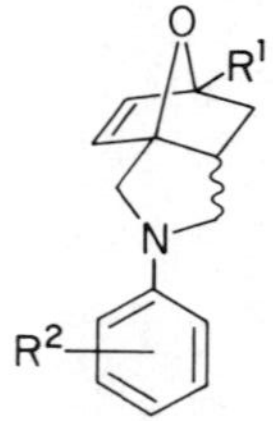 [a]	
1	H	H	Neat, 25°, 0.5–12 hr	(100)	163,55
2	H	p-Cl	CCl_4, 20°[b]	(—)	538
3	H	p-CH_3	Neat or CCl_4, 20°[b]	(100)	163,538
4	H	o-CH_3	CCl_4, 20°[b]	(—)	538
5	H	p-CH_3O	Neat or CCl_4, 20°[b]	(100)	163,538
6	H	p-CH_3O	CCl_4, 20°[b]	(—)	538
7	CH_3	H	CCl_4, 20°[b]	(—)	538
8	CH_3	p-Cl	CCl_4, 20°[b]	(—)	538
9	CH_3	p-CH_3	CCl_4, 20°[b]	(—)	538
10	CH_3	p-CH_3O	CCl_4, 20°[b]	(—)	538

No.	Reactant	Conditions	Product(s) and Yield(s) (%)	Refs.
	 N-COR (R below)		 N-COR, c	
	R			
11	CH_3	Neat or in C_2H_5OH, 90–95°, 12 hr	(95)	356,48,50
12	$ClCH_2$	"	(85)	356,50
13	CF_3	90–95°, 7 hr	(95)	48
14	CCl_3	Neat or in C_2H_5OH, 90–95°, 12 hr	(85)	356,50
15	C_2H_5	"	(85)	356,50
15a	t-C_4H_9	90–95°	(90)	48
16	2-Furyl	Neat or in C_2H_5OH, heat, 10–12 hr	(95)	358
17	3-Pyridyl	", ", "	(95)	358
18	C_6H_5	", 20°, 4–6 hr	(95)	358
19	p-BrC_6H_4*[d]	", 20°	(90)	358
20	p-$O_2NC_6H_4$*[d]	", "	(90)	358
	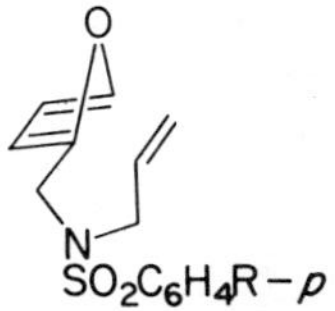 N-$SO_2C_6H_4R$-p (R below)	Neat or in C_2H_5OH, 90–95°, 50–60 hr	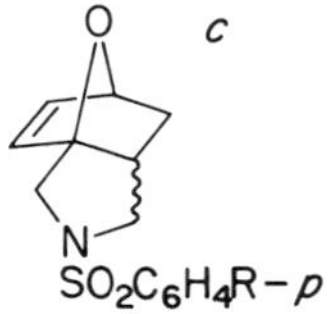 N-$SO_2C_6H_4R$-p, c	
	R			
21	H		(95)	539
22	CH_3		(90)	539
23	Cl		(95)	539

TABLE XVIIIb. FURANS; 3-ATOM CHAIN CONTAINING UNCHARGED NITROGEN (*Continued*)

Entry No.	Substrate 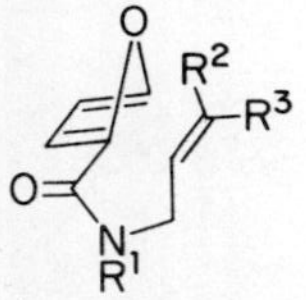					Conditions	Product(s) and Yield(s)(%) 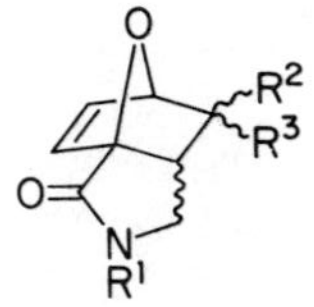	Refs.
	R^1	R^2	R^3					
24	H	H	H			DMF, 140°. 12 hr	(90)	47
25	CH_3	H	H			Benzene, reflux, 6 days	(100)	66
26	C_2H_5	H	H			DMF, 140°, 5 hr	(95)	47
27	CH_3	H	CH_3			Benzene, reflux, 6 days	(0)	66
28	CH_3	CH_3	H			", ", "	(0)	66
29	C_6H_5	H	H			DMF, 140°, 3 hr	(90)	47
30	p-$O_2NC_6H_4$ 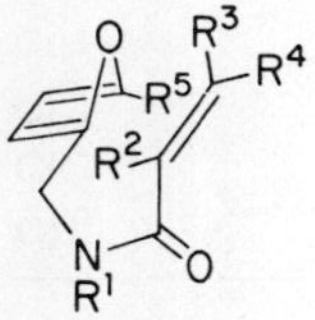	H	H			", ", 2 hr	(85) 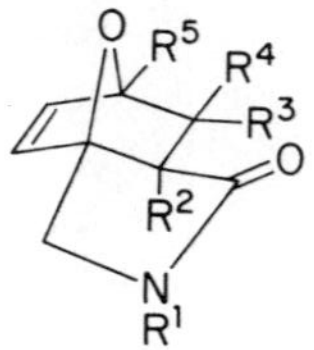	47
	R^1	R^2	R^3	R^4	R^5			
31	H	H	H	H	H	Benzene, reflux	(0)	66
32	CH_3	H	H	H	H	", ", 6 days	(100)[a]	66
33	H	H	H	HO_2C	H	25°	(0)	540
34	CH_3	CH_3	H	H	H	Benzene, reflux, 5 days	(100)[a]	66
35	CH_3	H	CH_3	H	H	", ", 14 days	(40)	66

36	CH_3	H	H	HO_2C	H*[f]	C_2H_5OH, 25°	(53)	540,541
37	CH_3	C_6H_5	H	H	H	C_2H_5OH, 80°	(61)[e]	224
						C_6D_6, 120°	(48)[e]	224
38	C_6H_5	H	H	HO_2C	H	C_2H_5OH, 25°, 16 hr	(96)	540,46,58
						Solid state, 115°	(—)	58
39	*o*-HOC_6H_4	CH_3	H	H	H	Benzene, 80°, 2.25 hr	(23)	67
40	*o*-$LiOC_6H_4$	CH_3	H	H	H	”, ”, ”	(45)	67
41	*o*-$NaOC_6H_4$	CH_3	H	H	H	”, ”, ”	(29)	67
42	*o*-KOC_6H_4	CH_3	H	H	H	”, ”, ”	(40)	67
43	*o*-$ClMgOC_6H_4$	CH_3	H	H	H	”, ”, ”	(79)	67
44	*o*-$BrMgOC_6H_4$	CH_3	H	H	H	”, ”, ”	(90)	67
45	*o*-$IMgOC_6H_4$	CH_3	H	H	H	”, ”, ”	(95)	67
46	$Mg(OC_6H_4\text{-}o)_2$	CH_3	H	H	H	”, ”, ”	(80)	67
47	$Zn(OC_6H_4\text{-}o)_2$	CH_3	H	H	H	”, ”, ”	(55)	67
48	C_6H_5	CH_3	H	H	O_2N	”, ”, 7 hr	(30)	68
49	*o*-HOC_6H_4	CH_3	H	H	O_2N	Toluene, reflux, 7 hr	(94)	68
50	*o*-$BrMgOC_6H_4$	CH_3	H	H	O_2N	Benzene, reflux, 7 hr	(57)	68
51	*o*-HOC_6H_4	H	CH_3	H	H	Toluene, 110°, 4.5 hr	(5)	67
52	*o*-$ClMgC_6H_4$	H	CH_3	H	H	”, ”, ”	(63)	67
53	*o*-$BrMgC_6H_4$	H	CH_3	H	H	”, ”, ”	(53)	67
54	*o*-$IMgC_6H_4$	H	CH_3	H	H	”, ”, ”	(46)	67
55	$Mg(OC_6H_4\text{-}o)_2$	H	CH_3	H	H	”, ”, ”	(43)	67
56	*o*-HOC_6H_4	H	CH_3	H	O_2N	”, ”, 9 hr	(93)	68
57	*o*-HOC_6H_4	H	CH_3	CH_3	H	”, ”, 7 hr	(0)	67
58	*o*-$BrMgC_6H_4$	H	CH_3	CH_3	H	”, ”, ”	(76)	67
59	*p*-$CH_3C_6H_4$	H	H	HO_2C	H	C_2H_5OH, 25°, 16 hr	(89)	540,58
60	CH_3	*m*-$CH_3OC_6H_4$	H	H	H	—	(—)	224
61	*o*-$CH_3OC_6H_4$	CH_3	H	H	O_2N	Benzene, reflux, 7 hr	(11)	68
62	*o*-HOC_6H_4	H	CH_3	CH_3	O_2N	Toluene, reflux, 9 hr	(73)	68
63	$C_6H_5CH_2$	H	H	HO_2C	H*[f]	Benzene, 25°	(99)	542

TABLE XVIIIb. FURANS; 3-ATOM CHAIN CONTAINING UNCHARGED NITROGEN (*Continued*)

Entry No.	Substrate R[1]	R[2]	R[3]	R[4]	R[5]	Conditions	Product(s) and Yield(s)(%)	Refs.
64	$C_6H_5(CH_2)_2$	H	H	HO_2C	H*[f]	Benzene, 25°	(94)	542
64a	C_6H_5	H	$C_2H_5O_2C$[m]	H	H	C_2H_5OH, 30–50°	(—)[a]	46
65	$HOCH_2CH(C_6H_5)$[g]	H	CH_3	H	H	Toluene, reflux, 3.25 hr	(57)[h]	69
66	$ClMgOCH_2CH(C_6H_5)$[g]	H	CH_3	H	H	", ", "	(84)[i]	69
67	$HOCH_2CH(C_6H_5)$[g]	H	CH_3	CH_3	H	", ", 50 hr	(44)[j]	69
68	$ClMgOCH_2CH(C_6H_5)$[g]	H	CH_3	CH_3	H	", ", "	(86)[k,l]	69
69	o-HOC_6H_4	—$(CH_2)_3$—		H	O_2N	", ", 7 hr	(96)	68
70	o-HOC_6H_4	—$(CH_2)_4$—		H	O_2N	", ", 9 hr	(30)	68
71	3-Indoleethyl	H	H	HO_2C	H*[f]	Benzene, 25°	(97)	542
72	3,4-$(CH_3O)_2C_6H_3(CH_2)_2$	H	H	HO_2C	H*[f]	", "	(92)	542
73	3-CH_3O, 4-$C_2H_5OC_6H_3(CH_2)_2$	H	H	HO_2C	H*[f]	", "	(85)	542
74	3-CH_3O, 4-$C_6H_5CH_2OC_6H_3(CH_2)_2$	H	H	HO_2C	H*[f]	", "	(95)	542
75	3-$C_6H_5CH_2O$, 4-$CH_3OC_6H_3(CH_2)_2$	H	H	HO_2C	H*[f]	", "	(58)	542

[a] The stereochemistry was not determined.
[b] For first-order rate constants at 20° and 50°, see ref. 538.
[c] The structures were assigned on the basis of elemental analysis and IR and/or UV spectra.
[d] The starting material cyclized too rapidly to be isolated.
[e] The yield shown is the equilibrium concentration; the remainder was starting material.
[f] The substrate was generated from the corresponding furanemethylamine and maleic anhydride.
[g] R[1] was derived from (−)-phenylglycinol.
[h] The ratio of diastereomers was 32:25.
[i] The ratio of diastereomers was 68:16.
[j] The ratio of diastereomers was 1:1.
[l] The BrMgO— and IMgO— salts gave the two diastereomers in yields of 68 and 9%, and 58 and 14%, respectively.
[m] It is not clear whether the dienophile is a fumaric or maleic acid derivative.
[k] The ratio of diastereomers was 74:12.
* This intermediate is hypothetical, or has not been isolated.

TABLE XVIIIc. FURANS; 3-ATOM CHAIN CONTAINING QUATERNARY NITROGEN

Entry No.	Substrate				Conditions	Product(s) and Yield(s) (%)	Refs.
	R^1	R^2	R^3	X		[a]	
1	CH_3	CH_3	H	Br	H_2O, 100°, 30 hr	(89)	158,51,543
2	C_2H_5	C_2H_5	H	Br	", ", "	(92)	158,51,52,543
3	$-(CH_2)_2O(CH_2)_2-$		H	Br	", ", "	(90)	158,51
4	$-(CH_2)_5-$		H	Br	", ", "	(93)	158,51
5	C_2H_5	C_2H_5	CH_3	Br	", ", 60 hr	(85)	158,52
6	C_2H_5	C_2H_5	$CH_2{=}CH$	Cl	", ", "	(85)	158
7	C_2H_5	C_2H_5	C_6H_5	Cl	H_2O, 100°, 60 hr	(87)	158
8	$-(CH_2)_5-$		C_6H_5	Cl	H_2O, 83°	(—)	51
9	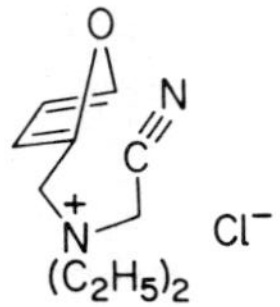 $(C_2H_5)_2$, Cl^-				H_2O, 100°, 60 hr	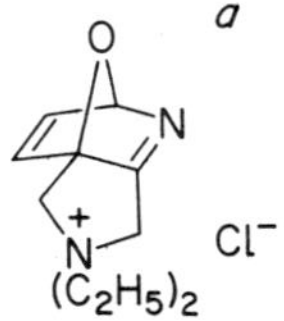 $(C_2H_5)_2$, Cl^- [a] (90)	158

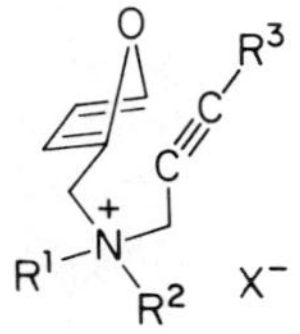

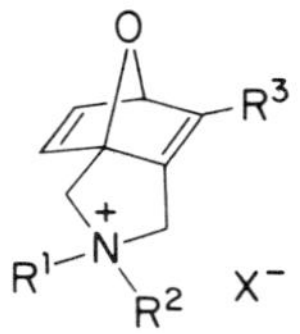

TABLE XVIIIc. FURANS; 3-ATOM CHAIN CONTAINING QUATERNARY NITROGEN (*Continued*)

Entry No.	Substrate R^1	R^2	R^3	R^4	R^5	X	Conditions	Product(s) and Yield(s)(%) [a]	Refs.
10	H	CH_3	H	H	H	Br	H_2O, 100°, 12 hr	(—)[c]	158
11	CH_3	CH_3	H	H	H	Br	", ", "	(96)[b]	158,51,543
12	CH_3	$NC(CH_2)_2$	H	H	H	Cl	", ", "	(90)[b]	158
13	CH_3	CH_3	H	CH_3	Cl	Cl	", 90°, 20 hr	(95)	544
14	CH_3	$CH_2{=}CHCH_2$	H	H	H	Cl	", 83°	(—)	51
15	C_2H_5	C_2H_5	H	H	H	Br	", 100°, 12 hr	(96)[b]	158,51,52,543,545
16	CH_3	CH_3	H	CH_3	CH_3	Cl	H_2O, 100°, 90 hr	(90)	158
17	$-(CH_2)_2O(CH_2)_2-$		H	H	H	Br	", ", 12 hr	(95)[b]	158,51
18	$-(CH_2)_5-$		H	H	H	Br	", ", "	(97)[b]	543,158
19	C_2H_5	C_2H_5	CH_3	H	H	Cl	", 100°, 60 hr	(92)	158,52
20	C_2H_5	C_2H_5	H	CH_3	H	Br	", ", 30 hr	(90)	158,52
21	C_2H_5	C_2H_5	H	CH_3	Cl	Cl	", 90°, 20 hr	(95)	544
22	$-(CH_2)_5-$		H	H	CH_3	Cl	", 83°	(—)	51
23	CH_3	C_6H_5	H	H	H	Br	", 100°, 12 hr	(93)[b]	158,51
24	$-(CH_2)_5-$		H	C_6H_5	H	Br	", ", 30 hr	(90)	158,51
25	$-(CH_2)_5-$		C_6H_5	H	H	Br	", 83°	(—)	51

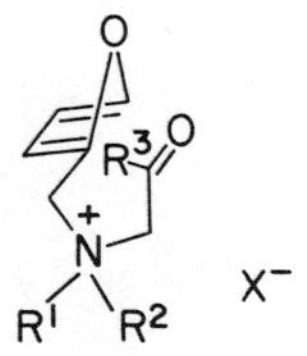

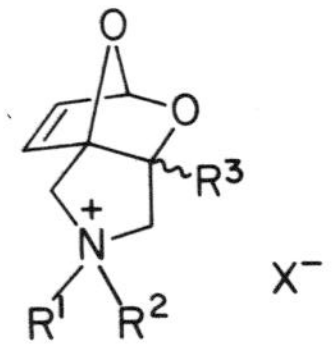

	R[1]	R[2]	R[3]	X			
26	$-(CH_2)_5-$		CH_3	—	H_2O, 83°	(—)[d]	545
27	C_2H_5	C_2H_5	CH_3O	Cl	", 100°, 90 hr	(85)[a]	158,545
28	$-(CH_2)_5-$		C_6H_5	—	", 83°	(—)[d]	545

[a] The structures were assigned on the basis of elemental analysis and IR spectra.
[b] Treatment of the products with hydrobromic acid gave the corresponding dihydroisoindolinium bromides.
[c] The yield was not reported; the corresponding tertiary amine did not cyclize.
[d] No structure proof was reported.

TABLE XVIIId. FURANS; 3-ATOM CHAIN NOT CONTAINING NITROGEN

Entry No.	Substrate	Conditions	Product(s) and Yield(s)(%)	Refs.
1	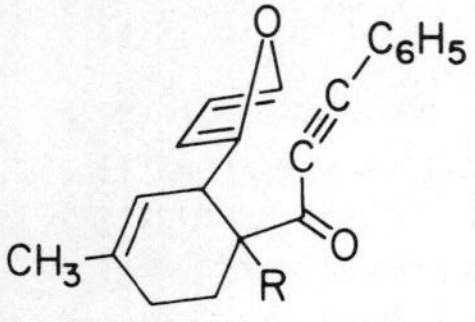 CH_3, R, O, C_6H_5; R = H, CH_3	—	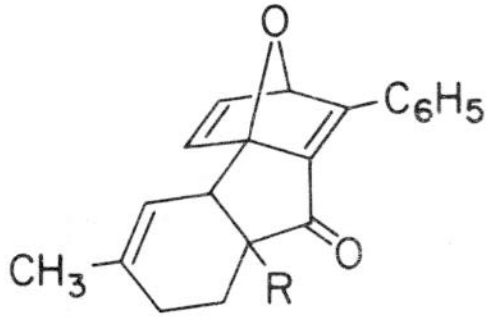 CH_3, R, O, C_6H_5 (0)	352
1a	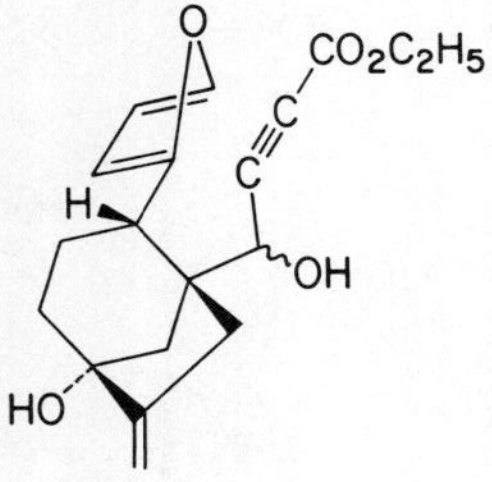 H, HO, OH, $CO_2C_2H_5$	Benzene, reflux, 24 hr	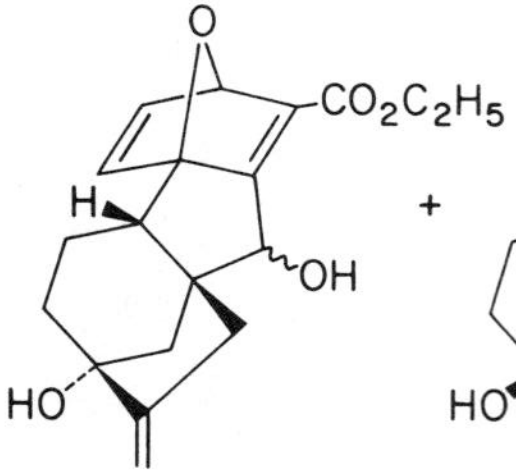 H, HO, OH, $CO_2C_2H_5$ (53) 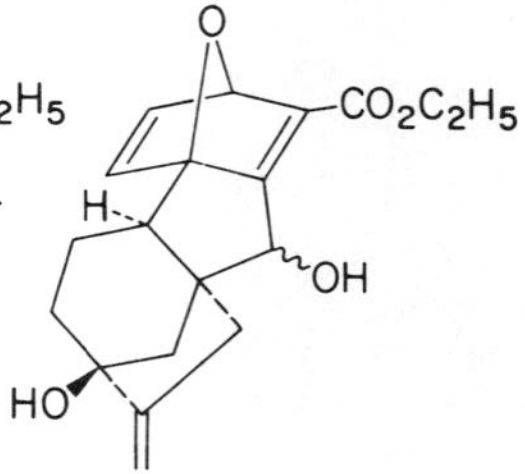 + H, HO, OH, $CO_2C_2H_5$ (17)	546

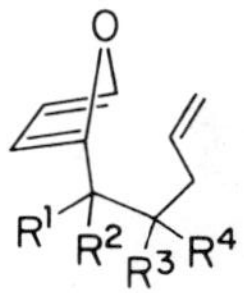

CH_3OH, 25°

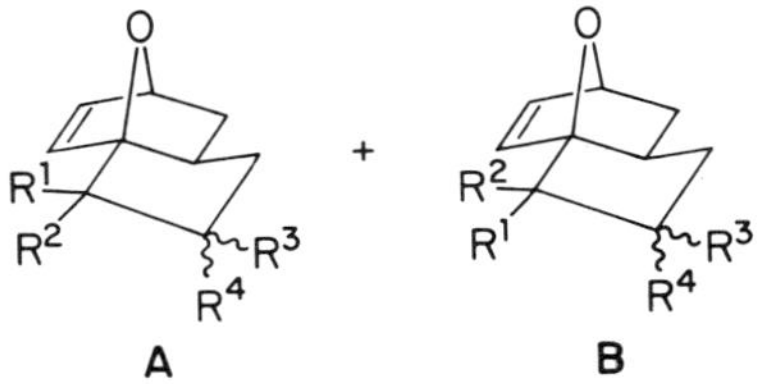

	R^1	R^2	R^3	R^4	Conditions	A	B	Refs.
2	O=		H	H	Benzene, reflux, 6 days	(0)	(0)	537
2a	HO	H	H	H	Benzene, reflux, 3 days	(0)	(0)	54,537
2b	HO	CH_3	H	H	", ", "	(0)	(0)	54
2c	HO	H	HO	CH_3	", ", "	(0)	(0)	54
2d	HO	H	C_2H_5O	C_2H_5O	", ", 1 day	(61)		55
2e	H	H	CH_3O_2C	CH_3O_2C	", "	(40)[a]		66
2f	HO	H	—$S(CH_2)_3S$—		", ", 3 days	(34)	(29)	54
					H_2O, β-cyclodextrin, 89°, 6 hr	(36)	(55)	55
2g	HO	CH_3	—$S(CH_2)_3S$—		", ", 2 days	(50)	(50)	54
2h	$C_6H_5CH_2O$	H	H	H	", ", 3 days	(0)	(0)	54
2i	HO	H	$C_6H_5CH_2O$	CH_3	", ", 2 weeks	(trace)		54
3	$C_6H_5CO_2$	H	—$S(CH_2)_3S$—		", ", 2 weeks	(39)	(23)	54
4	*[b]					(41–42)		547,548

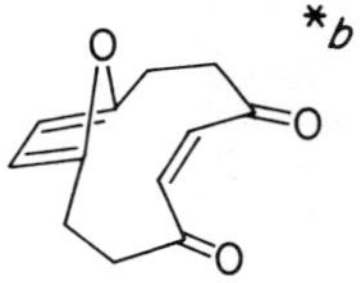

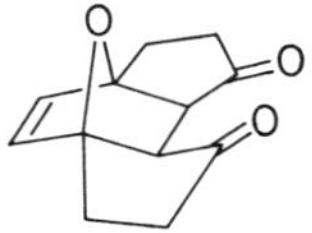

TABLE XVIIId. FURANS; 3-ATOM CHAIN NOT CONTAINING NITROGEN (*Continued*)

Entry No.	Substrate	Conditions	Product(s) and Yield(s)(%)	Refs.
5	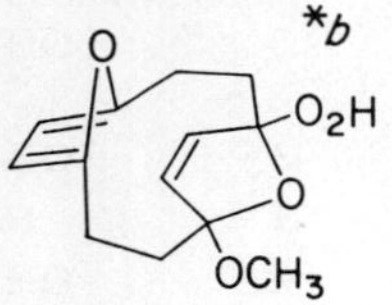 O; O_2H; O; OCH_3 *b	CH_3OH, 25°	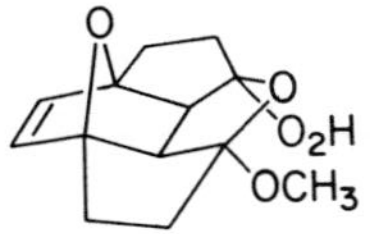 O; O; O_2H; OCH_3 (11)[c]	547
6	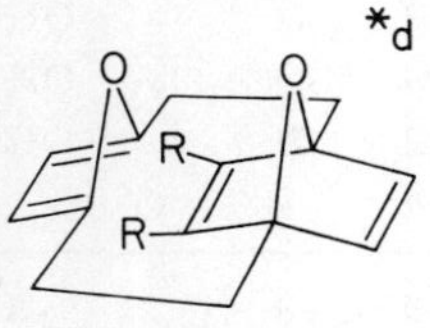 O; O; R; R *d R = CH_3O_2C	Benzene, 105°, 185 hr ⇌ 160°, 2 hr[e]	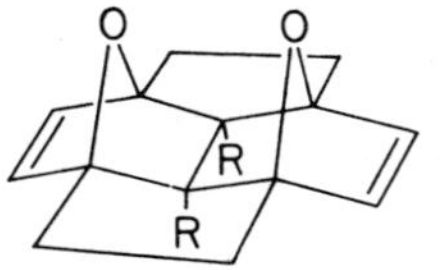 O; O; R; R (71)	549
7	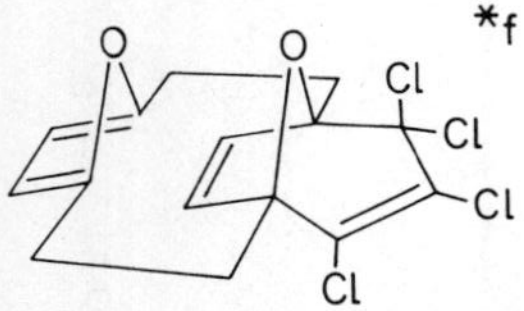 O; O; Cl; Cl; Cl; Cl *f	Benzene, reflux	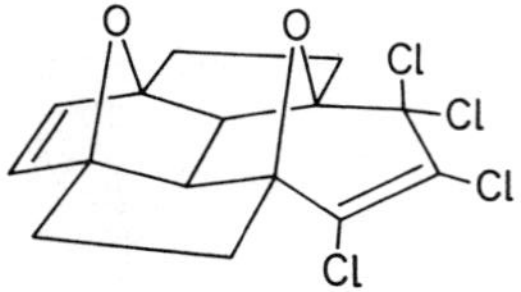 O; O; Cl; Cl; Cl; Cl (—)	550

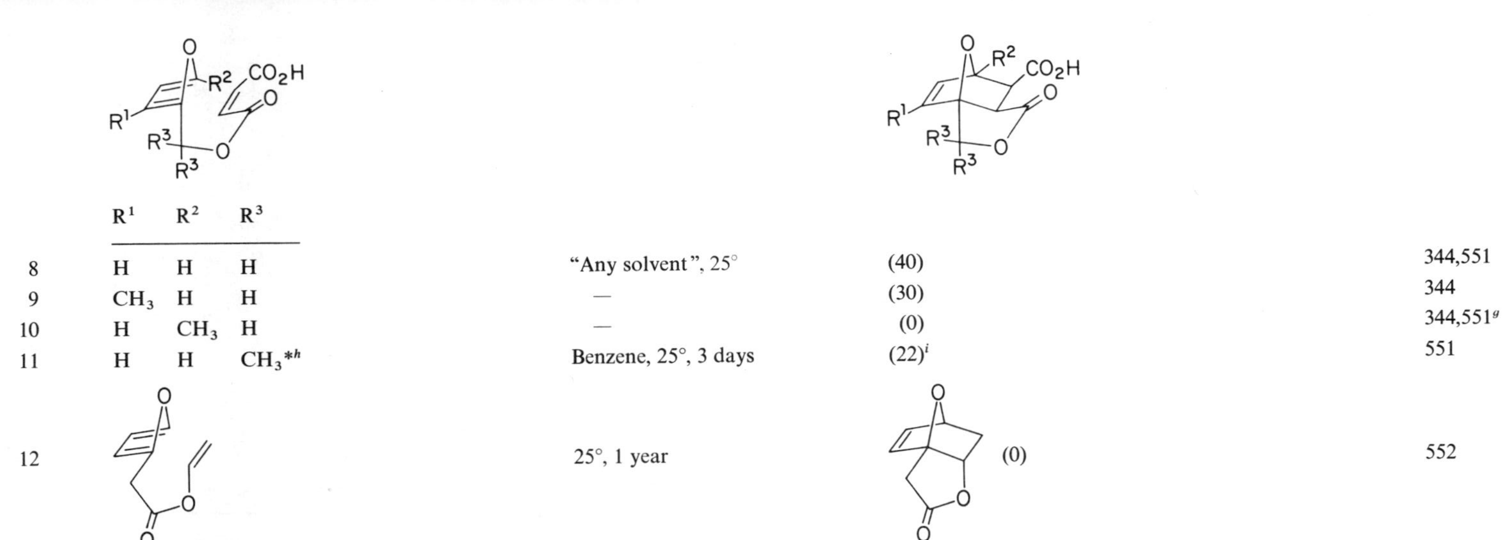

No.	R¹	R²	R³	Conditions	Product (%)	Refs.
8	H	H	H	"Any solvent", 25°	(40)	344,551
9	CH_3	H	H	—	(30)	344
10	H	CH_3	H	—	(0)	344,551[g]
11	H	H	CH_3*[h]	Benzene, 25°, 3 days	(22)[i]	551
12				25°, 1 year	(0)	552

[a] The stereochemistry was not established.

[b] The substrate was generated by photooxygenation of [2.2]2,5-furanophane in methanol.

[c] The structure was not established with certainty.

[d] The substrate was generated by addition of dimethyl acetylenedicarboxylate to furanophane.

[e] Under these conditions the product reverted to dimethyl acetylenedicarboxylate and furanophane.

[f] The substrate was generated by addition of tetrachlorocyclopropene to [2.2]2,5-furanophane.

[g] The products obtained by these authors by reaction of the appropriate furfuryl alcohols with maleic anhydride were later shown to be the intermolecular Diels–Alder adducts.[344]

[h] The substrate was generated from α,α-dimethylfurfuryl alcohol and maleic anhydride.

[i] The product may be the intermolecular Diels–Alder adduct; see footnote *g*.

* This intermediate is hypothetical, or has not been isolated.

TABLE XIX. ACYCLIC AND SEMICYCLIC DIENES CONTAINING HETEROATOMS

Entry No.	Substrate	Conditions	Product(s) and Yield(s)(%)	Refs.
1	CH_3 CH_3 O CH_3 CH_3	CH_3OH, *hν*, Pyrex	CH_3 O CH_3 CH_3 CH_3 (45)[a]	553
2	R R *b H N R R N R $R=CF_3$	Et_2O, −78 to 25°	R R H N R R N R (56)	554
3	R R c O R R R R $R=CF_3$	C_5H_{12}, 140°	R R O R (-) R R R	555

No.	Reactant	Conditions	Product(s) and Yield(s) (%)	Refs.
	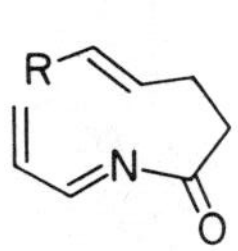 *d (R substituent)	Gas phase, heat		
	R			
4	H		(70–75)	225
5	C_6H_5		(74)	225
6	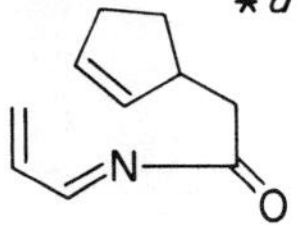 *d 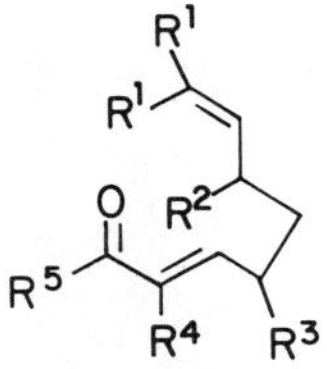	Gas phase, heat	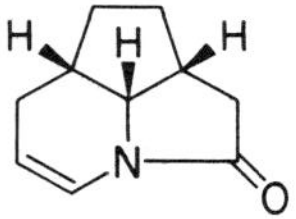 (69) 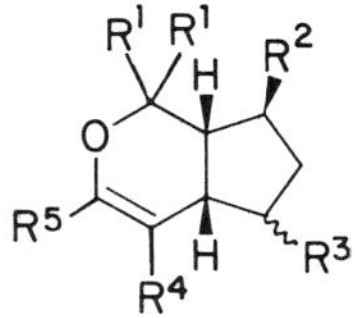	225

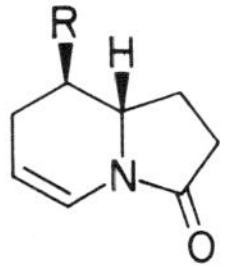

No.	R^1	R^2	R^3	R^4	R^5	Conditions	Yield (%)	Refs.
7	H	CH_3	H	CH_3	H[e]	Gas phase, 350°	(3)[f]	62
7a	CH_3	H	CH_3	H	CH_3	CH_2Cl_2, $(CH_3)_2AlCl$, 25°, 8 hr	(10)[g]	556

TABLE XIX. ACYCLIC AND SEMICYCLIC DIENES CONTAINING HETEROATOMS (*Continued*)

Entry No.	Substrate 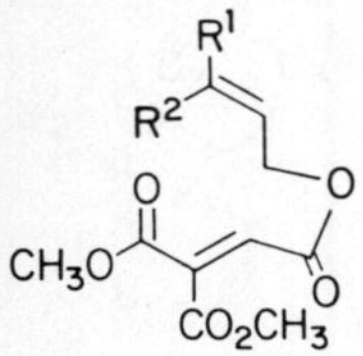		Conditions	Product(s) and Yield(s)(%) 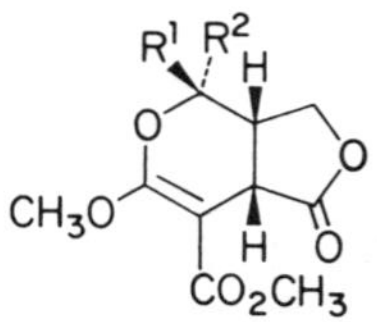	Refs.
	R^1	R^2			
8	CH_3	H	Benzene, 135°, 200 hr	(22)[h]	557
9	H	CH_3	", 137°, 200 hr	(—)	557
10	CH_3	CH_3	", 85°, 112 hr	(—)[h]	557
11	C_6H_5	H	", "	(33)[i]	557
12	$(CH_3)_2C{=}CH(CH_2)_2$	CH_3	", 125°, 6 hr	(—)[h]	557
13	CH_3	$(CH_3)_2C{=}CH(CH_2)_2$	", ", 6 hr	(—)[h]	557
14	*g 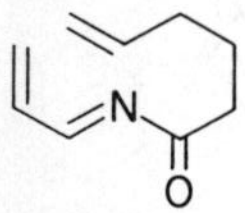		Gas phase, heat	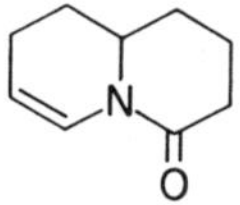 (70–75)	225

No.	Reactant	Conditions	Product(s) and Yield(s) (%)	Refs.
	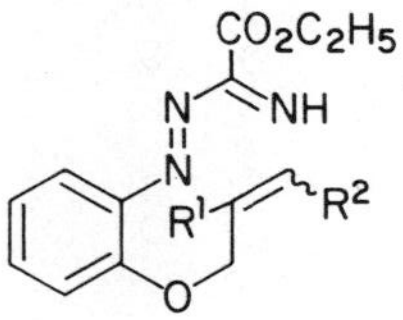 R^1 R^2 (*j)	Benzene, reflux, 2–3 hr 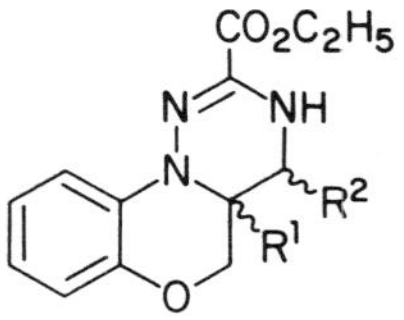		
15	H H		(63)	558
16	CH_3 H		(69)	558
17	H C_6H_5		(79)	558
18	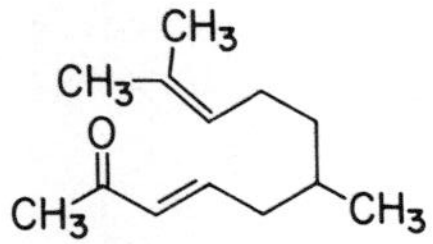	CH_2Cl_2, $(CH_3)_2AlCl$, 25°, 8 hr	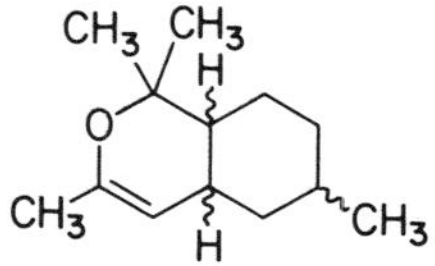 (58)[v]	556,559
	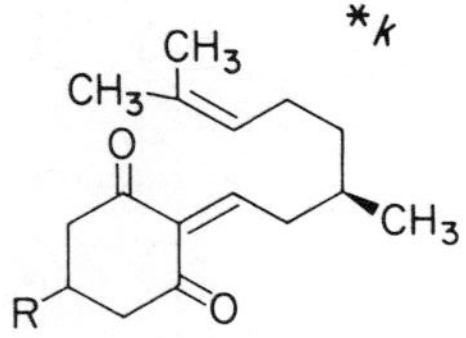 R (*k)	DMF, 100° 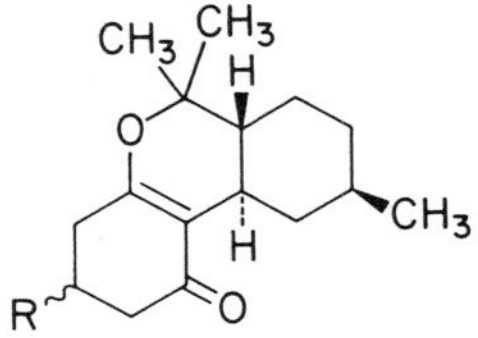		
18a	H		(64)	226,227
18b	n-C_5H_{11}		(65)	322

TABLE XIX. ACYCLIC AND SEMICYCLIC DIENES CONTAINING HETEROATOMS (*Continued*)

Entry No.	Substrate	Conditions	Product(s) and Yield(s)(%)	Refs.
	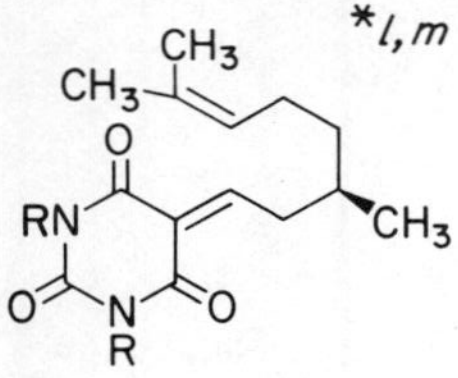 CH_3 CH_3 O RN CH_3 O N R O *l,m		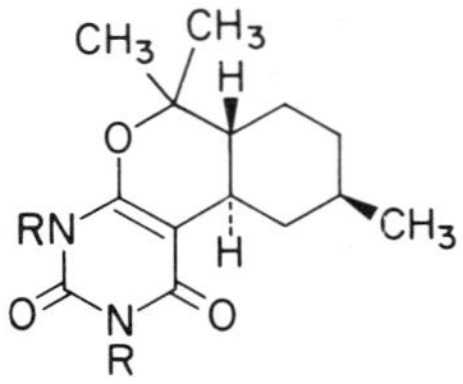 CH_3 CH_3 H O RN H CH_3 O N R O	
	R			
19	H	THF–pyridine–H_2O; reflux	(34)	321
20	CH_3	$H_2N(CH_2)_2NH_3^+CH_3CO_2^-$, 30°	(97)	321
21	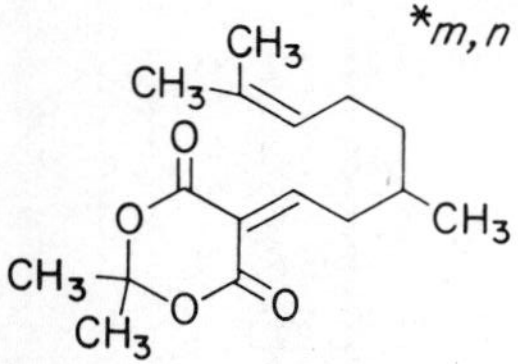 CH_3 CH_3 O O CH_3 CH_3 CH_3 O O *m,n	$H_2N(CH_2)_2NH_3^+CH_3CO_2^-$, 30°	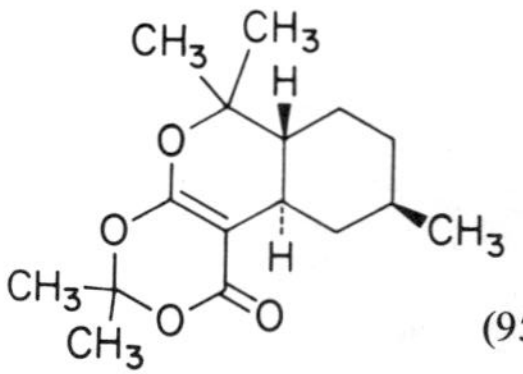 CH_3 CH_3 H O O H CH_3 CH_3 CH_3 O O (95)	321

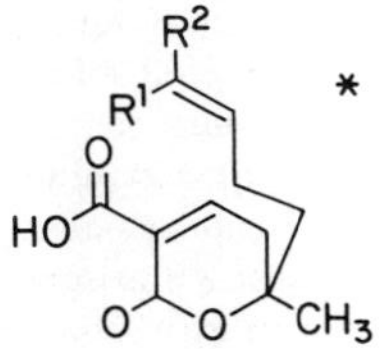

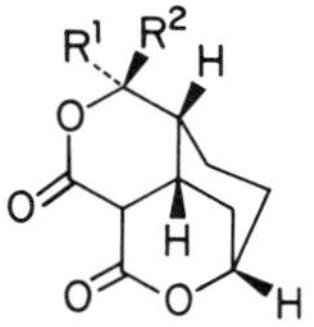

No.	R^1	R^2	Conditions	Product (Yield, %)	Refs.
22	CH_3	CH_3[o, p]	Pyridine, 110°, 15 hr	(35)	560
23	CH_3	$(CH_3)_2C{=}CH(CH_2)_2$[p, q]	", "	(—)	561
24	$(CH_3)_2C{=}CH(CH_2)_2$	CH_3[p, r]	", "	(—)	561

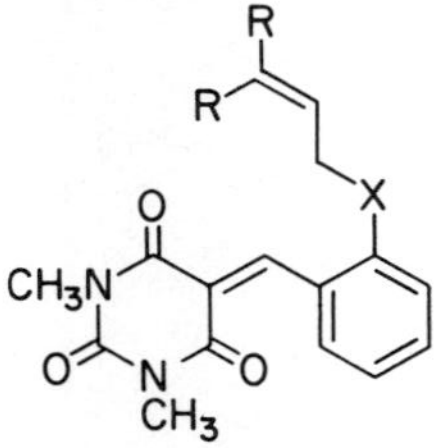

*[s]

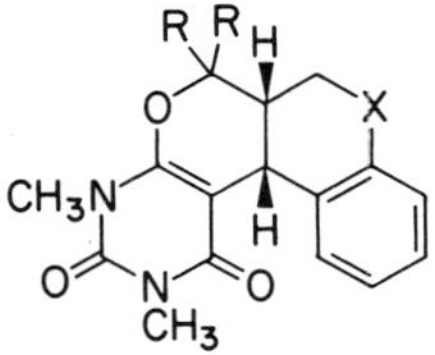

No.	R	X	Conditions	Product (Yield, %)	Refs.
25	H	O	CH_2Cl_2 or CH_3CN, 20°, 5 hr	(70)[t]	562
26	CH_3	O		(91)	562
27	CH_3	S		(33)	562
28	CH_3	CH_2		(69)	562
29	CH_3	CH_3N		(53)	562

TABLE XIX. ACYCLIC AND SEMICYCLIC DIENES CONTAINING HETEROATOMS (*Continued*)

Entry No.	Substrate	Conditions	Product(s) and Yield(s)(%)	Refs.
	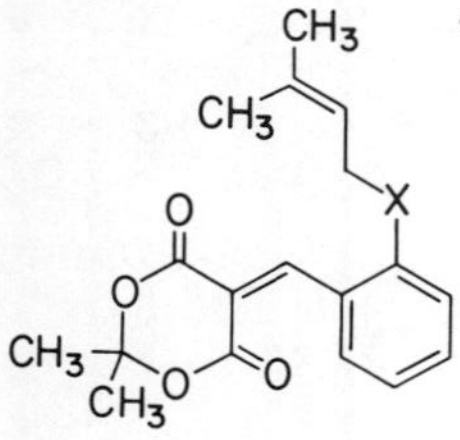 *u	CH_2Cl_2 or CH_3CN, 20°, 5 hr	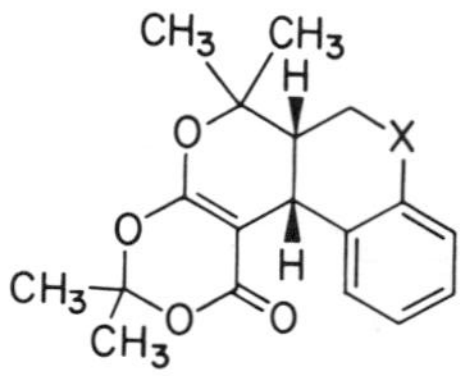	
	X			
30	O		(73)	562
31	CH_2		(65)	562

[a] The bridged product was formed in 5% yield (see Table XXVIc/21).

[b] The substrate was generated from tris(trifluoromethyl)cyclopropenyl trifluoromethyl ketone and trifluoroacetamidine in the presence of $TiCl_4$.

[c] The isomer with the *trans* diene did not cyclize.

[d] The substrate was generated from the *N*-acetoxy-*N*-allylamide by elimination of CH_3CO_2H.

[e] The stereochemistry of the reacting species is uncertain because *trans/cis* interconversion may occur at 350°.

[f] The number is the conversion; at 405° some bridged products were also formed (see Table XXVIc/24).

[g] The stereochemistry of the ring junction was not reported.

[h] The major products were those of an intramolecular ene reaction.

[i] At 115° the yield was 11%; the remainder at both temperatures was starting material.

[j] The substrate was generated from the corresponding $ArNHN{=}C(N_3)CO_2C_2H_5$; the mechanism of the cycloaddition has not been established.

[k] The substrate was generated from (*R*)- or (*S*)-citronellal and the appropriate 1,3-cyclohexanedione.

[l] The substrate was generated from (*R*)-citronellal and 1,3-di-R-barbituric acid.

[m] (*S*)-Citronellal gave the corresponding enantiomer in unspecified yields.

[n] The substrate was generated from (*R*)-citronellal and isopropylidene malonate.

[o] The substrate was generated from citral and malonic acid.

[p] The reaction was originally not formulated as an intramolecular Diels–Alder reaction, but such a mechanism was later proposed for related cyclizations.[232]

[q] The substrate was generated from (2*E*,6*E*)-farnesal and malonic acid.

[r] The substrate was generated from (2*E*,6*Z*)-farnesal and malonic acid.

[s] The substrate was generated from 1,3-dimethylbarbituric acid and the appropriate benzaldehyde in the presence of ethylenediammonium diacetate.

[t] The bridged product was formed in 4% yield; see Table XXVIc/25a.

[u] The substrate was generated from isopropylidene malonate and the appropriate benzaldehyde in the presence of ethylenediammonium diacetate; the substrate could be isolated in some cases.

[v] The reaction is reversible.

* This intermediate is hypothetical, or has not been isolated.

TABLE XX. *o*-QUINODIMETHANES CONTAINING HETEROATOMS IN THE DIENE SYSTEM

Entry No.	Substrate	Conditions	Product(s) and Yield(s)(%)	Refs.
1	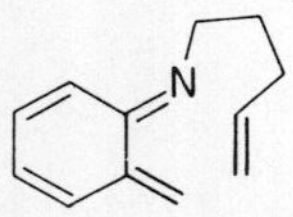 *a,v	CH_3CN, reflux, 2 hr	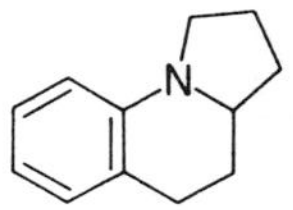 (53)	228,229
	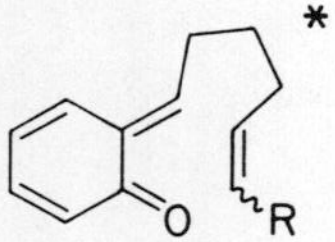 * (R)		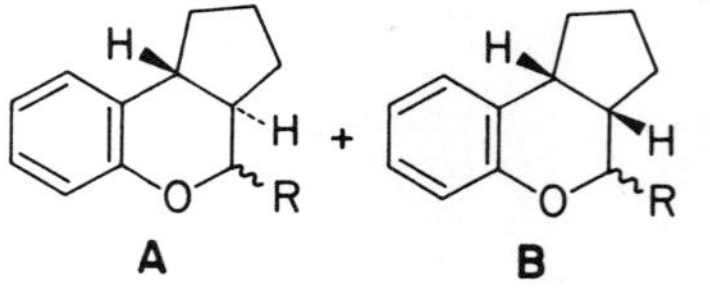 A + B	
	R		A B	
2	H[b]	$C_6H_4N(C_2H_5)_2$, 270°, 27 hr	(30)[c] (10)[c]	230
3	H[d]	Gas phase, 700°	(0) (12)	563
4	C_6H_5[e]	", 280–300°	(75)[f]	231
5	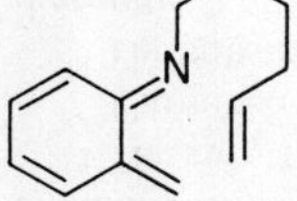 *a,v	CH_3CN, reflux, 2 hr	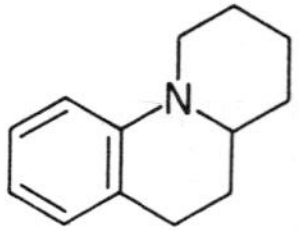 (58)	228,229
6	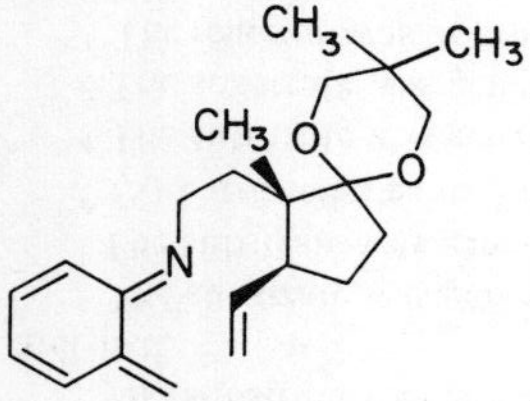 *a (CH_3, CH_3, CH_3)	CH_3CN, reflux, 2 hr	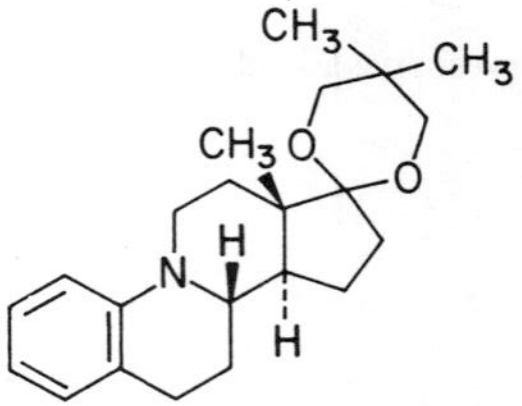 (CH_3, CH_3, CH_3) (60)	228

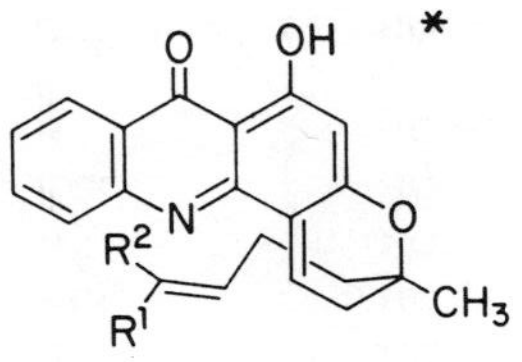

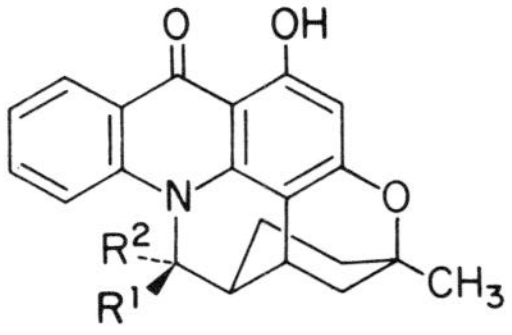

No.	Reactant	Conditions	Product(s) and Yield(s) (%)	Refs.
	*			
	R^1 / R^2			
7	CH_3 / CH_3[g]	Pyridine, 120°, 7 hr	(30)	564
8	$(CH_3)_2C{=}CH(CH_2)_2$ / CH_3[h]	", 150°, 24 hr	(2.5)	564
9	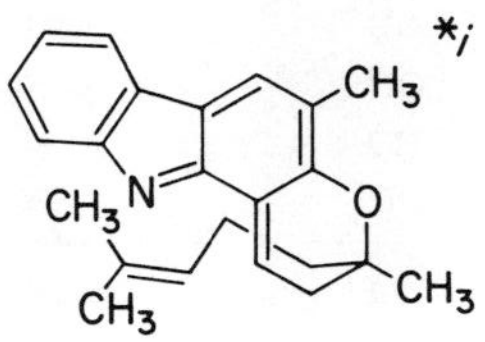 *[i]	Neat, 200°, 3 hr	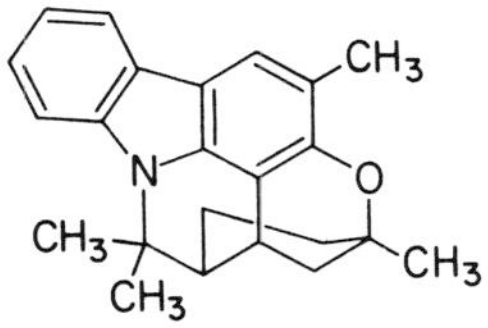 (3.5)	565
10	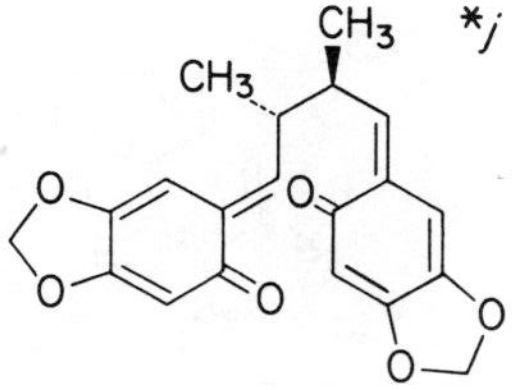 *[j]	CH_3OH, H_2O, 38°, 2 hr	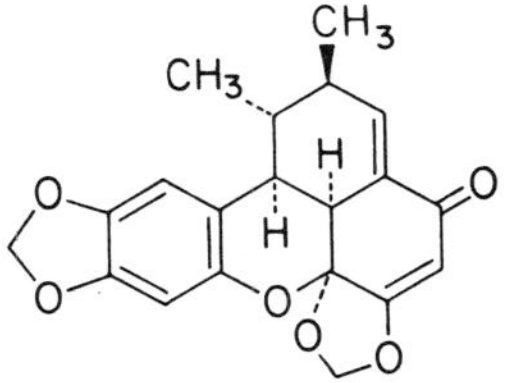 (46)	296

TABLE XX. *o*-Quinodimethanes Containing Heteroatoms in the Diene System (*Continued*)

Entry No.	Substrate					Conditions	Product(s) and Yield(s)(%)	Refs.
	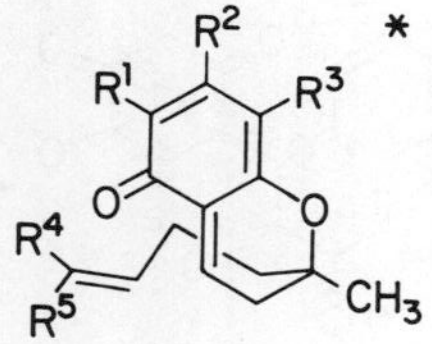 *						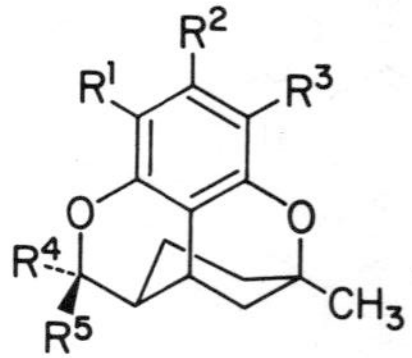	
	R^1	R^2	R^3	R^4	R^5			
11	H	HO	H	CH_3	CH_3[k]	Pyridine, 110°, 7 hr	(40)	233,234
12	H	CH_3	H	CH_3	CH_3[l]	", reflux, 5 hr	(—)	566
13	H	HO	HOC	CH_3	CH_3[m]	", 110°, 6 hr	(10)	567,568
14	HOC	HO	H	CH_3	CH_3[m]	", ", "	(59)	567,568
15	H	HO	CH_3CO	CH_3	CH_3[n]	", ", "	(8)	567,568
16	CH_3CO	HO	H	CH_3	CH_3[n]	", ", "	(68)	567,570
17	H	HO	H	CH_3	$(CH_3)_2C{=}CH(CH_2)_2$[o]	", "	(—)	561
18	H	HO	H	$(CH_3)_2C{=}CH(CH_2)_2$	CH_3[p]	Pyridine, 110°	(—)	561
19	H	HO	HOC	CH_3	$(CH_3)_2C{=}CH(CH_2)_2$[q]	", 50°	(—)	232
20	HOC	HO	H	CH_3	$(CH_3)_2C{=}CH(CH_2)_2$[q]	", "	(—)	232
21	H	HO	HOC	$(CH_3)_2C{=}CH(CH_2)_2$	CH_3[r]	", "	(—)	232
22	HOC	HO	H	$(CH_3)_2C{=}CH(CH_2)_2$	CH_3[r]	", "	(—)	232
23	H	n-C_5H_{11}	H	CH_3	CH_3[s]	", 110°, 7 hr	(26)	571,566
24	$C_6H_5CH{=}CHCO$	HO	H	CH_3	CH_3[t]	", 160°, 6 hr	(57)	572,573 574

	Substrate	Conditions	Product (Yield, %)	Refs.
25	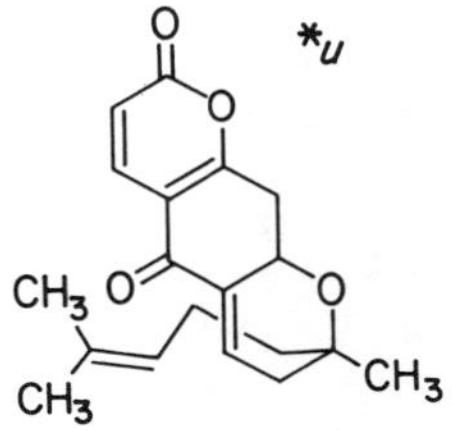 $*^{u}$	Pyridine, 110°, 9 hr	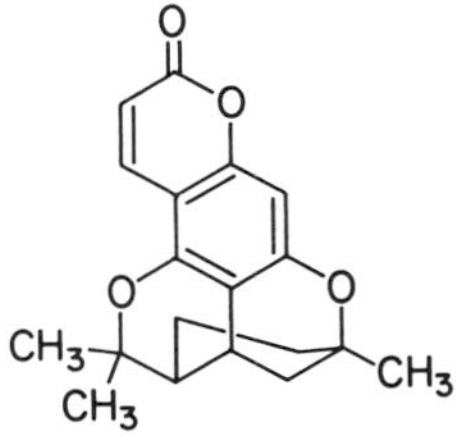 (10)	233,274,275

[a] The substrate was generated by treatment of $C_6H_4[N(R)Si(CH_3)_3][CH_2\overset{+}{N}(CH_3)_3]\ Br^-$ with CsF.

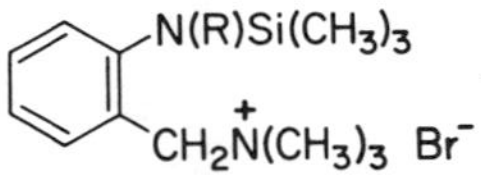

[b] The substrate was generated from 1-(2-hydroxyphenyl)-1,5-hexadiene by 1,5-hydrogen shift.
[c] The bridged product was formed in 29% yield (see Table XXVIc/26).
[d] The substrate was generated by dehydration of 1-(2-hydroxyphenyl)-5-hexenol.
[e] The substrate was generated by dehydration of 1-(2-hydroxyphenyl)-6-phenyl-5-hexenol.
[f] The stereochemistry was not reported.
[g] The substrate was generated from 1,3-dihydroxyacridone and citral.
[h] The substrate was generated from 1,3-dihydroxyacridone and farnesal.
[i] The substrate was generated from mohanimbine.
[j] The substrate was generated by $PdCl_2$-catalyzed dimerization of 2-hydroxy-β-methyl-4,5-methylenedioxystryene.
[k] The substrate was generated from phloroglucinol and citral.
[l] The substrate was generated from orcinol and citral.
[m] The substrate was generated from formylphloroglucinol and citral.
[n] The substrate was generated from acetylphloroglucinol and citral.
[o] The substrate was generated from phloroglucinol and (2*E*,6*E*)-farnesal.
[p] The substrate was generated from phloroglucinol and (2*E*,6*Z*)-farnesal.
[q] The substrate was generated from formylphloroglucinol and (6*E*)-farnesal.
[r] The substrate was generated from formylphloroglucinol and (6*Z*)-farnesal.
[s] The substrate was generated by reaction of olivetol with citral (ref. 566,571,572), dehydrogenation of cannabigerol with chloranil (ref. 573), or by heating of cannabichromene (ref. 571).
[t] The substrate was generated from pinocembrin and citral; the product is rubranine..
[u] The substrate was generated from 5,7-dihydroxycoumarin and citral.
[v] The substrate was generated by flash vacuum pyrolysis of a dihydrobenzoxazine.
* This intermediate is hypothetical, or has not been isolated.

TABLE XXI. DIHYDROPYRIDINES AND PYRIDONES

Entry No.	Substrate	Conditions	Product(s) and Yield(s)(%)	Refs.
1	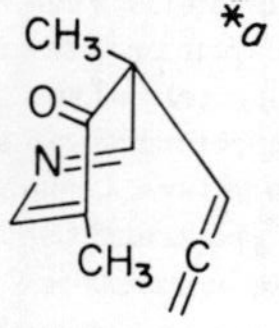 CH_3, O, N, CH_3, C *[a]	Benzene, flow system, 450°	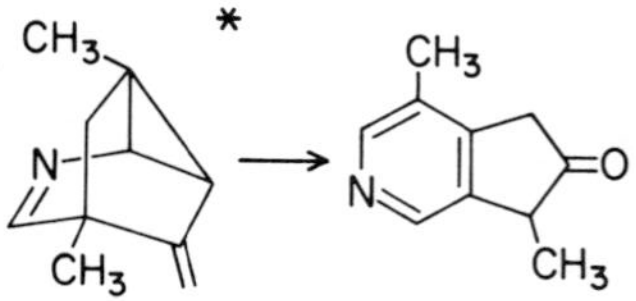 CH_3, N, CH_3 * → CH_3, N, O, CH_3 (55)	575
	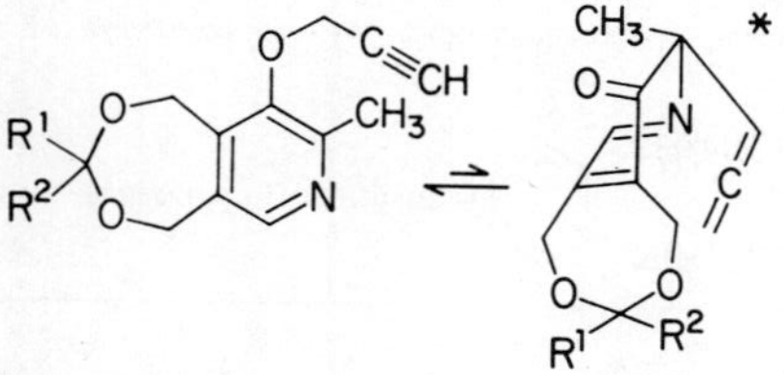 O, $C\equiv CH$, CH_3, N, R^1, R^2, O, O ⇌ CH_3, O, N, C, O, O, R^1, R^2 *		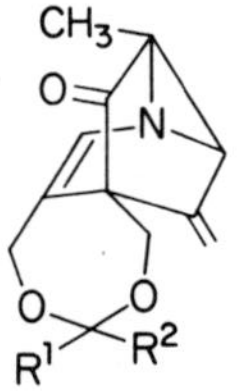 CH_3, O, N, O, O, R^1, R^2	
	R^1 R^2			
2	CH_3 CH_3	DMF, 190°, 10 hr	(35)	576
		n-$C_{10}H_{22}$, 188°, 7 hr	(21)	576
3	i-C_3H_7 H	DMF, 195°, 3 hr	(42)[b]	576
		n-$C_{10}H_{22}$, 208°, 3 hr	(38)[b]	576

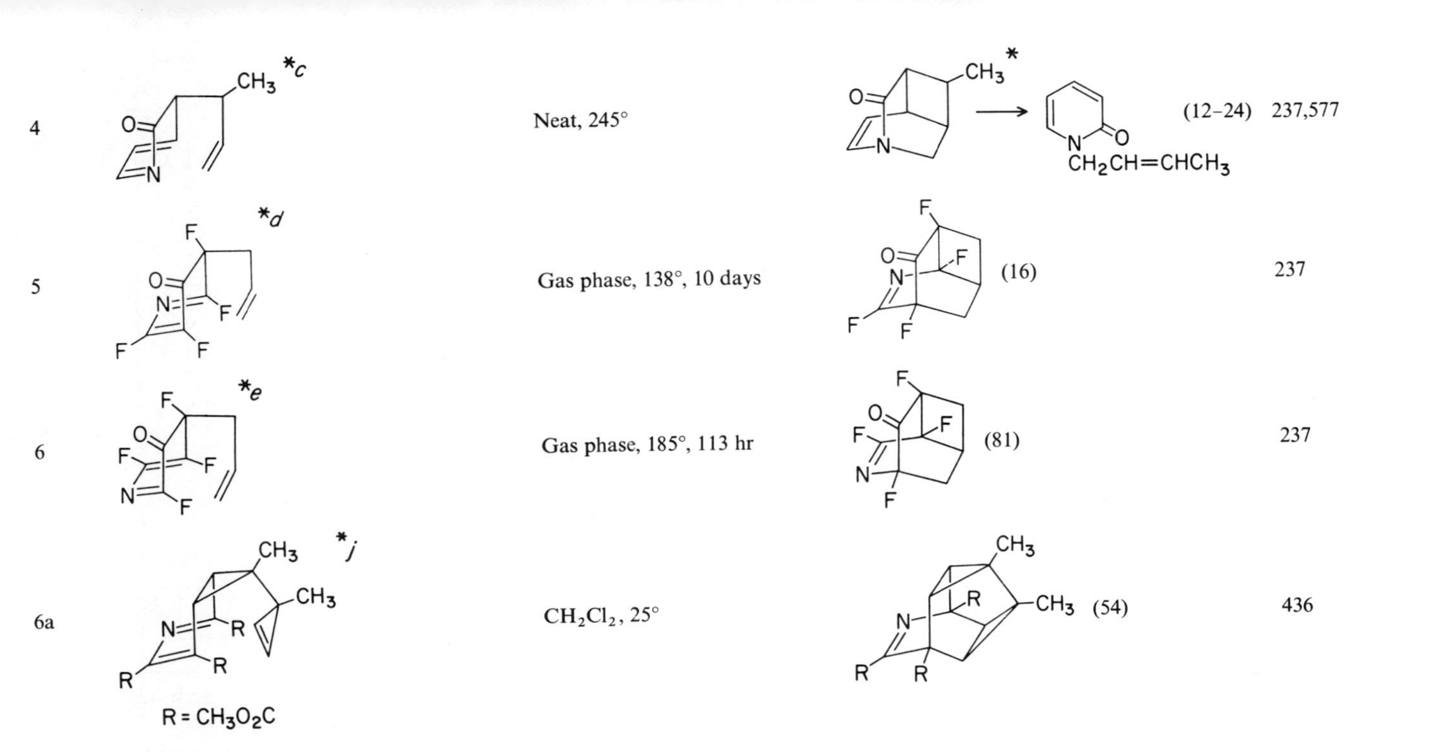

4
CH3 *c
O
N
Neat, 245°
CH3 *
O
N
CH2CH=CHCH3
(12–24)
237,577
5
*d
F
O
N
F
F
F
Gas phase, 138°, 10 days
(16)
237
6
*e
Gas phase, 185°, 113 hr
(81)
237
6a
CH3
*j
R
R = CH3O2C
CH2Cl2, 25°
(54)
436

TABLE XXI. DIHYDROPYRIDINES AND PYRIDONES (*Continued*)

Entry No.	Substrate 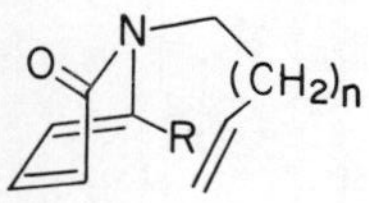	Conditions	Product(s) and Yield(s)(%) 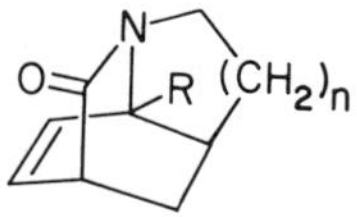	Refs.
	R n			
7	H 1	Heat, to 700°	(0)	350
7a	CH_3 2	", to 250°	(0)	578
7b	CH_3N … $Si(CH_3)_3$	Heat, to 250°	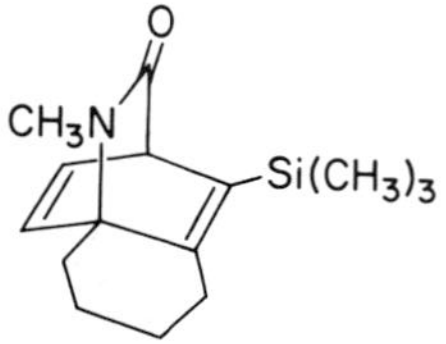 CH_3N … $Si(CH_3)_3$ (0)	578
7c	CH_3N	Heat, to 250°	CH_3N (0)	578

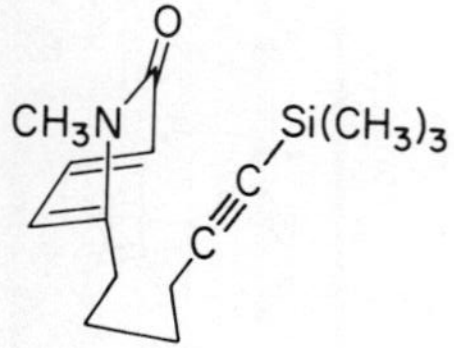

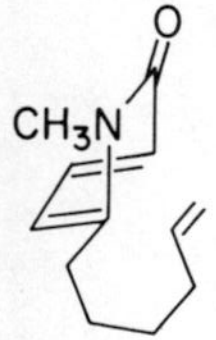

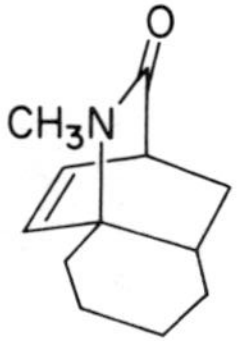

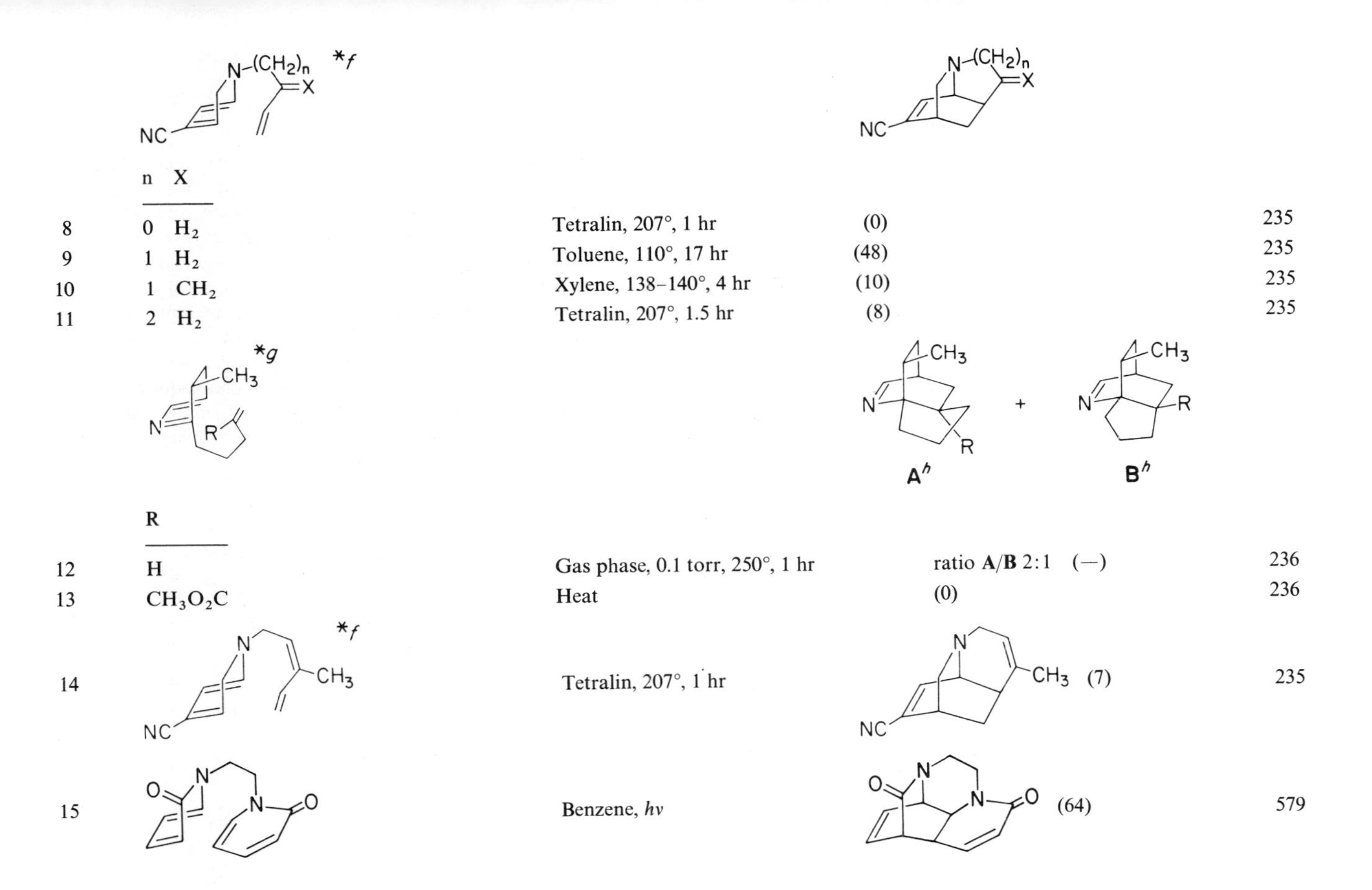

No.	Reactant	Conditions	Product(s) and Yield(s) (%)	Refs.
	*f (n, X)			
8	n = 0, X = H_2	Tetralin, 207°, 1 hr	(0)	235
9	n = 1, X = H_2	Toluene, 110°, 17 hr	(48)	235
10	n = 1, X = CH_2	Xylene, 138–140°, 4 hr	(10)	235
11	n = 2, X = H_2	Tetralin, 207°, 1.5 hr	(8)	235
	*g (R)		A[h] + B[h]	
12	R = H	Gas phase, 0.1 torr, 250°, 1 hr	ratio **A**/**B** 2:1 (—)	236
13	R = CH_3O_2C	Heat	(0)	236
14	*f	Tetralin, 207°, 1 hr	(7)	235
15		Benzene, *hν*	(64)	579

TABLE XXI. DIHYDROPYRIDINES AND PYRIDONES (*Continued*)

Entry No.	Substrate	Conditions	Product(s) and Yield(s)(%)	Refs.
16	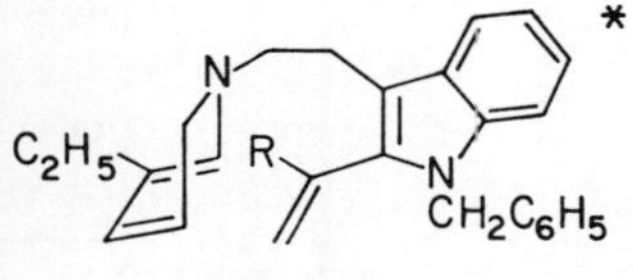 C_2H_5, R, N, $CH_2C_6H_5$ *[i] R = CH_3O_2C	$H_2N(CH_2)_2NH_2$, 25°, 25 hr	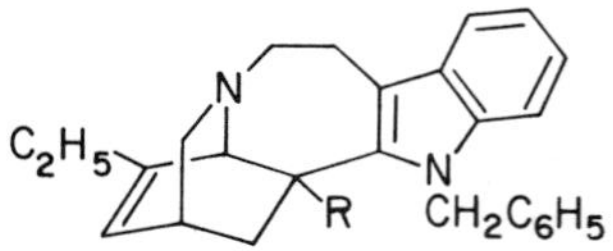 C_2H_5, N, R, $CH_2C_6H_5$ (1.5)	283

[a] The substrate was generated by Claisen rearrangement of 3,5-dimethyl-4-pyridyl propargyl ether.
[b] The product was a mixture of two isomers (R^1, R^2 *exo*/*endo*).
[c] The substrate was generated by Claisen rearrangement of 2-crotyloxypyridine; the mechanism was proposed by the authors of ref. 237 for a rearrangement reported in ref. 577.
[d] The substrate was generated by Claisen rearrangement of 2,3,5,6-tetrafluoro-4-pyridyl allyl ether.
[e] The substrate was generated by Claisen rearrangement of 2,3,5,6-tetrafluoro-3-pyridyl allyl ether.
[f] The substrate was generated by thermal cracking of the intermolecular Diels–Alder dimer.
[g] The substrate was generated from 1-[5(R)-5-hexenyl]-1,2-dihydropyridine.
[h] The structure assignments are tentative.
[i] The substrate was generated from the dihydropyridine chromium tricarbonyl complex.
[j] The substrate was generated by addition of 3-methyl-3-(3′-methyl-3′-cyclopropenyl)cyclopropene to trimethyl 1,2,4-triazine-2,5,6-tricarboxylate.
* This intermediate is hypothetical, or has not been isolated.

TABLE XXII. PYRIDAZINES

Entry No.	Substrate 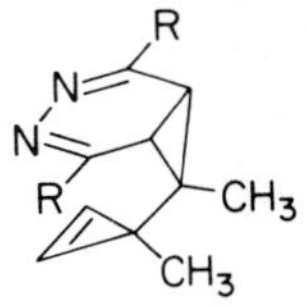	Conditions	Product(s) and Yield(s)(%) 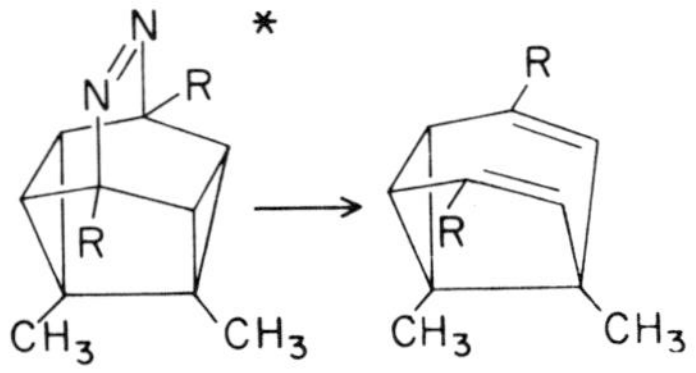	Refs.
	R		*	
1	CH_3	—	(38)	328
2	CH_3O_2C	—	(70)	328,436 579a
3	C_6H_5*[h]	$CHCl_3$, 50–60°	(74)	328,436
3a	*m*-ClC_6H_4*[h]	", "	(32)	436
3b	*p*-$CH_3C_6H_4$*[h]	", "	(84)	436
3c	*m*-$CH_3OC_6H_4$*[h]	", "	(33)	436
3d	*m*-$CF_3C_6H_4$*[h]	", "	(77)	436

TABLE XXII. PYRIDAZINES (*Continued*)

Entry No.	Substrate	Conditions	Product(s) and Yield(s)(%)		Refs.
	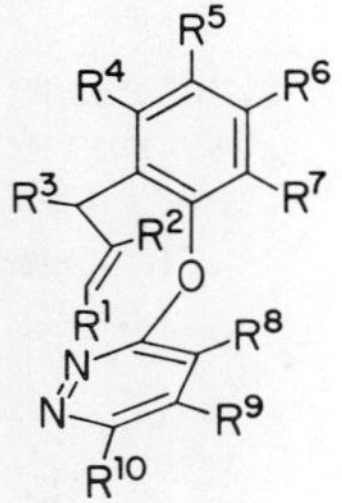		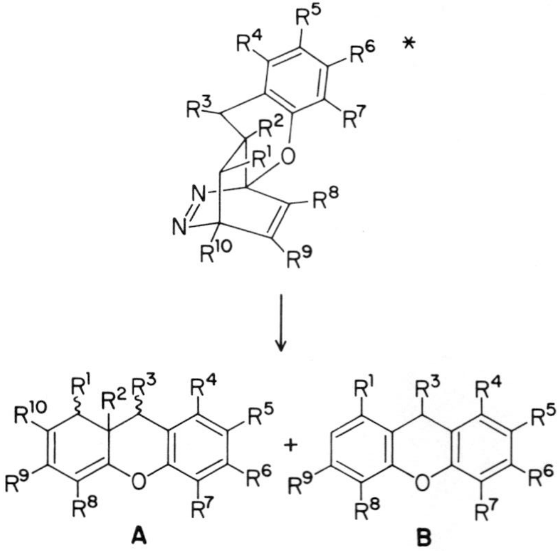		
	R^1 to R^9 = H; R^{10} = Cl except as noted below		A	B	
4		$C_6H_5N(C_2H_5)_2$ or tetralin, reflux, 2 hr	(0)	(80)	238
4a	R^5 = Cl	", reflux, 2 hr	(0)	(58)	238
4b	R^7 = Cl	", ", "	(0)	(75)	238
4c	R^7 = HO*[a]	", 210°, 3 hr	(0)	(30)	238
5	R^5 = CH_3; R^{10} = H	", reflux, 2 hr	(—)	(28)[b]	240
6	R^{10} = CH_3	", ", "	(—)	(35)[b,c]	240
7	R^1 = CH_3	", ", 12 hr	(0)	(20)	238

8	$R^2 = CH_3$	", ", 2 hr	(99)	(0)	239
9	$R^3 = CH_3$	", ", "	(0)	(98)	238
10	$R^5 = CH_3$	", ", "	(0)	(79)	238
11	$R^7 = CH_3$	", ", "	(0)	(64)	238
12	$R^8 = CH_3$	", 210–220°, 2 hr	(0)	(89)	238
13	$R^9 = CH_3$	", 200°, 2 hr	(0)	(65)	238
14	$R^3 = CH_3$; $R^5 = Cl$	$C_6H_5N(C_2H_5)_2$, reflux, 2 hr	(0)	(60)	238
15	$R^5 = CH_3O$	", ", "	(0)	(87)	238
16	$R^7 = CH_3O$	", ", "	(0)	(78)	238
17	$R^8 = CH_3O$	", 210–220°, 2 hr	(0)	(0)	238
18	$R^9 = CH_3O$	", reflux, 1 hr	(0)	(18)	238
19	$R^1 = CH_3$; $R^7 = HO^{*d}$	—	(0)	(—)	241
20	$R^3 = CH_3$; $R^7 = HO^{*e}$	$C_6H_5N(C_2H_5)_2$, 210°, 3 hr	(0)	(14)	241
21	R^2, $R^{10} = CH_3$	", reflux, 3 hr	(60)	(0)	239
22	R^2, $R^7 = CH_3$	", ", 2 hr	(94)	(0)	239
23	R^2, $R^5 = CH_3$	", ", 10 hr	(79)	(0)	239
24	R^3, $R^5 = CH_3$	", ", 2 hr	(0)	(66)	238
25	R^3, $R^7 = CH_3$	", ", "	(0)	(61)	238
26	R^4, $R^6 = CH_3$	", ", "	(0)	(76)	238
27	R^5, $R^7 = CH_3$	", ", "	(0)	(72)	238
28	$R^2 = CH_3$; $R^7 = CH_3O$	", ", "	(44)	(0)	239
29	$R^3 = CH_3$; $R^7 = CH_3O$	$C_6H_5N(C_2H_5)_2$, reflux, 3.5 hr	(0)	(100)	241
30	$R^7 = CH_3O_2C^{*f}$	Neat, K_2CO_3, 150–160°, 0.5 hr	(13)	(0)	240
31	$R^7 = i\text{-}C_3H_7$	$C_6H_5N(C_2H_5)_2$, reflux, 2 hr	(0)	(45)	238
32	$R^7 = CH_2{=}CHCH_2$	", ", "	(0)	(59)	238
33	$R^5 = C_2H_5O_2C^{*f}$	Neat, K_2CO_3, 150–160°, 1 hr	(13)	(0)	240
34	$R^5 = t\text{-}C_4H_9$	$C_6H_5N(C_2H_5)_2$, reflux, 2 hr	(0)	(70)	238
35	R^8–$R^9 = —(CH{=}CH)_2—$	", ", "	(0)	(64)	238
36	R^8–$R^9 = —(CH_2)_4—$	", ", "	(0)	(90)	238
37	$R^2 = CH_3$; $R^5 = C_2H_5O_2C$	Neat, 180–185°, 4 hr	(11)	(0)	239
38	$R^2 = CH_3$; R^4–$R^5 = —(CH{=}CH)_2—$	", ", 3 hr	(22)	(0)	239

TABLE XXII. PYRIDAZINES (*Continued*)

Entry No.	Substrate	Conditions	Product(s) and Yield(s)(%) A	B	Refs.
39	$R^5 = CH_3; R^8\text{–}R^9 = \text{—}(CH{=}CH)_2\text{—}$	$C_6H_5N(C_2H_5)_2$, reflux, 2 hr	(0)	(88)	238
40	$R^{10} = C_6H_5$	", ", 2.5 hr	(83)	(0)	240
41	$R^5 = C_6H_5$*[f]	Neat, K_2CO_3, 150–160°, 2 hr	(0)	(45)	240
42	$R^7 = C_6H_5$	$C_6H_5N(C_2H_5)_2$, reflux, 2 hr	(0)	(91)	238
43	$R^9 = C_6H_5$	", ", 1.5 hr	(0)	(98)	238
44	$R^5 = Cl; R^{10} = C_6H_5$*[f]	", "	(68)	(0)	240
45	$R^3 = C_6H_5; R^7 = HO$*[g]	", 210°, 3 hr	(0)	(26)	241
46	$R^5 = O_2N; R^{10} = C_6H_5$*[f]	Neat, 180°, 3.5 hr	(46)	(0)	240
47	$R^2 = CH_3; R^{10} = C_6H_5$	$C_6H_5N(C_2H_5)_2$, reflux, 3 hr	(88)	(0)	239
48	$R^3 = CH_3; R^{10} = C_6H_5$*[f]	", "	(73)	(0)	240
49	$R^5 = CH_3; R^{10} = C_6H_5$*[f]	", "	(100)	(0)	240
50	$R^2 = CH_3; R^7 = C_6H_5$	", ", 5 hr	(19)	(0)	239
51	$R^3 = C_6H_5; R^5 = CH_3$	", ", 2 hr	(0)	(65)	238
52	$R^3 = C_6H_5; R^7 = CH_3O$	", ", 3.5 hr	(0)	(87)	241
53	$R^5 = CH_3S; R^{10} = C_6H_5$*[f]	", "	(73)	(0)	240
54	$R^4, R^6 = CH_3; R^{10} = C_6H_5$	", "	(78)	(0)	240
55	$R^5 = C_6H_5C(CH_3)_2$	", ", 2 hr	(0)	(69)	238
56	$R^5 = C_2H_5O_2C; R^{10} = C_6H_5$	$C_6H_5N(C_2H_5)_2$, reflux	(100)	(0)	240
57	$R^8\text{–}R^9 = \text{—}(CH{=}CH)_2\text{—}; R^{10} = C_6H_5$	", ", 1.5 hr	(97)	(0)	240
58	$R^2 = CH_3; R^8\text{–}R^9 = \text{—}(CH{=}CH)_2\text{—}; R^{10} = C_6H_5$	", ", 6 hr	(60)	(0)	239

[a] The substrate was generated from 3-chloro-6-(2-allyloxyphenoxy)pyridazine by a formal *meta*-Claisen rearrangement.
[b] The yield is after oxidation with SeO_2.
[c] Product **B** retains the substituent R^{10}.
[d] The substrate was generated from 3-chloro-6-[2-(2-methallyloxy)phenoxy]pyridazine by a formal *meta*-Claisen rearrangement.
[e] The substrate was generated from 3-chloro-6-[2-(2-butenyloxy)phenoxy]pyridazine by a formal *meta*-Claisen rearrangement.
[f] The substrate was generated from 3,6-dichloropyridazine and the corresponding phenol.
[g] The substrate was generated from 3-chloro-6-[2-(3-phenylallyloxy)phenoxy]pyridazine by a formal *meta*-Claisen rearrangement.
[h] The substrate was generated by addition of 3-methyl-3-(3′-methyl-3′-cyclopropenyl)cyclopropene to the appropriate tetrazine.
* This intermediate is hypothetical, or has not been isolated.

TABLE XXIII. Pyrimidines and Pyrimidones

Entry No.	Substrate	Conditions	Product(s) and Yield(s)(%)	Refs.
	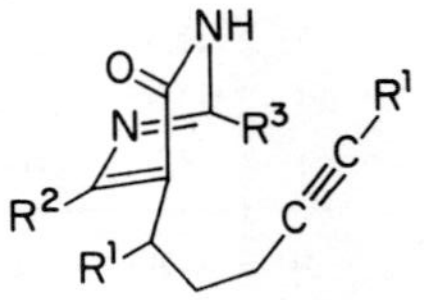		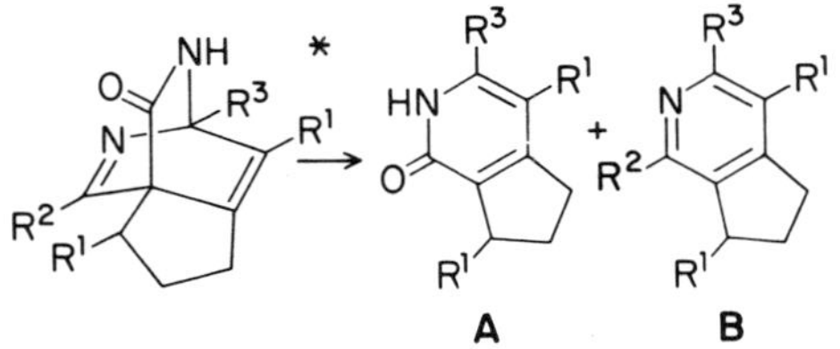	
	R^1 R^2 R^3		A B	
1	H HO CH_3	DMF, 200°, 6 hr	(46) (0)	289,243
2	H CH_3 CH_3	180°	(0) (60)	243
3	CH_3 HO H	Neat, 200°, 10 min	(87) (0)	289
4	H HO C_6H_5	DMF, 200°, 18 hr	(40) (0)	289
5	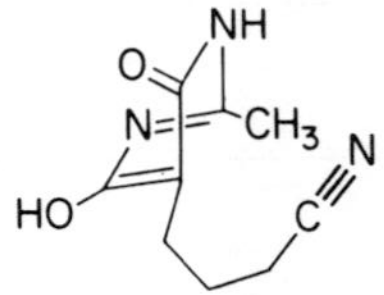	DMF, 198°, 20 hr	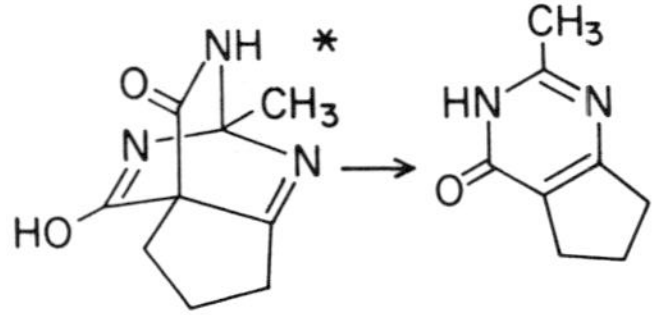 (—)	244,243

TABLE XXIII. PYRIMIDINES AND PYRIMIDONES (*Continued*)

Entry No.	Substrate			Conditions	Product(s) and Yield(s)(%)		Refs.
	(structure: R[1], R[2], $(CH_2)_n$)				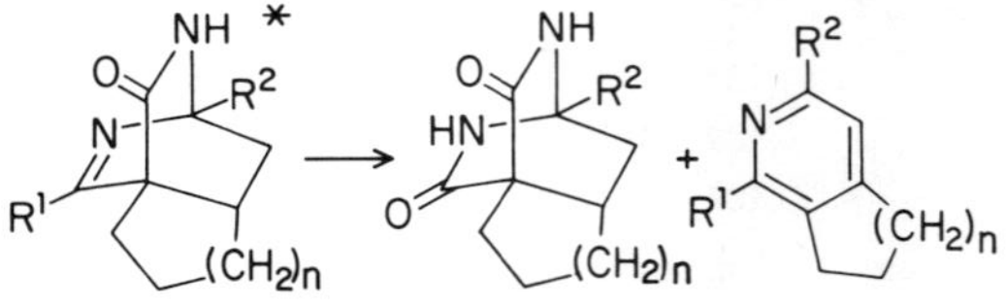 A * → B + C[a]		
	R^1	R^2	n		B	C	
6	HO	CH_3	1	DMF, 198°, 6 hr	(98)	(0)	244,245
7	HO	C_6H_5	1	198°, 48 hr	(59)	(0)	244
8	CH_3	CH_3	1	DMF or CH_3CN, 198°, 18 hr	(0)	(65)[b]	244,243
9	C_6H_5	CH_3	1	", ", 15 hr	(0)	(51)[b]	244,243
10	HO	CH_3	2	", 198°, 24 hr	(44)	(0)	244,245
11	(structure)			198°, 16 hr	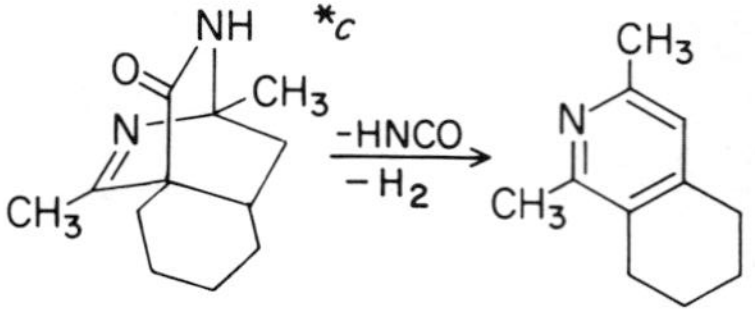 *[c] $\xrightarrow[-H_2]{-HNCO}$ (56)		244,243

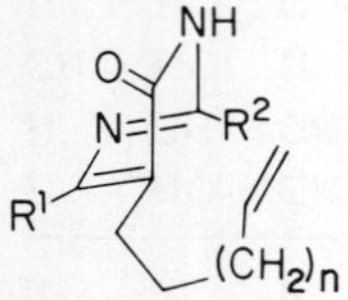

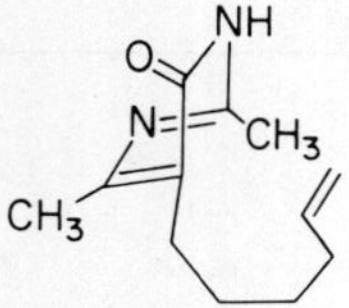

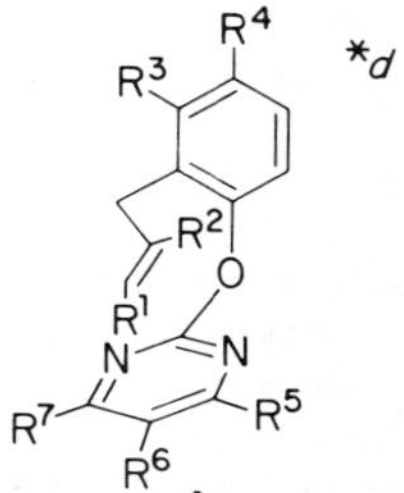

*[d]

A → B →(−R^7CN, −HR^6) C

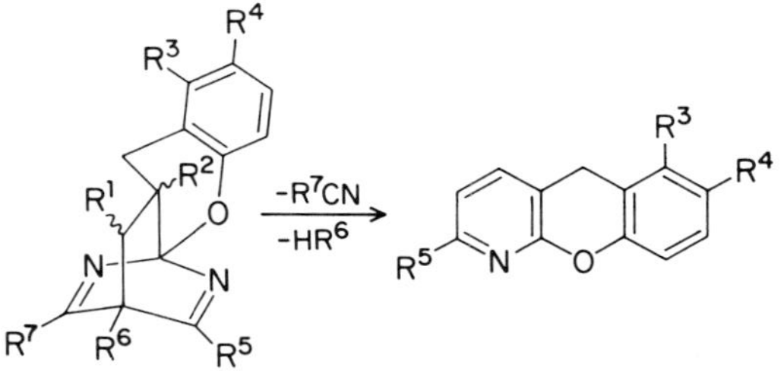

No.	R^1	R^2	R^3	R^4	R^5	R^6	R^7	Conditions	A	B	C	Refs.
12	H	H	H	H	H	Cl	H	Xylene, Na, reflux, 27 hr	(0)	(24)	(0)	242
13	H	H	H	H	H	Br	H	Neat, K_2CO_3, 150°, 2 hr	(14)	(0)	(27)	242
14	H	H	H	H	CH_3	H	CH_3	Xylene, Na, reflux, 7–13 hr	(78)	(5)	(0)	242
15	H	H	H	H	CH_3	Br	CH_3	", ", ", "	(39)	(25)	(0)	242
16	H	H	H	O_2N	CH_3	Br	CH_3	Neat, K_2CO_3, 150°, 1–7 hr	(17)	(24)	(0)	242
17	H	H	H	H	CH_3	CH_3	CH_3	Xylene, Na, reflux, 7–13 hr	(56)	(7)	(0)	242
18	H	H	H	CH_3	CH_3	Br	CH_3	", ", ", "	(40)	(19)	(0)	242
19	H	H	H	CH_3O	CH_3	Br	CH_3	", ", ", "	(67)	(9)	(0)	242
20	CH_3	H	H	H	CH_3	Br	CH_3[e]	Neat, 150°, 7 hr	(—)	(0)	(0)	242
21	H	CH_3	H	H	CH_3	Br	CH_3[e]	", ", "	(—)	(0)	(0)	242
22	H	H	—(CH=CH)$_2$—		CH_3	Br	CH_3	Xylene, Na, reflux, 1–7 hr	(16)	(43)	(0)	242

TABLE XXIII. PYRIMIDINES AND PYRIMIDONES (*Continued*)

Entry No.	Substrate	Conditions	Product(s) and Yield(s)(%)	Refs.
23	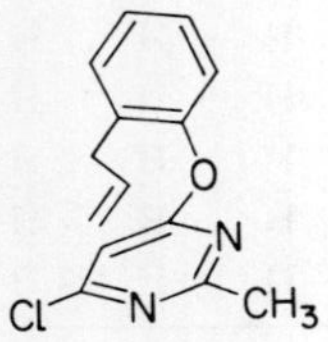 Cl, N, CH3, O	160°, 7 hr	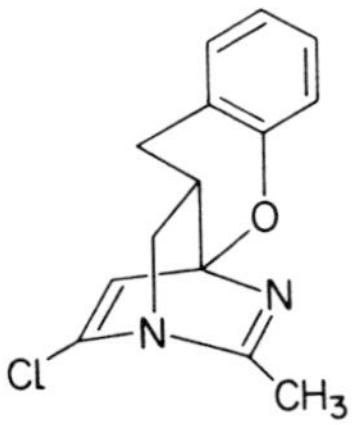 O, N, N, Cl, CH3 (0)	242

[a] The product **C** was formed from **A** by loss of HCNO and H_2.
[b] The intermediate **A** was detected by TLC.
[c] This intermediate could be detected by TLC.
[d] The substrate was generated *in situ* from the appropriate phenol and 2-chloro-R^5, R^6, R^7-pyrimidine.
[e] The reaction was carried out with preformed **A**.
* This intermediate is hypothetical, or has not been isolated.

TABLE XXIV. α-PYRONES

Entry No.	Substrate	Conditions	Product(s) and Yield(s)(%)	Refs.
	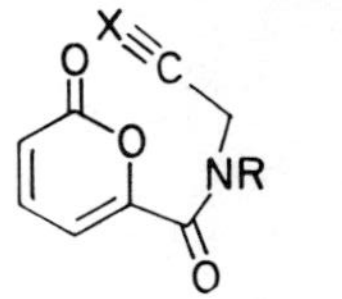		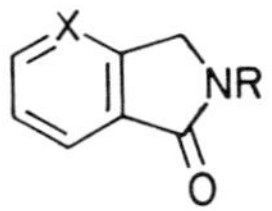	
	R / X			
1	$C_6H_5CH_2$ / CH	25° ($t_{1/2}$ = 4 days) or THF, reflux	(61)	167
2	CH_3 / N	Toluene, 250°, 4 hr	(—)	167
	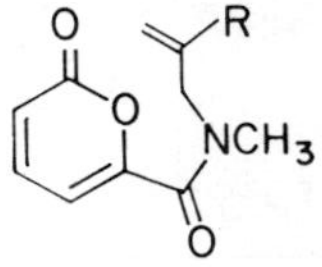		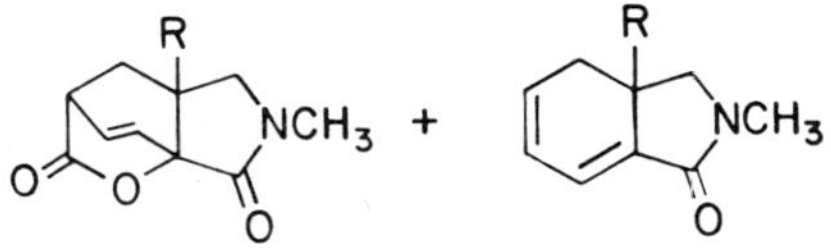 A B	
	R		A / B	
3	H	25° ($t_{1/2}$ = 12 hr)	(80)[a] / (0)	167
4	C_6H_5	p-Xylene, reflux, 25 min	(0) / (70)	167
5	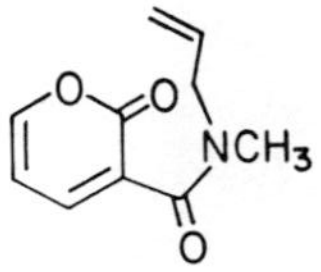	p-Xylene, 138°	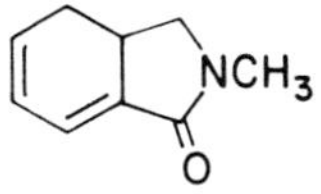 (—)	167

TABLE XXIV. α-PYRONES (*Continued*)

Entry No.	Substrate		Conditions	Product(s) and Yield(s)(%)	Refs.
6	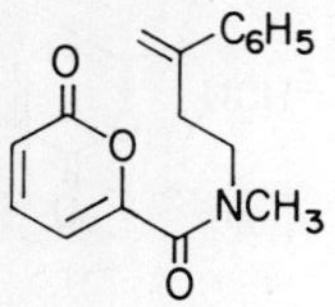		Toluene, 180°, 6 hr	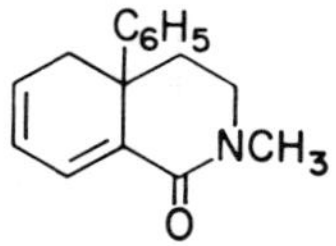 (48)	167
	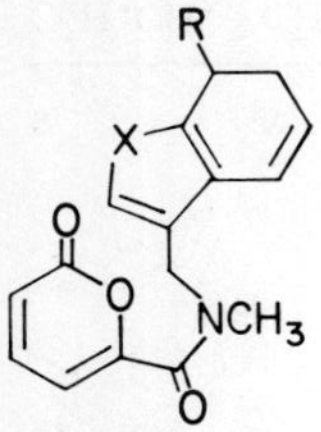			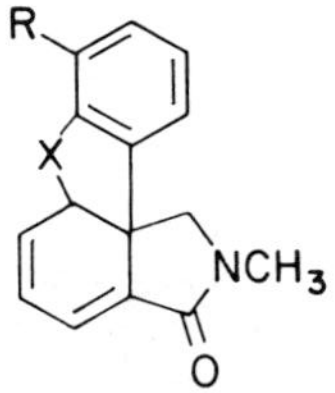	
	X	R			
7	o-C_6H_4	H	Toluene, 240°, 8 hr	(—)	167
8	$CH_2{=}CHCH_2N$	H	", 150°, 4 hr	(—)	167
9	O	CH_3O	o-$Cl_2C_6H_4$, 178°, 1 hr	(—)	167

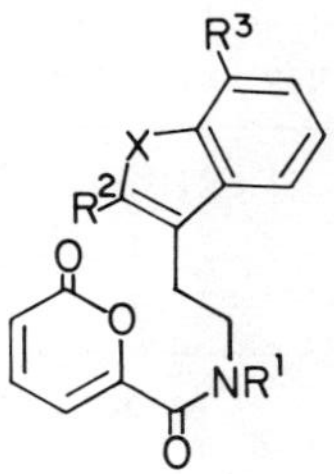

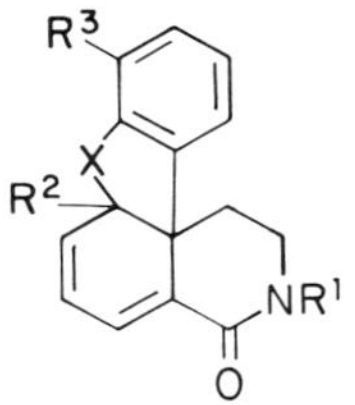

No.	X	R^1	R^2	R^3	Conditions	Product (Yield, %)	Refs.
10	CH_2	CH_3	H	H	Toluene, 210°, 6 hr	(—)	167
11	$(CH_2)_2$	CH_3	H	H	", 230°, 8 hr	(45)	167
12	CH_3N	CH_3	H	H	", ", 4 hr	(—)	167
13	O	CH_3	H	H	", 225°, 8 hr	(35)	580
14	O	CH_3	H	CH_3O	1,2,4-$Cl_3C_6H_3$, reflux, 10 hr	(53)	134,580
15	O	c-$C_3H_5CH_2$	H	CH_3O	", ", 7 hr	(67)	134,580
16	O	c-$C_3H_5CH_2$	CH_3	CH_3O	", ", 36 hr	(—)	167
17	O	$C_6H_5CH_2$	H	CH_3O	1,2,4-$Cl_3C_6H_3$, reflux, 5 hr	(50)	134,580
18	S	$C_6H_5CH_2$	H	H	Toluene, 270°, 4 hr	(—)	167
19					Toluene, 240°, 6 hr	* → (—)	167

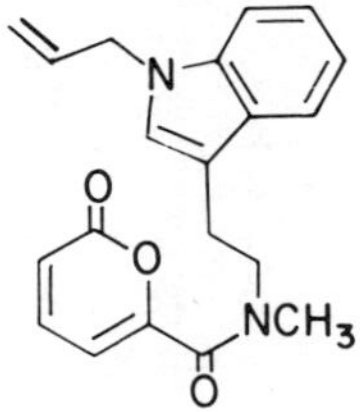

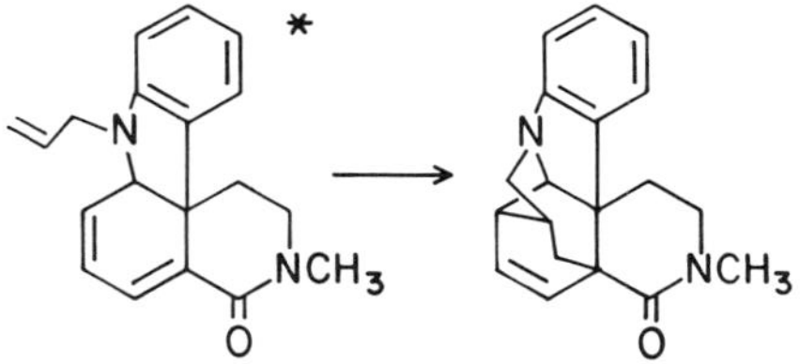

TABLE XXIV. α-PYRONES (*Continued*)

Entry No.	Substrate	Conditions	Product(s) and Yield(s)(%)	Refs.
	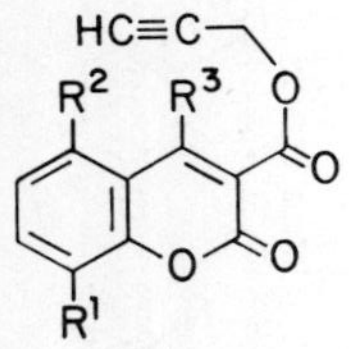	Toluene, 300°, 1.5–2 hr	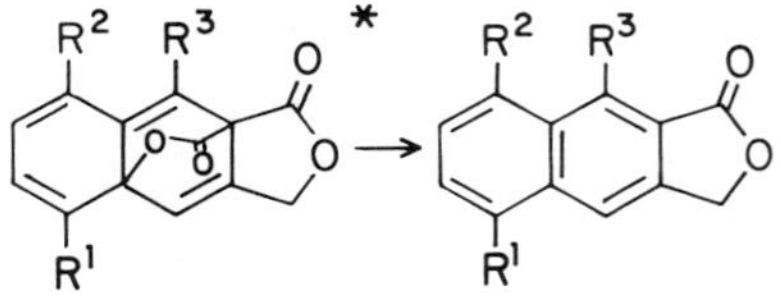	
	R^1 R^2 R^3			
20	H H H		(58)	246
21	H H CH_3		(47)	246
22	CH_3O H H		(61)	246
23	H CH_3O H		(51)	246
	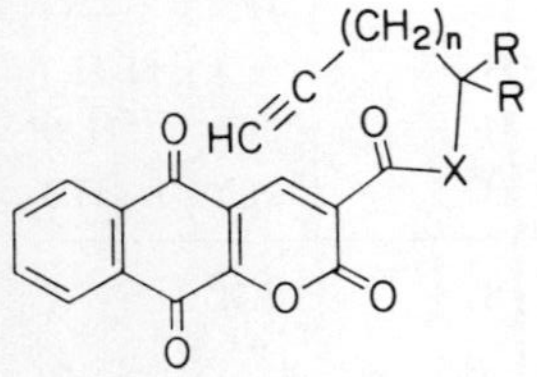		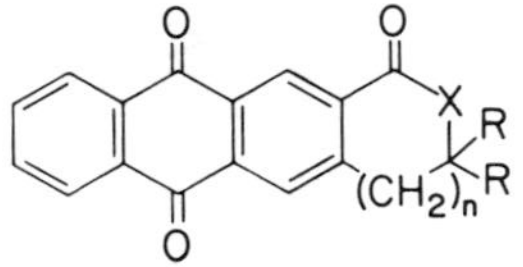	
	X R n			
24	O H 0[b]	Toluene, 230°, 1 hr	(100)	247
25	CH_2 H 1	Heat	(0)	247
26	CH_2 $C_2H_5O_2C$ 1[b]	Toluene, 230°, 2 hr	(100)	247

[a] The product was a single isomer of undetermined stereochemistry.
[b] The corresponding hydroquinone dimethyl ether could not be cyclized.
* This intermediate is hypothetical, or has not been isolated.

TABLE XXV. MISCELLANEOUS HETEROCYCLES[a]

Entry No.	Substrate	Conditions	Product(s) and Yield(s)(%)	Refs.
1	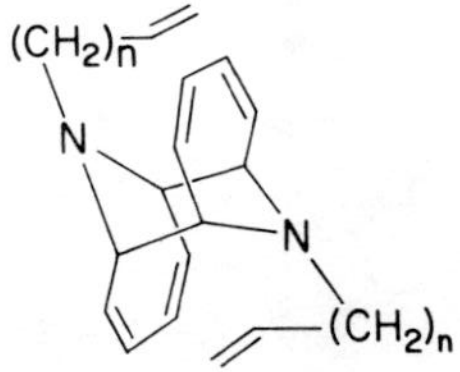 $(CH_2)_n$ … N … N … $(CH_2)_n$	n = 1 Neat, 180°, 2 hr	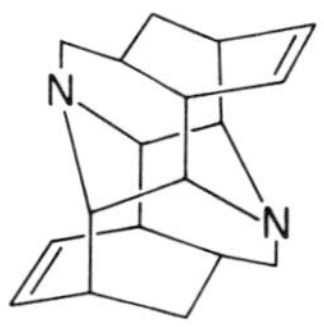 [b,c] (low to fair)	581
2		n = 2 Neat, 180°, 2 hr	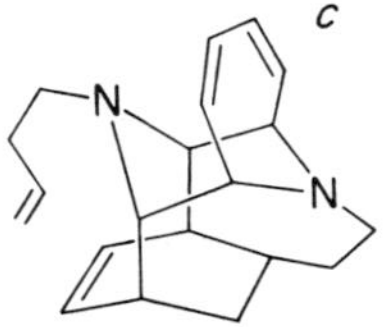 [c] (24)	581
3	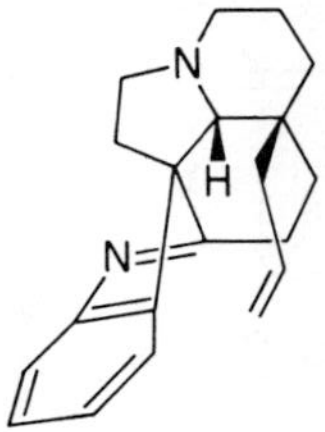	1% HCl, reflux, 15 min	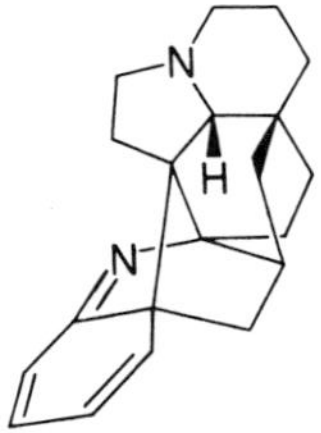 (100)	582

TABLE XXV. MISCELLANEOUS HETEROCYCLES (*Continued*)

Entry No.	Substrate	Conditions	Product(s) and Yield(s)(%)	Refs.
4	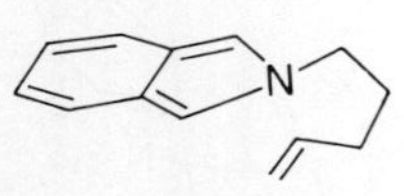 N; A	Slow distillation at 160–180° bath temperature (0.3 torr)	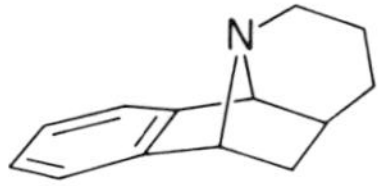 N; B (80)[d]	253
4a	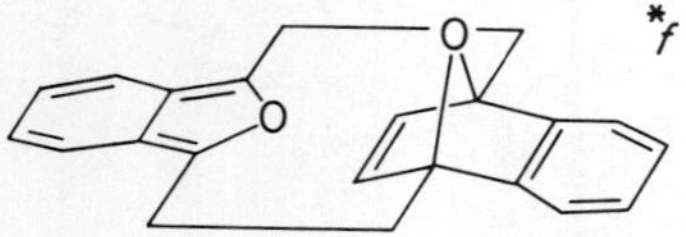 O; O *[f]	DMF, 150°	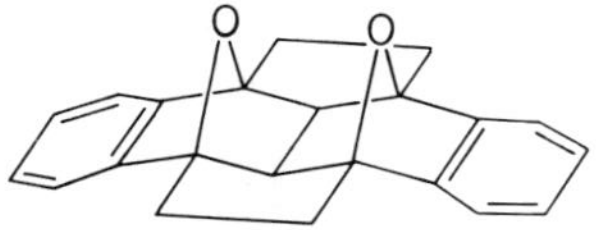 O; O (27)	583
4b	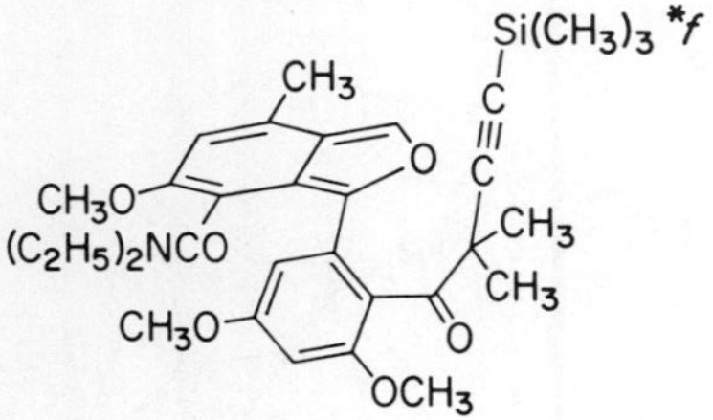 CH_3; O; CH_3O; $(C_2H_5)_2NCO$; $Si(CH_3)_3$ *[f]; C; C; CH_3; CH_3; O; CH_3O; OCH_3	Benzene, reflux, 8 hr	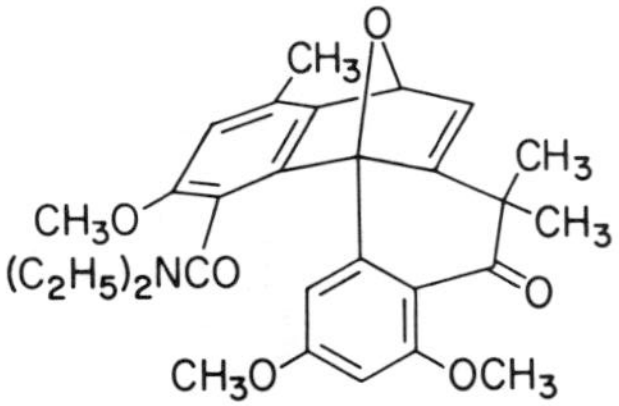 CH_3; O; CH_3; CH_3; CH_3O; $(C_2H_5)_2NCO$; O; CH_3O; OCH_3 (60)	323
	R^2; N; Br^-; R^1 — R^1 R^2		R^2; N; Br^-; R^1	
5	H H	Heat	(0)	150
5a	CH_3 H	CH_3CN, 145°, 3 hr	(46)	150,151
5b	CH_3 CH_3CONH	", reflux, 9 hr	(—)	150

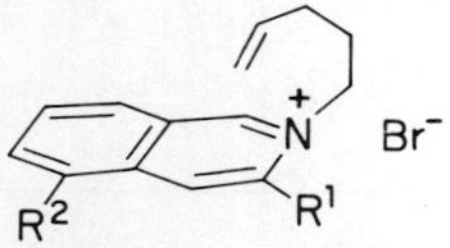

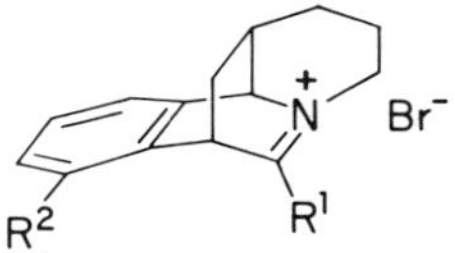

No.	Reactant	Conditions	Product (Yield, %)	Refs.
5c	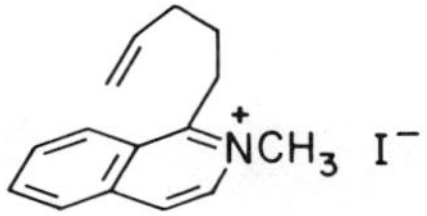 $\overset{+}{N}CH_3$ I^-	Heat	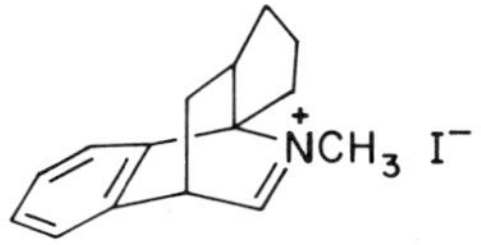 $\overset{+}{N}CH_3$ I^- (0)	150
5d	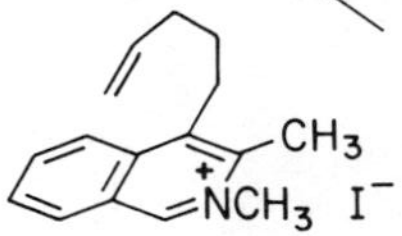 CH_3, $\overset{+}{N}CH_3$ I^-	CH_3CN, 145°, 3 hr	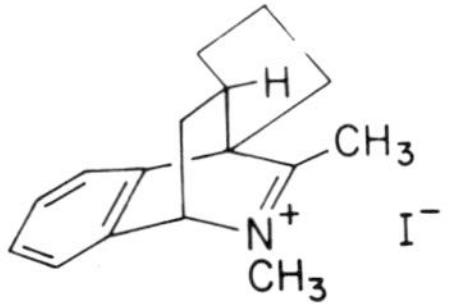 CH_3, N^+, CH_3, I^- (>84%)	150
6	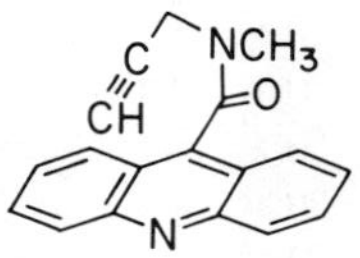 NCH_3, O, C≡CH, N	Toluene, 200°, 15 hr	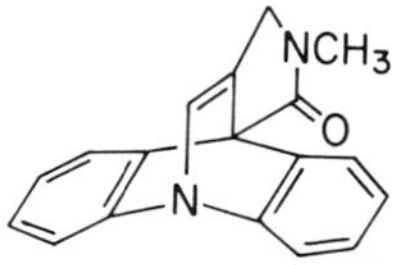 NCH_3, O, N (35)	45
7	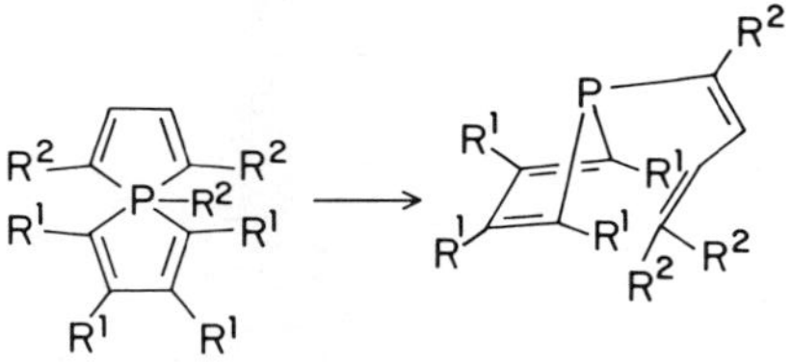 * $R^1 = CH_3O_2C$; $R^2 = C_6H_5$	$CHCl_3$, reflux, 4 hr	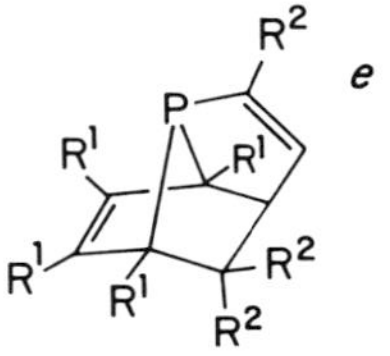 *e* (—)	584

TABLE XXV. MISCELLANEOUS HETEROCYCLES (*Continued*)

Entry No.	Substrate				Conditions	Product(s) and Yield(s)(%)	Refs.
	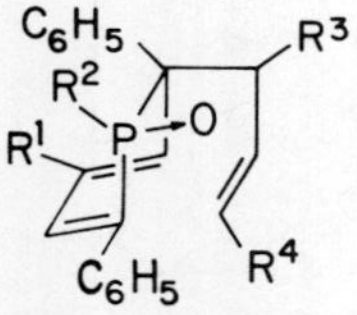 *f					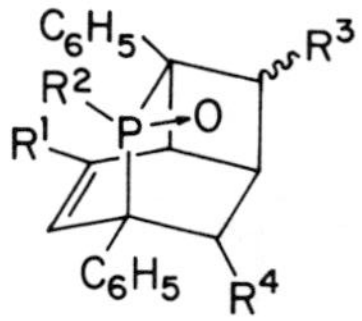	
	R^1	R^2	R^3	R^4			
8	C_6H_5	CH_3	H	H	Toluene, 100°, 6.5 hr	(60)	585
9	p-FC_6H_4	CH_3	H	H	", reflux, 4 hr	(11)	585
10	C_6H_5	CH_3	H	CH_3	", ", 1.5 hr	(8)	585
11	p-FC_6H_4	CH_3	CH_3	H	", 140°, 6 hr	(70)	585
12	p-$CF_3C_6H_4$	CH_3	H	H	", 100°, 6.5 hr	(25)	585
13	p-$CH_3OC_6H_4$	CH_3	H	H	", reflux, 3.5 hr	(15)	585
14	C_6H_5	2-CH_3-4-$O_2NC_6H_3$	H	H	Xylene, 136°, 3 hr	(26)	585
	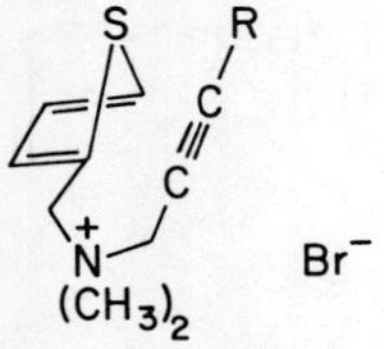 Br^-					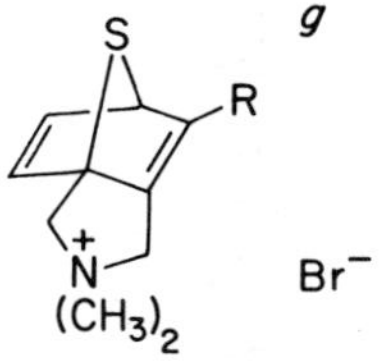 Br^- *g*	
	R						
15	H				DMF, 140°, 120 hr	(85)	586
16	CH_3				", ", 150 hr	(85)	586

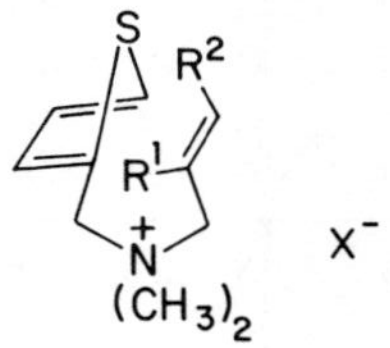

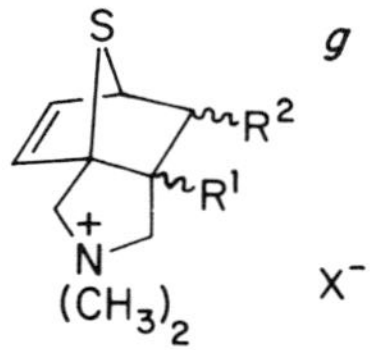

No.	R^1	R^2	X	Conditions	Product(s) (Yield, %)	Refs.
17	H	H	Br	DMF, 140°, 80 hr	(90)[h]	586
18	CH_3	H	Cl	", ", 100 hr	(85)	586
19	H	C_6H_5	Br	", ", "	(90)[h]	586

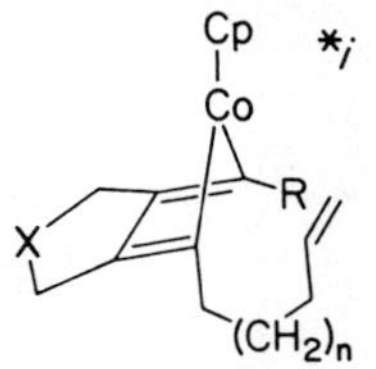

Reactant *[j]; products A[j] + B[j]

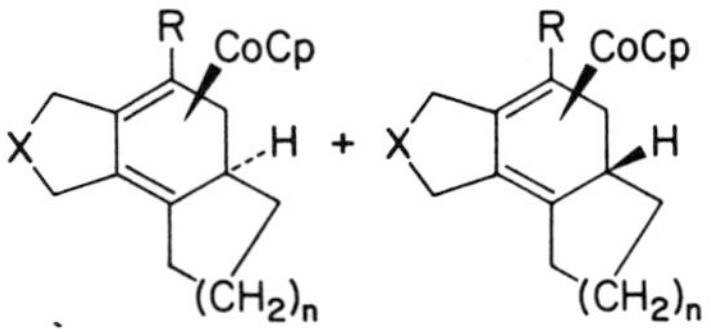

No.	X	R	n	Conditions	A	B	Refs.
				Isooctane, 4–5 days			
20	$(CH_2)_2$	H	1		(33)	(33)	248
21	$(CH_2)_2$	H	2		(38)	(38)[k]	248
22	O	$(CH_3)_3Si$	1		(35)	(0)	248
23	$(CH_2)_2$	$(CH_3)_3Si$	1		(85)	(0)	248
24	$(CH_2)_2$	$(CH_3)_3Si$	2		(47)	(47)	248

TABLE XXV. Miscellaneous Heterocycles (*Continued*)

Entry No.	Substrate	Conditions	Product(s) and Yield(s)(%)	Refs.
25	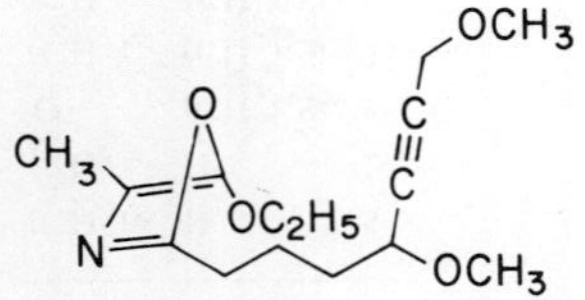	$C_6H_5C_2H_5$, reflux, 72 hr	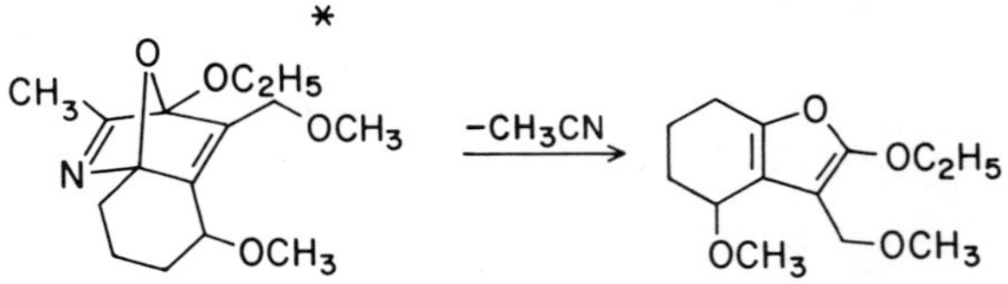 (80–90)	249
25a	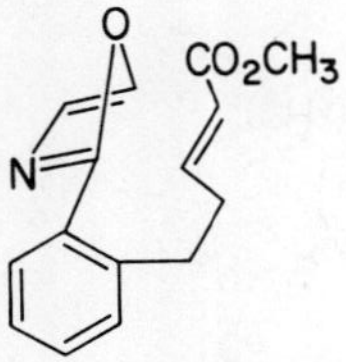	o-$Cl_2C_6H_4$, 1,5-Diazabicyclo[4.3.0]non-5-ene, heat	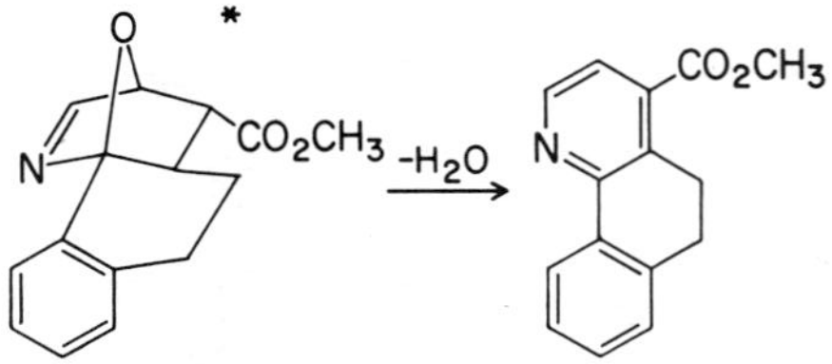 (76)	587
	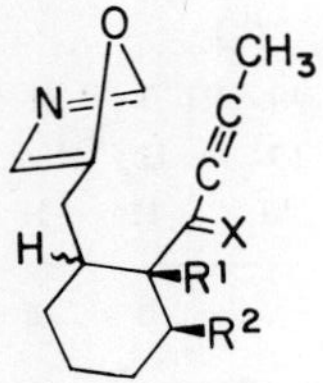		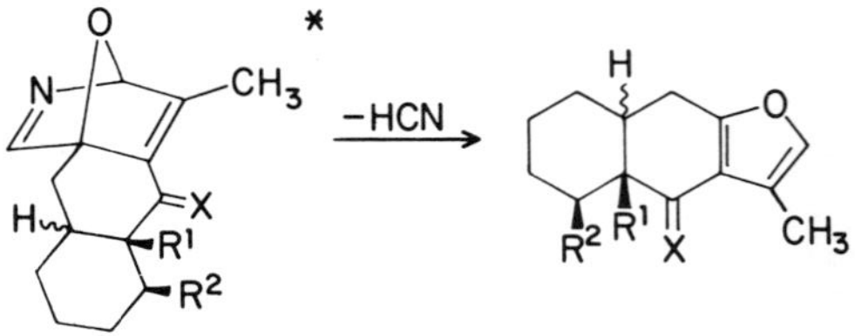	

Entry No.	R^1	R^2	X	Stereochemistry	Conditions	Product(s) and Yield(s)(%)	Refs.
26	H	H	O	*cis*	$C_6H_5C_2H_5$, reflux, 30 hr	(76)	154
27	H	H	O	*trans*	", ", "	(74)	154
28	CH_3	CH_3	O	*cis*	", ", 26 hr	(92)	164
29	CH_3	CH_3	H, OH	*cis*	", "	(84)[l]	164

Entry	Substrate	Conditions	Intermediate	Product(s) (% yield)	Ref.
30	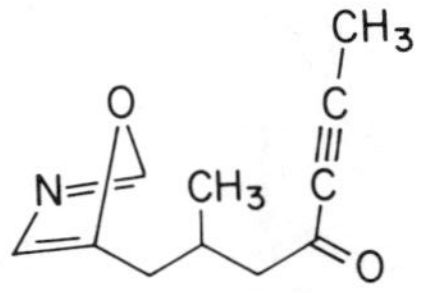 [structure]	$C_6H_5C_2H_5$, reflux, 96 hr	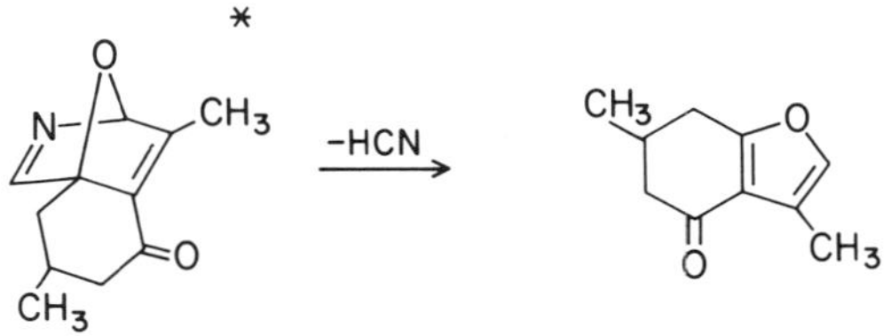 [structure] * $\xrightarrow{-HCN}$	[structure] (76)	154
31	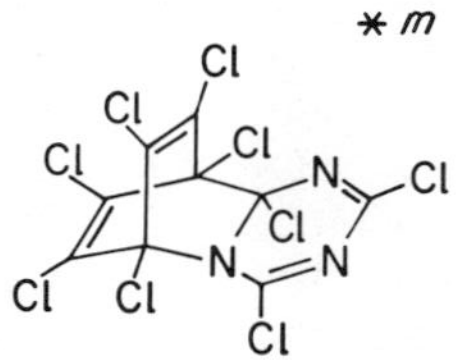 [structure] *[m]	Neat, 525°, 1 kbar ⇌	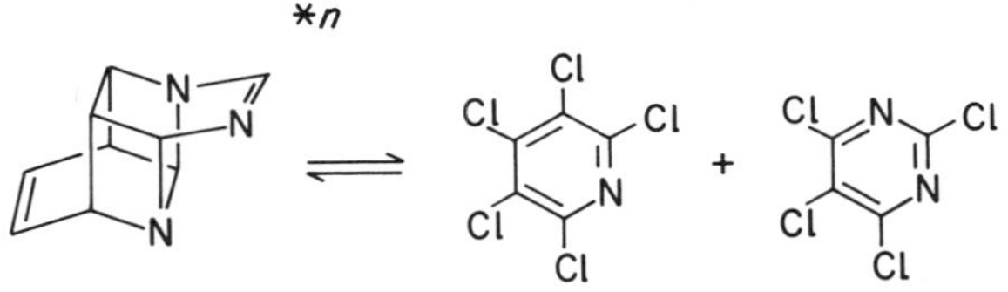 [structure] *[n] ⇌	[structure] (14)[o] + [structure] (14)[o]	588

[a] The entries are arranged in the order dihydroazepines, indolenines, isoindoles, isobenzofurans, isoquinolines, acridines, phosphacycloalkadienes, thiophenes, metallocycles, oxazoles, and triazines.

[b] The product was characterized as the tetrahydro derivative.

[c] The structures were assigned mostly on the basis of mass spactra and model considerations.

[d] **A** and **B** exist in equilibrium at elevated temperatures; the ratio **A**/**B** is 1 at 150° and 2.3 at 200°.

[e] The structure assignment is tentative.

[f] See original work for the precursors.

[g] The structures were assigned on the basis of elemental analysis and IR spectra.

[h] The product gave the corresponding dihydroisoindolinium salt on treatment with HBr.

[i] The substrate was generated from $RC{\equiv}CCH_2XCH_2C{\equiv}C(CH_2)_{(n+2)}CH{=}CH_2$ and $CpCo(CO)_2$.

[j] The ligands could be freed by treatment with $CuCl_2/(C_2H_5)_3N$.

[k] The yield includes a third isomer of unknown structure.

[l] The product was a single isomer with the hydroxyl group *cis* to $\mathbf{R}^1$.

[m] The substrate was generated from trichloro-1,3,5-triazine and hexachlorobenzene.

[n] The chlorine atoms have been omitted for clarity.

[o] The numbers are the conversions.

* This intermediate is hypothetical, or has not been isolated.

TABLE XXVIa. ACYCLIC 1,3,5-TRIENES GIVING BRIDGED PRODUCTS ($[_{\pi}4_a + {}_{\pi}2_a]$CYCLOADDITIONS)

Entry No.	Substrate						Conditions	Product(s) and Yield(s)(%)	Refs.
	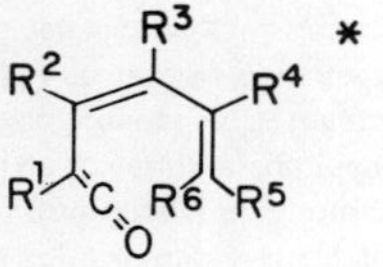 *							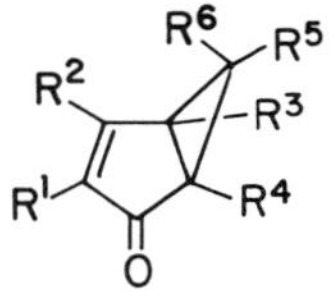	
	R^1	R^2	R^3	R^4	R^5	R^6			
1	CH_3	H	H	H	CH_3CO_2	CH_3[a]	CF_3CH_2OH, 25°	(—)	589
2	CH_3	CH_3	CH_3	H	CH_3	CH_3[a]	Et_2O, 25°	(70)	590
3	CH_3	H	CH_3	CH_3	CH_3	CH_3[a]	CH_3OH, "	(—)	590
4	CH_3	CH_3	CH_3	CH_3	CH_3	CH_3[a,b]	Et_2O, "	(50)	591,81
5	CH_3	CH_3	CH_3	CH_3	CH_3	CH_3CO[c]	CH_3OH, "	(Major product)	592
6	CH_3	CH_3	CH_3	CH_3	CH_3CO_2	CH_3[a]	Et_2O, "	(67)	593
7	CH_3	$ClCH_2$	CH_3	$ClCH_2$	CH_3CO_2	CH_3[a]	", "	(50)	594
8	C_2H_5	C_2H_5	C_2H_5	C_2H_5	C_2H_5	C_2H_5[a]	", "	(—)	591
9	CH_3	CH_3	CH_3	CH_3	$C_6H_5CH_2O$	CH_3[a]	", "	(30)	595
10	C_6H_5	H	C_6H_5	H	C_6H_5	CH_3CO_2[a]	Benzene, 25°	(83)	596
11	C_6H_5	H	C_6H_5	H	C_6H_5	$C_6H_5CO_2$[a]	Benzene, 25°	(75)	596
12	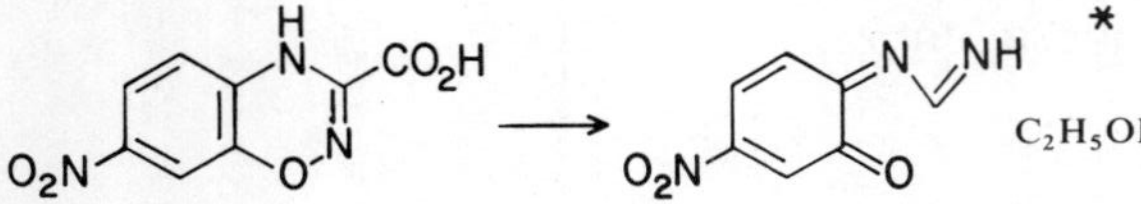 O_2N-benzoxadiazine-CO_2H → O_2N-quinone imine (N=CH–NH) *						C_2H_5OH, heat	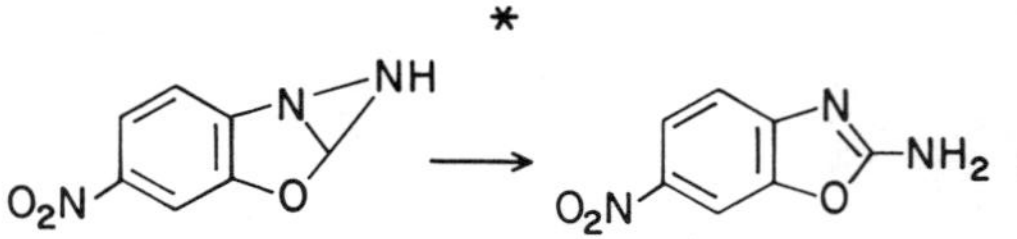 O_2N-(aziridino)benzoxazole * → O_2N-benzoxazole-NH_2 (—)	597,598

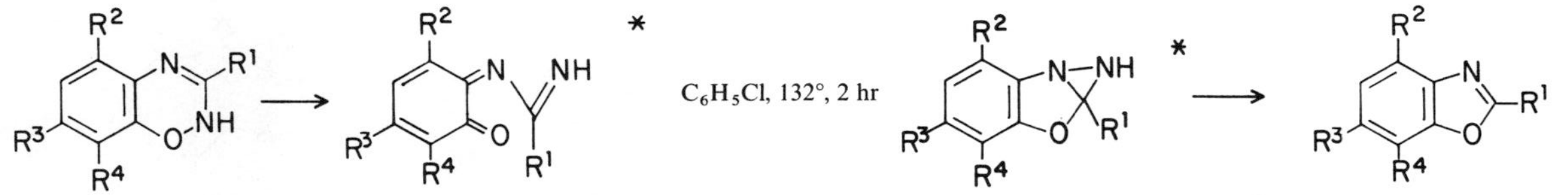

*

C_6H_5Cl, 132°, 2 hr

*

	R^1	R^2	R^3	R^4		
13	$C_2H_5O_2C$	H	O_2N	H	(56)	597
14	p-$CH_3C_6H_4$	H	H	H	(80)	597
15	p-$CH_3C_6H_4$	H	Cl	H	(80)	597
16	p-$CH_3C_6H_4$	Cl	H	O_2N	(74)	597
17	p-$CH_3C_6H_4$	H	O_2N	H	(33)	597
18	p-$CH_3C_6H_4$	C_6H_5	H	H	(77)	597

19 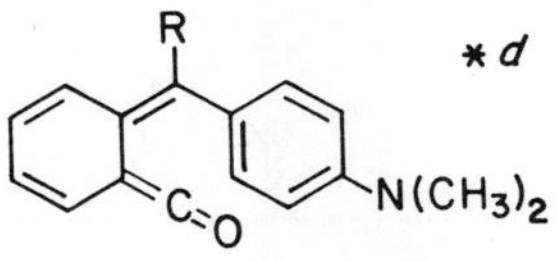 *d CH_2Cl_2, −78 to 10° * (≈100) 29

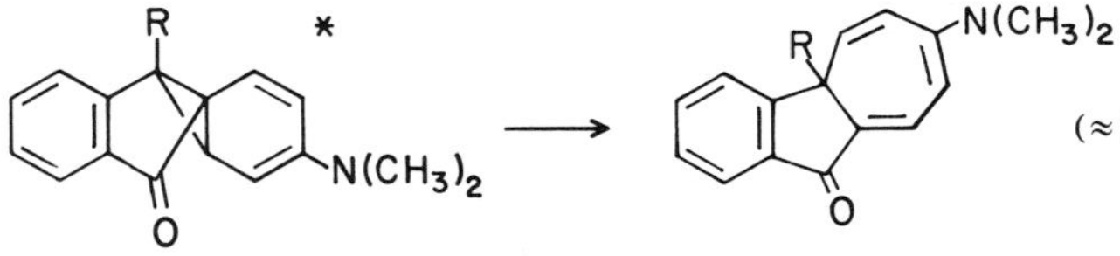

R = p-$(CH_3)_2NC_6H_4$

[a] The substrate was generated by irradiation of the corresponding 2,4-cyclohexadienone.

[b] The ketene was observed spectroscopically at −100°.

[c] The substrate was generated by irradiation of hexamethyl-8-oxabicyclo[3.2.1]octadienone.

[d] The substrate was generated by lead tetraacetate oxidation of 4,4-bis(p-dimethylaminophenyl)-3,4-dihydro-1(2H)-phthalazinone; see ref. 599 for examples of this cyclization involving substrates with other R groups.

* This intermediate is hypothetical, or has not been isolated.

TABLE XXVIb. Cyclic 1,3,5-Trienes Giving Bridged Products ($[_{\pi}4_a + {}_{\pi}2_a]$Cycloadditions)

Entry No.	Substrate	Conditions	Product(s) and Yield(s)(%)	Refs.
	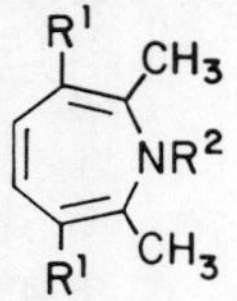		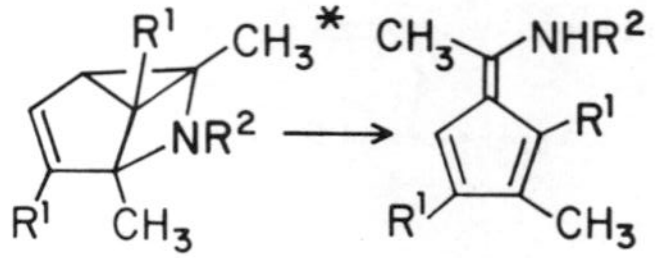	
	R^1 / R^2			
1	CH_3CO / CH_3*[a]	Mesitylene, reflux, 5–10 min	(49)	600
2	CH_3O_2C / CH_3	Benzene, reflux, 20 hr	(100)	600
3	CH_3O_2C / C_6H_5*[a]	160°, 0.1 torr	(10)	600
	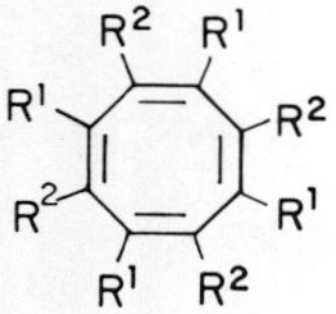		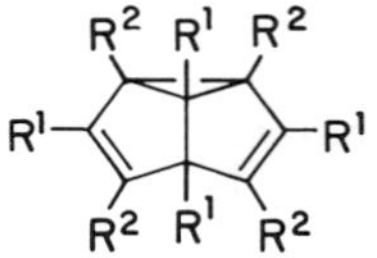	

No.	Reactant	Conditions	Product(s) and Yield(s) (%)	Refs.
	R^1 R^2			
4	CH_3 H	Nujol, 250°, 15 hr	(3)	601
5	CH_3 CH_3	C_2H_5OH, 1% C_2H_5ONa, 240°	(—)	602
6	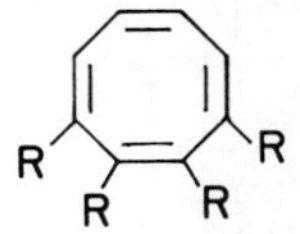 R = CF_3	C_5H_{12}, 170–180°, 6 days	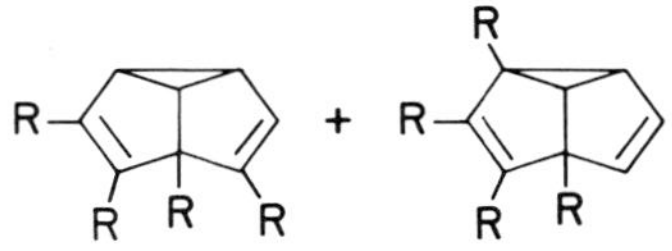 A (51)[b] + B (18)	603
7	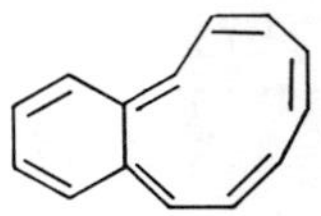 *[c]	CH_3CN, 40°	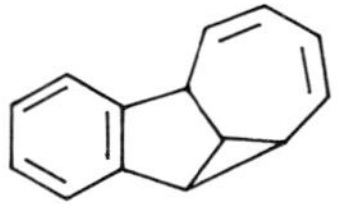 (20)	604
	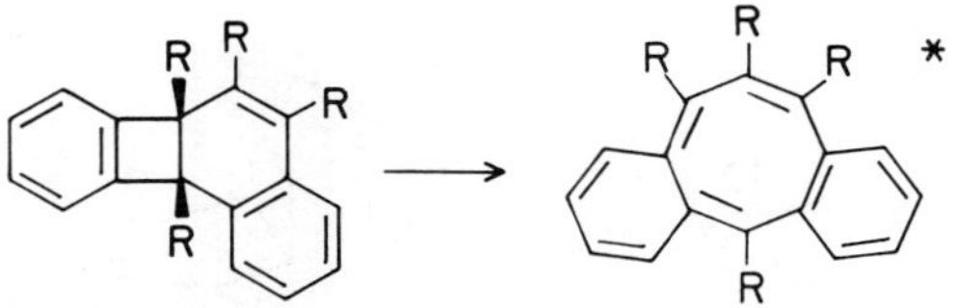 → *		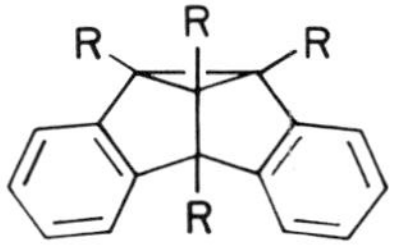	
	R			
8	H*[d]	—	(—)	605
9	CH_3	Neat, 250°	(75)	606

TABLE XXVIb. CYCLIC 1,3,5-TRIENES GIVING BRIDGED PRODUCTS ([$_{\pi}4_a + _{\pi}2_a$]CYCLOADDITIONS) (*Continued*)

Entry No.	Substrate			Conditions	Product(s) and Yield(s)(%)	Refs.
	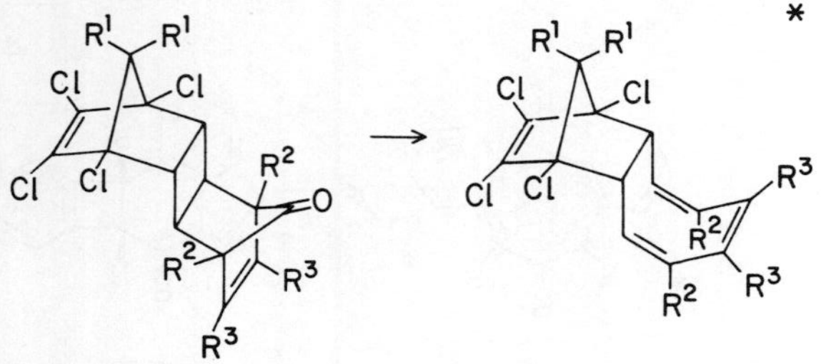 Cl	Cl			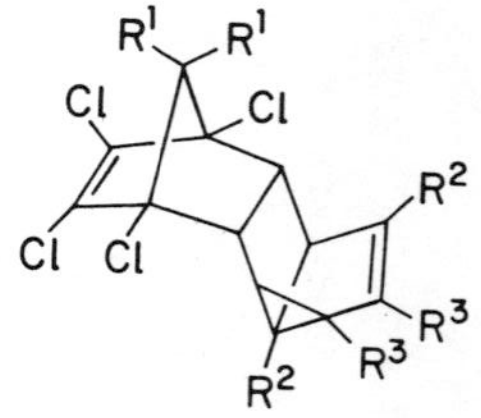 Cl	
	R^1	R^2	R^3			
10	Cl	CH_3	C_6H_5	Neat, 260°, 2 min[e]	(19)	31
11	Cl	C_6H_5	C_6H_5	o-$Cl_2C_6H_4$, 177–185°, 2 hr	(85)	31
12	CH_3O	C_6H_5	C_6H_5	Neat, 242°, 2 min	(55)	31
13	Cl 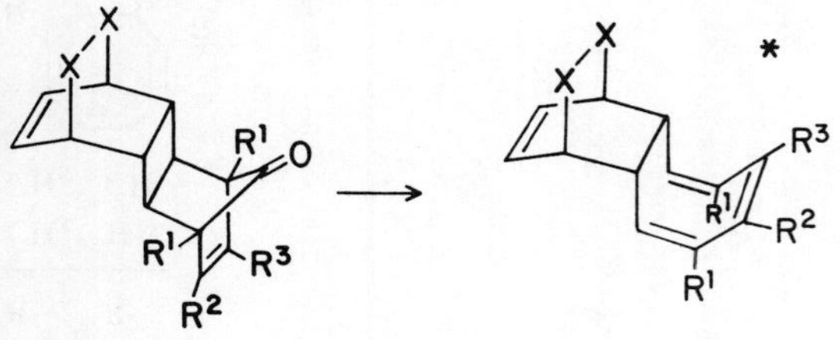	p-$CH_3OC_6H_4$	C_6H_5	Neat, 235°, 2 min	(14) 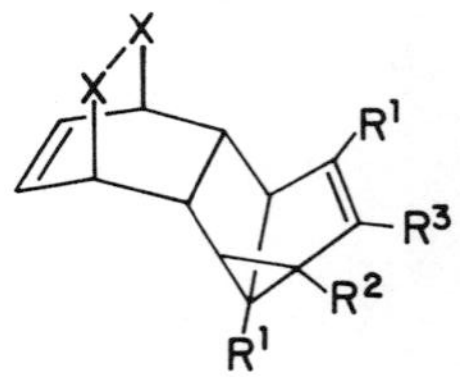	31

	R^1	R^2	R^3	X–X			
14	C_6H_5	C_6H_5	C_6H_5		Tetralin, reflux, 2 hr	(68)	607,608
15	C_6H_5	C_6H_5	C_6H_5	$C_2H_5O_2CN{-}NCO_2C_2H_5$	Neat, 210°	(62)	31,608
16	C_6H_5			$C_2H_5O_2CN{-}NCO_2C_2H_5$	Neat, 290°	(80)	31

[a] The substrate was generated from 4-chloromethyl-1,4-dihydro-2,6-dimethyl-3,5-R^1-1-R^2-pyridine and $BaCO_3$.
[b] **A** was converted completely into **B** on prolonged heating to 170°.
[c] The substrate was generated from benzyne and cyclooctatetraene.
[d] The substrate was generated by silver nitrate oxidation of (benzocyclobutadiene) iron tricarbonyl.
[e] At lower temperature, the intermediate cyclooctatriene could be isolated.
* This intermediate is hypothetical, or has not been isolated.

TABLE XXVIc. 1,3,n-TRIENES (n >

Entry No.	Substrate
1	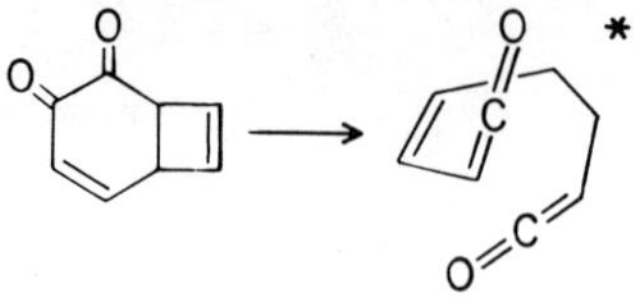
	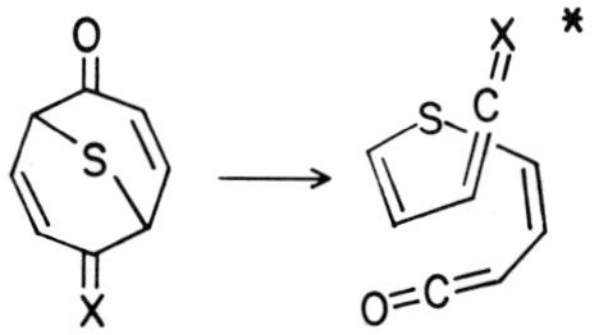
	X
2	O
3	H, CH_3CO_2
	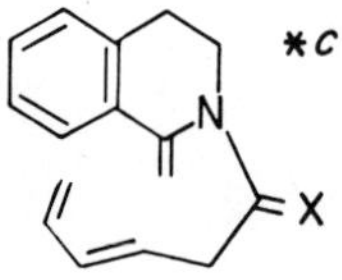
	X
4	H_2
5	O
	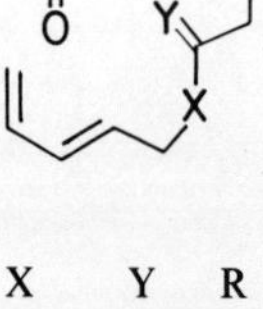

Entry No.	X	Y	R
6	O	O	H
7	CH_2	H_2	CH_3O_2C

…L TYPES) GIVING BRIDGED PRODUCTS[a]

…try ….	Conditions	Product(s) and Yield(s)(%)	Refs.
	Gas phase, 160°, or tetralin, 150°	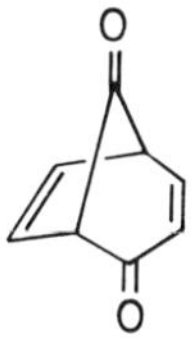 (11)	609
	510°	(43)	387
	520°	(19)[b]	387
	Xylene, reflux, 8 hr	(36)	22
	", ", 7 hr	(97)	22
	High dilution (slow addition to C_6H_5CN, 190°, 30 hr, Ar)	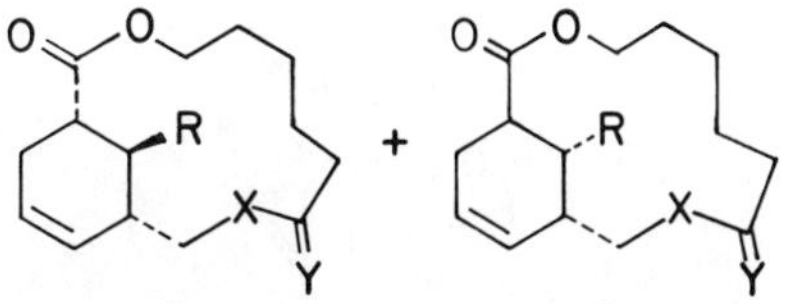 A + B	
		A (6) B (0)[d]	72
		A (21) B (5)[e]	72

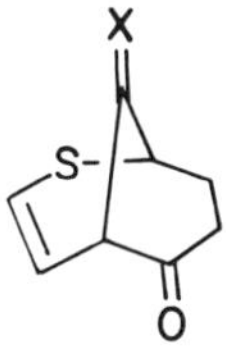

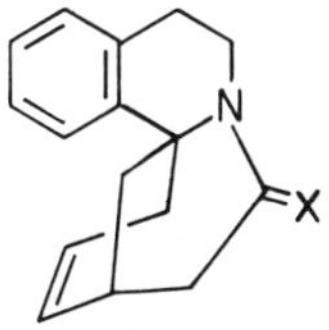

TABLE XXVIc. 1,3,n-TRIENES (n >

Entry No.	Substrate
8	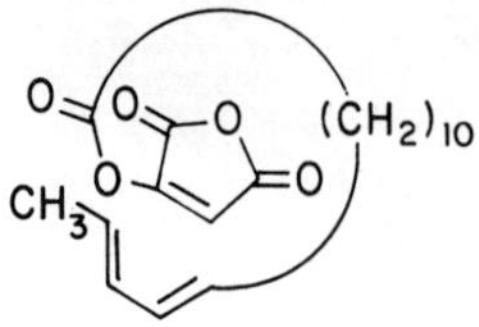 (structure: O, O, O, O, (CH2)10, CH3, O)
	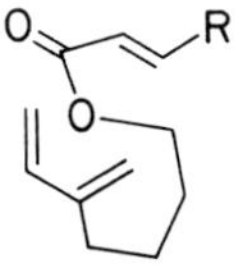 (structure: O, R, O)
	R
9a	H
9b	CH_3O_2C
10	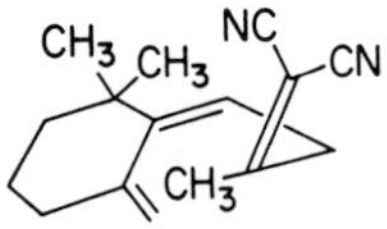 (structure: CH3, CH3, NC, CN, CH3)
11	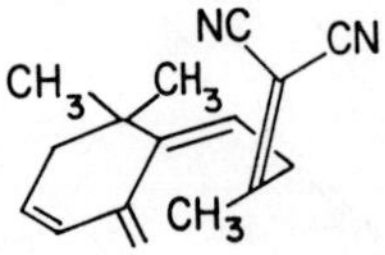 (structure: CH3, CH3, NC, CN, CH3)
	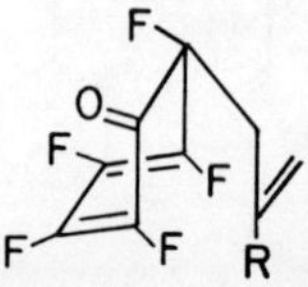 (structure: F, O, F, F, F, F, R, *h)
	R
12	H
13	CH_3

L TYPES) GIVING BRIDGED PRODUCTS[a] (*Continued*)

ry Conditions	Product(s) and Yield(s)(%)	Refs.
0.1 % in toluene, reflux, 96 hr	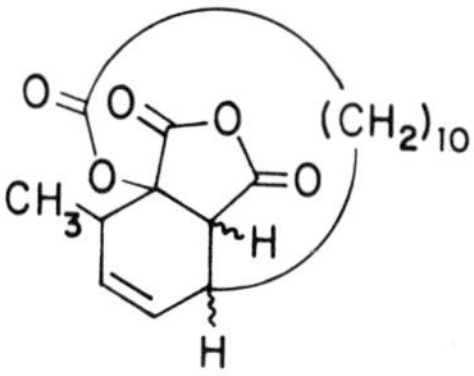 (5)[f]	314
Benzene, 190°, 9 hr	(4)[g]	83,82
Xylene, reflux, 16 hr	(—)[g]	83
100°, or *hv*	(35)	74
hv	(—)	74
	* *i* →	
Gas phase, 480°	(33)	430,429
Gas phase (0.05–0.1 torr), 410°	(22)	431

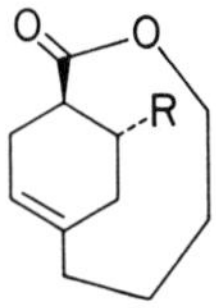

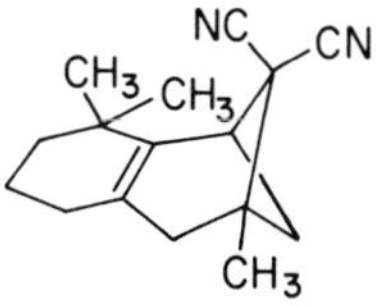

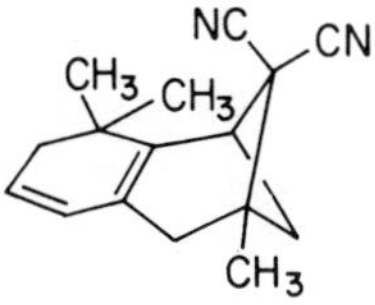

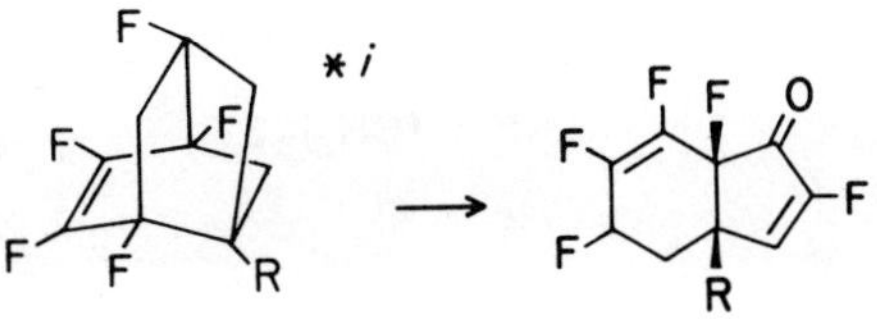

TABLE XXVIc. 1,3,n-TRIENES (n >

Entry No.	Substrate	
	R[1]	R[2]
14	CH_3O	H*[j]
15	CH_3O	H*[j]
15a	H	CH_3
	R[1]	R[2]
16	CH_3	H
17	H	CH_3
18	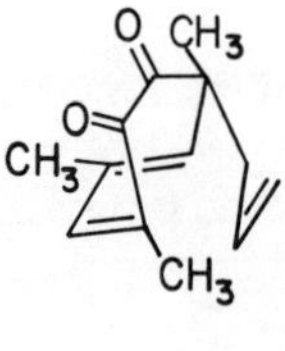 *o	
19	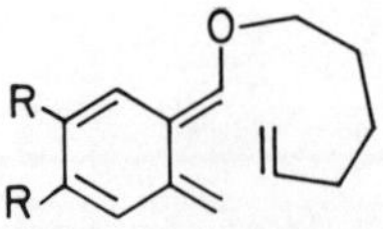 R = $(CH_3)_3Si$ *q	
20	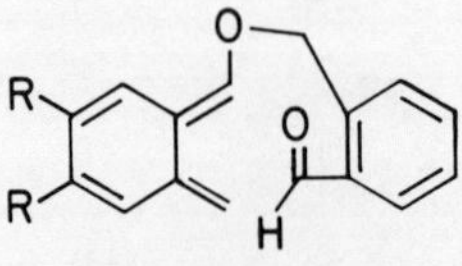 R = $(CH_3)_3Si$ *s	

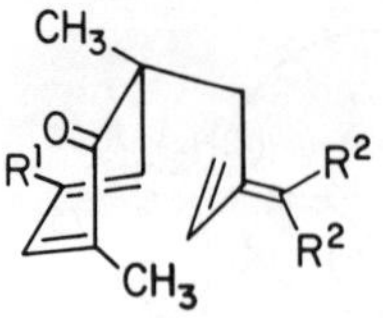

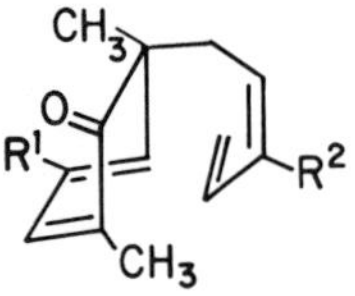

L TYPES) GIVING BRIDGED PRODUCTS[a] (*Continued*)

try .	Conditions	Product(s) and Yield(s)(%)	Refs.
		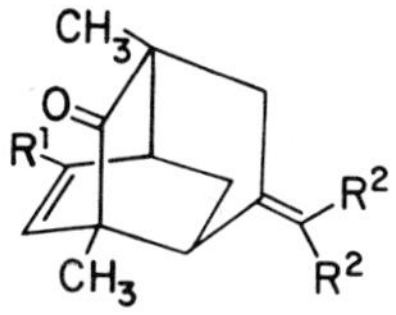	
	Benzene, reflux, 24 hr ($C_{10}H_{22}$, $C_6H_5N(C_2H_5)_2$ or sulfolane, 155–171°)	(44)[k]	78
	Benzene, reflux, 36 hr	(35)[l]	78
ı	", ", 8 hr	(14)[ee]	266
		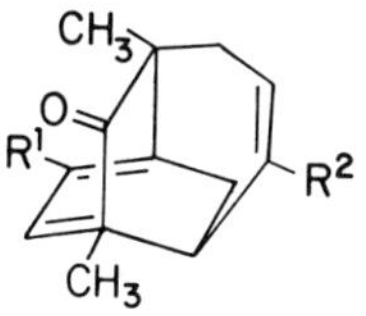	
	Benzene, reflux, 48 hr	(35)[m]	76
	", ", 6 hr	(44)[n]	152,76
	140°	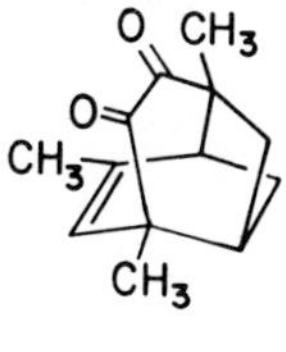 (—)[p]	440
	RC≡CR, reflux	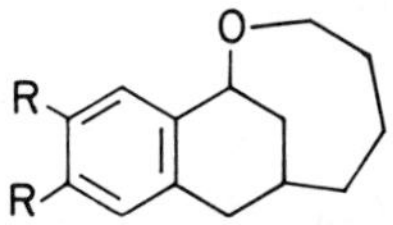 (65)[r]	73
	RC≡CR, reflux	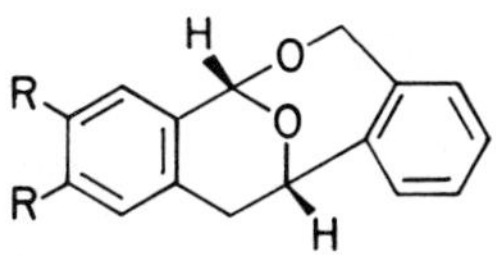 (49)	73

TABLE XXVIc. 1,3,n-TRIENES (n > 5,

Entry No.	Substrate
21	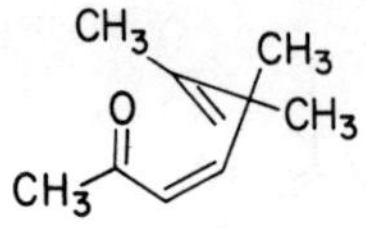

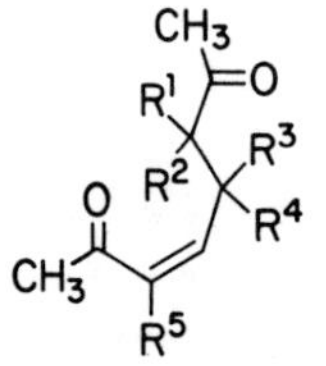

*

Entry No.	R^1	R^2	R^3	R^4	R^5
22	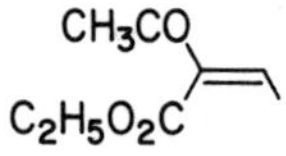 CH_3CO, $C_2H_5O_2C$	$C_2H_5O_2C$	CH_3CO	$C_2H_5O_2C$	$C_2H_5O_2C$[u]
23	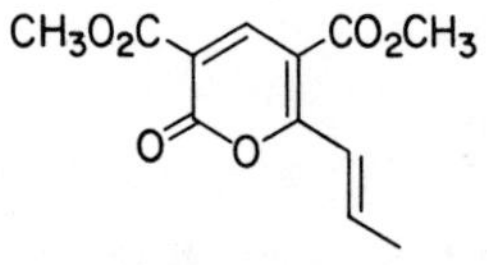 CH_3O_2C, CO_2CH_3	CH_3CO	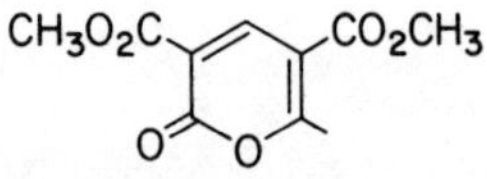 CH_3O_2C, CO_2CH_3	H	CH_3CO[w]

*

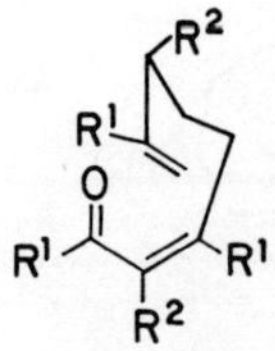

Entry No.	R^1	R^2
24	H	CH_3[x]
25	CH_3	H[z]

L TYPES) GIVING BRIDGED PRODUCTS[a] (*Continued*)

try .	Conditions	Product(s) and Yield(s)(%)	Refs.
	CH_3OH, *hν*, Pyrex	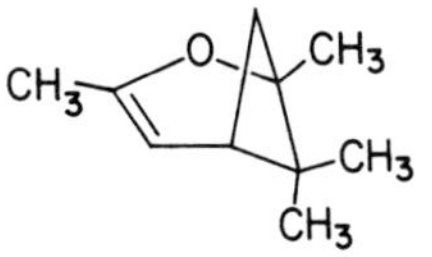 (5)[t]	553
		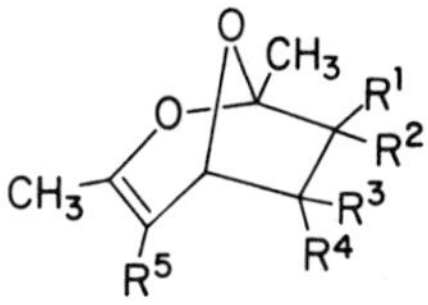	
	$CHCl_3$, 25°, 3 days	(1)[v]	64
	", ", 3 hr	(4)	64
		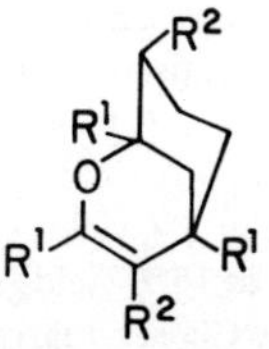	
	Benzene, 0.15 equiv BF_3/Et_2O, 25°, 4 hr	(49)	62
	", ", $AlCl_3$, 25°	(10)	62
	75% H_2SO_4, 0°, 1 min	(25)	62
	Gas phase, 405°	(—)[y]	62
	CH_2Cl_2, $AlCl_3$, 25°, 4 hr	(33)	610

TABLE XXVIc. 1,3,*n*-TRIENES (*n* >

Entry No	Substrate
	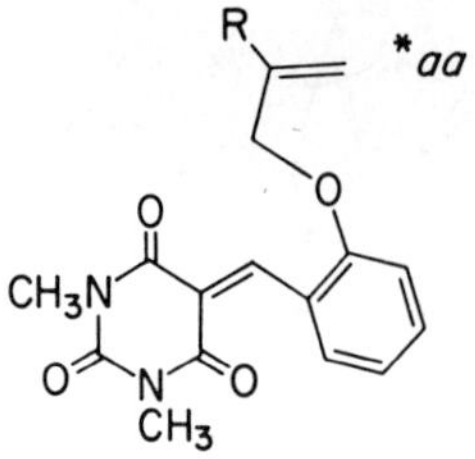 *aa
	R
25a	H
25b	CH_3
26	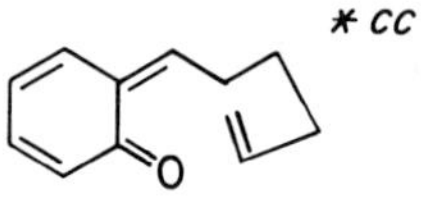 *cc
27	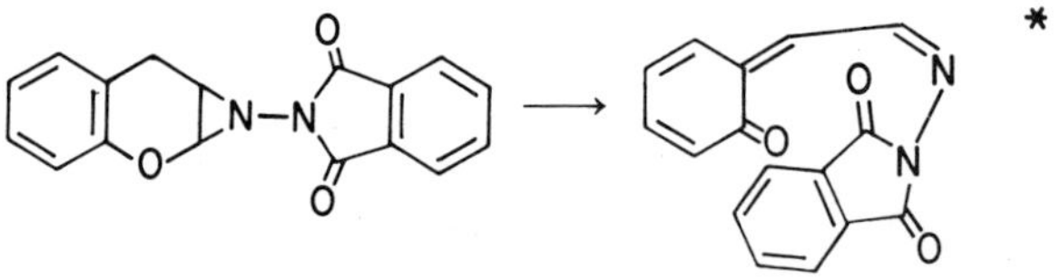 *

[a] This Table is arranged according to diene types in the same order as Tables I–XXV.
[b] The fused product was formed in 28% yield; see Table IIe/19.
[c] The substrate was generated from the diene-SO_2 adduct. This reaction is not a concerted [4 + 2] cyc addition; see Discussion section.
[d] Two fused products were formed in 71% combined yield; see Table IV/3.
[e] Two fused products were formed in 53% combined yield; see Table IV/4.
[f] The major product was the fused cycloadduct; see Table IV/5.
[g] The main product was the other regioisomer; see Tables V/15 and V/16, respectively.
[h] The substrate was generated by Claisen rearrangement from pentafluorophenyl 2-(*R*)-allyl ether.
[i] The fused product was also formed; see Table IXc/3,4.
[j] The substrate was generated by Claisen rearrangement from the corresponding phenyl allyl ether.
[k] The fused product was formed in 12% yield; see Table IXd/10.
[l] The fused product was formed in 9% yield; see Table IXd/11.
[m] The fused product was formed in 5% yield; see Table IXe/3.
[n] The fused product was formed in 15% yield; see Table IXe/4.
[o] The substrate was generated by Claisen rearrangement from the corresponding tropolone allyl ether.
[p] The fused product was also formed; see Table X/13.
[q] The substrate was generated by codimerization of bis(trimethylsilyl)acetylene with

$$HC{\equiv}CCH_2CH(C{\equiv}CH)O(CH_2)_4CH{=}CH_2$$

in the presence of $CpCo(CO)_2$ followed by ring opening of the resulting benzocyclobutene.

ALL TYPES) GIVING BRIDGED PRODUCTS[a] (*Continued*)

Entry No.	Conditions	Product(s) and Yield(s)(%)	Refs.
	CH_2Cl_2 or CH_3CN, 20°, 5 hr	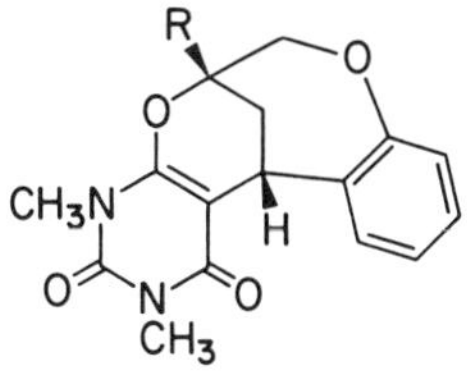	
25a		(4)[bb]	562
25b		(81)	562
26	$C_6H_5N(C_2H_5)_2$, 270°, 27 hr	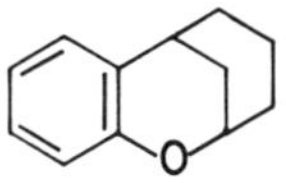 (29)[dd]	230
27	CH_3OH or toluene, reflux	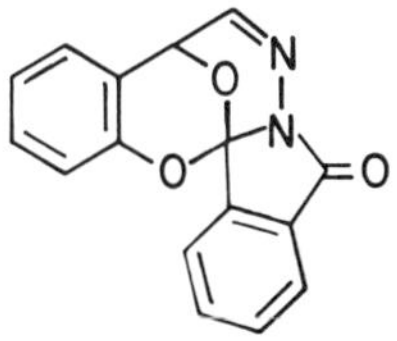 (—)	611

[r] The product was a mixture of fused and bridged isomers of unspecified stereochemistry; see Table XVIc/8.
[s] The substrate was generated by co-dimerization of bis(trimethylsilyl)acetylene with

$$HC{\equiv}CCH_2CH(C{\equiv}CH)OCH_2C_6H_4CHO\text{-}o$$

in the presence of $CpCo(CO)_2$ followed by ring opening of the resulting benzocyclobutene.
[t] The fused isomer was formed in 45% yield; see Table XIX/1.
[u] The substrate was generated by oxidative (MnO_2) dimerization of diethyl diacetylglutaconate.
[v] The stereochemistry was not determined.
[w] The substrate was generated by oxidative (MnO_2) dimerization of 3′,3′-diacetyl-3,5-bis(methoxycarbonyl)-xanthyrone.
[x] The substrate was generated from the corresponding *trans* diene.
[y] The isomer with the equatorial methyl group and the *cis*-fused isomer were also formed under these conditions; see Table XIX/7.
[z] The substrate was generated from 3-methyl-1,3-hexadien-5-one and isobutylene; mechanisms other than an intramolecular Diels–Alder reaction have been considered.
[aa] The substrate was generated from 1,3-dimethylbarbituric acid and the appropriate benzaldehyde in the presence of ethylenediammonium diacetate; the substrate could be isolated in some cases.
[bb] The fused product was formed in 70% yield; see Table XIX/25.
[cc] The substrate was generated from 1-(2-hydroxyphenyl)-1,5-hexadiene by 1,5-hydrogen shift.
[dd] The fused products were formed in 40% yield; see Table XX/2.
[ee] The fused product was formed in 42% yield; see Table IXd/14a.
* This intermediate is hypothetical, or has not been isolated.

XXVII. INTRAMOLECULAR HOMO DIELS-ALDER REACTIONS

Entry No.	Substrate	Conditions	Product(s) and Yield(s)(%)	Refs.
1	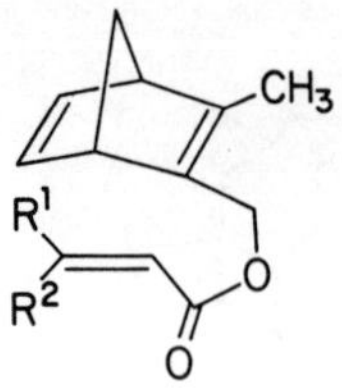 $R^1, R^2 = H$; $R^1 = H, R^2 = CH_3O_2C$; $R^1, R^2 = CH_3O_2C$	To 200°	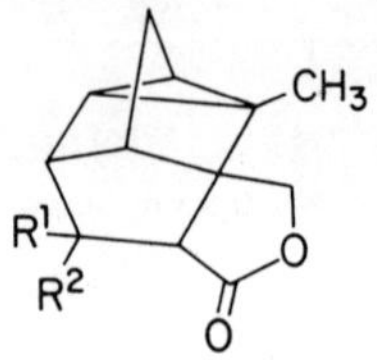 (0)[a]	251

[a] The substrate where $R^1 = R^2 = H$ polymerized. The diester $R^1 = R^2 = CH_3O_2C$ underwent an intramolecular ene reaction at room temperature.

TABLE XXVIIIa. REVERSE INTRAMOLECULAR DIELS-ALDER REACTIONS PRODUCING ACYCLIC DIENES[a]

Entry No.	Substrate	Conditions	Product(s) and Yield(s)(%)	Refs.
1	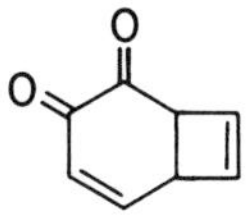	Gas phase, 160° or tetralin, 150°	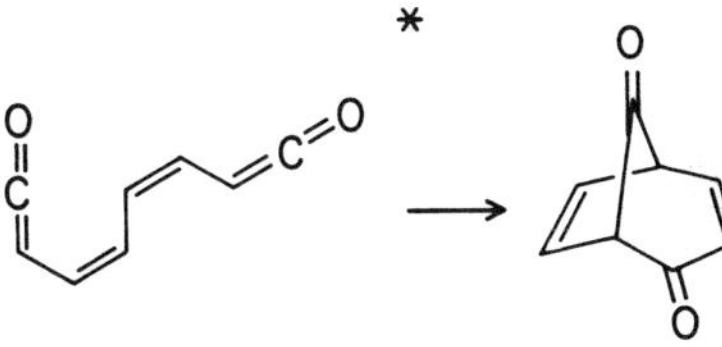 * → (11)	609
2	 *b	Gas phase, 300°	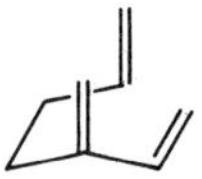 (31)	612
	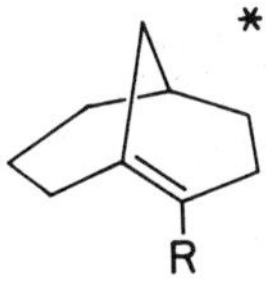 * R		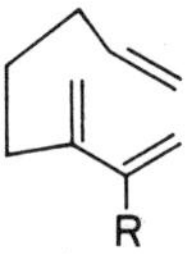 R	
	R			
3	H[c]	Gas phase, 375–400°, 12 torr	(—)	613
4	$(CH_3)_3SiO$[d]	", 550–750°	(1–64)	614

[a] For equilibria of this type, see Table V.

[b] The substrate was generated from the lithium salt of bicyclo[2.2.1]heptane-1-carboxaldehyde *p*-toluenesulfonylhydrazone by rearrangement of the initially formed carbene.

[c] The substrate was generated from the corresponding bridgehead acetate.

[d] The substrate was generated from

O

$CO_2Si(CH_3)_3$

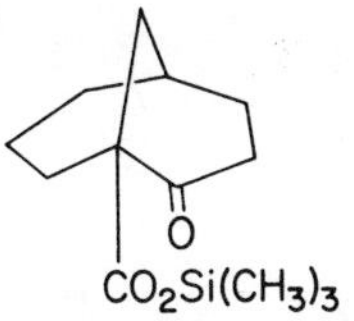

* This intermediate is hypothetical, or has not been isolated.

TABLE XXVIIIb. REVERSE INTRAMOLECULAR DIELS–ALDER REACTIONS PRODUCING CYCLOPENTADIENES, CYCLOHEXADIENES, AND CYCLOHEPTATRIENES[a]

Entry No.	Substrate	Conditions	Product(s) and Yield(s)(%)	Refs.
1	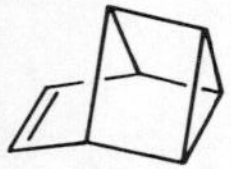 *	270–500°, flow system	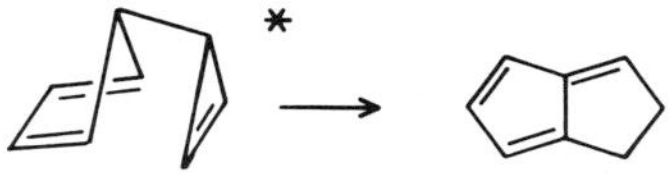 * → (—)	615,616
2	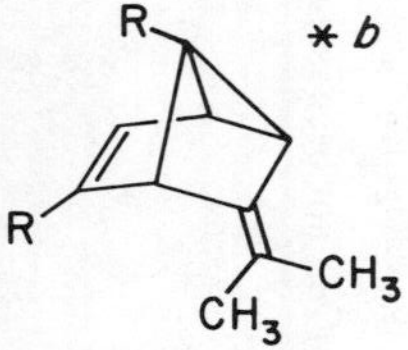 R, R, CH$_3$, CH$_3$ * b R = CH$_3$O$_2$C	CCl$_4$, reflux, 20 hr	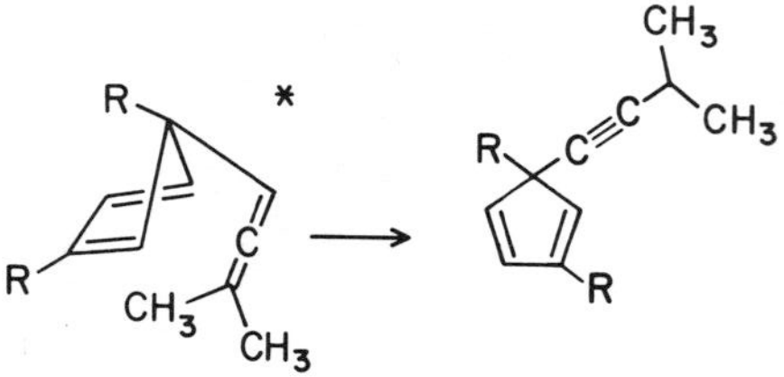 R, R, CH$_3$, CH$_3$ * → R, R, C≡C, CH$_3$, CH$_3$ (38)	403
3	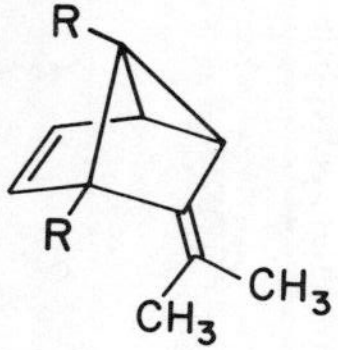 R, R, CH$_3$, CH$_3$	0°, weeks	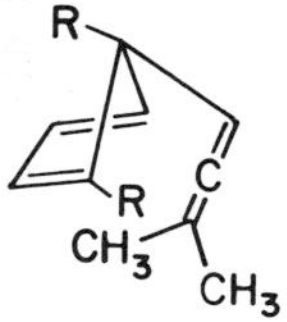 R, R, CH$_3$, CH$_3$ (—)	403
3a	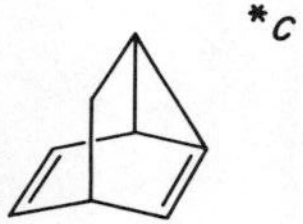 * c	Et$_2$O, 0°	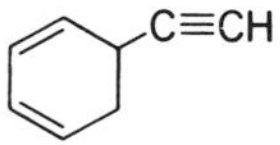 C≡CH (40)	617,618

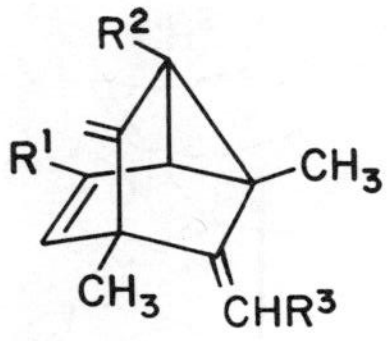

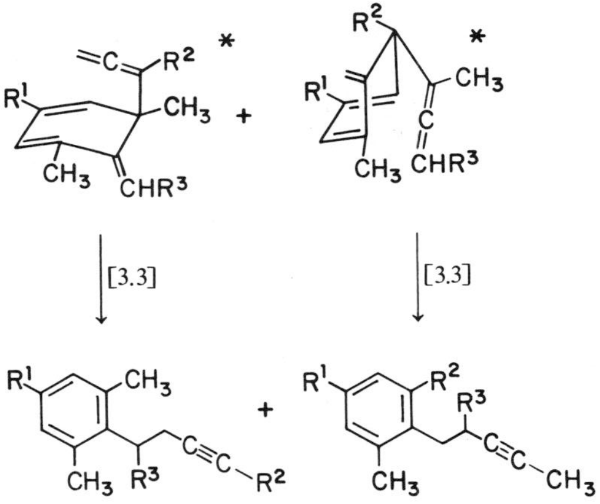

	R^1	R^2	R^3		A	B	
4	H	H	H	$C_{10}H_{22}$, 250°, 9 hr	(48)	(17)	619
5	CH_3	H	H	", 245°, 12 hr	(45)	(17)	619
6	H	H	CH_3	", 260°, 3.5 hr	(7)	(65)	619
7	CH_3	CH_3	H	", ", 10.5 hr	(0)	(28)	619

TABLE XXVIIIb. REVERSE INTRAMOLECULAR DIELS–ALDER REACTIONS PRODUCING CYCLOPENTADIENES, CYCLOHEXADIENES, AND CYCLOHEPTATRIENES (*Continued*)

Entry No.	Substrate			Conditions	Product(s) and Yield(s)(%)	Refs.
	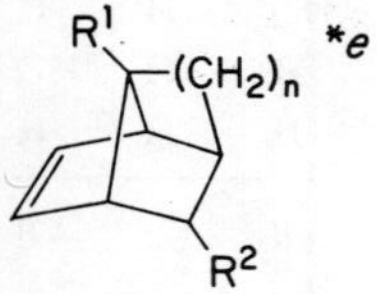 *e				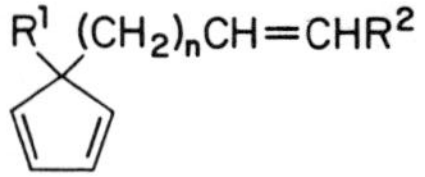 R^1 $(CH_2)_nCH{=}CHR^2$	
	R^1	R^2	n			
8	H	NC	0	C_2H_5OH, H_2O, 0°, 0.5 hr	(50)[f]	620
8a	CH_3	HO	0	Me_2CO, H_2O, 75°, 8 hr	(—)[g]	621
8b	NC	CF_3CH_2O	0	CF_3CH_2OH, 25°	(90)[h]	622
8c	CH_3	CH_3CO_2	1	Heat	(—)	623
8d	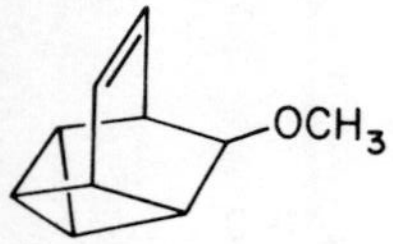 OCH_3			Heat	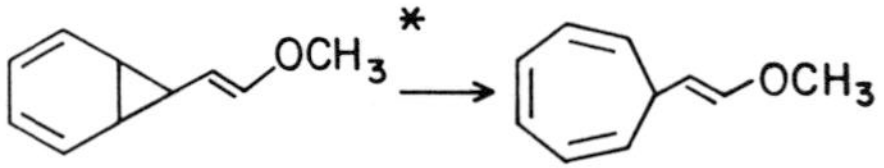 OCH_3 * → OCH_3 (—)	624
9	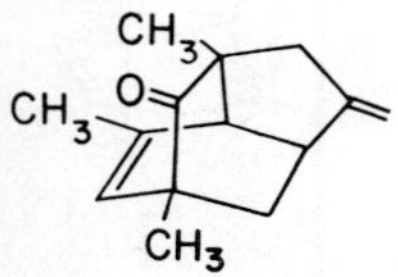 CH_3, CH_3, O, CH_3			n-$C_{10}H_{22}$, 176°, $t_{1/2}$ = 42 hr	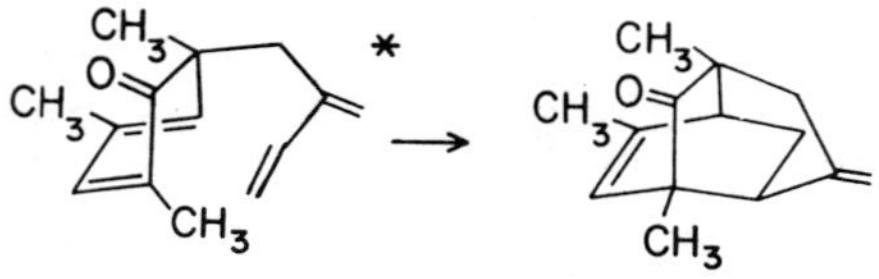 CH_3, CH_3, O, CH_3 * → CH_3, CH_3, O, CH_3 (—)	78

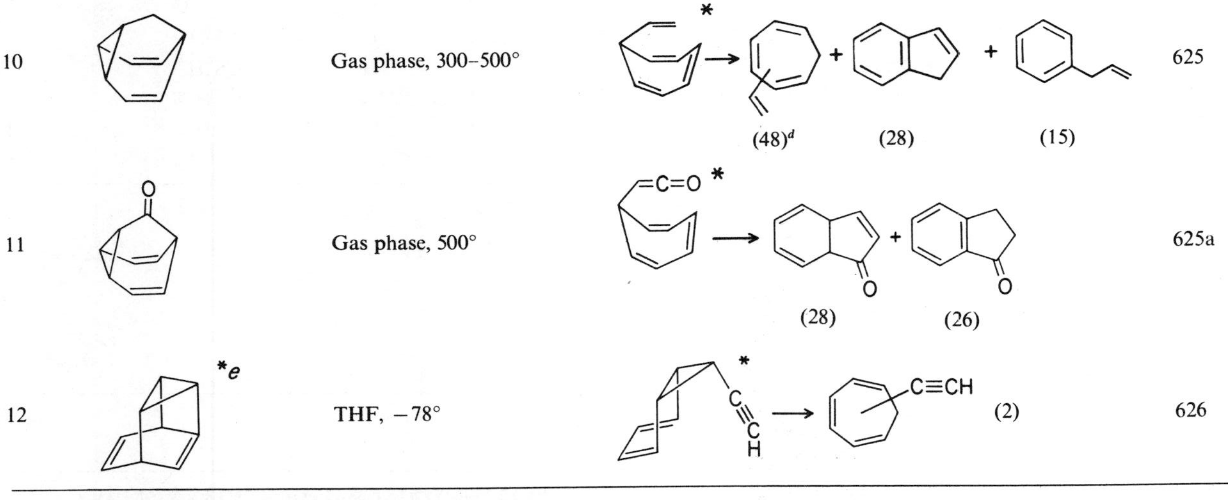

No.	Conditions	Ref.
10	Gas phase, 300–500°	625
11	Gas phase, 500°	625a
12	THF, −78°	626

[a] For equilibria of this type, see Tables IXb/1 and IXc/1,2.

[b] See Table VIIIb/1.

[c] The substrate was generated from 8,8-dibromotetracyclo$[5.1.0.0^{2,4}.0^{3,5}]$octane and methyllithium.

[d] The product composition varied with temperature; the values given are for 404°.

[e] See original work for the precursor.

[f] The product was isolated as the hexahydro derivative.

[g] The product isolated was the aldehyde.

[h] The product was isolated as the Diels–Alder dimer.

* This intermediate is hypothetical, or has not been isolated.

TABLE XXVIIIc. REVERSE INTRAMOLECULAR DIELS–ALDER REACTIONS PRODUCING 1,3,n-CYCLIC TRIENES (n > 5)[a]

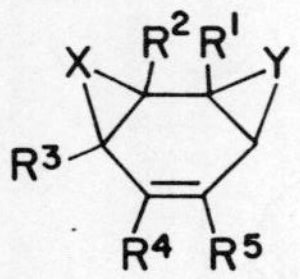

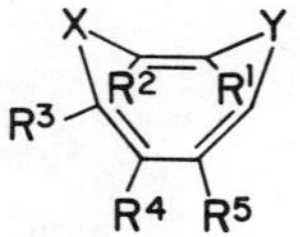

Entry No.	Substrate X	Y	R^1	R^2	R^3	R^4	R^5	Conditions	Product(s) and Yield(s)(%)	Refs.
1	CH_2	CH_2	CH_3O_2C	CH_3O_2C	H	H	H*[b]	THF, −78 to 25°	(60)	627
2	$(CH_3)_2C$	$(CH_3)_2C$	NC	H	H	NC	H	20°	(—)	628
3	$(CH_3)_2C$	$(CH_3)_2C$	CH_3O_2C	H	H	CH_3O_2C	H	C_6D_6	(—)	628
4	CH_2	CH_3SO_2N	H	H	CH_3O_2C	CH_3O_2C	CH_3	CH_3CN, 120°, 2–3 hr	(—)	629
5	CH_2	p-$CH_3C_6H_4SO_2N$	H	H	CH_3O_2C	CH_3O_2C	CH_3	Benzene, 120°, 2–3 hr	(—)	629
6	CH_2	O	H	C_6H_5	CH_3O_2C	CH_3O_2C	C_6H_5	30°($t_{1/2}$ = 1.5 hr)	(—)	147

7	CH_3N	CH_3N	H	H	H	H	H	Pyridine–CH_2Cl_2, 20°, 30 min	(59–74)	630
8	CH_3SO_2N	CH_3SO_2N	H	H	H	H	H	Me_2CO, heat, 2 hr	(94)	631
9	CH_3O_2CN	CH_3O_2CN	H	H	H	H	H*[c]	Pyridine–CH_2Cl_2, −10°	(60)	630
10	C_6H_5CON	C_6H_5CON	H	H	H	H	H*[c]	", ", 25°, 3 hr	(53)	630
11	p-$CH_3C_6H_4SO_2N$	p-$CH_3C_6H_4SO_2N$	H	H	H	H	H	$CHCl_3$, reflux, 4 hr	(100)	630
12	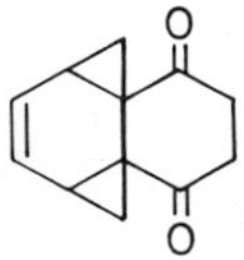 *[d]							t-C_4H_9OH	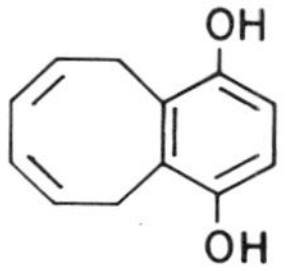 (—)	627

[a] For equilibria of this type, see Table XI/2–7.
[b] The substrate was generated from 1,2-dicarbomethoxy-1,4-cyclohexadiene-*cis*-3,6-dimethanol bis(methanesulfonate) and lithium naphthalene.
[c] The substrate was generated from the *N*-nitrosoaziridine by loss of N_2O.
[d] The substrate was generated from 2,3,4a,5,8,8a-hexahydro-1,4-naphthoquinone-*cis*-4a,8a-dimethanol bis(*p*-toluenesulfonate) and t-C_4H_9OK.
* This intermediate is hypothetical, or has not been isolated.

TABLE XXVIIId. REVERSE INTRAMOLECULAR DIELS–ALDER REACTIONS PRODUCING BRIDGED 1,3,6-CYCLOOCTATRIENES[a]

Entry No.	Substrate	Conditions	Product(s) and Yield(s)(%)	Refs.
1	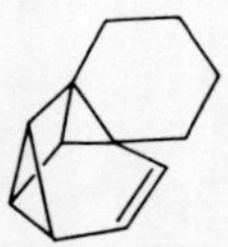 *[b]	Gas phase, flow system, 500°	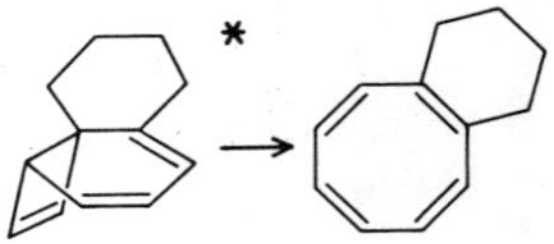 * → (69)	453
2	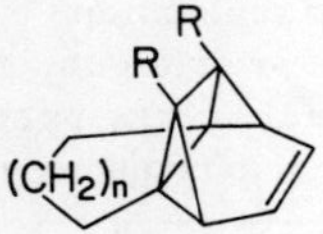 (CH$_2$)$_n$, R, R *[c]		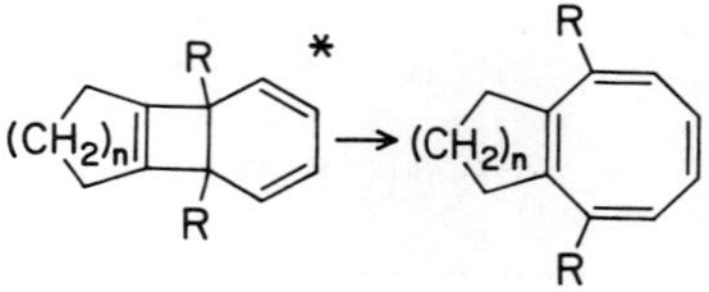 (CH$_2$)$_n$, R, R * → (CH$_2$)$_n$, R, R	
	n R			
3	1 H	Gas phase, 615°	(low)	454
4	2 H	", 470°	(60)	454
5	3 H	", "	(47)	454
6	2 CH_3	", "	(50)[d]	454
7	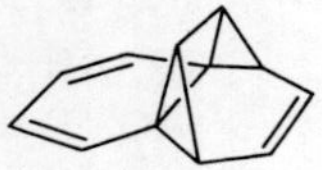 *[e]	Gas phase, 470°	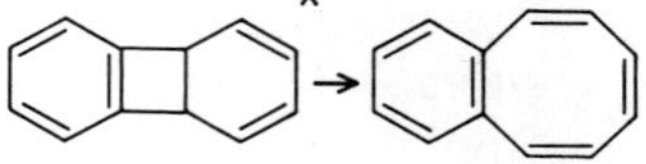 * → (53)	454

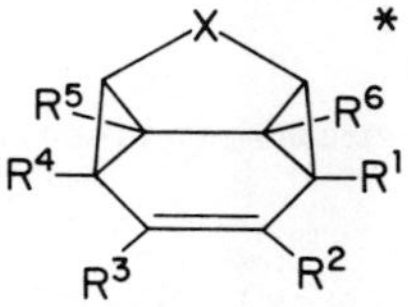

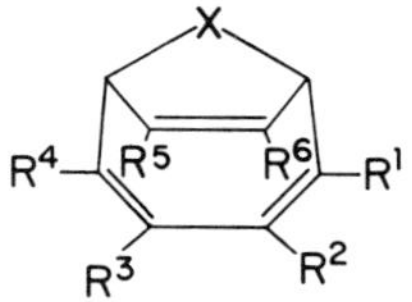

	R^1	R^2	R^3	R^4	R^5	R^6	X			
8	H	H	H	H	H	H	CO[f,g]	Me_2SO	(30)	632,633
9	H	H	H	H	CH_3	CH_3	CO[f]	"	(16)	632
10	H	H	H	H	CH_3	CH_3	SO_2[h]	"	(17)	632
11	H	H	H	H	$—(CH_2)_4—$		CO[f]	"	(24)	632
12	H	H	H	H	$—(CH_2)_4—$		SO_2[h]	"	(56)	632
13	H	H	H	H	$—(CH_2)_5—$		SO_2[h]	"	(56)	632
14	H	H	H	H	$—(CH{=}CH)_2—$		SO_2[h]	"	(68)	632
15	H	H	H	H	H	H	C_6H_5CH[i]	Diglyme, reflux	(40)	634
16	CH_3O_2C	CH_3O_2C	H	CH_3O_2C	H	H	$(CH_3)_2C{=}C$[j]	Me_2CO, $h\nu$	(80)	459
17	CH_3O_2C	H	H	CH_3O_2C	H	H	$(CH_3)_2C{=}C$[j]	$h\nu$	(4)	459
18	C_6H_5	CH_3O_2C	H	CH_3O_2C	H	H	$(CH_3)_2C{=}C$[j]	Me_2CO, $h\nu$	(55)	459

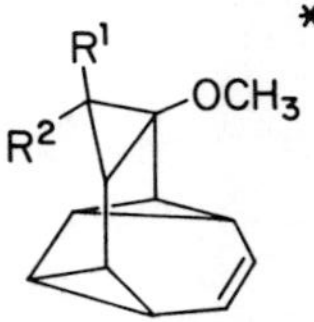

*[k]

A → 25° → B

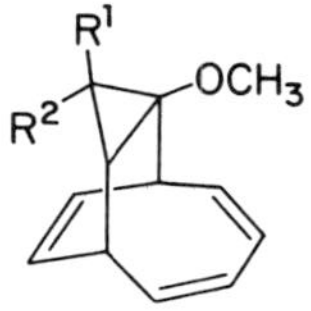

	R^1	R^2			
19	H	Br		(12)[l]	469
20	Br	H		(49)	469

TABLE XXVIIId. REVERSE INTRAMOLECULAR DIELS–ALDER REACTIONS PRODUCING BRIDGED 1,3,6-CYCLOOCTATRIENES (*Continued*)

Entry No.	Substrate	Conditions	Product(s) and Yield(s)(%)	Refs.
21	 * m	—	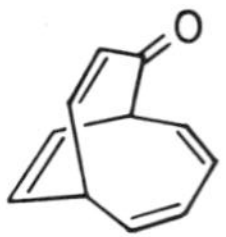 (—)	469

[a] For equilibria of this type, see Table XII/1, 2, 16–24, and 29–34.
[b] See Table XII/3.
[c] See Table XII/4–7.
[d] The number is a conversion.
[e] See Table XII/8.

[f] The substrate was generated by dehydrohalogenation of

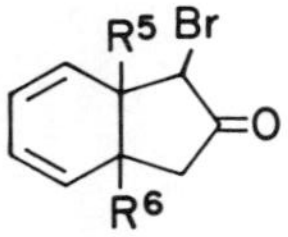

R^5 Br

R^6

[g] The substrate was generated from barbaralone by bromination followed by sodium amalgam reduction.

[h] The substrate was generated by dehydrohalogenation of

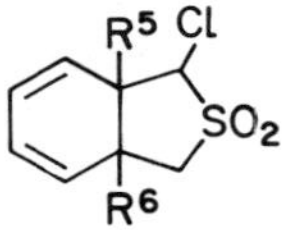

R5 Cl
SO2
R6

[i] The substrate was generated by irradiation of

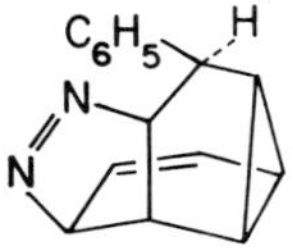

C6H5
H
N
N

[j] The substrate was generated by photolysis of

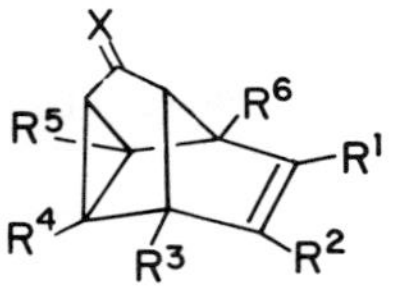

X
R5
R6
R1
R4
R3
R2

[k] The substrate was generated by reduction of the dibromocyclopropane ($R^1 = R^2 = Br$) with tributyltin hydride.
[l] The product was in equilibrium with isomer **A**, ratio **A**/**B** 71:29.
[m] The substrate was generated from the equilibrium mixture of Table XXVIIId/19 with silver perchlorate.
* This intermediate is hypothetical, or has not been isolated.

TABLE XXVIIIe. Reverse Intramolecular Diels–Alder Reactions Producing Bridged 1,3,n-Trienes Other Than 1,3,6-Cyclooctatrienes[a]

Entry No.	Substrate	Conditions	Product(s) and Yield(s) (%)	Refs.
	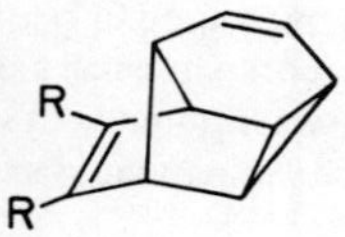		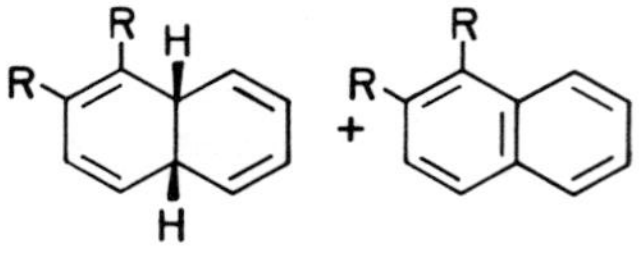 A B	
	R		A B	
1	H	CCl_4, 130°, 20 min	(50) (15)	411
2	NC	100°, 3.5 hr	(0) (63)	250
3	CH_3O_2C	150°, 4 hr	(0) (33)	250
	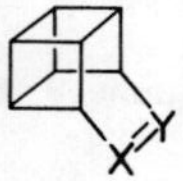		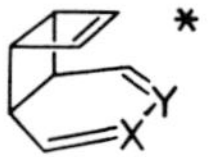	
	X Y			
4	CH CH	CCl_4, 65°, 2 days	(65)[b,c]	635–639
5	N N	Flash vacuum pyrolysis	(60)[d]	640
6	N $\overset{+}{N}-\overset{-}{O}$	$CHCl_3$, BF_3–Et_2O, reflux, 24 hr	(75)[e]	641

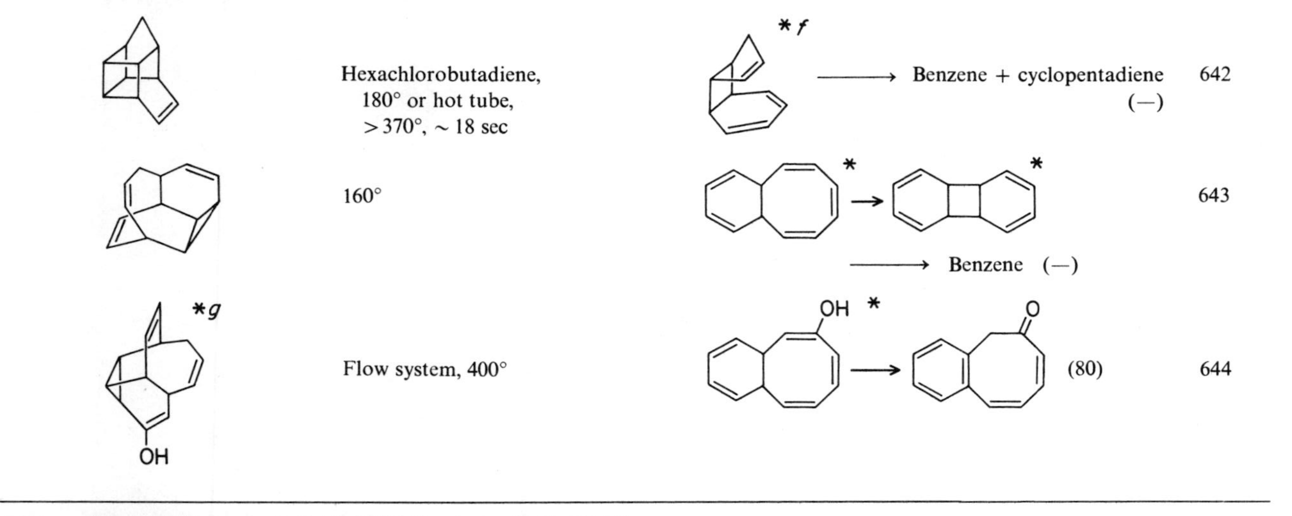

	Precursor	Conditions	Intermediate → Products (Yield %)	Ref.
7		Hexachlorobutadiene, 180° or hot tube, >370°, ~ 18 sec	*f → Benzene + cyclopentadiene (—)	642
8		160°	* → * → Benzene (—)	643
9	*g	Flow system, 400°	* → (80)	644

[a] For equilibria of this type, see Table XIII/14 and also Table XIII/2.

[b] The product was trapped as the adduct with maleic anhydride, tetracyanoethylene, or dimethyl azodicarboxylate or by irradiation to give hexacyclo[4.4.0.0^{2,4}.0^{3,9}.0^{5,8}.0^{7,10}]decane.

[c] The number is the yield of the maleic anhydride adduct.

[d] The product was azocine.

[e] The product was benzaldoxime.

[f] The intermediate was also trapped with maleic anhydride.

[g] See original work for the precursor.

* This intermediate is hypothetical, or has not been isolated.

TABLE XXVIIIf. REVERSE INTRAMOLECULAR DIELS–ALDER REACTIONS PRODUCING *o*-QUINODIMETHANES AND *o*-DIMETHYLENEHETEROCYCLES

Entry No.	Substrate	Conditions	Product(s) and Yield(s)(%)	Refs.
1		Gas phase, 400°	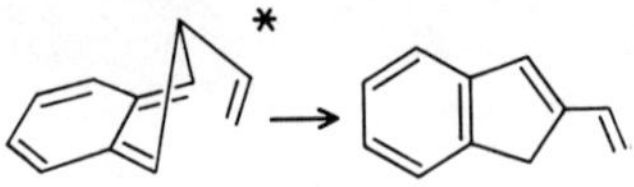 * → (100)	645
2	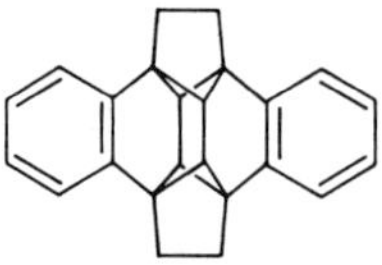	190–200°	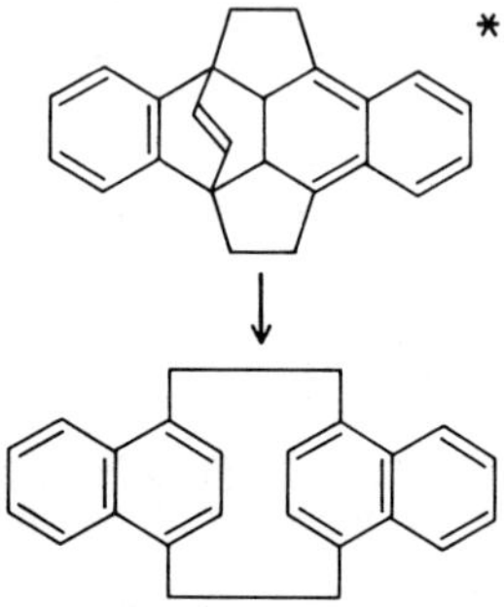 * ↓ (—)	523
3	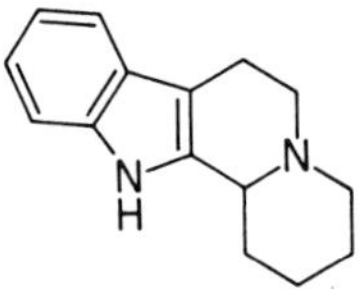 N, H, N	Mass spectrometer	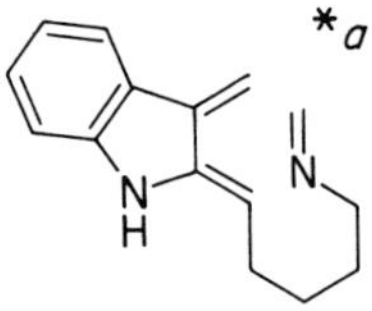 *[a] N, H, N (—)	646

[a] Various yohimbine alkaloids also fragment in this manner.

* This intermediate is hypothetical, or has not been isolated.

TABLE XXVIIIg. REVERSE INTRAMOLECULAR DIELS–ALDER REACTIONS PRODUCING AROMATICS[a]

Entry No.	Substrate		Conditions	Product(s) and Yield(s)(%)	Refs.
1	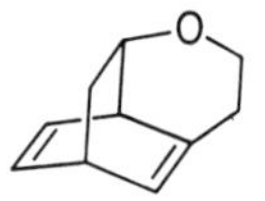		240°	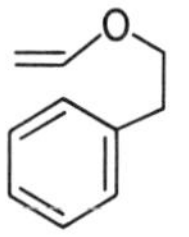 (—)	519
	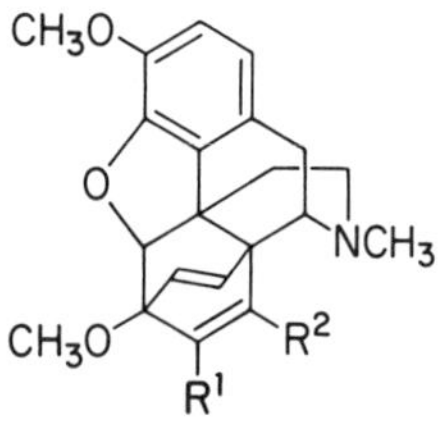 R^1	R^2	$(n\text{-}C_4H_9)_2O$, 141°, 10 min	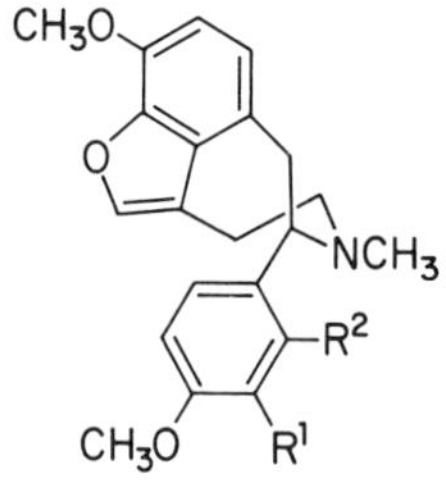	
2	$C_2H_5O_2C$	H		(90)	252
3	CH_3O_2C	CH_3O_2C		(87)	252
	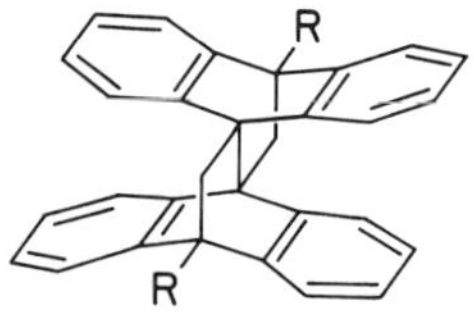 A			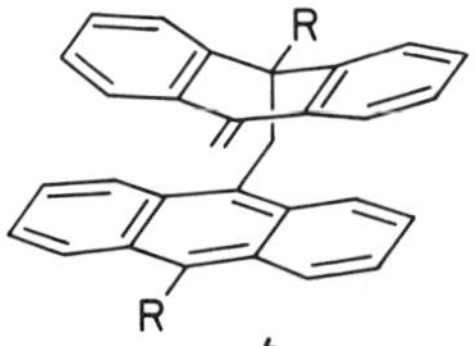 B[b]	
	R			$K_{A/B}$ (°C)	
4	H[c]			810 (21)	647,648
5	CH_3			$>10^7$ (25)	532
6	C_2H_5			$>2 \times 10^6$ (25)	532
7	$C_6H_5CH_2$			3×10^5 (25)	532

[a] For equilibria of this type, see Table XVIIa/4–9.

[b] See refs. 532 and 648 for the photochemically induced reverse intramolecular Diels–Alder reaction.

[c] See ref. 532 for the reverse intramolecular Diels–Alder reaction of a number of aromatic ring-chlorinated derivatives.

TABLE XXVIIIh. REVERSE INTRAMOLECULAR DIELS–ALDER REACTIONS PRODUCING FURANS[a]

Entry No.	Substrate		Conditions	Product(s) and Yield(s)(%)	Refs.
1	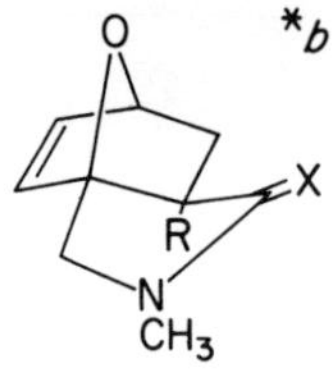 *[b]; $R=C_6H_5$; $X=H_2$		Et_2O, reflux	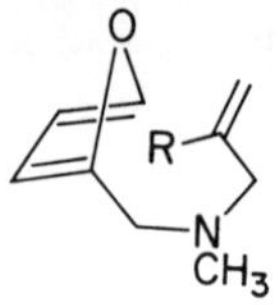 (100)	224
	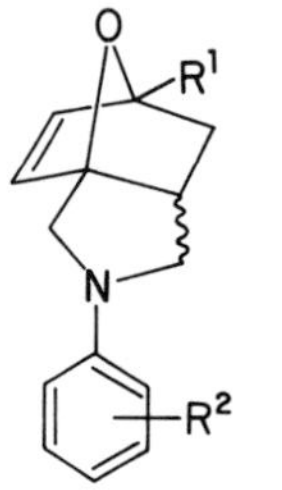 R^1	R^2		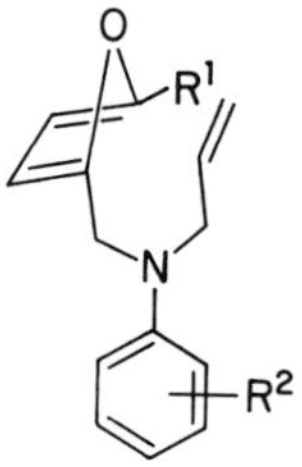	
2	H	H	Vacuum distillation, bp 170°/13 torr	(90)	163,649
3	H	H	Electron impact	(—)	649
4	H	p-CH_3	Vacuum distillation or electron impact	(—)	163,649
5	H	p-CH_3O	" " " "	(—)	163,649
6	H	o-CH_3O	Electron impact	(—)	649
7	CH_3	H	" "	(—)	649
7a	CH_3	p-Cl	" "	(—)	649
8	CH_3	p-CH_3	" "	(—)	649
9	CD_3	p-CH_3	" "	(—)	649
10	CH_3	p-CH_3O	" "	(—)	649
11	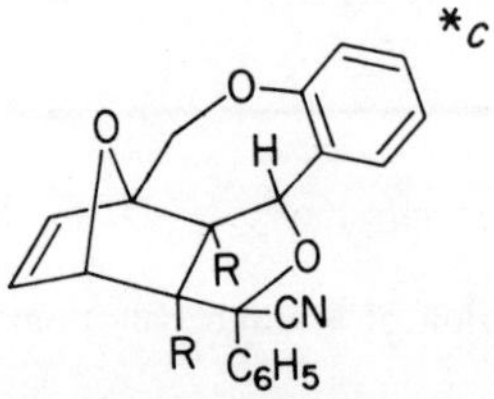 *[c]; $R=CH_3O_2C$		Isooctane, 120°, 6 hr	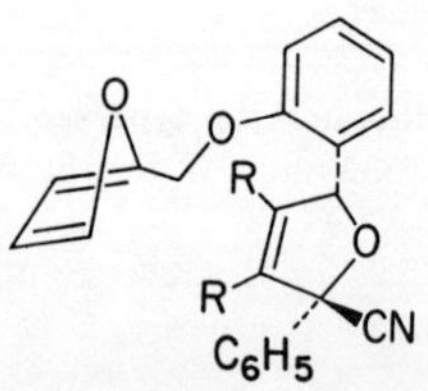 (86)[d]	650

[a] For equilibria of this type, see Tables XVIIIa/2, XVIIIb/37, and XVIIId/6.

[b] The substrate was generated by reduction of the lactam (X =O) with lithium aluminum hydride.

[c] The substrate was generated from dimethyl-1-{[2-(3-cyano-3-phenyloxiranyl)phenoxy]methyl}-7-oxabicyclo[2.2.1]hepta-2,5-diene-2,3-dicarboxylate by intramolecular 1,3-dipolar cycloaddition.

[d] The number is the conversion; the yield was 50%.

*This intermediate is hypothetical, or has not been isolated.

TABLE XXVIIIi. REVERSE INTRAMOLECULAR DIELS–ALDER REACTIONS PRODUCING DIENES CONTAINING HETEROATOMS

Entry No.	Substrate	Conditions	Product(s) and Yield(s)(%)	Refs.
	*a (R)		* (+CH_3CHO)	
1	2-Furyl	Neat, 270–280°, 10 min	(40)	651
2	C_6H_5	", 250°, "	(80)	651
3	p-BrC_6H_4	", 210°, "	(78)	651
4	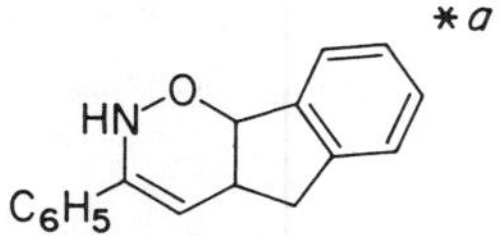 *a (C_6H_5)	Neat, 200°, 10 min	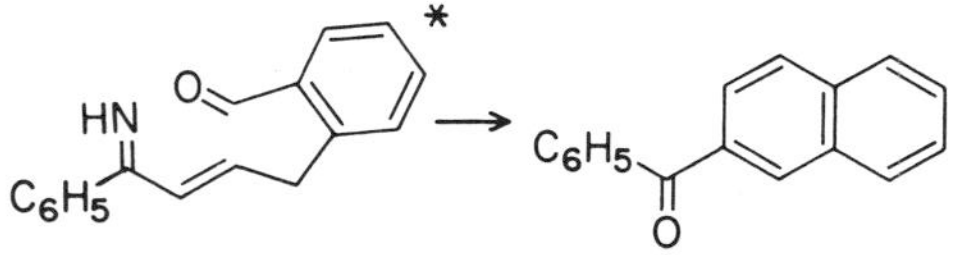 * (32)	651
	R^1 R^2 R^3			
5	CH_3 H H	<25°	(78)	652
6	H H CH_3	<25°	(92–97)	652

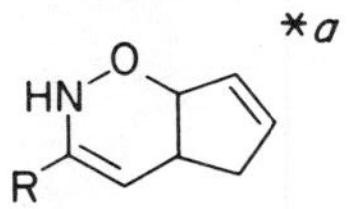

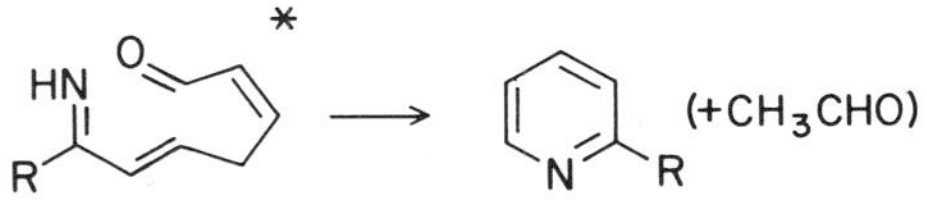

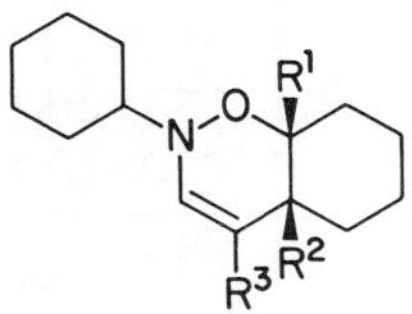

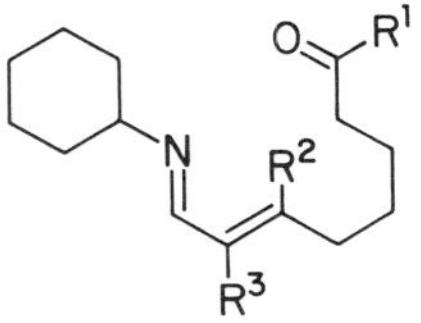

TABLE XXVIIIi. Reverse Intramolecular Diels–Alder Reactions Producing Dienes Containing Heteroatoms (*Continued*)

Entry No.	Substrate	Conditions	Product(s) and Yield(s)(%)	Refs.
	R^1 R^2 R^3			
7	CH_3 CH_3 H	80°, 2 hr	(91)	652
8	CH_3 CH_3 CH_3	80°, 2 hr	(95)	652
9	* [b]; R^1=H, CH_3; R^2=H, t—$C_4H_9(CH_3)_2SiOCH_2$.		(—)	653
		200°, Cu bronze	*	
10	R^1 = CH_3; R^2 = $(CH_3)_2C{=}CH(CH_2)_2$		(—)	232
11	R^1 = $(CH_3)_2C{=}CH(CH_2)_2$; R^2 = CH_3		(—)	232

[a] The substrate was generated by isomerization of the cyclic oxime.
[b] The substrate was generated by base treatment of the corresponding fused pyridinium tetrafluoroborate.
* This intermediate is hypothetical, or has not been isolated.

TABLE XXVIIIj. REVERSE INTRAMOLECULAR DIELS–ALDER REACTIONS PRODUCING MISCELLANEOUS HETEROCYCLES[a]

Entry No.	Substrate	Conditions	Product(s) and Yield(s) (%)	Refs.
1		0.2 torr, 400°	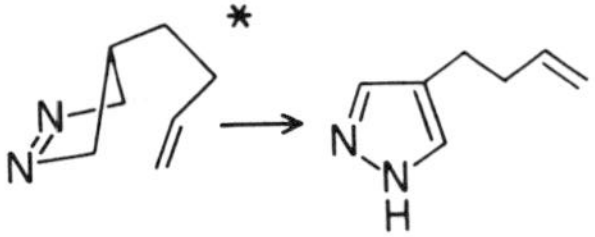 * → (98)	654
2	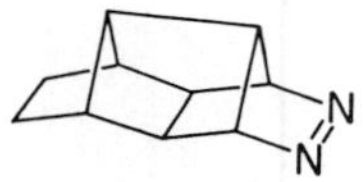	Gas phase, 380°	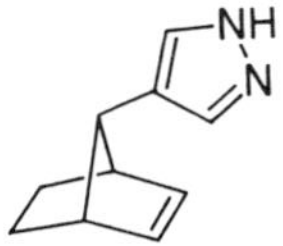 (95)	326
3	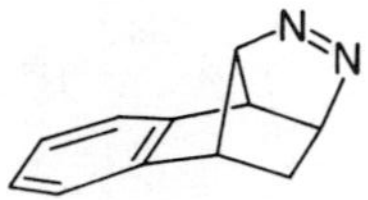	Gas phase, 400°	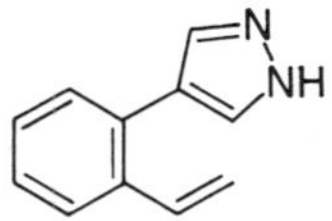 (100)	655

TABLE XXVIIIj. REVERSE INTRAMOLECULAR DIELS–ALDER REACTIONS PRODUCING MISCELLANEOUS HETEROCYCLES[a] (*Continued*)

Entry No.	Substrate	Conditions	Product(s) and Yield(s)(%)	Refs.
4	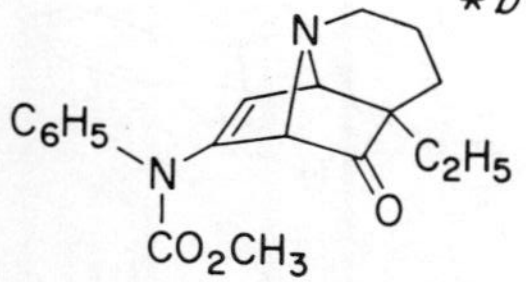 C_6H_5–N(CO_2CH_3)– ; N ; C_2H_5 ; O *b	CH_3OH, *hν*, 25°	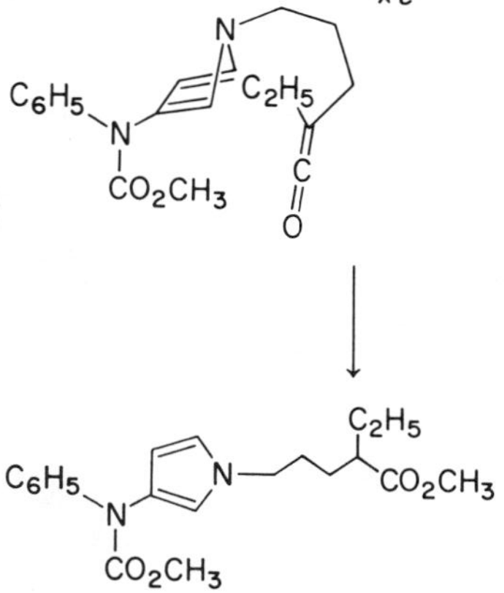 C_6H_5–N(CO_2CH_3)– ; N ; C_2H_5 ; C=O *c → C_6H_5–N(CO_2CH_3)–(pyrrol-1-yl)–$CH_2CH_2CH_2CH(C_2H_5)CO_2CH_3$ (75)	656

[a] For equilibria of this type, see Table XXV/4.

[b] The substrate is the proposed intermediate from the photolysis (Pyrex) of C_6H_5–N(CO_2CH_3)– ; N–N ; N ; H ; C_2H_5 ; O (75)

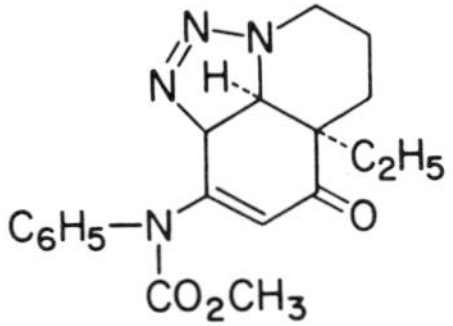

[c] In benzene, the ketene dimer was formed.

* This intermediate is hypothetical, or has not been isolated.

TABLE XXVIIIk. Reverse Intramolecular Diels–Alder Reactions of bridged Compounds Producing Acyclic 1,3,5-Trienes

Entry No.	Substrate					Conditions	Product(s) and Yield(s)(%)	Refs.
	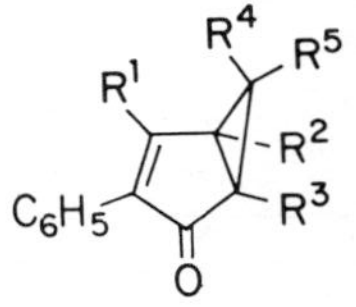						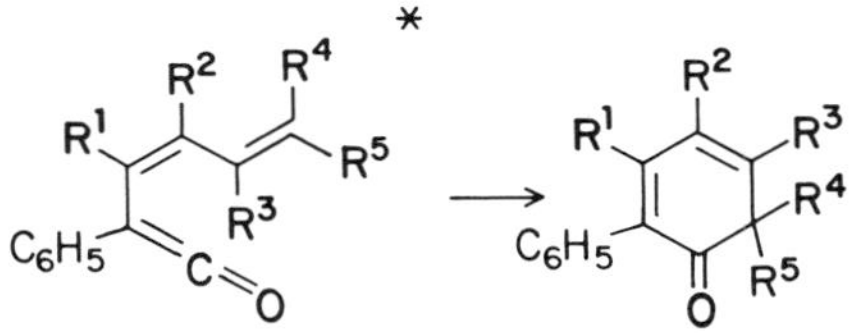	
	R^1	R^2	R^3	R^4	R^5			
1	H	H	H	C_6H_5	CH_3CO_2	C_6D_6, 129°, 2 hr	(—)	657
2	C_6H_5	H	H	C_6H_5	CH_3CO_2	", 142°, 17 hr	(—)[a]	657
3	H	C_6H_5	H	CH_3CO_2	C_6H_5	Toluene, reflux, 45 min	(90)	596
4	H	H	C_6H_5	C_6H_5	CH_3CO_2	C_6D_6, 145°, 41 hr		657
5	H	C_6H_5	H	$C_6H_5CO_2$	C_6H_5	Toluene, reflux, 45 min	(—)[a] (90)	596
	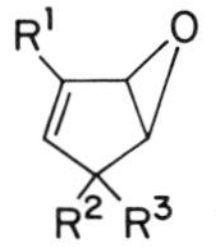						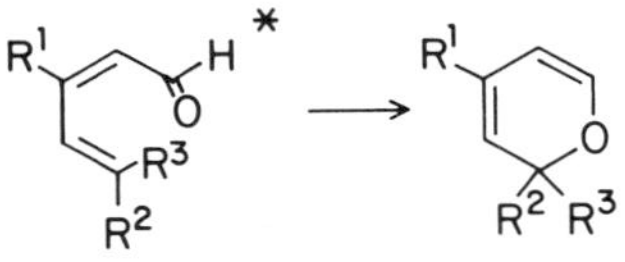 A B	
	R^1	R^2	R^3				A B	
6	H	H	H			14 torr, 400°	(9)*[b] (0)	658
7	HOC	O=*[c]				Vacuum, 550°	(0) (95)	659
8	$HOCH_2$	O=*[c]				", 540°	(0) (80)	659

TABLE XXVIIIk. REVERSE INTRAMOLECULAR DIELS–ALDER REACTIONS OF BRIDGED COMPOUNDS PRODUCING ACYCLIC 1,3,5-TRIENES (*Continued*)

Entry No.	Substrate R¹	R²	R³	Conditions	Product(s) and Yield(s) (%) A	B	Refs.
9	$CH_3CH{=}CH^d$	O=*[c]		Vacuum, 420°	(0)	(10)	660
10	$C_2H_5O_2C$	O=*[c]		", 510°	(0)	(90–95)	659
11	HOC	CH_3O	CH_3O*[c]	", 580°	(0)	(55)	659
12	$HOCH_2$	CH_3O	CH_3O*[c]	", 600°	(0)	(65)	659
13	$CH_3CH{=}CH^d$	CH_3O	CH_3O*[c]	", 450°	(0)	(—)	660
14	$C_2H_5O_2C$	CH_3O	CH_3O*[c]	", 520°	(0)	(60)	659

15

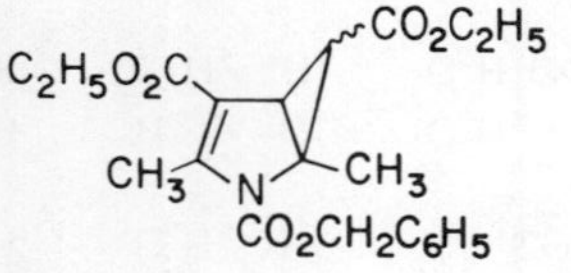

e

260°, 3 min

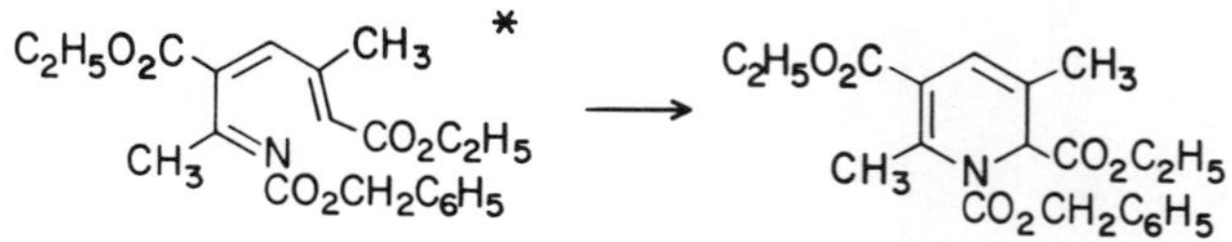

(100)

661

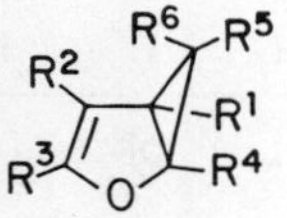

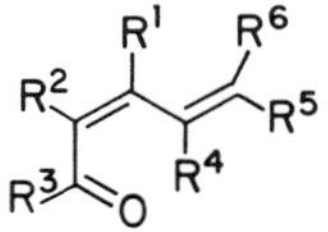

Entry No.	R¹	R²	R³	R⁴	R⁵	R⁶	Conditions	Product(s) and Yield(s) (%)	Refs.
16	H	H	H	H	—[f]	—*[g]	—	(13)[h]	662
17	H	H	H	H	CH_3CO	H*[i]	90°, 15 min	(43)	663
18	H	H	H	H	CH_3O_2C	H	Decalin, 160°, 1 hr	(90)	664
19	H	H	CH_3	H	CH_3CO	H*[i]	66°	(39)	663
20	H	H	H	—$(CH_2)_2$CO—		H*[i]	C_6H_{12}, reflux	(60)	665
21	H	H	CH_3	CH_3	CH_3CO	H*[i]	93°	(38)	663
22	H	H	CH_3	H	$C_2H_5O_2C$	H*[i]	66°	(48)	663

No.	R^1	R^2	R^3	R^4	R^5	R^6	Conditions	Product (Yield, %)	Ref.
23	H	H	CH_3	CH_3	$C_2H_5O_2C$	H		[illegible]	
24	H	H	H	H	$(CH_3)_2C{=}CH$	NC	60° ($t_{1/2}$ = 23 min)	(—)[k]	667,668
25	H	H	H	H	$(CH_3)_2C{=}CH$	CH_3CO	40°, 3 hr	(100)	668
26	C_6H_5	H	H	H	H	C_6H_5	SiO_2, 150°, 10 min	(—)[k]	669
27	H	H	H	H	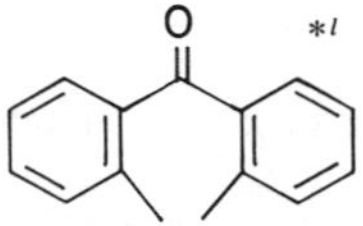 *[l]		16°	(44)[m]	670
28	C_6H_5	C_6H_5	C_6H_5	C_6H_5	C_6H_5	H	Vacuum, 180°, 30 min	(64)[n]	671
29	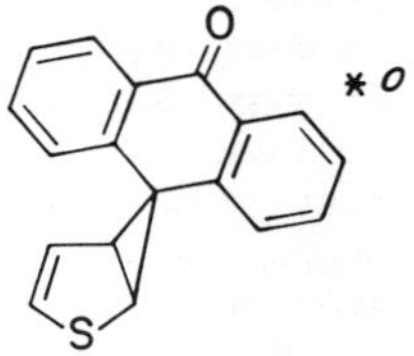 *[o]						16°	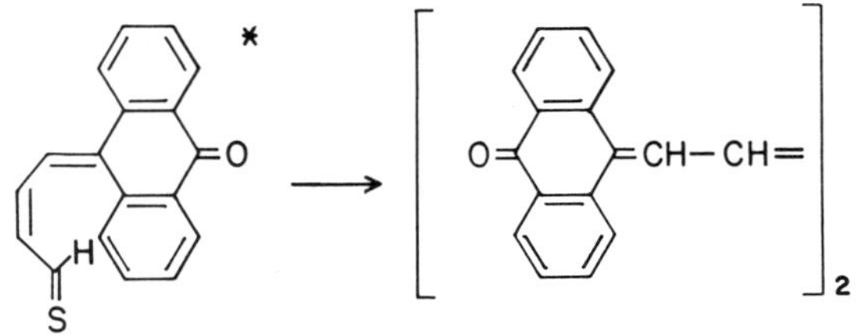 * (30)	672
29a	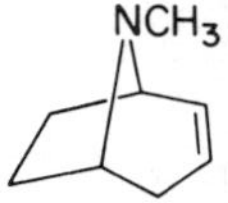 NCH_3						Pentane, $h\nu$	$CH_3N{=}CH(CH_2)_2CH{=}CHCH{=}CH_2$ (52)	673
30	CH_3	CH_3	CH_3	CH_3	CH_3	CH_3	Benzene, hot tube, 500°	* (38)[p]	674
							CH_3OH, $h\nu$	(—)[q]	592

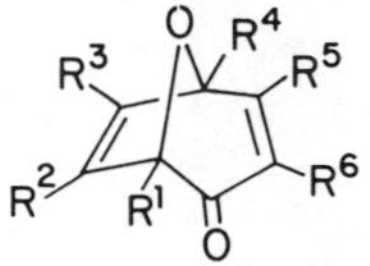

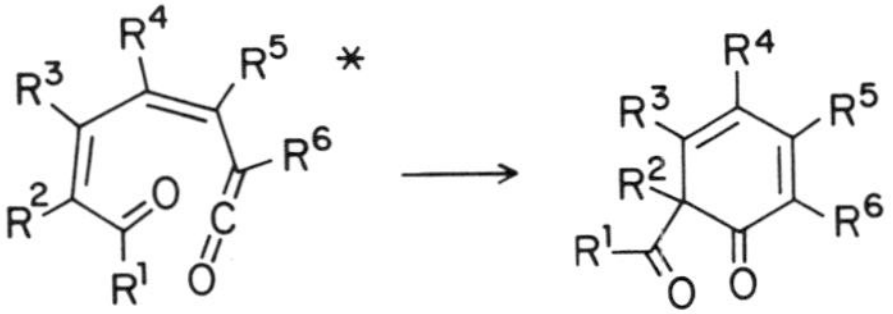

TABLE XXVIIIk. REVERSE INTRAMOLECULAR DIELS–ALDER REACTIONS OF BRIDGED COMPOUNDS PRODUCING ACYCLIC 1,3,5-TRIENES (*Continued*)

Entry No.	Substrate R^1	R^2	R^3	R^4	R^5	R^6	Conditions	Product(s) and Yield(s) (%)	Refs.
31	C_6H_5	CH_3O_2C	CH_3O_2C	C_6H_5	H	C_6H_5	Neat, 8 torr, 190°, 2 hr	(43)	675
32	C_6H_5	C_6H_5	C_6H_5	C_6H_5	H	C_6H_5	", ", 280°, 1 hr	(60)	675
33	C_6H_5	C_6H_5CO	C_6H_5CO	C_6H_5	H	C_6H_5	", 1 torr, 190°, 1 hr	(29)	675

[a] Entries 2 and 4 gave the same equilibrium mixture of quinone acetates interrelated by acyl shifts.
[b] The product was isolated as the 2,4-dinitrophenylhydrazone; the aldehyde polymerized at 25°.
[c] The substrate was generated from the formal Diels–Alder adduct with cyclopentadiene.
[d] The double bond is *trans*.
[e] Both *exo* and *endo* isomers gave the same product in identical yields.
[f] The substrate was the carbene.
[g] The substrate was generated by addition of atomic carbon to furan.
[h] The product was *cis*-2-penten-4-ynal.
[i] The substrate was generated by copper-catalyzed decomposition of the appropriate diazo compound in the appropriate furan.
[j] The rearrangement was catalyzed by Lewis acids.
[k] The product was a mixture of stereoisomers.
[l] The substrate was generated by photolysis of 10-diazoanthrone in furan.
[m] The *trans* aldehyde, which is a secondary photoproduct, was formed in 9% yield.
[n] $R^5 = H$; $R^6 = C_6H_5$.
[o] The substrate was generated by photolysis of 10-diazoanthrone in thiophene.
[p] The product was pentaphenylphenol.
[q] The product was 6-acetylpentamethylbicyclo[3.1.0]hexenone; see Table XXVIa/5.
* This intermediate is hypothetical, or has not been isolated.

TABLE XXVIIII. REVERSE INTRAMOLECULAR DIELS–ALDER REACTIONS OF BRIDGED COMPOUNDS PRODUCING CYCLIC 1,3,5-TRIENES

Entry No.	Substrate	Conditions	Product(s) and Yield(s)(%)	Refs.
	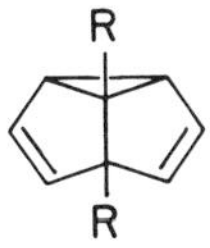 R		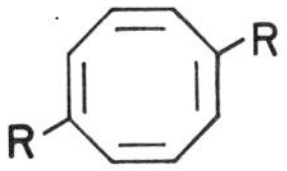	
1	H	Flow system, 427°	(56)	32,676
2	CH_3	", 390°	(76)	33,676
3	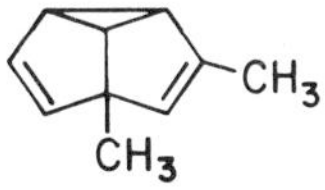 CH_3, CH_3	Flow system, 445°	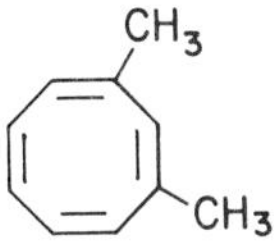 CH_3, CH_3 (69)	32
4	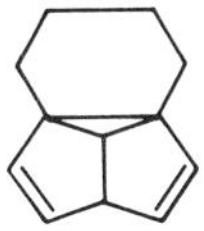	Flow system, 500°	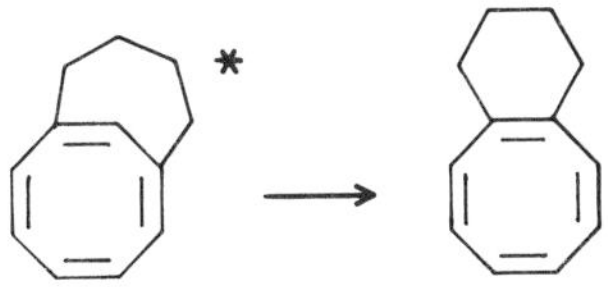 * → (69)	453
5	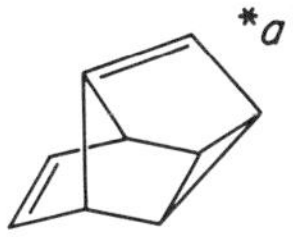 *[a]	THF, −78°	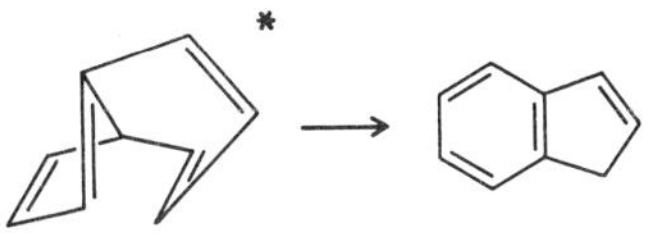 * → (15)	626

[a] See original work for the precursor.
*This intermediate is hypothetical, or has not been isolated.

TABLE XXVIIIm. REVERSE INTRAMOLECULAR HOMO DIELS–ALDER REACTIONS

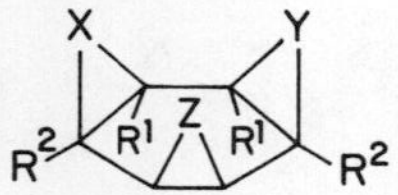

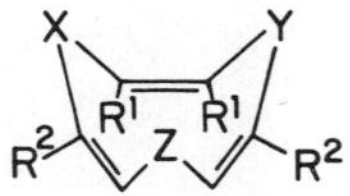

Entry No.	Substrate X	Y	Z	R^1	R^2	Conditions	Product(s) and Yield(s) (%)	Refs.
1	CH_2	CH_2	CH_2	$t\text{-}C_4H_9O_2C$	H*[a]	−72 to 25°	(74)	627
2	CH_2	CH_2	CH_3SO_2N	CH_3O_2C	H	Benzene, 90°, 2 hr	(65)	629
3	CH_2	CH_2	$p\text{-}CH_3C_6H_4SO_2N$	CH_3O_2C	H	", reflux, 2–3 hr	(100)	629
4	CH_3O_2CCH	CH_3O_2CCH	$p\text{-}CH_3C_6H_4SO_2N$	H	H	Me_2CO, 140°	(100)	677
5	NCCH	NCCH	O	H	H	Benzene, 140°	(100)	677
6	CH_2	CH_2	O	CH_3O_2C	H	", reflux, 2 hr	(100)	147
7	CH_3O_2CCH	CH_3O_2CCH	O	H	H	", 140°	(100)	677
8	CH_2	CH_2	O	CH_3O_2C	CH_3O_2C	$CHCl_3$, 65°, 1.5 hr	(96)	147
9	CH_2	CH_2	O	CH_3O_2C	C_6H_5	", 45°, 1 hr	(90)	147
10	CH_2	O	O	H	H	> 150°	(100)	678
11	CH_3N	CH_3N	CH_3N	H	H	C_6D_6, 90°, 10 min	(100)	679
12	CH_3SO_2N	CH_3SO_2N	CH_3SO_2N	H	H	$(CH_3)_2SO_2$, 200°, 10 min	(95)	679
13	CF_3SO_2N	CF_3SO_2N	CF_3SO_2N	H	H	", 170°, 5 hr	(66)	679

14	CH_3O_2CN	CH_3O_2CN	CH_3O_2CN	H	H	CH_3CN, 80°, 90 min	(—)	679
15	$NC(CH_2)_2N$	$NC(CH_2)_2N$	$NC(CH_2)_2N$	H	H	$CDCl_3$, 100°, 2.5 hr	(100)	679
16	$(CH_3)_2NCON$	$(CH_3)_2NCON$	$(CH_3)_2NCON$	H	H	Xylene, reflux, 3 hr	(93)	679
17	$CH_3O_2C(CH_2)_2N$	$CH_3O_2C(CH_2)_2N$	$CH_3O_2C(CH_2)_2N$	H	H	CD_3CN, 80°, 130 min	(100)	679
18	C_6H_5CON	C_6H_5CON	C_6H_5CON	H	H	$CHCl_3$, 80°, 5 hr	(93)	679
19	O	O	O	H	H	Gas phase, 400–470°	(95)	680,681
20	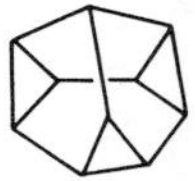					90°	 (—)	327
21	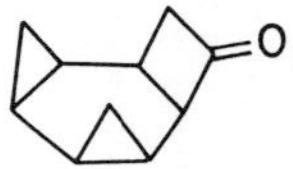					CCl_4, 150°	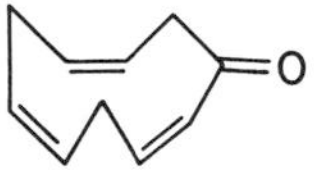 (—)	682
22	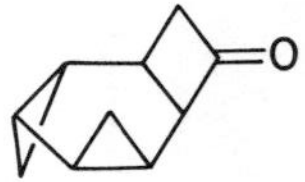					Gas phase, 300°	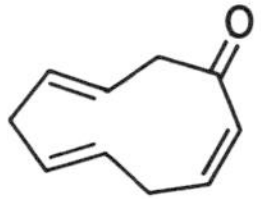 (44)	682

[a] The substrate was generated from 3,4-dicarbalkoxybicyclo[4.1.0]hept-3-ene-*cis*-2,5-dimethanol bis(methanesulfonate) and sodium naphthalene.
* This intermediate is hypothetical, or has not been isolated.

ADDENDA TO THE TABLES

The following is a list of references to recent papers that are not cited in the tables, although some are discussed in the text. The cut-off date is November 1983. The references are listed in the order of diene type and are keyed to the tables.

Table IIa	Refs. 140, 320a, 683, 684
Table IIb	Refs. 70b, 171, 172, 684a, 685, 686
Table IId	Ref. 687
Table IIe	Ref. 688
Table IIf	Refs. 178a,b
Table IIIa	Refs. 143, 144, 146, 154a,b, 191, 689–694
Table IIIb	Refs. 695, 696
Table IV	Ref. 57a, 151a
Table V	Refs. 71b, 697
Table VI	Refs. 166a, 698 (2,3-dimethylenecyclohexanone)
Table VIIIa	Refs. 699, 700
Table VIIIb	Refs. 71c, 700
Table IXd	Refs. 701–705
Table X	Refs. 706, 707
Table XI	Ref. 708
Table XII	Refs. 709, 710
Table XIII	Refs. 325, 704, 711, 712
Table XIVa	Refs. 172c,f
Table XIVb	Refs. 171, 172e (10-methyleneanthrone)
Table XIVd	Refs. 713, 714
Table XIVf	Ref. 715
Table XV	Refs. 280, 716
Table XVIa	Refs. 214, 717–722
Table XVIb	Refs. 723, 724 (benzo-*o*-quinodimethane)
Table XVId	Refs. 703, 719, 725
Table XVIIc	Refs. 172a,b
Table XVIIIb	Refs. 70a, 71a
Table XVIIId	Ref. 726
Table XIX	Refs. 172c,d,f, 727
Table XX	Refs. 728
Table XXI	Refs. 705, 730
Table XXII	Refs. 705, 731, 732
Table XXIII	Ref. 733
Table XXV	Refs. 734, 735 (metallocycles), 736 (oxazoles), 729
Table XXVIa	Ref. 737
Table XXVIc	Ref. 738
Table XXVIIIb	Refs. 708, 739
Table XXVIIIe	Ref. 740

Table XXVIIIg Ref. 172a
Table XXVIIIj Refs. 741, 742
Table XXVIIIk Ref. 743
Table XXVIIIm Refs. 744–746

REFERENCES

[1] For a recent discussion of the Diels–Alder reaction, including a list of references to reviews, see J. Sauer and R. Sustmann, *Angew. Chem., Int. Ed. Engl.*, **19**, 779 (1980).

[2] For a recent review of photochemical rearrangements of trienes, see W. G. Dauben, E. L. McInnes, and D. M. Michno in *Rearrangements in Ground and Excited States*, P. de Mayo, Ed., Vol. 3, Academic Press, New York, 1980.

[3] H. Wollweber, *Methoden Org. Chem. Houben Weyl*, 4th ed., 5/1c, 1113 (1970).

[4] R. G. Carlson, *Annu. Rep. Med. Chem.*, **9**, 270 (1974).

[5] G. Mehta, *J. Chem. Educ.*, **53**, 551 (1976).

[6] W. Oppolzer, *Angew. Chem., Int. Ed. Engl.*, **16**, 10 (1977).

[7] W. Oppolzer in *New Synthetic Methods*, Vol. 6, Verlag Chemie, Weinheim and New York, 1979, p. 1.

[8] G. Brieger and J. N. Bennett, *Chem. Rev.*, **80**, 63 (1980).

[8a] A. G. Fallis, *Can. J. Chem.*, **62**, 183 (1984).

[8b] D. F. Taber, in *Reactivity and Structure Concepts in Organic Chemistry*, Springer Verlag, Berlin and New York, 1984.

[9] W. Oppolzer, *Synthesis*, **1978**, 793.

[10] W. Oppolzer, *Heterocycles*, **14**, 1615 (1980).

[11] T. Kametani, *Pure Appl. Chem.*, **51**, 747 (1979).

[12] K. P. C. Vollhardt, *Ann. N. Y. Acad. Sci.*, **333**, 241 (1980).

[13] R. L. Funk and K. P. C. Vollhardt, *Chem. Soc. Rev.*, **9**, 41 (1980).

[14] T. Kametani and H. Nemoto, *Tetrahedron*, **37**, 3 (1981).

[15] J. L. Ripoll, A. Rouessac, and F. Rousseac, *Tetrahedron*, **34**, 19 (1978).

[16] R. B. Woodward and R. Hoffmann, *The Conservation of Orbital Symmetry*, Verlag Chemie Academic Press, New York, 1970, p. 65.

[17] R. F. Borch, A. J. Evans, and J. J. Wade, *J. Am. Chem. Soc.*, **99**, 1612 (1977); *ibid.*, **97**, 6282 (1975).

[18] G. E. Keck, E. Boden, and U. Sonnewald, *Tetrahedron Lett.*, **1981**, 2615.

[19] F. Näf, R. Decorzant, and W. Thommen, *Helv. Chim. Acta*, **62**, 114 (1979), *ibid.*, **65**, 2212 (1982).

[20] S. F. Martin, C. Tu, M. Kimura, and S. H. Simonsen, *J. Org. Chem.*, **47**, 3634 (1982).

[21] A. Krantz, *J. Am. Chem. Soc.*, **94**, 4020 (1972).

[22] S. F. Martin, C. Tu, and T. Chou, *J. Am. Chem. Soc.*, **102**, 5274 (1980); S. F. Martin, T. Chou, and C. Tu, *Tetrahedron Lett.*, **1979**, 3823.

[23] P. J. Garratt and S. B. Neoh, *J. Org. Chem.*, **44**, 2667 (1979).

[24] M. E. Kuehne, D. M. Roland, and R. Hafter, *J. Org. Chem.*, **43**, 3705 (1978).

[25] M. E. Kuehne, J. A. Huebner, and T. H. Matsko, *J. Org. Chem.*, **44**, 2477 (1979).

[26] M. E. Kuehne, T. H. Matsko, J. C. Bohnert, and C. L. Kirkemo, *J. Org. Chem.*, **44**, 1063 (1979).

[27] Ref. 16, pp. 80–82.

[28] H. Iwamura, *Tetrahedron Lett.*, **1973**, 369; H. Iwamura and K. Morio, *Bull. Chem. Soc. Jpn.*, **45**, 3599 (1972).

[29] M. Kuzuya, F. Miyake, and T. Okuda, *Tetrahedron Lett.*, **1980**, 1043.

[30] M. Kuzuya, F. Miyake, and T. Okuda, *Tetrahedron Lett.*, **1980**, 2185.

[31] W. P. Lay, K. Mackenzie, A. S. Miller, and D. L. Williams-Smith, *Tetrahedron*, **36**, 3021 (1980); G. E. Taylor, K. Mackenzie, and G. I. Fray, *Tetrahedron Lett.*, **1980**, 4935.

[32] L. A. Paquette, S. V. Ley, R. H. Meisinger, R. K. Russell, and M. Oku, *J. Am. Chem. Soc.*, **96**, 5806 (1974).

[33] R. B. Woodward and T. J. Katz, *Tetrahedron*, **5**, 70 (1959).

[34] K. N. Houk, *J. Am. Chem. Soc.*, **95**, 4092 (1973); M. J. S. Dewar and R. S. Pyron, *ibid.*, **92**, 3098 (1970); J. W. McIver, *Acc. Chem. Res.*, **7**, 72 (1974).

[35] I. Fleming, *Frontier Orbitals and Organic Chemical Reactions*, Wiley, New York, 1976.

[36] J. D. White and B. G. Sheldon, *J. Org. Chem.*, **46**, 2273 (1981).

[37] R. K. Boeckman, Jr. and S. S. Ko, *J. Am. Chem. Soc.*, **102**, 7146 (1980).

[38] W. R. Roush and S. M. Peseckis, *J. Am. Chem. Soc.*, **103**, 6696 (1981).

[39] D. F. Taber, C. Campbell, B. C. Gunn, and I.-C. Chiu, *Tetrahedron Lett.*, **1981**, 5141.

[40] W. R. Roush, A. I. Ko, and H. R. Gillis, *J. Org. Chem.*, **45**, 4267 (1980).

[41] W. R. Roush, H. R. Gillis, and A. I. Ko, *J. Am. Chem. Soc.*, **104**, 2269 (1982).

[42] K. N. Houk and R. W. Strozier, *J. Am. Chem. Soc.*, **95**, 4094 (1973); K. N. Houk, *Acc. Chem. Res.*, **8**, 361 (1975).

[43] H. W. Gschwend, A. O. Lee, and H.-P. Meier, *J. Org. Chem.*, **38**, 2169 (1973).

[44] H. W. Gschwend and H. P. Meier, *Angew. Chem., Int. Ed. Engl.*, **11**, 294 (1972).

[45] E. Ciganek, *J. Org. Chem.*, **45**, 1497 (1980).

[45a] K. Sakan and B. M. Craven, *J. Am. Chem. Soc.*, **105**, 3732 (1983).

[46] N. S. Isaacs and P. Van der Beeke, *Tetrahedron Lett.*, **1982**, 2147.

[47] T. R. Melikyan, G. O. Torosyan, R. S. Mkrtchyan, K. T. Tagmazyan, and A. T. Babayan, *Arm. Khim. Zh.*, **30**, 981 (1977) [*C.A.*, **89**, 75356 (1978)].

[48] R. S. Mkrtchyan, T. R. Melikyan, G. O. Torosyan, K. T. Tagmazyan, and A. T. Babayan, *Arm. Khim. Zh.*, **31**, 328 (1978) [*C.A.*, **89**, 108932 (1978)].

[49] G. O. Torosyan, K. T. Tagmazyan, and A. T. Babayan, *Arm. Khim. Zh.*, **27**, 752 (1974) [*C.A.*, **82**, 57110 (1975)].

[50] T. R. Melikyan, G. O. Torosyan, R. S. Mkrtchyan, K. T. Tagmazyan, and A. T. Babayan, *Arm. Khim. Zh.*, **30**, 138 (1977) [*C.A.*, **87**, 38422 (1977)].

[51] K. T. Tagmazyan, R. S. Mkrtchyan, and A. T. Babayan, *Arm. Khim. Zh.*, **27**, 957 (1974) [*C.A.*, **82**, 111234 (1975)].

[52] A. T. Babayan, K. T. Tagmazyan, A. I. Ioffe, R. S. Mkrtchyan, and G. O. Torosyan, *Dokl. Akad. Nauk Arm. SSR*, **58**, 275 (1974) [*C.A.*, **82**, 30640 (1975)].

[53] R. Breslow and D. C. Rideout, *J. Am. Chem. Soc.*, **102**, 7816 (1980).

[54] D. D. Sternbach and D. M. Rossana, *Tetrahedron Lett.*, **1982**, 303.

[55] D. D. Sternbach and D. M. Rossana, *J. Am. Chem. Soc.*, **104**, 5853 (1982).

[56] R. A. Firestone and M. A. Vitale, *J. Org. Chem.*, **46**, 2160 (1981).

[57] N. S. Isaacs, University of Reading, England, personal communication.

[57a] In a recent example, application of 11–13 kbar pressure permitted reduction of the cyclization temperature from 120° to 40° in the intramolecular Diels–Alder reaction of a substrate with an eleven-membered chain: S. A. Harkin and E. J. Thomas, *Tetrahedron Lett.*, **1983**, 5535.

[58] D. Bilović, *Croat. Chem. Acta*, **40**, 15 (1968) [*C.A.*, **69**, 86751 (1968)].

[59] J. W. Harder, Ph.D. Dissertation, University of California, Berkeley, 1978.

[60] A. P. Kozikowski, Ph.D. Dissertation, University of California, Berkeley, 1974.

[61] G. Stork, G. Clark, and C. S. Shiner, *J. Am. Chem. Soc.*, **103**, 4948 (1981).

[62] B. B. Snider and J. V. Dunčia, *J. Org. Chem.*, **45**, 3461 (1980).

[63] P. J. DeClerq and L. E. Van Royen, *Synth. Commun.*, **9**, 771 (1975).

[64] S. R. Baker, M. J. Begley, and L. Crombie, *J. Chem. Soc., Perkin Trans. 1*, **1981**, 190.

[65] O. P. Vig, I. H. Trehan, and R. Kumar, *Indian J. Chem., Sect. B.*, **15**, 319 (1977).

[66] K. A. Parker and M. R. Adamchuk, *Tetrahedron Lett.*, **1978**, 1689.

[67] T. Mukaiyama, T. Tsuji, and N. Iwasawa, *Chem. Lett.*, **1979**, 697.

[68] T. Mukaiyama and T. Takebayashi, *Chem. Lett.*, **1980**, 1013.

[69] T. Mukaiyama and N. Iwasawa, *Chem. Lett.*, **1981**, 29.

[70] T. Takebayashi, N. Iwasawa, and T. Mukaiyama, *Chem. Lett.*, **1982**, 579.

[70a] T. Takebayashi, N. Iwasawa, and T. Mukaiyama, *Bull. Chem. Soc. Jpn.*, **56**, 1107 (1983).

[70b] T. Takebayashi, N. Iwasawa, T. Mukaiyama, and T. Hata, *Bull. Chem. Soc. Jpn.*, **56**, 1669 (1983).

[71] F. Näf and G. Ohloff, *Helv. Chim. Acta*, **57**, 1868 (1974); F. Näf, R. Decorzant, W. Giersch, and G. Ohloff, *ibid.*, **64**, 1387 (1981).

[71a] M. S. Bailey, B. J. Brisdon, D. W. Brown, and K. M. Stark, *Tetrahedron Lett.*, **1983**, 3037.

[71b] K. J. Shea and J. W. Gilman, *Tetrahedron Lett.*, **1983**, 657.

[71c] S. K. Attah-Poku, G. Gallacher, A. S. Ng, L. E. B. Taylor, S. J. Alward, and A. G. Fallis, *Tetrahedron Lett.*, **1983**, 677.

[72] E. J. Corey and M. Petrzilka, *Tetrahedron Lett.*, **1975**, 2537.

[73] R. L. Funk and K. P. C. Vollhardt, *J. Am. Chem. Soc.*, **102**, 5245 (1980); *ibid.*, **98**, 6755 (1976).

[74] V. Ramamurthy and R. S. H. Liu, *J. Org. Chem.*, **39**, 3435 (1974).

[75] W. C. Ripka, E. I. duPont de Nemours & Co., Wilmington, DE, personal communication.

[76] H. Greuter, H. Schmid, and G. Fráter, *Helv. Chim. Acta*, **60**, 1701 (1977); H. Greuter, G. Fráter, and H. Schmid, *ibid.*, **55**, 526 (1972); G. Fráter, H. Greuter, and H. Schmid, *U.S. Pat.* 3925479 (1975) [*C.A.*, **85**, 46098 (1976)].

[77] A. Krantz and C. Y. Lin, *J. Am. Chem. Soc.*, **95**, 5662 (1973); *J. Chem. Soc., Chem. Commun.*, **1971**, 1287.

[78] H. Greuter and H. Schmid, *Helv. Chim. Acta*, **55**, 2382 (1972).

[79] R. L. Snowden, *Tetrahedron Lett.*, **1981**, 97, 101.

[80] C. A. Cupas, W. Schumann, and W. E. Heyd, *J. Am. Chem. Soc.*, **92**, 3237 (1970).

[80a] For a review of this type of cyclization in hetero-1,3,5-trienes, see M. V. George, A. Mitra, and K. B. Sukamaran, *Angew. Chem., Int. Ed. Engl.*, **19**, 973 (1980).

[81] J. Griffiths and H. Hart, *J. Am. Chem. Soc.*, **90**, 3297 (1968).

[82] K. J. Shea, P. S. Beauchamp, and R. S. Lind, *J. Am. Chem. Soc.*, **102**, 4544 (1980).

[83] P. S. Beauchamp, Ph.D. Dissertation, University of California, Irvine, 1981 [*Diss. Abstr. Int. B.*, **42**, 3684 (1982)].

[84] W. Oppolzer, C. Fehr, and J. Warneke, *Helv. Chim. Acta*, **60**, 48 (1977).

[85] R. K. Boeckman, Jr. and T. R. Alessi, *J. Am. Chem. Soc.*, **104**, 3216 (1982).

[86] S. G. Pyne, M. J. Hensel, and P. L. Fuchs, *J. Am. Chem. Soc.*, **104**, 5719 (1982).

[87] R. K. Boeckman, Jr. and S. S. Ko, *J. Am. Chem. Soc.*, **104**, 1033 (1982).

[88] M. E. Jung and K. M. Halweg, *Tetrahedron Lett.*, **1981**, 3929.

[89] J. J. S. Bajorek and J. K. Sutherland, *J. Chem. Soc., Perkin Trans. 1*, **1975**, 1559.

[90] A. Ichihara, R. Kimura, S. Yamada, and S. Sakamura, *J. Am. Chem. Soc.*, **102**, 6353 (1980).

[91] B. S. Joshi, N. Viswanathan, D. H. Gawad, V. Balakrishnan, and W. v. Philipsborn, *Helv. Chim. Acta*, **58**, 2295 (1975); B. S. Joshi, N. Viswanathan, D. H. Gawad, V. Balakrishnan, W. v. Philipsborn, and A. Quick, *Experientia*, **31**, 880 (1975).

[92] R. L. Funk and W. E. Zeller, *J. Org. Chem.*, **47**, 180 (1982).

[93] W. R. Roush and S. E. Hall, *J. Am. Chem. Soc.*, **103**, 5200 (1981).

[94] W. R. Roush and H. R. Gillis, *J. Org. Chem.*, **47**, 4825 (1982)

[95] S. R. Wilson and D. T. Mao, *J. Am. Chem. Soc.*, **100**, 6289 (1978).

[96] S. R. Wilson and D. T. Mao, *J. Org. Chem.*, **44**, 3093 (1979).

[97] W. Oppolzer and R. L. Snowden, *Tetrahedron Lett.*, **1976**, 4187; W. Oppolzer, R. L. Snowden, and D. P. Simmons, *Helv. Chim. Acta*, **64**, 2002 (1981).

[98] J.-L. Gras and M. Bertrand, *Tetrahedron Lett.*, **1979**, 4549.

[99] D. F. Taber and B. P. Gunn, Vanderbilt University, Nashville, Tenn., personal communication.

[100] E. A. Deutsch and B. B. Snider, *J. Org. Chem.*, **47**, 2682 (1982).

[101] S. F. Martin, University of Texas at Austin, personal communication; S. F. Martin, S. A. Williamson, R. P. Gist, and K. M. Smith, *J. Org. Chem.*, **48**, 5170 (1983).

[102] W. Oppolzer and W. Fröstl, *Helv. Chim. Acta*, **58**, 590 (1975).

[103] W. Oppolzer, W. Fröstl, and H. P. Weber, *Helv. Chim. Acta*, **58**, 593 (1975).

[104] W. Oppolzer and E. Flaskamp, *Helv. Chim. Acta*, **60**, 204 (1977).

[105] G. Stork and D. J. Morgans, Jr., *J. Am. Chem. Soc.*, **101**, 7110 (1979).

[106] R. Prewo, J. H. Bieri, U. Widmer, and H. Heimgartner, *Helv. Chim. Acta*, **64**, 1515 (1981).
[107] U. Widmer, H. Heimgartner, and H. Schmid, *Helv. Chim. Acta*, **61**, 815 (1978).
[108] H. W. Gschwend, *Helv. Chim. Acta*, **56**, 1763 (1973).
[109] S. F. Martin, S. R. Desai, G. W. Phillips, and A. C. Miller, *J. Am. Chem. Soc.*, **102**, 3294 (1980).
[110] S. F. Martin and C.-Y. Tu, *J. Org. Chem.*, **46**, 3763 (1981).
[111] D. J. Morgans, Jr., and G. Stork, *Tetrahedron Lett.*, **1979**, 1959.
[112] D. J. Morgans, Ph.D. Dissertation, Columbia University, 1979 [*Diss. Abstr. Int. B.*, **40**, 2195 (1979)].
[113] B. Nader, R. W. Franck, and S. M. Weinreb, *J. Am. Chem. Soc.*, **102**, 1153 (1980); B. Nader, T. R. Bailey, R. W. Franck, and S. M. Weinreb, *ibid.*, **103**, 7573 (1981).
[114] S. M. Weinreb and T. Bailey, Pennsylvania State University, State College, personal communication.
[115] G. Büchi and H. Wüest, unpublished results cited in ref. 116.
[116] G. Büchi, A. Hauser, and J. Limacher, *J. Org. Chem.*, **42**, 3323 (1977); G. H. Büchi and A. Hauser, *U.S. Pat.* 4124642 (1978) [*C.A.*, **90**, 168796 (1979)].
[117] R. F. Tavares and E. Katten, *Tetrahedron Lett.*, **1977**, 1713.
[118] L. Hodakowski and C. A. Cupas, *Tetrahedron Lett.*, **1973**, 1009.
[119] W. Oppolzer, R. Achini, E. Pfenninger, and H. P. Weber, *Helv. Chim. Acta*, **59**, 1186 (1976).
[120] R. Achini, W. Oppolzer, and E. Pfenninger, *Ger. Pat.*, 2348593 (1974) [*C.A.*, **81**, 13386 (1974)]; *Swiss Pat.* 611886 (1979) [*C.A.*, **91**, 140720 (1979)].
[121] L. H. Klemm, T. M. McGuire, and K. W. Gopinath, *J. Org. Chem.*, **41**, 2571 (1976).
[122] L. H. Klemm, V. T. Tran, and D. R. Olson, *J. Heterocycl. Chem.*, **13**, 741 (1976).
[123] M. T. Cox, *J. Chem. Soc., Chem. Commun.*, **1975**, 903.
[124] L. H. Klemm, K. W. Gopinath, D. H. Lee, F. W. Kelly, E. Trod, and T. M. McGuire, *Tetrahedron*, **22**, 1797 (1966).
[125] T. Kametani, H. Matsumoto, T. Honda, M. Nagai, and K. Fukumoto, *Tetrahedron*, **37**, 2555 (1981).
[126] K. C. Nicolaou, W. E. Barnette, and P. Ma, *J. Org. Chem.*, **45**, 1463 (1980); K. C. Nicolaou and W. E. Barnette, *J. Chem. Soc., Chem. Commun.*, **1979**, 1119.
[127] W. Oppolzer, D. A. Roberts, and T. G. C. Bird, *Helv. Chim. Acta*, **62**, 2017 (1979).
[128] W. Oppolzer, D. A. Roberts, and T. G. C. Bird, unpublished work cited in ref. 10.
[129] W. Oppolzer, K. Bättig, and M. Petrzilka, *Helv. Chim. Acta*, **61**, 1945 (1978).
[130] W. Oppolzer, M. Petrzilka, and K. Bättig, *Helv. Chim. Acta*, **60**, 2964 (1977).
[131] T. Kametani, M. Aizawa, and H. Nemoto, *Tetrahedron*, **37**, 2547 (1981).
[132] S. R. Wilson and J. C. Huffman, *J. Org. Chem.*, **45**, 560 (1980).
[133] S. E. Hall and W. R. Roush, *J. Org. Chem.*, **47**, 4611 (1982).
[134] E. Ciganek, *J. Am. Chem. Soc.*, **103**, 6261 (1981).
[135] H. F. Schmitthenner and S. M. Weinreb, *J. Org. Chem.*, **45**, 3372 (1980).
[136] N. A. Khatri, H. F. Schmitthenner, J. Shringarpure, and S. M. Weinreb, *J. Am. Chem. Soc.*, **103**, 6387 (1981).
[137] T. Schmidlin, W. Zürcher, and C. Tamm, *Helv. Chim. Acta*, **64**, 235 (1981); T. Schmidlin and C. Tamm, *ibid.*, **61**, 2096 (1978).
[138] K. C. Nicolaou and R. Magolda, *J. Org. Chem.*, **46**, 1506 (1981); K. C. Nicolaou, D. P. Papahatjis, D. A. Claremon, and R. E. Dolle, III, *J. Am. Chem. Soc.*, **103**, 6967 (1981).
[139] W. R. Roush and A. G. Myers, *J. Org. Chem.*, **46**, 1509 (1981).
[140] K. A. Parker and T. Iqbal, *J. Org. Chem.*, **47**, 337 (1982).
[141] D. F. Taber and B. P. Gunn, *J. Am. Chem. Soc.*, **101**, 3992 (1979).
[142] G. E. Keck and D. G. Nickell, *J. Am. Chem. Soc.*, **102**, 3632 (1980).
[143] S. R. Wilson, M. S. Hague, and R. N. Misra, *J. Org. Chem.*, **47**, 747 (1982).
[144] D. F. Taber and S. A. Saleh, *Tetrahedron Lett.*, **1982**, 2361.
[145] E. J. Corey and R. L. Danheiser, *Tetrahedron Lett.*, **1973**, 4477.
[146] T. Kametani, Y. Suzuki, H. Furuyama, and T. Honda, *J. Org. Chem.*, **48**, 31 (1983).
[147] H. Prinzbach, D. Stusche, M. Breuninger, and J. Markert, *Chem. Ber.*, **109**, 2823 (1976).

[148] A. Krantz, Ph.D. Dissertation, Yale University, 1967 [*Diss. Abstr. Int. B.*, **28**, 4067 (1968)].
[149] S. Wise, Ph.D. Dissertation, University of California, Irvine, 1979 [*Diss. Abstr. Int. B.*, **40**, 1184 (1979)].
[150] G. P. Gisby, P. G. Sammes, and R. A. Watt, *J. Chem. Soc., Perkin Trans. 1*, **1982**, 249.
[151] P. G. Sammes and R. A. Watt, *J. Chem. Soc., Chem. Commun.*, **1976**, 367.
[151a] G. Stork and E. Nakamura, *J. Am. Chem. Soc.*, **105**, 5510 (1983).
[152] G. Fráter, *Helv. Chim. Acta*, **57**, 172 (1974).
[153] T. Laird, W. D. Ollis, and I. O. Sutherland, *J. Chem. Soc., Perkin Trans. 1*, **1980**, 1477.
[154] P. A. Jacobi, D. G. Walker, and I. M. A. Odeh, *J. Org. Chem.*, **46**, 2065 (1981).
[154a] R. H. Schlessinger, J. L. Wood, A. J. Poss, R. A. Nugent, and W. H. Parsons, *J. Org. Chem.*, **48**, 1147 (1983).
[154b] J. M. Dewanckele, F. Zutterman, and M. Vandewalle, *Tetrahedron*, **39**, 3235 (1983).
[155] S. R. Wilson and D. T. Mao, unpublished results cited in ref. 156.
[156] S. R. Wilson and R. N. Misra, *J. Org. Chem.*, **45**, 5079 (1980).
[157] Sandoz Ltd., *Neth. Appl.* 75 03392 [*C.A.*, **85**, 5493c (1976)].
[158] K. T. Tagmazyan, R. S. Mkrtchyan, and A. T. Babayan, *Zh. Org. Khim.*, **10**, 1642 (1974); Engl. transl., p. 1657.
[159] R. K. Boeckman, Jr. and D. M. Demko, *J. Org. Chem.*, **47**, 1789 (1982).
[160] H. Kotsuki, A. Kawamura, M. Ochi, and T. Tokoroyama, *Chem. Lett.*, **1981**, 917.
[161] A. W. Johnson, *J. Chem. Soc.*, **1945**, 715.
[162] K. K. Light, E. J. Shuster, J. F. Vinals, and M. H. Vock, U.S. Pat. 3989760 (1976) [*C.A.*, **86**, 139501 (1977)].
[163] D. Bilović and V. Hahn, *Croat. Chem. Acta*, **39**, 189 (1967) [*C.A.*, **68**, 105055 (1968)]; D. Bilović, Ž. Stojanac, and V. Hahn, *Tetrahedron Lett.*, **1964**, 2071.
[164] P. A. Jacobi and D. G. Walker, *J. Am. Chem. Soc.*, **103**, 4611 (1981).
[165] M. E. Jung and K. M. Halweg, *Tetrahedron Lett.*, **1981**, 2735.
[166] S. D. Burke, University of South Carolina, Columbia, personal communication; see also ref. 166a.
[166a] S. D. Burke, S. M. S. Strickland, and T. H. Powner, *J. Org. Chem.*, **48**, 454 (1983).
[167] E. Ciganek, unpublished results.
[168] T. Shinmyozu, T. Inazu, and T. Yoshino, *Chem. Lett.*, **1978**, 405.
[169] E. L. Ghisalberti, P. R. Jefferies, and T. G. Payne, *Tetrahedron*, **30**, 3099 (1974).
[170] A. C. Bazan, J. M. Edwards, and U. Weiss, *Tetrahedron*, **34**, 3005 (1978); *Tetrahedron Lett.*, **1977**, 147.
[171] E. Schmitz, U. Heuck, H. Preuschhof, and E. Gründemann, *J. Prakt. Chem.*, **324**, 581 (1982).
[172] S. G. Pyne, D. C. Spellmeyer, S. Chen, and P. L. Fuchs, *J. Am. Chem. Soc.*, **104**, 5728 (1982).
[172a] H.-D. Becker, K. Sandros, and K. Andersson, *Angew. Chem., Int. Ed. Engl.*, **22**, 495 (1983).
[172b] H.-D. Becker and K. Andersson, *Tetrahedron Lett.*, **1983**, 3273.
[172c] O. Tsuge, S. Kanemasa, and S. Takenaka, *Bull. Chem. Soc. Jpn.*, **56**, 2073 (1983).
[172d] O. Tsuge and H. Shimoharada, *Chem. Pharm. Bull.*, **30**, 1903 (1982).
[172e] O. Tsuge, H. Shimoharada, M. Noguchi, and S. Kanemasa, *Chem. Lett.*, **1982**, 711.
[172f] O. Tsuge, S. Kanemasa, and S. Takenaka, *Chem. Lett.*, **1983**, 519.
[173] R. M. Harrison, J. D. Hobson, and A. W. Midgley, *J. Chem. Soc., Perkin Trans. 1*, **1973**, 1960.
[174] T. Miyashi, H. Kawamoto, T. Nakajo, and T. Mukai, *Tetrahedron Lett.*, **1979**, 155.
[175] H. R. Gillis, Ph.D. Dissertation, Massachusetts Institute of Technology, 1982.
[176] W. M. Best and D. Wege, *Tetrahedron Lett.*, **1981**, 4877.
[177] W. Oppolzer, *Angew. Chem., Int. Ed. Engl.*, **11**, 1031 (1972).
[178] D. A. Evans and L. McGee, California Institute of Technology, Pasadena, personal communication.
[178a] J. E. Baldwin and R. C. G. Lopez, *Tetrahedron*, **39**, 1487 (1983).
[178b] R. S. Garigipati and S. M. Weinreb, *J. Am. Chem. Soc.*, **105**, 4499 (1983).
[179] J.-L. Gras, *J. Org. Chem.*, **46**, 3738 (1981).

[180] N. Fukamiya, M. Kato, and A. Yoshikoshi, *J. Chem. Soc., Perkin Trans. 1*, **1973**, 1843; *J. Chem. Soc., Chem. Commun.*, **1971**, 1120.
[181] H.-D. Becker, K. Sandros, and A. Arvidsson, *J. Org. Chem.*, **44**, 1336 (1979); *c.f.* G. Felix, R. Lapouyade, A. Castellan, H. Bouas-Laurent, J. Gaultier, and C. Hauw, *Tetrahedron Lett.*, **1975**, 409.
[182] S. M. Weinreb, N. A. Khatri, and J. Shringarpure, *J. Am. Chem. Soc.*, **101**, 5073 (1979).
[183] R. E. Earl and K. P. C. Vollhardt, *Heterocycles*, **19**, 265 (1982).
[184] G. E. Keck, *Tetrahedron Lett.*, **1978**, 4767.
[185] H. O. House and T. H. Cronin, *J. Org. Chem.*, **30**, 1061 (1965).
[186] T. J. Brocksom and M. G. Constantino, *J. Org. Chem.*, **47**, 3450 (1982).
[187] G. Brieger and D. R. Anderson, *J. Org. Chem.*, **36**, 243 (1971).
[188] W. R. Roush, *J. Am. Chem. Soc.*, **102**, 1390 (1980); *ibid.*, **100**, 3599 (1978).
[189] G. Fráter, *Tetrahedron Lett.*, **1976**, 4517.
[190] E. Piers and M. Winter, *Justus Liebigs Ann. Chem.*, **1982**, 973.
[191] K. J. Shea and E. Wada, *Tetrahedron Lett.*, **1982**, 1523.
[192] A. T. Babayan, E. O. Chukhadzhyan, G. T. Babayan, El. O. Chukhadzhyan, and F. S. Kinoyan, *Arm. Khim. Zh.*, **23**, 149 (1970) [*C.A.*, **73**, 25225 (1970)].
[193] W. Oppolzer, E. Francotte, and K. Bättig, *Helv. Chim. Acta*, **64**, 478 (1981).
[194] R. H. Grubbs, T. A. Pancoast, and R. A. Grey, *Tetrahedron Lett.*, **1974**, 2425.
[195] E. J. Corey and R. S. Glass, *J. Am. Chem. Soc.*, **89**, 2600 (1967).
[196] F. Näf, R. Decorzant, and W. Thommen, *Helv. Chim. Acta*, **60**, 1196 (1977).
[197] D. W. Landry, Ph.D. Dissertation, Harvard University, 1979; *Tetrahedron*, **39**, 2761 (1983).
[198] T. Imagawa, T. Nakagawa, M. Kawanisi, and K. Sisido, *Bull. Chem. Soc. Jpn.*, **52**, 1506 (1979); T. Imagawa, M. Kawanisi, and K. Sisido, *J. Chem. Soc., Chem. Commun.*, **1971**, 1292.
[199] W. S. Trahanovsky and P. W. Mullen, *J. Am. Chem. Soc.*, **94**, 5911 (1972).
[200] P. Yates and H. Auksi, *Can. J. Chem.*, **57**, 2853 (1979); *J. Chem. Soc., Chem. Commun.*, **1976**, 1016.
[201] D. J. Bichan and P. Yates, *Can. J. Chem.*, **53**, 2054 (1975); *J. Am. Chem. Soc.*, **94**, 4773 (1972).
[202] Y. Kitahara, M. Oda, S. Miyakoshi, and S. Nakanishi, *Tetrahedron Lett.*, **1976**, 2149.
[203] W. G. Dauben, D. M. Michno, and E. G. Olsen, *J. Org. Chem.*, **46**, 687 (1981).
[204] T. S. Cantrell and H. Shechter, *J. Am. Chem. Soc.*, **89**, 5868 (1967).
[205] L. H. Klemm, R. A. Klemm, P. S. Santhanam, and D. V. White, *J. Org. Chem.*, **36**, 2169 (1971).
[206] L. H. Klemm and K. W. Gopinath, *J. Heterocycl. Chem.*, **2**, 225 (1965).
[207] W. Oppolzer, *J. Am. Chem. Soc.*, **93**, 3834 (1971).
[208] T. Kametani, M. Tsubuki, Y. Shiratori, Y. Kato, H. Nemoto, M. Ihara, K. Fukumoto, F. Satoh, and H. Inoue, *J. Org. Chem.*, **42**, 2672 (1977).
[209] T. Durst, M. Lancaster, and D. J. H. Smith, *J. Chem. Soc., Perkin Trans. 1*, **1981**, 1846.
[210] W. Oppolzer and B. Delpech, unpublished work cited in ref. 10.
[211] W. Oppolzer and K. Keller, *Angew. Chem., Int. Ed. Engl.*, **11**, 728 (1972).
[212] G. Quinkert, W.-D. Weber, U. Schwartz, and G. Dürner, *Angew. Chem., Int. Ed. Engl.*, **19**, 1027 (1980); E. Quinkert, *Chimia*, **31**, 225 (1977); G. Quinkert, W.-D. Weber, U. Schwartz, H. Stark, H. Baier, and G. Dürner, *Justus Liebigs Ann. Chem.*, **1981**, 2335; D. Leibfritz, E. Haupt, M. Feigel, W. E. Hull, and W.-D. Weber, *ibid.*, **1982**, 1971.
[213] J. M. Hornback and R. D. Barrows, *J. Org. Chem.*, **48**, 90 (1983).
[214] T. Tuschka, K. Naito, and B. Rickborn, *J. Org. Chem.*, **48**, 70 (1983).
[215] Y. Ito, M. Nakatsuka, and T. Saegusa, *J. Am. Chem. Soc.*, **102**, 863 (1980).
[216] S. Djuric, T. Sarkar, and P. Magnus, *J. Am. Chem. Soc.*, **102**, 6885 (1980).
[217] R. L. Funk and K. P. C. Vollhardt, *J. Am. Chem. Soc.*, **101**, 215 (1979).
[218] R. L. Funk and K. P. C. Vollhardt, *J. Am. Chem. Soc.*, **102**, 5253 (1980).
[219] W. Oppolzer, *J. Am. Chem. Soc.*, **93**, 3833 (1971).
[220] T. Gallagher and P. Magnus, *Tetrahedron*, **37**, 3889 (1981).
[221] D. T. Langone and J. A. Gladysz, *Tetrahedron Lett.*, **1976**, 4559.

[222] G. Himbert and L. Henn, *Angew. Chem., Int. Ed. Engl.*, **21**, 620 (1982); *Angew. Chem. Suppl.*, **1982**, 1472.
[223] E. Ciganek and M. A. Wuonola, unpublished results.
[224] H. W. Gschwend, M. J. Hillman, B. Kisis, and R. K. Rodebaugh, *J. Org. Chem.*, **41**, 104 (1976).
[225] Y.-S. Cheng, F. W. Fowler, and A. T. Lupo, Jr., *J. Am. Chem. Soc.*, **103**, 2090 (1981).
[226] L.-F. Tietze, G. v. Kiedrowski, and B. Berger, *Tetrahedron Lett.*, **1982**, 51.
[227] L.-F. Tietze, G. v. Kiedrowski, K. Harms, W. Clegg, and G. Sheldrick, *Angew. Chem., Int. Ed. Engl.*, **19**, 134 (1980).
[228] Y. Ito, S. Miyata, M. Nakatsuka, and T. Saegusa, *J. Am. Chem. Soc.*, **103**, 5250 (1981).
[229] R. D. Bowen, D. E. Davies, C. W. C. Fishwick, T. O. Glasbey, S. J. Noyce, and R. C. Storr, *Tetrahedron Lett.*, **1982**, 4501.
[230] R. Hug, H.-J. Hansen, and H. Schmid, *Helv. Chim. Acta*, **55**, 1675 (1972).
[231] A. M. Oude-Alink, A. W. K. Chan, and C. D. Gutsche, *J. Org. Chem.*, **38**, 1993 (1973).
[232] L. Crombie, S. D. Redshaw, and D. A. Whiting, *J. Chem. Soc., Chem. Commun.*, **1979**, 630.
[233] L. Crombie and R. Ponsford, *J. Chem. Soc. C*, **1971**, 788; *J. Chem. Soc., Chem. Commun.*, **1968**, 368; *Tetrahedron Lett.*, **1968**, 4557.
[234] V. V. Kane and R. K. Razdan, *Tetrahedron Lett.*, **1969**, 591.
[235] H. Greuter and H. Schmid, *Helv. Chim. Acta*, **57**, 1204 (1974).
[236] I. Hasan and F. W. Fowler, *J. Am. Chem. Soc.*, **100**, 6696 (1978).
[237] G. M. Brooke, R. S. Matthews, and N. S. Robson, *J. Chem. Soc., Perkin Trans. 1*, **1980**, 102.
[238] T. Jojima, H. Takeshiba, and T. Konotsune, *Chem. Pharm. Bull.*, **20**, 2191 (1972).
[239] T. Jojima, H. Takeshiba, and T. Kinoto, *Chem. Pharm. Bull.*, **24**, 1581 (1976).
[240] T. Jojima, H. Takeshiba, and T. Kinoto, *Chem. Pharm. Bull.*, **24**, 1588 (1976).
[241] T. Jojima, H. Takeshiba, and T. Kinoto, *Chem. Pharm. Bull.*, **28**, 198 (1980).
[242] T. Jojima, H. Takeshiba, and T. Kinoto, *Heterocycles*, **12**, 665 (1979).
[243] L. B. Davies, P. G. Sammes, and R. A. Watt, *J. Chem. Soc., Chem. Commun.*, **1977**, 663.
[244] L. B. Davies, O. A. Leci, P. G. Sammes, and R. A. Watt, *J. Chem. Soc., Perkin Trans. 1*, **1978**, 1293.
[245] P. G. Sammes and R. A. Watt, *J. Chem. Soc., Chem. Commun.*, **1975**, 502.
[246] G. A. Kraus, J. O. Pezzanite, and H. Sugimoto, *Tetrahedron Lett.*, **1979**, 853.
[247] G. A. Kraus and J. O. Pezzanite, *J. Org. Chem.*, **47**, 4337 (1982).
[248] E. D. Sternberg and K. P. C. Vollhardt, *J. Am. Chem. Soc.*, **102**, 4839 (1980); C.-A. Chang, J. A. King, Jr., and K. P. C. Vollhardt, *J. Chem. Soc., Chem. Commun.*, **1981**, 53.
[249] P. A. Jacobi and T. Craig, *J. Am. Chem. Soc.*, **100**, 7748 (1978).
[250] H. E. Zimmerman and G. L. Grunewald, *J. Am. Chem. Soc.*, **86**, 1434 (1964).
[251] T. R. Kelly, *Tetrahedron Lett.*, **1973**, 437.
[252] H. Rapoport and P. Sheldrick, *J. Am. Chem. Soc.*, **85**, 1636 (1963).
[253] E. Ciganek, *J. Org. Chem.*, **45**, 1512 (1980).
[254] O. P. Vig, I. R. Trehan, N. Malik, and R. Kumar, *Indian J. Chem., Sect. B.*, **16**, 449 (1978).
[255] P. A. Christenson, B. J. Willis, F. W. Wehrli, and S. Wehrli, *J. Org. Chem.*, **47**, 4786 (1982).
[256] E. Wenkert and K. Naemura, *Synth. Commun.*, **3**, 45 (1973).
[257] W. Oppolzer and R. L. Snowden, *Helv. Chim. Acta*, **64**, 2592 (1981).
[258] D. F. Taber and S. A. Saleh, *J. Am. Chem. Soc.*, **102**, 5085 (1980).
[259] W. Hofheinz and P. Schönholzer, *Helv. Chim. Acta*, **60**, 1367 (1977).
[260] E. J. Corey, R. L. Danheiser, S. Chandrasekaran, G. E. Keck, B. Gopalan, S. D. Larsen, P. Sirel, and J.-L. Gras, *J. Am. Chem. Soc.*, **100**, 8034 (1978); E. J. Corey, T. M. Brennan, and R. L. Carney, *ibid.*, **93**, 7316 (1971).
[261] E. G. Breitholle and A. G. Fallis, *J. Org. Chem.*, **43**, 1964 (1978); *Can. J. Chem.*, **54**, 1991 (1976).
[262] L. Givaudan & Cie, *Jpn. Kokai Tokkyo Koho*, 80 24193 [*C.A.*, **93**, 46885 (1980)].
[263] W. Oppolzer, *Eur. Pat. Appl.*, 8663 (1979).
[264] W. Oppolzer and R. L. Snowden, *Tetrahedron Lett.*, **1978**, 3505.

[265] F. Näf and G. Ohloff, *Ger. Pat.* 2537417 (1976) [*C.A.*, **85**, 46894 (1976)].
[266] G. Fráter and J. Wenger, *Helv. Chim. Acta*, **67**, in press (1984).
[267] G. A. Schiehser and J. D. White, *J. Org. Chem.*, **45**, 1864 (1980).
[268] H. Yamamoto and H. L. Sham, *J. Am. Chem. Soc.*, **101**, 1609 (1979).
[269] T. Kametani, Y. Hirai, Y. Shiratori, K. Fukumoto, and F. Satoh, *J. Am. Chem. Soc.*, **100**, 554 (1978); T. Kametani, Y. Hirai, F. Satoh, and K. Fukumoto, *J. Chem. Soc., Chem. Commun.*, **1977**, 16.
[270] T. Kametani, K. Suzuki, H. Nemoto, and K. Fukumoto, *J. Org. Chem.*, **44**, 1036 (1979); T. Kametani, H. Nemoto, and K. Fukumoto, *J. Chem. Soc. Chem. Commun.*, **1976**, 400.
[271] T. Kametani, M. Chihiro, T. Honda, and K. Fukumoto, *Chem. Pharm. Bull.*, **28**, 2468 (1980).
[272] T. Kametani, Y. Kato, T. Honda, and K. Fukumoto, *J. Am. Chem. Soc.*, **98**, 8185 (1976).
[273] T. Kametani, T. Honda, Y. Shiratori, and K. Fukumoto, *Tetrahedron Lett.*, **1980**, 1665; T. Kametani, T. Honda, Y. Shiratori, H. Matsumoto, and K. Fukumoto, *J. Chem. Soc., Perkin Trans. 1*, **1981**, 1386.
[274] M. J. Begley, L. Crombie, D. A. Slack, and D. A. Whiting, *J. Chem. Soc., Chem. Commun.*, **1976**, 140.
[275] M. J. Begley, L. Crombie, D. A. Slack, and D. A. Whiting, *J. Chem. Soc., Perkin Trans. 1*, **1977**, 2402.
[276] R. A. Gobao, M. L. Bremmer, and S. M. Weinreb, *J. Am. Chem. Soc.*, **104**, 7065 (1982).
[277] M. L. Bremmer and S. M. Weinreb, *Tetrahedron Lett.*, **1983**, 261; M. L. Bremmer, N. A. Khatri, and S. M. Weinreb, *J. Org. Chem.*, **48**, 3661 (1983).
[278] W. R. Roush and H. R. Gillis, *J. Org. Chem.*, **45**, 4283 (1980).
[279] M. E. Kuehne, C. L. Kirkemo, T. H. Matsko, and J. C. Bohnert, *J. Org. Chem.*, **45**, 3259 (1980).
[280] M. E. Kuehne, F. J. Okuniewicz, C. L. Kirkemo, and J. C. Bohnert, *J. Org. Chem.*, **47**, 1335 (1982).
[281] C. Kan-Fan, G. Massiot, A. Ahond, B. C. Das, H.-P. Husson, P. Potier, A. I. Scott, and C.-C. Wei, *J. Chem. Soc., Chem. Commun.*, **1974**, 164.
[282] A. I. Scott, *Bioorg. Chem.*, **3**, 398 (1974).
[283] J. B. Kutney, Y. Karton, N. Kawamura, and B. R. Worth, *Can. J. Chem.*, **60**, 1269 (1982).
[284] R. M. Wilson, R. A. Farr, and D. J. Burlett, *J. Org. Chem.*, **46**, 3293 (1981).
[285] W. Oppolzer and C. Robbiani, *Helv. Chim. Acta*, **66**, 1119 (1983).
[286] W. Oppolzer and K. Keller, *J. Am. Chem. Soc.*, **93**, 3836 (1971).
[287] T. Gallagher, P. Magnus, and J. C. Huffman, *J. Am. Chem. Soc.*, **104**, 1140 (1982).
[288] T. Kametani, Y. Kato, T. Honda, and K. Fukumoto, *Heterocycles*, **4**, 241 (1976).
[289] L. B. Davies, S. G. Greenberg, and P. G. Sammes, *J. Chem. Soc., Perkin Trans. 1*, **1981**, 1909.
[290] L. H. Klemm, D. R. Olson, and D. V. White, *J. Org. Chem.*, **36**, 3740 (1971).
[291] L. H. Klemm and K. W. Gopinath, *Tetrahedron Lett.*, **1963**, 1243.
[292] L. H. Klemm and P. S. Santhanam, *J. Heterocycl. Chem.*, **9**, 423 (1972).
[293] L. H. Klemm and P. S. Santhanam, *J. Org. Chem.*, **33**, 1268 (1968).
[294] B. S. Joshi, N. Viswanathan, V. Balakrishnan, D. H. Gawad, and K. R. Ravindranath, *Tetrahedron*, **35**, 1665 (1979); B. S. Joshi, K. R. Ravindranath, and N. Viswanathan, *Experientia*, **34**, 422 (1978).
[295] E. Block and R. Stevenson, *J. Org. Chem.*, **36**, 3453 (1971).
[296] O. L. Chapman, M. R. Engel, J. P. Springer, and J. C. Clardy, *J. Am. Chem. Soc.*, **93**, 6696 (1971); M. Matsumoto and K. Kuroda, *Tetrahedron Lett.*, **1981**, 4437; see also G. C. Brophy, J. Mohandas, M. Slaytor, S. Sternhell, T. R. Watson, and L. A. Wilson, *Tetrahedron Lett.*, **1969**, 5159.
[297] Y. Ito, M. Nakatsuka, and T. Saegusa, *J. Am. Chem. Soc.*, **103**, 476 (1981).
[298] T. Kametani, H. Matsumoto, H. Nemoto, and K. Fukumoto, *J. Am. Chem. Soc.*, **100**, 6218 (1978); *Tetrahedron Lett.*, **1978**, 2425.
[299] T. Kametani, H. Nemoto, M. Tsubuki, G.-E. Purvaneckas, M. Aizawa, and M. Nishiushi, *J. Chem. Soc., Perkin Trans. 1*, **1979**, 2830; T. Kametani, H. Nemoto, M. Tsubuki, and M. Nishiushi, *Tetrahedron Lett.*, **1979**, 27.

[300] T. Kametani, M. Aizawa, and H. Nemoto, *J. Chem. Soc., Perkin Trans. 1*, **1980**, 2793.
[301] W. Oppolzer and D. A. Roberts, *Helv. Chim. Acta*, **63**, 1703 (1980).
[302] G. Quinkert, U. Schwartz, H. Stark, W.-D. Weber, H. Baier, F. Adam, and G. Dürner, *Angew. Chem., Int. Ed. Engl.*, **19**, 1029 (1980); G. Quinkert, U. Schwartz, H. Stark, W.-D. Weber, F. Adam, H. Baier, G. Frank, and G. Dürner, *Justus Liebigs Ann. Chem.*, **1982**, 1999; see also G. Quinkert and H. Stark, *Angew. Chem., Int. Ed. Engl.*, **22**, 637 (1983).
[303] P. A. Grieco, T. Takigawa, and W. J. Schillinger, *J. Org. Chem.*, **45**, 2247 (1980).
[304] J. Tsuji, H. Okumoto, Y. Kobayashi, and T. Takahashi, *Tetrahedron Lett.*, **1981**, 1357.
[305] T. Kametani, H. Nemoto, H. Ishikawa, K. Shiroyama, H. Matsumoto, and K. Fukumoto, *J. Am. Chem. Soc.*, **98**, 3378 (1976); T. Kametani, H. Nemoto, H. Ishikawa, K. Shiroyama, H. Matsumoto, and K. Fukumoto, *ibid.*, **99**, 3461 (1977).
[306] T. Kametani, K. Suzuki, and H. Nemoto, *J. Org. Chem.*, **45**, 2204 (1980); *J. Chem. Soc., Chem. Commun.*, **1979**, 1127.
[307] T. Kametani, M. Tsubuki, and H. Nemoto, *J. Org. Chem.*, **45**, 4391 (1980).
[308] T. Kametani and H. Nemoto, *Tetrahedron Lett.*, **1979**, 3309.
[309] T. Kametani, H. Matsumoto, T. Honda, and K. Fukumoto, *Tetrahedron Lett.*, **1980**, 4847.
[310] T. Kametani, K. Suzuki, and H. Nemoto, *J. Am. Chem. Soc.*, **103**, 2890 (1981); *J. Org. Chem.*, **47**, 2331 (1982).
[311] G. Stork and D. H. Sherman, *J. Am. Chem. Soc.*, **104**, 3758 (1982).
[312] M. P. Edwards, S. V. Ley, and S. G. Lister, *Tetrahedron Lett.*, **1981**, 361.
[313] J. Auerbach and S. M. Weinreb, *J. Org. Chem.*, **40**, 3311 (1975).
[314] S. J. Bailey, E. J. Thomas, W. B. Turner, and J. A. J. Jarvis, *J. Chem. Soc., Chem. Commun.*, **1978**, 474; D. J. Williams, S. J. Bailey, E. J. Thomas, and J. A. J. Jarvis, *Tetrahedron*, **36**, 3571 (1980); S. J. Bailey, E. J. Thomas, S. M. Vather, and J. Wallis, *J. Chem. Soc., Perkin Trans. 1*, **1983**, 851.
[315] K. C. Nicolaou, N. A. Petasis, and R. E. Zipkin, *J. Am. Chem. Soc.*, **104**, 5560 (1982).
[316] K. C. Nicolaou, N. A. Petasis, R. E. Zipkin, and J. Uenishi, *J. Am. Chem. Soc.*, **104**, 5555 (1982).
[317] K. C. Nicolaou, N. A. Petasis, J. Uenishi, and R. E. Zipkin, *J. Am. Chem. Soc.*, **104**, 5557 (1982).
[318] W. M. Bandaranayake, J. E. Banfield, and D. St. C. Black, *J. Chem. Soc., Chem. Commun.*, **1980**, 902.
[319] S. Ito and Y. Hirata, *Tetrahedron Lett.*, **1972**, 2557.
[320] E. W. Thomas, Ph.D. Dissertation, Wayne State University, 1977 [*Diss. Abstr. Int. B.*, **39**, 5391 (1978)]; see also ref. 320a.
[320a] R. K. Boeckman, Jr., J. J. Napier, E. W. Thomas, and R. I. Sato, *J. Org. Chem.*, **48**, 4152 (1983).
[321] L.-F. Tietze and G. v. Kiedrowski, *Tetrahedron Lett.*, **1981**, 219.
[322] L.-F. Tietze, G. v. Kiedrowski, and B. Berger, *Angew. Chem., Int. Ed. Engl.*, **21**, 221 (1982).
[323] B. A. Keay and R. Rodrigo, *J. Am. Chem. Soc.*, **104**, 4725 (1982).
[324] C. A. Cupas and L. Hodakowski, *J. Am. Chem. Soc.*, **96**, 4668 (1974).
[325] D. P. G. Hamon and P. R. Spurr, *J. Chem. Soc., Chem. Commun.*, **1982**, 372; P. R. Spurr and D. P. G. Hamon, *J. Am. Chem. Soc.*, **105**, 4737 (1983).
[326] L. A. Paquette, M. J. Wyvratt, H. C. Berk, and R. E. Moerck, *J. Am. Chem. Soc.*, **100**, 5845 (1978); L. A. Poquette and M. J. Wyvratt, *J. Am. Chem. Soc.*, **96**, 4671 (1974).
[327] A. de Meijere, D. Kaufmann, and O. Schallner, *Angew. Chem., Int. Ed. Engl.*, **10**, 417 (1971); see also ref. 744.
[328] D. Paske, R. Ringshandl, I. Sellner, H. Sichert, and J. Sauer, *Angew. Chem., Int. Ed. Engl.*, **19**, 456 (1980).
[329] A. T. Babayan, E. O. Chukhadzhyan, and G. T. Babayan, *Zh. Org. Khim.*, **6**, 1161 (1970); Engl. transl., p. 1165.
[330] A. T. Babayan, E. O. Chukhadzhyan, and El. O. Chukhadzhyan, *Zh. Org. Khim.*, **7**, 470 (1971); Engl. transl., p. 476.
[331] A. T. Babayan, E. O. Chukhadzhyan, and El. O. Chukhadzhyan, *Zh. Org. Khim.*, **9**, 467 (1973); Engl. transl., p. 472.

[332] A. T. Babayan, E. O. Chukhadzhyan, El. O. Chukhadzhyan, and R. P. Babayan, *Zh. Org. Khim.*, **10**, 1638 (1974); Engl. transl., p. 1653.

[333] A. T. Babayan, E. O. Chukhadzhyan, El. O. Chukhadzhyan, and R. P. Babayan, *Zh. Org. Khim.*, **13**, 518 (1977); Engl. transl., p. 472.

[334] A. T. Babayan, E. O. Chukhadzhyan, and L. A. Manasyan, *Zh. Org. Khim.*, **15**, 942 (1979); Engl. transl. p. 842.

[335] A. T. Babayan, S. T. Kocharyan, D. V. Grigoryan, and P. S. Chobanyan, *Zh. Org. Khim.*, **7**, 2253 (1971); Engl. transl., p. 2344.

[336] A. T. Babayan, K. T. Tagmazyan, G. T. Babayan, and A. G. Oganesyan, *Zh. Org. Khim.*, **4**, 1323 (1968); Engl. transl., p. 1277.

[337] A. T. Babayan, K. T. Tagmazyan, and A. A. Cherkezyan, *Zh. Org. Khim.*, **9**, 1149 (1973); Engl. transl., p. 1178.

[338] A. T. Babayan, K. T. Tagmazyan, and G. O. Torosyan, *Zh. Org. Khim.*, **9**, 1156 (1973); Engl. transl., p. 1185.

[339] El. O. Chukhadzhyan, E. O. Chukhadzhyan, and A. T. Babayan, *Zh. Org. Khim.*, **10**, 46 (1974); Engl. transl., p. 43.

[340] E. O. Chukhadzhyan, G. L. Gabrielyan, and A. T. Babayan, *Zh. Org. Khim.*, **11**, 325 (1975); Engl. transl., p. 318.

[341] E. O. Chukhadzhyan, G. L. Gabrielyan, and A. T. Babayan, *Zh. Org. Khim.*, **14**, 2502 (1978); Engl. transl., p. 2305.

[342] K. T. Tagmazyan, L. P. Karapetyan, and A. T. Babayan, *Zh. Org. Khim.*, **10**, 740 (1974); Engl. transl., p. 745.

[343] K. T. Tagmazyan, G. O. Torosyan, and A. T. Babayan, *Zh. Org. Khim.*, **10**, 2082 (1974); Engl. transl., p. 2098.

[344] A. Pelter and B. Singaram, *Tetrahedron Lett.*, **1982**, 245; *J. Chem. Soc., Perkin Trans. 1*, **1983**, 1383.

[345] J. G. Cannon, T. Lee, F.-L. Hsu, J. P. Long, and J. R. Flynn, *J. Med. Chem.*, **23**, 502 (1980).

[346] H. Auksi and P. Yates, *Can. J. Chem.*, **59**, 2510 (1981).

[347] B. B. Snider, D. J. Rodini, M. Karras, T. C. Kirk, E. A. Deutsch, R. Cordova, and R. T. Price, *Tetrahedron*, **37**, 3927 (1981).

[348] B. B. Snider and G. B. Phillips, *J. Am. Chem. Soc.*, **104**, 1113 (1982); B. B. Snider, G. B. Phillips, and R. Cordova, *J. Org. Chem.*, **48**, 3003 (1983).

[349] K. J. Shea and S. Wise, *J. Am. Chem. Soc.*, **100**, 6519 (1978).

[350] S. R. Desai, Ph.D. Dissertation, University of Texas at Austin, 1980 [*Diss. Abstr. Int. B.*, **41**, 1369 (1980)].

[351] C. Bong, Ph.D. Dissertation, University of Cologne, 1952; cited by K. Alder and M. Schumacher, *Progr. Chem. Org. Nat. Products*, **10**, 66 (1953).

[352] G. A. Kraus and M. J. Taschner, *J. Am. Chem. Soc.*, **102**, 1974 (1980).

[353] W. R. Roush and H. R. Gillis, *J. Org. Chem.*, **45**, 4267 (1980).

[354] W. R. Roush, *J. Org. Chem.*, **44**, 4008 (1979).

[355] S. A. Bal and P. Helquist, *Tetrahedron Lett.*, **1981**, 3933.

[356] A. T. Babayan, T. R. Melikyan, G. H. Torosyan, R. S. Mkrtchyan, and K. T. Tagmazyan, *Arm. Khim. Zh.*, **29**, 388 (1976) [*C.A.*, **85**, 143012 (1976)].

[357] A. T. Babayan, K. T. Tagmazyan, G. O. Torosyan, and T. R. Melikyan, *Arm. Khim. Zh.*, **29**, 90 (1976) [*C.A.*, **85**, 32747 (1976)].

[358] G. O. Torosyan, R. S. Mkrtchyan, T. R. Melikyan, K. T. Tagmazyan, and A. T. Babayan, *Arm. Khim. Zh.*, **30**, 578 (1977) [*C.A.*, **88**, 37559 (1978)].

[359] S. G. Pyne, M. J. Hensel, S. R. Byrn, A. T. McKenzie, and P. L. Fuchs, *J. Am. Chem. Soc.*, **102**, 5960 (1980).

[360] A. T. Babayan, K. T. Tagmazyan, and G. O. Torosyan, *Arm. Khim. Zh.*, **24**, 1077 (1971) [*C.A.*, **76**, 153481 (1972)].

[361] A. T. Babayan, K. T. Tagmazyan, and A. A. Cherkezyan, *Arm. Khim. Zh.*, **27**, 43 (1974) [*C.A.*, **81**, 49510 (1974)].

[362] A. T. Babayan, K. T. Tagmazyan, and A. A. Cherkezyan, *Dokl. Akad. Nauk Arm. SSR*, **51**, 32 (1970) [*C.A.*, **74**, 53389 (1971)].
[363] G. O. Torosyan, K. T. Tagmazyan, and A. T. Babayan, *Arm. Khim. Zh.*, **29**, 350 (1976) [*C.A.*, **85**, 191799 (1976)].
[364] A. T. Babayan, K. T. Tagmazyan, and G. O. Torosyan, *Arm. Khim. Zh.*, **27**, 485 (1974) [*C.A.*, **81**, 105171 (1974)].
[365] I. A. Abramyan-Babayan and A. T. Babayan, *Arm. Khim. Zh.*, **25**, 19 (1972) [*C.A.*, **77**, 4546 (1972)].
[366] A. T. Babayan, S. T. Kocharyan, D. V. Grigoryan, T. L. Razina, and V. S. Voskanyan, *Arm. Khim. Zh.*, **27**, 213 (1974) [*C.A.*, **81**, 91292 (1974)].
[367] A. T. Babayan, S. T. Kocharyan, P. S. Chobanyan, and T. A. Azizyan, *Dokl. Akad. Nauk Arm. SSR*, **51**, 227 (1970) [*C.A.*, **75**, 5611 (1971)].
[368] E. O. Chukhadzhyan, G. L. Gabrielyan, and A. T. Babayan, *Arm. Khim. Zh.*, **29**, 71 (1976) [*C.A.*, **85**, 77965 (1975)].
[369] S. T. Kocharyan, D. V. Grigoryan, T. L. Razina, V. S. Voskanyan, and A. T. Babayan, *Arm. Khim. Zh.*, **27**, 861 (1974) [*C.A.*, **83**, 79023 (1975)].
[370] A. T. Babayan, E. O. Chukhadzhyan, El. O. Chukhadzhyan, and G. L. Gabrielyan, *Arm. Khim. Zh.*, **29**, 173 (1976) [*C.A.*, **85**, 62902 (1976)].
[371] A. T. Babayan, S. T. Kocharyan, T. L. Razina, V. S. Voskanyan, and S. M. Ogandzhanyan, *Arm. Khim. Zh.*, **28**, 903 (1975) [*C.A.*, **84**, 105330 (1976).
[372] A. T. Babayan, E. O. Chukhadzhyan, G. T. Babayan, and I. A. Abramyan, *Dokl. Akad. Nauk Arm. SSR*, **48**, 54 (1969) [*C.A.*, **71**, 124111 (1969)].
[373] A. T. Babayan, K. T. Tagmazyan, and G. T. Babayan, *Arm. Khim. Zh.*, **19**, 678 (1966) [*C.A.*, **66**, 104532 (1967)].
[374] K. T. Tagmazyan, L. P. Karapetyan, A. S. Melik-Ohandzhanyan, and A. T. Babayan, *Arm. Khim. Zh.*, **27**, 493 (1974) [*C.A.*, **81**, 120369 (1974)].
[375] A. T. Babayan, K. T. Tagmazyan, and G. T. Babayan, *Arm. Khim. Zh.*, **19**, 685 (1966) [*C.A.*, **66**, 104533 (1967)].
[376] A. T. Babayan, K. T. Tagmazyan, and G. T. Babayan, *Dokl. Akad. Nauk Arm. SSR*, **42**, 23 (1966) [*C.A.*, **65**, 2149 (1966)].
[377] A. T. Babayan, K. T. Tagmazyan, A. A. Cherkezyan, and L. V. Darbinyan, *Dokl. Akad. Nauk Arm. SSR*, **57**, 167 (1973) [*C.A.*, **80**, 70459 (1974)].
[378] E. O. Chukhadzhyan, G. L. Gabrielyan, and A. T. Babayan, *Arm. Khim. Zh.*, **29**, 452 (1976) [*C.A.*, **85**, 123836 (1976)].
[379] A. T. Babayan, K. T. Tagmazyan, L. P. Karapetyan, and A. V. Arutyunyan, *Dokl. Akad. Nauk Arm. SSR*, **61**, 40 (1975) [*C.A.*, **84**, 58806 (1976)].
[380] K. T. Tagmazyan, L. P. Karapetyan, A. A. Cherkezyan, and A. T. Babayan, *Mezhvuz. Sb. Nauchn. Tr.-Erevan. Politekh. Inst., Ser. 19*, **1974**, 178 [*C.A.*, **87**, 68011 (1977)].
[381] A. J. Bartlett, T. Laird, and W. D. Ollis, *J. Chem. Soc., Perkin Trans. 1*, **1975**, 1315; *J. Chem. Soc., Chem. Commun.*, **1974**, 496.
[382] S. Danishefsky and D. Dumas, Yale University, New Haven, CT, personal communication.
[383] B. G. Sheldon, Ph.D. Dissertation, Oregon State University, 1981 [*Diss. Abstr. Int. B.*, **41**, 3792 (1981)].
[384] J. D. White, B. G. Sheldon, B. A. Solheim, and J. Clardy, *Tetrahedron Lett.*, **1978**, 5189.
[385] W. K. Moberg, Ph.D. Dissertation, Harvard University, 1974 [*Diss. Abstr. Int. B.*, **36**, 3966 (1976)].
[386] J. C. Hubbs, Ph.D. Dissertation, Harvard University, 1981 [*Diss. Abstr. Int. B.*, **42**, 4424 (1982)].
[387] T. Miyashi, N. Suto, T. Yamaki, and T. Mukai, *Tetrahedron Lett.*, **1981**, 4421.
[388] Y. Butsugan, S. Yoshida, M. Muto, T. Bito, T. Matsuura, and R. Nakashima, *Tetrahedron Lett.*, **1971**, 1129.
[389] K.-H. Pförtner and W. E. Oberhänsli, *Helv. Chim. Acta*, **58**, 840 (1975).
[390] W. Oppolzer, R. L. Snowden, and P. H. Briner, *Helv. Chim. Acta*, **64**, 2022 (1981).

[390a] M. Katayama, S. Marumo, and H. Hattori, *Tetrahedron Lett.*, **1983**, 1703.

[391] W. Oppolzer, S. C. Burford, and F. Marazza, *Helv. Chim. Acta*, **63**, 555 (1980).

[392] D. T. Witiak, K. Tomita, R. J. Patch, and S. J. Enna, *J. Med. Chem.*, **24**, 788 (1981).

[393] J. H. Sellstedt, Ph.D. Dissertation, University of Minnesota, 1966 [*Diss. Abstr. Int. B.*, **27**, 7300 (1967)].

[394] R. N. Misra, Ph.D. Dissertation, Indiana University, 1980 [*Diss. Abstr. Int. B.*, **41**, 2185 (1980)].

[395] K. J. Shea, S. Wise, L. D. Burke, P. D. Davis, J. W. Gilman, and A. C. Greeley, *J. Am. Chem. Soc.*, **104**, 5708 (1982); K. J. Shea and S. Wise, *Tetrahedron Lett.*, **1979**, 1011.

[396] K. J. Shea and E. Wada, *J. Am. Chem. Soc.*, **104**, 5715 (1982).

[397] L. Mandell, D. E. Lee, and L. F. Courtney, *J. Org. Chem.*, **47**, 610 (1982).

[398] F. J. Jäggi and C. Ganter, *Helv. Chim. Acta*, **63**, 866 (1980).

[399] T. Olsson and O. Wennerstöm, *Tetrahedron Lett.*, **1979**, 1721.

[400] R. S. Glass, J. D. Herzog, and R. L. Sobczak, *J. Org. Chem.*, **43**, 3209 (1978).

[401] G. Brieger, *J. Am. Chem. Soc.*, **85**, 3783 (1963).

[402] J. N. Bennett, M. S. Dissertation, Oakland University, 1978; cited in ref. 8, p. 75.

[403] S. F. Nelsen, J. P. Gillespie, P. J. Hintz, and E. D. Seppanen, *J. Am. Chem. Soc.*, **95**, 8380 (1973).

[404] M. E. Jung and L. A. Light, *J. Org. Chem.*, **47**, 1084 (1982).

[405] S. A. Ismailov, M. S. Salakhov, and A. A. Bairamov, *Zh. Org. Khim.*, **18**, 1552 (1982); Engl. transl., p. 1361.

[406] J. C. L. Tam and P. Yates, *J. Chem. Soc., Chem. Commun.*, **1975**, 739.

[407] E. T. McBee, J. O. Stoffer, and H. P. Braendlin, *J. Am. Chem. Soc.*, **84**, 4540 (1962).

[408] M. A. Battiste and J. F. Timberlake, *J. Org. Chem.*, **42**, 176 (1977).

[409] D. McNeil, B. R. Vogt, J. J. Sudol, S. Theodoropulos, and E. Hedaya, *J. Am. Chem. Soc.*, **96**, 4673 (1974).

[409a] C. Bleasdale and D. W. Jones, *J. Chem. Soc., Chem. Commun.*, **1983**, 214.

[410] M. J. Wyvratt and L. A. Paquette, *Tetrahedron Lett.*, **1974**, 2433.

[411] M. Jones, Jr., S. D. Reich, and L. T. Scott, *J. Am. Chem. Soc.*, **92**, 3118 (1970).

[412] C. Moberg and M. Nilsson, *Tetrahedron Lett.*, **1974**, 4521.

[413] G. O. Torosyan, S. L. Paravyan, K. T. Tagmazyan, and A. T. Babayan, *Arm. Khim. Zh.*, **32**, 177 (1979) [*C.A.*, **92**, 58589 (1980)].

[414] T. Kametani, T. Honda, K. Fukumoto, M. Toyota, and M. Ihara, *Heterocycles*, **16**, 1673 (1981).

[415] W. v. E. Doering and W. R. Roth, *Tetrahedron*, **19**, 715 (1963).

[416] W. R. Roth and B. Peltzer, *Justus Liebigs Ann. Chem.*, **685**, 56 (1965); *Angew. Chem.*, **76**, 378 (1964).

[417] T. Imagawa, A. Haneda, T. Nakagawa, and M. Kawanisi, *Tetrahedron*, **34**, 1893 (1978).

[418] T. Imagawa, N. Sueda, and M. Kawanisi, *Tetrahedron*, **30**, 2227 (1974); *J. Chem. Soc., Chem. Commun.*, **1972**, 388.

[419] C. F. Huebner, E. Donoghue, L. Dorfman, F. A. Stuber, N. Danieli, and E. Wenkert, *Tetrahedron Lett.*, **1966**, 1185.

[420] K. Alder, F. Brochhagen, C. Kaiser, and W. Roth, *Justus Liebigs Ann. Chem.*, **593**, 1 (1955).

[421] P. Heimbach, K.-J. Ploner, and F. Thömel, *Angew. Chem., Int. Ed. Engl.*, **10**, 276 (1971).

[422] A. Haneda, H. Uenakai, T. Imagawa, and M. Kawanisi, *Synth. Commun.*, **6**, 141 (1976).

[423] E. N. Marvell and J. Seubert, *Tetrahedron Lett.*, **1969**, 1333; E. N. Marvell, J. Seubert, G. Vogt, G. Zimmer, G. Moy, and J. R. Siegmann, *Tetrahedron*, **34**, 1323 (1978).

[424] J. M. Riemann and W. S. Trahanovsky, *Tetrahedron Lett.*, **1977**, 1863.

[425] J. Zsindely and H. Schmid, *Helv. Chim. Acta*, **51**, 1510 (1968).

[426] P. Fahrni and H. Schmid, *Helv. Chim. Acta*, **42**, 1102 (1959).

[427] G. M. Brooke and D. H. Hall, *J. Fluorine Chem.*, **20**, 163 (1982).

[428] G. M. Brooke, D. H. Hall, and H. M. M. Shearer, *J. Chem. Soc., Perkin Trans. 1*, **1978**, 780.

[429] G. M. Brooke and D. H. Hall, *J. Chem. Soc., Perkin Trans. 1*, **1976**, 1463.

[430] G. M. Brooke, *J. Chem. Soc., Perkin Trans. 1*, **1974**, 233.

[431] G. M. Brooke and D. H. Hall, *J. Fluorine Chem.*, **10**, 495 (1977).
[432] M. S. Raasch, *J. Org. Chem.*, **45**, 856 (1980).
[433] G. Büchi and H. Wüest, *Helv. Chim. Acta*, **54**, 1767 (1971).
[434] D. G. Farnum, M. Ghandi, S. Raghu, and T. Reitz, *J. Org. Chem.*, **47**, 2598 (1982).
[435] W. L. Mock, C. M. Sprecher, R. F. Stewart, and M. G. Northolt, *J. Am. Chem. Soc.*, **94**, 2015 (1972).
[436] J. Sauer, University of Regensburg, Regensburg, Germany, personal communication.
[437] M. S. Raasch, *J. Org. Chem.*, **37**, 1347 (1972).
[438] S. D. Young and W. T. Borden, *J. Org. Chem.*, **45**, 724 (1980).
[439] A. Pryde, J. Zsindely, and H. Schmid, *Helv. Chim. Acta*, **57**, 1598 (1974).
[440] R. M. Harrison, J. D. Hobson, and M. M. Al Holly, *J. Chem. Soc. C*, **1971**, 3084.
[441] R. M. Harrison, J. D. Hobson, and A. W. Midgley, *J. Chem. Soc., Perkin Trans. 1*, **1976**, 2403.
[442] C. A. Cupas, M. S. Kong, M. Mullins, and W. E. Heyd, *Tetrahedron Lett.*, **1971**, 3157.
[443] C. A. Cupas, W. E. Heyd, and M.-S. Kong, *J. Am. Chem. Soc.*, **93**, 4623 (1971).
[444] J. Zirner and S. Winstein, *J. Chem. Soc., Chem. Commun.*, **1964**, 235.
[445] H.-J. Altenbach and E. Vogel, *Angew. Chem., Int. Ed. Engl.*, **11**, 937 (1972).
[446] E. Vogel, H.-J. Altenbach, and D. Cremer, *Angew. Chem., Int. Ed. Engl.*, **11**, 935 (1972).
[447] G. Kaupp and K. Rösch, *Angew. Chem., Int. Ed. Engl.*, **15**, 163 (1976).
[448] D. E. Minter, G. J. Fonken, and F. T. Cook, *Tetrahedron Lett.*, **1979**, 711.
[449] H. Antropiusová, K. Mach, V. Hanuš, F. Tureček, and P. Sedmera, *React. Kinet. Catal. Lett.*, **10**, 297 (1979); F. Tureček, V. Hanuš, P. Sedmera, H. Antropiusová, and K. Mach, *Tetrahedron*, **35**, 1463 (1979).
[450] L. A. Paquette and M. Oku, *J. Am. Chem. Soc.*, **96**, 1219 (1974).
[451] L. A. Paquette, M. Oku, W. E. Heyd, and R. H. Meisinger, *J. Am. Chem. Soc.*, **96**, 5815 (1974).
[452] L. A. Paquette, R. H. Meisinger, and R. E. Wingard, Jr., *J. Am. Chem. Soc.*, **94**, 9224 (1972).
[453] R. K. Russell, R. E. Wingard, Jr., and L. A. Paquette, *J. Am. Chem. Soc.*, **96**, 7483 (1974).
[454] L. A. Paquette, R. E. Wingard, Jr., and J. M. Photis, *J. Am. Chem. Soc.*, **96**, 5801 (1974); L. A. Paquette and R. E. Wingard, Jr., *ibid.*, **94**, 4398 (1972).
[455] T. A. Antkowiak, D. C. Sanders, G. B. Trimitsis, J. B. Press, and H. Shechter, *J. Am. Chem. Soc.*, **94**, 5366 (1972).
[456] J. A. Berson, R. R. Boetcher, and J. J. Vollmer, *J. Am. Chem. Soc.*, **93**, 1540 (1971); J. A. Berson and R. D. Bauer, *Nouv. J. Chim.*, **1**, 101 (1977).
[457] R. C. DeSelms, *J. Am. Chem. Soc.*, **96**, 1967 (1974).
[458] Y. Nomura, Y. Takeuchi, and S. Tomoda, *J. Chem. Soc., Chem. Commun.*, **1977**, 545 (1977).
[459] H. Prinzbach, W. Auge, and M. Basbudak, *Helv. Chim. Acta*, **54**, 759 (1971).
[460] W. v. E. Doering and J. W. Rosenthal, *Tetrahedron Lett.*, **1967**, 349.
[461] M. Jones, Jr., and B. Fairless, *Tetrahedron Lett.*, **1968**, 4881.
[462] W. Grimme, H. J. Riebel, and E. Vogel, *Angew. Chem., Int. Ed. Engl.*, **7**, 823 (1968); E. Vogel, W. Grimme, W. Meckel, H. J. Riebel, and J. F. M. Oth, *ibid.*, **5**, 590 (1966).
[463] R. T. Seidner, N. Nakatsuka, and S. Masamune, *Can. J. Chem.*, **48**, 187 (1970).
[464] S. Masamune, R. T. Seidner, H. Zenda, M. Wiesel, N. Nakatsuka, and G. Bigam, *J. Am. Chem. Soc.*, **90**, 5286 (1968).
[465] H. D. Carnadi, P. Hildenbrand, J. Richter, and G. Schröder, *Justus Liebigs Ann. Chem.*, **1978**, 2074.
[466] J. B. Press and H. Shechter, *J. Org. Chem.*, **40**, 2446 (1975).
[467] M. J. Goldstein and S. A. Kline, *Tetrahedron Lett.*, **1973**, 1085.
[468] L. A. Paquette, G. R. Krow, and J. R. Malpass, *J. Am. Chem. Soc.*, **91**, 5522 (1969).
[469] M. J. Goldstein and S. A. Kline, *Tetrahedron Lett.*, **1973**, 1089.
[470] J. Altman, E. Babad, D. Ginsburg, and M. B. Rubin, *Israel J. Chem.*, **7**, 435 (1969); J. Altman, E. Babad, M. B. Rubin, and D. Ginsburg, *Tetrahedron Lett.*, **1969**, 1125; E. Babad, D. Ginsburg, and M. B. Rubin, *Tetrahedron Lett.*, **1968**, 2361.
[471] W. v. Philipsborn, J. Altman, E. Babad, J. J. Bloomfield, D. Ginsburg, and M. B. Rubin, *Helv. Chim. Acta*, **53**, 725 (1970).

[472] J. S. McConaghy and J. J. Bloomfield, *Tetrahedron Lett.*, **1969**, 1121.

[473] M. J. Goldstein and S.-H. Dai, *Tetrahedron Lett.*, **1974**, 535.

[474] J. T. Groves and K. W. Ma, *J. Am. Chem. Soc.*, **97**, 4434 (1975); *ibid.*, **99**, 4076 (1977).

[474a] J. T. Groves and K. W. Ma, *Tetrahedron Lett.*, **1973**, 5225; K. W. Ma and J. T. Groves, *ibid.*, **1975**, 1141.

[475] M. J. Goldstein and D. P. Warren, *J. Am. Chem. Soc.*, **100**, 6539 (1978).

[476] E. Vedejs and W. R. Wilber, *Tetrahedron Lett.*, **1975**, 2679.

[477] A. Gilbert and R. Walsh, *J. Am. Chem. Soc.*, **98**, 1606 (1976).

[478] T. Sato and S. Ito, *Tetrahedron Lett.*, **1979**, 1051.

[479] I. A. Akhtar, G. I. Fray, and J. M. Yarrow, *J. Chem. Soc., C*, **1968**, 812; G. I. Fray and A. W. Oppenheimer, *J. Chem. Soc., Chem. Commun.*, **1967**, 599.

[480] M. Oda, H. Miyazaki, and Y. Kitahara, *Chem. Lett.*, **1976**, 1011.

[481] H. H. Wasserman, A. R. Doumaux, and R. E. Davis, *J. Am. Chem. Soc.*, **88**, 4517 (1966).

[482] J. A. Berson and R. F. Davis, *J. Am. Chem. Soc.*, **94**, 3658 (1972).

[483] K. Mach, H. Antropiusová, F. Tureček, V. Hanuš, and P. Sedmera, *Tetrahedron Lett.*, **1980**, 4879; F. Tureček, V. Hanuš, P. Sedmera, H. Antropiusová, and K. Mach, *Collect. Czech. Chem. Commun.*, **46**, 1474 (1981).

[484] I. Iwai and J. Ide, *Chem. Pharm. Bull.*, **12**, 1094 (1964).

[485] L. H. Klemm and T. M. McGuire, *J. Heterocycl. Chem.*, **9**, 1215 (1972).

[486] L. H. Klemm, Y. N. Hwang, and T. M. McGuire, *J. Org. Chem.*, **41**, 3813 (1976).

[487] I. Iwai and T. Hiroaka, *Chem. Pharm. Bull.*, **11**, 1564 (1963).

[488] A. V. Atomyan, N. P. Churkina, A. T. Babayan, I. S. Kislina, and M. I. Vinnik, *Arm. Khim. Zh.*, **34**, 389 (1981) [*C.A.*, **95**, 149484 (1981)].

[489] A. T. Banayan, E. O. Chukhadzhyan, and L. A. Manasyan, *Arm. Khim. Zh.*, **31**, 489 (1978) [*C.A.*, **89**, 215164 (1978)].

[490] I. A. Abramyan-Banayan, A. V. Atomyan, and A. T. Babayan, *Arm. Khim. Zh.*, **25**, 30 (1972) [*C.A.*, **77**, 18955 (1972)].

[491] S. T. Kocharyan, D. V. Grigoryan, and A. T. Babayan, *Arm. Khim. Zh.*, **27**, 757 (1974) [*C.A.*, **82**, 111899 (1975)].

[492] A. T. Babayan, E. O. Chukhadzhyan, and El. O. Chukhadzhyan, *Dokl. Akad. Nauk Arm. SSR*, **52**, 281 (1971) [*C.A.*, **76**, 14305 (1972)].

[493] I. A. Abramyan-Babayan, A. V. Atomyan, and A. T. Babayan, *Dokl. Akad. Nauk Arm. SSR*, **57**, 81 (1973) [*C.A.*, **80**, 69947 (1974)].

[494] A. T. Babyayan, E. O. Chukhadzhyan, El. O. Chukhadzhyan, G. L. Gabrielyan, V. G. Andrianov, A. A. Karapetyan, and Y. T. Struchkov, *Arm. Khim. Zh.*, **32**, 881 (1979) [*C.A.*, **93**, 46311 (1980)].

[495] T. Laird and W. D. Ollis, *J. Chem. Soc., Chem. Commun.*, **1972**, 557.

[496] L. H. Klemm, D. H. Lee, K. W. Gopinath, and C. E. Klopfenstein, *J. Org. Chem.*, **31**, 2376 (1966).

[497] L. Mandell, D. E. Lee, and L. F. Courtney, *J. Org. Chem.*, **47**, 731 (1982).

[498] M. E. Kuehne and J. C. Bohnert, *J. Org. Chem.*, **46**, 3443 (1981).

[499] B. Fraser-Ried and K. Mo Sun, University of Maryland, College Park, personal communication.

[500] J. V. Heck, Ph.D. Dissertation, Harvard University, 1976 [*Diss. Abstr. Int. B.*, **38**, 1207 (1977)].

[501] B. D. Gowland and T. Durst, *Can. J. Chem.*, **57**, 1462 (1979).

[502] T. Kametani, M. Tsubuki, H. Nemoto, and K. Fukumoto, *Chem. Pharm. Bull.*, **27**, 152 (1979).

[503] Y. N. Gupta, M. J. Doa, and K. N. Houk, *J. Am. Chem. Soc.*, **104**, 7336 (1982).

[504] W. Oppolzer and B. Delpech, unpublished work, cited in ref. 9; W. Oppolzer, *Chimia*, **32**, 32 (1978).

[505] W. Oppolzer and D. A. Roberts, unpublished work cited in ref. 10.

[506] T. Kametani, Y. Kato, F. Satoh, and K. Fukumoto, *J. Org. Chem.*, **42**, 1177 (1977).

[507] T. Kametani, H. Matsumoto, T. Honda, and K. Fukumoto, *Tetrahedron Lett.*, **1981**, 2379.

[508] R. L. Funk and K. P. C. Vollhardt, *J. Am. Chem. Soc.*, **99**, 5483 (1977).
[509] J. R. Cannon, I. A. McDonald, A. F. Sierakowski, A. B. White, and A. C. Willis, *Aust. J. Chem.*, **28**, 57 (1975).
[510] W. Oppolzer, *Ger. Pat.* 1963226 (1970) [*C.A.*, **73**, 77051 (1970)].
[511] H. J. J. Loozen, *Eur. Pat. Appl.* EP 50387 (1082) [*C.A.*, **97**, 127622 (1982)]; H. J. J. Loozen, F. T. L. Brands, and M. S. de Winter, *Recl.: J. R. Neth. Chem. Soc.*, **101**, 298 (1982).
[512] W. Oppolzer, *Tetrahedron Lett.*, **1974**, 1001.
[513] T. Kametani, T. Honda, and K. Fukumoto, *Heterocycles*, **14**, 419 (1980); T. Kametani, T. Honda, H. Matsumoto, and K. Fukumoto, *J. Chem. Soc., Perkin Trans. 1*, **1981**, 1383.
[514] C. Exon, T. Gallagher, and P. Magnus, *J. Chem. Soc., Chem. Commun.*, **1982**, 613.
[515] C. Exon, T. Gallagher, and P. Magnus, *J. Am. Chem. Soc.*, **105**, 4739 (1983).
[516] C. Kaneko, T. Naito, and M. Ito, *Tetrahedron Lett.*, **1980**, 1645.
[517] W. J. Houlihan, Y. Uike, and V. A. Parrino, *J. Org. Chem.*, **46**, 4515 (1981).
[518] R. F. C. Brown, F. W. Eastwood, N. Chaichit, B. M. Gatehouse, J. M. Pfeiffer, and D. Woodroffe, *Aust. J. Chem.*, **34**, 1467 (1981).
[519] A. Gilbert and G. N. Taylor, *J. Chem. Soc., Perkin Trans. 1*, **1980**, 1761; *J. Chem. Soc., Chem. Commun.*, **1978**, 129.
[520] H. H. Wasserman and P. M. Keehn, *Tetrahedron Lett.*, **1969**, 3227.
[521] H. H. Wasserman and R. Kitzing, *Tetrahedron Lett.*, **1969**, 3343.
[522] H. H. Wasserman and P. M. Keehn, *J. Am. Chem. Soc.*, **88**, 4522 (1966).
[523] H. H. Wasserman and P. M. Keehn, *J. Am. Chem. Soc.*, **89**, 2770 (1967).
[524] Y. Nakamura, J. Zsindely, and H. Schmid, *Helv. Chim. Acta*, **60**, 247 (1977).
[525] Y. Nakamura, R. Hollenstein, J. Zsindely, H. Schmid, and W. E. Oberhansli, *Helv. Chim. Acta*, **58**, 1949 (1975).
[526] T. Toyoda, A. Iwama, T. Otsubo, and S. Misumi, *Bull. Chem. Soc. Jpn.*, **49**, 3300 (1976); T. Toyoda, A. Iwama, Y. Sakata, and S. Misumi, *Tetrahedron Lett.*, **1975**, 3203.
[527] E. Ciganek, U.S. Pat. 4076830 (1978) [*C.A.*, **89**, 24136 (1978)].
[528] E. Ciganek, U.S. Pat. 4077977 (1978); *Ger. Pat.*, 2454634 (1975) [*C.A.*, **83**, 97015 (1975)].
[529] E. Ciganek, U.S. Pat. 4088772 (1978) [*C.A.*, **89**, 179853 (1978)].
[530] W. E. Hahn, W. Szalecki, and W. Boszczyk, *Pol. J. Chem.*, **52**, 2497 (1978).
[531] J. S. Meek and J. R. Dann, *J. Org. Chem.*, **21**, 968 (1956).
[532] H.-D. Becker, K. Andersson, and K. Sandros, *J. Org. Chem.*, **45**, 4549 (1980).
[533] E. Ciganek, *J. Org. Chem.*, **45**, 1505 (1980).
[534] H. Wynberg and R. Helder, *Tetrahedron Lett.*, **1971**, 4317.
[535] A. Iwama, T. Toyoda, T. Otsubo, and S. Misumi, *Chem. Lett.*, **1973**, 587.
[536] R. H. Martin, J. Jespers, and N. Defay, *Helv. Chim. Acta*, **58**, 776 (1975).
[537] L. A. Van Royen, R. Mijngheer, and P. J. de Clerq, *Tetrahedron Lett.*, **1982**, 3283.
[538] Ž. Klepo and K. Jakopčić, *Croat. Chem. Acta*, **47**, 45 (1975) [*C.A.*, **83**, 192105 (1975).].
[539] R. S. Mkrtchyan, G. O. Torosyan, T. R. Melikyan, K. T. Tagmazyan, and A. T. Babayan, *Arm. Khim. Zh.*, **30**, 573 (1977) [*C.A.*, **88**, 37558 (1978)].
[540] D. Bilović, *Croat. Chem. Acta*, **38**, 293 (1966) [*C.A.*, **66**, 55416 (1967)].
[541] W. Herz, *J. Am. Chem. Soc.*, **67**, 2272 (1945).
[542] S. Takano, Y. Oshima, F. Ito, and K. Ogasawara, *Yakagaku Zasshi*, **100**, 1194 (1980) [*C.A.*, **95**, 24871 (1981)].
[543] A. T. Babayan, K. T. Tagmazyan, and R. S. Mkrtchayan, *Dokl. Akad. Nauk Arm. SSR*, **55**, 224 (1972) [*C.A.*, **79**, 92058 (1973)].
[544] G. O. Torosyan, S. L. Paravyan, R. S. Mkrtchyan, K. T. Tagmazyan, and A. T. Babayan, *Arm. Khim. Zh.*, **32**, 182 (1979) [*C.A.*, **92**, 76234 (1980)].
[545] K. T. Tagmazyan, G. O. Torosyan, R. S. Mkrtchyan, and A. T. Babayan, *Arm. Khim. Zh.*, **29**, 352 (1976) [*C.A.*, **86**, 42746 (1977)].
[546] W. M. Grootaert and P. J. DeClerq, *Tetrahedron Lett.*, **1982**, 3291.
[547] T. J. Katz, V. Balogh, and J. Schulman, *J. Am. Chem. Soc.*, **90**, 734 (1968).
[548] H. H. Wasserman and A. R. Doumaux, Jr., *J. Am. Chem. Soc.*, **84**, 4611 (1962); H. H. Wasserman and R. Kitzing, *Tetrahedron Lett.*, **1969**, 5315.

[549] D. J. Cram, C. S. Montgomery, and G. R. Knox, *J. Am. Chem. Soc.*, **88**, 515 (1966); D. J. Cram and G. R. Knox, *ibid.*, **83**, 2204 (1961).
[550] M. A. Battiste, L. A. Kapicak, M. Mathew, and G. J. Palenik, *J. Chem. Soc., Chem. Commun.*, **1971**, 1536.
[551] R. Imagawa, T. Nakagawa, K. Matsuura, T. Akiyama, and M. Kawanisi, *Chem. Lett.*, **1981**, 903.
[552] D. A. Evans, California Institute of Technology, Pasadena, personal communication.
[553] R. A. Schneider and J. Meinwald, *J. Am. Chem. Soc.*, **89**, 2023 (1967).
[554] Y. Kobayashi, T. Nakano, M. Nakajima, and I. Kumadaki, *Tetrahedron Lett.*, **1981**, 1113.
[555] Y. Kobayashi, Y. Hanzawa, W. Miyashita, T. Kashiwagi, T. Nakano, and I. Kumadaki, *J. Am. Chem. Soc.*, **101**, 6445 (1979).
[556] B. B. Snider, M. Karras, R. T. Price, and D. J. Rodini, *J. Org. Chem.*, **47**, 4538 (1982).
[557] B. B. Snider, D. M. Roush, and T. A. Killinger, *J. Am. Chem. Soc.*, **101**, 6023 (1979).
[558] L. Garanti and G. Zecchi, *Tetrahedron Lett.*, **1980**, 559; L. Bruché, L. Garanti, and G. Zecchi, *J. Chem. Soc., Perkin Trans. 1*, **1982**, 755.
[559] Y.-R. Naves and P. Ardizio, *Bull. Soc. Chim. Fr.*, **1953**, 494.
[560] C. E. Berkoff and L. Crombie, *J. Chem. Soc.*, **1960**, 3734, and references cited therein.
[561] D. G. Clarke, L. Crombie, and D. A. Whiting, *J. Chem. Soc., Chem. Commun.*, **1973**, 582.
[562] L.-F. Tietze, H. Stegelmeier, K. Harms, and T. Brumby, *Angew. Chem., Int. Ed. Engl.*, **21**, 863 (1982).
[563] Y.-L. Mao and V. Boekelheide, *Proc. Natl. Acad. Sci.* (*USA*), **77**, 1732 (1980).
[564] W. M. Bandaranayake, M. J. Begley, B. O. Brown, D. G. Clarke, L. Crombie, and D. A. Whiting, *J. Chem. Soc., Perkin Trans. 1*, **1974**, 998.
[565] N. S. Narasimhan and S. L. Kelkar, *Indian J. Chem. Sect. B*, **14**, 430 (1976).
[566] V. V. Kane and R. K. Razdan, *J. Am. Chem. Soc.*, **90**, 6551 (1968).
[567] M. J. Begley, L. Crombie, R. W. King, D. A. Slack, and D. A. Whiting, *J. Chem. Soc., Perkin Trans 1*, **1977**, 2393; *J. Chem. Soc., Chem. Commun.*, **1976**, 138.
[568] L. Crombie, D. A. Slack, and D. A. Whiting, *J. Chem. Soc., Chem. Commun.*, **1976**, 139.
[569] V. V. Kane and T. L. Grayeck, *Tetrahedron Lett.*, **1971**, 3991.
[570] J. L. Montero and F. Winternitz, *Tetrahedron*, **29**, 1243 (1973).
[571] L. Crombie and R. Ponsford, *J. Chem. Soc., Chem. Commun.*, **1968**, 894.
[572] L. Crombie and R. Ponsford, *J. Chem. Soc., C*, **1971**, 796.
[573] R. Mechoulam, B. Yagnitinski, and Y. Gaoni, *J. Am. Chem. Soc.*, **90**, 2418 (1968).
[574] W. M. Bandaranayake, L. Crombie, and D. A. Whiting, *J. Chem. Soc., C*, **1971**, 804.
[575] M. Nitta, A. Sekiguchi, and H. Koba, *Chem. Lett.*, **1981**, 933.
[576] J. Bruhn, J. Zsindely, H. Schmid, and G. Fráter, *Helv. Chim. Acta*, **61**, 2542 (1978).
[577] F. J. Dinan and H. T. Tieckelmann, *J. Org. Chem.*, **29**, 892 (1964).
[578] G. P. Gisby, S. E. Royall, and P. G. Sammes, *J. Chem. Soc., Perkin Trans. 1*, **1982**, 169.
[579] Y. Nakamura, J. Zsindely, and H. Schmid, *Helv. Chim. Acta*, **59**, 2841 (1976).
[579a] I. Sellner, H. Schuster, H. Sichert, J. Sauer, and H. Nöth, *Chem. Ber.*, **116**, 3751 (1983).
[580] E. Ciganek, *Eur. Pat. Appl.* 9780 (1980) [*C.A.*, **93**, 220720 (1980)]; *U.S. Pat.* 4,243,668 (1981) [*C.A.*, **95**, 80924d (1981)].
[581] A. L. Johnson and H. E. Simmons, *J. Org. Chem.*, **34**, 1139 (1969).
[582] J. Hájiček and J. Trojánek, *Tetrahedron Lett.*, **1981**, 1823.
[583] A. Halverson and P. M. Keehn, *J. Am. Chem. Soc.*, **104**, 6125 (1982).
[584] A. N. Hughes and S. U. Uaboonkul, *Tetrahedron*, **24**, 3437 (1968).
[585] K. Dimroth, O. Schaffer, and G. Weiershäuser, *Chem. Ber.*, **114**, 1752 (1981).
[586] K. T. Tagmazyan, R. S. Mkrtchyan, and A. T. Babayan, *Arm. Khim. Zh.*, **27**, 587 (1974) [*C.A.*, **81**, 169451 (1974)].
[587] J. I. Levin and S. M. Weinreb, *J. Am. Chem. Soc.*, **105**, 1397 (1983).
[588] W. Mahler and T. Fukunaga, *J. Chem. Soc., Chem. Commun.*, **1977**, 307.
[589] G. Quinkert, F. Cech, E. Kleiner, and D. Rehm, *Angew. Chem., Int. Ed. Engl.*, **18**, 557 (1979).
[590] P. M. Collins and H. Hart, *J. Chem. Soc., C*, **1967**, 895.
[591] H. Hart, P. M. Collins, and A. J. Waring, *J. Am. Chem. Soc.*, **88**, 1005 (1966).

[592] H. Hart, S.-M. Chen, and M. Nitta, *Tetrahedron*, **37**, 3323 (1981).
[593] M. R. Morris and A. J. Waring, *J. Chem. Soc., C*, **1971**, 3269.
[594] M. R. Morris and A. J. Waring, *J. Chem. Soc., C*, **1971**, 3266.
[595] A. J. Waring, M. R. Morris, and M. M. Islam, *J. Chem. Soc., C*, **1971**, 3274.
[596] H. Perst and K. Dimroth, *Tetrahedron*, **24**, 5385 (1968).
[597] T. L. Gilchrist, C. J. Harris, M. E. Peek, and C. W. Rees, *J. Chem. Soc., Chem. Commun.*, **1975**, 962.
[598] L. Semper and L. Lichtenstadt, *Justus Liebigs Ann. Chem.*, **400**, 302 (1913); M. Z. Jovitschitsch, *Ber.*, **39**, 3821 (1906).
[599] M. Kuzuya, F. Miyake, and T. Okuda, *Chem. Lett.*, **1981**, 1593.
[600] R. F. Childs, R. Grigg, and A. W. Johnson, *J. Chem. Soc., C*, **1967**, 201.
[601] H. E. Zimmerman and H. Iwamura, *J. Am. Chem. Soc.*, **92**, 2015 (1970).
[602] R. Criegee and R. Askani, *Angew. Chem., Int. Ed. Engl.*, **7**, 537 (1968).
[603] Y. Kobayashi, A. Ando, K. Kawada, and I. Kumadaki, *J. Am. Chem. Soc.*, **103**, 3958 (1981).
[604] E. Vedejs, *Tetrahedron Lett.*, **1968**, 2633.
[605] G. F. Emerson, L. Watts, and R. Pettit, *J. Am. Chem. Soc.*, **87**, 131 (1965); W. Merk and R. Pettit, *ibid.*, **89**, 4787 (1967); for an alternate mechanism, see P. Warner, *Tetrahedron Lett.*, **1971**, 723; L. A. Paquette, *J. Chem. Soc., Chem. Commun.*, **1971**, 1076.
[606] H. Straub and J. Hambrecht, *Chem. Ber.*, **110**, 3221 (1977).
[607] I. A. Akhtar, R. J. Atkins, G. I. Fray, G. R. Geen, and T. J. King, *Tetrahedron*, **36**, 3033 (1980); G. I. Fray, D. P. Gale, and G. R. Geen, *ibid.*, **37**, 3101 (1981).
[608] G. I. Fray, W. P. Lay, K. Mackenzie, and A. S. Miller, *Tetrahedron Lett.*, **1979**, 2711.
[609] Y. Kitahara, M. Oda, and M. Oda, *J. Chem. Soc., Chem. Commun.*, **1976**, 446.
[610] R. C. Cookson and R. M. Tuddenham, *J. Chem. Soc., Perkin Trans. 1*, **1978**, 678.
[611] D. W. Jones, *J. Chem. Soc., Perkin Trans. 1*, **1972**, 225.
[612] A. D. Wolf and M. Jones, Jr., *J. Am. Chem. Soc.*, **95**, 8209 (1973).
[613] K. B. Becker and R. W. Pfluger, *Tetrahedron Lett.*, **1979**, 3713.
[614] R. Bloch, F. Boivin, and M. Bortolussi, *J. Chem. Soc., Chem. Commun.*, **1976**, 371.
[615] G. W. Klumpp and J. Stapersma, *J. Chem. Soc., Chem. Commun.*, **1980**, 670.
[616] J. Stapersma, I. D. C. Rood, and G. W.Klumpp, *Tetrahedron*, **38**, 2201 (1982).
[617] M. Christl and M. Lechner, *Angew. Chem., Int. Ed. Engl.*, **14**, 765 (1975).
[618] L. A. Paquette and R. T. Taylor, *Tetrahedron Lett.*, **1976**, 2745.
[619] P. Gilgen, J. Zsindely, and H. Schmid, *Helv. Chim. Acta*, **56**, 681 (1973).
[620] H. Tanida and Y. Hata, *J. Org. Chem.*, **30**, 977 (1965).
[621] R. K. Lustgarten, J. Lhomme, and S. Winstein, *J. Org. Chem.*, **37**, 1075 (1972).
[622] P. G. Gassman and J. J. Talley, *J. Am. Chem. Soc.*, **102**, 4138 (1980).
[623] R. S. Bly, R. K. Bly, and T. Shibata, *J. Org. Chem.*, **48**, 101 (1983).
[624] R. M. Magid and G. W. Whitehead, *Tetrahedron Lett.*, **1977**, 1951.
[625] H. Tsuruta, T. Kumagai, and T. Mukai, *Chem. Lett.*, **1972**, 981.
[625a] T. Mukai and K. Kurabayashi, *J. Am. Chem. Soc.*, **92**, 4493 (1970).
[626] J. Stapersma, I. D. C. Rood, and G. W. Klumpp, *Tetrahedron*, **38**, 3051 (1982).
[627] H. W. Whitlock, Jr., and P. F. Schatz, *J. Am. Chem. Soc.*, **93**, 3837 (1971).
[628] H. D. Fühlhuber, C. Gousetis, J. Sauer, and H. J. Lindner, *Tetrahedron Lett.*, **1979**, 1299.
[629] H. Prinzbach, D. Stusche, J. Markert, and H.-H. Limbach, *Chem. Ber.*, **109**, 3505 (1976).
[630] M. Breuninger, R. Schwesinger, B. Gallenkamp, K.-H. Müller, H. Fritz, D. Hunkler, and H. Prinzbach, *Chem. Ber.*, **113**, 3161 (1980).
[631] H.-J. Altenbach, H. Stegelmeier, M. Wilhelm, B. Voss, J. Lex, and E. Vogel, *Angew. Chem., Int. Ed. Engl.*, **18**, 962 (1979).
[632] L. A. Paquette, R. H. Meisinger, and R. E. Wingard, Jr., *J. Am. Chem. Soc.*, **95**, 2230 (1973).
[633] E. Vedejs, R. A. Gabel, and P. D. Weeks, *J. Am. Chem. Soc.*, **94**, 5842 (1972).
[634] H. Tsuruta, K. Kurabayashi, and T. Mukai, *J. Am. Chem. Soc.*, **90**, 7167 (1968).
[635] E. LeGoff and S. Oka, *J. Am. Chem. Soc.*, **91**, 5665 (1969).
[636] E. L. Allred and B. R. Beck, *J. Am. Chem. Soc.*, **95**, 2393 (1973).
[637] E. L. Allred and B. R. Beck, *Tetrahedron Lett.*, **1974**, 437.

[638] E. L. Allred, B. R. Beck, and N. A. Mumford, *J. Am. Chem. Soc.*, **99**. 2694 (1977).

[639] H. H. Westberg, E. N. Cain, and S. Masamune, *J. Am. Chem. Soc.*, **91**, 7512 (1969).

[640] D. W. McNeil, M. E. Kent, E. Hedaya, P. F. D'Angelo, and P. O. Schissel, *J. Am. Chem. Soc.*, **93**, 3817 (1971).

[641] J. P. Snyder, L. Lee, and D. G. Farnum, *J. Am. Chem. Soc.*, **93**, 3816 (1971).

[642] W. Mauer and W. Grimme, *Tetrahedron Lett.*, **1976**, 1835.

[643] E. Vedejs and R. A. Shepherd, *J. Org. Chem.*, **41**, 742 (1976).

[644] T. Miyashi, H. Kawamoto, and T. Mukai, *Tetrahedron Lett.*, **1977**, 4623; T. Miyashi, T. Nakajo, H. Kawamoto, K. Akiyama, and T. Mukai, *ibid.*, **1979**, 151.

[645] W. Adam and O. De Lucchi, *J. Org. Chem.*, **45**, 4167 (1980).

[646] L. D. Antonaccio, N. A. Pereira, B. Gilbert, H. Vorbrueggen, H. Budzikiewicz, J. M. Wilson, L. J. Durham, and C. Djerassi, *J. Am. Chem. Soc.*, **84**, 2161 (1962).

[647] H.-D. Becker and K. Sandros, *Chem. Phys. Lett.*, **55**, 498 (1978).

[648] H.-D. Becker, K. Sandros, and K. Andersson, *Chem. Phys. Lett.*, **77**, 246 (1981).

[649] M. Mintas, Ž. Klepo, K. Jakopčić, and L. Klasinc, *Org. Mass. Spectrom.*, **14**, 254 (1979).

[650] J. Brokatzky and W. Eberbach, *Chem. Ber.*, **114**, 384 (1981).

[651] R. Faragher and T. L. Gilchrist, *J. Chem. Soc., Perkin. Trans. 1*, **1979**, 258; *J. Chem. Soc., Chem. Commun.*, **1977**, 252.

[652] P. Gygax, T. K. Das Gupta, and A. Eschenmoser, *Helv. Chim. Acta*, **55**, 2205 (1972).

[653] M. Riediker and W. Graf, *Chimia*, **34**, 461 (1980).

[654] W. Adam, N. Carnalleira, and O. DeLucchi, *J. Am. Chem. Soc.*, **103**, 6406 (1981).

[655] W. Adam and O. De Lucchi, *J. Am. Chem. Soc.*, **102**, 2109 (1980); *J. Org. Chem.*, **46**, 4133 (1981).

[656] A. G. Schultz and C.-K. Sha, *J. Org. Chem.*, **45**, 2040 (1980).

[657] W. Dannenberg and H. Perst, *Justus Liebigs Ann. Chem.*, **1975**, 1873.

[658] P. Schiess and H. L. Chia, *Helv. Chim. Acta*, **53**, 485 (1970); P. Schiess and P. Radimerski, *ibid.*, **57**, 2583 (1974).

[659] A. J. H. Klunder, W. Bos, J. M. M. Verlaak, and B. Zwanenburg, *Tetrahedron Lett.*, **1981**, 4553.

[660] A. J. H. Klunder, W. Bos, and B. Zwanenburg, *Tetrahedron Lett.*, **1981**, 4557.

[661] J. F. Biellmann and M. P. Goeldner, *Tetrahedron*, **27**, 2957 (1971).

[662] S. F. Dyer and P. B. Shevlin, *J. Am. Chem. Soc.*, **101**, 1303 (1979).

[663] J. Novák and F. Šorm, *Collect. Czech. Chem. Commun.*, **23**, 1126 (1958).

[664] G. O. Schenck and R. Steinmetz, *Justus Liebigs Ann. Chem.*, **668**, 19 (1963).

[665] M. N. Nwaji and O. S. Onyiriuka, *Tetrahedron Lett.*, **1974**, 2255.

[666] E. Wenkert, M. L. F. Bakuzis, B. L. Buckwalter, and P. D. Woodgate, *Synth. Commun.*, **11**, 533 (1981).

[667] M. Franck-Neumann and C. Buchecker, *Angew. Chem., Int. Ed. Engl.*, **9**, 526 (1970).

[668] M. Franck-Neumann and C. Dietrich-Buchecker, *Tetrahedron*, **34**, 2797 (1978).

[669] D. Gravel, C. Leboeuf, and S. Caron, *Can. J. Chem.*, **55**, 2373 (1977).

[670] G. Cauquis, B. Divisia, M. Rastoldo, and G. Reverdy, *Bull. Soc. Chim. Fr.*, **1971**, 3022.

[671] J.-P. LeRoux, G. Letertre, P.-L. Desbene, and J.-J. Basselier, *Bull. Soc. Chim. Fr.*, **1971**, 4059.

[672] G. Cauquis, B. Divisia, and G. Reverdy, *Bull. Soc. Chim. Fr.*, **1971**, 3027.

[673] R. Srinivasan, J. Studebaker, and F. H. Brown, *Tetrahedron Lett.*, **1979**, 1955.

[674] H. Hart, S.-M. Chen, and S. Lee, *J. Org. Chem.*, **45**, 2096 (1980).

[675] K. T. Potts, A. J. Elliott, and M. Sorm, *J. Org. Chem.*, **37**, 3838 (1972).

[676] L. A. Paquette, R. K. Russell, and R. E. Wingard, Jr., *Tetrahedron Lett.*, **1973**, 1713.

[677] H. Prinzbach, H.-P. Böhm, S. Kagabu, V. Wessely, and H. V. Rivera, *Tetrahedron Lett.*, **1978**, 1243.

[678] E. Vogel, H.-J. Altenbach, and E. Schmidbauer, *Angew. Chem., Int. Ed. Engl.*, **12**, 838 (1973).

[679] R. Schwesinger, M. Breuninger, B. Gallenkamp, K.-H. Müller, D. Hunkler, and H. Prinzbach, *Chem. Ber.*, **113**, 3127 (1980).

[680] R. Schwesinger, H. Fritz, and H. Prinzbach, *Chem. Ber.*, **112**, 3318 (1979).

[681] E. Vogel, H.-J. Altenbach, and C.-D. Sommerfeld, *Angew. Chem., Int. Ed. Engl.*, **11**, 939 (1972).
[682] H. Prinzbach, H.-P. Schal, D. Hunkler, and H. Fritz, *Angew. Chem., Int. Ed. Engl.*, **19**, 567 (1980).
[683] W. R. Roush and S. M. Peseckis, *Tetrahedron Lett.*, **1982**, 4879.
[684] S. R. Wilson and M. S. Haque, *J. Org. Chem.*, **47**, 5411 (1982).
[684a] R. Brettle and I. A. Jafri, *J. Chem. Soc., Perkin Trans. 1*, **1983**, 387.
[685] G. Hwang and P. Magnus, *J. Chem. Soc., Chem. Commun.*, **1983**, 693.
[686] T. Kitahara, K. Tanida, and K. Mori, *Agric. Biol. Chem.*, **47**, 581 (1983) [*C.A.*, **99**, 71024 (1983)].
[687] E. O. Chukhadzhyan, El. O. Chukhadzhyan, L. A. Manasyan, and A. T. Babayan, *Arm. Khim. Zh.*, **34**, 46 (1981) [*C.A.*, **94**, 192048 (1981)].
[688] P. J. Duggan and J. L. Leng, *J. Chem. Soc., Perkin Trans. 1*, **1983**, 933.
[689] M. Hirama and M. Uei, *J. Am. Chem. Soc.*, **104**, 4251 (1982).
[690] A. S. Magee, Ph.D. Dissertation, University of Minnesota, **1982** [*Diss. Abstr. Int. B.*, **43**, 3599 (1983)]; T. R. Hoye and A. S. Magee, *Abstr. ACS Meeting, Kansas City, Mo.*, Sept. 1982.
[691] S. R. Wilson, A. Shedrinsky, and M. S. Haque, *Tetrahedron*, **39**, 895 (1983).
[692] S. D. Burke, T. H. Powner, and M. Kageyama, *Tetrahedron Lett.*, **1983**, 4529.
[693] E. A. Deutsch and B. B. Snider, *Tetrahedron Lett.*, **1983**, 3701.
[694] K. Takeda, M. Shinagawa, T. Koizumi, and E. Yoshii, *Chem. Pharm. Bull.*, **30**, 4000 (1982).
[695] M. Voyle, K. S. Kyler, S. Arseniyadis, N. K. Dunlap, and D. S. Watt., *J. Org. Chem.*, **48**, 470 (1983).
[696] D. C. Burdick, Ph.D. Dissertation, University of Michigan, 1982 [*Diss. Abstr. Int. B.*, **43**, 3239 (1983)].
[697] K. J. Shea and P. D. Davis, *Angew. Chem., Int. Ed. Engl.*, **22**, 419 (1983).
[698] A. J. Barker, M. J. Begley, A. M. Birch, and G. P. Pattenden, *J. Chem. Soc., Perkin Trans. 1*, **1983**, 1919.
[699] D. D. Sternbach, J. W. Hughes, D. F. Burdi, and R. M. Forstot, *Tetrahedron Lett.*, **1983**, 3295.
[700] O. Wallquist, M. Rey, and A. S. Dreiding, *Helv. Chim. Acta*, **66**, 1891 (1983).
[701] A. Gilbert, S. Krestonosich, and S. Wilson, *Tetrahedron Lett.*, **1982**, 4061.
[702] T. S. Macas and P. Yates, *Tetrahedron Lett.*, **1983**, 147.
[703] T. Gallagher and P. Magnus, *J. Am. Chem. Soc.*, **105**, 2086 (1983).
[704] K. Harano, M. Yasuda, and K. Kanematsu, *J. Org. Chem.*, **47**, 3736 (1982).
[705] H. Schuster and J. Sauer, *Tetrahedron Lett.*, **1983**, 4087.
[706] J. H. Rigby, *Tetrahedron Lett.*, **1982**, 1863.
[707] T. Toda, N. Shimazaki, H. Hotta, T. Hatakeyama, and T. Mukai, *Chem. Lett.*, **1983**, 523.
[708] H.-J. Altenbach, B. Voss, and E. Vogel, *Angew. Chem., Int. Ed. Engl.*, **22**, 410 (1983).
[709] H. Olsen, *Helv. Chim. Acta*, **65**, 1921 (1982).
[710] T. Tsuji and S. Nishida, *Tetrahedron Lett.*, **1983**, 1269.
[711] Y. Sekine and V. Boekelheide, *J. Am. Chem. Soc.*, **103**, 1777 (1981).
[712] K. Mach, H. Antropiusová, P. Sedmera, V. Hanuš, and F. Tureček, *J. Chem. Soc., Chem. Commun.*, **1983**, 805.
[713] A. V. Atomyan, N. P. Churkina, N. T. Babayan, I. S. Kislina, and M. I. Vinnik, *Izv. Akad. Nauk SSSR, Ser. Khim.*, **1981**, 525 [*C.A.*, **95**, 41786 (1981)].
[714] A. V. Atomyan, A. T. Babayan, I. S. Kislina, and M. I. Vinnik, *Arm. Khim. Zh.*, **34**, 398 (1981) [*C.A.*, **95**. 149485 (1981)].
[715] Tsumura Juntendo Co., Ltd., *Jpn. Kokai Tokkyo Koho* 5841874 (1983) [*C.A.*, **99**, 139653 (1983)].
[716] M. E. Kuehne, T. H. Matsko, J. C. Bohnert, L. Motyka, and D. Oliver-Smith, *J. Org. Chem.*, **46**, 2002 (1981).
[717] K. Fukumoto, M. Chihiro, Y. Shiratori, M. Ihara, T. Kametani, and T. Honda, *Tetrahedron Lett.*, **1982**, 2973; K. Fukumoto, M. Chihiro, M. Ihara, T. Kametani, and T. Honda, *J. Chem. Soc., Perkin Trans. 1*, **1983**, 2569.

[718] H. Nemoto, M. Hashimoto, K. Fukumoto, and T. Kametani, *J. Chem. Soc., Chem. Commun.*, **1982**, 699.

[719] Y. Ito, M. Nakatsuka, and T. Saegusa, *J. Am. Chem. Soc.*, **104**, 7609 (1982).

[720] K. Shishido, T. Matsuura, K. Fukumoto, and T. Kametani, *Chem. Pharm. Bull.*, **31**, 57 (1983).

[721] K. Shishido, T. Saitoh, K. Fukumoto, and T. Kametani, *J. Chem. Soc., Chem. Commun.*, **1983**, 852.

[722] Y. Ito, Y. Amino, M. Nakatsuka, and T. Saegusa, *J. Am. Chem. Soc.*, **105**, 1586 (1983).

[723] L. H. Klemm, Y. N. Hwang, and J. N. Louris, *J. Org. Chem.*, **48**, 1451 (1983).

[724] S. V. Kessar, I. R. Trehan, T. V. Singh, M. Narula, and N. P. Singh, *Tetrahedron Lett.*, **1982**, 4177.

[725] T. Gallagher, P. Magnus, and J. C. Huffman, *J. Am. Chem. Soc.*, **105**, 4750 (1983).

[726] L. A. Van Royen, R. Mijngheer, and P. J. DeClerq, *Tetrahedron Lett.*, **1983**, 3145.

[727] T. L. Gilchrist and P. Richards, *Synthesis*, **1983**, 153.

[728] Y. Ito, E. Nakajo, M. Nakatsuka, and T. Saegusa, *Tetrahedron Lett.*, **1983**, 2881.

[729] P. A. Jacobi, K. T. Weiss, and M. Egbertson, *Heterocycles*, **22**, 281 (1984).

[730] D. L. Comins, A. H. Abdullah, and R. K. Smith, *Tetrahedron Lett.*, **1983**, 2711.

[731] R. Askani and W. Schneider, *Chem. Ber.*, **116**, 2366 (1983).

[732] H. Schuster, H. Sichert, and J. Sauer, *Tetrahedron Lett.*, **1983**, 1485.

[733] J. F. W. Keana, P. J. Boyle, M. Erion, R. Hartling, J. R. Husman, J. E. Richman, R. B. Roman, and R. M. Wah, *J. Org. Chem.*, **48**, 3621 (1983).

[734] E. D. Sternberg and K. P. C. Vollhardt, *J. Org. Chem.*, **47**, 3447 (1982).

[735] T. R. Gadek and K. P. C. Vollhardt, *Angew. Chem., Int. Ed. Engl.*, **20**, 802 (1981).

[736] Daiichi Seiyaku Co., Ltd., *Jpn. Kokai Tokkyo Koho* 5888385 (1983) [*C.A.*, **99**, 139926 (1983)].

[737] S. Ramakanth, U. Rao, K. Narayanan, and K. K. Balasubramanian, *J. Chem. Soc., Chem. Commun.*, **1983**, 842.

[738] R. Appel, S. Korte, M. Halstenberg, and F. Knoch, *Chem. Ber.*, **115**, 3610 (1982).

[739] C. Rücker and H. Prinzbach, *Tetrahedron Lett.*, **1983**, 4099.

[740] H.-D. Martin, P. Pföhler, T. Urbanek, and R. Walsh, *Chem. Ber.*, **116**, 1415 (1983).

[741] A. G. Schultz, J. P. Dittami, S. O. Myong, and C.-K. Sha, *J. Am. Chem. Soc.*, **105**, 3273 (1983).

[742] A. G. Schultz and S. O. Myong, *J. Org. Chem.*, **48**, 2432 (1983).

[743] O. M. Nefedov, V. M. Shostakovski, and A. E. Vasilvizky, *Angew. Chem., Int. Ed. Engl.*, **16**, 646 (1977).

[744] D. Kaufmann, H.-H. Fick, O. Schallner, W. Spielmann, L.-U. Meyer, P. Gölitz, and A. de Meijere, *Chem. Ber.*, **116**, 587 (1983).

[745] M. Maas, M. Lutterbeck, D. Hunkler, and H. Prinzbach, *Tetrahedron Lett.*, **1983**, 2143.

[746] K.-H. Müller, C. Kaiser, M. Pillat, B. Zipperer, M. Froom, H. Fritz, D. Hunkler, and H. Prinzbach, *Chem. Ber.*, **116**, 2492 (1983).

CHAPTER 2

SYNTHESES USING ALKYNE-DERIVED ALKENYL- AND ALKYNYLALUMINUM COMPOUNDS

GEORGE ZWEIFEL AND JOSEPH A. MILLER

University of California, Davis, California

CONTENTS

ACKNOWLEDGMENT

We thank Mrs. Marilee Urban for typing the manuscript.

INTRODUCTION

The stereoselective elaboration of alkynes into substituted alkenes via vinylic organometallics derived from hydrometalation and carbometalation of the triple bond comprises one of the most powerful tools for olefinic synthesis. The discovery that mono- and disubstituted alkenylaluminum compounds can be directly synthesized via *cis* hydroalumination of alkynes[1,2] thus has played an important role in the development of stereoselective syntheses of functionally di- and tri-substituted olefins.[3–11] Also, the introduction of the zirconium-catalyzed carboalumination has made available alkenylaluminum compounds that can be elaborated into stereodefined trisubstituted olefins,[12] especially those of terpenoid origin. The fact that the hydroalumination of 1-alkynes not only provides alkenylalanes, but also can be controlled through the choice of solvent to afford 1-alkynylalanes and bis- and tris-aluminoalkanes greatly enhances the versatility of alkyne-derived organoaluminum derivatives. However, the chemistry of geminal aluminoalkanes remains to be delineated.

An important feature of alkenyl- and alkynylaluminum compounds is that they exhibit chemical properties that differ distinctly from those of the corresponding organolithium or organomagnesium compounds or of their organoboron congeners. For example, although unsaturated organoaluminum compounds undergo Grignard-like reactions with a variety of organic and inorganic electrophiles, they show some significant differences in reactivity and substrate selectivity as compared to organolithium and Grignard reagents. This results because the trigonal aluminum in organoalanes possesses a rather low intrinsic nucleophilicity. However, conversion of the organoalanes into the corresponding filled-octet aluminate species by treatment with bases markedly increases their carbanion character.[8] Also, transmetalation of alkenylaluminum compounds with transition metals produces stereodefined organometallics with reaction characteristics different from those of their progenitors, thus providing an additional tool for modifying the reactivity of unsaturated organoaluminum compounds.

For simplicity and clarity, organoaluminum compounds are referred to in this review as monomers even though they actually may be associated. Except for the 1-alkynylaluminum compounds, we have chosen to include in this chapter only those alkenylaluminum compounds whose preparations involve the direct formation of a carbon–aluminum bond. Those derived by exchange reactions are not covered.

PREPARATION OF ALKENYL- AND ALKYNYLALUMINUM COMPOUNDS

Alkenylaluminum Compounds via Hydroalumination

Dialkylaluminum hydrides, especially diisobutylaluminum hydride, and lithium aluminum hydride and its alkyl derivatives are the reagents most commonly employed for the hydroalumination of alkynes. Since these aluminum hydrides are capable of reducing a variety of functional groups, the presence of a functionality in an alkyne may interfere with the hydroalumination of the triple bond. Whether a functional group is compatible with the hydroalumination reaction depends on the nature of the aluminum hydride used, the reaction medium, and the reactivity of the triple bond toward a given reagent. For example, *trans* hydroaluminations of alkynes with the nucleophile lithium aluminum hydride tolerate the presence of acetal, ketal, and tetrahydropyranoxy groups. On the other hand, these functionalities may interfere in hydroaluminations with the electrophilic dialkylaluminum hydrides.

In the following discussion, the structure elucidations of the alkenylaluminum compounds prepared by the various methods are based on spectroscopic data or on derivatization of the vinyl carbon–aluminum bond by reactions that are known to proceed with retention of configuration.

***Cis* Hydroalumination of Monosubstituted Alkynes.** In 1956 Wilke and Müller[1] reported that monohydroalumination of aliphatic terminal acetylenes with dialkylaluminum hydrides, either neat or in hydrocarbon solvents, proceeds through a regio- and stereoselective *cis* addition of the aluminum–hydrogen bond to the triple bond to give (*E*)-1-alken-1-yldialkylalanes. However, the hydroalumination of acetylene itself is not clean.[2,13]

$$RC\equiv CH + R'_2AlH \longrightarrow (R)(H)C{=}C(H)(AlR'_2)$$

The hydroalumination of 1-alkynes with diisobutylaluminum hydride is accompanied by small amounts of metalation and bis-hydroalumination products (Eq. 1). Thus the reaction of 1-hexyne with diisobutylaluminum hydride in *n*-heptane solvent affords, after deuterolysis, 90% of (*E*)-1-^{2}H-1-hexene along with nearly equimolar amounts of deuterated 1-hexyne and *n*-hexane.[14] Since the NMR data clearly establish the *trans* structure for the intermediate alkenylalane, the formation of the *trans* deuterio-olefin is in agreement with retention of configuration in the deuterolysis step.

$$RC\equiv CH + (i\text{-}C_4H_9)_2AlH \xrightarrow[4\ \text{hr},\ 50^\circ]{n\text{-heptane}} (R)(H)C{=}C(H)(Al(C_4H_9\text{-}i)_2)$$

R	
n-C_4H_9	90%
C_6H_{11}	94%
t-C_4H_9	97%

$$+ RC\equiv CAl(C_4H_9\text{-}i)_2 + RCH_2CH[Al(C_4H_9\text{-}i)_2]_2 \qquad \text{(Eq. 1)}$$

R	$RC\equiv CAl(C_4H_9\text{-}i)_2$	$RCH_2CH[Al(C_4H_9\text{-}i)_2]_2$
n-C_4H_9	6%	4%
C_6H_{11}	2%	4%
t-C_4H_9	—	—

Secondary or tertiary alkyl-substituted 1-alkynes react more rapidly and provide higher yields of the (*E*)-alkenylalanes than do 1-alkynes with primary alkyl groups. For example, the hydroalumination of *tert*-butylacetylene is complete within 4 hours at 25°, yielding 97% of the alkenylalane.[14] By contrast, 1-hexyne requires 24 hours to proceed to completion (90%) under the same conditions.

The presence of an electron-withdrawing substituent on the triple bond, as in phenylacetylene[15–19] or conjugated enynes,[20] increases the acidity of the acetylenic hydrogen and hence leads to increased metalation during hydroalumination. Thus the reaction of phenylacetylene with one molar equivalent of diisobutylaluminum hydride in *n*-heptane solvent at 50° for 4 hours produces a complex mixture of products containing 29% of the metalated alkyne.[14] The hydroalumination of conjugated enynes such as cyclohexenylacetylene is chemoselective in that the aluminum–hydrogen addition occurs preferentially at the triple bond. However, metalation of the triple bond impairs the yield of dienylalane produced.

$$\text{(1-cyclohexenyl)}C{\equiv}CH + (i\text{-}C_4H_9)_2AlH \xrightarrow[4\ \text{hr}]{50^\circ} \text{(1-cyclohexenyl)}(H)C{=}C(H)Al(C_4H_9\text{-}i)_2\ (75\%) + \text{(1-cyclohexenyl)}C{\equiv}CAl(C_4H_9\text{-}i)_2\ (20\%)$$

It is important to note that the triple bonds of mono- and 1,2-disubstituted alkynes undergo monohydroalumination far more readily than do the double bonds of the corresponding alkenes. This differential reactivity provides the basis for selective hydroaluminations of triple bonds of nonconjugated enynes. Thus monohydroalumination of 4-hexen-1-yne with diisobutylaluminum hydride affords mainly the dienylalane.[21] Hydroalumination of 1-hexen-5-yne at 25°

$$CH_3CH{=}CHCH_2C{\equiv}CH \xrightarrow[4\ \text{hr},\ 50^\circ]{(i\text{-}C_4H_9)_2AlH} CH_3CH{=}CHCH_2(H)C{=}C(H)Al(C_4H_9\text{-}i)_2$$

produces, after protonolysis, 60% of 1,5-hexadiene along with methylcyclopentane and methylenecyclopentane.[22] The formation of the substituted cyclopentanes is the result of intramolecular cyclizations of the organoalanes initially formed. Interestingly, cyclization is the main reaction path when the hydroalumination is carried out in basic solvents (Eqs. 22 and 23).[22]

The high chemoselectivity in additions of the aluminum–hydrogen bond to the triple bond of nonconjugated enynes is further evidenced by the nearly exclusive formation of the dienylalane on treatment of 1-octen-7-yne with diisobutylaluminum hydride in hydrocarbon solvent.[23]

$$CH_2{=}CH(CH_2)_4C{\equiv}CH \xrightarrow{(i\text{-}C_4H_9)_2AlH} \begin{array}{c} CH_2{=}CH(CH_2)_4 \qquad H \\ \diagdown \quad \diagup \\ C{=}C \\ \diagup \quad \diagdown \\ H \qquad Al(C_4H_9\text{-}i)_2 \end{array}$$

Much effort has been devoted to the preparation of alkenylalanes derived from propargylic alcohols and their ether derivatives for use as precursors in prostaglandin syntheses. As pointed out above, the hydroalumination of functionally substituted alkynes is not without complications. Thus the reaction of diisobutylaluminum hydride with propargylic ethers may be accompanied by reductive cleavage of the carbon–oxygen bond. Moreover, depending on the nature of the hydroxyl-protecting group, the *cis*- rather than the *trans*-alkenylalane may be obtained.

A viable approach for the preparation of the alkenylalane derived from 1-octyn-3-ol involves sequential treatment of the alkyne with 2 molar equivalents of triisobutylaluminum followed by 1 molar equivalent of diisobutylaluminum hydride.[24] Direct hydroalumination of 1-octyn-3-ol with 2 equivalents of

$$n\text{-}C_5H_{11}\overset{\displaystyle OH}{\overset{|}{C}}HC{\equiv}CH \xrightarrow{(i\text{-}C_4H_9)_3Al} n\text{-}C_5H_{11}\overset{\displaystyle OAl(C_4H_9\text{-}i)_2}{\overset{|}{C}}HC{\equiv}CH$$

$$\xrightarrow[\text{2 hr, 60°}]{(i\text{-}C_4H_9)_2AlH} \begin{array}{c} OAl(C_4H_9\text{-}i)_2 \\ | \\ n\text{-}C_5H_{11}CH \qquad H \\ \diagdown \quad \diagup \\ C{=}C \\ \diagup \quad \diagdown \\ H \qquad Al(C_4H_9\text{-}i)_2 \end{array}$$

$(\sim 50\%)$

diisobutylaluminum hydride affords the alkenylalane in lower yield (25–40%).[24,25]

The preparation of *trans*-1-alkenylalanes via hydroalumination of protected 1-octyn-3-ol is critically dependent on the choice of the hydroxyl-blocking group. Thus treatment of the triphenylmethyl (Tr) ether derivative of the alkyne with diisobutylaluminum hydride affords the corresponding *trans*-alane (Eq. 2).[25] However, blocking the hydroxyl group of 1-octyn-3-ol as the *tert*-butyl ether leads to a net *trans* addition of the aluminum–hydrogen bond to the triple bond to produce the *cis*-octenylalane (Eq. 3).[25] The *trans* addition of diisobutylaluminum hydride to the triple bond is probably a consequence of a mechanistic change induced by prior coordination of the reagent with the ether oxygen.

$$n\text{-}C_5H_{11}\underset{\displaystyle|}{\overset{\displaystyle OTr}{C}}HC{\equiv}CH \xrightarrow{(i\text{-}C_4H_9)_2AlH} \begin{matrix} n\text{-}C_5H_{11}CH(OTr) & & H \\ & C{=}C & \\ H & & Al(C_4H_9\text{-}i)_2 \end{matrix} \quad \text{(Eq. 2)}$$

$$n\text{-}C_5H_{11}\underset{\displaystyle|}{\overset{\displaystyle OC_4H_9\text{-}t}{C}}HC{\equiv}CH \xrightarrow{(i\text{-}C_4H_9)_2AlH} \begin{matrix} n\text{-}C_5H_{11}CH(OC_4H_9\text{-}t) & & Al(C_4H_9\text{-}i)_2 \\ & C{=}C & \\ H & & H \end{matrix} \quad \text{(Eq. 3)}$$

The course of the reaction of dialkylaluminum hydrides with 1-alkynes is remarkably dependent on the choice of solvent.[14,19,22] Mono-, di-, or trihydroalumination products can be obtained by carrying out the reaction in hydrocarbon, ether, or tertiary amine solvents, respectively. Thus hydroalumination of 1-alkynes with diisobutylaluminum hydride in tetrahydrofuran (THF) produces a mixture of mono- and dihydroalumination products, unless two equivalents of the reagent are used.[26,27]

$$n\text{-}C_4H_9C{\equiv}CH + 2(i\text{-}C_4H_9)_2AlH \xrightarrow[70^\circ]{THF} n\text{-}C_5H_{11}CH[Al(C_4H_9\text{-}i)_2]_2$$

Bis-hydroalumination can also be achieved by heating the alkyne and hydroaluminating agent neat or in hydrocarbon solvent for prolonged periods of time.[2,13]

Whereas little metalation of alkyl-substituted 1-alkynes occurs during hydroalumination in hydrocarbon or ether solvents, reaction of one molar equivalent of dialkylaluminum hydrides with the alkynes in tertiary amine solvents results in the exclusive metalation of the triple bond.[28] Further treatment of the resultant alkynylalanes with two additional molar equivalents of dialkylaluminum hydride affords 1,1,1-tris(dialkylalumino)alkanes in high yields.[2,22,29]

$$RC{\equiv}CH \xrightarrow[(C_2H_5)_3N,\ 20^\circ\ (-H_2)]{R'_2AlH} RC{\equiv}CAlR'_2 \xrightarrow[70^\circ]{2R'_2AlH} RCH_2C(AlR'_2)_3$$

Interestingly, a high-yield conversion of alkynyldialkylalanes into the synthetically attractive 1,1-dialuminoalkenes via hydroalumination has not yet been realized.

***Cis* Hydroalumination of Disubstituted Acetylenes.** The hydroalumination of 1,2-disubstituted alkynes with dialkylaluminum hydrides neat or in hydrocarbon solvent also proceeds in a stereoselective *cis* manner.[2,13–15,30] The 1:1 reaction of internal alkynes with diisobutylaluminum hydride is conveniently carried out at 50–70° for several hours. The rate of hydroalumination is markedly lower in donor solvents such as diethyl ether or tetrahydrofuran. Diaddition of aluminum-hydrogen to disubstituted alkynes to give bis-alumino intermediates is usually negligible, especially when the reaction is carried out in a hydrocarbon solvent. However, the monohydroalumination of disubstituted acetylenes is frequently accompanied by the formation of small amounts of dienylaluminum compounds resulting from addition of the alkenylalane product to the triple bond of unreacted alkyne.[14] The portion of the dimerization product increases with higher temperatures and prolonged reaction times.

$$\mathrm{RC{\equiv}CR} \xrightarrow[\text{4 hr, 70°}]{(i\text{-}C_4H_9)_2AlH \text{ (1 eq)}} \underset{(75\text{–}85\%)}{(R)(H)C{=}C(R)Al(C_4H_9\text{-}i)_2}$$

$$+ \underset{(5\text{–}7\%)}{(R)(H)C{=}C(R){-}C(R){=}C(R)Al(C_4H_9\text{-}i)_2}$$

The rate of hydroalumination of internal alkynes with diisobutylaluminum hydride is pronouncedly decreased after 50% reaction because of complexation of the resulting alkenylalane with the dialkylaluminum hydride in the course of the reaction.[30,31] The 1:2 alkyne–hydride reaction circumvents the rate retardation problem and provides higher yields of alkenylalane with less dienylalane contamination of the product.

$$n\text{-}C_3H_7C{\equiv}CC_3H_7\text{-}n \xrightarrow[\text{6 hr, 50°}]{(i\text{-}C_4H_9)_2AlH \text{ (2 eq)}} \underset{(95\%)}{(n\text{-}C_3H_7)(H)C{=}C(C_3H_7\text{-}n)Al(C_4H_9\text{-}i)_2}$$

The hydroalumination of disubstituted alkynes is catalyzed by nickel acetylacetonate, making it possible to carry out the reaction at room temperature within 4 hours. However, the regio- and stereoselectivities are generally lower than in uncatalyzed reactions.[32]

The reaction of conjugated enynes having an internal triple bond as in pent-1-en-3-yne and hex-1-en-3-yne with dialkylaluminum hydrides leads mainly to polymeric products.[20] On the other hand, hept-2-en-4-yne, an enyne in which both unsaturations are internal, affords a 50% yield of the dienylalane on treatment with 1.1 molar equivalents of diisobutylaluminum hydride. The aluminum–

hydrogen addition is highly regioselective, placing the aluminum at the C-4 carbon of the triple bond.[33]

$$CH_3CH{=}CHC{\equiv}CC_2H_5 \xrightarrow[5\ hr,\ 50^\circ]{(i\text{-}C_4H_9)_2AlH} (i\text{-}C_4H_9)_2Al(CH_3CH{=}CH)C{=}C(C_2H_5)H \quad (50\%)$$

The hydroalumination of unsymmetrically 1,2-disubstituted acetylenes is only moderately regioselective, placing the aluminum preferentially at the least-hindered position of the triple bond.[14] This behavior is in contrast to the high

$$n\text{-}C_4H_9C{\equiv}CCH_3 \quad (33:67) \qquad C_6H_{11}C{\equiv}CCH_3 \quad (25:75) \qquad t\text{-}C_4H_9C{\equiv}CCH_3 \quad (15:85)$$

regioselectivity exhibited by dialkylboranes on addition to disubstituted alkynes.[34–36] More pronounced directive effects are encountered in hydroaluminations of alkylphenylacetylenes with diisobutylaluminum hydride, with the aluminum becoming attached preferentially alpha to the phenyl group.[37,38]

$$C_6H_5C{\equiv}CCH_3 \quad (82:18) \qquad C_6H_5C{\equiv}CC_4H_9\text{-}t \quad (100:0)$$

***Cis* Hydroalumination of Heteroatom-Substituted 1-Alkynes.** In contrast to the many reported studies of hydroaluminations of internal and terminal alkynes, relatively few examples of reactions of dialkylaluminum hydrides with heteroatom-substituted 1-alkynes have been reported. In view of the widely varying electronic and steric characteristics of the different heteroatoms, these substituents offer unique potential for affecting the regio- and stereochemical course of the aluminum-hydrogen addition to carbon–carbon triple bonds. Moreover, the resultant heteroatom-substituted alkenylaluminum compounds provide valuable synthons for further transformations.

Especially attractive from a synthetic point of view are those alkenylalanes containing another vinyl metal ligand. Unfortunately, monohydroalumination of alkynylalanes with dialkylaluminum hydrides in either hydrocarbon or ether solvents does not lead to the 1,1-dialuminum-1-alkenes but instead affords, besides starting material, the dihydroalumination product.[2,13,14,22,39]

$$RC{\equiv}CAlR'_2 \xrightarrow{R'_2AlH} \left[(R)(H)C{=}C(AlR'_2)_2 \right] \xrightarrow{R'_2AlH} RCH_2C(AlR'_2)_3$$

Apparently, the intermediate 1,1-dialumino-1-alkene formed is more reactive than the 1-alkynylalane toward remaining dialkylaluminum hydride. However, 1,1-dimetaloalkenes containing aluminum as well as titanium or zirconium have been synthesized by carbotitanation and carbozirconation of 1-alkynyldimethylalane.[40]

$$n\text{-}C_3H_7C{\equiv}CAl(CH_3)_2 \xrightarrow{(CH_3)_3Al\text{-}Cl_2Ti(C_5H_5)_2} n\text{-}C_3H_7(CH_3)C{=}C\begin{matrix}Al(CH_3)_2\\ Ti(C_5H_5)_2Cl\end{matrix}$$

$$n\text{-}C_3H_7C{\equiv}CAl(CH_3)_2 \xrightarrow{Cl(CH_3)Zr(C_5H_5)_2} n\text{-}C_3H_7(CH_3)C{=}C\begin{matrix}Al(CH_3)_2\\ Zr(C_5H_5)_2Cl\end{matrix}$$

The hydroalumination of (1-alkynyl)silanes with diisobutylaluminum hydride proceeds readily to the monoaddition stage. The reaction is highly regioselective in that the aluminum adds nearly exclusively to the carbon alpha to silicon to furnish 1-alumino-1-silyl-1-alkenes in high yields.[38] Also, the stereochemistry of the monohydroalumination products depends on the solvent used. Selective *cis* hydroalumination thus is observed when the reaction is carried out in a tertiary amine solvent (Eq. 4).[38] On the other hand, (*E*)-1-alumino-1-silyl-1-alkenes are obtained when the reaction medium is a hydrocarbon solvent (Eq. 5).[38]

$$RC{\equiv}CSi(CH_3)_3 \xrightarrow[R_3^1N,\ 60^\circ]{(i\text{-}C_4H_9)_2AlH} \begin{matrix}R\\ H\end{matrix}C{=}C\begin{matrix}Si(CH_3)_3\\ Al(C_4H_9\text{-}i)_2\end{matrix} \quad \text{(Eq. 4)}$$

$$RC{\equiv}CSi(CH_3)_3 \xrightarrow[\text{hydrocarbon},\ 25^\circ]{(i\text{-}C_4H_9)_2AlH} \begin{matrix}R\\ H\end{matrix}C{=}C\begin{matrix}Al(C_4H_9\text{-}i)_2\\ Si(CH_3)_3\end{matrix} \quad \text{(Eq. 5)}$$

It has been suggested that in nonpolar solvents, kinetically controlled *cis* addition of diisobutylaluminum hydride is followed by isomerization of the resultant (*Z*)-α-silylalkenylalane to the thermodynamically more stable *E* isomer.[41] The isomeric purities of the (*E*)-α-silylalkenylalanes obtained in hydrocarbon solvents depend on the substituent at the ethynylsilyl moiety. When R is $t\text{-}C_4H_9$, C_6H_5, or $CH_2{=}C(CH_3)$, therefore, hydroalumination of the corresponding (1-alkynyl)silanes affords essentially pure *E* monoaddition products. However, when R is a primary or a secondary alkyl group mixtures of *E* and *Z* isomers are obtained.[42]

The *cis* hydroalumination of (1-alkynyl)silanes with diisobutylaluminum hydride is most conveniently carried out in diethyl ether where the reaction proceeds in a stereo- and regioselective manner to produce [(Z)-1-alumino-1-alkenyl]silanes regardless of the steric requirements of the alkyl group at the β carbon of the (1-alkynyl)silanes.[42–44]

$$t\text{-}C_4H_9C{\equiv}CSi(CH_3)_3 \xrightarrow[\text{ether, }40^\circ]{(i\text{-}C_4H_9)_2AlH} (t\text{-}C_4H_9)(H)C{=}C(Si(CH_3)_3)(Al(C_4H_9\text{-}i)_2)$$

The hydroalumination of (1-alkynyl)silanes in ether is also highly chemoselective, permitting the selective reduction of the triple bond in enynes.[44]

$$\text{(1-cyclohexenyl)}C{\equiv}CSi(CH_3)_3 \xrightarrow[\text{ether, 1 hr, }40^\circ]{(i\text{-}C_4H_9)_2AlH} \text{(1-cyclohexenyl)}(H)C{=}C(Si(CH_3)_3)(Al(C_4H_9\text{-}i)_2)$$

The preference for aluminum–hydrogen additions to silicon-substituted triple bonds is also evidenced in the conversion of 1-silyl-1,4-diynes to the corresponding enynyl derivatives.[45]

$$n\text{-}C_6H_{13}C{\equiv}CCH_2C{\equiv}CSi(CH_3)_3 \xrightarrow[\text{ether, 2 hr, }40^\circ]{(i\text{-}C_4H_9)_2AlH} (n\text{-}C_6H_{13}C{\equiv}CCH_2)(H)C{=}C(Si(CH_3)_3)(Al(C_4H_9\text{-}i)_2)$$

Interestingly, the electron-withdrawing effect of a triple bond conjugated with the ethynyltrimethylsilyl moiety as in 1-silyl-1,3-diynes reduces its reactivity such that the hydroalumination proceeds very sluggishly.[45]

The outstanding feature of the hydroalumination of (1-alkynyl)silanes is that it provides a tool for converting those 1-alkynes into alkenylaluminum derivatives that cannot be obtained by the direct route. Thus prior silylation of the more acidic phenylacetylene and conjugated enynes circumvents the formation of the metalation byproducts encountered when these types of alkynes are directly subjected to hydroalumination. After elaboration of the vinyl carbon–aluminum bond, the trimethylsilyl group of the resultant vinylsilyl derivative can be removed by a variety of reagents with retention of configuration of the double bond. Also, masking the triple bond of tetrahydropyranyl (THP)-protected 1-alkyn-ols with a trimethylsilyl group enables their conversion into the corresponding alkenylaluminum compounds without adversely affecting the acetal protecting group. Hydroalumination of homopropargylic (tetrahydropyranyl-2′-oxy)-1-trimethylsilyl-1-alkynes with diisobutylaluminum hydride in diethyl ether thus affords the alkenylalanes in high yields.[46]

$$C_2H_5CH(OTHP)CH_2C\equiv CSi(CH_3)_3 \xrightarrow[\text{ether, 4 hr, }40^\circ]{(i\text{-}C_4H_9)_2AlH} (C_2H_5CH(OTHP)CH_2)(H)C{=}C(Si(CH_3)_3)(Al(C_4H_9\text{-}i)_2)$$

However, a propargylic THP group is partially cleaved under these reaction conditions. This problem can be obviated by carrying out the hydroalumination in ether by slowly warming the reaction mixture from 0 to 40° or in *n*-hexane containing 1.1 equivalents of tetrahydrofuran.[46]

$$CH_3CH(OTHP)C\equiv CSi(CH_3)_3 \xrightarrow[\text{THF (1.1 eq), 4 hr, }25^\circ]{(i\text{-}C_4H_9)_2AlH} (CH_3CH(OTHP))(H)C{=}C(Si(CH_3)_3)(Al(C_4H_9\text{-}i)_2)$$

The reaction of (phenylethynyl)trimethylgermane with diisobutylaluminum hydride proceeds analogously to the hydroalumination of the corresponding silanes, producing high yields of either the *cis* or the *trans* hydroalumination products depending on the nature of the solvent used.[38]

$$C_6H_5C\equiv CGe(CH_3)_3 \xrightarrow[R_3N]{(i\text{-}C_4H_9)_2AlH} (C_6H_5)(H)C{=}C(Ge(CH_3)_3)(Al(C_4H_9\text{-}i)_2)$$

$$C_6H_5C\equiv CGe(CH_3)_3 \xrightarrow[\text{hydrocarbon}]{(i\text{-}C_4H_9)_2AlH} (C_6H_5)(H)C{=}C(Al(C_4H_9\text{-}i)_2)(Ge(CH_3)_3)$$

Interestingly, treatment of the more polar (phenylethynyl)trimethylstannane with diisobutylaluminum hydride in the presence of a tertiary amine or in hydrocarbon solvent yields only phenylacetylene on hydrolysis.[38]

The hydroalumination of nitrogen-, phosphorus-, oxygen-, and sulfur-substituted 1-alkynes offers preparative possibilities for the stereoselective synthesis of the corresponding enamines, vinylic phosphines, ethers, and sulfides.

However, the scope of the hydroalumination of heterosubstituted 1-alkynes with dialkylaluminum hydrides and the synthetic utility of the resultant alkenylaluminum compounds have not yet been fully delineated.

Treatment of dimethyl(phenylethynyl)amine with diisobutylaluminum hydride in the presence of a tertiary amine yields the *trans* adduct, with the aluminum being attached to the vinylic carbon alpha to the phenyl group.[39] It appears that the alkyne undergoes initial *cis* hydroalumination and that the adduct formed isomerizes at room temperature to the *trans* adduct.

$$C_6H_5C{\equiv}CN(CH_3)_2 \xrightarrow[R_3N]{(i\text{-}C_4H_9)_2AlH} (C_6H_5)[(i\text{-}C_4H_9)_2Al]C{=}C[N(CH_3)_2](H)$$

$$\longrightarrow (C_6H_5)[(i\text{-}C_4H_9)_2Al]C{=}C(H)[N(CH_3)_2]$$

The hydroalumination of (phenylethynyl)dimethylphosphine using diisobutylaluminum hydride in hydrocarbon solvent with or without an added tertiary amine results in the exclusive *cis* addition of the aluminum-hydrogen bond to the triple bond but producing an 85:15 mixture of regioisomers.[38]

$$C_6H_5C{\equiv}CP(CH_3)_2$$

$$\xrightarrow[n\text{-heptane, 15 hr, 50°}]{(i\text{-}C_4H_9)_2AlH} \underset{(85\%)}{(C_6H_5)(H)C{=}C[P(CH_3)_2][Al(C_4H_9\text{-}i)_2]} + \underset{(15\%)}{(C_6H_5)[(i\text{-}C_4H_9)_2Al]C{=}C[P(CH_3)_2](H)}$$

In the absence of a tertiary amine, the reaction of 1-ethoxy-1-hexyne with diisobutylaluminum hydride results mainly in the cleavage of the alkynyl carbon–oxygen linkage, leading to a host of products.[39] On the other hand, regio- and stereoselective hydroalumination is observed as the predominant reaction path when the hydroalumination is carried out in the presence of a tertiary amine.[39]

$$n\text{-}C_4H_9C{\equiv}COC_2H_5$$

$$\xrightarrow[R_3N]{(i\text{-}C_4H_9)_2AlH} \underset{(70\%)}{(n\text{-}C_4H_9)[(i\text{-}C_4H_9)_2Al]C{=}C(OC_2H_5)(H)} + \underset{(30\%)}{n\text{-}C_4H_9C{\equiv}COC_2H_5}$$

Hydroalumination of ethyl phenylethynyl sulfide with diisobutylaluminum hydride proceeds to give exclusive *cis* addition of the aluminum–hydrogen bond to the triple bond in hydrocarbon solvent. The reaction is, however, only moderately regioselective.[39]

$$C_6H_5C{\equiv}CSC_2H_5 \xrightarrow[\text{cyclopentane, 1 hr, 25°}]{(i\text{-}C_4H_9)_2AlH} \underset{(83\%)}{(C_6H_5)(H)C{=}C(SC_2H_5)(Al(C_4H_9\text{-}i)_2)} + \underset{(17\%)}{(C_6H_5)((i\text{-}C_4H_9)_2Al)C{=}C(SC_2H_5)(H)}$$

The *cis* hydroalumination of alkynes has been postulated to involve a π complex between the electron-deficient trigonal aluminum of the dialkylaluminum hydride and the π orbital of the alkyne.[30,41] With unsymmetrically substituted acetylenes, collapse of the π complex to products is governed by both electronic and steric factors, the former usually predominating. Thus electrophilic attack by dialkylaluminum hydride on the triple bond of the substrate occurs preferentially to place the electron deficiency on the carbon best able to accommodate it. The regiochemistry depicted below should be favored when R′ is a better σ-electron donor than R″, or when R″ is a particularly effective π donor.[38] When electronic effects are less important, as in dialkylacetylenes, the regiochemistry depends mainly on the steric interaction between the substituents attached to the aluminum and to the alkyne. A mechanistic description of the hydroalumination of alkynes in nonpolar or weakly basic solvents must also take into account the association of dialkylaluminum hydrides and their formation of mixed alkenylhydride species in the presence of alkenylalanes.[31]

$$R'{-}\overset{\delta+}{C}{\equiv}\overset{\delta-}{C}{-}R'' \cdots \overset{\delta+}{AlR_2} \cdots \overset{\delta-}{H}$$

***Trans* Hydroalumination.** In contrast to the observed *cis* addition of dialkylaluminum hydrides to the triple bond of alkynes, lithium aluminum hydrides and lithium trialkylaluminum hydrides react with internal acetylenes in a stereoselective *trans* manner, thus making both *E* and *Z* alkenylaluminum compounds readily accessible. It should be noted, however, that the reactions of terminal acetylenes with these nucleophilic aluminum hydrides produce appreciable amounts of the metalated alkynes.

The *trans* hydroalumination of 1,2-dialkylacetylenes with lithium aluminum hydride in refluxing diethylene glycol dimethyl ether (diglyme)-tetrahydrofuran mixtures affords the lithium (*Z*)-alkenylalanates in high yields, as evidenced by their conversion into the corresponding *trans* alkenes.[47,48]

$$C_2H_5C{\equiv}CC_2H_5 \xrightarrow[\text{diglyme–THF, }120\text{–}150^\circ]{LiAlH_4} \left[\begin{matrix} C_2H_5 & & AlH_3Li \\ & C{=}C & \\ H & & C_2H_5 \end{matrix} \right] \xrightarrow{H_3O^+} \begin{matrix} C_2H_5 & & H \\ & C{=}C & \\ H & & C_2H_5 \end{matrix}$$

Lithium diisobutylmethylaluminum hydride in dimethoxyethane (DME), prepared from diisobutylaluminum hydride and methyllithium, is also effective for the *trans* hydroalumination of internal alkynes.[49]

$$(i\text{-}C_4H_9)_2AlH + CH_3Li \longrightarrow Li[AlH(C_4H_9\text{-}i)_2CH_3]$$

$$RC{\equiv}CR + Li[AlH(C_4H_9\text{-}i)_2CH_3] \xrightarrow[100\text{–}130^\circ]{DME} \begin{matrix} R & & Al(C_4H_9\text{-}i)_2CH_3Li \\ & C{=}C & \\ H & & R \end{matrix}$$

(60–80%)

An important feature of these alkenylalanates is that they react readily with a number of carbonyl reagents whereas those derived from lithium aluminum hydride do not.

The presence of an electron-withdrawing group on the triple bond has a marked effect on the rate and regiochemistry of the *trans* hydroalumination. For example, the reduction of 1-phenylpropyne with lithium aluminum hydride occurs at moderate temperatures, placing the aluminum moiety preferentially (95%) alpha to the phenyl group.[47,48]

$$C_6H_5C{\equiv}CCH_3 \xrightarrow[\text{THF, 13 hr, }66^\circ]{LiAlH_4} \begin{matrix} C_6H_5 & & H \\ & C{=}C & \\ LiH_3Al & & CH_3 \end{matrix}$$

This result is consistent with a nucleophilic attack on the triple bond by the hydride. Substituents that are better able to stabilize the developing negative charge of the incipient vinyl carbanion should increase the rate of reaction and favor attack of the aluminum adjacent to the more electronegative substituent (R″).

$$R'C{\equiv}CR'' + LiAlH_4 \xrightarrow[\text{(or } LiR_3AlH)]{} \left[\begin{matrix} R' & & \\ & C{=}\ddot{C}^{-} & \\ H & & R'' \end{matrix}\right] Li^+ + AlH_3$$

$$\downarrow$$

$$\begin{matrix} R' & & AlH_3Li \\ & C{=}C & \\ H & & R'' \end{matrix}$$

It has been proposed that dissolving metal reductions of disubstituted acetylenes generate vinyl carbanions possessing the *trans* stereochemistry.[50] The stereochemistry observed in the *trans* hydroalumination may be accounted for in terms of a similar reaction scheme.

The ethynyl group of conjugated diynes greatly facilitates addition of lithium hydroaluminates to the triple bond. Thus the hydroalumination of symmetrically substituted 1,3-diynes with lithium diisobutylmethylaluminum hydride in diglyme proceeds at room temperature. The reaction is highly stereoselective, as evidenced by the nearly exclusive formation of the corresponding *trans* enynes after hydrolysis of the intermediate enynylaluminates.[51] In addition to being stereoselective, the *trans* hydroalumination of symmetrically substituted conjugated diynes is also highly regioselective, placing the aluminum preferentially at the sterically less hindered position of the diyne system.[51]

$$t\text{-}C_4H_9C{\equiv}C{-}C{\equiv}CC_4H_9\text{-}t$$

$$\xrightarrow[\text{diglyme, 8 hr, } 25^\circ]{Li[AlH(C_4H_9\text{-}i)_2CH_3]} \begin{matrix} t\text{-}C_4H_9C{\equiv}C & & H \\ & C{=}C & \\ Li(i\text{-}C_4H_9)_2\underset{|}{Al} & & C_4H_9\text{-}t \\ CH_3 & & \end{matrix}$$

It is noteworthy that no di-reduction of the diyne is observed even when a 2:1 ratio of the reagents is used. Also, the hydroaluminating agent does not discriminate in its addition between the triple bonds of unsymmetrically alkyl-substituted conjugated diynes.

A remarkable selectivity is observed in reactions of 1-trimethylsilyl-1,3-diynes with lithium trialkylaluminum hydrides in 1,2-dimethoxyethane (DME). The aluminum-hydrogen addition to the diyne system is not only stereo- and regioselective, but is also highly chemoselective, placing the aluminum at the internal position of the alkyl-substituted triple bond (Eq. 6).[45]

$$n\text{-}C_6H_{13}C{\equiv}C{-}C{\equiv}CSi(CH_3)_3 \xrightarrow[\text{DME, 1 hr, 25°}]{Li[AlH(C_4H_9\text{-}i)_2(C_4H_9\text{-}n)]} (n\text{-}C_6H_{13})(H)C{=}C[Al(C_4H_9\text{-}i)_2(C_4H_9\text{-}n)Li](C{\equiv}CSi(CH_3)_3) \quad \text{(Eq. 6)}$$

These highly functionalized enynes are very attractive intermediates for further transformation into a host of organic derivatives by taking advantage of the differential reactivity of the vinyl carbon–aluminum and ethynyl carbon–silicon bonds toward organic and inorganic reagents.

Interestingly, treatment of 2-alkynenitriles with 0.5 equivalent of lithium aluminum hydride in ether does not affect the cyano group but results in the regio- and stereoselective hydroalumination of the triple bond. This is evidenced by the formation of α-deuterated *trans*-2-alkenylnitriles on deuterolysis of the intermediate alkenylalanates.[52]

$$C_2H_5C{\equiv}CCN \xrightarrow[\text{ether, }-60°\rightarrow 0°]{0.5\ LiAlH_4} (C_2H_5)(H)C{=}C(AlH_2Li)_{1/2}(CN) \xrightarrow{D_2O} (C_2H_5)(H)C{=}C(D)(CN)$$

The dominant influence of the cyano group on the regiochemistry of reduction is also evident in phenyl-substituted ethynylnitriles where the aluminum becomes attached alpha to the cyano moiety despite the competing stabilizing effect of the phenyl group.[52]

$$C_6H_5C{\equiv}CCN \xrightarrow[\text{ether}]{LiAlH_4} (C_6H_5)(H)C{=}C(AlH_2Li)_{1/2}(CN)$$

The subtle interplay of electronic factors exerted by substituents at the triple bond and the nature of hydroaluminating agent in determining the stereo- and regiochemical outcome of hydroalumination reactions is especially pronounced in alkynyl sulfides. The reaction of ethyl phenylethynyl sulfide with the electrophilic diisobutylaluminum hydride thus proceeds in a *cis* manner, producing a mixture of regioisomers.[39] Hydroalumination of methyl phenylethynyl sulfide with the nucleophile lithium aluminum hydride involves a *trans* addition of aluminum–hydrogen to the triple bond with the aluminum ending up alpha to the phenyl group.[53]

$$C_6H_5C{\equiv}CSCH_3 \xrightarrow[\text{THF}]{\text{LiAlH}_4} (C_6H_5)(LiH_3Al)C{=}C(H)(SCH_3)$$

Replacement of the phenyl group by an alkyl or alkenyl group results in a complete reversal of the regiochemistry on treatment of the alkyne with lithium aluminum hydride.[53]

$$CH_2{=}C(CH_3){-}C{\equiv}CSCH_3 \xrightarrow[\text{THF}]{\text{LiAlH}_4} (CH_2{=}C(CH_3))(H)C{=}C(AlH_3Li)(SCH_3)$$

Phenylthioacetylenes react with lithium aluminum deuteride by addition to the triple bond rather than by substitution of the methine hydrogen. Protonolysis of the reaction mixture furnishes the *cis*-phenyl vinyl sulfide-β-d$_1$.[54]

$$p\text{-}CH_3C_6H_4SC{\equiv}CH \xrightarrow[\text{THF, 1 hr, }25^\circ]{\text{LiAlD}_4} (p\text{-}CH_3C_6H_4S)(LiD_3Al)C{=}C(D)(H) \xrightarrow{H_2O} (p\text{-}CH_3C_6H_4S)(H)C{=}C(D)(H)$$

Attempts to convert 1-halo-1-alkynes into the synthetically attractive α-haloalkenylalanes via hydroalumination with diisobutylaluminum hydride have not been successful.[39] However, 1-chloro-1-alkynes containing primary and secondary alkyl groups attached at the triple bond undergo *trans* hydroalumination when reacted with an equimolar amount of lithium aluminum hydride in tetrahydrofuran (Eq. 7).[55]

$$RC{\equiv}CCl \xrightarrow[\text{THF, }-30^\circ \rightarrow 0^\circ]{\text{LiAlH}_4} \left[(R)(H)C{=}C(AlH_3Li)(Cl)\right] \xrightarrow{CH_3OH} (R)(H)C{=}C(H)(Cl) \qquad \text{(Eq. 7)}$$

These α-chlorovinylalanates are moderately stable at 0° and on methanolysis provide excellent yields of the corresponding (*E*)-1-chloro-1-alkenes. Hydroalumination of *tert*-butylchloroacetylene produces besides the α-chlorovinylalanate an appreciable amount of *tert*-butylacetylene (23%).[55]

The reduction of propargylic alcohols with lithium aluminum hydride is a standard procedure for preparing *trans* allylic alcohols.[56,57]

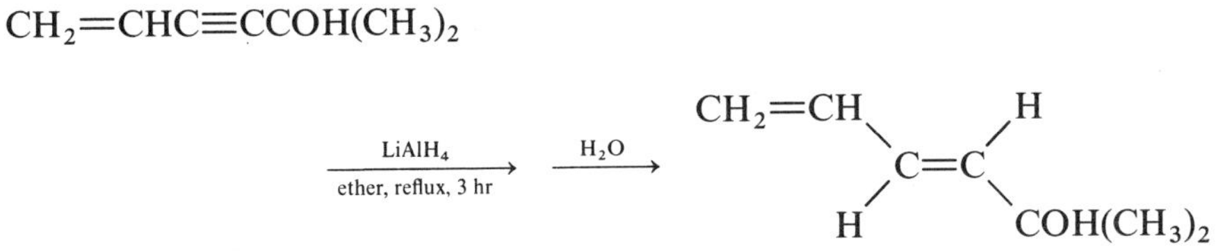

$$CH_2{=}CHC{\equiv}CCOH(CH_3)_2 \xrightarrow[\text{ether, reflux, 3 hr}]{LiAlH_4} \xrightarrow{H_2O} (CH_2{=}CH)(H)C{=}C(H)(COH(CH_3)_2)$$

Evidence that the reaction involves the intermediacy of alkenylaluminates resulting from *trans* addition of aluminum-hydrogen to the triple bond stems from the fact that they exhibit reaction characteristics similar to those observed with typical alkenylaluminum compounds (Eq. 8).

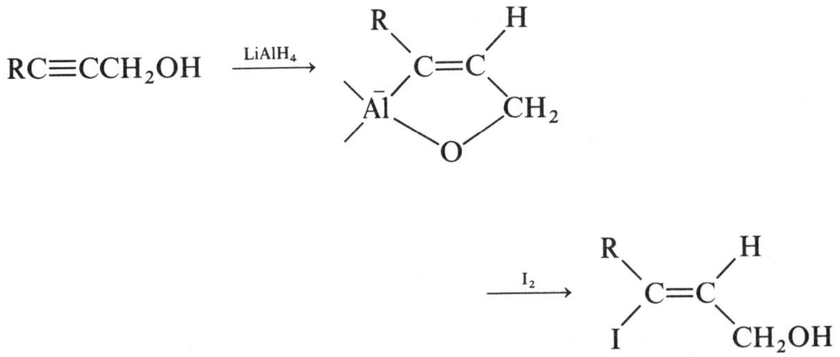

$$RC{\equiv}CCH_2OH \xrightarrow{LiAlH_4} \text{cyclic: } (R)C(\bar{Al}){=}C(H)CH_2O{-}\bar{Al} \xrightarrow{I_2} (R)(I)C{=}C(H)(CH_2OH) \qquad \text{(Eq. 8)}$$

If the hydroxyl group of an alkynol is not propargylic, reduction of the triple bond by lithium aluminum hydride occurs only at elevated temperature and at a rate comparable to that observed for the corresponding alkyne not containing a hydroxyl group.[58]

The nature of the solvent has a pronounced effect on the stereochemical course of hydroaluminations of alk-1-yn-3-ols.[59,60] In tetrahydrofuran, therefore, addition of lithium aluminum hydride to the triple bond occurs exclusively in a *trans* manner. On the other hand, in diisopropyl ether, *cis* addition predominates over *trans* addition. In general, a strong inverse correlation is observed between the Lewis basicity of the solvent and the extent of *cis* reduction.

The regiochemistry of hydride reductions of alk-2-yn-1-ols can be controlled by the proper choice of reagents. Thus attachment of aluminum occurs exclusively at C-3 of the propargylic alcohols using lithium aluminum hydride in the presence of excess sodium methoxide.[61] A complete reversal of the stereochemistry is observed in sequential treatment of the alk-2-yn-1-ols with *n*-butyllithium followed by diisobutylaluminum hydride (Eq. 9).[62]

$$RC{\equiv}CCH_2OH \xrightarrow[2.\ I_2]{1.\ LiAlH_4,\ NaOCH_3} (R)(I)C{=}C(H)(CH_2OH)$$

$$RC{\equiv}CCH_2OH \xrightarrow[2.\ (i\text{-}C_4H_9)_2AlH;\ 3.\ I_2]{1.\ n\text{-}C_4H_9Li} (R)(H)C{=}C(I)(CH_2OH) \qquad \text{(Eq. 9)}$$

The regiochemistry and stereochemistry of the intermediate alkenylaluminum compounds obtained follow from their conversion into the corresponding isomerically pure alkenyl iodides on treatment with iodine.[61,62]

Alkenylaluminum Compounds via Carboalumination

The hydroalumination of terminal alkynes followed by further elaboration of the resultant vinyl carbon–aluminum bond provides a convenient, stereoselective route to disubstituted olefins. However, its application to the synthesis of trisubstituted olefins is less satisfactory since the hydroalumination of unsymmetrically disubstituted alkynes is not very regioselective. An alternative approach to alkenylalanes that can be converted into trisubstituted olefins involves the controlled carboalumination of monosubstituted acetylenes by addition of the carbon–aluminum bond of trialkylalanes. Extension of the carboalumination reaction to disubstituted alkynes should provide alkenylalane precursors for the synthesis of tetrasubstituted olefins.

The reaction of trialkylalanes, such as triethyl- or triisobutylalane, with acetylene proceeds in a stereoselective manner and provides a simple route to *cis*-alkenyldialkylalanes.[2,13]

$$R_3Al + HC{\equiv}CH \xrightarrow{20\text{–}60^\circ} (H)(R_2Al)C{=}C(H)(R)$$

It should be noted that the synthetic utility of the *cis*-alkenylalanes has not been well delineated. The corresponding *trans*-alkenylalanes are available via *cis* hydroalumination of the appropriate monosubstituted acetylenes.

The reactions of trialkyl- and triarylalanes with monosubstituted alkynes produce appreciable amounts of metalated alkynes.[63]

$$(C_2H_5)_3Al + HC{\equiv}CC_4H_9\text{-}n \longrightarrow (C_2H_5)_2AlC{\equiv}CC_4H_9\text{-}n$$

The discovery that the methyl–aluminum moieties of trimethylalane add to terminal acetylenes in the presence of bis (cyclopentadienyl)zirconium dichloride, $Cl_2Zr(C_5H_5)_2$, has opened a unique, expeditious route to β-disubstituted alkenylalanes and via further reactions of these to a variety of trisubstituted olefins.[12] The reaction involves a zirconium-assisted direct carboalumination and proceeds in a stereo- and regioselective manner producing nearly exclusively the corresponding (E)-2-methylalkenylalane species.[64]

$$n\text{-}C_6H_{13}C{\equiv}CH + (CH_3)_3Al \xrightarrow[20\text{–}25^\circ]{Cl_2Zr(C_5H_5)_2} (n\text{-}C_6H_{13})(CH_3)C{=}C(H)[Al(CH_3)_2\cdot Cl_2Zr(C_5H_5)_2]$$

The methyl-alumination reaction can be extended to the more acidic phenylacetylene.[64]

Attempts to introduce an n-propyl group onto the triple bond of 1-octyne using $(n\text{-}C_3H_7)_3Al{-}Cl_2Zr(C_5H_5)_2$ result in a lower regioselectivity than that observed for the corresponding $(CH_3)_3Al{-}Cl_2Zr(C_5H_5)_2$ reaction.[64]

$$n\text{-}C_6H_{13}C{\equiv}CH + (n\text{-}C_3H_7)_3Al \xrightarrow{Cl_2Zr(C_5H_5)_2} (n\text{-}C_6H_{13})(n\text{-}C_3H_7)C{=}C(H)Al\langle \; + \; \rangle Al(n\text{-}C_6H_{13})C{=}C(H)(C_3H_7\text{-}n)$$

(80:20)

The regio- and stereoselective carboalumination of disubstituted alkynes with organoalanes is confined to a few favorable cases. Thus the *cis* addition of triphenylaluminum to 1-phenylpropyne proceeds to place the aluminum nearly exclusively at the phenyl-substituted carbon of the triple bond.[65]

$$C_6H_5C{\equiv}CCH_3 + (C_6H_5)_3Al \xrightarrow{90^\circ} (C_6H_5)[(C_6H_5)_2Al]C{=}C(CH_3)(C_6H_5)$$

The synthetic utility of the carboalumination of internal acetylenes with organoalanes is limited by the lack of regioselectivity and by competitive dimerization and polymerization.[6] For example, the 1-to-1 reaction of triethylalane with 3-hexyne does not produce the anticipated alkenylalane but results in dimerization to produce the dienylalane.[2,13]

$$C_2H_5C{\equiv}CC_2H_5 + (C_2H_5)_3Al \xrightarrow{80^\circ} (C_2H_5)(C_2H_5)C{=}C(C_2H_5)Al(C_2H_5)_2$$

$$\xrightarrow{C_2H_5C{\equiv}CC_2H_5} (C_2H_5)_2Al(C_2H_5)C{=}C(C_2H_5)C(C_2H_5){=}C(C_2H_5)_2$$

Internal acetylenes such as 5-decyne can, however, be stereoselectively carboaluminated with $(CH_3)_3Al{-}Cl_2Zr(C_5H_5)_2$ as evidenced by the formation of (*Z*)-5-methyl-5-decene on protonolysis of the intermediate organoalane.[64]

$$n\text{-}C_4H_9C{\equiv}CC_4H_9\text{-}n \xrightarrow[2.\ H_3O^+]{1.\ (CH_3)_3Al\text{-}Cl_2Zr(C_5H_5)_2} (n\text{-}C_4H_9)(CH_3)C{=}C(C_4H_9\text{-}n)H$$

The zirconium-assisted carboalumination can be extended to propargylic and homopropargylic alkynols. Thus the *syn* addition of trimethylalane to 1-octyn-3-ol places the aluminum exclusively at the 1 position of the triple bond.[66]

$$n\text{-}C_5H_{11}CHOHC{\equiv}CH \xrightarrow[2.\ H_3O^+]{1.\ (CH_3)_3Al\text{-}I_2Zr(C_5H_5)_2} (n\text{-}C_5H_{11}CHOH)(CH_3)C{=}C(H)Al(CH_3)_2$$

A reaction reminiscent of carboalumination is the *syn* addition of *tris*(trimethylsilyl)aluminum to 1-alkynes to afford 2-alumino-1-silyl-1-alkenes.[67]

$$HC{\equiv}CCH_2OTHP + Al[Si(CH_3)_3]_3 \xrightarrow{\text{ether}} (H)((CH_3)_3Si)C{=}C(CH_2OTHP)Al{<}$$

$$\xrightarrow{D_2O} (H)((CH_3)_3Si)C{=}C(CH_2OTHP)D$$

In contrast to the behavior observed with terminal acetylenes, the regio- and stereoselective silylation of the disubstituted triple bonds in 2-butyn-1-ol and 1-phenylpropyne requires activation by aluminum chloride.[67]

$$CH_3C{\equiv}CCH_2OH/AlCl_3 \xrightarrow[2.\ H_2O]{1.\ Al[Si(CH_3)_3]_3} (CH_3)_3Si(CH_3)C{=}C(H)CH_2OH \quad (78\%)$$

$$CH_3C{\equiv}CC_6H_5/AlCl_3 \xrightarrow[2.\ H_2O]{1.\ Al[Si(CH_3)_3]_3} (CH_3)_3Si(CH_3)C{=}C(H)C_6H_5 \quad (81\%)$$

Alkynylaluminum Compounds

A number of high-yield procedures are available for preparing 1-alkynylaluminum compounds.[6] These involve either metalation of 1-alkynes with the appropriate aluminum hydrides or transmetalation between readily available metal acetylides and aluminum chloride. An operationally simple route to 1-alkynyldialkylalanes is through the reaction of *tert*-amine complexes of dialkylaluminum hydrides with terminal acetylenes (Eq. 10).[28]

$$RC{\equiv}CH + R'_2AlH{\cdot}NR''_3 \xrightarrow{25^\circ} RC{\equiv}CAlR'_2{\cdot}NR''_3 + H_2 \quad \text{(Eq. 10)}$$

The reaction is essentially quantitative and also provides a convenient method for the determination of active hydride in solutions containing dialkylaluminum hydrides. Uncomplexed 1-alkynyldialkylalanes can be obtained by metalation of 1-alkynes with triorganoalanes.[6,68]

$$n\text{-}C_6H_{13}C{\equiv}CH + (CH_3)_3Al \xrightarrow{85^\circ} n\text{-}C_6H_{13}C{\equiv}CAl(CH_3)_2$$

A convenient preparative method for 1-alkynylalanes involves the exchange reaction of lithium or sodium acetylides with dialkylaluminum chlorides.[69]

$$C_2H_5OC{\equiv}CLi + (C_2H_5)_2AlCl \xrightarrow[\text{toluene}]{\text{hexane}} C_2H_5OC{\equiv}CAl(C_2H_5)_2 + LiCl$$

This procedure is especially valuable when nonsolvent complexed alkynylalanes are required.

Trialkynylalanes are obtainable by reacting alkynyllithium or alkynylsodium reagents with aluminum chloride in a 3:1 ratio in ether or in a hydrocarbon solvent.[70]

$$3\,n\text{-}C_4H_9C{\equiv}CLi + AlCl_3 \xrightarrow{n\text{-hexane}} (n\text{-}C_4H_9C{\equiv}C)_3Al + 3\,LiCl$$

The reaction of 1-alkynes with alkali metal aluminum hydrides in the presence of pyridine or ether solvents leads to metalation of the triple bond. The reactivity of the hydrides decreases in the order: $KAlH_4 > NaAlH_4 > LiAlH_4$. Treatment of the resultant alkali metal tetraalkynylalanates with aluminum chloride generates the corresponding trialkynylalanes.[71]

$$4RC{\equiv}CH + MAlH_4 \xrightarrow{\text{ether}} (RC{\equiv}C)_4AlM + 4H_2$$

$$3(RC{\equiv}C)_4AlM + AlCl_3 \xrightarrow{\text{ether}} 4(RC{\equiv}C)_3Al\cdot O(C_2H_5)_2$$

SYNTHESES VIA ALKENYL- AND ALKYNYLALUMINUM COMPOUNDS

The direct synthesis of stereodefined alkenylaluminum compounds via hydroalumination and carboalumination of alkynes provides a convenient source of potential vinyl carbanions. The intrinsic nucleophilicity of the aluminum–carbon bond of alkenylalanes, in which the aluminum possesses an empty *p* orbital, is rather low. Although alkenylalanes react readily with a number of inorganic electrophiles, such as water and halogens, their reactions with carbon electrophiles is less satisfactory. On the other hand, treatment of the alkenylalanes with organolithium reagents converts the aluminum into species with a completed octet. The resultant lithium alkenylalanates are now reasonably good nucleophiles whose reaction characteristics resemble those of the corresponding organolithium and Grignard reagents.[8,72]

$$(R)(H)C{=}C(R)(AlR'_2) \xrightarrow{R''Li} \left[(R)(H)C{=}C(R)(AlR'_2R'')\right]Li$$

The utility of alkenylaluminum compounds as intermediates stems from the observation that in reactions with electrophilic reagents the intermolecular transfer of the alkenyl moiety proceeds in the majority of cases with retention of configuration. This property coupled with their ready availability from alkynes makes the alkenylaluminum compounds useful synthons for elaboration into substituted alkenes, dienes, and enynes of predictable stereochemistry.

The chemistry of alkynylaluminum compounds is less well delineated. It has been shown that they exhibit some unique synthetic capabilities in additions to enones and epoxides. Also, their reaction with tertiary alkyl halides provides a valuable tool for introducing sterically bulky alkyl substituents onto terminal acetylenes.[70]

The following survey of organic syntheses involving alkenyl- and alkynylaluminum compounds will emphasize mainly those reactions that are of preparative interest.

Olefins

Via Protonolysis. The reactions of alkenylalanes and alkenylalanates with a variety of protic reagents are rapid and quantitative and proceed stereospecifically with retention of configuration yielding the corresponding olefins. Thus hydroalumination–protonolysis of alkynes provides a noncatalytic method for hydrogenation of triple to double bonds. The *cis* hydroalumination–protonolysis of internal alkynes provides pure *cis* olefins, whereas *trans* hydroalumination–protonolysis furnishes *trans* olefins, as exemplified in Eqs. 11–13:

$$n\text{-}C_4H_9C{\equiv}CC_4H_9\text{-}n \xrightarrow[\text{2. } H_3O^+]{\text{1. } 2(i\text{-}C_4H_9)_2AlH,\ n\text{-hexane, 6 hr, } 50^\circ} (n\text{-}C_4H_9)HC{=}CH(C_4H_9\text{-}n)\ (cis)\quad (89\%)$$ (Eq. 11)[73]

$$\text{cyclododecyne} \xrightarrow[\text{2. } CH_3OH\text{–}H_2SO_4]{\text{1. } 2(i\text{-}C_4H_9)_2AlH,\ 2\text{ hr, } 45^\circ} \text{cyclododecene}\quad (60\%)$$ (Eq. 12)[74]

$$C_6H_5C{\equiv}CCH_3 \xrightarrow[\text{2. } H_3O^+]{\text{1. } LiAlH_4,\ THF,\ 13\text{ hr, } 66^\circ} (C_6H_5)HC{=}CH(CH_3)\ (trans)\quad (99\%)$$ (Eq. 13)[47,48]

Carboalumination of symmetrically disubstituted alkynes with trimethylaluminum-bis(cyclopentadienyl)zirconium dichloride followed by protonolysis leads to a stereospecific synthesis of trisubstituted olefins.[64]

$$n\text{-}C_4H_9C{\equiv}CC_4H_9\text{-}n \xrightarrow{(CH_3)_3Al\text{–}Cl_2Zr(C_5H_5)_2} (n\text{-}C_4H_9)(CH_3)C{=}C(C_4H_9\text{-}n)H\quad (89\%)$$

Hydroalumination or carboalumination of alkynes followed by deuterolysis of the organoalane intermediates permits the stereospecific preparation of a variety of deuterated alkenes, as exemplified in Eqs. 14–16:

$$n\text{-}C_4H_9C{\equiv}CH \xrightarrow[\text{2. } D_2O]{\text{1. } (i\text{-}C_4H_9)_2AlH} (n\text{-}C_4H_9)(H)C{=}C(H)(D)$$ (Eq. 14)[1,75]

$$C_6H_5C{\equiv}CH \xrightarrow[\text{2. } D_2O]{\text{1. } (CH_3)_3Al\text{-}Cl_2Zr(C_5H_5)_2} (C_6H_5)(CH_3)C{=}C(H)(D)$$ (Eq. 15)[64]

$$C_2H_5C{\equiv}CC_2H_5 \xrightarrow[\text{2. } D_2O]{\text{1. } (i\text{-}C_4H_9)_2AlH} (C_2H_5)(H)C{=}C(C_2H_5)(D)$$ (Eq. 16)[1]

It should be noted that the *cis* hydroalumination of terminal acetylenes produces, besides the alkenylalanes, small amounts of metalation and dihydroalumination products, which on protonolysis give the corresponding 1-alkynes and alkanes, respectively, as byproducts (Eq. 1., p. 378).

Alkenylsilanes of defined stereochemistry are proving to be remarkably versatile starting materials for use in a variety of syntheses. The hydroalumination–protonolysis of the readily accessible 1-alkynylsilanes provides an operationally simple route to *cis*-1-alkenylsilanes in high yields and isomeric purities (Eq. 17). The corresponding *trans*-1-alkenylsilanes can be obtained through bromine-catalyzed isomerization of the *Z* isomers (Eq. 18).[76]

$$C_6H_{11}C{\equiv}CSi(CH_3)_3 \xrightarrow[\text{2. } H_3O^+]{\text{1. } (i\text{-}C_4H_9)_2AlH,\ \text{ether}} (C_6H_{11})(H)C{=}C(Si(CH_3)_3)(H)$$ (Eq. 17)

$$(C_6H_{11})(H)C{=}C(Si(CH_3)_3)(H) \xrightarrow[\text{pyridine, } h\nu]{N\text{-Bromosuccinimide}} (C_6H_{11})(H)C{=}C(H)(Si(CH_3)_3)\ (88\%)$$ (Eq. 18)

Hydroalumination of 2-alkynylnitriles, 1-alkynyl sulfides, and 1-chloro-1-alkynes with lithium aluminum hydride in ether solvents affords the corresponding vinylalanates from which the *trans*-2-alkenylnitriles (Eq. 19),[52] *trans*-1-alkenyl sulfides (Eq. 20),[53] and *trans*-1-chloro-1-alkenes (Eq. 21),[55] respectively, are obtained on protonolysis. This appears to be one of the most convenient methods for the preparation of these functionally substituted alkenes.

$$C_6H_5C{\equiv}CCN \xrightarrow[\text{2. } H_3O^+]{\text{1. LiAlH}_4\text{, ether, } -60 \text{ to } 0^\circ} (C_6H_5)HC{=}CH(CN) \; (98\%) \quad \text{(Eq. 19)}$$

$$t\text{-}C_4H_9C{\equiv}CSCH_3 \xrightarrow[\text{2. } H_3O^+]{\text{1. LiAlH}_4\text{, THF, } 50\text{–}60^\circ} (t\text{-}C_4H_9)HC{=}CH(SCH_3) \; (92\%) \quad \text{(Eq. 20)}$$

$$n\text{-}C_6H_{13}C{\equiv}CCl \xrightarrow[\text{2. } H_3O^+]{\text{1. LiAlH}_4\text{, THF, } -30 \text{ to } 0^\circ} (n\text{-}C_6H_{13})HC{=}CH(Cl) \; (80\%) \quad \text{(Eq. 21)}$$

Via Cyclization. In triethylamine solvent, 1-hexen-5-yne reacts with diisobutylaluminum hydride to give the metalated alkyne and hydrogen. Treatment of this with 1.2 molar equivalents of diisobutylaluminum hydride produces, after hydrolysis, 3-methylcyclopentene (Eq. 22).[22] The precursor for the cyclic alane appears to be the 1,1-dialuminohexa-1,5-diene.

$$CH_2{=}CHCH_2CH_2C{\equiv}CAlR_2 \xrightarrow[50^\circ]{R_2AlH} (R_2Al)_2C{=}CHCH_2CH_2CH{=}CH_2 \longrightarrow \text{1-}(CH_2AlR_2)\text{-2-}(AlR_2)\text{cyclopentene} \xrightarrow{H_3O^+} \text{3-methylcyclopentene} \; (79\%) \quad \text{(Eq. 22)}$$

Interestingly, the hydroalumination of 1-hexen-5-yne with 2 molar equivalents of diisobutylaluminum hydride in diethyl ether followed by hydrolysis of the reaction mixture affords methylcyclopentane.[22] The precursor for methylcyclopentane appears to be the bis-aluminoalkene resulting from dihydroalumination of the triple bond, which then undergoes intramolecular aluminum–carbon bond addition to the double bond (Eq. 23).

$$CH_2{=}CHCH_2CH_2C{\equiv}CH \xrightarrow[\text{2 hr, 40°}]{R_2AlH} \left[\begin{array}{c} R_2Al\text{—}CH(CH_2CH_2CH{=}CH_2) \\ R_2Al\text{—}CH_2\text{—}(H)C \end{array} \right] \longrightarrow \text{2-}(R_2AlCH_2)\text{-cyclopentyl-}AlR_2 \xrightarrow{H_3O^+} \text{methylcyclopentane } (80\%) \quad \text{(Eq. 23)}$$

Via Coupling with Alkyl Halides, Alkyl Sulfonates, and Aryl Halides. Coupling of alkenylaluminum compounds derived from terminal acetylenes with alkyl halides should provide a stereoselective route to alkenes. In fact, treatment of alkenylalanates with primary alkyl halides and alkyl sulfonates affords *trans* olefins (Eqs. 24–25).[43,77,78] On the other hand, the corresponding alkenylalanes do not react readily with these alkylating agents. Although the yields of disubstituted olefins derived from alkenylalanates are modest, the procedure for their preparation is operationally simple and does not require the isolation and/or purification of stereodefined precursors as in the corresponding coupling reactions involving alkenyllithium reagents.

$$n\text{-}C_5H_{11}C{\equiv}CH \xrightarrow[\text{2. } n\text{-}C_4H_9Li\text{; 3. } CH_3I]{\text{1. } (i\text{-}C_4H_9)_2AlH} (E)\text{-}n\text{-}C_5H_{11}CH{=}CHCH_3 \ (65\%) \quad \text{(Eq. 24)}^{77}$$

$$n\text{-}C_4H_9C{\equiv}CH \xrightarrow[\text{2. } n\text{-}C_4H_9Li\text{; 3. } n\text{-}C_8H_{17}OSO_2CH_3]{\text{1. } (i\text{-}C_4H_9)_2AlH} (E)\text{-}n\text{-}C_4H_9CH{=}CHC_8H_{17}\text{-}n \ (41\%) \quad \text{(Eq. 25)}^{77}$$

Interestingly, alkenylation of the activated chloromethyl ethyl ether does not require prior conversion of the *trans*-1-alkenyldiisobutylalane into the ate complex. Treatment of the vinylalanes in *n*-hexane solvent with chloromethyl ethyl ether thus affords the corresponding (*E*)-allyl ethyl ethers in good yields.[79]

$$n\text{-}C_4H_9C{\equiv}CH \xrightarrow[\text{2. } ClCH_2OC_2H_5 \text{ 3. } H_3O^+]{\text{1. } (i\text{-}C_4H_9)_2AlH} (n\text{-}C_4H_9)(H)C{=}C(H)(CH_2OC_2H_5) \quad (80\%)$$

Use of the vinylalanates derived from carboalumination of 1-alkynes instead of the vinylalanes for the alkenylation of chloromethyl methyl ether furnishes the corresponding (*E*)-3-methyl-2-alkenyl methyl ethers in higher purities.[77,80]

$$n\text{-}C_5H_{11}C{\equiv}CH \xrightarrow[\text{2. } n\text{-}C_4H_9Li \text{ 3. } ClCH_2OCH_3 \text{ 4. } H_3O^+]{\text{1. } (CH_3)_3Al\text{-}Cl_2Zr(C_5H_5)_2} (n\text{-}C_5H_{11})(CH_3)C{=}C(H)(CH_2OCH_3) \quad (79\%)$$

The reductive alkylation of symmetrical disubstituted alkylacetylenes via reaction of the corresponding alkenylalanates with alkyl halides to form trisubstituted olefins has not been reported. However, on sequential treatment with diisobutylaluminum hydride, methyllithium, and methyl iodide, 1-alkynylsilanes afford regio- and stereodefined disubstituted vinylsilanes.[43,78]

$$n\text{-}C_6H_{13}C{\equiv}CSi(CH_3)_3 \xrightarrow[\text{2. } CH_3Li \text{ 3. } CH_3I]{\text{1. } (i\text{-}C_4H_9)_2AlH} (n\text{-}C_6H_{13})(H)C{=}C(Si(CH_3)_3)(CH_3)$$

Arylation of alkenylalanates is of little synthetic value. However, alkenylalanes can be activated toward reaction with aryl bromides and iodides by nickel and palladium complexes to form arylated *trans*-alkenes.[81] The reaction is catalytic with respect to the nickel and palladium reagents.

$$(n\text{-}C_4H_9)(H)C{=}C(H)(Al(C_4H_9\text{-}i)_2) + \text{1-bromonaphthalene} \xrightarrow{Ni[P(C_6H_5)_3]_4} (n\text{-}C_4H_9)(H)C{=}C(H)(\text{1-naphthyl}) \quad (73\%)$$

The nickel- or palladium-catalyzed reactions of 1,2-disubstituted alkenylalanes with aryl halides either fail to give the cross-coupled products or produce them in only low yields.[82] However, arylation of these alkenylalanes is significantly improved by addition of zinc chloride. In a typical reaction the *tetrakis*-(triphenylphosphine)palladium-catalyzed cross-coupling of (*E*)-3-hexenyldiisobutylalane with *m*-iodotoluene, which fails to give the desired product even after one week, is complete within one hour at room temperature in the presence of 1 molar equivalent of zinc chloride.[82]

$$(E)\text{-}C_2H_5CH{=}C(C_2H_5)Al(C_4H_9\text{-}i)_2 + m\text{-}IC_6H_4CH_3 \xrightarrow[ZnCl_2]{Pd[P(C_6H_5)_3]_4} (E)\text{-}C_2H_5CH{=}C(C_2H_5)C_6H_4CH_3\text{-}m \quad (88\%)$$

Functionally Substituted Olefins

Alkenyl Halides (via Halogenation). A variety of stereochemically pure alkenyl halides can be prepared from alkenylaluminum compounds. Thus halogenation of alkenylalanes derived from the *cis*[83,84] or *trans*[49,61,62] hydroalumination of triple bonds results in the preferential cleavage of the vinyl carbon–aluminum bond to afford the corresponding alkenyl halides. The reaction proceeds with retention and offers a simple one-pot conversion of alkynes to stereodefined alkenyl halides.

It should be noted that the alkenyl halides derived from the *cis* hydroalumination of 1-alkynes may contain small amounts of the corresponding 1-halo-1-alkynes resulting from halogenation of the alkynylalane byproducts. Since certain alkenyl halides have a tendency to isomerize, it is recommended that solutions containing such alkenyl halides be treated with a few crystals of 2,6-di-*tert*-butyl-*p*-cresol (BHT) prior to removal of the solvents. Also, the isolated compounds should be stored over a few crystals of BHT.

The preparation of alkenyl bromides entails treatment of the vinylalane contained in ether with either bromine in the presence of pyridine[83] or with *N*-bromosuccinimide (NBS).[85] The latter procedure is the method of choice when the alkenylalane contains additional double or triple bonds. Iodinolysis of alkenylalanes derived from the *cis*-hydroalumination of alkynes proceeds readily in tetrahydrofuran solvent. *trans*-1-Chloro-1-alkenes, however, are best prepared from 1-chloro-1-alkynes as shown in Eq. 7. Some representative examples of alkenyl bromide and alkenyl iodide preparations are depicted in Eqs. 26–28.

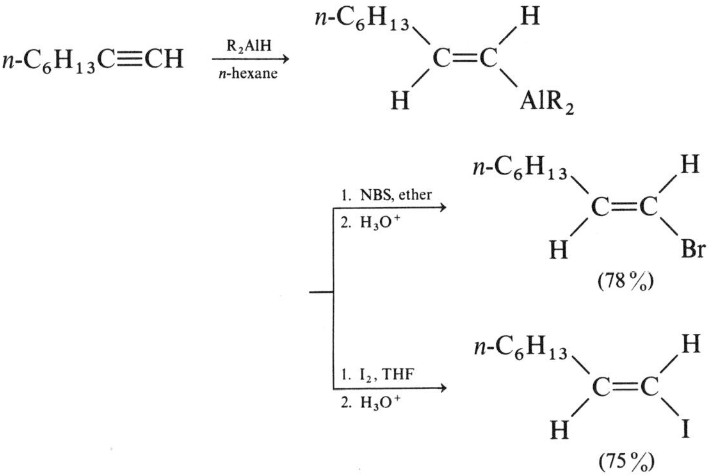

$$n\text{-}C_6H_{13}C{\equiv}CH \xrightarrow[n\text{-hexane}]{R_2AlH} (n\text{-}C_6H_{13})HC{=}CH(AlR_2)$$

$$\xrightarrow[\text{2. } H_3O^+]{\text{1. NBS, ether}} (n\text{-}C_6H_{13})HC{=}CHBr \quad (78\%)$$

$$\xrightarrow[\text{2. } H_3O^+]{\text{1. } I_2\text{, THF}} (n\text{-}C_6H_{13})HC{=}CHI \quad (75\%)$$

(Eq. 26)[85]

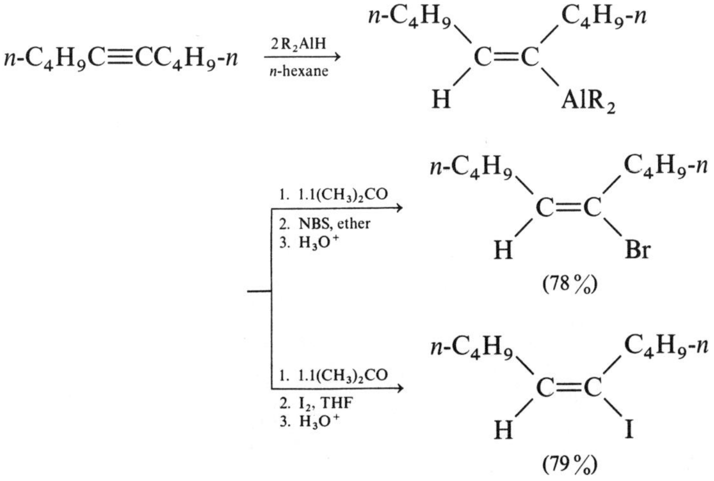

$$n\text{-}C_4H_9C{\equiv}CC_4H_9\text{-}n \xrightarrow[n\text{-hexane}]{2R_2AlH} (n\text{-}C_4H_9)HC{=}C(C_4H_9\text{-}n)(AlR_2)$$

$$\xrightarrow[\text{2. NBS, ether; 3. } H_3O^+]{\text{1. } 1.1(CH_3)_2CO} (n\text{-}C_4H_9)HC{=}C(C_4H_9\text{-}n)Br \quad (78\%)$$

$$\xrightarrow[\text{2. } I_2\text{, THF; 3. } H_3O^+]{\text{1. } 1.1(CH_3)_2CO} (n\text{-}C_4H_9)HC{=}C(C_4H_9\text{-}n)I \quad (79\%)$$

(Eq. 27)[85]

$$(C_2H_5)HC{=}C(C_2H_5){-}C(C_2H_5){=}C(C_2H_5)Al(C_4H_9\text{-}i)_2 \xrightarrow[\text{2. } H_3O^+]{\text{1. NBS, ether}} (C_2H_5)HC{=}C(C_2H_5){-}C(C_2H_5){=}C(C_2H_5)Br \quad (74\%)$$

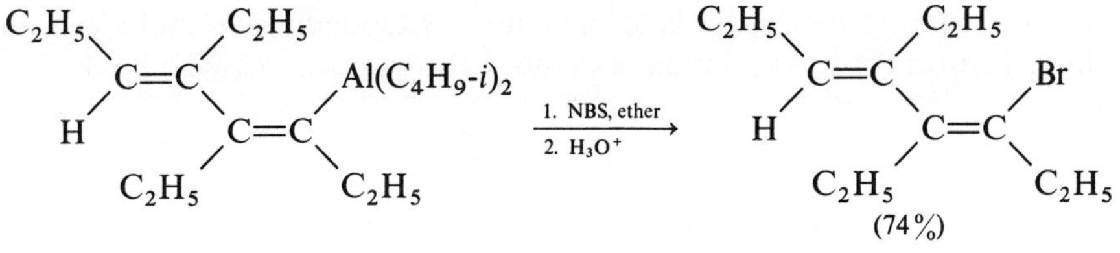

(Eq. 28)[85]

The *cis* hydroalumination–halogenation reaction when applied to alkynes containing hydroxyl or protected hydroxyl groups furnishes the corresponding alkenyl halides in only modest yields.[24]

$$n\text{-}C_5H_{11}CH(OH)C{\equiv}CH \xrightarrow[\text{2. }(i\text{-}C_4H_9)_2AlH\ \ \text{3. }I_2]{\text{1. }2(i\text{-}C_4H_9)_3Al} n\text{-}C_5H_{11}CH(OH)CH{=}CHI\ (47\%)$$

The trifunctional α-halovinylsilyl moiety of 1-halo-1-alkenylsilanes represents a uniquely constituted synthon for a variety of chemical transformations. An expeditious route to these compounds involves the halodealumination reaction of (*Z*)-1-alumino-1-alkenylsilanes derived by hydroalumination of 1-alkynylsilanes. Treatment of the α-silylalkenylalanes with *N*-chlorosuccinimide (NCS) (Eq. 29), bromine (Eq. 30), or iodine (Eq. 31) thus produces the corresponding (*E*)-1-halo-1-alkenylsilanes in high isomeric purities and yields.[44]

$$n\text{-}C_4H_9CH{=}C(Si(CH_3)_3)Al(C_4H_9\text{-}i)_2 \xrightarrow[\text{2. }H_3O^+]{\text{1. NCS}} n\text{-}C_4H_9CH{=}C(Si(CH_3)_3)Cl\ (84\%) \quad \text{(Eq. 29)}$$

$$\xrightarrow[\text{2. }H_3O^+]{\text{1. }Br_2\text{, pyridine}} n\text{-}C_4H_9CH{=}C(Si(CH_3)_3)Br\ (90\%) \quad \text{(Eq. 30)}$$

$$\xrightarrow[\text{2. }H_3O^+]{\text{1. }I_2} n\text{-}C_4H_9CH{=}C(Si(CH_3)_3)I\ (90\%) \quad \text{(Eq. 31)}$$

The conversion of dienylsilylalanes into the corresponding bromides can be achieved with a predried solution of cyanogen bromide in diethyl ether.[44]

$$\text{1-cyclohexenyl-}C{\equiv}CSi(CH_3)_3 \xrightarrow[\text{2. BrCN}]{\text{1. }(i\text{-}C_4H_9)_2AlH} \text{1-cyclohexenyl-}CH{=}C(Si(CH_3)_3)Br\ (96\%)$$

The [(*Z*)-1-chloro-, (*Z*)-1-bromo-, and (*Z*)-1-iodoalkenyl]silanes are also readily accessible by the reactions shown in Eqs. 32–34.

$$(Z)\text{-}C_6H_{11}CH{=}C(X)Si(CH_3)_3 \;(X = Cl, Br, I) \xrightarrow[h\nu]{Br_2} (Z)\text{-}C_6H_{11}CH{=}C(Cl)Si(CH_3)_3 \;(91\%) \quad \text{(Eq. 32)}^{44}$$

$$\xrightarrow[h\nu]{Br_2} (Z)\text{-}C_6H_{11}CH{=}C(Br)Si(CH_3)_3 \;(92\%) \quad \text{(Eq. 33)}^{44}$$

$$\xrightarrow[(5\text{ mol }\%)\; -65\text{ to }25^\circ]{t\text{-}C_4H_9Li} (Z)\text{-}C_6H_{11}CH{=}C(I)Si(CH_3)_3 \;(89\%) \quad \text{(Eq. 34)}^{86}$$

The α-haloalkenylsilanes not only provide a valuable starting point for stereospecific syntheses of trisubstituted olefins,[87] but their desilylation with methanolic sodium methoxide also offers a method for preparing either the (*E*)- or the corresponding (*Z*)-1-halo-1-alkenes from a single 1-alkynylsilane precursor. This procedure works not only with simple alkyl substituted 1-alkynylsilanes, but also with those containing phenyl or tetrahydropyranyl ether substituents.[46]

$$CH_3CH(OTHP)C{\equiv}CSi(CH_3)_3 \xrightarrow[\substack{2.\ ICl \\ 3.\ NaOCH_3\text{-}CH_3OH}]{\substack{1.\ (i\text{-}C_4H_9)_2AlH \\ n\text{-hexane–1.1 eq, THF}}} CH_3CH(OTHP)CH{=}CHI \;(68\%)$$

Iodination of (*E*)-2-methyl-1-alkenylalanes derived from zirconium-catalyzed carboalumination leads to β,β-dialkyl-substituted alkenyl iodides, which can be converted into a variety of terpenoid natural products (Eq. 35).[88] It is noteworthy

$$(CH_3)_2C{=}CH(CH_2)_2C{\equiv}CH \xrightarrow[2.\ I_2]{1.\ (CH_3)_3Al\text{–}Cl_2Zr(C_5H_5)_2} (CH_3)_2C{=}CH(CH_2)_2C(CH_3){=}CHI \;(72\%) \quad \text{(Eq. 35)}$$

that the carbometalation–iodination sequence is compatible with propargylic and homopropargylic alkynols (Eq. 36).[66]

$$HO(CH_2)_2C{\equiv}CH \xrightarrow[2.\ I_2]{1.\ (CH_3)_3Al\text{-}Cl_2Zr(C_5H_5)_2} (HO(CH_2)_2)(CH_3)C{=}C(H)(I)\ (62\%) \quad \text{(Eq. 36)}$$

The *trans* reduction of the triple bonds of alk-2-yn-1-ols with lithium aluminum hydride–sodium methoxide or diisobutylaluminum hydride followed by iodinolysis of the resultant vinylaluminum intermediates provides the basis for the stereoselective synthesis of functionally trisubstituted olefins (Eqs. 8 and 9).[61,62]

In contrast to the alkenyl halides, whose usefulness as synthons for preparing olefins and dienes has been clearly demonstrated, the 1,1-dihalo-1-ethenyl moiety has not yet played a major role in synthetic methodology. This is probably because most of the currently available methods for its synthesis are limited to the preparation of homo 1,1-dihaloolefins. The discovery that α-chloroalkenylalanates, derived from hydroalumination of 1-chloro-1-alkynes (Eq. 7, p. 392), undergo halogenolysis when treated with bromine or iodine monochloride has provided a simple route for preparing (*Z*)-1-bromo-1-chloro- and (*Z*)-1-iodo-1-chloro-1-alkenes, respectively (Eq. 37).[55] The (*E*)-1-bromo-1-chloro-1-alkenes

$$(C_6H_{11})(H)C{=}C(AlH_3Li)(Cl) \xrightarrow[(-H_2)]{(CH_3)_2CO\ (3.3\ eq)} (C_6H_{11})(H)C{=}C(Al(OC_3H_7\text{-}i)_3Li)(Cl)$$

$$\xrightarrow[-78^\circ]{Br_2} (C_6H_{11})(H)C{=}C(Br)(Cl)\ (87\%)$$

$$\xrightarrow[-30\ to\ 25^\circ]{ICl} (C_6H_{11})(H)C{=}C(I)(Cl)\ (89\%) \quad \text{(Eq. 37)}$$

can be obtained by either bromination–desilicobromination of (*E*)-α-chloroalkenylsilanes or chlorination–desilicochlorination of (*Z*)-α-bromoalkenylsilanes,[89] as typified in Eq. 38.

$$\begin{array}{c} C_6H_{11}(H)C{=}C(Cl)Si(CH_3)_3 \xrightarrow[-78 \text{ to } 0°]{Br_2} \\ C_6H_{11}(H)C{=}C(Br)Si(CH_3)_3 \xrightarrow[-60 \text{ to } 0°]{Cl_2} \end{array} \xrightarrow[CH_3OH]{NaOCH_3} C_6H_{11}(H)C{=}C(Cl)Br \qquad \text{(Eq. 38)}$$

An attractive, high-yield synthesis of β,β-dialkyl-substituted 1,1-diiodo-1-alkenes is the carbozirconation–iodination of 1-alkynylalanes.[40]

$$n\text{-}C_3H_7C{\equiv}CAl(CH_3)_2 \xrightarrow[2.\ I_2]{1.\ Cl(CH_3)Zr(C_5H_5)_2} n\text{-}C_3H_7(CH_3)C{=}CI_2 \quad (92\%)$$

Alkenyl Sulfides (via Sulfuridation). Treatment of alkenylalanates, derived from the *cis* or *trans* hydroalumination of mono- and disubstituted alkynes, respectively, with allyl thiosulfonates produces the corresponding isomerically pure allyl alkenyl sulfides.[90]

$$n\text{-}C_6H_{13}(H)C{=}C(H)Al(C_4H_9\text{-}i)_2(C_4H_9\text{-}n)Li \xrightarrow{CH_2{=}CHCH_2SSO_2C_6H_5} n\text{-}C_6H_{13}(H)C{=}C(H)SCH_2CH{=}CH_2 \quad (64\%)$$

$$H(n\text{-}C_3H_7)C{=}C(C_3H_7\text{-}n)Al(C_4H_9\text{-}i)_2(CH_3)Li \xrightarrow{CH_2{=}CHCH_2SSO_2C_6H_5} H(n\text{-}C_3H_7)C{=}C(C_3H_7\text{-}n)SCH_2CH{=}CH_2 \quad (51\%)$$

Allylic and Homoallylic Alcohols. The addition of the vinyl carbon–aluminum bond of *trans*-alkenylaluminum compounds to the carbonyl group of aldehydes and ketones proceeds with nearly complete retention, providing a stereoselective synthesis of *trans*-allylic alcohols.[84,91–93] Thus the one-carbon homologation of alkenylalanates with excess paraformaldehyde furnishes stereodefined 2-alken-1-ols (Eq. 39).[91] However, it should be noted that the reaction of alkenylalanes with paraformaldehyde produces the corresponding allylic carbinols in only modest yields.

$$n\text{-}C_4H_9C{\equiv}CH \xrightarrow[\text{2. } CH_3Li \quad \text{3. } 2(CH_2O)_n \quad \text{4. } H_3O^+]{\text{1. } (i\text{-}C_4H_9)_2AlH} (n\text{-}C_4H_9)(H)C{=}C(H)(CH_2OH) \quad (73\%) \qquad \text{(Eq. 39)}$$

The hydroxymethenylation of (*E*)-2-methylalkenylalanates offers a powerful tool for converting terminal acetylenes into (*E*)-3-methyl-2-alken-1-ols,[80,94,95] which contain a structural feature also found in a variety of natural products. This is illustrated by the conversion of 1-(3-butynyl)-2,6,6-trimethyl-1-cyclohexene into monocyclofarnesol.[94]

$$\text{(2,6,6-trimethylcyclohexenyl)}(CH_2)_2C{\equiv}CH \xrightarrow[\text{2. } n\text{-}C_4H_9Li \quad \text{3. } (CH_2O)_n \quad \text{4. } H_3O^+]{\text{1. } (CH_3)_3Al\text{–}Cl_2Zr(C_5H_5)_2} \text{(2,6,6-trimethylcyclohexenyl)}(CH_2)_2(CH_3)C{=}C(H)(CH_2OH) \quad (71\%)$$

Alkenylations of aldehydes other than paraformaldehyde do not require prior ate complex formation for achieving high yields of allylic alcohols.[92,93] The best results are obtained with straight-chain aldehydes and by adding the aldehydes to the alkenylalane contained in hydrocarbon[93] or ether[92] solvents.

$$n\text{-}C_4H_9C{\equiv}CH \xrightarrow[\text{2. } CH_3CHO,\ n\text{-hexane} \quad \text{3. } H_3O^+]{\text{1. } (i\text{-}C_4H_9)_2AlH} (n\text{-}C_4H_9)(H)C{=}C(H)(CHOHCH_3) \quad (67\%)$$

$$n\text{-}C_{13}H_{27}C{\equiv}CH \xrightarrow[\text{2. } C_6H_5CHO,\ \text{ether} \quad \text{3. } H_3O^+]{\text{1. } (i\text{-}C_4H_9)_2AlH} (n\text{-}C_{13}H_{27})(H)C{=}C(H)(CHOHC_6H_5) \quad (50\%)$$

An interesting application of the alkenylation of aldehydes is the conversion of aldehydes containing ester groups into vinyl lactones on sequential treatment with alkenylalanates followed by lactonization with trifluoroacetic acid.[96]

$n\text{-}C_5H_{11}(H)C{=}C(H)Al(C_4H_9\text{-}i)_2Li$ with CH_3 on Al

$\xrightarrow{OHC(CH_2)_3CO_2R}$ [intermediate: OR, OAl<, $C_5H_{11}\text{-}n$] $\xrightarrow{CF_3CO_2H}$ lactone with $C_5H_{11}\text{-}n$

The major byproducts in these reactions are the saturated alcohols resulting from reduction of the aldehydic carbonyl group by the isobutyl group of the organoalane.

$R(H)C{=}C(H)Al(i\text{-}C_4H_9)CH_2C(H)(CH_3)_2 + O{=}C(H)R' \longrightarrow R(H)C{=}C(H)Al(i\text{-}C_4H_9)OCH_2R' + (CH_3)_2C{=}CH_2$

The presence of an alkoxyl group in the resultant alkenylaluminum derivative decreases the reactivity of the vinyl carbon–aluminum bond toward further reaction with the aldehyde. With increasing steric requirements of the alkyl group attached to the aldehydic carbonyl group, reduction of the aldehyde becomes an increasingly important side reaction.[93]

Acyclic and cyclic ketones also react with alkenylalanes to afford tertiary alcohols of defined stereochemistry.[93] Again, reduction of the carbonyl group in hindered ketones competes seriously with alkenylation.

$n\text{-}C_4H_9C{\equiv}CH$ $\xrightarrow{1.\ (i\text{-}C_4H_9)_2AlH;\ 2.\ (CH_3)_2CO;\ 3.\ H_3O^+}$ $n\text{-}C_4H_9(H)C{=}C(H)COH(CH_3)_2$ (60%)

$n\text{-}C_4H_9C{\equiv}CH$ $\xrightarrow{1.\ (i\text{-}C_4H_9)_2AlH;\ 2.\ \text{cyclohexanone};\ 3.\ H_3O^+}$ $n\text{-}C_4H_9(H)C{=}C(H)$–1-hydroxycyclohexyl (HO) (60%)

Homoallylic alcohols are readily accessible by the reaction of alkenyl-aluminum compounds with not-too-hindered epoxides. Thus hydroxyethylation of *trans*-alkenyltrialkylalanates with ethylene oxide produces (*E*)-3-alken-1-ols.[23,97–99]

$$n\text{-}C_6H_{13}C\equiv CH \xrightarrow[\text{2. } CH_3Li;\ \text{3. } CH_2\text{—}CH_2\ (\text{O});\ \text{4. } H_3O^+]{\text{1. } (i\text{-}C_4H_9)_2AlH} (E)\text{-}n\text{-}C_6H_{13}CH{=}CH(CH_2)_2OH\ (81\%)$$

The reaction of *trans*-alkenylalanates with propylene oxide is regioselective, affording β-hydroxy-substituted alkenes[98] that are readily converted into the corresponding β,γ-unsaturated *trans*-alkenones.

$$C_6H_{11}C\equiv CH \xrightarrow[\text{2. } n\text{-}C_4H_9Li;\ \text{3. } CH_2\text{—}CHCH_3\ (\text{O});\ \text{4. } H_3O^+]{\text{1. } (i\text{-}C_4H_9)_2AlH} (E)\text{-}C_6H_{11}CH{=}CHCH_2CHOHCH_3\ (77\%)$$

Alkenylation of (*R*)-propylene oxide with the alkenylalanate derived from 1-nonen-8-yne proceeds with 99% *E*-stereoselectivity and with complete conservation of the chiral center.[23]

$$(E)\text{-}CH_2{=}CH(CH_2)_5CH{=}CH\text{—}Al(C_4H_9\text{-}i)_2Li\ (C_4H_9\text{-}n) \xrightarrow[\text{2. } H_3O^+]{\text{1. } (R)\text{-propylene oxide}} (E)\text{-}CH_2{=}CH(CH_2)_5CH{=}CH\text{—}CH_2\text{—}C(H)(CH_3)OH$$

On the other hand, styrene oxide gives an approximately 1:1 mixture of two possible regioisomers, and cyclohexene oxide fails to react with *trans*-alkenyl-alanates at a reasonable rate.[98]

A combination of the zirconium-catalyzed carboalumination of acetylenes and the reaction of alkenylalanates with epoxides provides (*E*)-4-methyl-3-alken-1-ols, which are useful intermediates for syntheses of terpenoids.[100]

$$(CH_3)_2C{=}CH(CH_2)_2(CH_3)C{=}CH{-}Al(CH_3)_2 \xrightarrow[2.\ CH_2{-}CHCH_3\ (\text{epoxide})]{1.\ n\text{-}C_4H_9Li} (CH_3)_2C{=}CH(CH_2)_2(CH_3)C{=}CH{-}CH_2CHOHCH_3\ (72\%)$$

It is noteworthy that when the alkenylalanes are employed instead of corresponding alkenylalanates the alkenols are produced in considerably lower yields.[98]

The reaction of *cis*-alkenylalanes with ethylene oxide or propylene oxide produces the corresponding *cis*-3-alken-1-ols, albeit in modest yields.[101]

$$C_2H_5(H)C{=}C(H)Al(C_2H_5)_2 \xrightarrow[\text{THF-benzene}]{CH_2{-}CH_2\ (\text{epoxide})} C_2H_5(H)C{=}C(H)(CH_2)_2OH$$

It is conceivable that conversion of the *cis*-alkenylalanes into the alanates prior to their reaction with epoxides might result in improved yields and purities of the homoallylic alcohols.

α,β-Unsaturated Acids, Esters, and Nitriles. The carbonation of alkenylaluminum compounds derived from *cis* or *trans* hydroalumination of alkynes is of considerable synthetic importance.[84,91,102] The reaction with carbon dioxide occurs preferentially at the vinyl carbon–aluminum bond and proceeds with retention of configuration, producing isomerically pure α,β-unsaturated acids.

Direct carbonation of *trans*-1-alkenyldiisobutylalanes derived from hydroalumination of terminal acetylenes in a hydrocarbon solvent affords *trans*-alkenoic acids only in modest yields (35–45%). However, conversion of the vinylalanes into the corresponding ate complexes prior to carbonation produces the (*E*)-α,β-unsaturated acids in good yields.[91,103]

$$n\text{-}C_4H_9C{\equiv}CH \xrightarrow[2.\ CH_3Li\quad 3.\ CO_2\quad 4.\ H_3O^+]{1.\ (i\text{-}C_4H_9)_2AlH} n\text{-}C_4H_9(H)C{=}C(H)CO_2H\ (78\%)$$

$$HO(CH_2)_7C{\equiv}CH \xrightarrow{(i\text{-}C_4H_9)_3Al} (i\text{-}C_4H_9)_2AlO(CH_2)_7C{\equiv}CH$$

$$\xrightarrow[2.\ CH_3Li\quad 3.\ CO_2\quad 4.\ H_3O^+]{1.\ (i\text{-}C_4H_9)_2AlH} HO(CH_2)_7(H)C{=}C(H)CO_2H\ (80\%)$$

Extension of the carbonation to β-methyl-substituted alkenylalanates derived from carboalumination of 1-alkynes provides a stereoselective synthesis of disubstituted alkenoic acids.[80]

$$n\text{-}C_5H_{11}C{\equiv}CH \xrightarrow[\substack{2.\ n\text{-}C_4H_9Li \\ 3.\ CO_2 \\ 4.\ H_3O^+}]{1.\ (CH_3)_3Al\text{-}Cl_2Zr(C_5H_5)_2} (n\text{-}C_5H_{11})(CH_3)C{=}C(H)(CO_2H) \quad (64\%)$$

Interestingly, the direct carbonation of alkenylalanes derived from monohydroalumination of disubstituted acetylenes proceeds quite readily to give the corresponding stereodefined alkenoic acids.[102,104]

$$C_6H_5C{\equiv}CC_6H_5 \xrightarrow[\substack{2.\ CO_2 \\ 3.\ H_3O^+}]{1.\ (i\text{-}C_4H_9)_2AlH} (C_6H_5)(H)C{=}C(C_6H_5)(CO_2H) \quad (90\%)$$

$$t\text{-}C_4H_9C{\equiv}CSi(CH_3)_3 \xrightarrow[\substack{2.\ CO_2,\ 60^\circ,\ heptane \\ 3.\ H_3O^+}]{1.\ (i\text{-}C_4H_9)_2AlH,\ ether} (t\text{-}C_4H_9)(H)C{=}C(Si(CH_3)_3)(CO_2H) \quad (85\%)$$

The (Z)-2-(trimethylsilyl)-2-butenoic acid obtained from sequential hydroalumination–carbonation of (1-propynyl)trimethylsilane has been used as starting material for the preparation of (trimethylsilyl)vinylketene, a stable vinylketene.[105]

$$CH_3C{\equiv}CSi(CH_3)_3 \xrightarrow[\substack{2.\ CH_3Li \\ 3.\ CO_2 \\ 4.\ H_3O^+}]{1.\ (i\text{-}C_4H_9)_2AlH} (CH_3)(H)C{=}C(Si(CH_3)_3)(CO_2H)$$

$$\xrightarrow[2.\ ClOCCOCl]{1.\ KH} (CH_3)(H)C{=}C(Si(CH_3)_3)(COCl) \xrightarrow{(C_2H_5)_3N} (CH_3)_3Si(CH_2{=}CH)C{=}C{=}O$$

Finally, it is possible to prepare either (E)- or (Z)-alkenoic acids from a common precursor by a slight modification in the experimental procedure.[49] Thus *cis* hydroalumination of 2-butyne with diisobutylaluminum hydride followed by sequential treatment of the resultant vinylalane with methyllithium and carbon dioxide affords tiglic acid (Eq. 40). On the other hand, carbonation of the *trans*-vinylalanate derived from *trans* hydroalumination of 2-butyne with lithium diisobutylmethylaluminum hydride yields angelic acid (Eq. 41).

$$CH_3C{\equiv}CCH_3 \xrightarrow[\substack{2.\ CH_3Li\\ 3.\ CO_2\\ 4.\ H_3O^+}]{1.\ (i\text{-}C_4H_9)_2AlH} (CH_3)(H)C{=}C(CH_3)(CO_2H) \quad \text{(Eq. 40)}$$

$$CH_3C{\equiv}CCH_3 \xrightarrow[\substack{2.\ CO_2\\ 3.\ H_3O^+}]{1.\ Li[(i\text{-}C_4H_9)_2(CH_3)AlH]} (CH_3)(H)C{=}C(CO_2H)(CH_3) \quad \text{(Eq. 41)}$$

trans-α,β-Unsaturated esters are readily accessible by hydroalumination of 1-alkynes followed by treatment of the resultant vinylalanes with the appropriate chloroformates.[80,106] This reaction does not require the intermediacy of the corresponding ate complexes.

$$n\text{-}C_6H_{13}C{\equiv}CH \xrightarrow[\substack{2.\ ClCO_2CH_3\\ 3.\ H_3O^+}]{1.\ (i\text{-}C_4H_9)_2AlH} (n\text{-}C_6H_{13})(H)C{=}C(H)(CO_2CH_3) \quad (65\%)$$

$$(CH_3)_2C{=}CH(CH_2)_2C{\equiv}CH \xrightarrow[\substack{2.\ ClCO_2C_2H_5\\ 3.\ H_3O^+}]{1.\ (CH_3)_3Al\text{-}Cl_2Zr(C_5H_5)_2} ((CH_3)_2C{=}CH(CH_2)_2)(CH_3)C{=}C(H)(CO_2C_2H_5) \quad (85\%)$$

It is noteworthy that the unsaturated esters are stable toward reduction both by unreacted vinylalanes and by the diisobutylaluminum chloride that is formed in the course of reaction. Extension of the alkoxycarbonylation to 1,2-disubstituted acetylenes such as 3-hexyne affords the corresponding esters in only modest yields.[106]

Vinylalanates derived from mono- and disubstituted alkynes react with cyanogen to afford α,β-unsaturated nitriles.[107]

$$C_2H_5C{\equiv}CC_2H_5 \xrightarrow[\substack{2.\ CH_3Li\\ 3.\ (CN)_2\\ 4.\ H_3O^+}]{1.\ (i\text{-}C_4H_9)_2AlH} (C_2H_5)(H)C{=}C(C_2H_5)(CN) \quad (76\%)$$

An alternative, operationally more convenient approach to monosubstituted *trans*-2-alkenenitriles is through *trans* hydroalumination–protonolysis of 2-alkynylnitriles (Eq. 19).[52]

Dienes

1,3-Dienes. Disubstituted alkynes undergo stereoselective vinylmetalation when treated with diisobutylaluminum hydride in a 2:1 ratio. The reaction involves addition of the vinyl carbon–aluminum groups of the initially formed alkenylalane to the unreacted alkyne to produce the dienylalane in high yield.[2,13,107] The dienylalane thus obtained can be further elaborated into stereodefined 1,3-dienes.

$$C_2H_5C{\equiv}CC_2H_5 + (i\text{-}C_4H_9)_2AlH \longrightarrow (C_2H_5)(H)C{=}C(C_2H_5)Al(C_4H_9\text{-}i)_2 \xrightarrow[70^\circ]{C_2H_5C{\equiv}CC_2H_5}$$

$$(C_2H_5)(H)C{=}C(C_2H_5){-}C(C_2H_5){=}C(C_2H_5)Al(C_4H_9\text{-}i)_2$$

$$\xrightarrow{H_3O^+} (C_2H_5)(H)C{=}C(C_2H_5){-}C(C_2H_5){=}C(H)(C_2H_5) \quad (75\%)$$

$$\xrightarrow{1.\ CH_3Li;\ 2.\ (CN)_2;\ 3.\ H_3O^+} (C_2H_5)(H)C{=}C(C_2H_5){-}C(C_2H_5){=}C(CN)(C_2H_5) \quad (63\%)$$

The dimerization reaction of 1-phenylpropyne has been found to proceed in a regioselective manner.[37]

$$C_6H_5C{\equiv}CCH_3 \xrightarrow[2.\ C_6H_5C{\equiv}CCH_3]{1.\ (i\text{-}C_4H_9)_2AlH} (C_6H_5)(H)C{=}C(CH_3){-}C(CH_3){=}C(C_6H_5)Al(C_4H_9\text{-}i)_2$$

$$\xrightarrow{H_3O^+} (C_6H_5)(H)C{=}C(CH_3){-}C(CH_3){=}C(H)(C_6H_5)$$

Disubstituted alkynes when heated at 140–150° in the presence of a small amount of diisobutylaluminum hydride undergo trimerization to give high yields of fully substituted benzenes.[2,13,37]

$$C_2H_5C{\equiv}CC_2H_5 \xrightarrow[48\text{ hr, }140°]{(i\text{-}C_4H_9)_2AlH} C_6(C_2H_5)_6 \quad (70\%)$$

The vinylmetalation reaction cannot be applied to 1-alkynes because of competing metalation of the acetylenes. Likewise, attempts to add the vinylalane derived from 3-hexyne and diisobutylaluminum hydride to 1-hexyne afford *cis*-3-hexene and diisobutyl(1-hexyn-1-yl)alane.[108] However, alkenylalanes derived from 1-alkynes are converted into *trans,trans*-1,3-dienes in the presence of copper(I) chloride in tetrahydrofuran.[108]

$$n\text{-}C_4H_9C{\equiv}CH \xrightarrow[\text{2. CuCl, THF; 3. }H_3O^+]{\text{1. }(i\text{-}C_4H_9)_2AlH} n\text{-}C_4H_9CH{=}CH{-}CH{=}CH{-}C_4H_9\text{-}n \quad (73\%)$$

This homocoupling reaction with cuprous chloride can be extended to vinylalanes derived from symmetrically disubstituted acetylenes, thus offering a convenient route for the preparation of symmetrical (*E*, *E*)-1,3-dienes.

A highly efficient and synthetically valuable procedure for the selective coupling of unlike groups is provided by the reaction of (*E*)-alkenylalanes with (*E*)- and (*Z*)-alkenyl halides in the presence of palladium or nickel complexes.[109] This cross-coupling reaction enables the synthesis of conjugated (*E*,*E*)- and (*E*,*Z*)-dienes in a chemo-, regio-, and stereoselective manner.

$$n\text{-}C_4H_9CH{=}CH{-}Al(C_4H_9\text{-}i)_2 + BrCH{=}C(CH_3)CO_2CH_3 \xrightarrow[\text{catalyst}]{\text{Pd}} n\text{-}C_4H_9CH{=}CH{-}CH{=}C(CH_3)CO_2CH_3$$

The reaction tolerates the presence of an ester group. However, low yields of dienes are obtained when the alkyl substituent of the alkenylalanes is sterically demanding. An important feature of the cross-coupling reaction is that it can be applied to alkenylalanes derived from the zirconium-catalyzed carboalumination of alkynes, thus providing a stereoselective, one-pot procedure for conversion of terminal alkynes into conjugated dienes.

$$(CH_3)_2C{=}CH(CH_2)_2C{\equiv}CH \xrightarrow[\substack{2.\ BrCH{=}CH_2,\ ZnCl_2,\\ Cl_2Pd[P(C_6H_5)_3]_2\\ +\ 2(i\text{-}C_4H_9)_2AlH}]{1.\ (CH_3)_3Al\text{-}Cl_2Zr(C_5H_5)_2} (CH_3)_2C{=}CH(CH_2)_2(CH_3)C{=}C(H)CH{=}CH_2$$

1,4-Dienes. The reaction of alkenylaluminum compounds with allylic or benzylic halides offers a synthetically attractive route to 1,4-dienes of predictable stereochemistry. Thus treatment of alkenylalanates with allyl bromide furnishes the corresponding 1,4-dienes (Eqs. 42–43).[43,77,78]

$$C_6H_5C{\equiv}CH \xrightarrow[\substack{2.\ n\text{-}C_4H_9Li\\ 3.\ BrCH_2CH{=}CH_2\\ 4.\ H_3O^+}]{1.\ (i\text{-}C_4H_9)_2AlH} C_6H_5(H)C{=}C(H)CH_2CH{=}CH_2\ (51\%) \qquad \text{(Eq. 42)}^{77}$$

$$n\text{-}C_6H_{13}C{\equiv}CSi(CH_3)_3 \xrightarrow[\substack{2.\ CH_3Li\\ 3.\ BrCH_2CH{=}CH_2}]{1.\ (i\text{-}C_4H_9)_2AlH} n\text{-}C_6H_{13}(H)C{=}C(Si(CH_3)_3)CH_2CH{=}CH_2\ (68\%) \qquad \text{(Eq. 43)}^{78}$$

On the other hand, alkenylalanes themselves fail to react with allyl bromide under these conditions. However, in the presence of copper(I) chloride or copper(I) iodide, alkenylalanes derived from mono- and disubstituted alkynes couple readily with certain allylic halides to yield isomerically pure (*E*)-1,4-dienes (Eq. 44).[110]

$$n\text{-}C_4H_9C{\equiv}CH \xrightarrow[\substack{2.\ \text{3-bromocyclohexene},\ CuI\\ THF}]{1.\ (i\text{-}C_4H_9)_2AlH} n\text{-}C_4H_9(H)C{=}C(H)\text{-(2-cyclohexenyl)}\ (67\%) \qquad \text{(Eq. 44)}$$

The scope of the 1,4-diene syntheses shown in Eqs. 42–44 is confined to allylic halides that can give only one product.

A highly selective synthesis of 1,4-dienes is the palladium-catalyzed cross-coupling of allylic halides or acetates with alkenylalanes derived by carbo-alumination of alkynes.[111] The reaction not only proceeds with essentially complete retention of the stereo- and regiochemistry of both the alkenyl and alkyl groups, but also is not accompanied by the formation of homocoupled products, as shown for the synthesis of α-farnesene.[111]

$$CH_2{=}CHC{\equiv}CH \xrightarrow{(CH_3)_3Al\text{-}Cl_2Zr(C_5H_5)_2} (CH_2{=}CH)(CH_3)C{=}C(H)(Al(CH_3)_2)$$

$$\xrightarrow[\text{THF}]{\text{Cl-geranyl},\ Pd[P(C_6H_5)_3]_4} \alpha\text{-farnesene}\ (86\%)$$

Enynes

Trans hydroalumination of disubstituted 1,3-diynes with lithium diisobutyl-methylaluminum hydride followed by protonolysis provides an operationally simple route to *trans*-enynes.[51]

$$C_6H_{11}C{\equiv}CC{\equiv}CC_6H_{11} \xrightarrow[2.\ H_3O^+]{1.\ Li[AlH(C_4H_9\text{-}i)_2CH_3]} C_6H_{11}C{\equiv}C{-}C(H){=}C(H)C_6H_{11}\quad (91\%)$$

The procedure is confined to the preparation of symmetrically substituted *trans*-enynes since the hydroaluminating agent does not discriminate between the triple bonds of unsymmetrically substituted conjugated diynes.[51]

A highly efficient method for preparing 3-en-1-ynes involves the regioselective reduction of 1-trimethylsilyl-1,3-diynes with lithium trialkylaluminum hydrides (Eq. 6). Depending on the mode of workup, the organoalanates formed can be converted either into enynes or into trimethylsilyl-protected enynes.[45]

$$C_6H_{11}C{\equiv}CC{\equiv}CSi(CH_3)_3 \xrightarrow[2.\ H_3O^+;\ 3.\ KF\cdot 2H_2O/DMF]{1.\ Li[AlH(C_4H_9\text{-}i)_2(C_4H_9\text{-}n)]} (C_6H_{11})(H)C{=}C(H)(C{\equiv}CH)$$

$$C_6H_{11}C{\equiv}CC{\equiv}CSi(CH_3)_3 \xrightarrow[2.\ H_3O^+]{1.\ Li[AlH(C_4H_9\text{-}i)_2(C_4H_9\text{-}n)]} (C_6H_{11})(H)C{=}C(H)(C{\equiv}CSi(CH_3)_3)$$

Alkenylalanes derived from the *cis* hydroalumination of 1-alkynes couple with 1-iodo-1-alkynes in the presence of nickel or palladium complexes and zinc chloride to produce unsymmetrically substituted conjugated enynes.[82]

$$n\text{-}C_5H_{11}C{\equiv}CH \xrightarrow[\text{2. } n\text{-}C_4H_9C{\equiv}CI,\ Ni^0\text{- or } Pd^0\text{-complex, } ZnCl_2]{\text{1. } (i\text{-}C_4H_9)_2AlH} (n\text{-}C_5H_{11})(H)C{=}C(H)(C{\equiv}CC_4H_9\text{-}n) \quad (92\%)$$

When employed in concert with carboalumination, this cross-coupling reaction permits the conversion of 1-alkynes into enynes containing a methyl substituent in the vinyl portion of the molecule.[82]

$$n\text{-}C_5H_{11}C{\equiv}CH \xrightarrow[\text{2. } n\text{-}C_4H_9C{\equiv}CI,\ Ni^0\text{- or } Pd^0\text{-complex, } ZnCl_2]{\text{1. } (CH_3)_3Al\text{–}Cl_2Zr(C_5H_5)_2} (n\text{-}C_5H_{11})(CH_3)C{=}C(H)(C{\equiv}CC_4H_9\text{-}n) \quad (90\%)$$

Finally, the reaction of alkenylalanates with propargyl bromide does not lead to 1,4-enynes but instead affords ene-allenes.[77]

$$n\text{-}C_4H_9C{\equiv}CH \xrightarrow[\text{2. } n\text{-}C_4H_9Li \quad \text{3. } BrCH_2C{\equiv}CH \quad \text{4. } H_3O^+]{\text{1. } (i\text{-}C_4H_9)_2AlH} (n\text{-}C_4H_9)(H)C{=}C(H)(CH{=}C{=}CH_2)$$

Acetylenes and β-Hydroxyacetylenes

The use of 1-alkynylaluminum compounds in organic synthesis is confined mainly to specific chemical transformations when the more readily available alkali metal acetylides fail to achieve the desired transformations. Thus alkynylalanes have proved to be valuable reagents for coupling tertiary alkyl–alkynyl groups, for opening of epoxides, and in conjugate additions to α,β-unsaturated carbonyl compounds.

Introduction of a quaternary carbon adjacent to an alkyne group cannot be achieved by the reaction of tertiary alkyl halides with alkynylmetals containing lithium, magnesium, or certain transition metals since elimination and other side reactions are predominant. However, trialkynylalanes derived from alkynyllithiums and anhydrous aluminum chloride undergo a remarkably clean reaction with tertiary alkyl bromides, chlorides, and sulfonates to produce cross-coupled products in high yields.[70]

$$(n\text{-}C_4H_9C{\equiv}C)_3{-}Al + \text{1-chloro-1-methylcyclohexane} \xrightarrow{ClCH_2CH_2Cl} \text{1-methyl-1-}(C{\equiv}CC_4H_9\text{-}n)\text{cyclohexane} \quad (71\%)$$

It should be noted that two of the three alkynyl groups on aluminum are not utilized, although they can be recovered nearly quantitatively if the reaction mixture is worked up soon after completion of the desired coupling. Also, secondary and primary alkyl halides do not readily couple with trialkynylalanes.[70]

Alkynylations of epoxides to produce β-hydroxyacetylenes is synthetically important. Unfortunately, the reaction of epoxides with alkali metal acetylides suffers from low yields even with substrates with moderate steric requirements. However, alkynylalanes open oxidocycloalkanes with a high degree of stereo-, regio-, and chemoselectivity. This has stimulated their applications in natural product research, especially in prostanoid syntheses.[112–114] The opening of unsubstituted cyclic epoxides by 1-alkynyldialkylalanes occurs in a stereoelectronically specific *trans*-diaxial manner.[69,115]

$$n\text{-}C_6H_{13}C{\equiv}CAl(C_2H_5)_2 + \text{cyclopentene oxide} \longrightarrow \textit{trans}\text{-2-}(C{\equiv}CC_6H_{13}\text{-}n)\text{cyclopentanol} \quad (77\%)$$

$$C_2H_5OC{\equiv}CAl(C_2H_5)_2 + \text{epoxy-decalin (H, H)} \longrightarrow C_2H_5OC{\equiv}C\text{-, HO-decalin (H, H)} \quad (80\%)$$

It should be noted that the ethoxyethynylalane represents an organometallic equivalent of $^-CH_2CO_2H$.[69,116]

trans-2-Ethynyl-substituted alcohols are accessible by treatment of epoxides with ether-stabilized dimethylethynylalane.[117]

$$\text{cyclohexene oxide} + HC{\equiv}CAl(CH_3)_2{\cdot}O(C_2H_5)_2 \longrightarrow \textit{trans}\text{-2-}(C{\equiv}CH)\text{cyclohexanol}$$

The regiospecificity of the alkynylalane–epoxide reaction is dependent on the composition of the organoalane reagent and on the presence of a suitably placed carbinol function elsewhere in the molecule.[113]

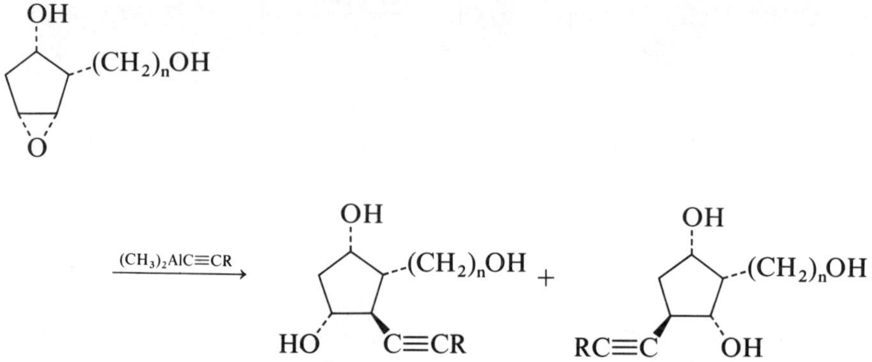

$$\xrightarrow{(CH_3)_2AlC{\equiv}CR}$$

$R = CHC_5H_{11}$-*n* (with OC_4H_9-*t* on the CH)

n	trans-alkynyl product	other regioisomer
2	60%	0%
3	50%	0%
7	10%	50%

In reactions of α-hydroxy epoxides and α-trimethylsilyloxy epoxides with diethylethoxyethynylalane, the direction of ring opening depends largely on the stereochemical relationship of the oxy function and the epoxide, and only slightly on the nature of the group on oxygen: H or $Si(CH_3)_3$.[118] Thus the stereochemistry of the product in Eq. 45 can be accounted for in terms of a *trans*-diaxial opening of the epoxide, assuming a preferential equatorial conformation of the neighboring oxygen substituent.[118]

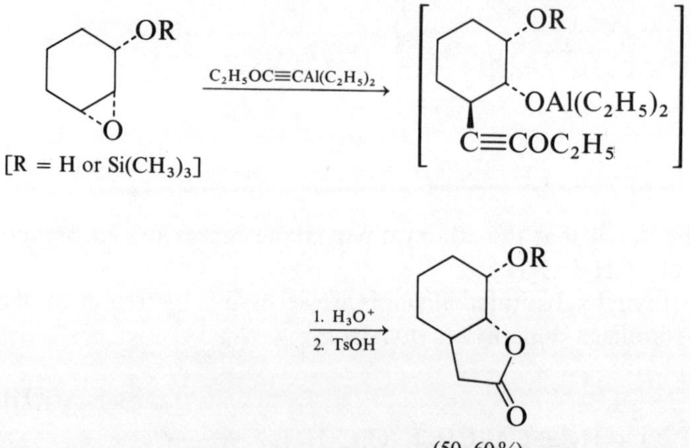

[R = H or $Si(CH_3)_3$]

$$\xrightarrow{C_2H_5OC{\equiv}CAl(C_2H_5)_2}$$

$OAl(C_2H_5)_2$, $C{\equiv}COC_2H_5$

$$\xrightarrow[2.\ TsOH]{1.\ H_3O^+}$$

(50–60%)

(Eq. 45)

Interestingly, treatment of 3,4-epoxycyclopentene with 1-hexynyldiethylalane gives two different products depending on the solvent used.[119]

n-$C_4H_9C{\equiv}CAl(C_2H_5)_2$

THF–toluene, −20° → $C{\equiv}CC_4H_9$-n, OH (53%)

toluene → [cyclopentenone] → OH, $C{\equiv}CC_4H_9$-n (53%)

These results can be explained by the strong affinity of aluminum for oxygen. In the absence of an ether solvent the organoaluminum reagent interacts with the epoxide oxygen and causes a highly selective rearrangement of the epoxide to the unsaturated ketone, which then reacts further with the alkynylalane.[119]

Finally, the reaction of epoxides with 1-alkynylalanes tolerates the presence of an ester group as illustrated by the conversion of ethyl *trans*-2,3-epoxybutanoate into the β-hydroxy acetylenic ester.[120]

OC_2H_5, O=C, H, C—C, H, O, CH_3 $\xrightarrow{CH_3C{\equiv}CAl(C_2H_5)_2}$ C_2H_5O, OH, $C{\equiv}CCH_3$, O, CH_3 (69%)

Alkenyl and Alkenyl Group Transfer to Conjugated Enones

The β-vinylation of conjugated enones via organocopper ate complexes provides a highly useful procedure for stereospecifically introducing alkenyl moieties onto acyclic and cyclic substrates.[121] Unfortunately, the preparation of the alkenylcopper reagents is somewhat tedious in that it entails the intermediacy of stereodefined alkenyllithium reagents. Since alkenylalanes of defined stereochemistry are readily available directly from acetylenes, the use of these organometallic reagents in 1,4 additions to conjugated enones is frequently the method of choice.

Thus a chemo- and stereospecific 1,4 delivery of the olefinic moiety is observed in reactions of *trans*-alkenyldiisobutylalanes with acyclic enones.[122]

$$(E)\text{-}n\text{-}C_4H_9CH{=}CHAl(C_4H_9\text{-}i)_2 + C_6H_5CH{=}CHCOCH_3 \longrightarrow (E)\text{-}n\text{-}C_4H_9CH{=}CHCH(C_6H_5)CH_2COCH_3 \; (67\%)$$

However, extension of the reaction to (*S*)-*trans* cyclic enones such as cyclohexenone[122] and (6-carbethoxyhexyl)cyclopent-2-en-1-one[123] affords either complex mixtures of products or products derived from 1,2 addition to the keto group.

Certain transoid enones containing a hydroxy group at a suitable position react with alkenylalanes *via* 1,4 addition.[123] The stereochemistry of the products strongly suggests that the hydroxy group participates in this reaction.

$$(E)\text{-}n\text{-}C_4H_9C(CH_3)(OSi(CH_3)_3)CH_2CH{=}CHAl(C_4H_9\text{-}i)_2 + \text{4-hydroxy-2-}[(CH_2)_2CO_2CH_3]\text{cyclopent-2-en-1-one} \longrightarrow$$

$$\text{3-hydroxy-2-}[(E)\text{-}CH{=}CHCH_2C(CH_3)(OSi(CH_3)_3)C_4H_9\text{-}n]\text{-1-oxo-cyclopentane-}(CH_2)_2CO_2CH_3 \; (35\%)$$

Interestingly, a chemo- and stereospecific conjugate addition to the transoid enone (6-carbethoxyhexyl)cyclopent-2-en-1-one takes place when using the alkenylalanate instead of the corresponding alkenylalane.[124]

Conjugate addition of alkynylmetallic reagents to α,β-unsaturated ketones provides γ,δ-acetylenic ketones that are readily converted into a variety of functional derivatives. It has been shown that alkynylcopper reagents are completely ineffective at promoting 1,4 addition of an acetylenic moiety. However, 1-alkynyldialkylalanes react with a variety of conjugated enones to give

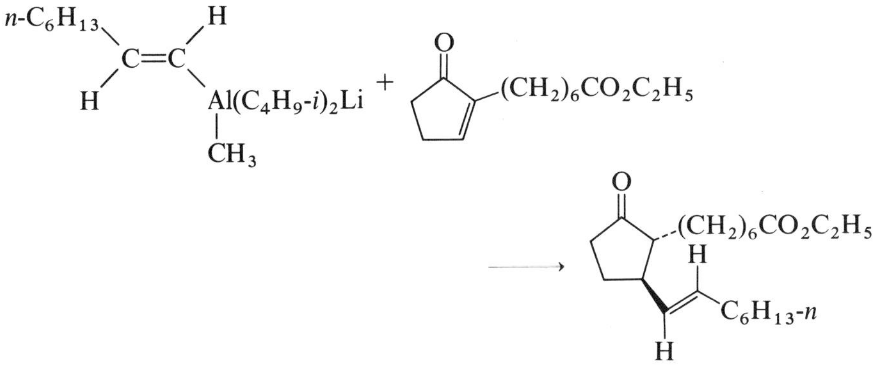

$$n\text{-}C_6H_{13}CH{=}CH\text{-}Al(C_4H_9\text{-}i)_2(CH_3)Li + \text{2-}[(CH_2)_6CO_2C_2H_5]\text{-2-cyclopentenone} \longrightarrow$$

1,4-addition products.[125,126] The success of the reaction depends critically on the proper selection of experimental conditions.

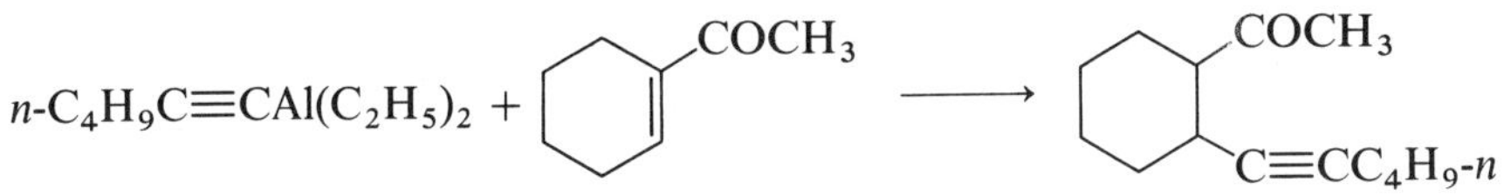

$$n\text{-}C_4H_9C{\equiv}CAl(C_2H_5)_2 + \text{1-acetylcyclohexene} \longrightarrow \text{1-}COCH_3\text{-2-}(C{\equiv}CC_4H_9\text{-}n)\text{cyclohexane}$$

The reaction is restricted to ketones capable of achieving an (*S*)-*cis* conformation; thus cyclic ketones in which the enone system is rigidly constrained in a transoid geometry, such as 2-cyclohexenone, react with the organoalane to provide the tertiary alcohol derived from 1,2 rather than 1,4 addition of the acetylenic unit.[125]

A plausible pathway for 1,4 addition involves intramolecular addition of the alkynyl group through a six-membered transition state. Thus 1,4 addition occurs when structural circumstances enable incorporation of the required *syn* geometry **A** (as in acyclic enones or acetylcyclohexene), but in the transoid enones **B** (cyclohexenone) geometric constraints operate to prohibit conjugate addition, and 1,2 addition supervenes.[125]

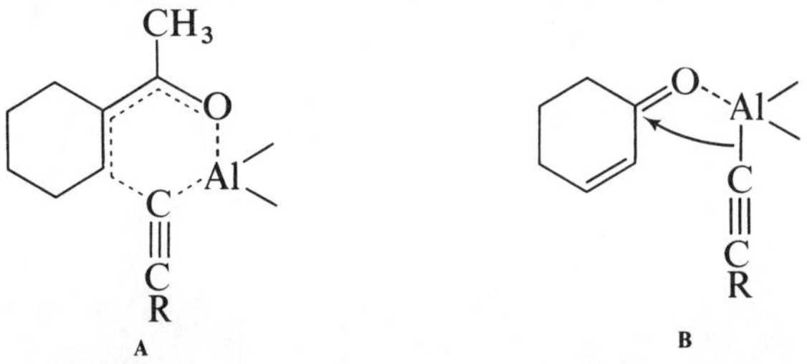

A B

However, 1,4 additions of 1-alkynylalanes to fixed (*S*)-*trans* enones do occur when the enone possesses a suitably located hydroxyl group (Eq. 46),[123,127] or if the reaction is carried out in the presence of a catalyst prepared from nickel acetylacetonate and diisobutylaluminum hydride.[128,129] The latter procedure represents a general approach for alkynylation of simple (*S*)-*trans* enones (Eq. 47).

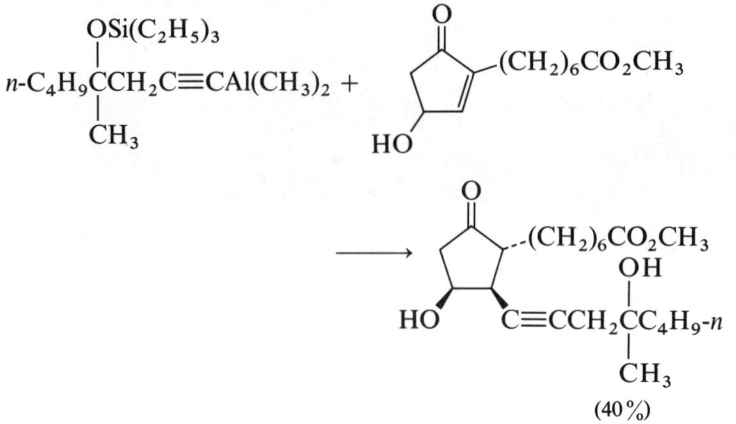

$$n\text{-}C_4H_9C(OSi(C_2H_5)_3)(CH_3)CH_2C\equiv CAl(CH_3)_2 + \text{4-hydroxy-2-[(CH}_2)_6CO_2CH_3\text{]cyclopent-2-enone}$$

$$\longrightarrow \text{3-hydroxy-2-(CH}_2)_6CO_2CH_3\text{-cyclopentanone with } C\equiv CCH_2C(OH)(CH_3)C_4H_9\text{-}n \quad (40\%)$$

(Eq. 46)[123]

$$t\text{-}C_4H_9C\equiv CAl(CH_3)_2 + \text{cyclohex-2-enone} \xrightarrow[(i\text{-}C_4H_9)_2AlH]{Ni(acac)2-} \text{3-}(C\equiv CC_4H_9\text{-}t)\text{cyclohexanone} \quad (72\%)$$

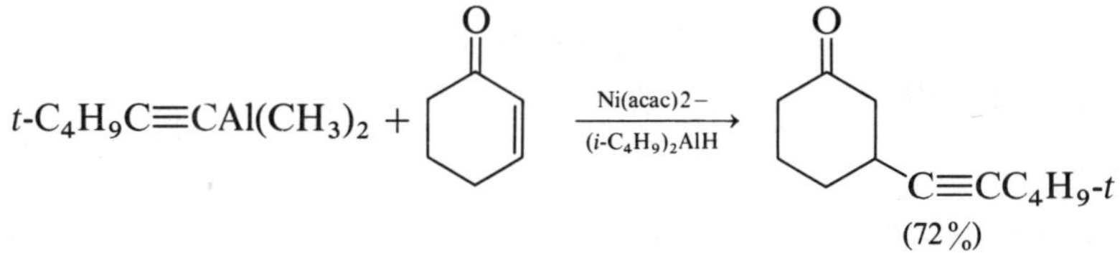

(Eq. 47)[128,129]

Cyclopropanation

The double bond of 1-alkenylalanes, derived from the *cis* hydroalumination of 1-alkynes with diisobutylaluminum hydride, reacts with methylene bromide in the presence of a zinc-copper couple to produce cyclopropylalanes.[130] These intermediates can be hydrolyzed to give alkylcyclopropanes or halogenated to afford *trans*-1-halo-2-alkylcyclopropanes.

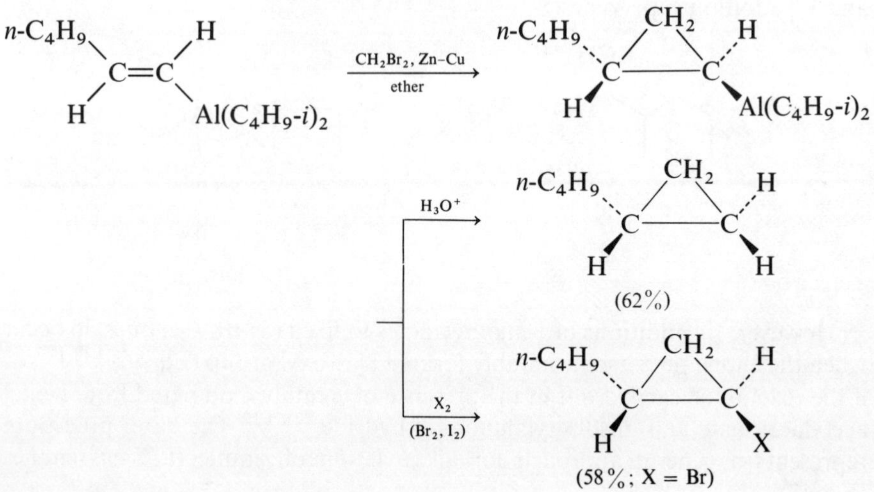

$$(E)\text{-}n\text{-}C_4H_9CH{=}CHAl(C_4H_9\text{-}i)_2 \xrightarrow[\text{ether}]{CH_2Br_2,\ Zn\text{-}Cu} trans\text{-}n\text{-}C_4H_9\text{-cyclopropyl-}Al(C_4H_9\text{-}i)_2$$

$$\xrightarrow{H_3O^+} n\text{-}C_4H_9\text{-cyclopropane} \quad (62\%)$$

$$\xrightarrow[(Br_2,\ I_2)]{X_2} trans\text{-}1\text{-}X\text{-}2\text{-}n\text{-}C_4H_9\text{-cyclopropane} \quad (58\%;\ X = Br)$$

However, extension of the methylenation reaction to alkenylalanes derived from disubstituted acetylenes produces complex mixtures of products.

Transmetalation

The vinyl carbon–aluminum bond undergoes facile transmetalation to stereospecifically form the corresponding alkenylboranes (Eq. 48),[131] alkenylzirconiums (Eq. 49),[131] and alkenylmercurials (Eq. 50),[132] thus greatly enhancing the scope of organoaluminum-mediated transformations.

$n\text{-}C_6H_{13}(CH_3)C{=}CH\text{-}Al(CH_3)_2$ $\xrightarrow{CH_3O\text{-}B\text{(9-BBN)}}$ $n\text{-}C_6H_{13}(CH_3)C{=}CH\text{-}B\text{(9-BBN)}$ $\xrightarrow{1.\ n\text{-}C_4H_9Li;\ 2.\ I_2}$ $n\text{-}C_6H_{13}(CH_3)C{=}CH\text{-}C_4H_9\text{-}n$ (77%) (Eq. 48)[131]

$n\text{-}C_6H_{13}(CH_3)C{=}CH\text{-}Al(CH_3)_2$ $\xrightarrow{1.\ n\text{-}C_4H_9Li;\ 2.\ Cl_2Zr(C_5H_5)_2}$ $n\text{-}C_6H_{13}(CH_3)C{=}CH\text{-}ZrCl(C_5H_5)_2$ $\xrightarrow{C_6H_5I,\ ZnCl_2,\ Pd[P(C_6H_5)_3]_4\ (5\ mol\ \%)}$ $n\text{-}C_6H_{13}(CH_3)C{=}CH\text{-}C_6H_5$ (89%) (Eq. 49)[131]

$n\text{-}C_6H_{13}(CH_3)C{=}CH\text{-}Al(CH_3)_2$ $\xrightarrow{HgCl_2}$ $n\text{-}C_6H_{13}(CH_3)C{=}CH\text{-}HgCl$ (83%) (Eq. 50)[132]

EXPERIMENTAL CONDITIONS

Organoaluminum compounds are highly reactive toward oxygen and moisture, and their preparation and handling have to be carried out in inert atmospheres.[5,7,9] Before working with organoaluminum reagents it is thus important to become familiar with the standard procedures employed for manipulating air-sensitive substances.[133–135]

For the majority of organic syntheses involving organoaluminum intermediates, working under a static pressure of nitrogen using benchtop techniques, common laboratory equipment, and magnetic stirring is satisfactory. Figure 1 illustrates a typical bench-top inert atmosphere apparatus suitable for most reactions involving organoaluminum compounds.

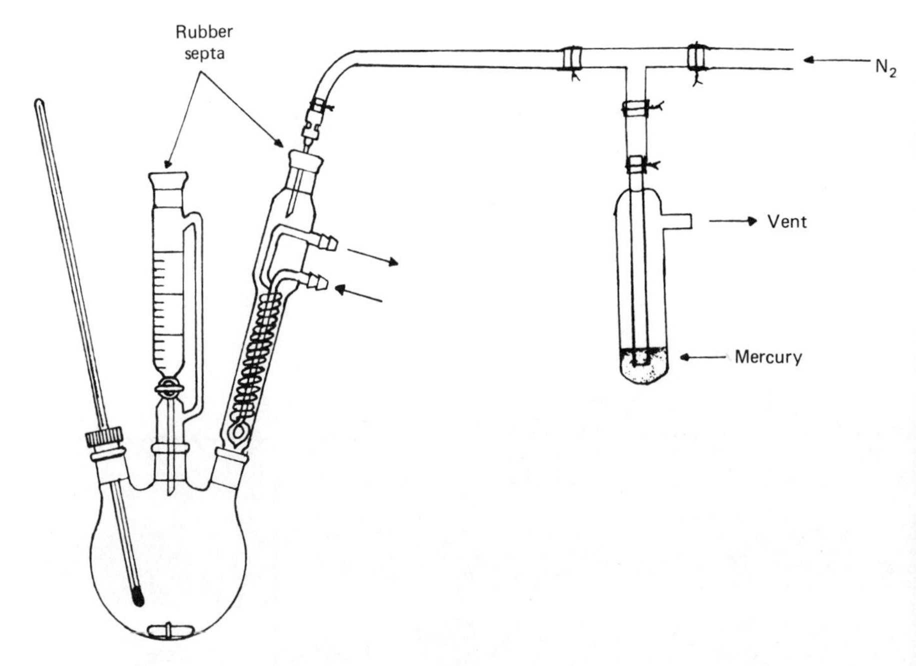

Figure 1.

All glassware for reactions involving organoaluminum reagents should be oven-dried at 150° for 6 hours, assembled hot, and cooled under a stream of pure nitrogen before use. Alternatively, the assembled apparatus can be flamed dry in a stream of dry nitrogen.

The solvents used in reactions of organoaluminum compounds must be meticulously dried as well as freed of impurities containing reactive sites (—OH, >NH, >CO, etc.). Procedures for obtaining solvents of high purities have been described in several monographs.[134–136]

For transferring organoaluminum reagents, it is advisable to use syringes with Luer–Lok fittings and equipped with 18-gauge needles. The syringes should be cleaned immediately after use by first drawing up a hydrocarbon such as *n*-pentane. The diluted residual organoaluminum reagent is then destroyed by adding it slowly to water. Further cleaning of the syringe is accomplished by successively drawing up dilute hydrochloric acid, water, acetone, and distilled water.

A number of synthetically important organoaluminum reagents are commercially available. Several can be obtained either neat or as solutions in a variety of solvents. In view of the pyrophoric nature of organoaluminum compounds, especially those with appreciable volatility such as trimethyl- and triethylaluminum, the instructions provided by the manufacturers must be observed.

Diisobutylaluminum hydride, which is an important reducing and hydroaluminating agent, can be obtained commercially either neat (5.4 *M*) or as a 1 *M* solution in hydrocarbon or in ether solvents. Especially convenient for use in hydroaluminations is the 1 *M* solution of the reagent in *n*-hexane. If desired, the solvent can be removed under reduced pressure to give neat diisobutylaluminum hydride. This can be used as such or can be diluted with an appropriate solvent while maintaining an inert atmosphere.

Solutions of diisobutylaluminum hydride are conveniently standardized as follows.[14] Dry cuprous chloride (10 g) is added to 50 mL of dry tetrahydrofuran contained in a 125-mL flask immersed into a water bath (25°) and connected to a gas buret. To this is added an appropriate volume (2–3 mmol) of the solution of diisobutylaluminum hydride by means of a syringe. The molarity of the reagent can be calculated from the amount of hydrogen evolved.[134] Usually the first measurement of hydrogen is low; thus two or three determinations are necessary to obtain the concentration accurately. Also, the reaction of *tert*-amine complexes of dialkylaluminum hydrides with 1-alkynes, as depicted in Eq. 10, provides an alternative method for the determination of active hydrides in solutions containing dialkylaluminum hydrides. Note that the usual hydrolysis procedure employed for the standardization of solutions of dialkylboranes[134] cannot be applied to dialkylalanes because they react with water or dilute acids to produce, in addition to hydrogen, the corresponding alkanes, which interfere with the hydrogen determination.[5]

TYPICAL HYDROALUMINATION PROCEDURE

A dry, nitrogen-flushed flask bearing a thermometer, a condenser (for volatile reagents), and an addition funnel capped with a rubber septum is equipped for magnetic strirring and for maintaining a static pressure of nitrogen throughout the reaction (Figure 1). The alkyne and the solvent are added to the flask through the addition funnel by means of a syringe. Then, to the stirred solution diisobutylaluminum hydride, either neat or as a solution, is introduced at 25° (water bath) by means of a syringe. The reaction mixture is stirred for 30 minutes at room temperature before being heated to an appropriate reaction temperature. The resultant organoaluminum intermediate is then treated with the desired organic or inorganic reagent. Hydrolysis of the reaction mixture is best accomplished by transferring the contents of the flask by means of a double-ended needle[134] into a mixture of 10% hydrochloric acid and ice. The two-phase mixture is shaken until the precipitate that forms dissolves and then is extracted with an appropriate solvent. For substrates containing acid-sensitive functionalities, the hydrolysis of the organoaluminum intermediate can be carried out with 6 *N* sodium hydroxide. At pH values intermediate between those present with the above acidic and basic workup procedures, the gelatinous aluminum hydroxide formed can seriously interfere with the isolation of the product.

EXPERIMENTAL PROCEDURES

(Z)-*β*-*tert*-Butylstyrene (Hydroalumination–Protonlysis of an Alkyne).[38] A solution of 4.40 g (28.0 mmol) of *tert*-butyl(phenyl)acetylene in 40 mL of *n*-heptane was treated at room temperature with 4.00 g (5.22 mL, 28.2 mmol) of diisobutylaluminum hydride and heated at 50° for 18 hours. The resulting mixture was then cooled to 0° and cautiously hydrolyzed with water. After filtering the suspended aluminum salts, the filtrate was concentrated and distilled to afford 3.76 g (84%) of the product, bp 73–75° (13 mm).

(*E*)-1-Bromo-1-octene (Hydroalumination–Bromination of an Alkyne).[85] To 2.76 g (25.0 mmol) of 1-octyne was added 25.0 mL of a 1.07 *M* solution of diisobutylaluminum hydride (26.8 mmol) in *n*-hexane while the temperature was maintained at 25–30° by means of a water bath. The solution was stirred at room temperature for 30 minutes and then was heated at 50° for 4 hours. The resultant alkenylalane was cooled to −30°, diluted with 15 mL of dry ether, and treated with 5.35 g (30.1 mmol) of *N*-bromosuccinimide while keeping the temperature below −15°. The reaction mixture was gradually warmed to room temperature and stirred for 1 hour before being poured slowly into a mixture of 6 *N* hydrochloric acid (50 mL), *n*-pentane (10 mL), and ice (10 g). The layers were separated, and the aqueous phase was extracted with pentane. The combined organic extract was washed successively with 1 *N* sodium hydroxide, 10% sodium sulfite, and saturated aqueous sodium chloride and then was treated with a few crystals of BHT to inhibit isomerization of the alkenyl bromide. After drying over magnesium sulfate, distillation afforded 3.72 g (78%) of (*E*)-1-bromo-1-octene, bp 67° (5 mm), n_D^{26} 1.4617. This compound, which contained 4% of 1-bromo-1-octyne, was stored over a few crystals of BHT.

(*E*)-1-Iodo-1-octene (Hydroalumination–Iodinolysis of an Alkyne).[85] To 2.76 g (25.0 mmol) of 1-octyne was added 27.5 mL of a 1.0 *M* solution of diisobutylaluminum hydride (27.5 mmol) in *n*-hexane while the temperature was maintained at 25–30° by means of a water bath. The resulting solution was stirred at room temperature for 30 minutes and then was heated at 50° for 4 hours. The volatiles were removed under reduced pressure from the alkenylalane and replaced at 0° with 20 mL of dry tetrahydrofuran. The resulting solution was cooled to −78° and treated with 10 mL of a solution of iodine (7.6 g, 30 mmol) in dry tetrahydrofuran at a rate such that the temperature was maintained below −60°. The reaction mixture was allowed to warm to 0° and then was hydrolyzed by transferring it by means of a double-ended needle into a stirred, ice-cooled mixture of 6 *N* hydrochloric acid (100 mL) and *n*-pentane (20 mL). The layers were separated, and the aqueous phase was extracted with *n*-pentane. The combined organic extract was washed successively with 1 *N* sodium hydroxide, 10% aqueous sodium sulfite, and aqueous sodium chloride, and then was treated with a few crystals of BHT to inhibit isomerization of the alkenyl iodide. After drying over magnesium sulfate, distillation yielded 4.10 g (75%) of the product, bp 85° (3 mm), n_D^{24} 1.5010. The compound was stored over a few crystals of BHT.

[(*E*)-1-Bromo-1-hexenyl]trimethylsilane [Hydroalumination–Bromination of a (1-Alkynyl)trimethylsilane.[44] A solution of 2.32 g (15.0 mmol) of (1-hexynyl)trimethylsilane in 7.5 mL of dry ether was treated with 3.00 mL of neat diisobutylaluminum hydride (16.5 mmol) at 25–30° and then was heated at 40° for 2 hours. The hydroalumination product formed was cooled to 0° and diluted with ether (15 mL) and pyridine (2.4 mL). To the resultant yellow reaction mixture was added at −70° a 1.50 *M* solution of bromine (22.5 mmol) in methylene chloride at such a rate as to maintain the temperature below −60°. The resulting yellow slurry was kept for an additional 15 minutes at −70° and then was poured slowly into a vigorously stirred mixture of 1 *N* sodium hydroxide (60 mL), ice (20 g), and *n*-pentane (15 mL). After the mixture was shaken until it became clear, it was extracted with *n*-pentane. The combined organic extract was washed successively with 1 *N* hydrochloric acid, a 20% aqueous solution of cadmium chloride (to remove small amounts of remaining pyridine), 1 *N* hydrochloric acid, and saturated aqueous sodium bicarbonate. To inhibit isomerization of the alkenyl bromide, a few crystals of BHT were added to the pentane extract prior to drying over magnesium sulfate. Distillation from a small amount of calcium carbonate afforded 3.17 g (90%) of the product, bp 48° (1 mm), n_D^{23} 1.4755. The compound was stored over a few crystals of BHT.[55]

(*Z*)-1-Bromo-1-chloro-1-hexene (Reduction of a 1-Chloro-1-alkyne with Lithium Aluminum Hydride and Reaction with Bromine).[55] A 1.0 *M* solution of lithium aluminum hydride (20 mmol) in dry tetrahydrofuran was cooled to −30° and treated with 2.33 g (20.0 mmol) of 1-chloro-1-hexyne while the temperature during the addition was maintained below −25°. After being stirred at −30° for an additional 15 minutes, the mixture was brought to 0° and stirred for 90 minutes. Dry acetone (66 mmol) was then added dropwise over 20 minutes while the temperature was maintained below 10°. After 1 hour, the reaction mixture was cooled to −78° and then was treated dropwise with a solution of bromine (22 mmol) in 10 mL of methylene chloride. The mixture was allowed to warm to room temperature in the dark and then was slowly poured into a mixture of 10% hydrochloric acid (80 mL), 10% aqueous sodium bisulfite (10 mL), *n*-pentane (20 mL), and ice (50 g). After extraction with *n*-pentane, the combined organic extract was washed with 10% hydrochloric acid and with saturated aqueous sodium bicarbonate and then was treated with a few crystals of BHT before drying over magnesium sulfate. Distillation from a small amount of calcium carbonate yielded 3.08 g (78%) of (*Z*)-1-bromo-1-chloro-1-hexene, bp 58–60° (9 mm), n_D^{25} 1.4790. To inhibit isomerization, it is important to treat the distilled compound immediately with a few crystals of BHT.

(*E*)-1-*tert*-Butylcinnamyl Alcohol (Reduction of a Propargylic Alcohol with Lithium Aluminum Hydride).[59] To a solution of 1.88 g (10.0 mmol) of 1-phenyl-4,4-dimethyl-1-pentyn-3-ol in 25 mL of tetrahydrofuran was added 0.42 g (11 mmol) of lithium aluminum hydride. The reaction mixture was heated at reflux for 6 hours before being hydrolyzed by cautious, sequential addition of water (0.42 mL), 15% aqueous sodium hydroxide (0.42 mL), and water (1.26 mL). The aluminum hydroxide precipitate was filtered and washed with ether. The filtrate was dried over anhydrous magnesium sulfate and distilled to afford 1.84 g (98%) of the product, mp as the benzoate ester 83–84°.

(*E*)-2-Hepten-1-ol (Reaction of a Lithium Alkenylalanate with an Aldehyde).[91] A solution of 8.21 g (0.100 mol) of 1-hexyne in 20 mL of *n*-heptane was treated at 25° with 18.5 mL of diisobutylaluminum hydride (0.10 mol) while maintaining the temperature at 25–30° by means of a water bath. The resulting solution was stirred at room temperature for 30 minutes, then was heated slowly to 50° and maintained at this temperature for 2 hours. The reaction mixture was cooled to 25° and 59 mL of a 1.7 *M* solution of methyllithium (0.10 mol) in ether was added. The vinylalanate formed was treated at room temperature with 6.6 g (0.22 mol) of dry paraformaldehyde at such a rate as to maintain a gentle reflux. The reaction mixture was refluxed for an additional hour before being poured slowly into a slurry of concentrated hydrochloric acid (50 mL) and ice (200 g). The aqueous phase was extracted with ether, and the combined ether extract was washed with 5% aqueous sodium bicarbonate and then dried over magnesium sulfate. Distillation afforded 8.3 g (73%) of (*E*)-2-hepten-1-ol, bp 43° (1 mm), n_D^{24} 1.4415.

(*E*)-5-(2,6,6-Trimethyl-1-cyclohexen-1-yl)-3-methyl-2-penten-1-ol (Zirconium-Catalyzed Carboalumination of an Alkyne and Reaction with an Aldehyde).[94] To a stirred slurry of 5.85 g (20.0 mmol) of bis(cyclopentadienyl) zirconium dichloride in 80 mL of 1,2-dichloroethane was added 2.88 g (40.0 mmol) of trimethylalane (*pyrophoric*) at 0°. The lemon-yellow solution was treated dropwise with 3.53 g (20.0 mmol) of 1-(3-butynyl)-2,6,6-trimethyl-1-cyclohexene in 20 mL of 1,2-dichloroethane at room temperature. The resulting mixture was stirred for 2–3 hours and the volatile compounds were removed at reduced pressure (maximum 50° and 0.3 mm Hg). The organic compounds were extracted with *n*-hexane, and the extract was transferred into another flask by means of a double-tipped needle.[134] To this was added 12.5 mL (20.0 mmol) of a 1.60 *M* solution of *n*-butyllithium in *n*-hexane. The precipitate that formed was dissolved in tetrahydrofuran and the resultant solution was added to a suspension of 1.80 g (60 mmol) of paraformaldehyde in tetrahydrofuran. The mixture was then stirred for several hours, quenched with 3 *N* hydrochloric acid and extracted with ether. The ether extract was washed with aqueous sodium bicarbonate, dried over magnesium sulfate and concentrated. After a simple column chromatographic purification (silica gel), 3.16 g (71 %) of the product was isolated, n_D^{27} 1.4984.

α-Farnesene (Zirconium-Catalyzed Carboalumination of an Alkyne followed by Palladium-Catalyzed Cross-Coupling with an Allylic Halide).[111] A solution of 3.84 mL (40.0 mmol) of trimethylalane (*pyrophoric*) and 2.34 g (8.00 mmol) of bis(cyclopentadienyl)zirconium dichloride in 30 mL of 1,2-dichloroethane was treated at room temperature under nitrogen with 1.04 g (20.0 mmol) of 1-buten-3-yne in xylene. After the reaction mixture was stirred for 12 hours at room temperature, 3.45 g (20.0 mmol) of geranyl chloride, 1.15 g (1.00 mmol) of tetrakis(triphenylphosphine)palladium, and 40 mL of THF were added at 0°. The reaction mixture was stirred for 3 hours at room temperature, treated with 30 mL of water, and extracted with pentane. The extract was washed with aqueous sodium bicarbonate and dried over anhydrous magnesium sulfate. After filtration and concentration, simple distillation yielded 3.50 g (86 %) of the product, bp 30–32° (0.15 mm), n_D^{23} 1.4977.

Methyl (*E,E*)-2-Methyl-2,4-nonadienoate [Palladium-Catalyzed Cross-Coupling of an (*E*)-Alkenylalane with an Alkenyl Halide].[109] To 0.74 g (1.0 mmol) of bis(triphenylphosphine)palladium dichloride suspended in 20 mL of tetrahydrofuran was added 0.37 mL (2.0 mmol) of neat diisobutylaluminum hydride at 25° over 10 minutes. To this mixture at 25° were added sequentially (*E*)-1-hexenyldiisobutylalane, prepared from 1.64 g (20.0 mmol) of 1-hexyne dissolved in 20 mL of *n*-hexane and 3.68 mL (20.0 mmol) of neat diisobutylaluminum hydride, and 3.58 g (20.0 mmol) of methyl (*E*)-3-bromo-2-methylpropenoate. The reaction mixture was refluxed for 15 minutes, treated with 3 *N* hydrochloric acid, and extracted with ether. The extract was dried over magnesium sulfate and then concentrated at reduced pressure and distilled to yield 2.2 g (61 %) of the product, bp 78–79° (1 mm).

(*E*)-1,4-Dicyclohexyl-1-buten-3-yne (Selective Reduction of a 1,3-Diyne with Lithium Diisobutylmethylaluminum Hydride).[51] To 16 mL of freshly distilled diglyme was added 5.6 mL (30 mmol) of neat diisobutylaluminum hydride followed by 19.5 mL (30.0 mmol) of methyllithium in ether while maintaining the temperature during the additions between 0° and 25°. The reaction mixture was then warmed to room temperature and the resultant milky lithium diisobutylmethylaluminum hydride slurry was treated at 25° with 4.29 g (20.0 mmol) of solid dicyclohexylbutadiyne. After being stirred at 25° for 8 hours the mixture was added to 50 mL of chilled 10% sulfuric acid by means of a double-ended needle.[134] The resulting mixture was diluted with an additional 25 mL of 10% sulfuric acid and extracted with *n*-pentane. The combined extract was washed sequentially with water, 10% hydrochloric acid, and a saturated solution of sodium bicarbonate and dried over magnesium sulfate. Distillation afforded 3.93 g (91%) of the enyne, bp 122° (1 mm), n_D^{21} 1.5223.

3,3-Dimethyl-4-nonyne (Reaction of a Trialkynylalane with a *tert*-Alkyl Halide).[70] To a solution of 1-hexyne (4.92 g, 60 mmol) in *n*-hexane (50 mL) was added 25 mL (60 mmol) of *n*-butyllithium in *n*-hexane at 0°. After the reaction mixture was stirred at this temperature for 30 minutes, 2.70 g (20 mmol) of aluminum chloride was added. The mixture was stirred for an additional 30 minutes at 0° and the *n*-hexane was removed under reduced pressure (~1 mm Hg). To the residue obtained were added sequentially 100 mL of 1,2-dichloroethane and 2.13 g (20 mmol) of 2-chloro-2-methylbutane at 0°. The reaction mixture was stirred for one hour at this temperature and then was poured into ice-cold aqueous 3 *N* hydrochloric acid. The organic phase was separated, and the aqueous layer was extracted with ether. The combined organic layers were washed with water, dried over magnesium sulfate, and concentrated. Distillation yielded 2.58 g (85%) of the product, bp 82–83° (40 mm), n_D^{20} 1.4314.

(*E*)-3-(3,3-Dimethyl-1-butyn-1-yl)-4-(cumyloxy)cyclopentanone(1,4 Addition of an 1-Alkynylalane to an Enone in the Presence of Nickel Acetylacetonate).[129] A solution of 0.09 g (0.36 mmol) of nickel (II) acetylacetonate in 15 mL of ether at 0° was treated with 0.60 mL (0.32 mmol) of a 0.53 *M* solution of diisobutylaluminum hydride in toluene. To the resulting mixture was added 8 mL (3.6 mmol) of a 0.45 *M* solution of dimethyl(3,3-dimethyl-1-butyn-1-yl)alane in ether, prepared by reaction of 1-lithio-3,3-dimethyl-1-butyne with dimethylaluminum chloride. The temperature of the reaction mixture was lowered to −5°, and 0.36 g (1.65 mmol) of 4-(cumyloxy)-2-cyclopentenone in 10 mL of ether was added dropwise over 15 minutes. The resulting mixture was allowed to stir at −5° for 1.5 hours and then was hydrolyzed with saturated potassium dihydrogen phosphate. Sufficient 10% aqueous sulfuric acid was added to dissolve the aluminum salts. The organic layer was separated and the aqueous phase was extracted with ether. The extract was washed with saturated aqueous sodium bicarbonate and saturated aqueous sodium chloride, dried over sodium sulfate, and concentrated. Liquid chromatographic separation (20% ethyl acetate, 80% hexane) yielded 0.42 g (85%) of the product.

TABULAR SURVEY

The 15 tables that follow include examples of significant, synthetically useful reactions of alkyne-derived alkenyl- and alkynylaluminum compounds published through December 1982 together with unpublished examples. The reactions in the tables appear in approximately the same order as reported in the text and are arranged according to increasing number of carbons. In entries where the product contains more than one functionality, the following sequence of priorities has been adopted: alkenyl halides, α,β-unsaturated acids, esters, nitriles, ketones, ethers; allylic- and homoallylic alcohols and ethers; enynes, dienes, alkenes.

An effort has been made to provide as complete an account as possible of experimental conditions and yields. However, products obtained in less than 10% yields are usually not shown. Yields that have not been specified in the reference cited are indicated by a dash in parentheses (—).

Standard abbreviations used throughout the tables are as follows:

acac	acetylacetonate
$(C_5H_5)_2TiCl_2$	bis(cyclopentadienyl)titanium dichloride
$(C_5H_5)_2ZrCl_2$	bis(cyclopentadienyl)zirconium dichloride
diglyme	diethylene glycol dimethyl ether
DME	1,2-dimethoxyethane
DMF	dimethylformamide
ether	diethyl ether
NBS	*N*-bromosuccinimide
NCS	*N*-chlorosuccinimide
THF	tetrahydrofuran
THP	tetrahydropyranyl
TsOH	*p*-toluenesulfonic acid

TABLE I. ALKENES

	Substrate	Reagents	Conditions	Product(s) and Yield(s)(%)	Refs.
C_2	$HC\equiv CH$	1. $(C_2H_5)_3Al$ 2. CH_2N_2 3. H_3O^+	Toluene–anisole (4:1), −78°	n-$C_3H_7CH{=}CH_2$ (—)	137
C_5	$CH_3C\equiv CC_2H_5$	1. $LiAlH_4$ 2. H_2O	THF, 125–130°, 2–3 hr	(E)-$CH_3CH{=}CHC_2H_5$ (92)	48
		1. $LiAlH_4$, H_2 2. H_3O^+	THF, 190°	$CH_3CH{=}CHC_2H_5$ (—) (E:Z = 91:9)	47
	$C_2H_5C\equiv CSCH_3$	1. $LiAlH_4$ (1 eq) 2. H_3O^+	THF, 50–60°, 1 hr	(E)-$C_2H_5CH{=}CHSCH_3$ (91)	53
C_6	n-$C_4H_9C\equiv CH$	1. (i-$C_4H_9)_2AlH$ (2 eq) 2. C_6H_5CHO	90°, 5 hr Heptane, 0°	$C_6H_5CH{=}CHC_5H_{11}$-n (25–30) (E:Z = 63.1:36.9)	27
		1. (i-$C_4H_9)_2AlH$ (2 eq) 2. n-$C_{11}H_{23}CHO$	90°, 5 hr Heptane, 0°	n-$C_5H_{11}CH{=}CHC_{11}H_{23}$-$n$ (15–20)	27
		1. (i-$C_4H_9)_2AlH$ (2 eq) 2. n-$C_5H_{11}COC_5H_{11}$-n	90°, 5 hr Heptane, 0°	(n-$C_5H_{11})_2C{=}CHC_5H_{11}$-$n$ (15–20)	27
		1. (i-$C_4H_9)_2AlH$ (2 eq) 2. cyclohexanone	90°, 5 hr Heptane, 0°	cyclohexylidene=CHC_5H_{11}-n (15–20)	27
		1. (i-$C_4H_9)_2AlH$ 2. $CH_3COC_6H_5$	90°, 5 hr Heptane, 0°	$C_6H_5C(CH_3){=}CHC_5H_{11}$-$n$ (15–20)	27
		1. (i-$C_4H_9)_2AlH$ (2 eq) 2. n-C_4H_9Li 3. CH_2O	THF or hexane	n-$C_4H_9CH_2CH{=}CH_2$ (70)	26
		1. (i-$C_4H_9)_2AlH$ 2. n-C_4H_9Li 3. $C_6H_5CH_2I$	Hexane, 50–55°, 4 hr Hexane, 25° THF, 40°, 18 hr	(E)-n-$C_4H_9CH{=}CHCH_2C_6H_5$ (51)	77

1. $(i\text{-}C_4H_9)_2AlH$ 2. $n\text{-}C_4H_9Li$ 3. $C_6H_5CH_2Br$	Hexane, 50–55°, 4 hr Hexane, 25° THF, reflux, 12 hr	" (46)	77
1. $(i\text{-}C_4H_9)_2AlH$ 2. $n\text{-}C_4H_9Li$ 3. $n\text{-}C_8H_{17}OSO_2CH_3$	Hexane, 50–55°, 4 hr Hexane, 25° Hexane, 25°, 24 hr	(*E*)-$n\text{-}C_4H_9CH{=}CHC_8H_{17}\text{-}n$ (41)	77
1. $(i\text{-}C_4H_9)_2AlH$ 2. $n\text{-}C_4H_9Li$ 3. $n\text{-}C_8H_{17}I$	Hexane, 50–55°, 4 hr Hexane, 25° THF, reflux, 110 hr	" (49)	77
1. $(i\text{-}C_4H_9)_2AlH$ 2. $Ni[P(C_6H_5)_3]_4$ (5 mol %) 3. $p\text{-}CH_3C_6H_4Br$ (0.5 eq)	Hexane Hexane–THF 25°, 24 hr	(*E*)-$n\text{-}C_4H_9CH{=}CHC_6H_4CH_3\text{-}p$ (84)	81
1. $(i\text{-}C_4H_9)_2AlH$ 2. $Ni[P(C_6H_5)_3]_4$ (5 mol %) 3. C_6H_5Br (0.5 eq)	Hexane Hexane–THF 50°, 3 hr	(*E*)-$n\text{-}C_4H_9CH{=}CHC_6H_5$ (85)	81
1. $(i\text{-}C_4H_9)_2AlH$ 2. $Ni[P(C_6H_5)_3]_4$ (5 mol %) 3. C_6H_5I (0.5 eq)	Hexane Hexane–THF 25°, 12 hr	" (89)	81
1. $(i\text{-}C_4H_9)_2AlH$ 2. $Ni[P(C_6H_5)_3]_4$ (5 mol %) 3. 1-Bromonaphthalene (structure: Br on naphthalene) (0.5 eq)	Hexane Hexane–THF 50°, 3 hr	(*E*)-$n\text{-}C_4H_9CH{=}CH$–(1-naphthyl, structure) (93)	81
1. $(i\text{-}C_4H_9)_2AlH$ 2. $Ni[P(C_6H_5)_3]_4$ (5 mol %) 3. $p\text{-}NCC_6H_4Br$ (1.0 eq)	Hexane Hexane–THF 25°, 1 hr	(*E*)-$n\text{-}C_4H_9CH{=}CHC_6H_4CN\text{-}p$ (64)	81
1. $(C_2H_5)_2AlH$ 2. D_2O	35–45°, 4.5 hr	(*E*)-$n\text{-}C_4H_9CH{=}CHD$ (79), $n\text{-}C_4H_9C{\equiv}CD$ (21)	75
1. $(C_2H_5)_2AlH$ 2. D_2O	40°, 1 hr	(*E*)-$n\text{-}C_4H_9CH{=}CHD$ (—)	13

TABLE I. ALKENES (*Continued*)

Substrate	Reagents	Conditions	Product(s) and Yield(s)(%)	Refs.
C_6 n-$C_4H_4C{\equiv}CH$ (*cont'd*)	1. $(C_2H_5)_2AlD$ 2. H_2O	40°, 1 hr	n-$C_4H_9CD{=}CH_2$ (—)	13
	1. $(C_2H_5)_2AlD$ 2. D_2O	40°, 1 hr	(E)-n-$C_4H_9CD{=}CHD$ (—)	13
n-$C_4H_9C{\equiv}CD$	1. $(C_2H_5)_2AlH$ 2. H_2O	10–20°, 3 hr	(Z)-n-$C_4H_9CH{=}CHD$ (45)	75
$C_2H_5C{\equiv}CC_2H_5$	1. $(i\text{-}C_4H_9)_2AlH$ 2. H_2O	40–50°, 3–5 hr	(Z)-$C_2H_5CH{=}CHC_2H_5$ (—)	2
	1. $LiAlH_4$ 2. H_2O	THF, 125°, 1.7 hr	(E)-$C_2H_5CH{=}CHC_2H_5$ (66)	48
	1. $(i\text{-}C_4H_9)_2AlH$ 2. $Pd[P(C_6H_5)_3]_4$ (5 mol %) 3. $ZnCl_2$ 4. m-$CH_3C_6H_4I$	Hexane 20–25°	(E)-$C_2H_5CH{=}C(C_2H_5)C_6H_4CH_3$-$m$ (88)	82
	1. $(i\text{-}C_4H_9)_2AlH$ 2. D_2O	45°, 3–4 hr	(Z)-$C_2H_5CH{=}CDC_2H_5$ (—)	1,13
	1. $(C_2H_5)_2AlD$ 2. D_2O	40–45°, 8 hr	(Z)-$C_2H_5CD{=}CDC_2H_5$ (—)	13
$CH_2{=}CH(CH_2)_2C{\equiv}CH$	1. $(i\text{-}C_4H_9)_2AlH$ (1 eq) 2. $(i\text{-}C_4H_9)_2AlH$ (1.2 eq) 3. H_3O^+	Octane–$(C_2H_5)_3N$, 25° 50°, 8 hr	3-methylcyclopentene (structure: cyclopentene ring with CH_3) (77)	22
C_7 n-$C_5H_{11}C{\equiv}CH$	1. $(i\text{-}C_4H_9)_2AlH$ 2. n-C_4H_9Li 3. CH_3I	Hexane, 50–55°, 4 hr Hexane, 25° THF, 25°, 24 hr	(E)-n-$C_5H_{11}CH{=}CHCH_3$ (65)	77
	1. $(i\text{-}C_4H_9)_2AlH$ 2. n-C_4H_9Li 3. n-$C_3H_7OSO_2CH_3$	Hexane, 50–55°, 4 hr Hexane, 25° THF, 25°, 24 hr	(E)-n-$C_5H_{11}CH{=}CHC_3H_7$-n (44)	77

	Reactant	Reagents	Conditions	Product(s) (Yield, %)	Refs.
	t-$C_4H_9C{\equiv}CSCH_3$	1. $LiAlH_4$ (1 eq) 2. H_3O^+	THF, 50–60°, 1 hr	(E)-t-$C_4H_9CH{=}CHSCH_3$ (92)	53
	n-$C_3H_7C{\equiv}CAl(CH_3)_2$	1. $(CH_3)_3Al$, $Cl_2Ti(C_5H_5)_2$ 2. D_2O	CH_2Cl_2	n-$C_3H_7C(CH_3){=}CD_2$ (85–95)	40
	$HC{\equiv}CCH_2CH(CH_3)CH{=}CH_2$	1. $(C_2H_5)_2AlH$ 2. H_2O	Hexane, reflux	CH_3–[ring] (—)	138
C_8	n-$C_6H_{13}C{\equiv}CH$	1. $(CH_3)_3Al$ (2 eq), $Cl_2Zr(C_5H_5)_2$ (0.4 eq) 2. $Pd[P(C_6H_5)_3]_4$ (5 mol %) 3. $C_6H_5CH_2Cl$	$ClCH_2CH_2Cl$, 25° THF THF, 25°, 3 hr	(E)-n-$C_6H_{13}C(CH_3){=}CHCH_2C_6H_5$ (92)	139
		1. $(CH_3)_3Al$ (2 eq), $Cl_2Zr(C_5H_5)_2$ (0.4 eq) 2. $Pd[P(C_6H_5)_3]_4$ (5 mol %) 3. $C_6H_5CH_2Br$	$ClCH_2CH_2Cl$, 25° THF THF, 25°, 3 hr	" (91)	139
		1. $(i$-$C_4H_9)_2AlH$ 2. i-$C_4H_9OCH_2N$[ring]	Hexane 55°, 2 hr	(E)-n-$C_6H_{13}CH{=}CHCH_2N$[ring] (68)	140
		1. $(i$-$C_4H_9)_2AlH$ 2. CH_3Li 3. CH_3I	Hexane Ether–THF	(E)-n-$C_6H_{13}CH{=}CHCH_3$ (60)	78
		1. $(n$-$C_3H_7)_3Al$ (2 eq), $Cl_2Zr(C_5H_5)_2$ (0.4 eq) 2. H_2O	$ClCH_2CH_2Cl$, 0°, 1 hr	n-$C_6H_{13}C(C_3H_7$-$n){=}CH_2$ I (E)-n-$C_6H_{13}CH{=}CHC_3H_7$-n II (75) (I:II = 80:20)	64
		1. $(CH_3)_3Al$ (2 eq), $Cl_2Zr(C_5H_5)_2$ (0.4 eq) 2. H_3O^+	$ClCH_2CH_2Cl$, 20–25°, 3 hr	n-$C_6H_{13}C(CH_3){=}CH_2$ I, (E)-n-$C_6H_{13}CH{=}CHCH_3$ II (100) (I:II = 95:5)	64
		1. $(i$-$C_4H_9)_2AlH$ 2. n-C_4H_9Li 3. $C_6H_5CH_2SSO_2C_6H_5$	Heptane, 50°, 4 hr Heptane–toluene, −70 to 25°	(E)-n-$C_6H_{13}CH{=}CHSCH_2C_6H_5$ (56)	90

TABLE I. ALKENES (*Continued*)

Substrate	Reagents	Conditions	Product(s) and Yield(s)(%)	Refs.
C_8 n-$C_6H_{13}C{\equiv}CH$ (*cont'd*)	1. $(i\text{-}C_4H_9)_2AlH$ (2 eq), CuBr (1 eq), LiCl (2 eq) 2. H_3O^+	THF, 20°, 5 hr	n-$C_6H_{13}CH{=}CH_2$ (85)	141
	1. $(i\text{-}C_4H_9)_2AlH$ (2 eq), $CH_3Cu{-}MgBrCl$ (1 eq), LiCl (2 eq) 2. H_3O^+	THF, 20°, 1.25 hr	" (100)	141
	1. $(i\text{-}C_4H_9)_2AlH$ (2 eq), $(CH_3)_2CuMgBr{-}MgCl_2$ (1 eq), LiCl (2 eq) 2. H_3O^+	THF, 20°, 5 hr	" (90)	141
	1. $(CH_3)_3Al$ (2 eq), $Cl_2Zr(C_5H_5)_2$ (0.4 eq) 2. *B*-methoxy-9-borabicyclo[3.3.1]nonane 3. n-C_4H_9Li 4. I_2	$ClCH_2CH_2Cl$, 25° Hexane, 25°, 1 hr THF THF, −78 to 25°	(*E*)-n-$C_6H_{13}C(CH_3){=}CHC_4H_9$-$n$ (77)	131
	1. $(CH_3)_3Al$ (2 eq), $Cl_2Zr(C_5H_5)_2$ (0.4 eq) 2. n-C_4H_9Li 3. $Cl_2Zr(C_5H_5)_2$ 4. C_6H_5I, $Pd[P(C_6H_5)_3]_4$ (5 mol %), $ZnCl_2$ (1 eq)	$ClCH_2CH_2Cl$, 25° THF, −78 to 25° THF, 25°	(*E*)-n-$C_6H_{13}C(CH_3){=}CHC_6H_5$ (89)	131
	1. $(i\text{-}C_4H_9)_2AlH$ 2. $HgCl_2$ (1.5 eq)	Hexane THF, 0°	(*E*)-n-$C_6H_{13}CH{=}CHHgCl$ (82)	132
	1. $(CH_3)_3Al$ (2 eq), $Cl_2Zr(C_5H_5)_2$ (0.2 eq) 2. 2-bromopyridine, $Pd[P(C_6H_5)_3]_4$ (3 mol %)	$ClCH_2CH_2Cl$, 25° THF, 25°, 5 hr	(*E*)-2-pyridyl-$CH{=}C(CH_3)C_6H_{13}$-n (82)	142

	1. $(CH_3)_3Al$ (1.7 eq), $Cl_2Zr(C_5H_5)_2$ (1.0 eq) 2. $HgCl_2$ (1.5 eq)	$ClCH_2CH_2Cl$, 25°, 2 hr THF, 0°	(*E*)-*n*-$C_6H_{13}C(CH_3){=}CHHgCl$ (83)	132
$C_6H_{11}C{\equiv}CH$	1. $(i\text{-}C_4H_9)_2AlH$ 2. $Ni[P(C_6H_5)_3]_4$ (5 mol %) 3. *p*-$CH_3C_6H_4Br$ (0.5 eq)	Hexane THF–hexane 50°, 24 hr	(*E*)-$C_6H_{11}CH{=}CHC_6H_4CH_3$-*p* (75)	81
n-$C_3H_7C{\equiv}CC_3H_7$-*n*	1. $(i\text{-}C_4H_9)_2AlH$ (2 eq) 2. H_3O^+	Hexane, 70°, 2 hr	(*Z*)-*n*-$C_3H_7CH{=}CHC_3H_7$-*n* (100)	73
$C_6H_5C{\equiv}CH$	1. $Al[Si(CH_3)_3]_3$ (0.33 eq) 2. CH_3OH	Ether, 25°, 1 hr	(*E*)-$C_6H_5CH{=}CHSi(CH_3)_3$ (92)	67
	1. $Al[Si(CH_3)_3]_3$ (0.33 eq) 2. D_2O	Ether, 25°, 1 hr	(*E*)-$C_6H_5CD{=}CHSi(CH_3)_3$ (—)	67
	1. $(CH_3)_3Al$ 2. H_3O^+	120°, 80 min	$C_6H_5C(CH_3){=}CH_2$ (9), (*E*)-$C_6H_5CH{=}CHCH_3$ (7)	17
	1. $(C_2H_5)_3Al$ 2. H_3O^+	120°, 80 min	(*E*)-$C_6H_5CH{=}CHC_2H_5$ (30)	17
	1. $(n\text{-}C_3H_7)_3Al$ 2. H_3O^+	120°, 80 min	(*E*)-$C_6H_5CH{=}CHC_3H_7$-*n* (25)	17
	1. $(CH_3)_3Al$ (2 eq), $Cl_2Zr(C_5H_5)_2$ (0.4 eq) 2. D_2O	$ClCH_2CH_2Cl$, 20–25°, 3 hr	(*E*)-$C_6H_5C(CH_3){=}CHD$ (95), (*E*)-$C_6H_5CD{=}CHCH_3$ (5)	64
	1. [structure: CH_3, CH_3, H, D, D, $Al(C_4H_9\text{-}i)_2$, $\ddot{O}(C_2H_5)_2$] 2. H_2O	110°	(*E*)-[structure: CH_3, CH_3, H, D, D, $CH{=}CHC_6H_5$] (—)	143
	1. $(CH_3)_3Al$ (1.7 eq), $Cl_2Zr(C_5H_5)_2$ (1.0 eq) 2. $HgCl_2$ (1.5 eq)	$ClCH_2CH_2Cl$, 25° THF, 0°	(*E*)-$C_6H_5C(CH_3){=}CHHgCl$ (81)	132

TABLE I. ALKENES (*Continued*)

	Substrate	Reagents	Conditions	Product(s) and Yield(s)(%)	Refs.
C_8 (*cont'd*)	$C_6H_5C \equiv CH$	1. $(i\text{-}C_4H_9)_2AlH$ (2 eq), $CH_3Cu\text{-}MgBrCl$ (1 eq), LiCl (2 eq) 2. H_3O^+	THF, 20°, 1.25 hr	$C_6H_5CH{=}CH_2$ (82)	141
	$C_6H_5C \equiv CD$	1. $(CH_3)_3Al$ (2 eq), $Cl_2Zr(C_5H_5)_2$ (0.4 eq) 2. H_2O	$ClCH_2CH_2Cl$, 20–25°, 3 hr	(*Z*)-$C_6H_5C(CH_3){=}CHD$ (95), (*E*)-$C_6H_5CH{=}CDCH_3$ (5)	64
	$C_6H_5SC \equiv CH$	1. $LiAlH_4$ 2. H_3O^+	THF, 25°, 1 hr	$C_6H_5SCH{=}CH_2$ (90)	54
		1. $LiAlD_4$ 2. H_3O^+	THF, 25°, 1 hr	(*Z*)-$C_6H_5SCH{=}CHD$ (93)	54
	$C_6H_5SC \equiv CD$	1. $LiAlH_4$ 2. H_3O^+	THF, 25°, 1 hr	(*E*)-$C_6H_5SCH{=}CHD$ (93)	54
	p-$ClC_6H_4SC \equiv CH$	1. $LiAlD_4$ 2. H_3O^+	THF, 25°, 1 hr	(*Z*)-*p*-$ClC_6H_4SCH{=}CHD$ (72)	54
	p-$ClC_6H_4SC \equiv CD$	1. $LiAlH_4$ 2. H_3O^+	THF, 25°, 1 hr	*p*-$ClC_6H_4SCH{=}CHD$ (66) ($E:Z$ = 87:13)	54
	$CH_2{=}C(CH_3)C \equiv CSi(CH_3)_3$	1. $(i\text{-}C_4H_9)_2AlH$ 2. H_3O^+	Hexane, 25°, 4 days	(*E*)-$CH_2{=}C(CH_3)CH{=}CHSi(CH_3)_3$ (67)	38
	$Cl(CH_2)_3C \equiv CSi(CH_3)_3$	1. $(i\text{-}C_4H_9)_2AlH$ 2. H_3O^+ 3. NBS (3 × 5 mol %)	Ether, 40°, 1 hr Ether–pyridine, *hv*, 45 min	(*E*)-$Cl(CH_2)_3CH{=}CHSi(CH_3)_3$ (82)	76
	i-$C_3H_7C \equiv CSi(CH_3)_3$	1. $(i\text{-}C_4H_9)_2AlH$ 2. H_3O^+ 3. NBS (3 × 5 mol %)	Ether, 40°, 1 hr Ether-pyridine, *hv*, 45 min	(*E*)-*i*-$C_3H_7CH{=}CHSi(CH_3)_3$ (76)	76

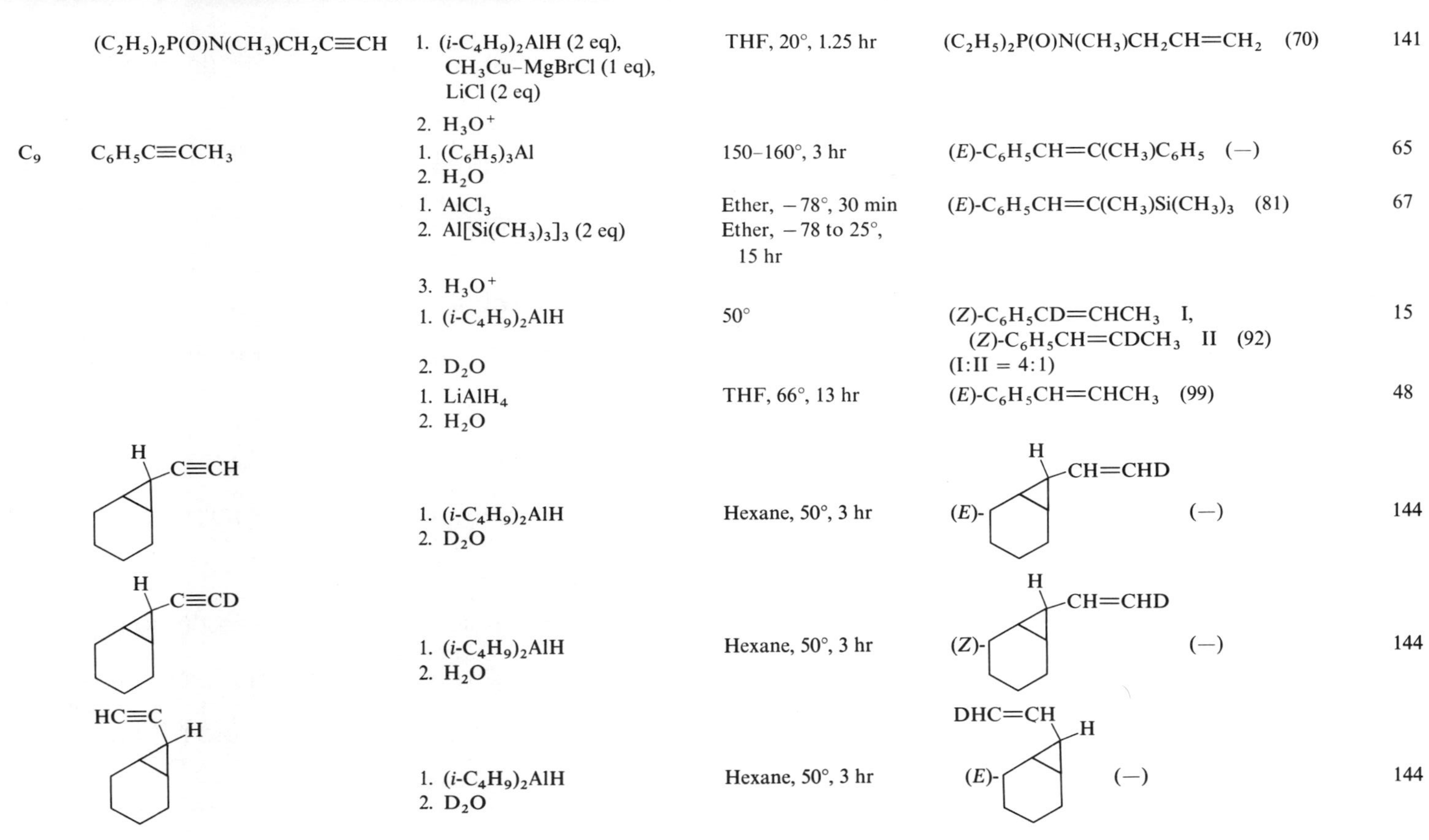

	Reactant	Reagent(s)	Conditions	Product(s) and Yield(s) (%)	Refs.
	$(C_2H_5)_2P(O)N(CH_3)CH_2C{\equiv}CH$	1. $(i\text{-}C_4H_9)_2AlH$ (2 eq), $CH_3Cu\text{–}MgBrCl$ (1 eq), LiCl (2 eq) 2. H_3O^+	THF, 20°, 1.25 hr	$(C_2H_5)_2P(O)N(CH_3)CH_2CH{=}CH_2$ (70)	141
C_9	$C_6H_5C{\equiv}CCH_3$	1. $(C_6H_5)_3Al$ 2. H_2O	150–160°, 3 hr	(*E*)-$C_6H_5CH{=}C(CH_3)C_6H_5$ (—)	65
		1. $AlCl_3$ 2. $Al[Si(CH_3)_3]_3$ (2 eq) 3. H_3O^+	Ether, −78°, 30 min Ether, −78 to 25°, 15 hr	(*E*)-$C_6H_5CH{=}C(CH_3)Si(CH_3)_3$ (81)	67
		1. $(i\text{-}C_4H_9)_2AlH$ 2. D_2O	50°	(*Z*)-$C_6H_5CD{=}CHCH_3$ I, (*Z*)-$C_6H_5CH{=}CDCH_3$ II (92) (I:II = 4:1)	15
		1. $LiAlH_4$ 2. H_2O	THF, 66°, 13 hr	(*E*)-$C_6H_5CH{=}CHCH_3$ (99)	48
	H, C≡CH	1. $(i\text{-}C_4H_9)_2AlH$ 2. D_2O	Hexane, 50°, 3 hr	(*E*)- H, CH=CHD (—)	144
	H, C≡CD	1. $(i\text{-}C_4H_9)_2AlH$ 2. H_2O	Hexane, 50°, 3 hr	(*Z*)- H, CH=CHD (—)	144
	HC≡C, H	1. $(i\text{-}C_4H_9)_2AlH$ 2. D_2O	Hexane, 50°, 3 hr	(*E*)- DHC=CH, H (—)	144

TABLE I. ALKENES (*Continued*)

Substrate	Reagents	Conditions	Product(s) and Yield(s)(%)	Refs.
C_9 *p*-$CH_3C_6H_4SC{\equiv}CD$ (*cont'd*)	1. $LiAlH_4$ 2. H_3O^+	THF, 25°, 1 hr	(*E*)-*p*-$CH_3C_6H_4SCH{=}CHD$ (81)	54
p-$CH_3C_6H_4SC{\equiv}CH$	1. $LiAlD_4$ 2. H_3O^+	THF, 25°, 1 hr	(*Z*)-*p*-$CH_3C_6H_4SCH{=}CHD$ (85)	54
p-$CH_3OC_6H_4C{\equiv}CH$	1. $Al[Si(CH_3)_3]_3$ (0.33 eq) 2. CH_3OH	Ether, 25°, 1 hr	(*E*)-*p*-$CH_3OC_6H_4CH{=}CHSi(CH_3)_3$ (89)	67
	1. $Al[Si(CH_3)_3]_3$ (0.33 eq) 2. D_2O	Ether, 25°, 1 hr	(*E*)-*p*-$CH_3OC_6H_4CD{=}CHSi(CH_3)_3$ (—)	67
n-$C_4H_9C{\equiv}CSi(CH_3)_3$	1. $(i\text{-}C_4H_9)_2AlH$ 2. H_3O^+	Ether, 25°, 7 hr	(*Z*)-*n*-$C_4H_9CH{=}CHSi(CH_3)_3$ (94)	145
	1. $(i\text{-}C_4H_9)_2AlH$ 2. H_3O^+ 3. NBS (3 × 5 mol %)	Ether, 40°, 1 hr Ether–pyridine, *hν*, 45 min	(*E*)-*n*-$C_4H_9CH{=}CHSi(CH_3)_3$ (77)	76
t-$C_4H_9C{\equiv}CSi(CH_3)_3$	1. $(i\text{-}C_4H_9)_2AlH$ 2. H_3O^+	Hexane	(*E*)-*t*-$C_4H_9CH{=}CHSi(CH_3)_3$ (65)	38
	1. $(i\text{-}C_4H_9)_2AlH$ 2. H_3O^+	*N*-Methylpyrrolidine, heptane, 60°, 48 hr	(*Z*)-*t*-$C_4H_9CH{=}CHSi(CH_3)_3$ (72)	38
	1. $(i\text{-}C_4H_9)_2AlH$ 2. H_3O^+	Ether, 25°, 7 hr	" (92)	145
	1. $(i\text{-}C_4H_9)_2AlH$ 2. H_3O^+ 3. NBS (3 × 5 mol %)	Ether, 40°, 1 hr Ether–pyridine, *hν*, 45 min	(*E*)-*t*-$C_4H_9CH{=}CHSi(CH_3)_3$ (82)	76
$C_6H_5C{\equiv}CSCH_3$	1. $LiAlH_4$ (1 eq) 2. H_3O^+	THF, 50–60°, 1 hr	(*E*)-$C_6H_5CH{=}CHSCH_3$ (90)	53

	$C_6H_5SC{\equiv}CCH_3$	1. $(i\text{-}C_4H_9)_2AlH$ (2 eq), $CH_3Cu\text{–}MgBrCl$ (1 eq), LiCl (2 eq) 2. H_3O^+	THF, 20°, 1.25 hr	(*Z*)-$C_6H_5SCH{=}CHCH_3$ (70)	141
C_{10}	[structure]	1. $(i\text{-}C_4H_9)_2AlH$ (1.5 eq) 2. H_3O^+	40°, 12 hr	[structure] (—)	2
	$(CH_3)_2PC{\equiv}CC_6H_5$	1. $(i\text{-}C_4H_9)_2AlH$ 2. H_3O^+	Heptane, 45–50°, 15 hr	(*Z*)-$(CH_3)_2PCH{=}CHC_6H_5$ (70)	38
	$(CH_3)_2NC{\equiv}CC_6H_5$	1. $(i\text{-}C_4H_9)_2AlH$ 2. H_3O^+	*N*-Methylpyrrolidine, heptane, 30°	(*E*)-$(CH_3)_2NCH{=}CHC_6H_5$ (—)	39
	$C_2H_5SC{\equiv}CC_6H_5$	1. $(i\text{-}C_4H_9)_2AlH$ 2. H_3O^+	Cyclopentane, 25°, 12 hr	(*Z*)-$C_2H_5SCH{=}CHC_6H_5$ (100)	39
	$n\text{-}C_4H_9C{\equiv}CC_4H_9\text{-}n$	1. $(CH_3)_3Al$ (2 eq), $Cl_2Zr(C_5H_5)_2$ (0.4 eq) 2. H_2O	$ClCH_2CH_2Cl$, 50°, 6 hr	(*Z*)-$n\text{-}C_4H_9C(CH_3){=}CHC_4H_9\text{-}n$ (89)	64
		1. $(i\text{-}C_4H_9)_2AlH$ (2 eq) 2. H_3O^+	Hexane, 50°, 6 hr	(*Z*)-$n\text{-}C_4H_9CH{=}CHC_4H_9\text{-}n$ (89)	73
C_{11}	$n\text{-}C_6H_{13}C{\equiv}CSi(CH_3)_3$	1. $(CH_3)_3Al$ (2 eq), $Cl_2Zr(C_5H_5)_2$ (1 eq) 2. H_2O	Not specified	$n\text{-}C_6H_{13}C(CH_3){=}CHSi(CH_3)_3$ (—) (*E*:*Z* = 1:1)	146
		1. $(i\text{-}C_4H_9)_2AlH$ 2. H_3O^+	Ether, 40°, 1 hr	(*Z*)-$n\text{-}C_6H_{13}CH{=}CHSi(CH_3)_3$ (95)	73
		1. $(i\text{-}C_4H_9)_2AlH$ 2. CH_3Li 3. CH_3I	Heptane–ether, 25°, 17 hr 0°, 20 min 25°, 20 hr	(*Z*)-$n\text{-}C_6H_{13}CH{=}C(CH_3)Si(CH_3)_3$ (88)	43
	$C_6H_{11}C{\equiv}CSi(CH_3)_3$	1. $(i\text{-}C_4H_9)_2AlH$ 2. H_3O^+	Ether, 25°, 7 hr	(*Z*)-$C_6H_{11}CH{=}CHSi(CH_3)_3$ (91)	145

TABLE I. ALKENES (*Continued*)

Substrate	Reagents	Conditions	Product(s) and Yield(s)(%)	Refs.
C_{11} $C_6H_{11}C{\equiv}CSi(CH_3)_3$ (*cont'd*)	1. $(i\text{-}C_4H_9)_2AlH$ 2. H_3O^+ 3. NBS (3 × 5 mol %)	Ether, 40°, 1 hr Ether–pyridine, *hv*, 45 min	(*E*)-$C_6H_{11}CH{=}CHSi(CH_3)_3$ (88)	76
$C_6H_5C{\equiv}CSi(CH_3)_3$	1. $(i\text{-}C_4H_9)_2AlH$ 2. H_3O^+	Ether, 25°, 7 hr	(*Z*)-$C_6H_5CH{=}CHSi(CH_3)_3$ (96)	145
	1. $(i\text{-}C_4H_9)_2AlH$ 2. H_3O^+	Pentane, 25°, 15 hr	(*E*)-$C_6H_5CH{=}CHSi(CH_3)_3$ (96)	38
	1. $(i\text{-}C_4H_9)_2AlH$ 2. H_3O^+	*N*-Methylpyrrolidine, heptane, 55–60°, 24 hr	(*Z*)-$C_6H_5CH{=}CHSi(CH_3)_3$ (96)	38
$C_6H_5C{\equiv}CGe(CH_3)_3$	1. $(i\text{-}C_4H_9)_2AlH$ 2. H_3O^+	Pentane, 50°	(*E*)-$C_6H_5CH{=}CHGe(CH_3)_3$ (90–93)	38
	1. $(i\text{-}C_4H_9)_2AlH$ 2. H_3O^+	*N*-Methylpyrrolidine, heptane, 60°, 15 hr, then reflux 5 hr	(*Z*)-$C_6H_5CH{=}CHGe(CH_3)_3$ (90–94)	38
$n\text{-}C_8H_{17}C{\equiv}CCH_3$	1. $(i\text{-}C_4H_9)_2AlH$ 2. CH_3OH	65°	(*Z*)-$n\text{-}C_8H_{17}CH{=}CHCH_3$ (—)	147
$n\text{-}C_7H_{15}C{\equiv}CC_2H_5$	1. $(i\text{-}C_4H_9)_2AlH$ 2. CH_3OH	65°	(*Z*)-$n\text{-}C_7H_{15}CH{=}CHC_2H_5$ (—)	147
$n\text{-}C_6H_{13}C{\equiv}CC_3H_7\text{-}n$	1. $(i\text{-}C_4H_9)_2AlH$ 2. CH_3OH	65°	(*Z*)-$n\text{-}C_6H_{13}CH{=}CHC_3H_7\text{-}n$ (—)	147
$n\text{-}C_5H_{11}C{\equiv}CC_4H_9\text{-}n$	1. $(i\text{-}C_4H_9)_2AlH$ 2. CH_3OH	65°	(*Z*)-$n\text{-}C_5H_{11}CH{=}CHC_4H_9\text{-}n$ (—)	147

	Reactant	Reagents	Conditions	Product(s) (Yield, %)	Refs.
	$C_6H_5S(CH_2)_3C{\equiv}CH$	1. $(i\text{-}C_4H_9)_2AlH$ (2 eq), $CH_3Cu\text{–}MgBrCl$ (1 eq), LiCl (2 eq) 2. H_3O^+	THF, 20°, 1.25 hr	$C_6H_5S(CH_2)_3CH{=}CH_2$ (85)	141
	$(CH_3)_3SiC(CH_3)_2C{\equiv}CSi(CH_3)_3$	1. $(i\text{-}C_4H_9)_2AlH$ 2. H_2O	Not specified	(*Z*)-$(CH_3)_3SiC(CH_3)_2CH{=}CHSi(CH_3)_3$ (100)	148
C_{12}	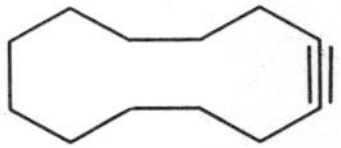	1. $(i\text{-}C_4H_9)_2AlH$ (2 eq) 2. CH_3OH	45°, 3–4 hr	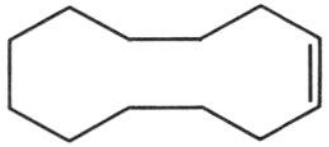 (60)	74
	$C_6H_5C{\equiv}CC_4H_9\text{-}t$	1. $(i\text{-}C_4H_9)_2AlH$ 2. H_3O^+	Heptane, 50°, 18 hr	(*Z*)-$C_6H_5CH{=}CHC_4H_9\text{-}t$ (84)	38
		1. $(C_6H_5)_3Al$ 2. H_2O	Toluene, 90°, 4 days, then reflux 24 hr	$(C_6H_5)_2C{=}CHC_4H_9\text{-}t$ (75)	149
	$n\text{-}C_4H_9C{\equiv}C(CH_2)_6OH$	1. $LiAlH_4$ 2. H_2O	Diglyme, THF, 140°, 48–55 hr	(*E*)-$n\text{-}C_4H_9CH{=}CH(CH_2)_6OH$ (93)	58
C_{13}	$(C_2H_5)_2NCH_2C{\equiv}CC_6H_5$	1. $(i\text{-}C_4H_9)_2AlH$ (2.5 eq) 2. H_2O	Toluene, 40°, 3 hr	(*E*)-$(C_2H_5)_2NCH_2CH{=}CHC_6H_5$ (84)	150
C_{14}	$n\text{-}C_6H_{13}C{\equiv}CSi(C_2H_5)_3$	1. $(i\text{-}C_4H_9)_2AlH$ 2. CH_3Li 3. CH_3I	Heptane, 25°, 20 hr Heptane–ether, 0°	$n\text{-}C_6H_{13}CH{=}C(CH_3)Si(C_2H_5)_3$ (68) ($E{:}Z = 93{:}7$)	43
	$C_6H_5C{\equiv}CGe(C_2H_5)_3$	1. $(i\text{-}C_4H_9)_2AlH$ 2. H_3O^+	Pentane, 50°	(*E*)-$C_6H_5CH{=}CHGe(C_2H_5)_3$ (90–93)	38
	$C_6H_5C{\equiv}CC_6H_5$	1. $(C_6H_5CH_2)_3Al$ 2. H_2O	Xylene, reflux, 48 hr	(*Z*)-$C_6H_5CH{=}CH(C_6H_5)CH_2C_6H_5$ (80)	151
		1. $(C_2H_5)_3Al$ 2. H_2O	85°, 20 hr	(*Z*)-$C_6H_5C(C_2H_5){=}CHC_6H_5$ (60)	2
		1. $(C_6H_5)_3Al$ 2. H_2O	$C_6H_5OC_6H_5$, 200°, 12 hr	$(C_6H_5)_2C{=}CHC_6H_5$ (68)	65

TABLE I. ALKENES (*Continued*)

Substrate	Reagents	Conditions	Product(s) and Yield(s) (%)	Refs.
C_{14} $C_6H_5C\equiv CC_6H_5$ (*cont'd*)	1. $LiAlH_4$ 2. H_2O	66°, 2 hr	(*E*)-$C_6H_5CH=CHC_6H_5$ (100)	48
	1. $(i\text{-}C_4H_9)_2AlH$ 2. H_3O^+	45°, 12 hr	(*Z*)-$C_6H_5CH=CHC_6H_5$ (73)	1
	1. $[(C_6H_5)_3Si]_2Al(C_2H_5)$, 2 LiBr, 2 THF 2. H_2O	Toluene, 120°, 3 hr	$[(C_6H_5)_3Si]C(C_6H_5)=CHC_6H_5$ (96)	152
C_{16} $n\text{-}C_8H_{17}C\equiv C(CH_2)_6OH$	1. $LiAlH_4$ 2. H_2O	Diglyme, THF, 140°, 48–55 hr	(*E*)-$n\text{-}C_8H_{17}CH=CH(CH_2)_6OH$ (94)	58
C_{17} $sec\text{-}C_4H_9(CH_2)_4C\equiv C(CH_2)_7OH$	1. $LiAlH_4$ 2. H_2O	Diglyme, THF, 140°, 48–55 hr	(*E*)-$sec\text{-}C_4H_9(CH_2)_4CH=CH(CH_2)_7OH$ (85)	58
C_{27} $C_6H_5C\equiv CC(C_6H_5)_3$	1. $(i\text{-}C_4H_9)_2AlH$ 2. H_3O^+	Toluene, 110°	(*Z*)-$C_6H_5CH=CHC(C_6H_5)_3$ (40)	15
C_{36} $n\text{-}C_{18}H_{37}C\equiv C(CH_2)_{16}OH$	1. $LiAlH_4$ 2. H_2O	Diglyme, 125°, 12 days	(*E*)-$n\text{-}C_{18}H_{37}CH=CH(CH_2)_{16}OH$ (85)	153

TABLE II. ALKENYL HALIDES

	Substrate	Reagents	Conditions	Product(s) and Yield(s)(%)	Refs.
C_2	$HC{\equiv}CH$	1. $(i\text{-}C_4H_9)_3Al$ 2. Br_2	Hexane–ether, -50 to 25°	$i\text{-}C_4H_9CH{=}CHBr$ (78)	84
		1. $(i\text{-}C_4H_9)_3Al$ 2. I_2	Ether, 0 to 25°	$i\text{-}C_4H_9CH{=}CHI$ (90)	84
C_3	$HOCH_2C{\equiv}CBr$	1. $LiAlH_4$, $AlCl_3$ 2. H_3O^+	Ether, reflux, 2 hr	(E)-$HOCH_2CH{=}CHBr$ (50)	154
C_4	$CH_3OCH_2C{\equiv}CBr$	1. $LiAlH_4$, $AlCl_3$ 2. H_3O^+	Ether, reflux, 2 hr	(E)-$CH_3OCH_2CH{=}CHBr$ (39)	154
	$HO(CH_2)_2C{\equiv}CH$	1. $(CH_3)_3Al$ (3 eq), $Cl_2Zr(C_5H_5)_2$ (0.25 eq) 2. I_2	$ClCH_2CH_2Cl$, 25° THF	(E)-$HO(CH_2)_2C(CH_3){=}CHI$ (62)	66
		1. $(CH_3)_3Al$ (3 eq), $Cl_2Zr(C_5H_5)_2$ (0.25 eq) 2. I_2 3. $(CH_3)_2(t\text{-}C_4H_9)SiCl$, $(C_2H_5)_3N$, 4-(dimethylamino)pyridine [structure: pyridine ring bearing $N(CH_3)_2$]	$ClCH_2CH_2Cl$, 25° THF CH_2Cl_2	(E)-$(CH_3)_2(t\text{-}C_4H_9)SiO(CH_2)_2C(CH_3){=}CHI$ (87)	66
	$I(CH_2)_2C{\equiv}CH$	1. $(CH_3)_3Al$ (3 eq), $Cl_2Zr(C_5H_5)_2$ (1 eq) 2. I_2	$ClCH_2CH_2Cl$ THF	(E)-$I(CH_2)_2C(CH_3){=}CHI$ (60)	66
C_5	$CH_2{=}C(CH_3)C{\equiv}CH$	1. $(CH_3)_3Al$ (2 eq), $Cl_2Zr(C_5H_5)_2$ (0.4 eq) 2. I_2	$ClCH_2CH_2Cl$, 25°, 24 hr THF–$ClCH_2CH_2Cl$, 0°	(E)-$CH_2{=}C(CH_3)C(CH_3){=}CHI$ (70)	88
	$CH_2{=}C(CH_3)C{\equiv}CBr$	1. $LiAlH_4$, $AlCl_3$ 2. H_3O^+	Ether, reflux, 3 hr	(E)-$CH_2{=}C(CH_3)CH{=}CHBr$ (—)	154

TABLE II. ALKENYL HALIDES (*Continued*)

	Substrate	Reagents	Conditions	Product(s) and Yield(s)(%)	Refs.
C_6	n-$C_4H_9C{\equiv}CH$	1. $(i\text{-}C_4H_9)_2AlH$ 2. I_2 (1.2 eq)	Hexane, 50°, 4 hr THF, −78 to 0°	(E)-n-$C_4H_9CH{=}CHI$ (69)	85
		1. $(i\text{-}C_4H_9)_2AlH$ 2. Br_2	Hexane–ether, −50 to 25°	n-$C_4H_9CH{=}CHBr$ (77)	84
		1. $(i\text{-}C_4H_9)_2AlH$ 2. I_2	Ether, 0 to 25°	n-$C_4H_9CH{=}CHI$ (89)	84
		1. $(CH_3)_3Al$ (2 eq), $Cl_2Zr(C_5H_5)_2$ (0.4 eq) 2. I_2	$ClCH_2CH_2Cl$, 25°, 24 hr THF–$ClCH_2CH_2Cl$, 0°	(E)-n-$C_4H_9C(CH_3){=}CHI$ (85)	88
		1. $(i\text{-}C_4H_9)_2AlH$ 2. I_2	Heptane, 50°, 2 hr THF, −50 to 25°	(E)-n-$C_4H_9CH{=}CHI$ (94)	83
		1. $(i\text{-}C_4H_9)_2AlH$ 2. ICl	Heptane, 50°, 2 hr THF, −50 to 25°	" (76)	83
		1. $(i\text{-}C_4H_9)_2AlH$ 2. Br_2	Heptane, 50°, 2 hr THF, −50 to 25°	(E)-n-$C_4H_9CH{=}CHBr$ (72)	84
	t-$C_4H_9C{\equiv}CH$	1. $(i\text{-}C_4H_9)_2AlH$ 2. I_2 (1.2 eq)	Hexane, 25°, 4 hr THF, −78 to 0°	(E)-t-$C_4H_9CH{=}CHI$ (82)	85
	$C_2H_5C{\equiv}CC_2H_5$	1. $(i\text{-}C_4H_9)_2AlH$ 2. Br_2	Hexane–ether, −50 to 25°	$C_2H_5CH{=}C(Br)C_2H_5$ (69)	84
		1. $(i\text{-}C_4H_9)_2AlH$ 2. I_2	Ether, 0 to 25°	$C_2H_5CH{=}CIC_2H_5$ (85)	84
		1. $(i\text{-}C_4H_9)_2AlH$ 2. I_2	Heptane, 50°, 2 hr THF, −50 to 25°	(E)-$C_2H_5CH{=}CIC_2H_5$ (72)	83
		1. $(i\text{-}C_4H_9)_2AlH$ 2. Br_2	Heptane, 50°, 2 hr THF, −50 to 25°	(E)-$C_2H_5CH{=}CBrC_2H_5$ (42)	83

		1. $(i\text{-}C_4H_9)_2AlH$ (0.5 eq) 2. NBS	Hexane, 70°, 12 hr Ether, −30 to 25°, 1 hr	$C_2H_5CH{=}C(C_2H_5)C(C_2H_5){=}C(Br)C_2H_5$ (74)	85
		1. $(i\text{-}C_4H_9)_2(CH_3)AlHLi$ (2 eq) 2. I_2	DME, 100–130°, 6 hr	(Z)-$C_2H_5CH{=}CIC_2H_5$ (60)	49
	$CH_2{=}CHCOH(CH_3)C{\equiv}CBr$	1. $LiAlH_4$, $AlCl_3$ 2. H_3O^+	Ether, reflux, 2 hr	(E)-$CH_2{=}CHCOH(CH_3)CH{=}CHBr$ (58)	154
	$n\text{-}C_4H_9C{\equiv}CCl$	1. $LiAlH_4$ (1.0 eq) 2. CH_3OH	THF, −30 to 0°, 90 min	(E)-$n\text{-}C_4H_9CH{=}CHCl$ (96)	55
	$t\text{-}C_4H_9C{\equiv}CCl$	1. $LiAlH_4$ (1.0 eq) 2. CH_3OH	THF, −30 to 0°, 90 min	(E)-$t\text{-}C_4H_9CH{=}CHCl$ (76)	55
	$HC{\equiv}CCH_2Si(CH_3)_3$	1. $(CH_3)_3Al$ (2 eq), $Cl_2Zr(C_5H_5)_2$ (0.1 eq) 2. I_2	$ClCH_2CH_2Cl$, 25°, 2 hr THF, −30°	(E)-$ICH{=}C(CH_3)CH_2Si(CH_3)_3$ (63)	155
C_7	$n\text{-}C_5H_{11}C{\equiv}CH$	1. $(CH_3)_3Al$ (2 eq), $Cl_2Zr(C_5H_5)_2$ (0.4 eq) 2. I_2	$ClCH_2CH_2Cl$, 25°, 24 hr THF–$ClCH_2CH_2Cl$, 0°	(E)-$n\text{-}C_5H_{11}C(CH_3){=}CHI$ (83)	88
	$(C_2H_5)_2NCH_2C{\equiv}CBr$	1. $LiAlH_4$, $AlCl_3$ 2. H_3O^+	Ether, reflux	(E)-$(C_2H_5)_2NCH_2CH{=}CHBr$ (47)	154
	$CH_2{=}CH(CH_2)_3C{\equiv}CH$	1. $(i\text{-}C_4H_9)_2AlH$ 2. I_2	Not specified	(E)-$CH_2{=}CH(CH_2)_3CH{=}CHI$ (—)	156
C_8	$n\text{-}C_6H_{13}C{\equiv}CH$	1. $(i\text{-}C_4H_9)_2AlH$ 2. Br_2	Hexane, 50°, 2 hr THF, −50 to 25°	(E)-$n\text{-}C_6H_{13}CH{=}CHBr$ (75)	157
		1. $(i\text{-}C_4H_9)_2AlH$ 2. NCS (1.2 eq)	Hexane, 50°, 4 hr CH_2Cl_2, −78 to 0°, 1 hr	(E)-$n\text{-}C_6H_{13}CH{=}CHCl$ (67), $n\text{-}C_6H_{13}C{\equiv}CCl$ (5)	85

TABLE II. ALKENYL HALIDES (*Continued*)

Substrate	Reagents	Conditions	Product(s) and Yield(s)(%)	Refs.
C_8 n-$C_6H_{13}C{\equiv}CH$ (*cont'd*)	1. $(i$-$C_4H_9)_2AlH$ 2. NBS (1.2 eq)	Hexane, 50°, 4 hr Ether, −30 to 25°, 1 hr	(E)-n-$C_6H_{13}CH{=}CHBr$ (78), n-$C_6H_{13}C{\equiv}CBr$ (4)	85
	1. $(i$-$C_4H_9)_2AlH$ 2. I_2 (1.2 eq)	Hexane, 50°, 4 hr THF, −78 to 0°	(E)-n-$C_6H_{13}CH{=}CHI$ (75), n-$C_6H_{13}C{\equiv}CI$ (1)	85
	1. $(i$-$C_4H_9)_2AlH$ 2. I_2	Heptane, 50°, 2 hr THF, −50 to 25°	(E)-n-$C_6H_{13}CH{=}CHI$ (86)	24, 158
	1. $(i$-$C_4H_9)_2AlH$ 2. n-C_4H_9Li 3. $CH_2{=}CClCH_2SSO_2C_6H_5$	Heptane, 50°, 4 hr Heptane–toluene, −70 to 25°	(E)-n-$C_6H_{13}CH{=}CHSCH_2CCl{=}CH_2$ (55)	90
n-$C_6H_{13}C{\equiv}CCl$	1. $LiAlH_4$ (1.0 eq) 2. CH_3OH	THF, −30 to 0°, 90 min	(E)-n-$C_6H_{13}CH{=}CHCl$ (80)	55
$C_6H_{11}C{\equiv}CH$	1. $(i$-$C_4H_9)_2AlH$ 2. NCS (1.2 eq)	Hexane, 50°, 2 hr Hexane–ether, −78 to 25°	(E)-$C_6H_{11}CH{=}CHCl$ (70), $C_6H_{11}C{\equiv}CCl$ (5)	85
	1. $(i$-$C_4H_9)_2AlH$ 2. NBS (1.2 eq)	Hexane, 50°, 2 hr Ether, −30 to 25°, 1 hr	(E)-$C_6H_{11}CH{=}CHBr$ (78), $C_6H_{11}C{\equiv}CBr$ (1)	85
	1. $(i$-$C_4H_9)_2AlH$ 2. I_2	Hexane, 50°, 2 hr THF, −78 to 0°	(E)-$C_6H_{11}CH{=}CHI$ (76)	85
$C_6H_5C{\equiv}CH$	1. $(CH_3)_3Al$ (2 eq), $Cl_2Zr(C_5H_5)_2$ (0.4 eq) 2. I_2	$ClCH_2CH_2Cl$, 25°, 24 hr THF–$ClCH_2CH_2Cl$, 0°	(E)-$C_6H_5C(CH_3){=}CHI$ (73)	64
$C_6H_5C{\equiv}CBr$	1. $LiAlH_4$, $AlCl_3$ 2. H_3O^+	Ether, reflux, 10 hr	(E)-$C_6H_5CH{=}CHBr$ (45)	154
$C_6H_{11}C{\equiv}CCl$	1. $LiAlH_4$ (1.0 eq) 2. CH_3OH	THF, −30 to 0°, 90 min	(E)-$C_6H_{11}CH{=}CHCl$ (93)	55

(*S*)-*n*-C_5H_{11}CHOHC≡CH	1. $(i\text{-}C_4H_9)_3Al$ (2 eq) 2. $(i\text{-}C_4H_9)_2AlH$ 3. I_2 (3 eq)	Heptane, 10–20° 56–58°, 2 hr THF, −50 to 25°	(*S*)-(*E*)-*n*-C_5H_{11}CHOHCH=CHI (47)	24, 159, 160
n-C_3H_7C≡CC_3H_7-*n*	1. $(i\text{-}C_4H_9)_2AlH$ (2 eq) 2. CH_3COCH_3 (1 eq) 3. NBS (1.2 eq)	Hexane, 70°, 2 hr Ether, −78 to −30° Ether, −30 to 25°, 1 hr	(*E*)-*n*-C_3H_7CH=CBrC_3H_7-*n* (78)	85
	1. $(i\text{-}C_4H_9)_2AlH$ (2 eq) 2. CH_3COCH_3 (1 eq) 3. I_2 (1.2 eq)	Hexane, 70°, 2 hr THF, −78° THF, −78 to 0°	(*E*)-*n*-C_3H_7CH=CIC_3H_7-*n* (79)	85
n-C_5H_{11}CHOHC≡CH	1. $(CH_3)_2Al$ (3 eq), $I_2Zr(C_5H_5)_2$ (1 eq) 2. I_2	$ClCH_2CH_2Cl$ THF	(*E*)-*n*-C_5H_{11}CHOHC(CH_3)=CHI (60)	66
$(CH_3)_2C$=CH$(CH_2)_2$C≡CH	1. $(CH_3)_3Al$ (2 eq), $Cl_2Zr(C_5H_5)_2$ (0.4 eq) 2. I_2	$ClCH_2CH_2Cl$, 25°, 24 hr THF–$ClCH_2CH_2Cl$, 0°	(*E*)-$(CH_3)_2C$=CH$(CH_2)_2$C(CH_3)=CHI (72)	88
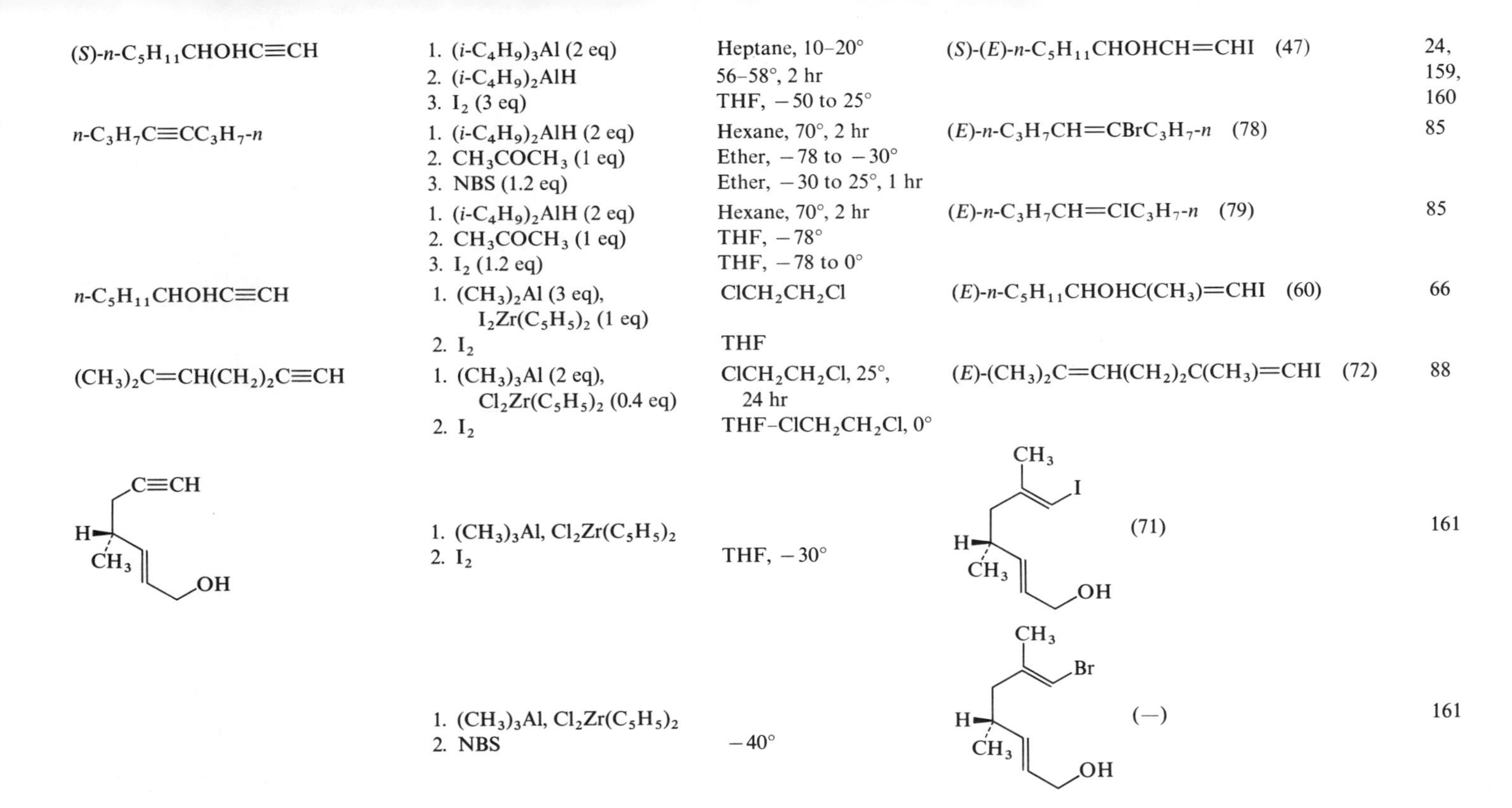	1. $(CH_3)_3Al$, $Cl_2Zr(C_5H_5)_2$ 2. I_2	THF, −30°	(71)	161
	1. $(CH_3)_3Al$, $Cl_2Zr(C_5H_5)_2$ 2. NBS	−40°	(—)	161

TABLE II. ALKENYL HALIDES (*Continued*)

	Substrate	Reagents	Conditions	Product(s) and Yield(s)(%)	Refs.
C_9	$n\text{-}C_4H_9C{\equiv}CSi(CH_3)_3$	1. $(i\text{-}C_4H_9)_2AlH$ 2. I_2 3. $t\text{-}C_4H_9Li$ (15 mol %) 4. $NaOCH_3$	Ether, 40°, 1 hr Ether, −78 to 0° Pentane, −70 to 25° THF–CH_3OH, 25°, 4 hr	(*Z*)-$n\text{-}C_4H_9CH{=}CHI$ (69)	46
		1. $(i\text{-}C_4H_9)_2AlH$ 2. I_2 3. $NaOCH_3$	Ether, 40°, 1 hr Ether, −78 to 0° CH_3OH, 40°, 2 hr	(*E*)-$n\text{-}C_4H_9CH{=}CHI$ (75)	46
		1. $(i\text{-}C_4H_9)_2AlH$ 2. Br_2 (1.5 eq) 3. $NaOCH_3$	Ether, 40°, 1 hr Ether–pyridine, −78° CH_3OH, 40°, 2 hr	(*E*)-$n\text{-}C_4H_9CH{=}CHBr$ (76)	46
	$t\text{-}C_4H_9C{\equiv}CSi(CH_3)_3$	1. $(i\text{-}C_4H_9)_2AlH$ 2. Br_2 (1.5 eq) 3. $NaOCH_3$	Ether, 40°, 1 hr Ether–pyridine, −78° CH_3OH, 40°, 2 hr	(*E*)-$t\text{-}C_4H_9CH{=}CHBr$ (93)	46
	$C_6H_5SCH_2C{\equiv}CH$	1. $(CH_3)_3Al$ (3 eq), $Cl_2Zr(C_5H_5)_2$ (1 eq) 2. I_2	$ClCH_2CH_2Cl$, 25° THF	(*E*)-$C_6H_5SCH_2C(CH_3){=}CHI$ (75)	88 66
C_{10}	[structure: C_2H_5-substituted alkene chain ending in $C{\equiv}CCH_2OH$]	1. $LiAlH_4$ (2 eq), $NaOCH_3$ (4 eq) 2. I_2 (excess)	THF, reflux, 45 min −60°	[structure: C_2H_5-substituted diene bearing I, ending in CH_2OH] (65)	162
	$(CH_3O)_2CH(CH_2)_4C{\equiv}CCH_2OLi$	1. $(i\text{-}C_4H_9)_2AlH$ 2. I_2	Not specified	(*Z*)-$(CH_3O)_2CH(CH_2)_4CH{=}CICH_2OH$ (38)	163
	$t\text{-}C_4H_9Si(CH_3)_2O(CH_2)_2C{\equiv}CH$	1. $(CH_3)_3Al$ (3 eq), $Cl_2Zr(C_5H_5)_2$ (1 eq) 2. I_2, THF	$ClCH_2CH_2Cl$	(*E*)-$t\text{-}C_4H_9Si(CH_3)_2O(CH_2)_2C(CH_3){=}CHI$ (52)	66

	Reactant	Reagents	Conditions	Product(s) and Yield(s) (%)	Refs.
	n-$C_7H_{15}C{\equiv}CCH_2OH$	1. n-C_4H_9Li	Ether, $-20°$	(Z)-n-$C_7H_{15}CH{=}CICH_2OH$ (67)	62
		2. (i-$C_4H_9)_2AlH$ (3 eq)	Ether, -20 to $35°$, 48 hr		
		3. $CH_3CO_2C_2H_5$	$0°$		
		4. I_2 (9 eq)	Ether, $-78°$, 10 min		
		1. n-C_4H_9Li	Ether, $-20°$	(Z)-n-$C_7H_{15}CH{=}CBrCH_2OH$ (75)	62
		2. ($iC_4H_9)_2AlH$ (3 eq)	Ether, -20 to $35°$, 48 hr		
		3. $CH_3CO_2C_2H_5$	$0°$		
		4. Br_2 (9 eq)	Ether, $-78°$		
	$THPO(CH_2)_3C{\equiv}CCl$	1. $LiAlH_4$ (1.0 eq)	THF, -30 to $0°$, 90 min	(E)-$THPO(CH_2)_3CH{=}CHCl$ (95)	55
		2. CH_3OH			
C_{11}	n-$C_6H_{13}C{\equiv}CSi(CH_3)_3$	1. (i-$C_4H_9)_2AlH$	Heptane–ether, $25°$, 17 hr	(Z)-n-$C_6H_{13}CH{=}CSi(CH_3)_3CH_2CCl{=}CH_2$ (77)	43
		2. CH_3Li	Heptane–ether, $0°$, 20 min		
		3. $ClCH_2CCl{=}CH_2$	0 to $25°$, 20 hr		
		1. (i-$C_4H_9)_2AlH$	Ether, $40°$, 1 hr	(Z)-n-$C_6H_{13}CH{=}CHBr$ (69)	46
		2. Br_2 (1.5 eq)	Ether–pyridine, $-78°$		
		3. Br_2 (15 mol %)	Ether–pyridine, $h\nu$		
		4. $NaOCH_3$	CH_3OH, $40°$, 4 hr		
		1. (i-$C_4H_9)_2AlH$	Ether, $40°$, 1 hr	(Z)-n-$C_6H_{13}CH{=}CHCl$ (76)	46
		2. NCS	Ether, -25 to $0°$		
		3. Br_2 (15 mol %)	Ether–pyridine, $h\nu$		
		4. $NaOCH_3$	CH_3OH, $65°$, 4 hr		
		1. (i-$C_4H_9)_2AlH$	Ether, $40°$, 1 hr	(E)-n-$C_6H_{13}CH{=}CHCl$ (84)	46
		2. NCS	Ether, -25 to $0°$		
		3. $NaOCH_3$	CH_3OH, $40°$, 4 hr		
	$C_6H_{11}C{\equiv}CSi(CH_3)_3$	1. (i-$C_4H_9)_2AlH$	Ether, $40°$, 1 hr	(E)-$C_6H_{11}CH{=}CHBr$ (93)	46
		2. Br_2 (1.5 eq)	Ether–pyridine, $-78°$		
		3. $NaOCH_3$	CH_3OH, $40°$, 2 hr		

TABLE II. ALKENYL HALIDES (*Continued*)

Substrate	Reagents	Conditions	Product(s) and Yield(s) (%)	Refs.
C_{11} $C_6H_{11}C{\equiv}CSi(CH_3)_3$ (*cont'd*)	1. $(i\text{-}C_4H_9)_2AlH$ 2. Br_2 (1.5 eq) 3. Br_2 (15 mol %) 4. $NaOCH_3$	Ether, 40°, 1 hr Ether–pyridine, −78° Ether–pyridine, *hv* CH_3OH, 40°, 4 hr	(*Z*)-$C_6H_{11}CH{=}CHBr$ (74)	46
$C_6H_5C{\equiv}CSi(CH_3)_3$	1. $(i\text{-}C_4H_9)_2AlH$ 2. Br_2 (1.5 eq) 3. $NaOCH_3$	Ether, 40°, 4 hr Ether–pyridine, −78° CH_3OH, 40°, 2 hr	(*E*)-$C_6H_5CH = CHBr$ (88)	46
	1. $(i\text{-}C_4H_9)_2AlH$ 2. Br_2 (1.5 eq) 3. Br_2 (15 mol %) 4. $NaOCH_3$	Ether, 40°, 4 hr Ether–pyridine, −78° Ether–pyridine, *hv* CH_3OH, 40°, 4 hr	(*Z*)-$C_6H_5CH{=}CHBr$ (71)	46
C_{12} $CH_3CH(OTHP)C{\equiv}CSi(CH_3)_3$	1. $(i\text{-}C_4H_9)_2AlH$ 2. Br_2 (1.5 eq) 3. $NaOCH_3$	Hexane–THF (1.1 eq), 0°, 15 min to 25°, 4 hr Hexane–THF, −78° CH_3OH, 40°, 4 hr	(*E*)-$CH_3CH(OTHP)CH{=}CHBr$ (82)	46
	1. $(i\text{-}C_4H_9)_2AlH$ 2. Br_2 (1.5 eq) 3. Br_2 (15 mol %) 4. $NaOCH_3$	Hexane–THF (1.1 eq), 0°, 15 min; 25°, 4 hr Hexane–THF, −78° Ether–pyridine, *hv* CH_3OH, 40°, 4 hr	(*Z*)-$CH_3CH(OTHP)CH{=}CHBr$ (75)	46
	1. $(i\text{-}C_4H_9)_2AlH$ 2. ICl (1.5 eq) 3. $NaOCH_3$	Hexane–THF (1.1 eq), 0°, 15 min; 25°, 4 hr CH_3OH, 40°, 2 hr	(*E*)-$CH_3CH(OTHP)CH{=}CHI$ (68)	46
n-$C_5H_{11}CH(OC_4H_9\text{-}t)C{\equiv}CH$	1. $(i\text{-}C_4H_9)_2AlH$ 2. Br_2	C_6H_6–toluene THF	(*Z*)-*n*-$C_5H_{11}CH(OC_4H_9\text{-}t)CH{=}CHBr$ (—)	25

	Reactant	Reagent(s)	Conditions	Product(s) and Yield(s) (%)	Refs.
	n-$C_4H_9C{\equiv}CC{\equiv}CC_4H_9$-$n$	1. $(i\text{-}C_4H_9)_2(CH_3)AlHLi$ 2. CNBr (2 eq)	Diglyme–ether, 45°, 2 hr −78 to 25°	(Z)-n-$C_4H_9CH{=}CBrC{\equiv}CC_4H_9$-$n$ (88)	45
C_{13}	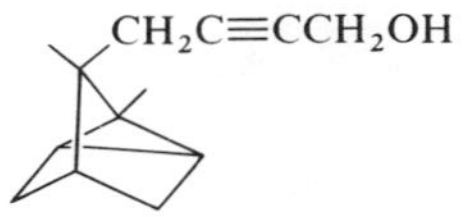 [bicyclic structure]-$CH_2C{\equiv}CCH_2OH$	1. n-C_4H_9Li 2. $(i\text{-}C_4H_9)_2AlH$ 3. $CH_3CO_2C_2H_5$ 4. I_2 (9 eq)	Ether, −20° Ether, −20 to 35°, 48 hr 0° Ether, −78°	(Z)-[bicyclic structure]-$CH_2CH{=}CICH_2OH$ (70) 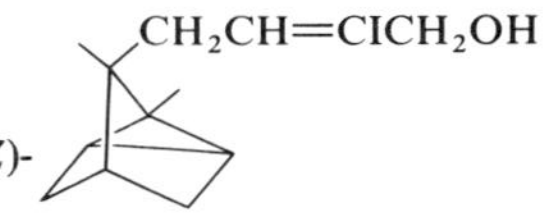	62
	n-$C_6H_{13}C{\equiv}CC{\equiv}CSi(CH_3)_3$	1. $(i\text{-}C_4H_9)_2(n\text{-}C_4H_9)AlHLi$ 2. CNBr (2 eq)	DME–hexane, 25°, 1 hr −78 to 25°	(Z)-n-$C_6H_{13}CH{=}CBrC{\equiv}CSi(CH_3)_3$ (90)	45
C_{14}	$C_6H_5C{\equiv}CC_6H_5$	1. $(C_6H_5)_3Al$ 2. I_2	$C_6H_5OC_6H_5$, 200°, 12 hr C_6H_6	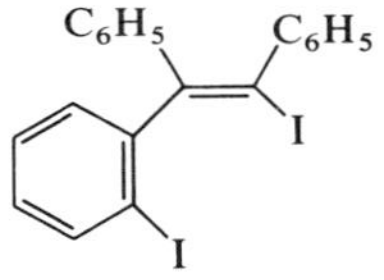 2-I-$C_6H_4C(C_6H_5){=}C(I)C_6H_5$ [structure] (72)	65
		1. $(i\text{-}C_4H_9)_2AlH$ 2. Br_2	Hexane–ether, −50 to 25°	$C_6H_5CH{=}CBrC_6H_5$ (98.8)	84
		1. $(i\text{-}C_4H_9)_2AlH$ 2. I_2	Ether, 0 to 25°	$C_6H_5CH{=}CIC_6H_5$ (100)	84
	$CH_2{=}CH(CH_2)_9C{\equiv}CCH_2OH$	1. n-C_4H_9Li 2. $(i\text{-}C_4H_9)_2AlH$ 3. $CH_3CO_2C_2H_5$ 4. I_2	Ether, −20° Ether, −20 to 35°, 48 hr 0° Ether, −78°	(Z)-$CH_2{=}CH(CH_2)_9CH{=}CICH_2OH$ (50)	164
	$C_2H_5CH(OTHP)CH_2C{\equiv}CSi(CH_3)_3$	1. $(i\text{-}C_4H_9)_2AlH$ 2. Br_2 (1.5 eq) 3. $NaOCH_3$	Ether, 40°, 4 hr Ether–pyridine, −78° CH_3OH, 40°, 4 hr	(E)-$C_2H_5CH(OTHP)CH_2CH{=}CHBr$ (78)	46

TABLE II. ALKENYL HALIDES (*Continued*)

Substrate 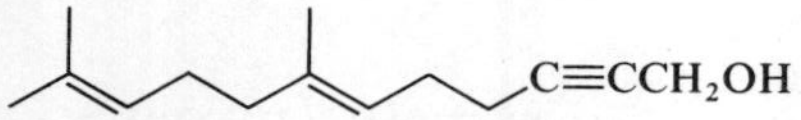	Reagents	Conditions	Product(s) and Yield(s)(%)	Refs.
C≡CCH$_2$OH	1. LiAlH$_4$, AlCl$_3$ 2. I$_2$	THF, reflux, 3 hr −78°	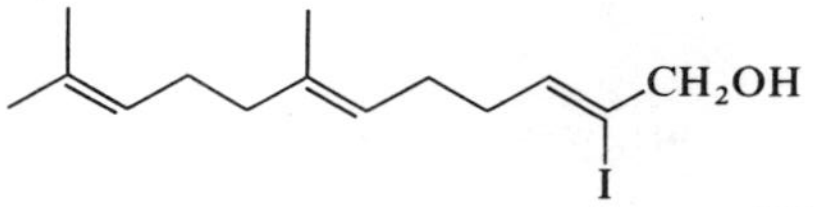 CH$_2$OH, I (60–75)	61
	1. LiAlH$_4$, NaOCH$_3$ 2. I$_2$	THF, reflux, 3 hr −78°	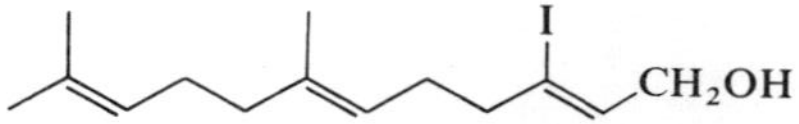 I, CH$_2$OH (60–75)	61
C$_{16}$ 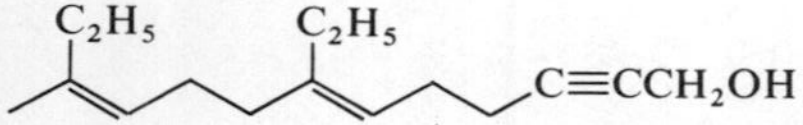 C$_2$H$_5$, C$_2$H$_5$, C≡CCH$_2$OH	1. LiAlH$_4$, NaOCH$_3$ 2. I$_2$ (excess) 3. H$_3$O$^+$	THF, reflux, 3 hr −78°	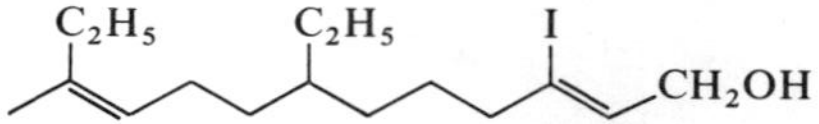 C$_2$H$_5$, C$_2$H$_5$, I, CH$_2$OH (53)	162

TABLE III. 1-HALO-1-ALKENYLSILANES

	Substrate	Reagents	Conditions	Product(s) and Yield(s) (%)	Refs.
C_9	$n\text{-}C_4H_9C\equiv CSi(CH_3)_3$	1. $(i\text{-}C_4H_9)_2AlH$ 2. NCS	Ether, 40°, 1 hr Ether, −25 to 0°	$(E)\text{-}n\text{-}C_4H_9CH{=}CClSi(CH_3)_3$ (84)	44
		1. $(i\text{-}C_4H_9)_2AlH$ 2. NCS 3. Br_2 (15 mol %)	Ether, 40°, 1 hr Ether, −25 to 0° Ether–pyridine, $h\nu$	$(Z)\text{-}n\text{-}C_4H_9CH{=}CClSi(CH_3)_3$ (81)	44
		1. $(i\text{-}C_4H_9)_2AlH$ 2. Br_2 (1.5 eq)	Ether, 40°, 1 hr Ether–pyridine, −78°	$(E)\text{-}n\text{-}C_4H_9CH{=}CBrSi(CH_3)_3$ (90)	44
		1. $(i\text{-}C_4H_9)_2AlH$ 2. Br_2 (1.5 eq) 3. Br_2 (15 mol %)	Ether, 40°, 1 hr Ether–pyridine, −78° Ether–pyridine, $h\nu$	$(Z)\text{-}n\text{-}C_4H_9CH{=}CBrSi(CH_3)_3$ (81)	44
		1. $(i\text{-}C_4H_9)_2AlH$ 2. I_2	Ether, 40°, 1 hr Ether, −70 to 0°	$(E)\text{-}n\text{-}C_4H_9CH{=}CISi(CH_3)_3$ (90)	44
		1. $(i\text{-}C_4H_9)_2AlH$ 2. I_2 3. $t\text{-}C_4H_9Li$ (15 mol %)	Ether, 40°, 1 hr Ether, −70 to 0° Pentane–ether −78 to 25°	$(Z)\text{-}n\text{-}C_4H_9CH{=}CISi(CH_3)_3$ (72)	86
	$t\text{-}C_4H_9C\equiv CSi(CH_3)_3$	1. $(i\text{-}C_4H_9)_2AlH$ 2. Br_2 (1.5 eq)	Ether, 40°, 1 hr Ether–pyridine, −78°	$(E)\text{-}t\text{-}C_4H_9CH{=}CBrSi(CH_3)_3$ (86)	44
		1. $(i\text{-}C_4H_9)_2AlH$ 2. Br_2 (1.5 eq) 3. Br_2 (15 mol %)	Ether, 40°, 1 hr Ether–pyridine, −78° Ether–pyridine, $h\nu$	$(Z)\text{-}t\text{-}C_4H_9CH{=}CBrSi(CH_3)_3$ (89)	44

TABLE III. 1-HALO-1-ALKENYLSILANES (*Continued*)

Substrate	Reagents	Conditions	Product(s) and Yield(s)(%)	Refs.
C_9 $t\text{-}C_4H_9C{\equiv}CSi(CH_3)_3$ (*cont'd*)	1. $(i\text{-}C_4H_9)_2AlH$ 2. I_2	Ether, 40°, 1 hr Ether, −78 to 0°	(*E*)-$t\text{-}C_4H_9CH{=}CISi(CH_3)_3$ (85)	44
	1. $(i\text{-}C_4H_9)_2AlH$ 2. I_2 3. $t\text{-}C_4H_9Li$ (15 mol %)	Ether, 40°, 1 hr Ether, −78 to 0° Pentane–ether −70 to 25°	(*Z*)-$t\text{-}C_4H_9CH{=}CISi(CH_3)_3$ (83)	86
C_{10} $n\text{-}C_5H_{11}C{\equiv}CSi(CH_3)_3$	1. $(i\text{-}C_4H_9)_2AlH$ 2. I_2	Hexane, ether	(*E*)-$n\text{-}C_5H_{11}CH{=}CISi(CH_3)_3$ (70)	165
C_{11} $C_6H_{11}C{\equiv}CSi(CH_3)_3$	1. $(i\text{-}C_4H_9)_2AlH$ 2. NCS	Ether, 40°, 1 hr Ether, −25 to 0°	(*E*)-$C_6H_{11}CH{=}CClSi(CH_3)_3$ (87)	44
	1. $(i\text{-}C_4H_9)_2AlH$ 2. NCS 3. Br_2 (15 mol %)	Ether, 40°, 1 hr Ether, −25 to 0° Ether-pyridine, *hv*	(*Z*)-$C_6H_{11}CH{=}CClSi(CH_3)_3$ (91)	44
	1. $(i\text{-}C_4H_9)_2AlH$ 2. Br_2 (1.5 eq)	Ether, 40°, 1 hr Ether–pyridine, −78°	(*E*)-$C_6H_{11}CH{=}CBrSi(CH_3)_3$ (96)	44
	1. $(i\text{-}C_4H_9)_2AlH$ 2. Br_2 (1.5 eq) 3. Br_2 (15 mol %)	Ether, 40°, 1 hr Ether–pyridine, −78° Ether–pyridine, *hv*	(*Z*)-$C_6H_{11}CH{=}CBrSi(CH_3)_3$ (92)	44
	1. $(i\text{-}C_4H_9)_2AlH$ 2. I_2	Ether, 40°, 1 hr Ether, −78 to 0°	(*E*)-$C_6H_{11}CH{=}CISi(CH_3)_3$ (90)	44
	1. $(i\text{-}C_4H_9)_2AlH$ 2. I_2 3. $t\text{-}C_4H_9Li$ (15 mol %)	Ether, 40°, 1 hr Ether, −78 to 0° Pentane–ether −70 to 25°	(*Z*)-$C_6H_{11}CH{=}CISi(CH_3)_3$ (82)	86

	Substrate	Reagent(s)	Conditions	Product(s) (Yield, %)	Refs.
	n-$C_6H_{13}C{\equiv}CSi(CH_3)_3$	1. $(i$-$C_4H_9)_2AlH$ 2. I_2	Hexane	(Z)-n-$C_6H_{13}CH{=}CISi(CH_3)_3$ (78)	165
	$C_6H_5C{\equiv}CSi(CH_3)_3$	1. $(i$-$C_4H_9)_2AlH$ 2. Br_2 (1.5 eq)	Ether, 40°, 4 hr Ether–pyridine, −78°	(E)-$C_6H_5CH{=}CBrSi(CH_3)_3$ (64)	73
	1-cyclohexenyl-$C{\equiv}CSi(CH_3)_3$	1. $(i$-$C_4H_9)_2AlH$ 2. BrCN (1.1 eq)	Ether, 40°, 1 hr Ether–pyridine, 0°	(E)-1-cyclohexenyl-$CH{=}CBrSi(CH_3)_3$ (84)	44
		1. $(i$-$C_4H_9)_2AlH$ 2. BrCN (1.1 eq)	Hexane, 25°, 30 min Ether, 25°, 1 hr	(Z)-1-cyclohexenyl-$CH{=}CBrSi(CH_3)_3$ (84)	44
C_{12}	n-$C_4H_9C{\equiv}CSi(C_2H_5)_3$	1. $(i$-$C_4H_9)_2AlH$ 2. Br_2 (1.5 eq)	Ether, 40°, 1 hr Ether–pyridine, −78°	(E)-n-$C_4H_9CH{=}CBrSi(C_2H_5)_3$ (85)	73
		1. $(i$-$C_4H_9)_2AlH$ 2. Br_2 (1.5 eq) 3. Br_2 (15 mol %)	Ether, 40°, 1 hr Ether–pyridine, −78° Ether–pyridine, $h\nu$	(Z)-n-$C_4H_9CH{=}CBrSi(C_2H_5)_3$ (81)	73
C_{15}	(S)-$C_6H_5CH_2OCH_2CH(CH_3)C{\equiv}CSi(CH_3)_3$	1. $(i$-$C_4H_9)_2AlH$ 2. I_2	Heptane–ether, 25° −78°	(E)-(S)-$C_6H_5CH_2OCH_2CH(CH_3)CH{=}CISi(CH_3)_3$ (89)	166

TABLE IV. 1,1-DIHALO-1-ALKENES

	Substrate	Reagents	Conditions	Product(s) and Yield(s)(%)	Refs.
C_6	n-$C_4H_9C{\equiv}CCl$	1. $LiAlH_4$ (1.0 eq) 2. CH_3COCH_3 (3.3 eq) 3. Br_2 (1.1 eq)	THF, −30 to 0°, 90 min 0°, 1 hr −78 to 25°	(Z)-n-$C_4H_9CH{=}CBrCl$ (78)	55
		1. $LiAlH_4$ (1.0 eq) 2. CH_3COCH_3 (3.3 eq) 3. ICl (1.1 eq)	THF, −30 to °0, 90 min 0°, 1 hr −30°, 1 hr to 25°, 2 hr	(Z)-n-$C_4H_9CH{=}CICl$ (85)	55
	t-$C_4H_9C{\equiv}CCl$	1. $LiAlH_4$ (1.0 eq) 2. CH_3COCH_3 (3.3 eq) 3. Br_2 (1.1 eq)	THF, −30 to 0°, 90 min 0°, 1 hr −78 to 25°	(Z)-t-$C_4H_9CH{=}CBrCl$ (61)	55
C_7	n-$C_3H_7C{\equiv}CAl(CH_3)_2$	1. $Cl(CH_3)Zr(C_5H_5)_2$ 2. I_2 (2 eq)	Not specified	n-$C_3H_7C(CH_3){=}CI_2$ (92)	40
C_8	n-$C_6H_{13}C{\equiv}CCl$	1. $LiAlH_4$ (1.0 eq) 2. CH_3COCH_3 (3.3 eq) 3. Br_2 (1.1 eq)	THF, −30 to 0°, 90 min 0°, 1 hr −78 to 25°	(Z)-n-$C_6H_{13}CH{=}CBrCl$ (80)	55
		1. $LiAlH_4$ (1.0 eq) 2. CH_3COCH_3 (3.3 eq) 3. ICl (1.1 eq)	THF, −30 to 0°, 90 min 0°, 1 hr −30°, 1 hr; 25°, 2 hr	(Z)-n-$C_6H_{13}CH{=}CICl$ (86)	55
	$C_6H_{11}C{\equiv}CCl$	1. $LiAlH_4$ (1.0 eq) 2. CH_3COCH_3 (3.3 eq) 3. Br_2 (1.1 eq)	THF, −30 to 0°, 90 min 0°, 1 hr −78 to 25°	(Z)-$C_6H_{11}CH{=}CBrCl$ (87)	55
		1. $LiAlH_4$ (1.0 eq) 2. CH_3COCH_3 (3.3 eq) 3. ICl (1.1 eq)	THF, −30 to 0°, 90 min 0°, 1 hr −30°, 1 hr; 25°, 2 hr	(Z)-$C_6H_{11}CH{=}CICl$ (89)	55
C_{10}	$THPO(CH_2)_3C{\equiv}CCl$	1. $LiAlH_4$ (1.0 eq) 2. CH_3COCH_3 (3.3 eq) 3. Br_2 (1.1 eq)	THF, −30 to 0°, 90 min 0°, 1 hr −78 to 25°	(Z)-$THPO(CH_2)_3CH{=}CBrCl$ (86)	55

TABLE V. ALLYLIC ALCOHOLS

	Substrate	Reagents	Conditions	Product(s) and Yield(s)(%)	Refs.
C_2	$HC{\equiv}CH$	1. $(i\text{-}C_4H_9)_3Al$ 2. CH_3CHO 3. H_2O	Heptane, 50°, 4 hr	$i\text{-}C_4H_9CH{=}CHCHOHCH_3$ (72)	84
C_3	$HOCH_2C{\equiv}CH$	1. $(CH_3)_3Al$ (3 eq), $Cl_2Zr(C_5H_5)_2$ (1 eq) 2. D_2O	$ClCH_2CH_2Cl$, 25°	$(E)\text{-}HOCH_2C(CH_3){=}CHD$ (41)	66
		1. $Al[Si(CH_3)_3]_3$ (0.33 eq) 2. CH_3OH	Ether, 25°, 1 hr	$(E)\text{-}HOCH_2CH{=}CHSi(CH_3)_3$ (84)	67
		1. $LiAlH_4$ 2. H_3O^+	THF, reflux, 22 hr	$HOCH_2CH{=}CH_2$ (100)	167
C_4	$HOCH_2C{\equiv}CCH_3$	1. $AlCl_3$ 2. $Al[Si(CH_3)_3]_3$ (2 eq) 3. H_3O^+	Ether, −78°, 30 min Ether, −78 to 25°, 15 hr	$(E)\text{-}HOCH_2CH{=}C(CH_3)Si(CH_3)_3$ (78)	67
	$CH_3CHOHC{\equiv}CH$	1. $LiAlD_4$ 2. D_2O	THF, 25°, 3 hr	$CH_3CHOHCD{=}CHD$ (—)	60,168
	$CH_3CHOHC{\equiv}CD$	1. $LiAlH_4$ 2. D_2O	THF, 25°, 3 hr	$CH_3CHOHCH{=}CD_2$ (—)	60
C_5	$HOCH_2C{\equiv}CC{\equiv}CH$	1. $LiAlH_4$ 2. H_2O	Ether, reflux, 4 hr	$(E)\text{-}HOCH_2CH{=}CHC{\equiv}CH$ (—)	56
	$CH_2{=}C(CH_3)C{\equiv}CH$	1. $(i\text{-}C_4H_9)_2AlH$ 2. CH_3Li 3. $(CH_2O)_n$ (0.6 eq) 4. H_2O	Hexane, 50°, 2 hr Hexane, ether, −30° Hexane, ether, −30°, 1 hr; 25°, 14 hr	$(E)\text{-}CH_2{=}C(CH_3)CH{=}CHCH_2OH$ (42)	169
		1. $(i\text{-}C_4H_9)_2AlH$ 2. CH_3Li 3. $n\text{-}C_5H_{11}CHO$ (0.6 eq) 4. H_2O	Hexane, 50°, 2 hr Hexane, ether, −30° Hexane, ether, −30°, 1 hr, 25°, 14 hr	$(E)\text{-}CH_2{=}C(CH_3)CH{=}CHCHOHC_5H_{11}\text{-}n$ (63)	169

TABLE V. ALLYLIC ALCOHOLS (*Continued*)

	Substrate	Reagents	Conditions	Product(s) and Yield(s)(%)	Refs.
C_5 (*cont'd*)	$CH_2{=}C(CH_3)C{\equiv}CH$	1. $(i\text{-}C_4H_9)_2AlH$ 2. CH_3Li 3. OHC-(cyclohex-3-enyl) (0.5 eq) 4. H_2O	Hexane, 50°, 2 hr Hexane, ether, −30° Hexane, ether, −30°, 1 hr, 25°, 14 hr	(E)-$CH_2{=}C(CH_3)CH{=}CHCHOH$-(cyclohex-3-enyl) (76)	169
		1. $(i\text{-}C_4H_9)_2AlH$ 2. CH_3Li 3. C_6H_5CHO (0.5 eq) 4. H_2O	Hexane, 50°, 2 hr Hexane, ether, −30° Hexane, ether, −30°, 1 hr, 25°, 14 hr	(E)-$CH_2{=}C(CH_3)CH{=}CHCHOHC_6H_5$ (66)	169
C_6	$n\text{-}C_4H_9C{\equiv}CH$	1. $(i\text{-}C_4H_9)_2AlH$ 2. CH_3CHO 3. H_2O	Heptane, 50°, 4 hr	$n\text{-}C_4H_9CH{=}CHCHOHCH_3$ (76)	84
		1. $(i\text{-}C_4H_9)_2AlH$ 2. $(C_2H_5)_2CO$ 3. H_2O	Hexane, 50°, 4 hr Hexane, −20 to 25°	(E)-$n\text{-}C_4H_9CH{=}CHCOH(C_2H_5)_2$ (36)	93
		1. $(i\text{-}C_4H_9)_2AlH$ 2. $t\text{-}C_4H_9COCH_3$ 3. H_2O	Hexane, 50°, 4 hr Hexane, −20 to 25°	(E)-$n\text{-}C_4H_9CH{=}CHCOH(CH_3)C_4H_9\text{-}t$ (25)	93
		1. $(i\text{-}C_4H_9)_2AlH$ 2. CH_3CHO 3. H_3O^+	Hexane, 50°, 4 hr Hexane, 0 to 25°	(E)-$n\text{-}C_4H_9CH{=}CHCHOHCH_3$ (67)	93
		1. $(i\text{-}C_4H_9)_2AlH$ 2. CH_3Li 3. CH_3CHO 4. H_3O^+	Heptane, 50°, 2 hr Ether–heptane, 25° 25°	" (68)	91
		1. $(i\text{-}C_4H_9)_2AlH$ 2. $n\text{-}C_4H_9CHO$ 3. H_3O^+	Hexane, 50°, 4 hr Hexane, 0 to 25°	(E)-$n\text{-}C_4H_9CH{=}CHCHOHC_4H_9\text{-}n$ (61)	93

	1. $(i\text{-}C_4H_9)_2AlH$ 2. $C_6H_{11}CHO$ 3. H_3O^+	Hexane, 50°, 4 hr Hexane, 0 to 25°	$(E)\text{-}n\text{-}C_4H_9CH{=}CHCHOHC_6H_{11}$ (44)	93
	1. $(i\text{-}C_4H_9)_2AlH$ 2. $t\text{-}C_4H_9CHO$ 3. H_3O^+	Hexane, 50°, 4 hr Hexane, 0 to 25°	$(E)\text{-}n\text{-}C_4H_9CH{=}CHCHOHC_4H_9\text{-}t$ (28)	93
	1. $(i\text{-}C_4H_9)_2AlH$ 2. cyclohexanone 3. H_3O^+	Hexane, 50°, 4 hr Hexane, −20 to 25°	$(E)\text{-}n\text{-}C_4H_9CH{=}CH$–1-hydroxycyclohexyl (HO) (60)	93
	1. $(i\text{-}C_4H_9)_2AlH$ 2. cyclohexanone 3. H_3O^+	Hexane, 50°, 2 hr Ether-hexane, 0°	" (40)	92
	1. $(i\text{-}C_4H_9)_2AlH$ 2. 2-methylcyclohexanone 3. H_3O^+	Hexane, 50°, 4 hr Hexane, −20 to 25°	$(E)\text{-}n\text{-}C_4H_9CH{=}CH$–1-hydroxy-2-methylcyclohexyl (HO) (48)	93
	1. $(i\text{-}C_4H_9)_2AlH$ 2. CH_3COCH_3 3. H_3O^+	Hexane, 50°, 4 hr Hexane, −20 to 25°	$(E)\text{-}n\text{-}C_4H_9CH{=}CHCOH(CH_3)_2$ (60)	93
	1. $(i\text{-}C_4H_9)_2AlH$ 2. CH_3COCH_3 3. H_3O^+	Hexane, 50°, 2 hr Ether–hexane, 0°	" (42)	92
	1. $(i\text{-}C_4H_9)_2AlH$ 2. CH_3Li 3. $(CH_2O)_n$ 4. H_3O^+	Heptane, 50°, 2 hr Ether–heptane, 25° 35°	$(E)\text{-}n\text{-}C_4H_9CH{=}CHCH_2OH$ (73)	91
$C_2H_5C{\equiv}CC_2H_5$	1. $(i\text{-}C_4H_9)_2AlH$ 2. CH_3CHO 3. H_2O	Heptane, 50°, 4 hr	$C_2H_5CH{=}C(C_2H_5)CHOHCH_3$ (72)	84

TABLE V. ALLYLIC ALCOHOLS (*Continued*)

	Substrate	Reagents	Conditions	Product(s) and Yield(s)(%)	Refs.
C_6 (*cont'd*)	$C_2H_5C{\equiv}CC_2H_5$	1. $(i\text{-}C_4H_9)_2AlH$ 2. CH_3Li 3. HCHO 4. H_3O^+	Heptane, 50°, 2 hr Heptane–ether	(*E*)-$C_2H_5CH{=}C(C_2H_5)CH_2OH$ (73)	49
		1. $(i\text{-}C_4H_9)_2(CH_3)AlHLi$ 2. HCHO 3. H_3O^+	DME, 100–130°, 6 hr	(*Z*)-$C_2H_5CH{=}C(C_2H_5)CH_2OH$ (68)	49
	$HOCH_2C{\equiv}CSi(CH_3)_3$	1. $LiAlH_4$ 2. H_2O	Not specified	(*E*)-$HOCH_2CH{=}CHSi(CH_3)_3$ (—)	170,171
		1. $NaAlH_2(OCH_2CH_2OCH_3)_2$ 2. H_3O^+	Ether, toluene, 25°, 1 hr	" (85)	172
	$C_2H_5CHOHC{\equiv}CCH_3$	1. $LiAlH_4$ 2. H_3O^+	THF, reflux, 1 hr	(*E*)-$C_2H_5CHOHCH{=}CHCH_3$ (100)	167
	$HOCH_2C{\equiv}CC{\equiv}CCH_2OH$	1. $LiAlH_4$ 2. $CH_3CO_2C_2H_5$ 3. H_3O^+	Ether, reflux	(*E,E*)-$HOCH_2CH{=}CHCH{=}CHCH_2OH$ (27)	56
	$CH_3CHOHC{\equiv}CC{\equiv}CH$	1. $LiAlH_4$ 2. H_2O	Ether, reflux, 4 hr	(*E*)-$CH_3CHOHCH{=}CHC{\equiv}CH$ (—)	56
C_7	$n\text{-}C_5H_{11}C{\equiv}CH$	1. $(CH_3)_3Al$ (2 eq), $Cl_2Zr(C_5H_5)_2$ (0.4 eq) 2. $n\text{-}C_4H_9Li$ 3. $(CH_2O)_n$ 4. H_3O^+	$ClCH_2CH_2Cl$, 25°, 2 hr Hexane THF, 25°, 3 hr	(*E*)-$n\text{-}C_5H_{11}C(CH_3){=}CHCH_2OH$ (82)	80
		1. $(i\text{-}C_4H_9)_2AlH$ 2. C_2H_5CHO 3. H_3O^+	Hexane, 50°, 2 hr Ether–hexane, 0°	(*E*)-$n\text{-}C_5H_{11}CH{=}CHCHOHC_2H_5$ (30)	92
	$n\text{-}C_4H_9CHOHC{\equiv}CH$	1. $LiAlD_4$ 2. H_3O^+	THF, 25°, 3 hr	$n\text{-}C_4H_9CHOHCD{=}CH_2$ (85)	60
	$n\text{-}C_4H_9CHOHC{\equiv}CD$	1. $LiAlH_4$ 2. D_2O	THF, 25°, 3 hr	$n\text{-}C_4H_9CHOHCH{=}CD_2$ (—)	60

	$(CH_3)_2COHC \equiv CCH{=}CH_2$	1. $LiAlH_4$ 2. H_2O	Ether, reflux, 3 hr	(*E*)-$(CH_3)_2COHCH{=}CHCH{=}CH_2$ (88)	56
C_8	n-$C_6H_{13}C \equiv CH$	1. $(i\text{-}C_4H_9)_2AlH$ 2. CH_3Li 3. C_2H_5CHO 4. H_3O^+	Hexane, 50°, 4 hr Hexane–ether, 0–20° −60 to 25°	(*E*)-n-$C_6H_{13}CH{=}CHCHOHC_2H_5$ (56)	93
		1. $(i\text{-}C_4H_9)_2AlH$ 2. CH_3Li 3. n-C_4H_9CHO 4. H_3O^+	Hexane, 50°, 4 hr Hexane–ether, 0–20° −60° to 25°	(*E*)-n-$C_6H_{13}CH{=}CHCHOHC_4H_9$-$n$ (56)	93
	$C_6H_{10}(OH)C \equiv CH$	1. $LiAlH_4$ 2. H_3O^+	THF, reflux, 22 hr	$C_6H_{10}(OH)CH{=}CH_2$ (67)	56,167
	$(CH_3)_2COHC \equiv CSi(CH_3)_3$	1. $LiAlH_4$ 2. H_3O^+	THF	(*E*)-$(CH_3)_2COHCH{=}CHSi(CH_3)_3$ (—)	173
	$(CH_3)_2C{=}CH(CH_2)_2C \equiv CH$	1. $(CH_3)_3Al$ (2 eq), $Cl_2Zr(C_5H_5)_2$ (0.4 eq) 2. n-C_4H_9Li 3. $(CH_2O)_n$ 4. H_3O^+	$ClCH_2CH_2Cl$, 25°, 2 hr Hexane THF, 25°, 3 hr	(*E*)-$(CH_3)_2C{=}CH(CH_2)_2C(CH_3){=}CHCH_2OH$ (87)	80
	n-$C_5H_{11}C \equiv CCH_2OH$	1. $LiAlH_4$ 2. H_3O^+	THF, reflux, 1 hr	(*E*)-n-$C_5H_{11}CH{=}CHCH_2OH$ (100)	167
	n-$C_5H_{11}CHOHC \equiv CH$	1. $LiAlH_4$ 2. H_3O^+	THF, reflux, 22 hr	n-$C_5H_{11}CHOHCH{=}CH_2$ (96)	167
C_9	$C_6H_5SCH_2C \equiv CH$	1. $(CH_3)_3Al$ (2 eq), $Cl_2Zr(C_5H_5)_2$ (1 eq) 2. n-C_4H_9Li 3. $(CH_2O)_n$	$ClCH_2CH_2Cl$, 25°	(*E*)-$C_6H_5SCH_2C(CH_3){=}CHCH_2OH$ (78)	66
	$C_6H_{10}(OH)C \equiv CCH_3$	1. $LiAlH_4$ 2. H_3O^+	THF, reflux, 6 hr	(*E*)-$C_6H_{10}(OH)CH{=}CHCH_3$ (100)	167
	$C_6H_5C \equiv CCH_2OH$	1. $LiAlH_4$ 2. H_3O^+	Ether, reflux, 4 hr	(*E*)-$C_6H_5CH{=}CHCH_2OH$ (94)	56

TABLE V. ALLYLIC ALCOHOLS (*Continued*)

	Substrate	Reagents	Conditions	Product(s) and Yield(s)(%)	Refs.
C_{10}	$C_6H_5S(CH_2)_2C\equiv CH$	1. $(CH_3)_3Al$ (2 eq), $Cl_2Zr(C_5H_5)_2$ (1 eq) 2. n-C_4H_9Li 3. $(CH_2O)_n$	$ClCH_2CH_2Cl$, 25°	(E)-$C_6H_5S(CH_2)_2C(CH_3)=CHCH_2OH$ (78)	66
	$(CH_3)_2COH(C\equiv C)_2COH(CH_3)_2$	1. $LiAlH_4$ 2. $CH_3CO_2C_2H_5$ 3. H_3O^+	Ether, reflux	(E,E)-$(CH_3)_2COH(CH=CH)_2COH(CH_3)_2$ (—)	56
	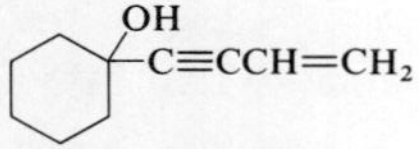 OH, $C\equiv CCH=CH_2$	1. $LiAlH_4$ 2. H_3O^+	Ether, reflux, 3 hr	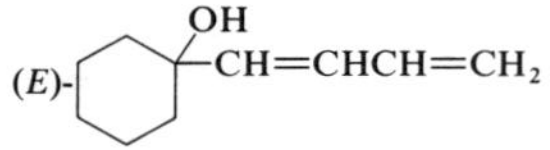 (E)- OH, $CH=CHCH=CH_2$ (—)	56
C_{11}	n-$C_5H_{11}CHOHC\equiv CSi(CH_3)_3$	1. $LiAlH_4$ 2. H_3O^+	THF	(E)-n-$C_5H_{11}CHOHCH=CHSi(CH_3)_3$ (—)	173
	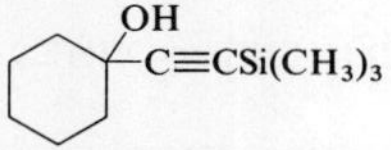 OH, $C\equiv CSi(CH_3)_3$	1. $LiAlH_4$ 2. H_3O^+	THF	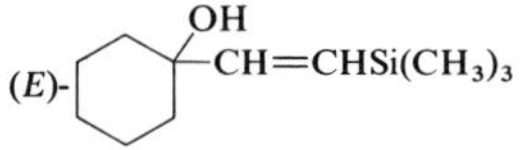 (E)- OH, $CH=CHSi(CH_3)_3$ (—)	173
	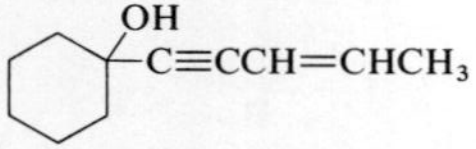 OH, $C\equiv CCH=CHCH_3$	1. $LiAlH_4$ 2. H_2O	Ether, reflux	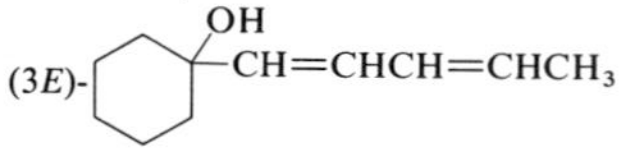 (3E)- OH, $CH=CHCH=CHCH_3$ (—)	56
C_{12}	$(CH_3)_2COH(C\equiv C)_3COH(CH_3)_2$	1. $LiAlH_4$ 2. $CH_3CO_2C_2H_5$ 3. H_3O^+	Ether, reflux	(E,E)-$(CH_3)_2COHCH=CHC\equiv CCH=CHCOH(CH_3)_2$ (44)	56
C_{13}	$(CH_3)_2C=CH(CH_2)_2COH(CH_3)C\equiv CSi(CH_3)_3$	1. $LiAlH_4$ 2. H_3O^+	THF	(E)-$(CH_3)_2C=CH(CH_2)_2COH(CH_3)CH=CHSi(CH_3)_3$ (—)	173
	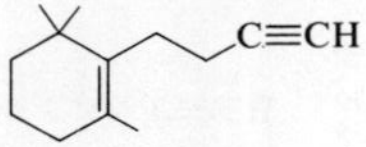 $C\equiv CH$	1. $(CH_3)_3Al$ (2 eq), $Cl_2Zr(C_5H_5)_2$ (0.4 eq) 2. n-C_4H_9Li 3. $(CH_2O)_n$	$ClCH_2CH_2Cl$, 25°, 2–3 hr THF	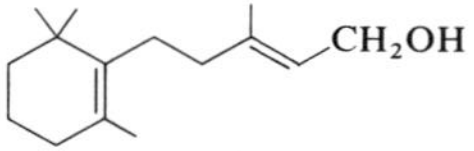 CH_2OH (71)	94

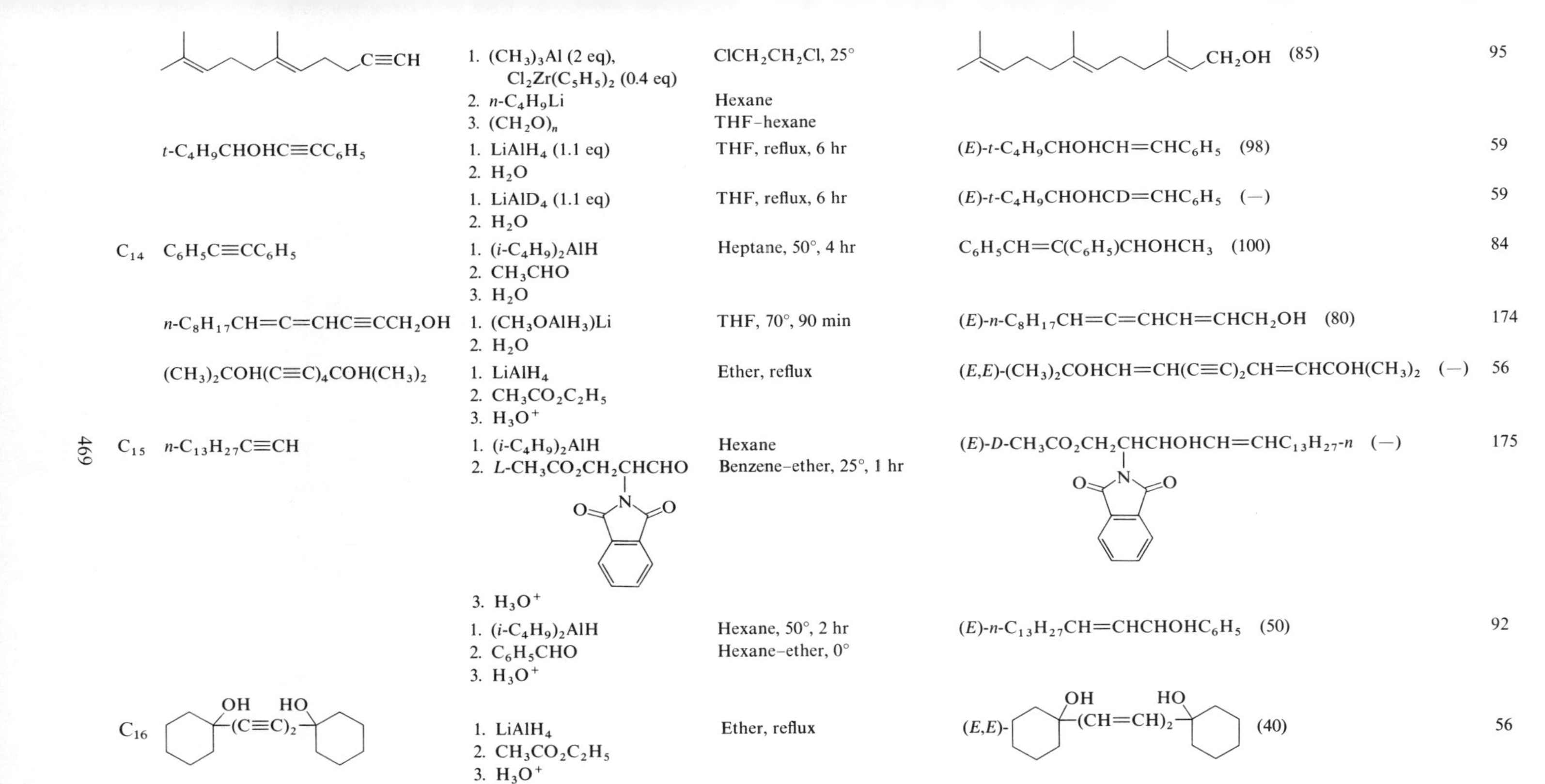

	$C{\equiv}CH$	1. $(CH_3)_3Al$ (2 eq), $Cl_2Zr(C_5H_5)_2$ (0.4 eq) 2. n-C_4H_9Li 3. $(CH_2O)_n$	$ClCH_2CH_2Cl$, 25° Hexane THF–hexane	CH_2OH (85)	95
	t-$C_4H_9CHOHC{\equiv}CC_6H_5$	1. $LiAlH_4$ (1.1 eq) 2. H_2O	THF, reflux, 6 hr	(E)-t-$C_4H_9CHOHCH{=}CHC_6H_5$ (98)	59
		1. $LiAlD_4$ (1.1 eq) 2. H_2O	THF, reflux, 6 hr	(E)-t-$C_4H_9CHOHCD{=}CHC_6H_5$ (—)	59
C_{14}	$C_6H_5C{\equiv}CC_6H_5$	1. $(i$-$C_4H_9)_2AlH$ 2. CH_3CHO 3. H_2O	Heptane, 50°, 4 hr	$C_6H_5CH{=}C(C_6H_5)CHOHCH_3$ (100)	84
	n-$C_8H_{17}CH{=}C{=}CHC{\equiv}CCH_2OH$	1. $(CH_3OAlH_3)Li$ 2. H_2O	THF, 70°, 90 min	(E)-n-$C_8H_{17}CH{=}C{=}CHCH{=}CHCH_2OH$ (80)	174
	$(CH_3)_2COH(C{\equiv}C)_4COH(CH_3)_2$	1. $LiAlH_4$ 2. $CH_3CO_2C_2H_5$ 3. H_3O^+	Ether, reflux	(E,E)-$(CH_3)_2COHCH{=}CH(C{\equiv}C)_2CH{=}CHCOH(CH_3)_2$ (—)	56
C_{15}	n-$C_{13}H_{27}C{\equiv}CH$	1. $(i$-$C_4H_9)_2AlH$ 2. L-$CH_3CO_2CH_2CHCHO$ (N-phthalimido) 3. H_3O^+	Hexane Benzene–ether, 25°, 1 hr	(E)-D-$CH_3CO_2CH_2CHCHOHCH{=}CHC_{13}H_{27}$-$n$ (N-phthalimido) (—)	175
		1. $(i$-$C_4H_9)_2AlH$ 2. C_6H_5CHO 3. H_3O^+	Hexane, 50°, 2 hr Hexane–ether, 0°	(E)-n-$C_{13}H_{27}CH{=}CHCHOHC_6H_5$ (50)	92
C_{16}	OH HO $(C{\equiv}C)_2$	1. $LiAlH_4$ 2. $CH_3CO_2C_2H_5$ 3. H_3O^+	Ether, reflux	(E,E)- OH HO $(CH{=}CH)_2$ (40)	56

TABLE V. ALLYLIC ALCOHOLS (*Continued*)

	Substrate	Reagents	Conditions	Product(s) and Yield(s)(%)	Refs.
C_{23}	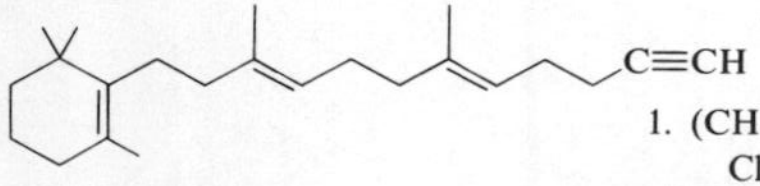 $C \equiv CH$	1. $(CH_3)_3Al$ (2 eq), $Cl_2Zr(C_5H_5)_2$ (0.4 eq) 2. n-C_4H_9Li 3. $(CH_2O)_n$ 4. H_2O	$ClCH_2CH_2Cl$, 25° Hexane THF–hexane	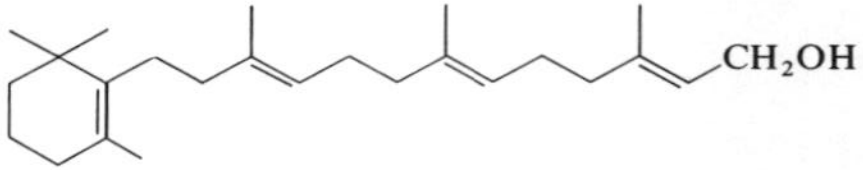 CH_2OH (61)	95
C_{30}	$(C_6H_5)_2COH(C\equiv C)_2COH(C_6H_5)_2$	1. $LiAlH_4$ 2. $CH_3CO_2C_2H_5$ 3. H_3O^+	Ether, reflux	(E,E)-$(C_6H_5)_2COH(CH{=}CH)_2COH(C_6H_5)_2$ (50)	56
C_{32}	$(C_6H_5)_2COH(C\equiv C)_3COH(C_6H_5)_2$	1. $LiAlH_4$ 2. $CH_3CO_2C_2H_5$ 3. H_3O^+	Ether, reflux	(E,E)-$(C_6H_5)_2COHCH{=}CHC\equiv CCH{=}CHCOH(C_6H_5)_2$ (58)	56

TABLE VI. HOMOALLYLIC ALCOHOLS

	Substrate	Reagents	Conditions	Product(s) and Yield(s)(%)	Refs.
C_2	$HC{\equiv}CH$	1. $(C_2H_5)_3Al$ 2. $\overset{O}{CH_2{-}CH_2}$ (ethylene oxide) 3. H_3O^+	Hexane, 49°, 1 hr autoclave	(*Z*)-$C_2H_5CH{=}CH(CH_2)_2OH$ (66)	101
		1. $(C_2H_5)_3Al$ 2. $\overset{O}{CH_3CH{-}CH_2}$ (propylene oxide) 3. H_3O^+	30–60°	(*Z*)-$C_2H_5CH{=}CHCH_2CHOHCH_3$ (33)	101
C_4	$C_2H_5C{\equiv}CH$	1. $(i\text{-}C_4H_9)_2AlH$ 2. $\overset{O}{CH_2{-}CH_2}$, $(n\text{-}C_4H_9)_3N$ 3. H_3O^+	Octane 15°	(*E*)-$C_2H_5CH{=}CH(CH_2)_2OH$ (54)	97
		1. $(i\text{-}C_4H_9)_2AlH$ 2. CH_3Li 3. $\overset{O}{CH_2{-}CH_2}$, $(n\text{-}C_4H_9)_3N$ 4. H_3O^+	Octane Ether 15°	" (84)	97
	$HO(CH_2)_2C{\equiv}CH$	1. $(C_2H_5)_6Al_2$, $Cl_2Zr(C_5H_5)_2$ (0.5 eq) 2. H_2O	20–25°, 8 hr	(*E*)-$C_2H_5CH{=}CH(CH_2)_2OH$ I, (67) $CH_2{=}C(C_2H_5)CH_2CH_2OH$ II I:II = 63:37	176
		1. $(CH_3)_6Al_2$, $Cl_2Zr(C_5H_5)_2$ (0.5 eq) 2. H_2O	20–25°, 14 hr	(*E*)-$CH_3CH{=}CH(CH_2)_2OH$ I, (66) $CH_2{=}C(CH_3)(CH_2)_2OH$ II I:II = 91:9	176
C_5	$n\text{-}C_3H_7C{\equiv}CH$	1. $(i\text{-}C_4H_9)_2AlH$ 2. $\overset{O}{CH_2{-}CH_2}$, $(n\text{-}C_4H_9)_3N$ 3. H_2O^+	Octane −15°	(*E*)-$n\text{-}C_3H_7CH{=}CH(CH_2)_2OH$ (58)	97

TABLE VI. HOMOALLYLIC ALCOHOLS (*Continued*)

Substrate	Reagents	Conditions	Product(s) and Yield(s)(%)	Refs.
C_5 n-$C_3H_7C{\equiv}CH$ (*cont'd*)	1. $(i\text{-}C_4H_9)_2AlH$ 2. CH_3Li 3. $CH_2{-}CH_2$ (epoxide, O-bridged), $(n\text{-}C_4H_9)_3N$ 4. H_3O^+	Octane Ether $-15°$	(E)-n-$C_3H_7CH{=}CH(CH_2)_2OH$ (88)	97
C_6 n-$C_4H_9C{\equiv}CH$	1. $(i\text{-}C_4H_9)_2AlH$ 2. $CH_2{-}CH_2$ (epoxide, O-bridged), $(n\text{-}C_4H_9)_3N$ 3. H_3O^+	Octane $-15°$	(E)-n-$C_4H_9CH{=}CH(CH_2)_2OH$ (58)	97
	1. $(i\text{-}C_4H_9)_2AlH$ 2. CH_3Li 3. $CH_2{-}CH_2$ (epoxide, O-bridged), $(n\text{-}C_4H_9)_3N$ 4. H_3O^+	Octane Ether $-15°$	" (81)	97
	1. $(i\text{-}C_4H_9)_2AlH$ 2. $CH_2{-}CH_2$ (epoxide, O-bridged) 3. H_3O^+	Hexane, 50°, 2 hr 25°	" (47)	98
	1. $(i\text{-}C_4H_9)_2AlH$ 2. n-C_4H_9Li 3. $CH_2{-}CH_2$ (epoxide, O-bridged) 4. H_3O^+	Hexane, 50°, 2 hr Hexane, 25° 25°	" (75)	98
	1. $(i\text{-}C_4H_9)_2AlH$ 2. $CH_3CH{-}CH_2$ (epoxide, O-bridged) 3. H_3O^+	Hexane, 50°, 2 hr 25°	(E)-n-$C_4H_9CH{=}CHCH_2CHOHCH_3$ (33)	98

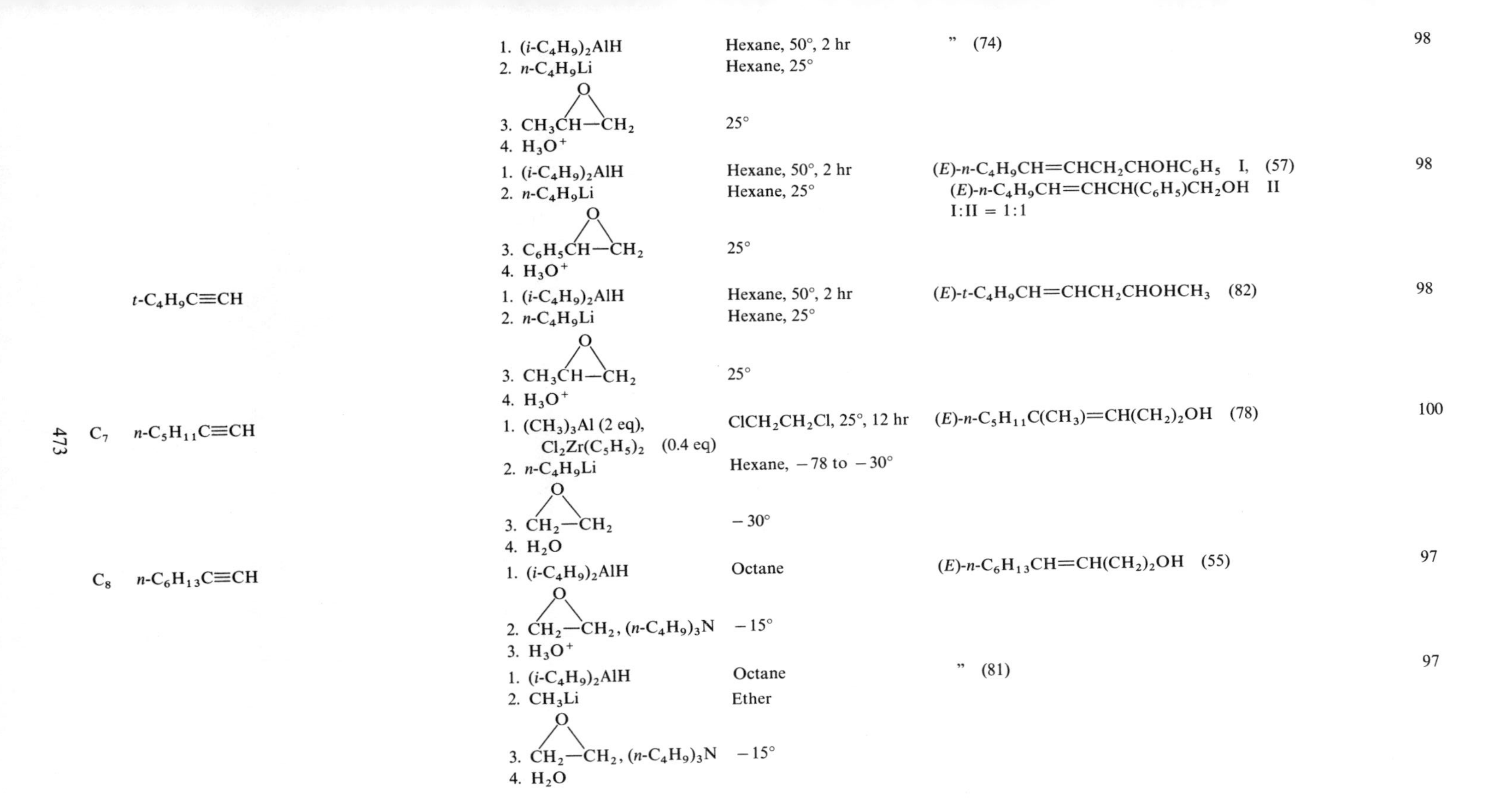

		1. $(i\text{-}C_4H_9)_2AlH$	Hexane, 50°, 2 hr	" (74)	98
		2. $n\text{-}C_4H_9Li$	Hexane, 25°		
		3. $CH_3\overset{O}{\widehat{CH{-}CH_2}}$	25°		
		4. H_3O^+			
		1. $(i\text{-}C_4H_9)_2AlH$	Hexane, 50°, 2 hr	(*E*)-$n\text{-}C_4H_9CH{=}CHCH_2CHOHC_6H_5$ I, (57)	98
		2. $n\text{-}C_4H_9Li$	Hexane, 25°	(*E*)-$n\text{-}C_4H_9CH{=}CHCH(C_6H_5)CH_2OH$ II	
				I:II = 1:1	
		3. $C_6H_5\overset{O}{\widehat{CH{-}CH_2}}$	25°		
		4. H_3O^+			
	$t\text{-}C_4H_9C{\equiv}CH$	1. $(i\text{-}C_4H_9)_2AlH$	Hexane, 50°, 2 hr	(*E*)-$t\text{-}C_4H_9CH{=}CHCH_2CHOHCH_3$ (82)	98
		2. $n\text{-}C_4H_9Li$	Hexane, 25°		
		3. $CH_3\overset{O}{\widehat{CH{-}CH_2}}$	25°		
		4. H_3O^+			
C_7	$n\text{-}C_5H_{11}C{\equiv}CH$	1. $(CH_3)_3Al$ (2 eq), $Cl_2Zr(C_5H_5)_2$ (0.4 eq)	$ClCH_2CH_2Cl$, 25°, 12 hr	(*E*)-$n\text{-}C_5H_{11}C(CH_3){=}CH(CH_2)_2OH$ (78)	100
		2. $n\text{-}C_4H_9Li$	Hexane, −78 to −30°		
		3. $\overset{O}{\widehat{CH_2{-}CH_2}}$	−30°		
		4. H_2O			
C_8	$n\text{-}C_6H_{13}C{\equiv}CH$	1. $(i\text{-}C_4H_9)_2AlH$	Octane	(*E*)-$n\text{-}C_6H_{13}CH{=}CH(CH_2)_2OH$ (55)	97
		2. $\overset{O}{\widehat{CH_2{-}CH_2}}$, $(n\text{-}C_4H_9)_3N$	−15°		
		3. H_3O^+			
		1. $(i\text{-}C_4H_9)_2AlH$	Octane	" (81)	97
		2. CH_3Li	Ether		
		3. $\overset{O}{\widehat{CH_2{-}CH_2}}$, $(n\text{-}C_4H_9)_3N$	−15°		
		4. H_2O			

TABLE VI. HOMOALLYLIC ALCOHOLS (*Continued*)

Substrate	Reagents	Conditions	Product(s) and Yield(s)(%)	Refs.
C_8 n-$C_6H_{13}C\equiv CH$ (*cont'd*)	1. $(CH_3)_3Al$ (2 eq), $Cl_2Zr(C_5H_5)_2$ (0.4 eq)	$ClCH_2CH_2Cl$, 25°, 12 hr	(E)-n-$C_6H_{13}C(CH_3){=}CHCH_2CHOHCH_3$ (87)	100
	2. n-C_4H_9Li	Hexane, −78 to −30°		
	3. $CH_3CH{-}CH_2$ (epoxide, O bridging) (2 eq)	−30°		
	4. H_2O			
$C_6H_{11}C\equiv CH$	1. (i-$C_4H_9)_2AlH$	Hexane, 50°, 2 hr	(E)-$C_6H_{11}CH{=}CHCH_2CHOHCH_3$ (77)	98
	2. n-C_4H_9Li	Hexane, 25°		
	3. $CH_3CH{-}CH_2$ (epoxide, O bridging)	25°		
	4. H_3O^+			
$(CH_3)_2C{=}CH(CH_2)_2C\equiv CH$	1. $(CH_3)_3Al$ (2 eq), $Cl_2Zr(C_5H_5)_2$ (0.4 eq)	$ClCH_2CH_2Cl$, 25°, 12 hr	(E)-$(CH_3)_2C{=}CH(CH_2)_2C(CH_3){=}CH(CH_2)_2OH$ (71)	100
	2. n-C_4H_9Li	Hexane, −78 to −30°		
	3. $CH_2{-}CH_2$ (epoxide, O bridging)	−30°		
	4. H_2O			
	1. $(CH_3)_3Al$ (2 eq), $Cl_2Zr(C_5H_5)_2$ (0.4 eq)	$ClCH_2CH_2Cl$, 25°, 12 hr	(E)-$(CH_3)_2C{=}CH(CH_2)_2C(CH_3){=}CHCH_2CHOHCH_3$ (72)	100
	2. n-C_4H_9Li	Hexane, −78 to −30°		
	3. $CH_3CH{-}CH_2$ (epoxide, O bridging) (2 eq)	−30°		
	4. H_2O			

C_9	$CH_2{=}CH(CH_2)_5C{\equiv}CH$	1. $(i\text{-}C_4H_9)_2AlH$ 2. $n\text{-}C_4H_9Li$ 3. (R)-propylene oxide (H, O, CH_3) 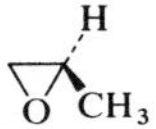 4. H_3O^+	Heptane, 50°, 4 hr 25°, 30 min 25°, 24 hr	$(E)\text{-}(R)\text{-}CH_2{=}CH(CH_2)_5CH{=}CHCH_2CHOHCH_3$ (55)	23
C_{10}	$n\text{-}C_6H_{13}C{\equiv}C(CH_2)_2OH$	1. $LiAlH_4$ 2. H_2O	Diglyme, THF, 140°, 48–55 hr	$(E)\text{-}n\text{-}C_6H_{13}CH{=}CH(CH_2)_2OH$ (89)	58
C_{23}	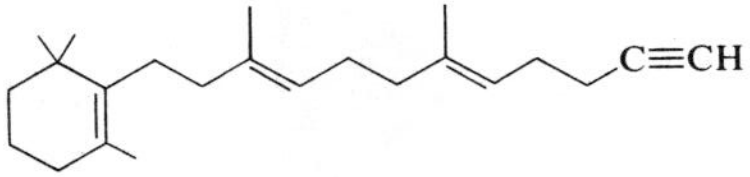 [structure] $C{\equiv}CH$	1. $(CH_3)_3Al$ (2 eq), $Cl_2Zr(C_5H_5)_2$ 2. $n\text{-}C_4H_9Li$ 3. 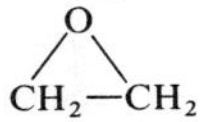 $CH_2{-}CH_2$ (O) (ethylene oxide) 4. H_3O^+	$ClCH_2CH_2Cl$, 25° Hexane, −78 to −30° −30°	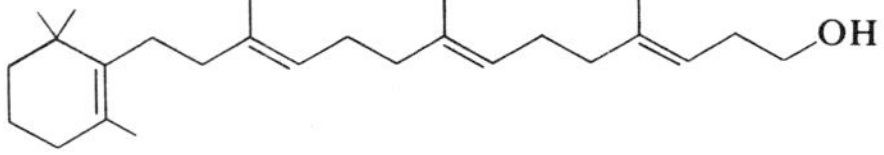 [structure] OH (62)	177

TABLE VII. ALLYLIC AND HOMOALLYLIC ETHERS

	Substrate	Reagents	Conditions	Product(s) and Yield(s)(%)	Refs.
C_4	$C_2H_5C{\equiv}CH$	1. $(i\text{-}C_4H_9)_2AlH$ 2. CH_3Li 3. $OHC(CH_2)_2CH(CH_3)CO_2CH_3$ 4. F_3CCO_2H	Not specified	CH_3, H; (*E*)- [lactone ring: O, O]; H, $CH{=}CHC_2H_5$ (—)	96
C_5	$CH_2{=}C(CH_3)C{\equiv}CH$	1. $(i\text{-}C_4H_9)_2AlH$ 2. $(C_2H_5O)(CH_3O)CHCl$	Hexane, 50°, 2 hr	(*E*)-$CH_2{=}C(CH_3)CH{=}CHCH(OCH_3)(OC_2H_5)$ (72)	169
C_6	$n\text{-}C_4H_9C{\equiv}CH$	1. $(i\text{-}C_4H_9)_2AlH$ 2. $ClCH_2OC_2H_5$	Hexane, 50°, 4 hr Hexane, 25°, 1 hr	(*E*)-$n\text{-}C_4H_9CH{=}CHCH_2OC_2H_5$ (80)	79
		1. $(i\text{-}C_4H_9)_2AlH$ 2. $n\text{-}C_4H_9Li$ 3. $ClCH_2OCH_3$	Hexane, 50°, 4 hr Hexane, 25° THF, 25°	(*E*)-$n\text{-}C_4H_9CH{=}CHCH_2OCH_3$ (51)	77
	$t\text{-}C_4H_9C{\equiv}CH$	1. $(i\text{-}C_4H_9)_2AlH$ 2. $ClCH_2OC_2H_5$	Hexane, 50°, 4 hr Hexane, 25°, 1 hr	(*E*)-$t\text{-}C_4H_9CH{=}CHCH_2OC_2H_5$ (75)	79
C_7	$n\text{-}C_5H_{11}C{\equiv}CH$	1. $(CH_3)_3Al$ (2 eq), $Cl_2Zr(C_5H_5)_2$ (0.4 eq) 2. $n\text{-}C_4H_9Li$ 3. $ClCH_2OCH_3$ (3 eq)	$ClCH_2CH_2Cl$, 25°, 2 hr Hexane Hexane, 0°, 2 hr	(*E*)-$n\text{-}C_5H_{11}C(CH_3){=}CHCH_2OCH_3$ (79)	80
		1. $(i\text{-}C_4H_9)_2AlH$ 2. CH_3Li 3. $OHC(CH_2)_3CO_2CH_3$ 4. F_3CCO_2H	Not specified	(*E*)- [lactone ring: O, O]; $CH{=}CHC_5H_{11}\text{-}n$ (—)	96
C_8	$THPOCH_2C{\equiv}CH$	1. $Al[Si(CH_3)_3]_3$ (0.33 eq) 2. CH_3OH	Ether, 25°, 1 hr	(*E*)-$THPOCH_2CH{=}CHSi(CH_3)_3$ (86)	67
		1. $Al[Si(CH_3)_3]_3$ (0.33 eq) 2. D_2O	Ether, 25°, 1 hr	(*E*)-$THPOCH_2CD{=}CHSi(CH_3)_3$ (—)	67
	$C_6H_{11}C{\equiv}CH$	1. $(i\text{-}C_4H_9)_2AlH$ 2. $ClCH_2OC_2H_5$	Hexane, 50°, 4 hr Hexane, 25°, 1 hr	(*E*)-$C_6H_{11}CH{=}CHCH_2OC_2H_5$ (72)	79

	Substrate	Reagents	Conditions	Product(s) and Yield(s) (%)	Refs.
	$HC{\equiv}CCH_2CH(CH_2OCH_3)CH{=}CH_2$	1. $(C_2H_5)_2AlH$ 2. H_2O	Hexane, reflux	CH_3OCH_2-(2-methylenecyclopentyl) (—)	138
C_9	n-$C_4H_9C{\equiv}CSi(CH_3)_3$	1. $(i$-$C_4H_9)_2AlH$ 2. CH_3Li 3. (pyrrolidine) $NCO_2CH_2C_6H_5$, H CH_3, O (epoxide) 4. H_3O^+	Ether Ether Ether, reflux	(Z)- (pyrrolo-oxazolone) N, O, O, H, CH_3, $CH_2\overset{Si(CH_3)_3}{C}{=}CHC_4H_9$-$n$ (35–50)	99
		1. $(i$-$C_4H_9)_2AlH$ 2. CH_3Li 3. (pyrrolidine) $NCO_2CH_2C_6H_5$, H CH_3, O (epoxide) 4. H_3O^+	Ether Ether Ether, reflux	(Z)- (pyrrolo-oxazolone) N, O, O, H, CH_3, $CH_2\overset{Si(CH_3)_3}{C}{=}CHC_4H_9$-$n$ (35–50)	99
	n-$C_5H_{11}CH(OCH_3)C{\equiv}CH$	1. $(i$-$C_4H_9)_2AlH$ 2. CH_3CO_2D	Heptane $-5°$	(Z)-n-$C_5H_{11}CH(OCH_3)CH{=}CHD$ (—)	25
C_{11}	(R)-n-$C_4H_9CHCH_3$ $(CH_3)_3SiC{\equiv}C$	1. $(i$-$C_4H_9)_2AlH$ 2. CH_3Li 3. (pyrrolidine) $NCO_2CH_2C_6H_5$, H CH_3, O (epoxide) 4. H_3O^+	Ether Ether Ether, reflux	(Z)-(R)- (pyrrolo-oxazolone) N, O, O, H, CH_3, $CH_2\overset{Si(CH_3)_3}{C}{=}CHCH(CH_3)C_4H_9$-$n$ (41)	99
C_{12}	$CH_3CH(OTHP)C{\equiv}CSi(CH_3)_3$	1. $(i$-$C_4H_9)_2AlH$ 2. 6N NaOH	Hexane, 1.1 eq THF, 0 to 25°	(Z)-$CH_3CH(OTHP)CH{=}CHSi(CH_3)_3$ (85)	73
	n-$C_5H_{11}CH(OC_4H_9$-$t)C{\equiv}CH$	1. $(i$-$C_4H_9)_2AlH$ 2. D_2O	C_6H_6, toluene	(Z)-n-$C_5H_{11}CH(OC_4H_9$-$t)CH{=}CHD$ (—)	25
C_{14}	$C_2H_5CH(OTHP)CH_2C{\equiv}CSi(CH_3)_3$	1. $(i$-$C_4H_9)_2AlH$ 2. 6N NaOH	Ether, 40°, 6 hr	(Z)-$C_2H_5CH(OTHP)CH_2CH{=}CHSi(CH_3)_3$ (79)	73
	n-$C_3H_7CH(OTHP)C{\equiv}CSi(CH_3)_3$	1. $(i$-$C_4H_9)_2AlH$ 2. 6N NaOH	Hexane, 1.1 eq THF, 0 to 25°	(Z)-n-$C_3H_7CH(OTHP)CH{=}CHSi(CH_3)_3$ (93)	73

TABLE VII. ALLYLIC AND HOMOALLYLIC ETHERS (*Continued*)

	Substrate	Reagents	Conditions	Product(s) and Yield(s)(%)	Refs.
C_{15}	1-methylcyclohexyl-$OCH(C_5H_{11}\text{-}n)C\equiv CH$	1. $(i\text{-}C_4H_9)_2AlH$ 2. CH_3CO_2D	Heptane $-5°$	(*Z*)-1-methylcyclohexyl-$OCH(C_5H_{11}\text{-}n)CH{=}CHD$ (70)	25
C_{27}	$n\text{-}C_5H_{11}CH(OC(C_6H_5)_3)C\equiv CH$	1. $(i\text{-}C_4H_9)_2AlH$ 2. CH_3Li 3. D_2O	C_6H_6–hexane Ether–hexane	(*E*)-$n\text{-}C_5H_{11}CH(OC(C_6H_5)_3)CH{=}CHD$ (40)	25

TABLE VIII. α,β-UNSATURATED ACIDS, ESTERS, NITRILES, KETONES, AND ETHERS

	Substrate	Reagents	Conditions	Product(s) and Yield(s)(%)	Refs.
C_2	$HC\equiv CH$	1. $(i\text{-}C_4H_9)_3Al$ 2. CO_2 3. H_3O^+	Heptane, 25°, 4 hr	$i\text{-}C_4H_9CH{=}CHCO_2H$ (94)	84
C_4	$HO(CH_2)_2C\equiv CH$	1. $(CH_3)_3Al$ (3 eq), $Cl_2Zr(C_5H_5)_2$ (0.25 eq) 2. $ClCO_2CH_3$	$ClCH_2CH_2Cl$	(E)-$HO(CH_2)_2C(CH_3){=}CHCO_2CH_3$ (63)	66
	$I(CH_2)_2C\equiv CH$	1. $(CH_3)_3Al$ (3 eq), $Cl_2Zr(C_5H_5)_2$ (1 eq) 2. $ClCO_2C_2H_5$	$ClCH_2CH_2Cl$	(E)-$I(CH_2)_2C(CH_3){=}CHCO_2C_2H_5$ (74)	66
	$CH_3C\equiv CCH_3$	1. $(i\text{-}C_4H_9)_2(CH_3)AlHLi$ 2. CO_2 3. H_3O^+	DME, 100–130°, 6 hr	(Z)-$CH_3CH{=}C(CH_3)CO_2H$ (72)	49
		1. $(i\text{-}C_4H_9)_2AlH$ 2. CH_3Li 3. CO_2 4. H_3O^+	Heptane Heptane–ether, 25° −30 to −10°	(E)-$CH_3CH{=}C(CH_3)CO_2H$ (76)	91
C_5	$i\text{-}C_3H_7C\equiv CH$	1. $(i\text{-}C_4H_9)_2AlH$ 2. $ClCO_2C_2H_5$	Hexane, 50°, 4 hr 25°, 1 hr	(E)-$i\text{-}C_3H_7CH{=}CHCO_2C_2H_5$ (64)	106
	$C_2H_5C\equiv CCN$	1. $LiAlH_4$ (0.5 eq) 2. H_3O^+	Ether, reflux, 30 min	(E)-$C_2H_5CH{=}CHCN$ (70)	52
C_6	$n\text{-}C_4H_9C\equiv CH$	1. $(i\text{-}C_4H_9)_2AlH$ 2. CH_3Li 3. CNCl	Hexane, 50°, 2 hr Ether–THF, −40 to −50° −40 to 25°, 15 min	(E)-$n\text{-}C_4H_9CH{=}CHCN$ (62)	178
		1. $(i\text{-}C_4H_9)_2AlH$ 2. CH_3Li 3. $(CN)_2$	Hexane, 50°, 2 hr Ether–hexane, −20° −10 to 10°	" (87)	107

TABLE VIII. α,β-Unsaturated Acids, Esters, Nitriles, Ketones, and Ethers (*Continued*)

Substrate	Reagents	Conditions	Product(s) and Yield(s)(%)	Refs.
C_6 n-$C_4H_9C{\equiv}CH$ (*cont'd*)	1. $(i\text{-}C_4H_9)_2AlH$ 2. $ClCO_2C_2H_5$	Hexane, 50°, 4 hr 25°, 1 hr	(E)-n-$C_4H_9CH{=}CHCO_2C_2H_5$ (64)	106
	1. $(i\text{-}C_4H_9)_2AlH$ 2. CO_2 3. H_3O^+	Heptane, 25°, 4 hr	(E)-n-$C_4H_9CH{=}CHCO_2H$ (76)	84
	1. $(i\text{-}C_4H_9)_2AlH$ 2. CH_3Li 3. CO_2 4. H_3O^+	Heptane, 50°, 2 hr Heptane–ether, 25° −30 to −10°	" (78)	91
	1. $(i\text{-}C_4H_9)_2AlH$ 2. $Cl_2Pd[P(C_6H_5)_3]_2$–$(i\text{-}C_4H_9)_2AlH$ 3. (E)-$BrCH{=}C(CH_3)CO_2CH_3$	Hexane THF–hexane THF–hexane, 25°, reflux 15 min	n-C_4H_9(H)C=C(H)–C(H)=C(CH_3)(CO_2CH_3) (75)	109
	1. $(i\text{-}C_4H_9)_2AlH$ 2. CO_2 3. H_3O^+	Heptane 25°	(E)-n-$C_4H_9CH{=}CHCO_2H$ (35)	102
t-$C_4H_9C{\equiv}CH$	1. $(i\text{-}C_4H_9)_2AlH$ 2. C_6H_5OCN	Hexane, 50°, 4 hr Hexane, −78°, 1 hr	(E)-t-$C_4H_9CH{=}CHCN$ (72)	179
	1. $(i\text{-}C_4H_9)_2AlH$ 2. $ClCO_2C_2H_5$	Hexane, 50°, 4 hr	(E)-t-$C_4H_9CH{=}CHCO_2C_2H_5$ (72)	106
$C_2H_5C{\equiv}CC_2H_5$	1. $(i\text{-}C_4H_9)_2(CH_3)AlHLi$ 2. $(CN)_2$	DME, 100–130°, 6 hr	(Z)-$C_2H_5CH{=}C(CN)C_2H_5$ (65)	107
	1. $(i\text{-}C_4H_9)_2AlH$ 2. CH_3Li 3. $(CN)_2$	Hexane, 50°, 2 hr Ether–hexane, −20° −10 to 10°	(E)-$C_2H_5CH{=}C(CN)C_2H_5$ (76)	107

	1. $(i\text{-}C_4H_9)_2AlH$ 2. CO_2 3. H_3O^+	Heptane, 25°, 4 hr	$C_2H_5CH{=}C(CO_2H)C_2H_5$ (89)	84
	1. $(i\text{-}C_4H_9)_2(CH_3)AlHLi$ 2. CO_2 3. H_3O^+	DME, 100–130°, 6 hr	(Z)-$C_2H_5CH{=}C(CO_2H)C_2H_5$ (67)	49
	1. $(i\text{-}C_4H_9)_2AlH$ 2. CO_2 3. H_3O^+	Heptane 25°	(E)-$C_2H_5CH{=}C(CO_2H)C_2H_5$ (60)	102
	1. $(i\text{-}C_4H_9)_2AlH$ 2. CH_3Li 3. CO_2 4. H_3O^+	Heptane Heptane–ether, 25° -30 to $-10°$	" (78)	91
	1. $(i\text{-}C_4H_9)_2AlH$ (0.5 eq) 2. $n\text{-}C_4H_9Li$ 3. $(CN)_2$	Hexane, 70°	C_2H_5 C_2H_5 CN H C_2H_5 C_2H_5 (63)	107
	1. $(i\text{-}C_4H_9)_2AlH$ 2. $Pd[P(C_6H_5)_3]_4$ (5 mol %) 3. $ZnCl_2$ 4. (E)-$BrCH{=}C(CH_3)CO_2CH_3$	20–25°	C_2H_5 C_2H_5 CH_3 H H CO_2CH_3 (65)	82
$CH_3C{\equiv}CSi(CH_3)_3$	1. $(i\text{-}C_4H_9)_2AlH$ 2. CH_3Li 3. CO_2 4. H_3O^+	Ether–hexane, 25°, 21 hr 0°, 30 min	(Z)-$CH_3CH{=}C(CO_2H)Si(CH_3)_3$ (68)	105
$CH_2{=}C(CH_3)C{\equiv}CCN$	1. $LiAlH_4$ (0.5 eq) 2. H_3O^+	Ether, -60 to 0°, 1 hr	(E)-$CH_2{=}C(CH_3)CH{=}CHCN$ (85)	52

TABLE VIII. α,β-UNSATURATED ACIDS, ESTERS, NITRILES, KETONES, AND ETHERS (*Continued*)

Substrate	Reagents	Conditions	Product(s) and Yield(s)(%)	Refs.
C_6 $HC{\equiv}CCH_2Si(CH_3)_3$ (*cont'd*)	1. $(CH_3)_3Al$ (2 eq), $Cl_2Zr(C_5H_5)_2$ (0.1 eq) 2. $ClCO_2CH_3$	$ClCH_2CH_2Cl$, 25°, 2 hr	(*E*)-$CH_3O_2CCH{=}C(CH_3)CH_2Si(CH_3)_3$ (57)	155
C_7 *n*-$C_5H_{11}C{\equiv}CH$	1. $(CH_3)_3Al$ (2 eq), $Cl_2Zr(C_5H_5)_2$ (0.4 eq) 2. $ClCO_2C_2H_5$ (3 eq)	$ClCH_2CH_2Cl$, 25°, 2 hr Hexane, 25°, 1 hr	(*E*)-*n*-$C_5H_{11}C(CH_3){=}CHCO_2C_2H_5$ (86)	80
	1. $(CH_3)_3Al$ (2 eq), $Cl_2Zr(C_5H_5)_2$ (0.4 eq) 2. *n*-C_4H_9Li 3. CO_2 4. H_3O^+	$ClCH_2CH_2Cl$, 25°, 2 hr Hexane 25°, 3 hr	(*E*)-*n*-$C_5H_{11}C(CH_3){=}CHCO_2H$ (64)	80
n-$C_4H_9C{\equiv}CCN$	1. $LiAlH_4$ (0.5 eq) 2. H_3O^+	Ether, reflux, 30 min	(*E*)-*n*-$C_4H_9CH{=}CHCN$ (90)	52
n-$C_3H_7C{\equiv}CAl(CH_3)_2$	1. $Cl(CH_3)Zr(C_5H_5)_2$ 2. CH_3COCl–$AlCl_3$	Not specified	*n*-$C_3H_7C(CH_3){=}CHCOCH_3$ (61) ($Z{:}E$ = 92:8)	40
C_8 *n*-$C_6H_{13}C{\equiv}CH$	1. (*i*-$C_4H_9)_2AlH$ 2. C_6H_5OCN	Hexane, 50°, 4 hr Hexane, −78°, 1 hr	(*E*)-*n*-$C_6H_{13}CH{=}CHCN$ (71)	179
	1. (*i*-$C_4H_9)_2AlH$ 2. $ClCO_2CH_3$	Hexane, 50°, 4 hr 25°, 1 hr	(*E*)-*n*-$C_6H_{13}CH{=}CHCO_2CH_3$ (65)	106
$C_6H_{11}C{\equiv}CH$	1. (*i*-$C_4H_9)_2AlH$ 2. C_6H_5OCN	Hexane, 50°, 4 hr Hexane, −78°, 1 hr	(*E*)-$C_6H_{11}CH{=}CHCN$ (78)	179
	1. (*i*-$C_4H_9)_2AlH$ 2. CH_3Li 3. $(CN)_2$	Hexane, 50°, 2 hr Ether–hexane, −20° −10 to 10°	" (78)	107
	1. (*i*-$C_4H_9)_2AlH$ 2. $ClCO_2C_2H_5$	Hexane, 50°, 4 hr 25°, 1 hr	(*E*)-$C_6H_{11}CH{=}CHCO_2C_2H_5$ (74)	106

	Reactant	Reagents	Conditions	Product (Yield, %)	Refs.
		1. $(i\text{-}C_4H_9)_2AlH$ 2. CH_3Li 3. CO_2 4. H_3O^+	Heptane, 50°, 2 hr Heptane–ether, 25° −30 to −10°	(*E*)-$C_6H_{11}CH{=}CHCO_2H$ (72)	91
	[1-cyclohexenyl]$C{\equiv}CH$	1. $(i\text{-}C_4H_9)_2AlH$ 2. $n\text{-}C_4H_9Li$ 3. $(CN)_2$	Hexane, 50°, 2 hr Hexane, −20° −10 to 10°	(*E*)-[1-cyclohexenyl]$CH{=}CHCN$ (62)	107
	$C_6H_5C{\equiv}CH$	1. $(i\text{-}C_4H_9)_2AlH$ 2. $n\text{-}C_4H_9Li$ 3. $(CN)_2$	Hexane, 50°, 2 hr Hexane, −20° −10 to 10°	(*E*)-$C_6H_5CH{=}CHCN$ (64)	107
	$n\text{-}C_4H_9C{\equiv}COC_2H_5$	1. $(i\text{-}C_4H_9)_2AlH$ 2. H_2O	Pentane, *N*-methylpyrrolidine, 25°, 10 hr	(*Z*)-$n\text{-}C_4H_9CH{=}CHOC_2H_5$ (70)	39
	$(CH_3)_2C{=}CH(CH_2)_2C{\equiv}CH$	1. $(CH_3)_3Al$ (2 eq), $Cl_2Zr(C_5H_5)_2$ (0.4 eq) 2. $ClCO_2C_2H_5$ (3 eq)	$ClCH_2CH_2Cl$, 25°, 2 hr Hexane, 25°, 1 hr	(*E*)-$(CH_3)_2C{=}CH(CH_2)_2C(CH_3){=}CHCO_2C_2H_5$ (85)	80
C_9	$C_6H_5C{\equiv}CCN$	1. $LiAlH_4$ (0.5 eq) 2. H_3O^+	Ether, −60°, 0°, 1 hr	(*E*)-$C_6H_5CH{=}CHCN$ (98)	52
	[1-cyclohexenyl]$C{\equiv}CCN$	1. $LiAlH_4$ (0.5 eq) 2. H_3O^+	Ether, −60°, 0°, 1 hr	(*E*)-[1-cyclohexenyl]$CH{=}CHCN$ (75)	52
	$HO(CH_2)_7C{\equiv}CH$	1. $(i\text{-}C_4H_9)_3Al$ 2. $(i\text{-}C_4H_9)_2AlH$ 3. CH_3Li 4. CO_2 5. H_3O^+	Toluene, 25°, 1 hr 25°, 12 hr 5–10°, 0.5 hr −30 to −40°, 5 hr 0–5°	(*E*)-$HO(CH_2)_7CH{=}CHCO_2H$ (80)	103
	$n\text{-}C_4H_9C{\equiv}CSi(CH_3)_3$	1. $(i\text{-}C_4H_9)_2AlH$ 2. CO_2 3. H_3O^+	Ether, 40°, 1 hr Hexane, 60°, 2 hr	(*Z*)-$n\text{-}C_4H_9CH{=}C(CO_2H)Si(CH_3)_3$ (82)	104

TABLE VIII. α,β-UNSATURATED ACIDS, ESTERS, NITRILES, KETONES, AND ETHERS (*Continued*)

Substrate	Reagents	Conditions	Product(s) and Yield(s)(%)	Refs.
C_9 *n*-$C_4H_9C{\equiv}CSi(CH_3)_3$ (*cont'd*)	1. (*i*-$C_4H_9)_2AlH$ 2. CH_3Li 3. CO_2 4. H_3O^+	Ether, 40°, 1 hr Ether, 25°, 15 min Ether, −30 to 25°, 1 hr	(*Z*)-*n*-$C_4H_9CH{=}C(CO_2H)Si(CH_3)_3$ (88)	104
	1. (*i*-$C_4H_9)_2AlH$ 2. Br_2 (1.5 eq) 3. Br_2 (15 mol %) 4. *n*-C_4H_9Li 5. CO_2 6. H_3O^+	Ether, 40°, 1 hr Ether–pyridine, −78° Ether–pyridine, *hν* THF, −78°, 2 hr −78 to 25°	(*E*)-*n*-$C_4H_9CH{=}C(CO_2H)Si(CH_3)_3$ (73)	104
t-$C_4H_9C{\equiv}CSi(CH_3)_3$	1. (*i*-$C_4H_9)AlH$ 2. CO_2 3. H_3O^+	Ether, 40°, 1 hr Hexane, 60°, 2 hr	(*Z*)-*t*-$C_4H_9CH{=}C(CO_2H)Si(CH_3)_3$ (85)	104
	1. (*i*-$C_4H_9)_2AlH$ 2. Br_2 (1.5 eq) 3. Br_2 (15 mol %) 4. *n*-C_4H_9Li 5. CO_2 6. H_3O^+	Ether, 40°, 1 hr Ether–pyridine, −78° Ether–pyridine, *hν* THF, −78°, 2 hr −78 to 25°	(*E*)-*t*-$C_4H_9CH{=}C(CO_2H)Si(CH_3)_3$ (83)	104

C_{10}	$C_6H_5S(CH_2)_2C{\equiv}CH$	1. $(CH_3)_3Al$ (3 eq), $Cl_2Zr(C_5H_5)_2$ (1 eq) 2. n-C_4H_9Li 3. CO_2 4. H_3O^+	$ClCH_2CH_2Cl$, 25°	(E)-$C_6H_5S(CH_2)_2C(CH_3){=}CHCO_2H$ (62)	66
C_{11}	$C_6H_{11}C{\equiv}CSi(CH_3)_3$	1. $(i$-$C_4H_9)_2AlH$ 2. CO_2 3. H_3O^+	Ether, 40°, 1 hr Hexane, 60° 2 hr	(Z)-$C_6H_{11}CH{=}C(CO_2H)Si(CH_3)_3$ (97)	104
		1. $(i$-$C_4H_9)_2AlH$ 2. Br_2 (1.5 eq) 3. Br_2 (15 mol %) 4. n-C_4H_9Li 5. CO_2 6. H_3O^+	Ether, 40°, 1 hr Ether–pyridine, −78° Ether–pyridine, $h\nu$ THF, −78°, 2 hr −78 to 25°	(E)-$C_6H_{11}CH{=}C(CO_2H)Si(CH_3)_3$ (100)	104
C_{14}	$C_6H_5C{\equiv}CC_6H_5$	1. $(i$-$C_4H_9)_2AlH$ 2. CO_2 3. H_3O^+	Heptane, 25°, 4 hr	$C_6H_5CH{=}C(CO_2H)C_6H_5$ (100)	84
		1. $(i$-$C_4H_9)_2AlH$ 2. CO_2 3. H_3O^+	Heptane, 25°, 24 hr 50°, 4 hr	(E)-$C_6H_5CH{=}C(CO_2H)C_6H_5$ (96)	102

TABLE IX. DIENES

	Substrate	Reagents	Conditions	Product(s) and Yield(s)(%)	Refs.
C_4	$CH_2{=}CHC{\equiv}CH$	1. $(CH_3)_3Al$ (2 eq), $Cl_2Zr(C_5H_5)_2$(0.4 eq) 2. $Pd[P(C_6H_5)_3]_4$ (5 mol %)	$ClCH_2CH_2Cl$, 25°, 12 hr	(86)	111
		3. Cl	THF–$ClCH_2CH_2Cl$, 0 to 25°, 3 hr		
		1. $(CH_3)_3Al$ (2 eq), $Cl_2Zr(C_5H_5)_2$ (0.4 eq) 2. $Pd[P(C_6H_5)_3]_4$ (5 mol %)	$ClCH_2CH_2Cl$, 25°, 12 hr	(77)	111
		3. Cl	THF–$ClCH_2CH_2Cl$, 0 to 25°, 3 hr		
C_5	$CH_3C{\equiv}CSCH{=}CH_2$	1. $LiAlH_4$ (1 eq) 2. H_3O^+	THF, 50–60°, 1 hr	(*E*)-$CH_3CH{=}CHSCH{=}CH_2$ (92)	53
	$CH_2{=}C(CH_3)C{\equiv}CH$	1. $(CH_3)_3Al$ (1.7 eq), $Cl_2Zr(C_5H_5)_2$ (1.0 eq) 2. $HgCl_2$ (1.5 eq)	$ClCH_2CH_2Cl$, 25°, 2 hr THF, 0°	(*E*)-$CH_2{=}C(CH_3)C(CH_3){=}CHHgCl$ (75)	132
C_6	n-$C_4H_9C{\equiv}CH$	1. $(i\text{-}C_4H_9)_2AlH$ 2. n-C_4H_9Li 3. $BrCH_2C{\equiv}CH$	Hexane, 50–55°, 4 hr Hexane THF, 25°, 6 hr	(*E*)-n-$C_4H_9CH{=}CHCH{=}C{=}CH_2$ (60)	77
		1. $(i\text{-}C_4H_9)_2AlH$ (2 eq) 2. $C_6H_5CH{=}CHCHO$	90°, 5 hr Heptane, 0°	$C_6H_5CH{=}CHCH{=}CHC_5H_{11}$-$n$ (25–30)	27
		1. $(CH_3)_3Al$ (2 eq), $Cl_2Zr(C_5H_5)_2$ (0.4 eq) 2. $Pd[P(C_6H_5)_3]_4$ (5 mol %) 3. Cl	$ClCH_2CH_2Cl$, 25° THF, 25°, 1 hr	n-C_4H_9, CH_3 (100)	111, 180
		1. $(CH_3)_3Al$ (2 eq), $Cl_2Zr(C_5H_5)_2$ (0.4 eq) 2. $Pd[P(C_6H_5)_3]_4$ (5 mol %) 3. O_2CCH_3	$ClCH_2CH_2Cl$, 25° THF, 25°, 1 hr	" (100)	180

Reagents	Conditions	Product (Yield, %)	Refs.
1. $(CH_3)_3Al$ (2 eq), $Cl_2Zr(C_5H_5)_2$ (0.4 eq) 2. $Pd[P(C_6H_5)_3]_4$ (5 mol %) 3. $OAl(CH_3)_2$	$ClCH_2CH_2Cl$, 25° THF, 25°, 48 hr	" (100)	180
1. $(CH_3)_3Al$ (2 eq), $Cl_2Zr(C_5H_5)_2$ (0.4 eq) 2. $Pd[P(C_6H_5)_3]_4$ (5 mol %) 3. $OPO(OC_2H_5)_2$	$ClCH_2CH_2Cl$, 25° THF, 25°, 6 hr	" (43)	180
1. $(CH_3)_3Al$ (2 eq), $Cl_2Zr(C_5H_5)_2$ (0.4 eq) 2. $Pd[P(C_6H_5)_3]_4$ (5 mol %) 3. $OSi(CH_3)_3$	$ClCH_2CH_2Cl$, 25° THF, 25°, 48 hr	" (94)	180
1. $(CH_3)_3Al$ (2 eq), $Cl_2Zr(C_5H_5)_2$ (0.4 eq) 2. $Pd[P(C_6H_5)_3]_4$ (5 mol %) 3. $OSi(CH_3)_2(C_4H_9\text{-}t)$	$ClCH_2CH_2Cl$, 25° THF, 25°, 48 hr	" (46)	180
1. $(i\text{-}C_4H_9)_2AlH$ 2. $n\text{-}C_4H_9Li$ 3. $BrCH_2CH{=}CH_2$	Hexane, 50–55°, 2–4 hr Hexane THF, 25°, 3 hr	$(E)\text{-}n\text{-}C_4H_9CH{=}CHCH_2CH{=}CH_2$ (73)	77
1. $(i\text{-}C_4H_9)_2AlH$ 2. CuCl	Hexane, 50°, 4 hr THF, 25°, 1 hr	$(E,E)\text{-}n\text{-}C_4H_9CH{=}CHCH{=}CHC_4H_9\text{-}n$ (73)	108
1. $(i\text{-}C_4H_9)_2AlH$ 2. $Cl_2Pd[P(C_6H_5)_3]_2$ (5 mol %), $(i\text{-}C_4H_9)_2AlH$ 3. $(E)\text{-}t\text{-}C_4H_9CH{=}CHI$	Hexane THF–hexane, 25°	$(E,E)\text{-}n\text{-}C_4H_9CH{=}CHCH{=}CHC_4H_9\text{-}t$ (82)	109
1. $(i\text{-}C_4H_9)_2AlH$ 2. $Ni(acac)_2$ (5 mol %), $(C_6H_5)_3P$, $(i\text{-}C_4H_9)_2AlH$ 3. $(E)\text{-}t\text{-}C_4H_9CH{=}CHI$	Hexane Ether–hexane Ether–hexane, 25°	" (48)	109
1. $(i\text{-}C_4H_9)_2AlH$ 2. Br, CuI	Hexane THF, 25°, 4 hr	$n\text{-}C_4H_9$, H, H (67)	110

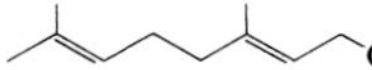

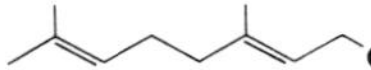

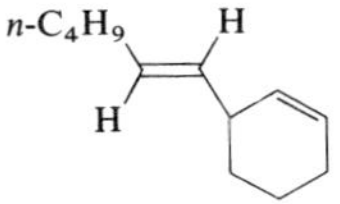

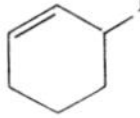

TABLE IX. DIENES (*Continued*)

Substrate	Reagents	Conditions	Product(s) and Yield(s)(%)	Refs.
C_6 n-$C_4H_9C{\equiv}CH$ (*cont'd*)	1. $(CH_3)_3Al$ (2 eq), $Cl_2Zr(C_5H_5)_2$ (0.4 eq) 2. 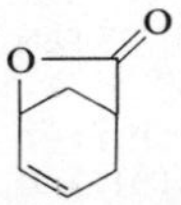 [bicyclic lactone structure], $Pd[P(C_6H_5)_3]_4$ (5 mol %), 3. H_3O^+	$ClCH_2CH_2Cl$ THF, 25°, 3 hr	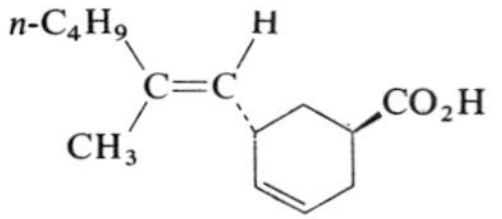 n-$C_4H_9(CH_3)C{=}CH$-(cyclohexenyl)-CO_2H (90)	181
	1. $(CH_3)_3Al$ (2 eq), $Cl_2Zr(C_5H_5)_2$ (0.4 eq) 2. 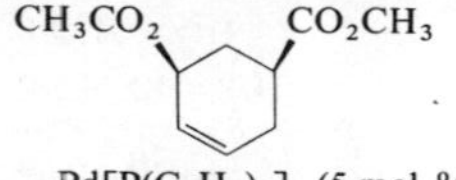 CH_3CO_2-(cyclohexenyl)-CO_2CH_3, $Pd[P(C_6H_5)_3]_4$ (5 mol %)	$ClCH_2CH_2Cl$ THF	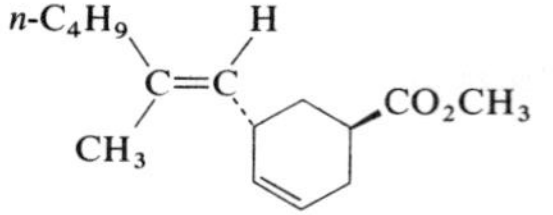 n-$C_4H_9(CH_3)C{=}CH$-(cyclohexenyl)-CO_2CH_3 (86)	181
t-$C_4H_9C{\equiv}CH$	1. $(i$-$C_4H_9)_2AlH$ 2. n-C_4H_9Li 3. $BrCH_2CH{=}CH_2$	Hexane, 50–55°, 2–4 hr Hexane THF, 25°, 3 hr	(E)-t-$C_4H_9CH{=}CHCH_2CH{=}CH_2$ (79)	77
	1. $(i$-$C_4H_9)_2AlH$ 2. $CH_2{=}CHCH_2Br$, CuCl	Hexane, 50°, 4 hr Hexane, 25°, 1 hr	" (67)	110
	1. $(i$-$C_4H_9)_2AlH$ 2. CuCl	Hexane, 50°, 4 hr THF, 25°, 1 hr	(E,E)-t-$C_4H_9CH{=}CHCH{=}CHC_4H_9$-$t$ (67)	108
	1. $(i$-$C_4H_9)_2AlH$ 2. $Ni(acac)_2$ (5 mol %), $(C_6H_5)_3P$, $(i$-$C_4H_9)_2AlH$ 3. (E)-n-$C_4H_9CH{=}CHI$	Hexane Ether–hexane Ether–hexane, 25°	(E,E)-t-$C_4H_9CH{=}CHCH{=}CHC_4H_9$-$n$ (15)	109
	1. $(i$-$C_4H_9)_2AlH$ 2. $Cl_2Pd[P(C_6H_5)_3]_2$ (5 mol %), $(i$-$C_4H_9)_2AlH$ 3. (Z)-n-$C_4H_9CH{=}CHI$	Hexane THF–hexane THF–hexane, 25°	$(3E,5Z)$-t-$C_4H_9CH{=}CHCH{=}CHC_4H_9$-$n$ (36)	109

	$C_2H_5C{\equiv}CC_2H_5$	1. $(i\text{-}C_4H_9)_3Al$ (0.16 eq) 2. H_2O	82°, 16 hr	$C_2H_5CH{=}C(C_2H_5)C(C_2H_5){=}CHC_2H_5$ (—)	2
		1. $(i\text{-}C_4H_9)_2AlH$ 2. $C_2H_5C{\equiv}CC_2H_5$ 3. CH_3OH	 60 to 70°, 24 hr	" (75)	2
		1. $(C_2H_5)_3Al$ (0.5 eq) 2. $C_2H_5C{\equiv}CC_2H_5$ 3. H_2O	80°, 20 hr	" (68)	2
		1. $(i\text{-}C_4H_9)_2AlH$ 2. CuCl	Hexane, 70°, 4 hr THF, 25°, 1 hr	" (71)	108
	$CH_2{=}C(CH_3)C{\equiv}CSCH_3$	1. $LiAlH_4$ (1 eq) 2. H_3O^+	THF, 50 to 60°, 1 hr	(*E*)-$CH_2{=}C(CH_3)CH{=}CHSCH_3$ (91)	53
C_7	$n\text{-}C_5H_{11}C{\equiv}CH$	1. $(CH_3)_3Al$ (2 eq), $Cl_2Zr(C_5H_5)_2$ (0.4 eq) 2. $Pd[P(C_6H_5)_3]_4$ (5 mol %) 3. $ZnCl_2$ 4. $CH_2{=}CHBr$	$ClCH_2CH_2Cl$, 25° 20 to 25°	(*E*)-$n\text{-}C_5H_{11}C(CH_3){=}CHCH{=}CH_2$ (73)	82
		1. $(CH_3)_3Al$ (2 eq), $Cl_2Zr(C_5H_5)_2$ (0.4 eq) 2. $Pd[P(C_6H_5)_3]_4$ (5 mol %) 3. $ZnCl_2$ 4. (*E*)-$n\text{-}C_4H_9CH{=}CHI$	$ClCH_2CH_2Cl$, 25° 20 to 25°	(*E,E*)-$n\text{-}C_5H_{11}C(CH_3){=}CHCH{=}CHC_4H_9\text{-}n$ (65)	82
		1. $(i\text{-}C_4H_9)_2AlH$ 2. $Cl_2Pd[P(C_6H_5)_3]_2$ (5 mol %), $(i\text{-}C_4H_9)_2AlH$ 3. (*E*)-$n\text{-}C_4H_9CH{=}CHI$	Hexane Ether–hexane Ether–hexane, 25°	(*E,E*)-$n\text{-}C_5H_{11}CH{=}CHCH{=}CHC_4H_9\text{-}n$ (74)	109
		1. $(i\text{-}C_4H_9)_2AlH$ 2. $Ni(acac)_2$ (5 mol %), $(C_6H_5)_3P$, $(i\text{-}C_4H_9)_2AlH$ 3. (*E*)-$n\text{-}C_4H_9CH{=}CHI$	Hexane Ether–hexane Ether–hexane, 25°	" (70)	109
		1. $(i\text{-}C_4H_9)_2AlH$ 2. $Cl_2Pd[P(C_6H_5)_3]_2$ (5 mol %), $(i\text{-}C_4H_9)_2AlH$ 3. (*Z*)-$n\text{-}C_4H_9CH{=}CHI$	Hexane Ether–hexane Ether–hexane, 25°	(5*Z*,7*E*)-$n\text{-}C_4H_9CH{=}CHCH{=}CHC_5H_{11}\text{-}n$ (55)	109

TABLE IX. DIENES (*Continued*)

Substrate	Reagents	Conditions	Product(s) and Yield(s)(%)	Refs.
C_7 n-$C_5H_{11}C{\equiv}CH$ (*cont'd*)	1. $(i\text{-}C_4H_9)_2AlH$ 2. $Ni(acac)_2$ (5 mol %), $(C_6H_5)_3P$, $(i\text{-}C_4H_9)_2AlH$ 3. (Z)-n-$C_4H_9CH{=}CHI$	Hexane Ether–hexane Ether–hexane, 25°	$(5Z,7E)$-n-$C_4H_9CH{=}CHCH{=}CHC_5H_{11}$-$n$ (55)	109
(E)-$CH_3CH{=}CHC{\equiv}CCHDCH_3$	1. $(i\text{-}C_4H_9)_2AlH$ 2. H_3O^+	Not specified	$(2E,4Z)$-$CH_3CH{=}CHCH{=}CHCHDCH_3$ (—)	33
(Z)-$CH_3CH{=}CHC{\equiv}CCHDCH_3$	1. $(i\text{-}C_4H_9)_2AlH$ 2. H_3O^+	Not specified	(Z,Z)-$CH_3CH{=}CHCH{=}CHCHDCH_3$ (—)	33
n-$C_3H_7C{\equiv}CAl(CH_3)_2$	1. $Cl(CH_3)Ti(C_5H_5)_2$ 2. cyclohexanone	0°, 30 min	n-$C_3H_7C(CH_3){=}C{=}$cyclohexylidene (83)	40
	1. $Cl(CH_3)Ti(C_5H_5)_2$ 2. C_6H_5CHO	−30°, 30 min	n-$C_3H_7C(CH_3){=}C{=}CHC_6H_5$ (67)	40
$C_5H_9C{\equiv}CH$	1. $(i\text{-}C_4H_9)_2AlH$ 2. CuCl	Hexane, 50°, 4 hr THF, 25°, 1 hr	(E,E)-$C_5H_9CH{=}CHCH{=}CHC_5H_9$ (68)	108
C_8 n-$C_6H_{13}C{\equiv}CH$	1. $(CH_3)_3Al$ (2 eq), $Cl_2Zr(C_5H_5)_2$ (0.4 eq) 2. $Pd[P(C_6H_5)_3]_4$ (5 mol %) 3. $CH_2{=}CHCH_2Br$	$ClCH_2CH_2Cl$, 25° THF THF	(E)-n-$C_6H_{13}C(CH_3){=}CHCH_2CH{=}CH_2$ (90)	111
	1. $(CH_3)_3Al$ (2 eq), $Cl_2Zr(C_5H_5)_2$ (0.4 eq) 2. $Pd[P(C_6H_5)_3]_4$ (5 mol %) 3. $CH_2{=}CHCH_2O_2CCH_3$	$ClCH_2CH_2Cl$, 25° THF THF	" (84)	111
	1. $(CH_3)_3Al$ (2 eq), $Cl_2Zr(C_5H_5)_2$ (0.4 eq) 2. $Pd[P(C_6H_5)_3]_4$ (5 mol %) 3. $(CH_3)_2C{=}CHCH_2Cl$	$ClCH_2CH_2Cl$, 25° THF THF	(E)-n-$C_6H_{13}C(CH_3){=}CHCH_2CH{=}C(CH_3)_2$ (98)	111
	1. $(i\text{-}C_4H_9)_2AlH$ 2. CH_3Li 3. $CH_2{=}CHCH_2Br$	Hexane THF–ether–hexane Reflux, 12 hr	(68)	78
	1. $(i\text{-}C_4H_9)_2AlH$ 2. $CH_2{=}CHCH_2Br$, CuCl	Hexane, 50°, 4 hr Hexane, 25°, 1 hr	" (70)	110

Formula	Reactant	Reagents	Conditions	Product(s) and Yield(s) (%)	Refs.
		1. $(i\text{-}C_4H_9)_2AlH$ 2. $n\text{-}C_4H_9Li$ 3. $CH_2{=}CCH_2SSO_2C_6H_5$	Heptane, 50°, 4 hr Heptane–toluene, −70 to 25°	$(E)\text{-}n\text{-}C_6H_{13}CH{=}CHSCH_2CH{=}CH_2$ (64)	90
		1. $(i\text{-}C_4H_9)_2AlH$ 2. $n\text{-}C_4H_9Li$ 3. $CH_2{=}C(CH_3)CH_2SSO_2C_6H_5$	Heptane, 50°, 4 hr Heptane–toluene −70 to 25°	$(E)\text{-}n\text{-}C_6H_{13}CH{=}CHSCH_2C(CH_3){=}CH_2$ (52)	90
		1. $(i\text{-}C_4H_9)_2AlH$ 2. $n\text{-}C_4H_9Li$ 3. $(E)\text{-}CH_3CH{=}CHCH_2SSO_2C_6H_5$	Heptane, 50°, 4 hr Heptane–toluene −70 to 25°	$(E,E)\text{-}n\text{-}C_6H_{13}CH{=}CHSCH_2CH{=}CHCH_3$ (65)	90
	$C_6H_{11}C{\equiv}CH$	1. $(i\text{-}C_4H_9)_2AlH$ 2. $CH_2{=}CHCH_2Br$, CuCl	Hexane Hexane, 25°, 1 hr	$(E)\text{-}C_6H_{11}CH{=}CHCH_2CH{=}CH_2$ (66)	110
	$n\text{-}C_3H_7C{\equiv}CC_3H_7\text{-}n$	1. $(i\text{-}C_4H_9)_2AlH$ 2. $CH_2{=}CHCH_2Br$, CuCl	Hexane, 70°, 4 hr Hexane, 25°, 1 hr	$(E)\text{-}n\text{-}C_3H_7CH{=}C(C_3H_7\text{-}n)CH_2CH{=}CH_2$ (70)	110
		1. $(i\text{-}C_4H_9)_2(CH_3)AlHLi$ 2. $CH_2{=}CHCH_2SSO_2C_6H_5$	Not specified	$(Z)\text{-}n\text{-}C_3H_7CH{=}C(C_3H_7\text{-}n)SCH_2CH{=}CH_2$ (51)	90
	$CH_2{=}C(CH_3)C{\equiv}CSi(CH_3)_3$	1. $(i\text{-}C_4H_9)_2AlH$ 2. H_3O^+	N-methylpyrrolidine, heptane, 60°, 50 hr	$(Z)\text{-}CH_2{=}C(CH_3)CH{=}CHSi(CH_3)_3$ (78)	38
		1. $(i\text{-}C_4H_9)_2AlH$ 2. H_3O^+	20–25°, 4 days	$(E)\text{-}CH_2{=}C(CH_3)CH{=}CHSi(CH_3)_3$ (67)	38
	$(CH_3)_2C{=}CH(CH_2)_2C{\equiv}CH$	1. $(CH_3)_3Al$ (2 eq), $Cl_2Zr(C_5H_5)_2$ (0.4 eq) 2. $Cl_2Pd[P(C_6H_5)_3]_2$ (5 mol %), $(i\text{-}C_4H_9)_2AlH$ 3. $ZnCl_2$ 4. $CH_2{=}CHBr$	$ClCH_2CH_2Cl$, 25° 20 to 25°	$(E)\text{-}(CH_3)_2C{=}CH(CH_2)_2C(CH_3){=}CHCH{=}CH_2$ (70)	82
C_9	$C_6H_5C{\equiv}CCH_3$	1. $(i\text{-}C_4H_9)_2AlH$ 2. H_2O	50 to 80°, 60 hr	$C_6H_5CH{=}C(CH_3)C(CH_3){=}CHC_6H_5$ (Major) + $C_6H_5CH{=}C(CH_3)C(C_6H_5){=}CHCH_3$ (Minor) (—)	37

TABLE IX. DIENES (*Continued*)

	Substrate	Reagents	Conditions	Product(s) and Yield(s)(%)	Refs.
C_{10}	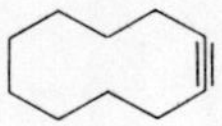	1. $(i\text{-}C_4H_9)_2AlH$ (0.5 eq) 2. H_2O	100°, 18 hr	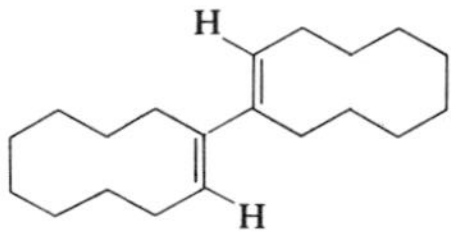 (60)	2
	(*E*)-$(CH_3)_3SiCH{=}CHC{\equiv}CSi(CH_3)_3$	1. $(i\text{-}C_4H_9)_2AlH$ 2. 3 *N* NaOH	Ether, 40°, 4 hr	(*E,Z*)-$(CH_3)_3SiCH{=}CHCH{=}CHSi(CH_3)_3$ (93)	45
C_{11}	*n*-$C_6H_{13}C{\equiv}CSi(CH_3)_3$	1. $(i\text{-}C_4H_9)_2AlH$ 2. CH_3Li 3. $(CH_3)_2C{=}CHCH_2Cl$	Heptane–ether, 25°, 17 hr Heptane–ether, 0°, 20 min 0 to 25°, 20 hr	(*Z*)-*n*-$C_6H_{13}CH{=}C[Si(CH_3)_3]CH_2CH{=}C(CH_3)_2$ (77)	43
		1. $(i\text{-}C_4H_9)_2AlH$ 2. CH_3Li 3. $CH_2{=}C(CH_3)CH_2Cl$	Heptane–ether, 25°, 17 hr Heptane–ether, 0°, 20 min 0 to 25°, 20 hr	(*Z*)-*n*-$C_6H_{13}CH{=}C[Si(CH_3)_3]CH_2C(CH_3){=}CH_2$ (84)	43
	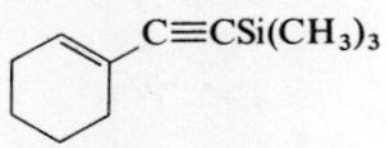 $C{\equiv}CSi(CH_3)_3$	1. $(i\text{-}C_4H_9)_2AlH$ 2. H_3O^+	Ether, 40°, 1 hr	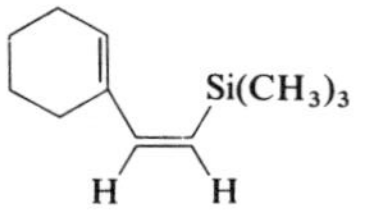 $Si(CH_3)_3$ (96)	42
	$C_6H_5C{\equiv}CSi(CH_3)_3$	1. $(i\text{-}C_4H_9)_2AlH$ 2. CH_3Li 3. $CH_2{=}CHCH_2Br$	Hexane, 25°, 4 hr THF–ether–hexane Reflux, 12 hr	(*E*)-$C_6H_5CH{=}C[Si(CH_3)_3]CH_2CH{=}CH_2$ (90)	78
		1. $(i\text{-}C_4H_9)_2AlH$ 2. CH_3Li 3. $CH_2{=}CHCH_2I$	*N*-methylpyrrolidine, hexane, 55° Ether–THF–hexane–*N*-methylpyrrolidine 25°, 36 hr	(*Z*)-$C_6H_5CH{=}C[Si(CH_3)_3]CH_2CH{=}CH_2$ (40)	78
	(*E*)-*t*-$C_4H_9CH{=}CHC{\equiv}CSi(CH_3)_3$	1. $(i\text{-}C_4H_9)_2AlH$ 2. 3 *N* NaOH	Ether, 40°, 4 hr	(1*Z*,3*E*)-*t*-$C_4H_9CH{=}CHCH{=}CHSi(CH_3)_3$ (91)	45
C_{12}	(*E*)-*n*-$C_4H_9CH{=}CHC{\equiv}CC_4H_9$-*n*	1. $(i\text{-}C_4H_9)_2(CH_3)AlHLi$ 2. H_3O^+	Diglyme, 90°, 4 hr	(*E,E*)-*n*-$C_4H_9CH{=}CHCH{=}CHC_4H_9$-*n* (85)	179

	Substrate	Reagents	Conditions	Product(s) and Yield(s) (%)	Refs.
C_{13}	(*E*)-*n*-$C_6H_{13}CH{=}CHC{\equiv}CSi(CH_3)_3$	1. $(i\text{-}C_4H_9)_2AlH$ 2. 3 *N* NaOH	Ether, 40°, 4 hr	(1*Z*,3*E*)-*n*-$C_6H_{13}CH{=}CHCH{=}CHSi(CH_3)_3$ (97)	45
		1. $(i\text{-}C_4H_9)_2(n\text{-}C_4H_9)AlHLi$ 2. H_3O^+	Diglyme, 25°, 6 hr	*n*-$C_6H_{13}CH_2CH{=}C{=}CHSi(CH_3)_3$ I, *n*-$C_6H_{13}CH_2CH_2C{\equiv}CSi(CH_3)_3$ II (86) I:II = 52:48	45
	(*E*)-$C_6H_{11}CH{=}CHC{\equiv}CSi(CH_3)_3$	1. $(i\text{-}C_4H_9)_2AlH$ 2. 3 *N* NaOH	Ether, 40°, 4 hr	(1*Z*,3*E*)-$C_6H_{11}CH{=}CHCH{=}CHSi(CH_3)_3$ (96)	45
	$(C_2H_5)_2NCH_2C{\equiv}C$–(cyclohexenyl)	1. $(i\text{-}C_4H_9)_2AlH$ (2.5 eq) 2. H_2O	Toluene, 40°, 3 hr	(*E*)-$(C_2H_5)_2NCH_2CH{=}CH$–(cyclohexenyl) (94)	150
		1. $(i\text{-}C_4H_9)_2AlH$ (2.5 eq) 2. D_2O	Toluene, 40°, 3 hr	(*E*)-$(C_2H_5)_2NCH_2CD{=}CH$–(cyclohexenyl) (86)	150
C_{14}	$C_6H_5C{\equiv}CC_6H_5$	1. $(i\text{-}C_4H_9)_3Al$ (0.16 eq) 2. H_2O	82°, 12 hr	$C_6H_5CH{=}C(C_6H_5)C(C_6H_5){=}CHC_6H_5$ (75–80)	2
		1. $(i\text{-}C_4H_9)_2AlH$ 2. H_2O	85°, 4 hr	" (77)	2
C_{17}	$C_6H_5CH_2N(CH_3)CH_2C{\equiv}C$–(cyclohexenyl)	1. $(i\text{-}C_4H_9)_2AlH$ (2.5 eq) 2. H_2O	Toluene, 40°, 3 hr	(*E*)-$C_6H_5CH_2N(CH_3)CH_2CH{=}CH$–(cyclohexenyl) (92)	150
	$C_6H_5CH_2N(CH_3)CH_2C{\equiv}C{-}C{\equiv}CC_4H_9$-*n*	1. $(i\text{-}C_4H_9)_2AlH$ (5 eq) 2. H_2O	Toluene, 40°, 24 hr	(*E*)-$C_6H_5CH_2N(CH_3)CH_2CH{=}CHC{\equiv}CC_4H_9$-*n* (46), (2*E*,4*Z*)-$C_6H_5CH_2N(CH_3)CH_2CH{=}CHCH{=}CHC_4H_9$-*n* (51)	150
	$CH_3(CH_2)_5C{\equiv}CCH_2{-}C{\equiv}C(CH_2)_6Cl$	1. $(i\text{-}C_4H_9)_2AlH$ (3 eq) 2. CH_3OH-petroleum ether 3. H_3O^+	Heptane, 25°, 15 hr	$CH_3(CH_2)_5CH{=}CHCH_2{-}CH{=}CH(CH_2)_6Cl$ (65)	188
C_{23}	*n*-$C_3H_7CH{=}C(SC_6H_5)OP(O)(OC_6H_5)_2$	(*E*)-*n*-$C_5H_{11}CH{=}CHAl(C_4H_9\text{-}i)_2$ (2 eq), $Pd[P(C_6H_5)_3]_4$ (0.1 eq)	C_6H_6, 25°	*n*-$C_3H_7CH{=}C(SC_6H_5)CH{=}CHC_5H_{11}$-*n* (62)	182
C_{24}	*n*-$C_{10}H_{21}C(OP(O)(OC_6H_5)_2){=}CH_2$	(*E*)-*n*-$C_5H_{11}CH{=}CHAl(C_4H_9\text{-}i)_2$ (2 eq), $Pd[P(C_6H_5)_3]_4$ (0.1 eq)	$ClCH_2CH_2Cl$–hexane, 25°, 4 hr	(*E*)-*n*-$C_5H_{11}CH{=}CHC(C_{10}H_{21}\text{-}n){=}CH_2$ (66)	183

TABLE X. ENYNES

	Substrate	Reagents	Conditions	Product(s) and Yield(s)(%)	Refs.
C_7	n-$C_5H_{11}C{\equiv}CH$	1. $(i\text{-}C_4H_9)_2AlH$ 2. $Pd[P(C_6H_5)_3]_4$ (5 mol %) 3. $ZnCl_2$ 4. n-$C_4H_9C{\equiv}CI$	Hexane 20 to 25°	(E)-n-$C_5H_{11}CH{=}CHC{\equiv}CC_4H_9$-$n$ (92)	82
		1. $(CH_3)_3Al$ (2 eq), $Cl_2Zr(C_2H_5)_2$ (0.4 eq) 2. $Pd[P(C_6H_5)_3]_4$ (5 mol %) 3. $ZnCl_2$ 4. n-$C_4H_9C{\equiv}CI$	$ClCH_2CH_2Cl$ 20 to 25°	(E)-n-$C_5H_{11}C(CH_3){=}CHC{\equiv}CC_4H_9$-$n$ (90)	82
C_{10}	i-$C_3H_7C{\equiv}CC{\equiv}CC_3H_7$-$i$	1. $(i\text{-}C_4H_9)_2(CH_3)AlHLi$ 2. H_3O^+	Diglyme–ether, 25°, 8 hr	(E)-i-$C_3H_7CH{=}CHC{\equiv}CC_3H_7$-$i$ (89)	51
	$(CH_3)_3SiC{\equiv}CC{\equiv}CSi(CH_3)_3$	1. $(i\text{-}C_4H_9)_2(n\text{-}C_4H_9)AlHLi$ 2. H_3O^+	DME–hexane, 25°, 18 hr	(E)-$(CH_3)_3SiCH{=}CHC{\equiv}CSi(CH_3)_3$ (91)	45
	n-$C_5H_{11}C{\equiv}CCH_2C{\equiv}CH$	1. $(i\text{-}C_4H_9)_2AlH$ (2 eq), CH_3Cu–$MgBrCl$ (1 eq), $LiCl$ (2 eq) 2. H_3O^+	THF, 20°, 1.25 hr	n-$C_5H_{11}C{\equiv}CCH_2CH{=}CH_2$ (45)	141
C_{11}	t-$C_4H_9C{\equiv}CC{\equiv}CSi(CH_3)_3$	1. $(i\text{-}C_4H_9)_2(n\text{-}C_4H_9)AlHLi$ 2. H_3O^+	DME–hexane, 25°, 1 hr	(E)-t-$C_4H_9CH{=}CHC{\equiv}CSi(CH_3)_3$ (90)	45
C_{12}	n-$C_4H_9C{\equiv}CC{\equiv}CC_4H_9$-$n$	1. $(i\text{-}C_4H_9)_2(CH_3)AlHLi$ 2. H_3O^+	Diglyme–ether, 25°, 8 hr	(E)-n-$C_4H_9CH{=}CHC{\equiv}CC_4H_9$-$n$ (92)	51
	t-$C_4H_9C{\equiv}CC{\equiv}CC_4H_9$-$t$	1. $(i\text{-}C_4H_9)_2(CH_3)AlHLi$ 2. H_3O^+	Diglyme–ether, 25°, 8 hr	(E)-t-$C_4H_9CH{=}CHC{\equiv}CC_4H_9$-$t$ (80)	51
	t-$C_4H_9C{\equiv}CC{\equiv}CC_4H_9$-$n$	1. $(i\text{-}C_4H_9)_2(CH_3)AlHLi$ 2. H_3O^+	Diglyme–ether, 25°, 8 hr	(E)-t-$C_4H_9C{\equiv}CCH{=}CHC_4H_9$-$n$ (58), (E)-t-$C_4H_9CH{=}CHC{\equiv}CC_4H_9$-$n$ (42)	51

C_{13}	$C_6H_{11}C{\equiv}CC{\equiv}CSi(CH_3)_3$	1. $(i\text{-}C_4H_9)_2(n\text{-}C_4H_9)AlHLi$ 2. H_3O^+	DME–hexane, 25°, 1 hr	(*E*)-$C_6H_{11}CH{=}CHC{\equiv}CSi(CH_3)_3$ (93)	45
	$n\text{-}C_6H_{13}C{\equiv}CC{\equiv}CSi(CH_3)_3$	1. $(i\text{-}C_4H_9)_2(n\text{-}C_4H_9)AlHLi$ 2. H_3O^+	DME–hexane, 25°, 1 hr	(*E*)-$n\text{-}C_6H_{13}CH{=}CHC{\equiv}CSi(CH_3)_3$ (95)	45
		1. $(i\text{-}C_4H_9)_2(n\text{-}C_4H_9)AlHLi$ 2. H_3O^+ 3. $KF \cdot 2H_2O$	DME–hexane, 25°, 1 hr DMF, 25°, 30 min	(*E*)-$n\text{-}C_6H_{13}CH{=}CHC{\equiv}CH$ (93)	45
		1. $(i\text{-}C_4H_9)_2(n\text{-}C_4H_9)AlHLi$ 2. H_3O^+ 3. $KF \cdot 2H_2O$ 4. $n\text{-}C_4H_9Li$ 5. $n\text{-}C_5H_{11}Br$	DME–hexane, 25°, 1 hr DMF, 25°, 30 min Diglyme–hexane, −78 to 25° Diglyme, 80°, 18 hr	(*E*)-$n\text{-}C_6H_{13}CH{=}CHC{\equiv}CC_5H_{11}\text{-}n$ (80)	45
		1. $(i\text{-}C_4H_9)_2(n\text{-}C_4H_9)AlHLi$ 2. $BrCH_2CH{=}CH_2$	DME, 25°, 1 hr DME, 25°, 24 hr	(*E*)-$n\text{-}C_6H_{13}CH{=}C(CH_2CH{=}CH_2)C{\equiv}CSi(CH_3)_3$ (94)	45
		1. $(i\text{-}C_4H_9)_2(n\text{-}C_4H_9)AlHLi$ 2. CH_3I (2 eq), CuI (1.1 eq)	DME, 25°, 1 hr DME, −35°, 24 hr	(*E*)-$n\text{-}C_6H_{13}CH{=}C(CH_3)C{\equiv}CSi(CH_3)_3$ (81)	45
C_{14}	$n\text{-}C_6H_{13}C{\equiv}CCH_2C{\equiv}CSi(CH_3)_3$	1. $(i\text{-}C_4H_9)_2AlH$ 2. H_3O^+	Ether, 40°, 2 hr	(*Z*)-$n\text{-}C_6H_{13}C{\equiv}CCH_2CH{=}CHSi(CH_3)_3$ (92)	45
C_{16}	$C_6H_{11}C{\equiv}CC{\equiv}CC_6H_{11}$	1. $(i\text{-}C_4H_9)_2(CH_3)AlHLi$ 2. H_3O^+	Diglyme–ether, 25°, 8 hr	(*E*)-$C_6H_{11}CH{=}CHC{\equiv}CC_6H_{11}$ (91)	51
	$t\text{-}C_4H_9(C{\equiv}C)_4C_4H_9\text{-}t$	1. $LiAlH_4$ 2. $CH_3OH\text{–}HCl$	Ether	$t\text{-}C_4H_9(C{\equiv}C)_3CH{=}CHC_4H_9\text{-}t$ (84)	184
C_{18}	$t\text{-}C_4H_9(C{\equiv}C)_5C_4H_9\text{-}t$	1. $LiAlH_4$ 2. H_2O	Ether	$t\text{-}C_4H_9(C{\equiv}C)_4CH{=}CHC_4H_9\text{-}t$ (—)	184
	$t\text{-}C_4H_9(C{\equiv}C)_4CH{=}CHC_4H_9\text{-}t$	1. $LiAlH_4$ 2. H_2O	Ether	$t\text{-}C_4H_9CH{=}CH(C{\equiv}C)_3CH{=}CHC_4H_9\text{-}t$ (—)	184
C_{20}	$C_6H_5C(OP(O)(OC_6H_5)_2){=}CH_2$	$C_6H_5C{\equiv}CAl(C_2H_5)_2$ (2 eq), $Pd[P(C_6H_5)_3]_4$ (0.1 eq)	$ClCH_2CH_2Cl$–hexane, 25°, 3 hr	$C_6H_5C{\equiv}CC(C_6H_5){=}CH_2$ (67)	183
C_{21}	$THPO(CH_2)_9C{\equiv}CC{\equiv}CSi(CH_3)_3$	1. $(i\text{-}C_4H_9)_2(n\text{-}C_4H_9)AlHLi$ 2. H_2O	DME, 25°, 1 hr	(*E*)-$THPO(CH_2)_9CH{=}CHC{\equiv}CSi(CH_3)_3$ (—)	45

TABLE X. ENYNES (*Continued*)

	Substrate	Reagents	Conditions	Product(s) and Yield(s)(%)	Refs.
C_{22}	4-t-C_4H_9-cyclohexenyl-1-$OP(O)(OC_6H_5)_2$	$C_6H_5C\equiv CAl(C_2H_5)_2$ (2 eq), $Pd[P(C_6H_5)_3]_4$ (0.1 eq)	$ClCH_2CH_2Cl$–hexane, 25°, 6 hr	4-t-C_4H_9-cyclohexenyl-1-$C\equiv CC_6H_5$ (70)	183
C_{23}	n-$C_3H_7CH{=}C(OP(O)(OC_6H_5)_2)SC_6H_5$	$C_6H_5C\equiv CAl(C_2H_5)_2$ (2 eq), $Pd[P(C_6H_5)_3]_4$ (0.1 eq)	C_6H_6, 25°	n-$C_3H_7CH{=}C(SC_6H_5)C\equiv CC_6H_5$ (83)	182
C_{24}	n-$C_{10}H_{21}C(OP(O)(OC_6H_5)_2){=}CH_2$	$C_6H_5C\equiv CAl(C_2H_5)_2$ (2 eq), $Pd[P(C_6H_5)_3]_4$ (0.1 eq)	$ClCH_2CH_2Cl$–hexane, 25°, 2 hr	$C_6H_5C\equiv CC(C_{10}H_{21}$-$n){=}CH_2$ (82)	183
		n-$C_5H_{11}C\equiv CAl(C_2H_5)_2$ (2 eq), $Pd[P(C_6H_5)_3]_4$ (0.1 eq)	$ClCH_2CH_2Cl$–hexane, 25°, 3 hr	n-$C_5H_{11}C\equiv CC(C_{10}H_{21}$-$n){=}CH_2$ (57)	183

TABLE XI. ACETYLENES AND HYDROXYACETYLENES

	Substrate	Reagents	Conditions	Product(s) and Yield(s)(%)	Refs.
C_2	$HC\equiv CNa$	1. $(CH_3)_3Al_2Cl_3$ 2.  [cyclopropene oxide] (0.7 eq)	Ether, 25°, 5 hr Toluene, 25°, 40 min	 [trans-2-ethynylcyclopropanol] (36)	117
		1. $(CH_3)_3Al_2Cl_3$ 2. 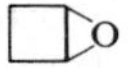 [cyclobutene oxide] (0.7 eq)	Ether, 25°, 5 hr Toluene, 25°, 40 min	 [trans-2-ethynylcyclobutanol] (53)	117
		1. $(CH_3)_3Al_2Cl_3$ 2. 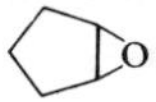 [cyclopentene oxide] (0.7 eq)	Ether, 25°, 5 hr Toluene, 25°, 2 hr	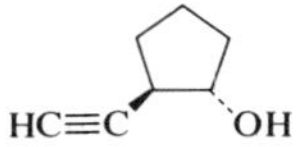 [trans-2-ethynylcyclopentanol] (20)	117
		1. $(CH_3)_3Al_2Cl_3$ 2. 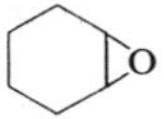 [cyclohexene oxide] (0.7 eq)	Ether, 25°, 5 hr Toluene, 50°, 24 hr	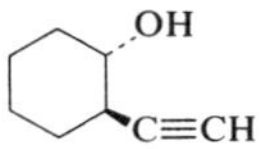 [trans-2-ethynylcyclohexanol] (4)	117
		1. $(CH_3)_3Al_2Cl_3$ 2.  [norbornene oxide]	Ether, 25°, 5 hr Toluene, 25°, 2 hr	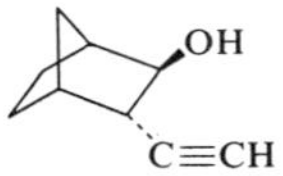 [ethynylnorbornanol] (44)	117
C_3	$CH_3C\equiv CNa$	1. $(C_2H_5)_2AlBr$ 2. 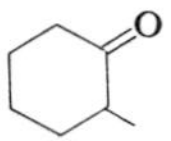 [2-methylcyclohexanone]	THF, reflux, 1 hr	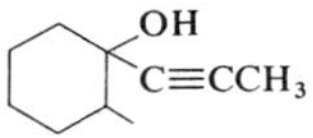 [1-(1-propynyl)-2-methylcyclohexanol] (96)	185
		1. $(C_2H_5)_2AlBr$ 2. CH_3COCH_3 (3 eq)	Ether, reflux, 1 hr	$CH_3COH(CH_3)C\equiv CCH_3$ (45)	185
C_4	$C_2H_5OC\equiv CH$	1. n-C_4H_9Li 2. $(C_2H_5)_2AlCl$ 3. 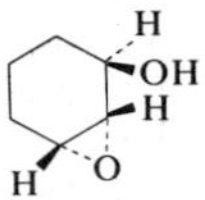 [epoxy cyclohexanol] 4. H_3O^+ 5. TsOH	Hexane Toluene–hexane, −40° Toluene, −40 to 25° C_6H_6, reflux, 16 hr	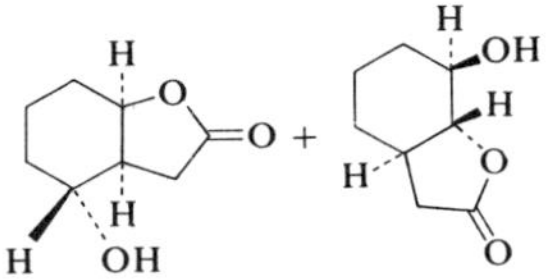 [hydroxy bicyclic lactone] + [hydroxy bicyclic lactone] (50–60)[a] 1.3:1	118

TABLE XI. ACETYLENES AND HYDROXYACETYLENES (*Continued*)

Substrate	Reagents	Conditions	Product(s) and Yield(s)(%)	Refs.
C_4 $C_2H_5OC{\equiv}CH$ (*cont'd*)	1. n-C_4H_9Li 2. $(C_2H_5)_2AlCl$ 3. 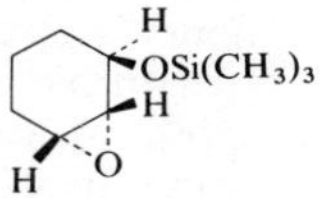 H, $OSi(CH_3)_3$, H, H, O 4. H_3O^+ 5. TsOH	Hexane Toluene–hexane, −40° Toluene, −40 to 25° C_6H_6, reflux, 16 hr	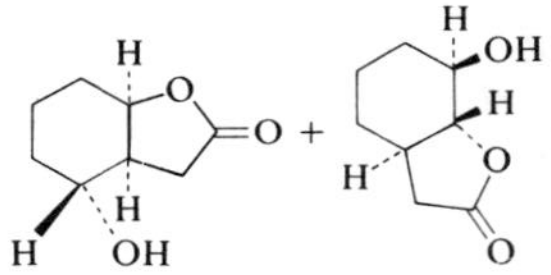 H, O, =O + H, OH, H, H, O, O; H, H, OH (50–60)[a] 2.7 : 1	118
	1. n-C_4H_9Li 2. $(C_2H_5)_2AlCl$ 3. 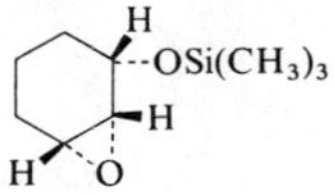 H, $OSi(CH_3)_3$, H, H, O 4. H_3O^+ 5. TsOH	Hexane Toluene–hexane, −40° Toluene, −40 to 25°	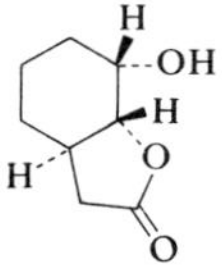 H, OH, H, O, H, O (50–60)[a]	118
	1. n-C_4H_9Li 2. $(C_2H_5)_2AlCl$ 3. 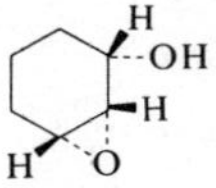 H, OH, H, H, O 4. H_3O^+ 5. TsOH	Hexane Toluene–hexane, −40° Toluene, −40 to 25° Benzene, reflux, 16 hr	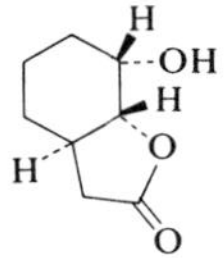 H, OH, H, O, H, O (50–60)[a]	118
	1. n-C_4H_9Li 2. $(C_2H_5)_2AlCl$ 3. 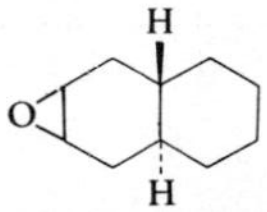 H, O, H	Hexane Toluene–hexane, −40° Toluene–hexane, −40°, 25° 15 hr	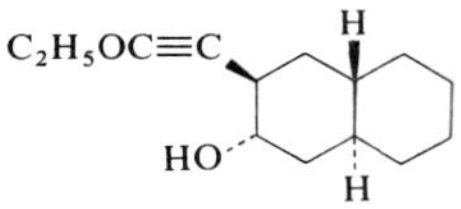 $C_2H_5OC{\equiv}C$, H, HO, H (80)	69

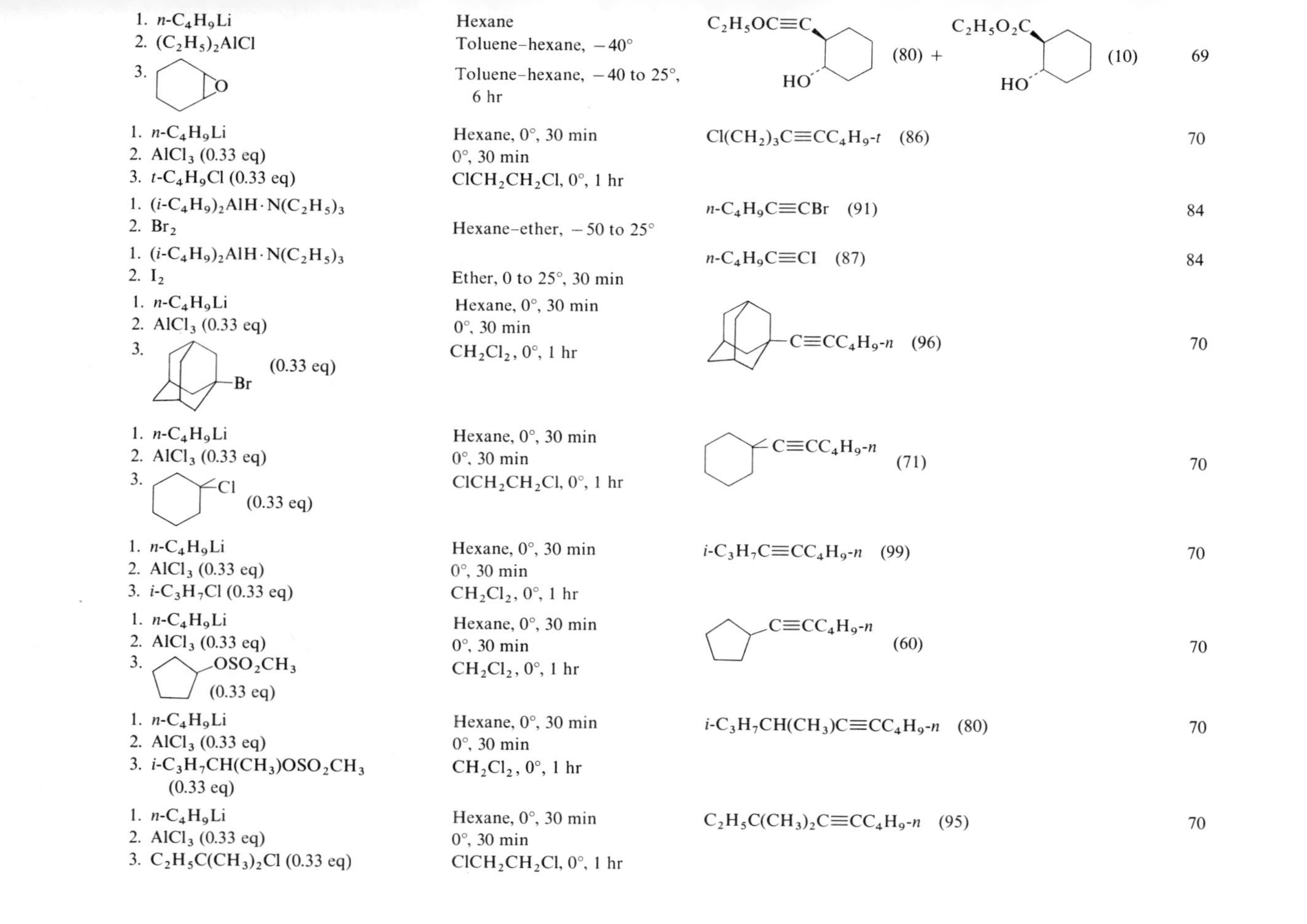

		1. $n\text{-}C_4H_9Li$ 2. $(C_2H_5)_2AlCl$ 3. O	Hexane Toluene–hexane, −40° Toluene–hexane, −40 to 25°, 6 hr	$C_2H_5OC{\equiv}C$, HO (80) + $C_2H_5O_2C$, HO (10)	69
C_5	$Cl(CH_2)_3C{\equiv}CH$	1. $n\text{-}C_4H_9Li$ 2. $AlCl_3$ (0.33 eq) 3. $t\text{-}C_4H_9Cl$ (0.33 eq)	Hexane, 0°, 30 min 0°, 30 min $ClCH_2CH_2Cl$, 0°, 1 hr	$Cl(CH_2)_3C{\equiv}CC_4H_9\text{-}t$ (86)	70
C_6	$n\text{-}C_4H_9C{\equiv}CH$	1. $(i\text{-}C_4H_9)_2AlH \cdot N(C_2H_5)_3$ 2. Br_2	Hexane–ether, −50 to 25°	$n\text{-}C_4H_9C{\equiv}CBr$ (91)	84
		1. $(i\text{-}C_4H_9)_2AlH \cdot N(C_2H_5)_3$ 2. I_2	Ether, 0 to 25°, 30 min	$n\text{-}C_4H_9C{\equiv}CI$ (87)	84
		1. $n\text{-}C_4H_9Li$ 2. $AlCl_3$ (0.33 eq) 3. Br (0.33 eq)	Hexane, 0°, 30 min 0°, 30 min CH_2Cl_2, 0°, 1 hr	$C{\equiv}CC_4H_9\text{-}n$ (96)	70
		1. $n\text{-}C_4H_9Li$ 2. $AlCl_3$ (0.33 eq) 3. Cl (0.33 eq)	Hexane, 0°, 30 min 0°, 30 min $ClCH_2CH_2Cl$, 0°, 1 hr	$C{\equiv}CC_4H_9\text{-}n$ (71)	70
		1. $n\text{-}C_4H_9Li$ 2. $AlCl_3$ (0.33 eq) 3. $i\text{-}C_3H_7Cl$ (0.33 eq)	Hexane, 0°, 30 min 0°, 30 min CH_2Cl_2, 0°, 1 hr	$i\text{-}C_3H_7C{\equiv}CC_4H_9\text{-}n$ (99)	70
		1. $n\text{-}C_4H_9Li$ 2. $AlCl_3$ (0.33 eq) 3. OSO_2CH_3 (0.33 eq)	Hexane, 0°, 30 min 0°, 30 min CH_2Cl_2, 0°, 1 hr	$C{\equiv}CC_4H_9\text{-}n$ (60)	70
		1. $n\text{-}C_4H_9Li$ 2. $AlCl_3$ (0.33 eq) 3. $i\text{-}C_3H_7CH(CH_3)OSO_2CH_3$ (0.33 eq)	Hexane, 0°, 30 min 0°, 30 min CH_2Cl_2, 0°, 1 hr	$i\text{-}C_3H_7CH(CH_3)C{\equiv}CC_4H_9\text{-}n$ (80)	70
		1. $n\text{-}C_4H_9Li$ 2. $AlCl_3$ (0.33 eq) 3. $C_2H_5C(CH_3)_2Cl$ (0.33 eq)	Hexane, 0°, 30 min 0°, 30 min $ClCH_2CH_2Cl$, 0°, 1 hr	$C_2H_5C(CH_3)_2C{\equiv}CC_4H_9\text{-}n$ (95)	70

TABLE XI. ACETYLENES AND HYDROXYACETYLENES (*Continued*)

Substrate	Reagents	Conditions	Product(s) and Yield(s)(%)	Refs.
C_6 $n\text{-}C_4H_9C\equiv CH$ (*cont'd*)	1. $n\text{-}C_4H_9Li$ 2. $AlCl_3$ (0.33 eq) 3. $t\text{-}C_4H_9Cl$ (0.33 eq)	Hexane, 0°, 30 min 0°, 30 min CH_2Cl_2, 0°, 1 hr	$t\text{-}C_4H_9C\equiv CC_4H_9\text{-}n$ (98)	70
	1. $n\text{-}C_4H_9Li$ 2. $(C_2H_5)_2AlCl$ 3. 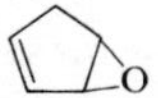 (0.5 eq)	Hexane–THF, 0° Toluene–THF–hexane, 0 to 25°, 90 min −20°	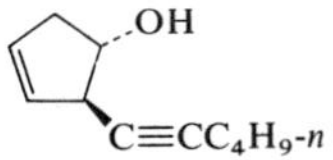 OH, $C\equiv CC_4H_9\text{-}n$ (53)	119
	1. $n\text{-}C_4H_9Li$ 2. $(C_2H_5)_2AlCl$ 3. 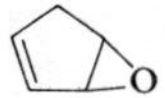 (0.5 eq)	Hexane–THF, 0° Toluene–THF–hexane, 0° to 25°, 90 min −25°, 2 hr, 25°, 12 hr	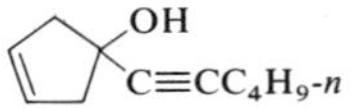 OH, $C\equiv CC_4H_9\text{-}n$ (53)	119
C_7 $CH_3C\equiv CAl(C_2H_5)_2$	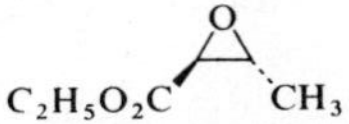 $C_2H_5O_2C$, O, CH_3	Not specified	$CH_3C\equiv CCH(CH_3)CHOHCO_2C_2H_5$ (69)	120
C_8 $n\text{-}C_6H_{13}C\equiv CH$	1. $n\text{-}C_4H_9Li$ 2. $(C_2H_5)_2AlCl$ 3. 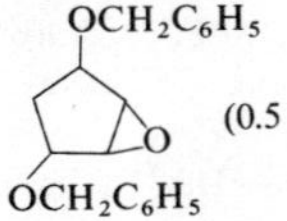 $OCH_2C_6H_5$, O, $OCH_2C_6H_5$ (0.5 eq)	Toluene–hexane Toluene–hexane Toluene, 25°, 20 hr	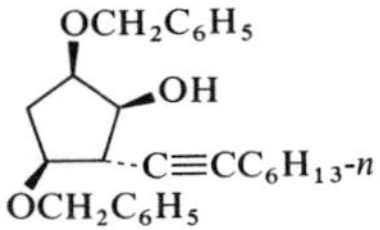 $OCH_2C_6H_5$, OH, $C\equiv CC_6H_{13}\text{-}n$, $OCH_2C_6H_5$ (78)	115
	1. $n\text{-}C_4H_9Li$ 2. $(C_2H_5)_2AlCl$ 3. 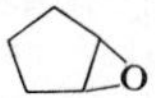 O (0.5 eq)	Toluene–hexane Toluene–hexane 25°, 18 hr	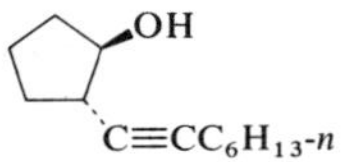 OH, $C\equiv CC_6H_{13}\text{-}n$ (77)	115
	1. $n\text{-}C_4H_9Li$ 2. $(C_2H_5)_2AlCl$ 3. 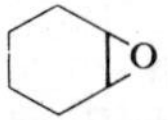 O (0.5 eq)	Toluene–hexane Toluene–hexane 25°, 18 hr	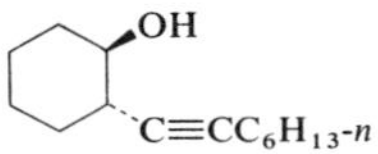 OH, $C\equiv CC_6H_{13}\text{-}n$ (98)	115

	$C_6H_{11}C{\equiv}CH$	1. n-C_4H_9Li 2. $AlCl_3$ (0.33 eq) 3. t-C_4H_9Cl (0.33 eq)	Hexane, 0°, 30 min 0°, 30 min CH_2Cl_2, 0°, 1 hr	$C_6H_{11}C{\equiv}CC_4H_9$-$t$ (90)	70
C_{11}	(R)-n-$C_5H_{11}CH(OSi(CH_3)_3)C{\equiv}CH$	1. n-C_4H_9Li 2. $(CH_3)_2AlCl$ 3. cyclopentene oxide bearing $OSi(CH_3)_3$ and $CH_2CH(OCH_2)_2$ (0.1 eq) 4. 5% K_2CO_3	Toluene, −35°, 10 min Toluene, 0°, 1 hr 40°, 2 hr	HO, OH-cyclopentane bearing $CH_2CH(OCH_2)_2$ and $C{\equiv}CCHOHC_5H_{11}$-n (26) + OH, OH-cyclopentane bearing $CH_2CH(OCH_2)_2$ and n-$C_5H_{11}CHOHC{\equiv}C$ (42)	112
C_{12}	(R)-n-$C_5H_{11}CH(OC_4H_9$-$t)C{\equiv}CH$	1. n-C_4H_9Li 2. $(CH_3)_2AlCl$ 3. cyclopentene oxide bearing OH and $(CH_2)_2OH$ (0.1 eq)	Toluene, −35°, 10 min Toluene, 0°, 1 hr 60°, 2 hr	HO, OH-cyclopentane bearing $(CH_2)_2OH$ and $C{\equiv}CCH(OC_4H_9$-$t)C_5H_{11}$-n (60)	113
		1. n-C_4H_9Li 2. $(CH_3)(OCH_3)AlCl$ 3. cyclopentene oxide bearing OH and $(CH_2)_2OH$ (0.1 eq)	Toluene, 0°, 15 min Toluene, 0°, 50 min Toluene, 60°, 2 hr	HO, HO-cyclopentane bearing $(CH_2)_2OH$ and $C{\equiv}CCH(OC_4H_9$-$t)C_5H_{11}$-n (62)	114
C_{13}	n-$C_5H_{11}CH(OTHP)C{\equiv}CH$	1. n-C_4H_9Li 2. $(C_2H_5)_2AlCl$ 3. cyclopentene oxide (0.5 eq)	Toluene–hexane Toluene–hexane Toluene, reflux, 18 hr	2-($C{\equiv}CCH(OTHP)C_5H_{11}$-$n$)cyclopentanol (OH) (27)	115
		1. n-C_4H_9Li 2. $(C_2H_5)_2AlCl$ 3. cyclohexene oxide (0.5 eq)	Toluene–hexane Toluene–hexane Toluene, reflux, 18 hr	2-($C{\equiv}CCH(OTHP)C_5H_{11}$-$n$)cyclohexanol (OH) (20)	115

TABLE XI. ACETYLENES AND HYDROXYACETYLENES (*Continued*)

Substrate	Reagents	Conditions	Product(s) and Yield(s)(%)	Refs.
C_{13} n-$C_5H_{11}CH(OTHP)C{\equiv}CH$ (*cont'd*)	1. n-C_4H_9Li 2. $(C_2H_5)_2AlCl$ 3. 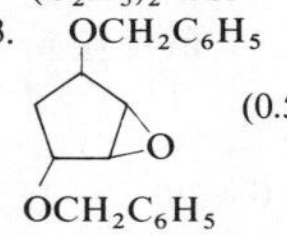 $OCH_2C_6H_5$, $OCH_2C_6H_5$ (0.5 eq)	Toluene–hexane Toluene–hexane Toluene, reflux, 140 hr	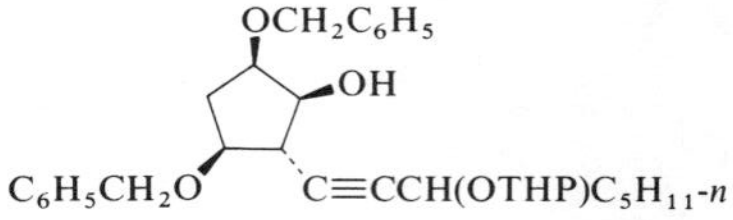 $OCH_2C_6H_5$, OH, $C_6H_5CH_2O$, $C{\equiv}CCH(OTHP)C_5H_{11}$-$n$ (28)	115
C_{15} n-$C_5H_{11}CH(OCH_2C_6H_5)C{\equiv}CH$	1. n-C_4H_9Li 2. $(C_2H_5)_2AlCl$ 3.  (0.5 eq)	Toluene–hexane Toluene–hexane Toluene, 85°, 72 hr	 OH, $C{\equiv}CCH(OCH_2C_6H_5)C_5H_{11}$-$n$ (30)	115
	1. n-C_4H_9Li 2. $(C_2H_5)_2AlCl$ 3 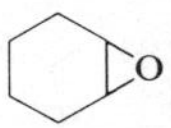 (0.5 eq)	Toluene–hexane Toluene–hexane Toluene, 90°, 72 hr	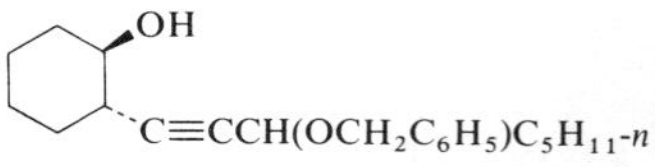 OH, $C{\equiv}CCH(OCH_2C_6H_5)C_5H_{11}$-$n$ (38)	115
	1. n-C_4H_9Li 2. $(C_2H_5)_2AlCl$ 3. 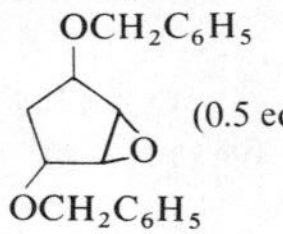 $OCH_2C_6H_5$, $OCH_2C_6H_5$ (0.5 eq)	Toluene–hexane Toluene–hexane Toluene, 85°, 72 hr	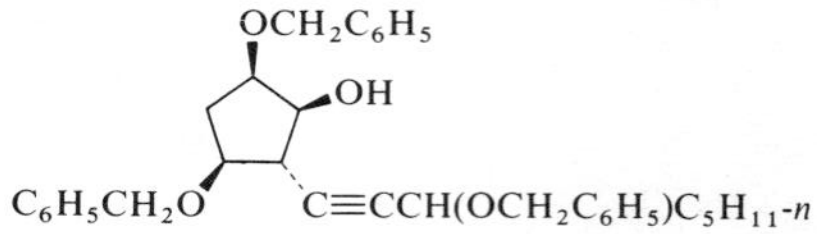 $OCH_2C_6H_5$, OH, $C_6H_5CH_2O$, $C{\equiv}CCH(OCH_2C_6H_5)C_5H_{11}$-$n$ (59)	115

[a] The intermediate ethoxy acetylene was subjected to hydrolysis and lactonization.

TABLE XII. γ-KETOALKENES

Substrate	Reagents	Conditions	Product(s) and Yield(s)(%)	Refs.
C_6 n-$C_4H_9C{\equiv}CH$	1. $(i\text{-}C_4H_9)_2AlH$ 2. 4-hydroxy-1-acetylcyclopentene (HO, $COCH_3$)	Not specified	(E)-HO-cyclopentane-$COCH_3$, $CH{=}CHC_4H_9$-n I (—), (E)-HO-cyclopentane-$COCH_3$, $CH{=}CHC_4H_9$-n II (—) I:II = 5:1	186
	1. $(i\text{-}C_4H_9)_2AlH$ 2. $C_6H_5CH{=}CHCOCH_3$	Ligroin, 50°, 2 hr Ligroin–C_6H_6, 25°, 1.25 hr	(E)-n-$C_4H_9CH{=}CHCH(C_6H_5)CH_2COCH_3$ (67)	122
	1. $(i\text{-}C_4H_9)_2AlH$ 2. $C_6H_5CH{=}CHCOC_6H_5$	Ligroin, 50°, 2 hr Ligroin–C_6H_6, 25°, 1.25 hr	(E)-n-$C_4H_9CH{=}CHCH(C_6H_5)CH_2COC_6H_5$ (60)	122
	1. $(i\text{-}C_4H_9)_2AlH$ 2. 1-acetylcyclohexene ($COCH_3$)	Ligroin, 50°, 2 hr Ligroin–C_6H_6, 25°, 1.25 hr	(E)-cyclohexane-$COCH_3$, $CH{=}CHC_4H_9$-n (35)	122
	1. $(i\text{-}C_4H_9)_2AlH$ 2. $CH_2{=}CHCOCH_3$	Ligroin, 50°, 2 hr Ligroin–ether (1:1)	(E)-n-$C_4H_9CH{=}CH(CH_2)_2COCH_3$ (30)	122
C_8 n-$C_6H_{13}C{\equiv}CH$	1. $(i\text{-}C_4H_9)_2AlH$ 2. CH_3Li 3. 2-$(CH_2)_6CO_2C_2H_5$-cyclopent-2-enone (O)	Heptane Heptane–ether 25°	(E)-cyclopentanone (O), $(CH_2)_6CO_2C_2H_5$, $CH{=}CHC_6H_{13}$-n (76)	124
	1. $(i\text{-}C_4H_9)_2AlH$ 2. CH_3Li 3. 2-$(CH_2)_6CO_2C_2H_5$-cyclopent-2-enone (O)	Hexane, 50°, 2 hr Ether–hexane, 0°, 20 min 25°, 16 hr	" (76)	25

TABLE XII. γ-KETOALKENES (*Continued*)

Substrate	Reagents	Conditions	Product(s) and Yield(s)(%)	Refs.
C_8 n-$C_6H_{13}C{\equiv}CH$ (*cont'd*)	1. $(i\text{-}C_4H_9)_2AlH$ 2. CH_3Li 3. 2-$(CH_2)_4CO_2C_2H_5$-2-cyclopenten-1-one	Heptane Heptane–ether 25°	(*E*)-2-$(CH_2)_4CO_2C_2H_5$-3-$(CH{=}CHC_6H_{13}\text{-}n)$cyclopentanone (66)	124
	1. $(i\text{-}C_4H_9)_2AlH$ 2. CH_3Li 3. 2-$CH_2CO_2C_2H_5$-2-cyclopenten-1-one	Heptane Heptane–ether 25°	(*E*)-2-$CH_2CO_2C_2H_5$-3-$(CH{=}CHC_6H_{13}\text{-}n)$cyclopentanone (57)	124
	1. $(i\text{-}C_4H_9)_2AlH$ 2. n-C_4H_9Li 3. 4-CH_3CO_2-2-$(CH_2)_6CO_2C_2H_5$-2-cyclopenten-1-one	Hexane, 50°, 2 hr Hexane	(*E*)-4-CH_3CO_2-2-$(CH_2)_6CO_2C_2H_5$-3-$(CH{=}CHC_6H_{13}\text{-}n)$cyclopentanone (53)	25
	1. $(CH_3)_3Al$ (2 eq), $Cl_2Zr(C_5H_5)_2$ (0.4 eq) 2. *B*-Methoxy-9-borabicyclo[3.3.1]nonane 3. $CH_2{=}CHCOCH_3$ (3 eq)	$ClCH_2CH_2Cl$, 25° Hexane, 25°, 1 hr 0 to 25°, 6–8 min	(*E*)-n-$C_6H_{13}C(CH_3){=}CH(CH_2)_2COCH_3$ (75)	131
C_{15} n-$C_4H_9C(OSi(C_2H_5)_3)(CH_3)CH_2C{\equiv}CH$	1. $(i\text{-}C_4H_9)_2AlH$ 2. 4-HO-2-$(CH_2)_6CO_2CH_3$-2-cyclopenten-1-one 3. CH_3CO_2H–THF–H_2O	 60°	(*E*)-(±)-4-HO-2-$(CH_2)_6CO_2CH_3$-3-$(CH{=}CHCH_2COH(CH_3)C_4H_9\text{-}n)$cyclopentanone (35)	123

No. of C Atoms	Alkyne	Reagents	Conditions	Product(s) and Yield(s) (%)	Refs.
C_{26}	$CH_3CH(OC(C_6H_5)_3)(CH_2)_3C\equiv CH$	1. $(i\text{-}C_4H_9)_2AlH$ 2. 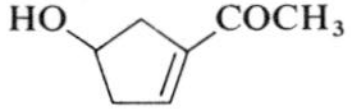 4-HO-1-$COCH_3$-cyclopentene	Hexane, 50°, 5 hr THF–ether–hexane, – 10 to 0°, 45 min	(E)-HO-cyclopentyl-$COCH_3$, $CH{=}CH(CH_2)_3CH(OC(C_6H_5)_3)CH_3$ (—) 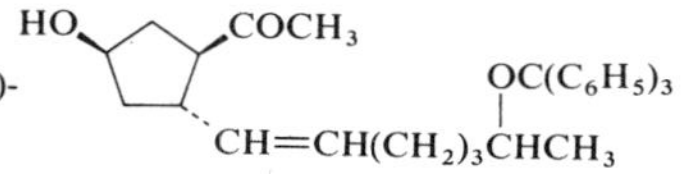	186
C_{27}	$n\text{-}C_5H_{11}CH(OC(C_6H_5)_3)C\equiv CH$	1. $(i\text{-}C_4H_9)_2AlH$ 2. CH_3Li 3. 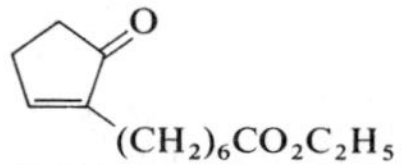 2-$(CH_2)_6CO_2C_2H_5$-cyclopent-2-enone 4. H_3O^+	C_6H_6–hexane Ether–hexane 25°, 18 hr	(E)-cyclopentanone, $(CH_2)_6CO_2C_2H_5$, $CH{=}CHCHOHC_5H_{11}\text{-}n$ (20) 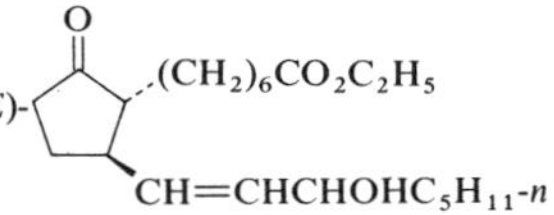	25
		1. $(i\text{-}C_4H_9)_2AlH$ 2. CH_3Li 3. 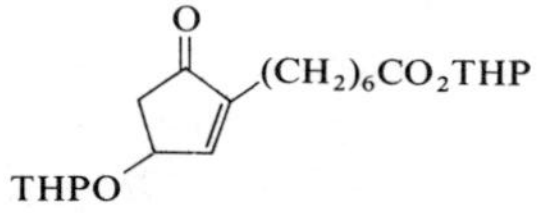 4-THPO-2-$(CH_2)_6CO_2THP$-cyclopent-2-enone 4. H_3O^+	C_6H_6–hexane Ether–hexane, 0°	(E)-HO-cyclopentanone, $(CH_2)_6CO_2H$, $CH{=}CHCHOHC_5H_{11}\text{-}n$ (12) 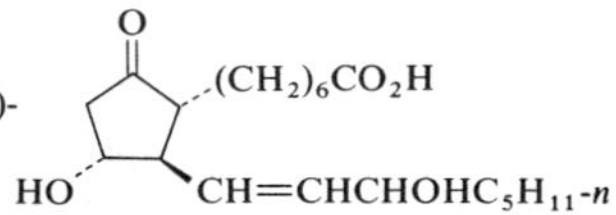	25

TABLE XIII. γ-KETOACETYLENES

	Substrate	Reagents	Conditions	Product(s) and Yield(s) (%)	Refs.
C_2	$HC{\equiv}CLi$	1. $(CH_3)_2AlCl$ 2. $Ni(acac)_2$–$(i\text{-}C_4H_9)_2AlH$ 3. [structure]	Ether Ether, 0° Ether, −5°	[structure] C≡CH (15)	128,129
C_5	$(CH_3)_3SiC{\equiv}CLi$	1. $(CH_3)_2AlCl$ 2. $Ni(acac)_2$–$(i\text{-}C_4H_9)_2AlH$ 3. [structure]	Ether Ether, 0° Ether, −5°	[structure] $C{\equiv}CSi(CH_3)_3$ (80)	128,129
		1. $(CH_3)_2AlCl$ 2. $Ni(acac)_2$–$(i\text{-}C_4H_9)_2AlH$ 3. [structure]	Ether Ether, 0° Ether, −5°	[structure] H, $C{\equiv}CSi(CH_3)_3$ (70)	128,129
		1. $(CH_3)_2AlCl$ 2. $Ni(acac)_2$–$(i\text{-}C_4H_9)_2AlH$ 3. $C_6H_5C(CH_3)_2O$–[structure]	Ether Ether, 0° Ether, −5°	$C_6H_5C(CH_3)_2O$–[structure]–$C{\equiv}CSi(CH_3)_3$ (50)	128,129
C_6	$n\text{-}C_4H_9C{\equiv}CLi$	1. $(C_2H_5)_2AlCl$ 2. [structure] $COCH_3$	Ether–ligroin Ether–ligroin, 0°, 1 hr	[structure] $COCH_3$, $C{\equiv}CC_4H_9\text{-}n$ (79)	125
		1. $(C_2H_5)_2AlCl$ 2. $C_6H_5CH{=}CHCOC_6H_5$	Ether–ligroin Ether–ligroin, 5°, 0.75 hr	$n\text{-}C_4H_9C{\equiv}CCH(C_6H_5)CH_2COC_6H_5$ (65)	125

	1. $(C_2H_5)_2AlCl$ 2. $CH_2{=}CHCOCH_3$	Ether–ligroin Ether–ligroin, $-15°$, 1.5 hr	n-$C_4H_9C{\equiv}C(CH_2)_2COCH_3$ (48)	125
	1. $(C_2H_5)_2AlCl$ 2. $(CH_3)_3C{=}CHCOCH_3$	Ether–ligroin Ether–ligroin, 25°, 4 hr	n-$C_4H_9C{\equiv}CC(CH_3)_2CH_2COCH_3$ (30), n-$C_4H_9C{\equiv}CCOH(CH_3)CH{=}C(CH_3)_2$ (40)	125
	1. $(CH_3)_2AlCl$ 2. $Ni(acac)_2$–$(i$-$C_4H_9)_2AlH$ 3. 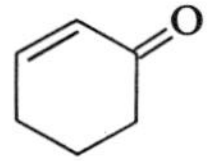 O	Ether Ether, 0° Ether, $-5°$	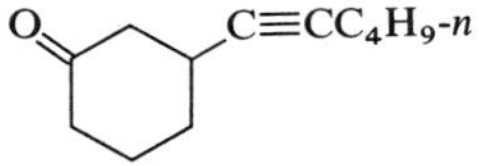 O, $C{\equiv}CC_4H_9$-n (71)	128,129
t-$C_4H_9C{\equiv}CLi$	1. $(CH_3)_2AlCl$ 2. $Ni(acac)_2$–$(i$-$C_4H_9)_2AlH$ 3.  O	Ether Ether, 0° Ether, $-5°$	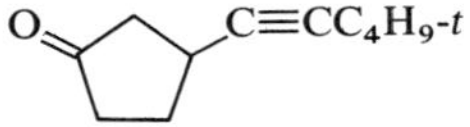 O, $C{\equiv}CC_4H_9$-t (60)	128,129
	1. $(CH_3)_2AlCl$ 2. $Ni(acac)_2$–$(i$-$C_4H_9)_2AlH$ 3. 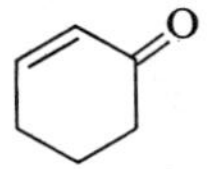 O	Ether Ether, 0° Ether, $-5°$	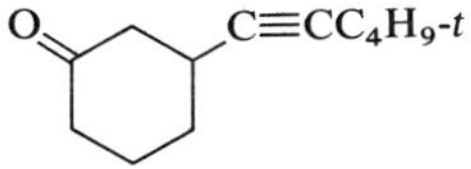 O, $C{\equiv}CC_4H_9$-t (72)	128,129
	1. $(CH_3)_2AlCl$ 2. $Ni(acac)_2$–$(i$-$C_4H_9)_2AlH$ 3. 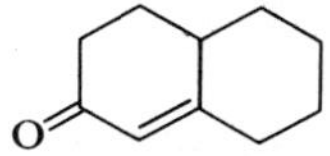 O	Ether Ether, 0° Ether, $-5°$	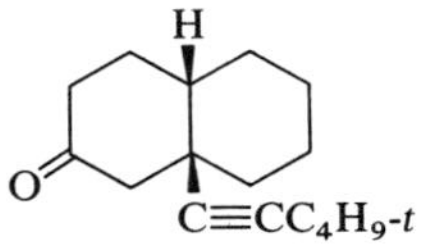 H, O, $C{\equiv}CC_4H_9$-t (49)	128,129
	1. $(CH_3)_2AlCl$ 2. $Ni(acac)_2$–$(i$-$C_4H_9)_2AlH$ 3. 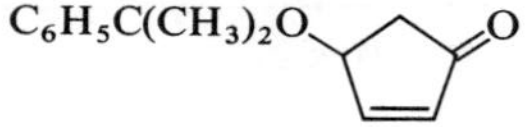 $C_6H_5C(CH_3)_2O$, O	Ether Ether, 0° Ether, $-5°$	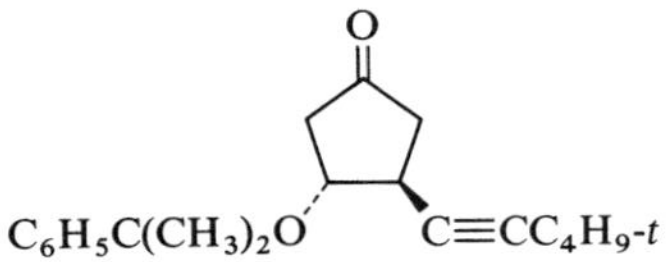 O, $C_6H_5C(CH_3)_2O$, $C{\equiv}CC_4H_9$-t (85)	128,129

TABLE XIII. γ-KETOACETYLENES (*Continued*)

	Substrate	Reagents	Conditions	Product(s) and Yield(s)(%)	Refs.
C_7	$Cl(CH_2)_3C{\equiv}CAl(CH_3)_2$	$CH_3COCH{=}CH_2$	Not specified	$CH_3CO(CH_2)_2C{\equiv}C(CH_2)_3Cl$ I, $CH_2{=}CHCOH(CH_3)C{\equiv}C(CH_2)_3Cl$ II (—) I:II = 1:2	126
C_8	n-$C_6H_{13}C{\equiv}CLi$	1. $AlCl_3$ 2. 4-hydroxy-2-[$(CH_2)_6CO_2CH_3$]cyclopent-2-en-1-one	Ether, 25°	3-HO-2-($C{\equiv}CC_6H_{13}$-n)-1-[$(CH_2)_6CO_2CH_3$]cyclopentan-one I + II (stereoisomer) (—) I:II = 1:2	127
	$C_6H_5C{\equiv}CLi$	1. $(C_2H_5)_2AlCl$ 2. 1-acetylcyclohexene	Ether–ligroin Ether–ligroin, −15°, 1.5 hr	1-$COCH_3$-2-($C{\equiv}CC_6H_5$)cyclohexane (94)	125
		1. $(C_2H_5)_2AlCl$ 2. $C_6H_5CH{=}CHCOC_6H_5$	Ether–ligroin Ether–ligroin, −10°, 1.5 hr	$C_6H_5C{\equiv}CCH(C_6H_5)CH_2COC_6H_5$ (81)	125
		1. $(C_2H_5)_2AlCl$ 2. $C_6H_5CH{=}CHCOCH_3$	Ether–ligroin Ether–ligroin, −10°, 1.5 hr	$C_6H_5C{\equiv}CCH(C_6H_5)CH_2COCH_3$ (95)	125
		1. $(C_2H_5)_2AlCl$ 2. $CH_3COC(CH_3){=}CH_2$	Ether–ligroin Ether–ligroin, −15°, 1.5 hr	$C_6H_5C{\equiv}CCH_2CH(CH_3)COCH_3$ (50)	125
		1. $(C_2H_5)_2AlCl$ 2. $CH_3COCH{=}CHCH_3$	Ether–ligroin Ether–ligroin, −10°, 1 hr	$C_6H_5C{\equiv}CCH(CH_3)CH_2COCH_3$ (54)	125

C_{13}	$n\text{-}C_5H_{11}CH(OTHP)C{\equiv}CLi$	1. $AlCl_3$ 2. 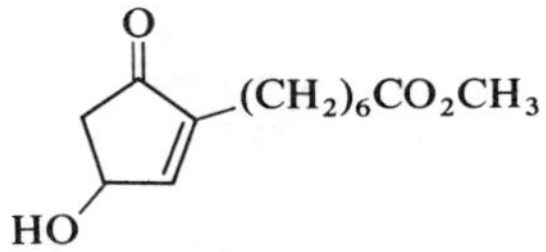 [cyclopentenone: O, $(CH_2)_6CO_2CH_3$, HO] 3. CH_3CO_2H, H_2O, THF	Ether, 25°	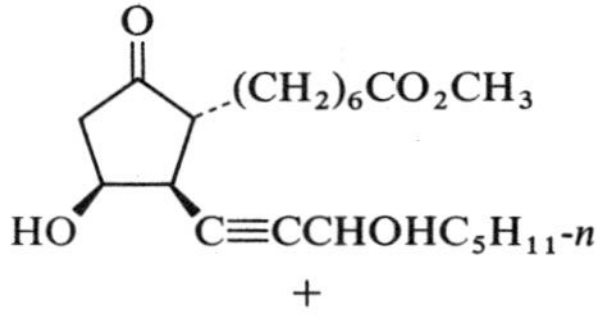 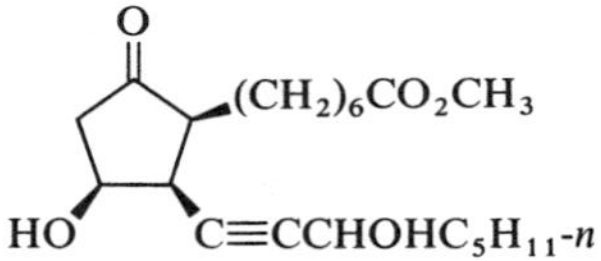 [cyclopentanone: O, $(CH_2)_6CO_2CH_3$, HO, $C{\equiv}CCHOHC_5H_{11}\text{-}n$] + [cyclopentanone: O, $(CH_2)_6CO_2CH_3$, HO, $C{\equiv}CCHOHC_5H_{11}\text{-}n$]	(—) (—)	127
C_{15}	$n\text{-}C_4H_9C(OSi(C_2H_5)_3)(CH_3)CH_2C{\equiv}CLi$	1. $(CH_3)_2AlCl$ 2. 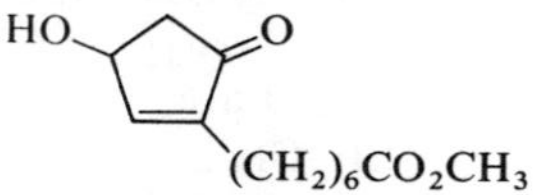 [cyclopentenone: HO, O, $(CH_2)_6CO_2CH_3$] 3. CH_3CO_2H, H_2O, THF	25°	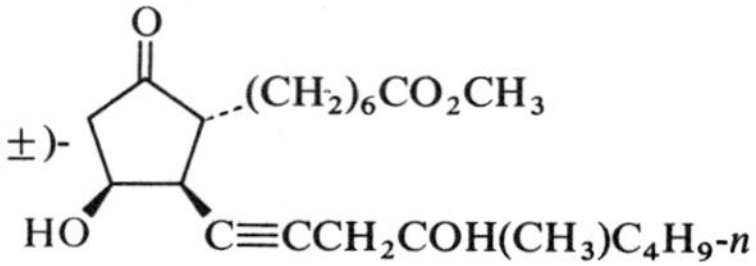 (±)-[cyclopentanone: O, $(CH_2)_6CO_2CH_3$, HO, $C{\equiv}CCH_2COH(CH_3)C_4H_9\text{-}n$]	(40)	123

TABLE XIV. CYCLOPROPANES

	Substrate	Reagents	Conditions	Product(s) and Yield(s)(%)	Refs.
C_6	$n\text{-}C_4H_9C{\equiv}CH$	1. $(i\text{-}C_4H_9)_2AlH$ 2. Zn–Cu, CH_2Br_2 3. H_3O^+	n-Hexane, 50°, 4 hr Ether, reflux, 24 hr	CH_2 $n\text{-}C_4H_9CH{-}CH_2$ (62)	130
		1. $(i\text{-}C_4H_9)_2AlH$ 2. Zn–Cu, CH_2Br_2 3. I_2 (3 eq)	n-Hexane, 50°, 4 hr Ether, reflux, 24 hr THF–ether, −65°, 3 hr, 25°	$n\text{-}C_4H_9$ CH_2 H C—C H I (47)	130
		1. $(i\text{-}C_4H_9)_2AlH$ 2. Zn–Cu, CH_2Br_2 3. Br_2 (3 eq)	n-Hexane, 50°, 4 hr Ether, reflux, 24 hr Ether–CH_2Cl_2, −65°, 3 hr	$n\text{-}C_4H_9$ CH_2 H C—C H Br (58)	130
	$t\text{-}C_4H_9C{\equiv}CH$	1. $(i\text{-}C_4H_9)_2AlH$ 2. Zn–Cu, CH_2Br_2 3. H_3O^+	n-Hexane, 50°, 2 hr Ether, reflux, 24 hr	CH_2 $t\text{-}C_4H_9CH{-}CH_2$ (51)	130
		1. $(i\text{-}C_4H_9)_2AlH$ 2. Zn–Cu, CH_2Br_2 3. Br_2 (3 eq)	n-Hexane, 50°, 2 hr Ether, reflux, 24 hr Ether–CH_2Cl_2, −65°, 3 hr	$t\text{-}C_4H_9$ CH_2 H C—C H Br (51)	130
C_8	$C_6H_{11}C{\equiv}CH$	1. $(i\text{-}C_4H_9)_2AlH$ 2. Zn–Cu, CH_2Br_2 3. H_3O^+	n-Hexane, 50°, 2 hr Ether, reflux, 24 hr	CH_2 $C_6H_{11}CH{-}CH_2$ (58)	130
		1. $(i\text{-}C_4H_9)_2AlH$ 2. Zn–Cu, CH_2Br_2 3. I_2 (3 eq)	n-Hexane, 50°, 2 hr Ether, reflux, 24 hr THF–ether, −65°, 3 hr, 25°	C_6H_{11} CH_2 H C—C H I (53)	130
		1. $(i\text{-}C_4H_9)_2AlH$ 2. Zn–Cu, CH_2Br_2 3. Br_2 (3 eq)	n-Hexane, 50°, 2 hr Ether, reflux, 24 hr Ether–CH_2Cl_2, −65°, 3 hr	C_6H_{11} CH_2 H C—C H Br (51)	130

TABLE XV. MISCELLANEOUS

	Substrate	Reagents	Conditions	Product(s) and Yield(s)(%)	Refs.
C_6	n-$C_4H_9C{\equiv}CH$	1. $(i\text{-}C_4H_9)_2AlH$ (2 eq) 2. D_2O	Neat	n-$C_5H_{11}CHD_2$ (—)	2
		1. $(i\text{-}C_4H_9)_2AlH$ (2 eq) 2. O_2 3. H_2O	Neat	n-$C_5H_{11}CHO$ (—)	2
		1. $(i\text{-}C_4H_9)_2AlH$ (2 eq) 2. n-C_4H_9Li 3. CH_3I (3 eq) 4. H_3O^+	Hexane–THF Hexane–THF	n-C_7H_{16} (74)	26
		1. $(i\text{-}C_4H_9)_2AlH$ (2 eq) 2. n-C_4H_9Li 3. CH_3I (3 eq) 4. O_2 5. H_3O^+	Hexane–THF Hexane–THF	n-$C_5H_{11}CHOHCH_3$ (90)	26
		1. $(i\text{-}C_4H_9)_2AlH$ (2 eq) 2. n-C_4H_9Li 3. CO_2 4. H_3O^+	Hexane–THF Hexane–THF	n-$C_5H_{11}CH(CO_2H)_2$ (73)	26
	$C_2H_5C{\equiv}CC_2H_5$	$(i\text{-}C_4H_9)_2AlH$	140°, 48 hr	Hexaethylbenzene ($C_6(C_2H_5)_6$) (—)	2
	$CH_2{=}CH(CH_2)_2C{\equiv}CH$	1. $(i\text{-}C_4H_9)_2AlH$ (2 eq) 2. H_3O^+	Ether, 37°, 4 hr	Methylcyclopentane (CH_3-cyclopentane) (80)	22

TABLE XV. MISCELLANEOUS (*Continued*)

	Substrate	Reagents	Conditions	Product(s) and Yield(s)(%)	Refs.
C_6 (*cont'd*)	$CH_2{=}CH(CH_2)_2C{\equiv}CH$	1. $(i\text{-}C_4H_9)_2AlH$ (2 eq) 2. O_2 3. H_3O^+	Ether, 37°, 4 hr	CH_2OH, OH (—)	22
C_7	$CH_2{=}CH(CH_2)_3C{\equiv}CH$	1. $(C_2H_5)_2AlH$ 2. H_2O	Ether	CH_3 (—)	187
	$CH_2{=}CHCH(CH_3)CH_2C{\equiv}CH$	1. $(C_2H_5)_2AlH$ 2. H_2O	Hexane, ether, reflux	CH_3, CH_3 (69)	138
C_8	$CH_2{=}CHCH(CH_3)CH_2CH_2C{\equiv}CH$	1. $(C_2H_5)_2AlH$ 2. H_2O	Ether	CH_3, CH_3 (—)	187
	$CH_2{=}CHCH(CH_2OCH_3)CH_2C{\equiv}CH$	1. $(C_2H_5)_2AlH$ 2. H_2O	Hexane, ether, reflux	CH_3, CH_2OCH_3 (76)	138
C_{14}	$C_6H_5C{\equiv}CC_6H_5$	$(i\text{-}C_4H_9)_2AlH$	140°, 50 hr	C_6H_5, C_6H_5, C_6H_5, C_6H_5, C_6H_5, C_6H_5 (—)	2

REFERENCES

[1] G. Wilke and H. Müller, *Chem. Ber.*, **89**, 444 (1956).

[2] G. Wilke and H. Müller, *Justus Liebigs Ann. Chem.*, **629**, 222 (1960).

[3] R. Köster and P. Binger, *Adv. Inorg. Chem. Radiochem.*, **7**, 263 (1965).

[4] H. Reinheckel, K. Hoage, and D. Jahnke, *Organometal. Chem. Rev.* (*A*), **4**, 47 (1969).

[5] H. Lehmkuhl and K. Ziegler, Organische Aluminum-Verbindungen, *Houben-Weyl Methoden der Organischen Chemie*, **13**(4), 9 (1970).

[6] T. Mole and E. A. Jeffrey, *Organoaluminum Compounds*, Elsevier, Amsterdam, 1972.

[7] K. L. Henold and J. P. Oliver, *Organometallic Reactions*, Vol. 5, Wiley-Interscience, New York, 1975, p. 387.

[8] E. Negishi, *J. Organometal. Chem. Lib.*, **1**, 93 (1976); E. Negishi, *Organometallics in Organic Synthesis*, Wiley, New York, 1980.

[9] G. Bruno, *The Use of Aluminum Alkyls in Organic Synthesis*, Ethyl Corporation, **1970**, **1973**, **1977**.

[10] G. Zweifel, *Comprehensive Organic Chemistry*, D. H. R. Barton and W. D. Ollis, Eds., Vol. 3, Pergamon Press, Oxford, 1979, p. 1013.

[11] J. J. Eisch, *Comprehensive Organometallic Chemistry*, Vol. 1, G. Wilkinson, Ed., Pergamon Press, Oxford, 1982.

[12] E. Negishi, *Pure Appl. Chem.*, **53**, 2333 (1981).

[13] G. Wilke and H. Müller, *Justus Liebigs Ann. Chem.*, **618**, 267 (1958).

[14] R. L. Miller and G. Zweifel, University of California, Davis, unpublished results.

[15] J. J. Eisch and W. C. Kaska, *J. Am. Chem. Soc.*, **85**, 2165 (1963).

[16] J. J. Eisch and W. C. Kaska, *J. Organomet. Chem.*, **2**, 184 (1964).

[17] T. Mole and J. R. Surtees, *Aust. J. Chem.*, **17**, 1229 (1964).

[18] J. R. Surtees, *Aust. J. Chem.*, **18**, 14 (1965).

[19] V. V. Gavrilenko, B. A. Palei, and L. I. Zakharkin, *Izv. Akad. Nauk SSSR, Ser. Khim.*, **1968**, 910 [*C.A.*, **69**, 77310*h* (1968)].

[20] V. V. Markova, V. A. Korma, and A. A. Petrov, *Zh. Obshch. Khim.*, **37**, 226 (1967) [*C.A.*, **66**, 95106 p. (1967)].

[21] G. M. Clark and G. Zweifel, University of California, Davis, unpublished results.

[22] G. Zweifel, G. M. Clark, and R. A. Lynd, *J. Chem. Soc., Chem. Commun.*, **1971**, 1593.

[23] K. Utimoto, K. Uchida, M. Yamaya, and H. Nozaki, *Tetrahedron Lett.*, **1977**, 3641.

[24] C. J. Sih, R. G. Salomon, P. Price, R. Sood, and G. Peruzzotti, *J. Am. Chem. Soc.*, **97**, 857 (1975).

[25] K. F. Bernady, M. B. Floyd, J. F. Poletto, and M. J. Weiss, *J. Org. Chem.*, **44**, 1438 (1979).

[26] G. Zweifel and R. B. Steele, *Tetrahedron Lett.*, **1966**, 6021.

[27] G. Cainelli, F. Bertini, P. Grasseli, and G. Zubiania, *Tetrahedron Lett.*, **1967**, 1581.

[28] P. Binger, *Angew. Chem., Int. Ed. Engl.*, **2**, 686 (1963).

[29] G. Wilke and W. Schneider, *Bull. Soc. Chim. Fr.*, **1963**, 1462.

[30] J. J. Eisch and S. G. Rhee, *J. Am. Chem. Soc.*, **96**, 7276 (1974).

[31] G. M. Clark and G. Zweifel, *J. Am. Chem. Soc.*, **93**, 527 (1971).

[32] J. J. Eisch and M. W. Foxton, *J. Organomet. Chem.*, **12**, P33 (1968).

[33] I. Knox, S.-C. Chang, and A. H. Andrist, *J. Org. Chem.*, **42**, 3981 (1977).

[34] H. C. Brown, C. G. Scouten, and R. Liotta, *J. Am. Chem. Soc.*, **101**, 96 (1979).

[35] P. Teisseire, B. Corbier, and M. Plattier, *Recherches* (*Paris*), **16**, 5 (1967).

[36] G. Zweifel, G. M. Clark, and N. L. Polston, *J. Am. Chem. Soc.*, **93**, 3395 (1971).

[37] J. J. Eisch and W. C. Kaska, *J. Am. Chem. Soc.*, **88**, 2213 (1966).

[38] J. J. Eisch and M. W. Foxton, *J. Org. Chem.*, **36**, 3520 (1971).

[39] J. J. Eisch, H. Gopal, and S.-G. Rhee, *J. Org. Chem.*, **40**, 2064 (1975).

[40] T. Yoshida and E. Negishi, *J. Am. Chem. Soc.*, **103**, 1276 (1981).

[41] J. J. Eisch and S. G. Rhee, *J. Am. Chem. Soc.*, **97**, 4673 (1975).

[42] W. Lewis and G. Zweifel, University of California, Davis, unpublished results.

[43] K. Uchida, K. Utimoto, and H. Nozaki, *J. Org. Chem.*, **41**, 2215 (1976).

[44] G. Zweifel and W. Lewis, *J. Org. Chem.*, **43**, 2739 (1978).
[45] J. A. Miller and G. Zweifel, *J. Am. Chem. Soc.*, **105**, 1383 (1983), and unpublished results.
[46] H. P. On, W. Lewis, and G. Zweifel, *Synthesis*, **1981**, 999.
[47] L. H. Slaugh, *Tetrahedron*, **22**, 1741 (1966).
[48] E. F. Magoon and L. H. Slaugh, *Tetrahedron*, **23**, 4509 (1967).
[49] G. Zweifel and R. B. Steele, *J. Am. Chem. Soc.*, **89**, 5085 (1967).
[50] H. O. House, *Modern Synthetic Reactions*, 2nd ed., Benjamin/Cummings, Menlo Park, CA, 1972.
[51] G. Zweifel, R. A. Lynd, and R. E. Murray, *Synthesis*, **1977**, 52.
[52] H. Westmijze, H. Kleijn, and P. Vermeer, *Synthesis*, **1979**, 430.
[53] P. Vermeer, J. Meijer, C. Eylander, and L. Brandsma, *Recl. Trav. Chim. Pays-Bas*, **95**, 25 (1976).
[54] M. Hojo, R. Masuda, and S. Takagi, *Synthesis*, **1978**, 284.
[55] G. Zweifel, W. Lewis, and H. P. On, *J. Am. Chem. Soc.*, **101**, 5101 (1979).
[56] A. B. Bates, E. R. H. Jones, and M. C. Whiting, *J. Chem. Soc.*, **1954**, 1854.
[57] R. A. Raphael, *Acetylenic Compounds in Organic Synthesis*, Butterworths, London, 1955. 1955.
[58] R. Rossi and A. Carpita, *Synthesis*, **1977**, 561.
[59] W. J. Borden, *J. Am. Chem. Soc.*, **92**, 4898 (1970).
[60] B. Grant and C. Djerassi, *J. Org. Chem.*, **39**, 968 (1974).
[61] E. J. Corey, J. A. Katzenellenbogen, and G. H. Posner, *J. Am. Chem. Soc.*, **89**, 4245 (1967).
[62] E. J. Corey, H. A. Kirst, and J. A. Katzenellenbogen, *J. Am. Chem. Soc.*, **92**, 6314 (1970).
[63] R. Rienäcker and D. Schwenger, *Justus Liebigs Ann. Chem.*, **737**, 183 (1970).
[64] D. E. Van Horn and E. Negishi, *J. Am. Chem. Soc.*, **100**, 2252 (1978).
[65] J. J. Eisch and W. C. Kaska, *J. Am. Chem. Soc.*, **88**, 2976 (1966).
[66] C. L. Rand, D. E. Van Horn, M. W. Moore, and E. Negishi, *J. Org. Chem.*, **46**, 4093 (1981).
[67] G. Altnau and L. Rösch, *Tetrahedron Lett.*, **1980**, 4069.
[68] D. E. Van Horn, L. F. Valente, M. J. Idacavage, and E. Negishi, *J. Organomet. Chem.*, **156**, C20 (1978).
[69] S. Danishefsky, T. Kitahara, M. Tsai, and J. Dynak, *J. Org. Chem.*, **41**, 1669 (1976).
[70] E. Negishi and S. Baba, *J. Am. Chem. Soc.*, **97**, 7385 (1975).
[71] L. L. Ivanov, V. V. Gavrilenko, and L. I. Zakharkin, *Izv. Akad. Nauk SSSR, Ser. Khim.*, **1964**, 1989 [*C.A.*, **62**, 7660*f* (1965)].
[72] E. Negishi, *Organometallics in Organic Synthesis*, Wiley, New York, 1980.
[73] H. P. On and G. Zweifel, University of California, Davis, unpublished results.
[74] W. Ziegenbein and W. M. Schneider, *Chem. Ber.*, **98**, 824 (1965).
[75] P. S. Skell and P. K. Freeman, *J. Org. Chem.*, **29**, 2524 (1964).
[76] G. Zweifel and H. P. On, *Synthesis*, **1980**, 803.
[77] S. Baba, D. E. Van Horn, and E. Negishi, *Tetrahedron Lett.*, **1976**, 1927.
[78] J. J. Eisch and G. A. Damasevitz, *J. Org. Chem.*, **41**, 2214 (1976).
[79] G. Zweifel and R. A. Lynd, *Synthesis*, **1976**, 816.
[80] N. Okukado and E. Negishi, *Tetrahedron Lett.*, **1978**, 2357.
[81] E. Negishi and S. Baba, *J. Chem. Soc., Chem. Commun.*, **1976**, 596.
[82] E. Negishi, N. Okukado, A. O. King, D. E. Van Horn, and B. I. Spiegel, *J. Am. Chem. Soc.*, **100**, 2254 (1978).
[83] G. Zweifel and C. C. Whitney, *J. Am. Chem. Soc.*, **89**, 2753 (1967).
[84] B. A. Palei, V. V. Gavrilenko, and L. I. Zakharkin, *Izv. Akad. Nauk SSSR, Ser. Khim.*, **1969**, 2760 [*C.A.*, **72**, 79143*s* (1970)].
[85] H. P. On, J. A. Werner, and G. Zweifel, University of California, Davis, unpublished results.
[86] G. Zweifel, R. E. Murray, and H. P. On, *J. Org. Chem.*, **46**, 1292 (1981).
[87] R. B. Miller and G. McGarvey, *J. Org. Chem.*, **44**, 4623 (1979).
[88] E. Negishi, D. E. Van Horn, A. O. King, and N. Okukado, *Synthesis*, **1979**, 501.
[89] R. P. Fisher, H. P. On, J. T. Snow, and G. Zweifel, *Synthesis*, **1982**, 127.
[90] A. P. Kozikowski, A. Ames, and H. Wetter, *J. Organomet. Chem.*, **164**, C33 (1979).

[91] G. Zweifel and R. B. Steele, *J. Am. Chem. Soc.*, **89**, 2754 (1967).
[92] H. Newman, *Tetrahedron Lett.*, **1971**, 4571.
[93] R. A. Lynd and G. Zweifel, University of California, Davis, unpublished results.
[94] E. Negishi, A. O. King, W. Klima, W. Patterson, and A. Silveira, Jr., *J. Org. Chem.*, **45**, 2526 (1980).
[95] E. Negishi, L. F. Valente, and M. Kobayashi, *J. Am. Chem. Soc.*, **102**, 3298 (1980).
[96] S. Danishefsky, R. L. Funk, and J. F. Kerwin, Jr., *J. Am. Chem. Soc.*, **102**, 6889 (1980).
[97] S. Warwel, G. Schmitt, and B. Ahlfaenger, *Synthesis*, **1975**, 632.
[98] E. Negishi, S. Baba, and A. O. King, *J. Chem. Soc., Chem. Commun.*, **1976**, 17.
[99] L. E. Overman and K. L. Bell, *J. Am. Chem. Soc.*, **103**, 1851 (1981).
[100] M. Kobayashi, L. F. Valente, E. Negishi, W. Patterson, and A. Silveira, Jr., *Synthesis*, **1980**, 1034.
[101] D. B. Malpass, J. C. Watson, and G. S. Yeargin, *J. Org. Chem.*, **42**, 2712 (1977).
[102] J. J. Eisch and M. W. Foxton, *J. Organomet. Chem.*, **11**, P7 (1963).
[103] V. M. Bulina, L. L. Ivanov, and Y. B. Pyatnova, *Zh. Org. Khim.*, **9**, 491 (1973) [*C.A.*, **78**, 158856*d* (1973)].
[104] W. Lewis, M. Musumeci, H. P. On, and G. Zweifel, University of California, Davis, unpublished results.
[105] R. L. Dansheiser and H. Sard, *J. Org. Chem.*, **45**, 4810 (1980).
[106] G. Zweifel and R. A. Lynd, *Synthesis*, **1976**, 625.
[107] G. Zweifel, J. T. Snow, and C. C. Whitney, *J. Am. Chem. Soc.*, **90**, 7139 (1968).
[108] G. Zweifel and R. L. Miller, *J. Am. Chem. Soc.*, **92**, 6678 (1970).
[109] S. Baba and E. Negishi, *J. Am. Chem. Soc.*, **98**, 6729 (1976).
[110] R. A. Lynd and G. Zweifel, *Synthesis*, **1974**, 658.
[111] H. Matsushita and E. Negishi, *J. Am. Chem. Soc.*, **103**, 2882 (1981).
[112] J. Fried, C.-H. Lin, J. C. Sih, P. Dalven, and G. F. Cooper, *J. Am. Chem. Soc.*, **94**, 4342 (1972).
[113] J. Fried, J. C. Sih, C.-H. Lin, and P. Dalven, *J. Am. Chem. Soc.*, **94**, 4343 (1972).
[114] J. Fried and J. C. Sih, *Tetrahedron Lett.*, **1973**, 3899.
[115] J. Fried, C.-H. Lin, and S. H. Ford, *Tetrahedron Lett.*, **1969**, 1379.
[116] J. Fried, C.-H. Lin, M. Mehra, W. Kao, and P. Dalven, *Ann. N.Y. Acad. Sci.*, **180**, 36 (1971).
[117] T. F. Murray, V. Verma, and J. R. Norton, *J. Chem. Soc., Chem. Commun.*, **1976**, 907.
[118] S. Danishefsky, M.-Y. Tsai, and T. Kitanara, *J. Org. Chem.*, **42**, 394 (1977).
[119] G. A. Crosby and R. A. Stephenson, *J. Chem. Soc., Chem. Commun.*, **1975**, 287.
[120] P. A. Bartlett and J. Myerson, *J. Am. Chem. Soc.*, **100**, 3950 (1978).
[121] G. H. Posner, *Org. React.*, **19**, 1 (1972).
[122] J. Hooz and R. B. Layton, *Can. J. Chem.*, **51**, 2098 (1973).
[123] P. W. Collins, E. Z. Dajani, M. S. Bruhn, C. H. Brown, J. R. Palmer, and R. Pappo, *Tetrahedron Lett.*, **1975**, 4217.
[124] K. F. Bernady and M. J. Weiss, *Tetrahedron Lett.*, **1972**, 4083.
[125] J. Hooz and R. B. Layton, *J. Am. Chem. Soc.*, **93**, 7320 (1971).
[126] M. Bruhn, C. H. Brown, P. W. Collins, J. R. Palmer, E. Z. Dajani, and R. Pappo, *Tetrahedron Lett.*, **1976**, 235.
[127] R. Pappo and P. W. Collins, *Tetrahedron Lett.*, **1972**, 2627.
[128] R. T. Hansen, D. B. Carr, and J. Schwartz, *J. Am. Chem. Soc.*, **100**, 2244 (1978).
[129] J. Schwartz, D. B. Carr, R. T. Hansen, and F. M. Dayrit, *J. Org. Chem.*, **45**, 3053 (1980).
[130] G. Zweifel, G. M. Clark, and C. C. Whitney, *J. Am. Chem. Soc.*, **93**, 1305 (1971).
[131] E. Negishi and L. D. Boardman, *Tetrahedron Lett.*, **1982**, 3327.
[132] E. Negishi, K. P. Jadhav, and N. Daotien, *Tetrahedron Lett.*, **1982**, 2085.
[133] D. F. Shriver, *The Manipulation of Air-Sensitive Compounds*, McGraw-Hill, New York, 1969.
[134] H. C. Brown, *Organic Syntheses* via *Boranes*, Wiley, New York, 1975.
[135] J. J. Eisch, *Organometallic Syntheses. Non-Transition Metal Compounds*, Vol. 2, Academic Press, New York, 1981.

[136] G. Zweifel and H. C. Brown, *Org. React.*, **13**, 1 (1963).
[137] H. Hoberg, *Angew. Chem., Int. Ed. Engl.*, **5**, 513 (1966).
[138] M. J. Smith and S. E. Wilson, *Tetrahedron Lett.*, **1982**, 5013.
[139] E. Negishi, H. Matsushita, and N. Okukado, *Tetrahedron Lett.*, **1981**, 2715.
[140] C. Germon, A. Alexakis, and J. F. Normant, *Tetrahedron Lett.*, **1980**, 3763.
[141] D. Masure, Ph. Coutrot, and J. F. Normant, *J. Organomet. Chem.*, **226**, C55 (1982).
[142] E. Negishi, F.-T. Luo, R. Frisbee, and H. Matsushita, *Heterocycles*, **18**, 117 (1982).
[143] J. J. Eisch and K. C. Fichter, *J. Am. Chem. Soc.*, **96**, 6815 (1974).
[144] R. Salomon, M. F. Salomon, and J. L. C. Kachinski, *J. Am. Chem. Soc.*, **99**, 1043 (1977).
[145] R. B. Miller and G. McGarvey, *J. Org. Chem.*, **43**, 4424 (1978).
[146] B. B. Snider and M. Karras, *J. Organomet. Chem.*, **179**, C37 (1979).
[147] F. Asinger, B. Fell, and G. Steffan, *Chem. Ber.*, **97**, 1555 (1964).
[148] B. Bennetau, J.-P. Pillot, J. Dunogues, and R. Calas, *J. Chem. Soc., Chem. Commun.*, **1981**, 1094.
[149] J. J. Eisch and R. Amtmann, *J. Org. Chem.*, **37**, 3410 (1972).
[150] W. Granitzer and A. Stütz, *Tetrahedron Lett.*, **1979**, 3145.
[151] J. J. Eisch and J.-M. Biedermann, *J. Organomet. Chem.*, **30**, 167 (1971).
[152] G. A. Razuvaev, I. V. Lomakova, and L. P. Stepovik, *Zh. Obshch. Khim.*, **43**, 2416 (1973) [*C.A.*, **80**, 59991*d* (1974)].
[153] H. C. Huang, J. K. Rehmann, and G. R. Gray, *J. Org. Chem.*, **47**, 4018 (1982).
[154] R. I. Kruglikova, L. P. Kravets, and B. V. Unkovskii, *Zh. Org. Khim.*, **11**, 263 (1975) [*C.A.*, **82**, 124634*x* (1975)].
[155] E. Negishi, F.-T. Luo, and C. L. Rand, *Tetrahedron Lett.*, **1982**, 27.
[156] K. A. M. Kremer, G.-H. Kuo, E. J. O'Connor, P. Helquist, and R. C. Kerber, *J. Am. Chem. Soc.*, **104**, 6119 (1982).
[157] G. H. Posner and P.-W. Tang, *J. Org. Chem.*, **43**, 4131 (1978).
[158] C. J. Sih, R. G. Salomon, P. Price, G. Peruzzotti, and R. Sood, *J. Chem. Soc., Chem. Commun.*, **1972**, 240.
[159] C. J. Sih, J. B. Heather, G. P. Peruzzotti, P. Price, R. Sood, and L.-F. H. Lee, *J. Am. Chem. Soc.*, **95**, 1676 (1973).
[160] C. J. Sih, P. Price, R. Sood, R. G. Salomon, G. Peruzzotti, and M. Casey, *J. Am. Chem. Soc.*, **94**, 3643 (1972).
[161] D. R. Williams, B. A. Barner, K. Nishitani, and J. G. Phillips, *J. Am. Chem. Soc.*, **104**, 4708 (1982).
[162] E. J. Corey, J. A. Katzenellenbogen, N. W. Gilman, S. A. Roman, and B. W. Erickson, *J. Am. Chem. Soc.*, **90**, 5618 (1968).
[163] M. F. Semmelhack and E. S. C. Wu, *J. Am. Chem. Soc.*, **98**, 3384 (1976).
[164] S. W. Rollinson, R. A. Amos, and J. A. Katzenellenbogen, *J. Am. Chem. Soc.*, **103**, 4114 (1981).
[165] H. Tomioka, T. Suzuki, K. Oshima, and H. Nozaki, *Tetrahedron Lett.*, **1982**, 3387.
[166] I. Hasan and Y. Kishi, *Tetrahedron Lett.*, **1980**, 4229.
[167] B. B. Molloy and K. L. Hauser, *J. Chem. Soc., Chem. Commun.*, **1968**, 1017.
[168] H. Kwart and T. J. George, *J. Am. Chem. Soc.*, **99**, 5214 (1977).
[169] A. P. Kozikowski and Y. Kitigawa, *Tetrahedron Lett.*, **1982**, 2087.
[170] R. F. Cunico and H. M. Lee, *J. Am. Chem. Soc.*, **99**, 7613 (1977).
[171] G. Stork, M. E. Jung, E. Colvin, and Y. Noel, *J. Am. Chem. Soc.*, **96**, 3684 (1974).
[172] S. E. Denmark and T. K. Jones, *J. Org. Chem.*, **47**, 4595 (1982).
[173] I. Cutting and P. J. Parsons, *Tetrahedron Lett.*, **1981**, 2021.
[174] R. Baudouy and J. Gore, *Synthesis*, **1974**, 573.
[175] H. Newman, *J. Am. Chem. Soc.*, **95**, 4098 (1973).
[176] D. C. Brown, S. A. Nichols, A. B. Gilpin, and D. W. Thompson, *J. Org. Chem.*, **44**, 3457 (1979).
[177] M. Kobayashi and E. Negishi, *J. Org. Chem.*, **45**, 5223 (1980).
[178] C. C. Whitney and G. Zweifel, University of California, Davis, unpublished results.

[179] R. E. Murray and G. Zweifel, University of California, Davis, unpublished results.
[180] E. Negishi, S. Chatterjee, and H. Matsushita, *Tetrahedron Lett.*, **1981**, 3737.
[181] H. Matsushita and E. Negishi, *J. Chem. Soc., Chem. Commun.*, **1982**, 160.
[182] M. Sato, K. Takai, K. Oshima, and H. Nozaki, *Tetrahedron Lett.*, **1981**, 1609.
[183] K. Takai, K. Oshima, and H. Nozaki, *Tetrahedron Lett.*, **1980**, 2531.
[184] F. Bohlmann, E. Inhoffen, and J. Politt, *Justus Liebigs Ann. Chem.*, **604**, 207 (1957).
[185] H. Demarne and P. Cadiot, *Bull. Soc. Chim. Fr.*, **1968**, 205.
[186] P. A. Bartlett and F. R. Green, *J. Am. Chem. Soc.*, **100**, 4858 (1978).
[187] R. Rienäcker and D. Schwengers, *Justus Liebigs Ann. Chem.*, **1977**, 1633.
[188] W. J. Gensler and J. J. Bruno, *J. Org. Chem.*, **28**, 1254 (1963).

AUTHOR INDEX, VOLUMES 1–32

CHAPTER AND TOPIC INDEX, VOLUMES 1–32

Many chapters contain brief discussions of reactions and comparisons of alternative synthetic methods related to the reaction that is the subject of the chapter. These related reactions and alternative methods are not usually listed in this index. In this index the volume number is in **BOLDFACE,** the chapter number in ordinary type.

SUBJECT INDEX, VOLUME 32

Since the table of contents provides a quite complete index, only those items not readily found from the contents page are listed here. Numbers in **BOLDFACE** refer to experimental procedures.